The Peptides *Volume 2*

THE PEPTIDES

Analysis, Synthesis, Biology

Treatise Editors

E. GROSS AND J. MEIENHOFER

Volume 1
Major Methods of Peptide Bond Formation

Volume 2
Special Methods in Peptide Synthesis, Part A

The Peptides

Analysis, Synthesis, Biology

VOLUME 2 Special Methods in Peptide Synthesis, Part A

Edited by

ERHARD GROSS

National Institutes of Health
Bethesda, Maryland

JOHANNES MEIENHOFER

Chemical Research Department
Hoffmann-La Roche Inc.
Nutley, New Jersey

1980

ACADEMIC PRESS

A Subsidiary of Harcourt Brace Jovanovich, Publishers

New York London Toronto Sydney San Francisco

ACADEMIC PRESS, INC.
111 Fifth Avenue, New York, New York 10003

United Kingdom Edition published by
ACADEMIC PRESS, INC. (LONDON) LTD.
24/28 Oval Road, London NW1 7DX

Library of Congress Cataloging in Publication Data
Main entry under title:

The Peptides.
Includes bibliographical references and index.
CONTENTS: v. 1. Major methods of peptide bond Formation.--[etc.]--v. 3. Protection of functional groups in peptide synthesis.
1. Peptides. I. Gross, Erhard. II. Meienhofer, Johannes. [DNLM: 1. Peptides. QU68 P424]
QP552.P4P47 574.1'9245 78-31958
ISBN 0-12-304202-X (v. 2)

PRINTED IN THE UNITED STATES OF AMERICA

79 80 81 82 9 8 7 6 5 4 3 2 1

Contents

Chapter 3 *Polymeric Reagents in Peptide Synthesis*

Mati Fridkin

Chapter 4 *The Four Component Synthesis*

Ivar Ugi

Chapter 5 *The Oxidation–Reduction Condensation*

Teruaki Mukaiyama, Rei Matsueda, and Masaaki Ueki

Chapter 6 *Repetitive Methods in Solution*

Lajos Kisfaludy

List of Contributors

Numbers in parentheses indicate the pages on which the authors' contributions begin.

George Barany (1), The Rockefeller University, New York, New York 10021

Ernst Bayer (285), Institut für Organische Chemie der Universität Tübingen, Auf der Morgenstelle 18, 74—Tübingen, Germany

Mati Fridkin (333), Department of Organic Chemistry, The Weizmann Institute of Science, Rehovot, Israel

Lajos Kisfaludy (417), Chemical Works of Gedeon Richter Ltd. Budapest, Hungary

József Kovács (485), Department of Chemistry, St. John's University, New York, New York 11439

*Rei Matsueda** (383), Central Research Laboratories, Sankyo Company, Ltd., Hiromachi, Shinagawa-ku, Tokyo, Japan

R. B. Merrifield (1), The Rockefeller University, New York, New York 10021

Teruaki Mukaiyama (383), Department of Chemistry, Faculty of Science, The University of Tokyo, Bunkyo-ku, Japan

Manfred Mutter (285), Institut für Organische Chemie der Universität Tübingen, Auf der Morgenstelle 18, 74—Tübingen, Germany

Robert C. Sheppard (441), MRC Laboratory of Molecular Biology, Cambridge CB2 2QH, England

Masaaki Ueki (383), Department of Chemistry, Science University of Tokyo, Kagurazaka, Shinjuku-ku, Tokyo, Japan

Ivar Ugi (365), Organisch-Chemisches Institut, Technische Universitat München. Lichenbergstrasse 4, 8046 Garching, Germany

* *Present address:* Department of Chemistry, University of Chicago, Chicago, Illinois 60637

Preface

"The Peptides" is an open-ended treatise providing comprehensive and critical reviews of important developments in all areas of peptide research including analysis, synthesis, and biology. These reviews are intended as a reference for the specialist, a guide for the novice, and a forum for investigators in all areas of research concerned with peptides and proteins.

The purpose of the volumes on synthesis is to present the current state of methodology. Peptide synthesis has been variously described as being either very difficult or merely routine "cook-book" chemistry. The important aspects and considerations are the polymeric nature of peptides and proteins and their wide range of properties. The polymeric character mandates the use of special approaches to synthesis. The spectrum of peptide properties—from very basic to very acidic, from highly hydrophilic to totally hydrophobic, from easily soluble to completely insoluble—invariably present surprises and/or obstacles during synthesis of these molecules. For these reasons, in the Pierce Award Lecture at the Fifth American Peptide Symposium in 1977, M. Bodanszky stated that peptide synthesis remains "an undiminished challenge."

In Volume 1, the "Major Methods of Peptide Bond Formation" were described. This second volume deals with special approaches to peptide synthesis that have generally contributed in major ways to rapid progress and specifically to increased efficiency, process simplifications, and time saving.

The first chapter by G. Barany and R. B. Merrifield presents a comprehensive up-to-date account of the steadily growing efficacy, scope, and application of the solid-phase method. All pertinent aspects of the liquid-phase procedure are discussed in Chapter 2 by M. Mutter and E. Bayer. In Chapter 3, M. Fridkin describes the application of polymeric reagents to peptide synthesis. The continuing exploration and refinement of the four component condensation, reviewed by I. Ugi in Chapter 4, is leading to more frequent application of this ingenious new approach to practical peptide synthesis. In Chapter 5, T. Mukaiyama and collaborators present convincing evidence that the potential of the oxidation–reduction condensation for considerably improved peptide bond formation are not being fully utilized. Repetitive methods of peptide synthesis in solution, reviewed in Chapter 6 by L. Kisfaludy, may frequently achieve time savings that approach those obtained with the solid-phase method. In Chapter 7, R. C. Sheppard presents a lucid account of the relatively young endeavor of partial synthesis of peptides and proteins. The final Chapter (8), by J. Kovács, is devoted to novel approaches for quantitation of racemization in model systems and to prediction of racemization dangers in actual synthesis.

Other special methods will be compiled in a later volume; protection of functional groups in peptide synthesis will be presented in Volume 3.

We are much indebted to the authors for their contributions and cooperation during the editorial process, and to the staff of Academic Press for their efficient preparation of this volume.

Erhard Gross
Johannes Meienhofer

Nomenclature and Abbreviations

In several chapters of this volume the peptide-size nomenclature proposed by M. Bodanszky (1977) at the Fifth American Peptide Symposium has been used. Greek prefixes are retained for di- to decapeptides only. For peptides with more than 10 amino acid residues arabic numerals are used, for example, "15-peptide," "31-peptide," or corticotropin-(1–24), β-LPH-(61–91).

Peptide *segment* and *segment condensation* have been used instead of *fragment* or *fragment condensation*. Amino acids are of the L configuration unless otherwise indicated. In peptide structures, hyphens between D and the amino acid symbol are omitted, e.g., H-Gly-DPhe-Ala-OH.

The current *Chemical Abstracts* term 5(4*H*)-oxazolone is used for azlactone [viz. *Chem. Abstr.* **83**, 3633 CS(1975)], since it is unambiguous in contrast to other (earlier) terms in the literature (see Volume 1, p. 39, footnote).

Abbreviations are kept as short as possible, preferably three letters (four or more letters only if unavoidable). Many abbreviations are indeed those recommended by the IUPAC–IUB Commission on Biochemical Nomenclature, e.g., in *J. Biol. Chem.* **242**, 6489–6497 (1970); **250**, 3215–3216 (1975). However, the system proposed by the Commission in 1972 (viz. *J. Biol. Chem.* **247**, 997–982) of complex symbols frequently exceeding five spaces and including Greek letters and subscript numerals will not be used. These symbols are impractical for typesetting and are generally of infrequent use.

Symbols used in this volume are listed below (except those of the common amino acids).

Abbreviations

AII	angiotensin II
AA	amino acid
Ac	acetyl
Acm	acetamidomethyl
AcOH	acetic acid

ACTH	corticotropin
Adc	adamantyloxycarbonyl
Aib	α-aminoisobutyric acid
Alk	alkyl
Aoc	*tert*-amyloxycarbonyl
Azoc	2-(4-phenylazophenyl)propyl(2)oxycarbonyl
Bic	5-benzisoxazolylmethyleneoxycarbonyl
Bmv	2-benzoyl-1-methylvinyl
Boc	*tert*-butyloxycarbonyl
Bpoc	2-(4-biphenylyl)propyl(2)oxycarbonyl
BPTI	bovine pancreatic trypsin inhibitor
Btf	2,2,2-trifluoro-1-*tert*-butyloxycarbonylaminoethyl
Btm	benzylthiomethyl
Bz	benzoyl
Bzl	benzyl
4CC	four component condensation (method)
cd	circular dichroism
CDI	*N,N'*-carbonyldiimidazole
*c*Hex	cyclohexyl
Cpase B	carboxypeptidase B
2Cz	2-chlorobenzyloxycarbonyl
DBN	1,5-diazabicyclo[4.3.0]nonene-5
Dcb	2,6-dichlorobenzyl
DCC	dicyclohexylcarbodiimide
DCU	*N,N'*-dicyclohexylurea
Ddz	2-(3,5-dimethyloxyphenyl)propyl(2)oxycarbonyl
DEA	diethylamine
DEPC	diethylphosphorocyanidate
Dmb	3,4-dimethylbenzyl
Dmc	diisopropylmethyloxycarbonyl
DMF	dimethylformamide
Dmob	2,4-dimethyloxybenzyl
DMSO	dimethyl sulfoxide
Dnp	2,4-dinitrophenyl
Dns	5-dimethylamino-1-naphthalenesulfonyl (dansyl)
Dpm	diphenylmethyl (benzhydryl)
Dpp	diphenylphosphinoyl
Dts	dithiasuccinoyl
DVB	divinylbenzene

Ec	ethylcarbamoyl
EEDQ	*N*-ethyloxycarbonyl-2-ethyloxy-1,2-dihydroquinoline
Eoc	ethyloxycarbonyl
esr	electron spin resonance
Et	ethyl
Et_3N	triethylamine
Et_2NH	diethylamine
EtOH	ethanol
Fmoc	9-fluorenylmethyloxycarbonyl
Foc	furfuryloxycarbonyl
For	formyl
GIP	gastric inhibitory polypeptide
HA	acid
HGH	human growth hormone (somatotropin)
HOBt	1-hydroxybenzotriazole
HONb	*N*-hydroxy-5-norbornene-2, 3-dicarboximide
HOSu	*N*-hydroxysuccinimide
HPLC	high-performance liquid chromatography
Hse	homoserine
*i*Bom	isobutyloxymethyl
IIDQ	*N*-isobutyloxycarbonyl-2-isobutyloxy-1,2-dihyroquinoline
ir	infrared (spectroscopy)
LHRH	luliberin
Maq	methyleneanthraquinone
Mbs	4-methyloxybenzenesulfonyl
Mcb	1-methylcyclobutyloxycarbonyl
Me	methyl
Meb	4-methylbenzyl
MeOH	methanol
Met (O)	methionine sulfoxide
Mob	4-methyloxybenzyl
Moz	4-methyloxybenzyloxycarbonyl
Mpt	dimethylphosphinothioyl
Msc	methylsulfonylethyloxycarbonyl
β-MSH	β-melanotropin

N^g	guanidine nitrogen
N^i	indole nitrogen
N^{im}	imidazole nitrogen
N^π	*pros*-imidazole nitrogen
N^τ	*tele*-imidazole nitrogen
NCA	*N*-carboxyanhydride
NEPIS	*N*-ethyl-5-phenylisoxazolium-3′-sulfonate
Nle	norleucine
nmr	nuclear magnetic resonance (spectroscopy)
NO_2	nitro
Nps	2-nitrophenylsulfenyl
NTA	*N*-thiocarboxyanhydride (1,3-thiazolidine-2,5-dione)
Nu	nucleophile
4Nz	4-nitrobenzyloxycarbonyl
OBt	benzotriazolyl ester
OBzl	benzyl ester
ODnp	2,4*-dinitrophenyl ester
OEt	ethyl ester
OMe	methyl ester
ONb	4-nitrobenzyl ester
ONo	2-nitrophenyl ester
ONp	4-nitrophenyl ester
OPa	phenacyl ester
OPcp	pentachlorophenyl ester
OPfp	pentafluorophenyl ester
OPh	phenyl ester
OPic	4-picolyl ester
OQ	8-quinolyl ester
OR	ester (general)
OSu	*N*-succinimidoyl ester
OTbp	2,4,6-tribromophenyl ester
O*t*Bu	*tert*-butyl ester
OTce	β,β,β-trichloroethyl ester
OTcp	2,4,5*-trichlorophenyl ester
Pac	phenacyl
Pam	phenylacetamidomethyl
Pc[CoI]	cobalt I phthalocyanine

* Unless a different position-isomer is indicated.

PCP	*o*-phenylene chlorophosphite
PEG*	polyethylene glycol
pGlu	pyroglutamic acid (2-pyrrolidone-5-carboxylic acid)
Ph	phenyl
[PHMI]	cross-linked copoly(ethylene-*N*-hydroxymaleimide)
[PHNB]	(4-hydroxy-3-nitro)benzylated polystyrene
Phol	L-phenylalaninol
Ph_3P	triphenylphosphine
Pht	pthtaloyl
Picoc	4-picolyloxycarbonyl
Pms	4-toluylmethylsulfonyl
[PNP]	poly(4-hydroxy-3-nitrostyrene)
Poc	2-phenylpropyl(2)oxycarbonyl
Ppt	diphenylphosphinothioyl
PS	polystyrene
Py	pyridine
$(Pys)_2$	2,2′-dipyridyl disulfide
REMA	repetitive excess mixed anhydride (method)
–Ⓡ, –Ⓟ	resin support
Scm	*S*-carboxymethylsulfenyl
Smc	9-(2-sulfo)fluorenylmethyloxycarbonyl (Sulfmoc)
SPS	solid-phase synthesis
STI	soybean trypsin inhibitor
Tac	4-toluenesulfonylaminocarbonyl
*t*Bu	*tert*-butyl
Tce	β,β,β-trichloroethyl
TFA	trifluoroacetic acid
Tfa	trifluoroacetyl
TFE	trifluoroethanol
THF	tetrahydrofuran
Thp	tetrahydropyranyl
Thz	thiazolidine-4-carboxylic acid
tlc	thin-layer chromatography
Tos	*p*-toluencesulfonyl (tosyl)
Trt	triphenylmethyl (trityl)

* The number following PEG signifies average molecular weight.

uv	ultraviolet (spectroscopy)
VIP	vasointestinal polypeptide
Xan	9-xanthenyl
Z	benzyloxycarbonyl
Ztf	2,2,2-trifluoro-1-benzyloxycarbonylaminoethyl

Contents of Volume 1

MAJOR METHODS OF PEPTIDE BOND FORMATION

ERHARD GROSS AND JOHANNES MEIENHOFER

Chapter 1

Solid-Phase Peptide Synthesis

GEORGE BARANY and R.B. MERRIFIELD

THE PEPTIDES, VOLUME 2

ISBN 0-12-304202-X

I. INTRODUCTION

The science and art of peptide synthesis was founded at the beginning of the twentieth century by Emil Fischer and Theodor Curtius, and has steadily been developed into a discipline of great power and sophistication. The classical methodology has been reviewed by Greenstein and Winitz (1961), Schröder and Lübke (1965), Wünsch (1974), Bodanszky *et al.* (1976), Finn and Hofmann (1976), and in these volumes of "The Peptides." However, as recently as 15 years ago, the successful synthesis of biologically active materials still required considerable investments of time, effort, and manpower. Perhaps the most intimidating aspect of the field has been the need to purify, by discriminating methods, all intermediates of lengthy multistep routes, a non-routine and demanding task. The problem is made all the more challenging by the unpredictable solubility characteristics of the various intermediates.

Solid-phase peptide synthesis was conceived in 1959, at a time when chromatography on insoluble resins was being applied with great success to the analysis of amino acids, peptides, and proteins. It was reasoned that by attaching a growing peptide chain to an insoluble polymeric support, excess reagents and by-products from the synthetic cycles could be removed by simple filtration and washing steps. Thus, the purification problem of classical methods of peptide synthesis ought to be readily circumvented, while most of the highly honed chemistry of coupling and functional group protection was expected to be readily adaptable to the new strategy. As an added benefit, the entire solid-phase procedure was predicted to be amenable to automation. The feasibility of these ideas was first shown by the synthesis of the crystalline tetrapeptide L-leucyl-L-alanyl-glycyl-L-valine (Merrifield, 1962, 1963). At about the same time, Letsinger and Kornet (1963) reported the polymer-supported synthesis of L-leucyl-glycine, using a different chemical approach. Further developments followed soon thereafter in several laboratories, and a useful and practical set of methods has evolved that will be the subject of this review.

The solid-phase principle has had three main areas of achievement and influence.

1. At a time when there is an explosion of interest in the biochemistry and pharmacology of peptides, the nominal simplicity of the solid-phase approach has made peptide synthesis accessible for the first time to nonchemically oriented laboratories. Extensive structure–function studies through the preparation of analogs have contributed greatly to the understanding of the action of peptide hormones, toxins, and antibiotics (e.g., Stewart, 1968; Manning *et al.*, 1970; Wieland *et al.*, 1972b; Hruby *et al.*, 1973; Khosla *et al.*, 1974; Rivier *et al.*, 1975a; Rosenblatt *et al.*, 1976; van Batenburg *et al.*, 1976a; Gisin *et al.*, 1978; Granier *et al.*, 1978; Meyers *et al.*, 1978; Yamashiro and Li, 1978; Yamashiro *et al.*, 1978; also see Meienhofer, 1973, pp. 180–216, for an essentially complete tabulation of the pre-1972 literature).

2. The opportunity to functionally select desired chemical species based on their covalent attachment to polymeric phases of different physical properties has been elegantly applied in pure chemistry as a way to discern reaction mechanisms (e.g., Blackburn *et al.*, 1969; Rebek and Gaviña, 1975b) or to prepare elusive molecules of theoretical interest (e.g., "hooplanes," Harrison and Harrison, 1967). Similar ideas guided the evolution of such staples of biochemical technique as affinity chromatography (Lerman, 1953; Cuatrecasas and Anfinsen, 1971), enzyme immobilization (Axén *et al.*, 1967; Katchalski *et al.*, 1971; Mosbach, 1976) and solid-phase Edman degradation (Laursen, 1971, 1975). Solid-phase methodology has also been explored for the synthesis of oligonucleotides and oligosaccharides (e.g., Letsinger and Mahadevan, 1966; Hayatsu and Khorana, 1967). In the field of peptide synthesis, the use of soluble polymeric carriers (Shemyakin *et al.*, 1965; reviewed by Mutter and Bayer, this volume, Chapter 2) is an interesting offshoot of the original idea, and the application of polymeric reagents (reviewed by Fridkin, this volume, Chapter 3) has been considered as the reciprocal of supported synthesis. Finally, some of the objectives of the solid-phase principle, including rapid repetitive methods (Medzihradszky and Rádóczy, 1966; reviewed by Kisfaludy, this volume, Chapter 6) and facilitated purification of intermediates (use of 4-picolyl esters as "handles" by Young and co-workers, e.g., Camble *et al.*, 1968), have been achieved in solution without relying on polymeric supports.

3. A major goal of peptide chemistry is the synthesis of proteins with biologically specific activities. At the present time, such undertakings are far from routine and require careful planning and skillful execution, regardless of whether solid-phase, solution, or semisynthetic strategies are selected (overview by Merrifield, 1973a). Nonetheless, the initially encouraging results (Gutte and Merrifield, 1969, 1971; Hirschmann *et al.*, 1969; Ontjes and Anfinsen, 1969a) indicate the promise of chemical synthesis for contributing

to our understanding of the mechanisms of protein folding and structural organization, of the binding of proteins to organic ligands, peptides or other proteins, and of enzyme action.

Solid-phase peptide synthesis is by no means a panacea; a number of practical as well as theoretical limitations are well appreciated. It seems safe to assert that with current methods, the solid-phase assembly of moderate-sized peptides containing up to 40 amino acid residues can generally be reliably and conveniently carried out; in several instances, larger molecules have been made. Careful purification and analytical characterization leads in most cases to good products that permit meaningful conclusions to be drawn in subsequent biochemical studies.

Solid-phase peptide synthesis has been thoroughly reviewed by Merrifield (1969, 1973b), Meienhofer (1973), and most recently by Erickson and Merrifield (1976). The latter authors also listed (pp. 258–259) earlier reviews of the subject. Treatments appearing since then, with varying points of emphasis, include those of Atherton *et al.* (1979b), Birr (1978), Doscher (1977), Manning (1976), Marnett *et al.* (1976), Merrifield *et al.* (1977b), Sheppard (1977), and Stewart (1976). A book by Stewart and Young (1969) dealing with laboratory aspects of the technique continues to be very useful, and an updated edition would be most welcome.

In this chapter, we propose to discuss critically the scope and limitations of the best available procedures for tackling the synthesis of target peptides by the solid-phase or allied approaches. At the same time, we wish to mention important new developments and trends that foreshadow future levels of development. Following the editorial policy of this treatise, the references are weighted whenever possible toward detailed reports with full experimental procedures.

II. THE SOLID-PHASE PRINCIPLE

The key ideas and features of the solid-phase principle are fourfold: (1) the peptide is synthesized while it is covalently attached to a polymeric support; this allows ready separation of the product from by-products; (2) reactions of the polymer-supported peptide chains can be driven to completion through the use of excess soluble reagents; (3) mechanical losses are avoided by retaining the peptide–polymer beads in a single reaction vessel throughout the synthesis; and (4) physical manipulations are easy, rapid, and amenable to automation.

Specific steps in carrying out solid-phase peptide synthesis are outlined below in connection with Fig. 1, taken from the first publication on the subject. The *polymeric support* chosen was a polystyrene (2% crosslinked with divinylbenzene) resin bead that had been first chloromethylated (*functionalization*), and further modified by nitration (line 1). The carboxyl terminal residue (R^1) was esterified (*anchoring*) to the support (line 2), and an acid step (*deprotection*) served to liberate a free amino group (line 3) which had been temporarily blocked by the benzyloxycarbonyl (Z) group of Bergmann and Zervas (1932). The next N^α-benzyloxycarbonyl-protected

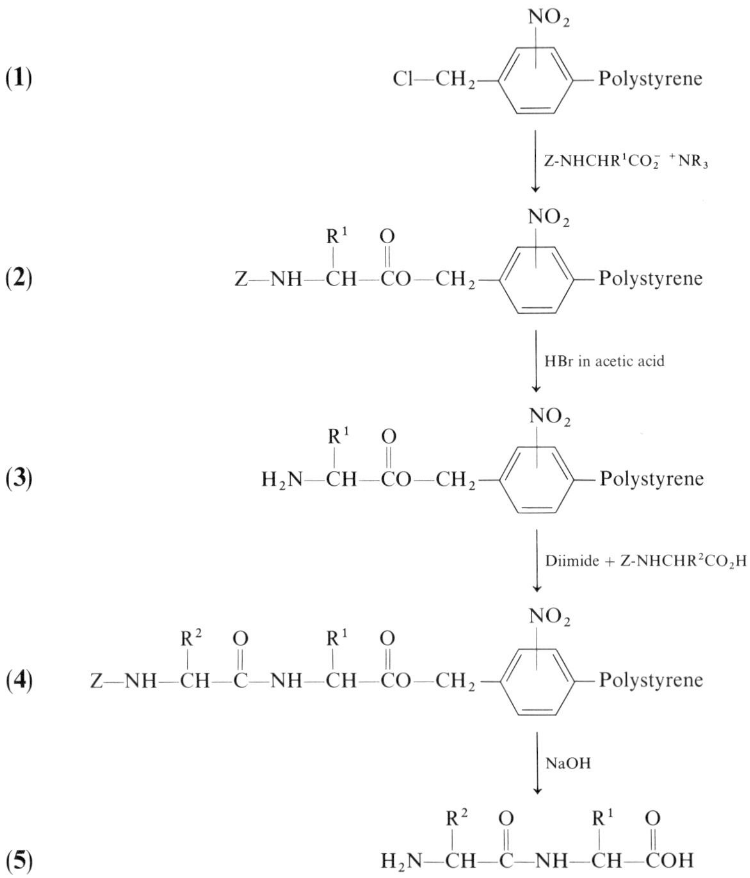

Figure 1. The original scheme for solid-phase peptide synthesis, redrawn from Merrifield (1963). Note that some of the intermediate chemical steps have not been shown. By the following year (Merrifield, 1964a), the α-amino protecting group was changed from Z to Boc, the deprotecting reagent was changed from HBr in acetic acid to HCl in acetic acid, the nitrobenzyl ester anchoring linkage was replaced by a simple *p*-alkylbenzyl ester, and the reagent for cleavage from the support was changed from NaOH to HBr in TFA. This *latter* scheme (not the early one illustrated) is referred to as "standard," and with certain variations and improvements to be described in this chapter, remains the most widely used up to the present day.

amino acid residue (R^2) was incorporated (line 4) with the aid of dicyclohexylcarbodiimide (*coupling*). To ensure that the amino group of R^1 would be completely covered, these soluble reactants were both used in excess and were readily removed after completion of the reaction by filtration and washing of the Z-dipeptidyl-resin. Finally, the free peptide was liberated into solution (line 5) by saponification (*cleavage*). Further amino acid residues could be introduced into the growing polypeptide chain by repetition of the deprotection and coupling steps (one *cycle*).

Two specific problems were observed in the application of this original scheme. They typify some of the practical difficulties that can arise in solid-phase peptide synthesis, and much of the developmental research effort of the field has been aimed at eradicating or minimizing their occurrence.

1. The coupling step could not be driven to completion under the conditions used. Since renewed chain growth at a temporarily unreactive site at later stages of the synthesis would lead to the undesirable appearance of deletion peptides, an acetylation step was added to cover the free amino groups remaining after coupling. The acetylated peptides were easily separated during purification of the crude product mixture by ion-exchange chromatography. In later work using improved chemical procedures and polymeric supports, coupling yields were found to be nearly quantitative, and the acetylation step was omitted.
2. The nitrobenzyl ester-anchoring bond was stable to acid and required excessively harsh treatment with alkali for cleavage of the final product from the support at the end of the synthesis. The given anchor was selected because exploratory experiments had shown that the acid-labile bromobenzyl or unsubstituted benzyl ester linkages were not compatible with the N^{α}-benzyloxycarbonyl group and gave rise to premature cleavage of peptides from the support. These problems were circumvented by a frame-shift of the *protection scheme*, whereby the simple *p*-alkylbenzyl ester anchor, cleavable by HBr in acetic acid, could be employed in conjunction with the more acid-labile *tert*-butyloxycarbonyl (Boc) group, removable by HCl in acetic acid (Merrifield, 1964a; see Fig. 1 legend). The selection of the *tert*-butyloxycarbonyl–benzyl combination of protecting groups and anchoring linkages was the first step in the evolution of the technique. Improved combinations of labile temporary protecting groups and reversible anchoring linkages continue to be developed in the search for greater selectivity in the solid-phase synthesis of peptides.

A. Generalized Solid-Phase Strategy

As has often been pointed out (e.g., Marshall and Merrifield, 1971), the solid-phase principle (Fig. 1) is suited not only for the synthesis of peptides,

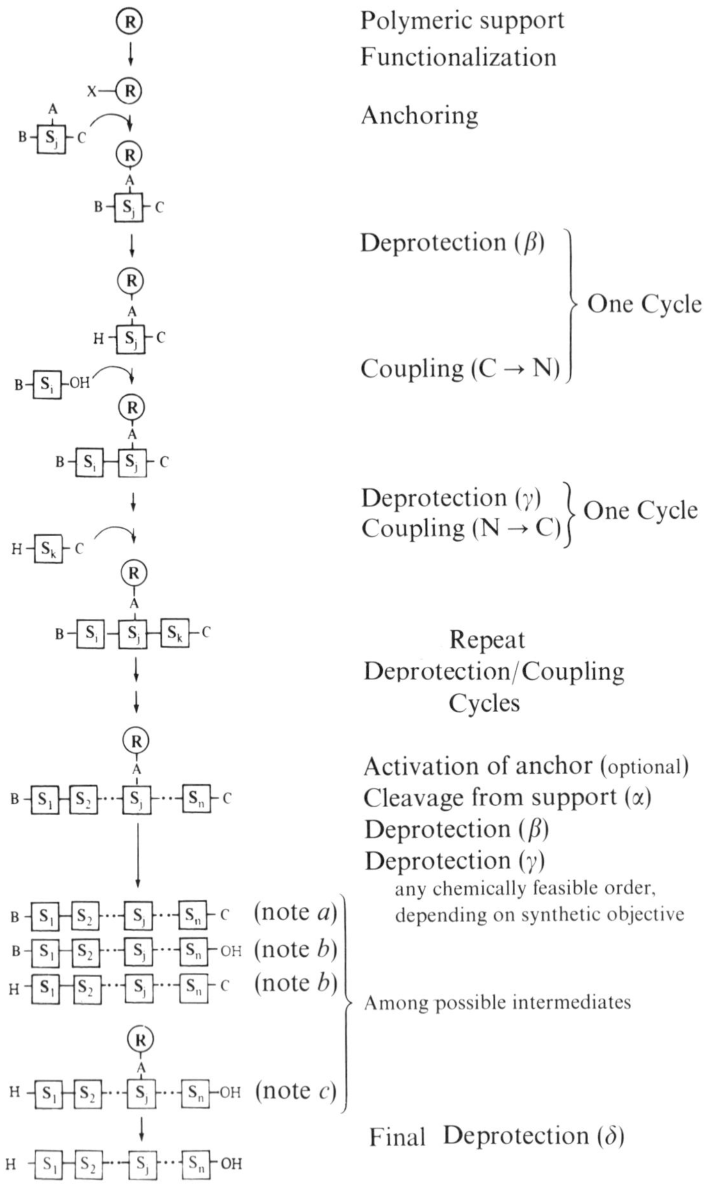

Figure 2. A general bidirectional polymer-supported synthesis of linear peptides. For this presentation, a synthetic unit S_τ is defined as an amino acid or peptide segment residue [..(NHCHR)$_x$..]. $S_1, S_2, \ldots, S_i, S_j, S_k, \ldots, S_n$ = set of n synthetic units, in their exact final order (NH_2- to COOH-terminal); boldface means that each such unit has a full complement of permanent protecting groups removed only at the end of the synthesis; R = support; A = anchoring group; B = temporary amino protecting group; C = carboxyl protecting group; H– = free amino group; –OH = free carboxyl group; $\alpha, \beta, \gamma, \delta$ = selective cleavage conditions for removal, respectively, of A, B, C, or all of the permanent protecting groups.

[a] Fully protected peptide.

[b] Partially protected peptide suitable for segment condensation.

[c] Anchored peptide, suitable for cyclization.

but for the synthesis of a number of other classes of oligomeric compounds as well. Much of what will be stated in the following about peptides is also applicable to oligonucleotides, oligosaccharides, polyamides, and polyesters. The solid-phase principle has also been adapted to a variety of single and multistep synthetic organic reactions. However, more specific consideration of these other synthetic targets is beyond the scope of this chapter (see Erickson and Merrifield, 1976, pp. 453–476, for a recent discussion).

A number of variations have been schematically diagrammed and discussed, with specific illustrative examples (Merrifield, 1973b). Some of these ideas have reached advanced levels of practical implementation, while others have only been demonstrated for simple models. In the present treatment, we show the interrelationships of the different strategies that have been explored. Thus, a general scheme for the polymer-supported synthesis of linear peptides is presented in Fig. 2. Details of the chemical implementation of the scheme will be described in subsequent sections.

1. The Polymeric Support

A central feature of the solid-phase principle is the opportunity to override the unpredictable physical characteristics of synthetic intermediates by covalent attachment to a polymeric carrier. Excess reagents and by-products may thus be separated from the desired peptide chain on the basis of the physical properties of the polymer. The term *solid-phase* refers to the most widely applied variation of the general idea, in which insolubility of the polymer is the basis of the separations, usually achieved by filtration or centrifugation. Beginning with Shemyakin *et al.* (1965), several authors have suggested, and experimentally examined, the potential benefits of conducting polymer-supported peptide synthesis in a homogeneous, rather than a heterogeneous system. These *liquid-phase* approaches rely on separation techniques such as membrane ultrafiltration, precipitation, or crystallization for the removal of excess low-molecular-weight reagents. Finally, two polymeric supports may be used in combination, to great advantage in the simultaneous synthesis of two peptides under essentially identical experimental conditions (Tregear, 1972; see van Rietschoten *et al.*, 1975a, and Hugli and Erickson, 1977, for applications, and Rebek *et al.*, 1975, for mechanistic studies with the same theme). At the end of the synthesis, separations may be achieved on the basis of differences in density (solvent flotation, e.g., Tregear, 1972; Rebek and Gaviña, 1975b); size (screening, e.g., Rebek *et al.*, 1975); solubility (Frank and Hagenmaier, 1975; Jung *et al.*, 1975); or phase immiscibility (Albertsson, 1971; not applied to peptide synthesis to date). Regardless of the choice of polymeric support and the attendant physical strategy for accomplishing separations, the remaining chemical considerations for successful applications to peptide synthesis are similar.

2. Functionalization and Anchoring

The introduction of functional groups suitable for further chemical elaboration is usually accomplished by derivatization of preformed polymers, although, in a number of cases, a functionalized monomeric unit may be copolymerized in the course of preparation of the support.

A variety of anchoring strategies have been used. In the formulation of Fig. 2, a given protected amino acid or peptide S_j eventually becomes linked to the support via an anchoring group A. As drawn, A is first connected to S_j, and then reacted with the functionalized polymer; alternatively, the anchoring group may have its origin on the support. Similarly, upon cleavage α, the anchoring group A may either leave with the assembled peptide chain or remain on the support. The requirement for complete stability of the anchor is not as stringent as that for the permanent protecting groups, since any prematurely cleaved peptide is released into solution and does not contaminate the final product. However, such processes lead to decreased yields, and in addition several low-level side reactions might result from such losses due to the liberation of new functional sites (see p. 33).

Since the initially anchored unit S_j is the limiting reactant for all subsequent reactions of an n-step solid-phase synthesis (Fig. 2), its quantitative amount is often used as a basis for calculations of overall yield.*

3. The Repetitive Deprotection/Coupling Cycle

The characteristic feature of the solid-phase approach is the attachment of a starting synthetic unit (S_j in Fig. 2) onto a polymeric support, as outlined above. Once this has been accomplished, the peptide chain can be extended by selective removal of the appropriate temporary protecting groups, followed by coupling of suitably protected additional synthetic units. Much of the wealth and range of classical methods of peptide chemistry can be drawn on at these stages. On the other hand, the idea of driving reactions to completion by the use of excess reagents is one to which the solid-phase method is uniquely suited. Furthermore, due to the repetitive nature of these steps, the solid-phase method readily lends itself to automation.

* Following Bodanszky (1970), the *efficiency* E of a solid-phase synthesis may be defined as

$$E = \frac{M(\mathrm{Pr}) \cdot 100}{\frac{1}{n} \sum_{\tau=1}^{n} M(S_\tau)}$$

where the function $M(\mathrm{X})$ represents the number of moles of substance X, and Pr indicates isolated product. The formula for E will reflect the excesses of reagents routinely employed to drive reactions to completion; these excesses become an important consideration for certain synthetic applications including industrial scale-up or incorporation of valuable isotopically

4. Cleavage from the Support

Once a protected peptide chain has been satisfactorily assembled on the polymeric support, the anchoring linkage must be cleaved (α in Fig. 2) to release the crude product into solution. Normally, the peptide–resin is treated directly with an appropriate cleavage reagent, but occasionally a "safety-catch" option may be exploited. In the latter case, an anchor that is entirely stable throughout all normal operations of the synthesis becomes activated, via a separate chemical reaction, into a labile and cleavable form. The protection scheme, and the actual cleavage reagent used, will determine whether the peptide is obtained in the fully or partially protected, or free form, and whether the COOH-terminal residue is the free acid or, for example, the amide, ester, or hydrazide. Finally, since the cleavage from the support is only performed once, at the end of a solid-phase synthesis, it is not essential that this step proceed in quantitative yield.

5. Selection of the Protection Scheme

The flexibility of stepwise schemes of solid-phase peptide synthesis as generalized in Fig. 2 depends on the nature of the protecting groups selected, along with the corresponding chemical methods for their removal. In its simplest formulation (monodirectional chain extension, as exemplified in Fig. 1), two levels of protecting group stability are required. Quantitative and selective cleavage (β) of the temporary group B must be repeatedly achieved under conditions of reagent, solvent, temperature, and time that leave all of the permanent protecting groups, as well as the anchor A, unaffected. These latter groups may then all be removed under the same set of conditions (α, γ, and δ of Fig. 2 combined into a single step).

In practice, the straightforward strategy outlined above is by far the most commonly applied. Furthermore, the same chemical mechanism (usually acidolysis) is often exploited to remove both classes of protecting groups. Therefore, selectivity must be achieved by kinetic "fine-tuning," i.e., a dependence on quantitative rate differences in mechanistically similar reactions. An often-limiting consequence of this approach is the relative severity of the cleavage conditions required for the last step.

---

labeled residues. However, since some current methods allow the recovery of unreacted reagents, we suggest the modified expression

$$E^* = \frac{M(\mathrm{Pr}) \cdot 100}{\frac{1}{n}\sum_{\tau=1}^{n} [M(S_\tau) - R(S_\tau)]}$$

where $R(\mathrm{X})$ represents the number of moles of substance X recovered from coupling steps, and suitable for reuse.

A more general strategy would make use of two or more mechanistically different types of chemical reactions for removal of the various protecting groups. Thus, an *orthogonal system* has been defined as a set of completely independent classes of protecting groups, such that each class of groups can be removed in any order and in the presence of all other classes (Barany and Merrifield, 1977). An orthogonal protection scheme will, in general, permit the use of substantially milder deprotective conditions, since selectivity can now be attained on the basis of differences in chemistry rather than in reaction rates. Similar approaches to the design and selection of protecting groups have been discussed, either implicitly or explicitly, by numerous authors (e.g., Greenstein and Winitz, 1961; Schröder and Lübke, 1965; Veber *et al.*, 1972a; Kemp *et al.*, 1975). Among the reagents that can be considered as candidates to mediate orthogonal modes of deprotection are (1) acids (anhydrous or hydrolytic), (2) bases (anhydrous or hydrolytic), (3) reducing agents (including hydrogen over a catalyst, metals and their hydrides, and thiols), (4) oxidizing agents, (5) nucleophiles, (6) electrophiles, (7) ions, (8) electrons, (9) light, (10) solvents, and (11) enzymes. Any given organic functionality may be susceptible to several of these under appropriate conditions. In order for a proposed protection scheme to be of value for peptide synthesis, the conditions of removal for the various classes of protecting groups need to be specific, mild, and nondestructive to the polypeptide backbone and side chains. Although an entirely orthogonal system, as ideally required for the full generality of Fig. 2, has not yet been applied to the solid-phase synthesis of a complex peptide, the development of such combinations of protecting groups is underway (Barany and Merrifield, 1977; Merrifield *et al.*, 1977b; Atherton *et al.*, 1978a,b, 1979a; Chang and Meienhofer, 1978).

B. Specific Variations of Solid-Phase Strategy

Table I summarizes the variations of the solid-phase principle that have been proposed and tested since the development of the initial concepts. Representative peptides are listed to suggest the scope of present practical applications of the different strategies. Each variation in Table I represents a special case of the generalized scheme of Fig. 2. Recent discussions with a different emphasis are those of Merrifield (1973b) and Erickson and Merrifield (1976, pp. 262–269).

1. Stepwise Condensation

In the stepwise mode the COOH-terminal (or NH_2-terminal) amino acid of the desired peptide is anchored to the support, through its carboxyl (or amino) group, and monodirectional chain growth proceeds by repeated coupling of successive protected amino acid units. This strategy was the first one examined (Fig. 1) and remains by far the most frequently used. The major theoretical objection to this approach lies in the possibility of creating closely related (and hence, difficult to separate) peptide by-products that remain attached to the polymeric support, along with the desired peptide, until the end of a synthesis.

2. *Segment Condensation*

The solid-phase coupling of protected peptide segments* has recently been successfully applied to a number of significant synthetic problems (Table I). The overall advantages and disadvantages of such segment condensation strategies are well known from the field of peptide synthesis by solution methods (reviewed by Finn and Hofmann, 1976, pp. 127–128). Chief among the positive attributes are a simplification of the subsequent analytical and purification problems; difficulties that occasionally arise include sparing solubilities of intermediates, marginal yields, rearrangements, and the risk of partial racemization during coupling steps. It seems possible that as improved experimental procedures become available, segment condensation may become the method of choice for the preparation of moderate- to large-sized peptides by solid-phase techniques.

3. *Preparation of Protected Peptide Segments*

Protected peptide segments are required for segment condensation. All four possible permutations of synthetic strategy have been demonstrated; i.e., protected peptides prepared in solution have been further coupled either in solution or on a solid phase; and protected peptides prepared by solid-phase methodology have been purified and subsequently condensed either in solution and on a solid phase. Practical methods currently exist for obtaining any of the variety of partially protected intermediates shown in the penultimate lines of Fig. 2. Their preparation requires at least three classes of selectively removable protecting groups, so that both the temporary and permanent groups remain unaffected upon cleavage of the protected peptide from the support.

4. *Side-Chain Anchoring*

Both protected amino acids and peptides have been attached to solid supports via side-chain functional groups. Schemes based on this idea

* We concur with the aptly descriptive term "segment" introduced by Pettit (1976) and recommended by Bodanszky (1977) to describe oligopeptide intermediates. The standard peptide synthetic literature to date makes use of the term "fragment" to express the same meaning.

Table I. Variations of the Solid-Phase Principle[a]

No.	Name	Case	Remarks	Examples[b]	References
1.	Stepwise condensation	All S_τ = amino acids			
1a.	C → N	j = n, A = C	Original, and most commonly useful scheme	Leu-Ala-Gly-Val (>99%[c]) Ribonuclease A (124, 17%[d])	Merrifield (1963) Gutte and Merrifield (1971)
1b.	N → C	j = 1, A = B	Usually less preferred than No. 1a for specific chemical reasons; see, however, No. 2b	Leu-Gly (31%) Leu-Ala-Gly-Val (30%)	Letsinger *et al.* (1964) Felix and Merrifield (1970)
2.	Segment condensation	All S_τ = peptides	Protected peptides prepared in solution or by No. 3; coupling reagents or strategies required that minimize the risk of racemization		
2a.	C → N	j = n, A = C		Valinomycin (12 = 2 + 2 + 2 + 2 + 2 + 2, ~100%[e], after solution cyclization, 33%[f]) Trypsin inhibitor (58 = 2 + 19 + 9 + 16 + 12, ~80%[e], 8%) Glucagon (27 = 1 + 1 + 1 + 5 + 6 + 9 + 6, >95%[e], 17%[f])	Gisin *et al.* (1969) Yajima and Kiso (1974) Protein Synthesis Group (1975);
2b.	N → C	j = 1, A = B		ACTH (1–24) (24 = 2 + 5 + 3 + 4 + 5 + 5, >98%[e], 25%)	Maruyama *et al.* (1976)
3.	Preparation of protected peptide segments	Selectivity $\beta < \alpha < \delta$[g] (ideally *orthogonal*)	For purification and further coupling: (i) in solution, or (ii) on solid-phase as in No. 2; See also No. 5a (safety-catch)	Z-Phe-Val-Ala-Leu-NHNH$_2$ (89%, 76%[f]) Linear protected decapeptide hydrazide (98%, final yield of cyclic Gramicidin S = 50%[f]) Boc-Ser(Bzl)-Gln-Arg(NO$_2$)-His(Bzl)-Gly-OH (85%, 67%[f])	Wang and Merrifield (1969b) Ohno *et al.* (1971) Barton *et al.* (1973)
4.	Side-chain anchoring	A ≠ B or C			
4a.	Monodirectional	j = 1 or n	Anchoring residue in NH$_2$- or COOH-terminal segment	[Lys[8]]-Vasopressin (9, 25%)	Meienhofer and Trzeciak (1971)

4b. Bidirectional	$1 < j < n$	In principle, most flexible approach	*p*Glu-His-Pro-NH_2 (49%) Oxytocin (9, 15%)	Glass *et al.* (1972, 1974)
4c. Precursor of cyclic peptides	$\beta, \gamma < \alpha^g$	Potentially valuable, compare to No. 5b	[Gly^5, Gly^{10}]-Gramicidin S (cyclic-10, 20%)	Sklyarov and Shashkova (1969)
5. Safety-catch anchoring				Rudinger (1963)
5a. For linear peptides	Activation prerequisite to cleavage α	Useful for preparation of protected peptide segments	Boc-Met-Leu-Gly-OH (92%, 75%[f])	Kenner *et al.* (1971)
5b. For cyclic peptides	Activation prerequisite to β or γ	(i) Peptides prepared in solution anchored onto support and cyclized concomitant with cleavage off support	c-$(Ala)_4$ (65%)	Fridkin *et al.* (1965); Patchornik *et al.* (1967)
		(ii) Peptides synthesized on unactivated support by normal procedures, then cleaved by activation and cyclization.	c-(Gly-Val-Ala-Phe-Ala-Gly) (50%[f])	Flanigan and Marshall (1970)
		Racemization may be a problem with both these approaches		
6. Dual anchoring	Both components to be coupled are anchored onto polymeric supports	Excess amino and carboxyl components removed in separate steps	Luteinizing hormone releasing hormone (9, 26%)	Frank *et al.* (1977a,b)

[a] Nomenclature of Fig. 2.
[b] Values in parentheses refer to number of amino acid residues (if not self-evident), followed by overall isolated yields.
[c] By analytical ion-exchange chromatography of crude product mixture, Kent *et al.* (1978).
[d] Yield based on resin-bound protein remaining after completion of assembly of the chain (123 deprotection/coupling cycles). Overall yield based on first amino acid residue = 3% (specific activity 78% of native enzyme).
[e] Yields of individual segment coupling steps.
[f] Isolated as crystalline material.
[g] Notation $\beta < \alpha$ means that protecting groups **B** can be removed by reaction β under conditions that do not cleave anchor A.

provide a considerably greater amount of flexibility in the planning and execution of syntheses, by comparison to standard stepwise strategies. Side-chain anchoring is especially suited, in principle, to the preparation of cyclic peptides (Fig. 2, note c). In their complete generality, the subvariations 4a–c of Table I require a more substantial range of selectively removable protecting groups than is presently available.

5. Safety-Catch Anchoring

The principles of this approach have already been described (p. 11). As shown in Table I, it provides a promising method for the preparation of a number of protected peptide segments or cyclic peptides.

6. Dual Anchoring

Frank and Hagenmaier (1975) introduced an *alternating liquid–solid phase* method of peptide synthesis. In the context of Fig. 2, two polymeric supports, I and II, are utilized satisfying the following criteria: (1) support I (the liquid phase) and support II (the solid phase) are physically separable from one another; (2) each molecule of support I serves as the carrier for a *single* growing peptide chain,* anchored through the group A; (3) the incoming synthetic unit is linked to support II, through its protecting group B. Then, after carrying out the coupling step, the ensuing fully protected peptide will be attached to both supports (much like a hammock swinging between a tree and a house); and (4) the "cross-linked" species behaves physically like support II.

Removal of excess support I by filtration, followed by cleavage (β) of the protecting group linked to support II, allows unambiguous isolation solely of the extended peptide, anchored onto support I. The overall result is removal of the excesses of both the amino and carboxyl components in separate steps. Kemp *et al.* (1975) have discussed progress toward the same goal, termed "affinity purification," in solution. These approaches circumvent the need for quantitative coupling yields, which is a characteristic of stepwise condensation strategies; nonetheless, high ($>95\%$) yields are still desirable to ensure that a reasonable amount of product may still be isolated after several repetitions of coupling cycles.

* If more than one growing peptide chain is anchored onto a single molecule of support I, the advantage of step (3), described in the text, is defeated because as soon as one attachment point is established, other chains will be carried along, irrespective of whether the coupling step was successful. The alternating liquid–solid phase method is viable so long as support I (including the growing peptide chain) can effectively penetrate the matrix of support II, thereby permitting reactions to proceed with minimal diffusion barriers. It is worth remarking that sufficiently large, protected peptide molecules could, in principle, serve as their own carriers (pointed out in other contexts; e.g., Bodanszky, 1960; Bodanszky *et al.*, 1967; Morley, 1967).

C. Conceptual Ramifications

Various features of solid-phase peptide synthesis are summarized in Table II. A number of advantages were anticipated from the inception of the idea; however, each of these is accompanied by an often-demanding set of requirements which must be satisfied before the method may be reasonably expected to give good results. Significant difficulties have been encountered sufficiently often to conclude that no synthetic effort can automatically be assumed to have provided the correct product free of all by-products. For this reason, careful analytical control is strongly recommended. In our judgment, the scope of the solid-phase method is primarily defined by the flexibility of chemical methods that can be brought to bear on a problem, rather than by physical–chemical constraints resulting from the carrying out of reactions on a polymeric support.

In the remainder of this section, we explore the consequences, and evaluate the experimental status, of efforts to satisfy some of the requirements for solid-phase peptide synthesis. Meienhofer (1973) presented a thorough theoretical discussion of these questions under the heading "intrinsic problems", which, however, ought not be taken to imply that such problems cannot be solved.

1. *Physical Chemistry of the Solid Phase*

Individual polymeric particles used as supports for solid-phase synthesis may each contain as many as 10^{15} sites for peptide chain growth; this is in marked contrast to some carriers for liquid-phase synthesis (including those for a dual anchoring strategy), which may contain only one site. The larger loadings are a consequence of the mechanical methods used to effect solid-phase separation, which generally require particle sizes $\geqslant 16\ \mu m$ (Stark, 1971 p. 187). Furthermore, a simple calculation (p. 24) shows that peptide chain growth cannot be restricted to the actual spherical surface area of even the smallest particles, because of the very low capacity that would result.

To achieve adequate reaction rates, it is required that all functional sites at which elaboration of peptide chains takes place be located in well-solvated, noncrowded, and accessible regions of the solid support, where diffusion or steric barriers are expected to be minimal. The static and dynamic microstructure of the polymeric matrix becomes critical in this regard. Related aspects of the problem are the capacity and potential for interaction of polymer-supported sites.

a. Swelling. Interaction of the dry polymeric matrix with appropriate solvents often leads to dramatically expanded structures. The extent of swelling may be quantitated by a variety of techniques, including: (1) volumetric (Heitz and Platt, 1969; Wieland *et al.*, 1969a); (2) gravimetric after centrifugation or rapid filtration (Pepper *et al.*, 1952; Gregor *et al.*, 1965);

Table II. Features of Solid-Phase Peptide Synthesis

Advantages	Requirements[a]	Auxiliary[b]
No solubility problems for intermediates; no manipulative losses and a controlled environment due to single reaction vessel	Adequate functionalized polymeric support: (a) insoluble in all solvents (b) chemically and physically stable (c) all reactive sites accessible to solvents and reagents (d) reasonable capacity	Model syntheses to test supports (o)
Reactions may be driven to completion with excess soluble reagents, which are removable by filtration and washing	Quantitative yields for all repetitive steps, i.e., deprotection/coupling cycles; reaction rates should be comparable to those observed in homogeneous phase	Monitoring of reaction rates and yields (o, r)
Chemical side reactions giving soluble or permanently bound by-products do not affect purity of synthetic product. Novel reagents for deprotection or coupling may be uniquely compatible with solid-phase methodology	Minimal side reactions (affecting structure of peptide backbone and side chains): (a) racemization-free coupling (b) compatible protecting groups (c) minimal cleavage, modification, or rearrangements, especially during final deprotection steps	Thorough purification of crude product from end of synthesis (n); test final product by battery of analytical methods (n)
	Purity of: (a) starting amino acid derivatives (chemically and optically) (b) reagents and solvents	Analytical control (n) and purification if necessary
Repetitive nature of method lends itself to automation		Feedback control of synthesis (o)

[a] When these requirements cannot be satisfied, they are disadvantages.
[b] Code: (o) = optional, (r) = recommended, especially for trouble-shooting, (n) = necessary.

(3) direct microscopic examination (Gutte and Merrifield, 1971); and (4) exclusion of high-molecular-weight substances (Rudinger and Buetzer, 1975).

Rudinger and Buetzer (1975) have noted that polymer-bound reactants in solid-phase synthesis are in reality confined to the gel phase or a part of it. Thus (as a first approximation), the effective concentrations of reactants can be estimated with respect to the amount of solvent taken up by the support on swelling. The rate of disappearance of a soluble reactant from the bulk solution is then given by the product of the "true" second-order rate constant, the effective concentrations of each of the soluble and polymer-bound reactants, and the ratio of volumes of the reaction phase and the bulk solvent. When partition effects are neglected (see the following), this expression reduces to the intuitive "apparent" second-order rate law in which the volume of the total bulk solvent is used to calculate concentrations of both soluble and polymer-bound reactants.

b. Partition. The macroscopic phenomenon of "partition" has long been recognized (Davies and Owen, 1956; Gut and Rudinger, 1968; R. B. Merrifield, unpublished observations). Experimentally, the concentration of a given solute in the external bulk solvent phase is determined before and after mixing with a dry, chemically inert polymer. Differences in either direction have been observed under suitable conditions.

A distribution coefficient, K_{AXL}, may be defined according to Rudinger and Buetzer (1975) as c_{AX}/c_{AL}, where A is the solute, c is concentration, X is the reaction phase, and L is the remaining bulk solution. A coefficient <1 implies selective exclusion of solute from the polymer relative to solvent; >1 implies adsorption. The reported values ranged most typically from 0.7 to 1.5, and were a function of the resin, solute, and solvent. At low concentrations ($\leqslant 1.5$ mM), certain solutes showed markedly decreased values of the partition coefficient.

As a practical application of the partition phenomenon, the effective concentrations of soluble reactants in solid-phase peptide synthesis may often be different from the nominal bulk-phase concentrations, leading to corresponding discrepancies between the true and apparent kinetic patterns (see p. 30). The problem may be accentuated when two soluble reactants (e.g., protected amino acid and *in situ* activating agent) are employed, if the relative stoichiometry is critical.

c. Morphology of Polymeric Particles. A variety of structural forms result from the application of different polymerization protocols in the preparation of solid-phase supports (reviewed, with specific documentation, by Merrifield, 1969; pp. 226–233; Erickson and Merrifield, 1976, pp. 269–281). These are summarized in Table III. Specific supports of demonstrated or potential usefulness for solid-phase peptide synthesis are listed in Table IV.

Table III. Morphologies of Solid Supports

Name	Permeability[a]	Swelling[b]	Capacity	Remarks	References
Gel (microporous)	+ + +	+ + +	+ + +	Sites uniformly distributed throughout polymeric matrix; $\rho_0 \simeq \rho_s$ (note *c*)	Altgelt and Segal (1971)
Macroporous	+ +	±	+	Sites restricted to the substantially increased surface area of otherwise rigid structure; $\rho_0 < \rho_s$ (note *c*)	Kunin *et al.* (1962); Millar *et al.* (1963); Haller (1965); Seidl *et al.* (1967)
Popcorn	+ +[a]	±	+ +	Expanded rigid network structure with many voids	Letsinger *et al.* (1964); Breitenbach (1968)
Pellicular				Impermeable core, sites restricted to mobile shell which is still thick on molecular scale	Tregear *et al.* (1966); Horvath *et al.* (1967)
Core	–	–	–		
Shell	+ + + +	+ + +[d]	+ +		
Ghost[e]	+ +	?	+	"Inverted" macroporous, with shell, but no core	Meyers and Royer (1977)

[a] The term "permeability" is used loosely to refer to a variety of related phenomena, including accessibility and mobility (vs. rigidity) of chains. In general, permeability is decreased by increased cross-linking of polymeric matrix; the very low cross-linked popcorn polymers are an exception.

[b] Swelling in compatible solvents; decreased by increased cross-linking.

[c] ρ_s = true skeletal density (helium displacement), independent of macrostructure. ρ_0 = apparent density (mercury displacement). Note, for statistical hexagonal packing of uniform spheres, the dry volume of 1 gm resin beads $\sim 1.6/\rho_0$.

[d] The shell of a pellicular support is assumed to be well solvated, but conventional volume expansion experiments to measure swelling in this system must be interpreted with caution.

[e] Not yet applied to peptide synthesis.

Table IV. Selected Polymeric Supports for Solid-Phase Peptide Synthesis[a]

No.	Chemical class	Cross-linkage[b]	Morphology[c]	Dimensions	Suppliers/ trade name[d]	Remarks[e]	References
1.	Polystyrene	2% Divinylbenzene	Microporous gel (suspension copolymer)	50 μm Beads	Dow Chemical[f] Bio-Rad/ Biobeads S-X2 and several other companies	First successful support	Merrifield (1963)
2.	Polystyrene	1% Divinylbenzene	Microporous gel	150 μm Beads	Bio-Rad/Biobeads S-X1; chloromethylated form also from Lab Systems, Pierce, Fluka, etc.	Support of choice for large majority of applications	Gutte and Merrifield (1971)
3.	Polystyrene	0.5% Divinylbenzene	Microporous gel	Not specified "fine" beads	X	High swelling fragile beads require a special centrifugal reactor	Birr (1973); see also Birr and Lochinger (1971)
4.	Polyacrylamide	5% *N*,*N*′-methylene-bis-acrylamide	Microporous gel	100 μm Beads	Bio-Rad/Biogel P	N, but used for enzyme immobilization	Inman and Dintzis (1969); Hjertén (1971)
5.	Polydimethylacrylamide-co-Boc-βAla-*N*′-acryloyl-hexamethylenediamine (0.4 mmol βAla/gm)	14% *N*,*N*′-ethylene-bis-acrylamide	Microporous gel (emulsion copolymer)[g]	Beads or amorphous	X	Support suggested to have solvation properties similar to peptide chains	Altherton *et al.* (1975); see also Arshady *et al.* (1979)[h]
6.	Poly(*N*-acryloylpyrrolidine)-co-*N*-acryloylhexamethyl-enediamine (11 mol %)	5% *N*,*N*′-ethylene-bis-acrylamide	Microporous gel (reverse-phase suspension copolymer)	Entirely beaded, 150 μm	X	Favorable solvation in solvents over wide range of polarity, including water	Smith *et al.* (1979)

Table IV. *(Continued)*

No.	Chemical class	Cross-linkage[b]	Morphology[c]	Dimensions	Suppliers/ trade name[d]	Remarks[e]	References
7.	Polydimethylacrylamide	2% *N,N'*-dimethyl-*N,N'*-ethylene-bis-acrylamide	Microporous gel	100 μm Beads	X	N, but applied to solid-phase nucleotide synthesis	Köster and Heidmann (1976)
8.	Polydextran (hydroxypropylated)	Varied, with epichlorhydrin	Microporous gel	50 μm Beads	Pharmacia/ Sephadex LH-20	Limited experience summarized	see Erickson and Merrifield (1976, pp. 277–279)
9.	Polystyrene–polyethylene glycol graft	Divinylbenzene	Microporous gel	50 μm Beads	X	N, "cosolvent" features have been discussed	Regen and Dulak (1977)
10.	Polystyrene	~20% Divinylbenzene	Macroporous, average pore diameter = 200 Å	400 μm Beads surface area = 100 m^2/gm	Rohm & Haas/ Amberlite; improved supports now available	Early unsatisfactory results with similar resins reviewed	Merrifield (1969, pp. 227–228)
11.	Polystyrene	8% Divinylbenzene	Macroporous, average pore diameter = 1630 Å	200 μm Beads surface area = 20 m^2/gm	Not specified	Satisfactory results reported for small peptides	Losse (1973); see also Sano *et al.* (1971)
12.	Brush on Silica (controlled pore glass)	None	Macroporous, average pore diameter = 1150 Å	120 μm Beads surface area = 15 m^2/gm	Waters/Porasil-E; also Pierce (formerly Corning)/CPG	Solvent exchange within the glass pores is difficult	Parr *et al.* (1974)
13.	Polystyrene	0.2% Divinylbenzene	Popcorn	20 μg fragments (after grinding in Waring Blender)	X	Used in early studies	Letsinger *et al.* (1964)
14.	Polystyrene on glass (1:200 w/w)	0.5% Divinylbenzene	Pellicular	100 μm glass core beads 1 μm shell	Northgate Labs	Suitable for column operation; low capacity and occasional peeling	Scott *et al.* (1971)

No.	Support	Cross-linking	Type	Size	Source	Comments	Reference
15.	Polystyrene on polystyrene	Divinylbenzene: low crosslinked shell on high crosslinked core	Pellicular	40 μm core; 5–6 μm thick shell	No experimental details or specifications	Used in successful segment condensation synthesis of glucagon	Protein Synthesis Group (1975)
16.	Polystyrene radiation grafted onto Kel-F (5–15% w/w)	Linear polystyrene shell	Pellicular	120 μm core of density 2.2; 20 μm thick shell, measured for 7.5% w/w of polystyrene[i]	ICI, Australia	Results[i] comparable to standard resins (Nos. 1 and 2); useful in a double-resin strategy with standard resins (see p. 9)	Tregear *et al.* (1966); Tregar (1972); see also Kent and Merrifield (1978).
17.	Polyethyleneimine (initially adsorbed onto macroporous alumina)	Glutaraldehyde	Ghost	?, original core dissolved out with mineral acid.	X	Large surface to volume ratio of potential interest for synthesis	Meyers and Royer (1977)

[a] The supports as listed are generally functionalized at a later stage of the overall solid-phase methodology.

[b] The nominal percentage cross-linkage is the amount by weight of cross-linking agent relative to the monomer in the polymerization mixture. The actual number of cross-links may be quite different, and in general will not be uniform throughout the copolymer.

[c] See Table III.

[d] Resins from different sources, or even different lots, are not necessarily of comparable quality. X = not commercially available, see reference for procedure.

[e] N = not yet applied to solid-phase peptide synthesis.

[f] The original polystyrene-divinylbenzene resins used in the development of solid-phase peptide synthesis were obtained as gifts from the Dow Chemical Company.

[g] Since, in the described polymerization procedure, an aqueous mixture of monomer and cross-linking agent was dispersed into a nonpolar solvent, the product microporous gel should be referred to as a *reverse-phase suspension copolymer*.

[h] Arshady *et al.* (1979) made a polydimethylacrylamide-co-acryloylsarcosine methyl ester (0.35 mmol Sar/gm) resin cross-linked with 8% *N,N'*-ethylene bis-acrylamide. This support, which is more easily prepared (as is No. 6) than No. 5, is now preferred in the Cambridge laboratories (R. C. Sheppard, personal communication, 1979).

[i] The "apparent" density (mass volume) of the polystyrene shell portion is readily calculated as 0.13, in contrast to the true "skeletal" density for polystyrene of 1.05. Also the respective linear and volume expansions of the shell portion only upon swelling in CH_2Cl_2 are 1.4 × and 1.7 ×, compared to 1.7 × and 5 × for normal polystyrene. Thus, polystyrene chains of the pellicular graft resin are rather loosely packed.

The best practical results have been obtained with microporous gels in good swelling solvents. Examples include suspension copolymers of styrene and divinylbenzene, and a variety of carbohydrates or polyamides (Table IV, numbers 1–8). In optimal cases (appropriate solvent and lowest amount of cross-linking consistent with mechanical stability), the solvent may occupy 80% or more of the volume of the swollen gel. On a macroscopic level, the overwhelming majority (>99.9%) of the functional sites are found in the interior of the polymeric matrix, and they are randomly distributed as shown by radioautography (Merrifield and Littau, 1968).

No data are available on the precise molecular distribution of functional sites. One might expect the microstructure of the solid support to be related both to the intimate chemistry and mechanics (reagents, reactivities, rates, concentrations, solubilities, mixing, initiation conditions) of the original polymerization and subsequent functionalization processes. Some early reports suggested hypothetical "local clustering" or heterogeneity of site distribution. However, from an empirical point of view, *effective* site homogeneity (equivalent to random distribution) is indicated by the uniformity of rate data (p. 29).

Some calculations initially performed by Merrifield and Littau (1968) may be extended to provide some instructive insights. We shall assume the recommended poly(styrene-co-1% divinylbenzene) resin bead of average dry diameter 150 μm, which swells to five times its packed volume in dichloromethane (Gutte and Merrifield, 1971). Only about one out of 4×10^4 aromatic rings is found on the true spherical surface of beads this size. The commonly useful loading of 0.2 mmol sites/gm resin corresponds to substitution of 1 out of 50 aromatic rings. A single bead at this loading contains about 2×10^{14} sites, each in an average solvated volume of 3.3×10^4 Å^3, which means that the effective concentration of polymer-supported reactants is 0.05 *M*. Assuming a uniformly distributed cubic lattice (and including the skeletal volume of the polymer in the calculation), the average distance between sites is 34 Å.

Supports that possess a rigid or nonswellable polymeric matrix structure (Table III; Table IV, numbers 10–13) have been examined. "Controlled-pore" glasses and extensively cross-linked polystyrenes (termed macroporous or macroreticular) prepared in the presence of suitable diluents fall into this category. Functional sites must of necessity be restricted to the effective surface areas of the support, i.e., the free faces of the various pores. These pores have been found to be approximately evenly distributed throughout the internal structure of the polymeric particles (Burlitch and Winterton, 1975). Even in the most favorable situations, capacities of these classes of supports are quite low. For example, the average site separation on a support with a surface area of 100 m^2/gm (an atypically high value) is 34 Å for a loading of 0.015 mmol/gm. The effective concentration of polymer-supported reactants may be estimated at 0.015 *M* for this case, based on the assumption that the reaction phase is the pore volume of typically 1 ml/gm. These calculations ought to be contrasted with those presented in the previous paragraph dealing with microporous gel supports. It should be noted that, recently, several new macroporous polymers have been prepared with a relatively low cross-linking. While these supports still have low capacities based on the surface area, they also appear to have a gel phase that permits ready diffusion.

Pellicular supports are composed of two layers: an impervious core to provide physical and mechanical stability, and a mobile and accessible layer near the surface of the particles to which the actual chemical sites for synthesis are restricted. Suitable core materials include glass, extensively cross-linked polystyrene, and Kel-F (Table IV, numbers 14–16). The surface layer may be expected to possess physical-chemical characteristics analogous or possibly even superior to those of the microporous gels previously discussed. The local surface swelling will not be reflected in a substantial overall macroscopic volume expansion; and, hence, pellicular supports can be safely packed for column operation. The capacity per weight of these supports is decreased due to "dilution" with the core. To compensate, more total pellicular support can be used, but a large increase in loading of the surface polymeric layer should be avoided to prevent excessive crowding (pp. 31–32).

d. Mobility of Polymeric Chains. Nitroxide spin labels have been used as "molecular reporters" to delineate the internal flexibility of several supports used for solid-phase synthesis (Regen, 1974, 1975a,b). The rotational correlation times calculated from the electron spin resonance (esr) spectra were shown to vary inversely with the degree of swelling of the polymer in various solvents, although the precise mathematical relationship is unknown. When the solvent was held constant, the average mobility measured by this technique increased with decreasing cross-linking (improved swelling characteristics) of microporous gel supports, both when the label was attached to the polymer backbone or was imbibed in solvent channels of the matrix. The results asymptotically approached those measured on noncross-linked polymers or on spin labels in free solution, respectively, and could be usefully expressed in terms of the "internal viscosity" of the polymer (Table V). The esr technique has also been applied by Sparrow (1976) for the qualitative assessment of the use of spacer linkages for anchoring peptides to the solid phase.

There is now a large body of evidence to show that under appropriate conditions appreciable amounts of intersite* reactions may occur. These are

* Two or more sites on the same polymeric bead.

Table V. Motion within Solvent Channels of Polymeric Matrices[a]

Resin[b]	$10^{10}\ \tau$, sec[c] Resin-bound[f]	Soluble[g]	q[d]	η_i[e]
Soluble	0.9	0.19		0.6
S-X1	2.9	0.22	5.4	0.7
S-X2	3.1	0.31	3.7	1.0
S-X4	5.5	0.49	2.8	1.5
S-X8	15.0	1.3	2.0	4.1

[a] Reprinted from Regen (1974, 1975b). Copyright by the American Chemical Society. Reprinted by permission of the copyright owner.

[b] Soluble polystyrene, or polystyrene with indicated extent of cross-linking with divinyl benzene.

[c] Rotational correlation time, by esr.

[d] Swelling volume (in benzene)/dry volume.

[e] "Internal viscosity", in cP.

[f] 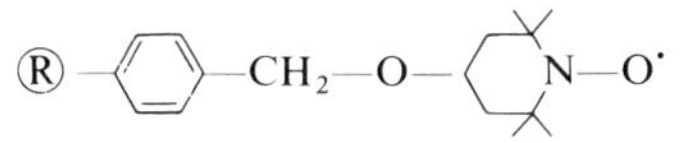

~0.002 mmol/gm, swollen in benzene.

[g] 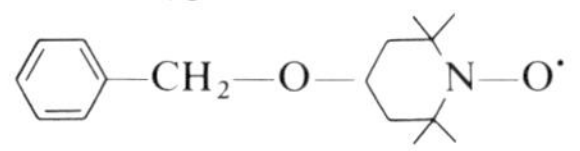

$10^{-4}\ M$ in benzene; imbibed in underivatized resin.

observed with: (1) sufficiently high loadings of sites; (2) long reaction times or high temperatures; (3) appropriate distances of sites from polymer backbone; and (4) suitable swelling solvents. Two sites that react may in principle be close, relatively near, or distant from one another in the linear covalent structure of the polymer. In the intermediate case, flexibility and mobility of the polymeric chains is required, and there are reasons to believe that this is often the principal route of interaction. When extensive levels of cross-linkage occur between linearly distant but geometrically adjacent sites, substantial changes in measurable physical properties of the support, for example swelling, may be expected. Such changes have occasionally been observed (pp. 56–57). The influence of support morphology on the occurrence of intersite reactions has been considered by several authors, but no clear-cut conclusions may be drawn from the presently available data.

It is important to stress that under the normal conditions of solid-phase synthesis, the known side reactions involving interactions of different sites can usually be safely ignored or at least, once identified, entirely circumvented by modifications in methodology. The major well-documented examples from the peptide literature are (1) nearly quantitative oxidation, under vigorous conditions, of resin-bound cysteine peptides to form intersite disulfides (Lunkenheimer and Zahn, 1970); (2) low levels of intermolecular aminolysis of a benzyl ester linkage to a resin, leading to chain doubling (Beyerman *et al.*, 1972; Rothe and Mazânek, 1974); however, this reaction was not detectable in another comparable case at 20-fold increased sensitivity (Merrifield *et al.*, 1974; Kent and Merrifield, 1978); (3) significant cyclodimerization (and higher cyclo-oligomerization) during attempted unsuccessful cyclization of the pentapeptide semi-gramicidin off an activated safety-catch support (Rothe *et al.*, 1977); (4) partial N-alkylation of proline residues of a model peptide by excess chloromethyl sites (Schou *et al.*, 1976; see Fig. 4); and (5) nucleophilic transfer of trifluoroacetyl groups from hydroxy to amino sites (Kent *et al.*, 1979). The interaction of separate polymeric sites has also been concluded from a variety of chemical studies: (6) multiple (usually two, but in a special case, three) polymer-bound chelation sites of certain transition metal complexes (Collman *et al.*, 1972); (7) intersite anhydride formation* (Crowley *et al.*, 1973; Rebek and Gaviña, 1975a; Scott *et al.*, 1977); (8) mixed and self-condensation reactions of esters attached to a common polymeric backbone (Kraus and Patchornik, 1974); (9) quaternization of chloromethylated resins with bifunctional tertiary amines (Regen and Lee, 1974); (10) alkylation of chloromethylated resins with a dithiol (Farrall and Fréchet, 1978), and (11) intermolecular glycyl transfer within a glycine *p*-nitrophenyl ester-resin, as detected by subsequent intramolecular diketopiperazine formation (Rebek and Trend, 1979).

Among the intersite reactions just catalogued that result in the creation of cross-links (multiple attachment points), all but numbers (1) and (6) could not be driven to completion even under the most extreme conditions. Control experiments showed that those sites that had not been multiply attacked either remained accessible to monofunctional reagents [e.g., number (9)] or had undergone reactions by alternate chemical pathways [e.g., number (7)]. Such results have occasionally been interpreted to mean that certain regions of the polymeric matrix retain some measure of rigidity.

e. Effective Site Isolation. The intersite reactions, as documented in the preceding section, can often be minimized. In the ideal case of zero contact between sites, a situation approaching infinite dilution of reactants

* Further chemical characterization of these reactions would be helpful. Factors which might warrant study include (1) equilibration of anhydrides through exchange reactions, (2) the extent of side reactions during anhydride synthesis, (3) reactivity of anhydrides with nucleophiles, and (4) whether the apparently extensive cross-linking that was observed has any significant effect on swelling of the resin beads.

exists, favoring only intramolecular processes (Fridkin *et al.*, 1965). Several types of observations suggest that effective site isolation on the solid support can be obtained under appropriate conditions.

1. Letsinger *et al.* (1964) noted absorbances in the infrared spectra of carboxyl-group-containing polymers due both to the free and hydrogen-bonded (associated) state; in solution models, only the latter absorbances are found.
2. Even under heavily overloaded conditions, carbodiimide-mediated phosphodiester formation from soluble alcohols and polymer-supported phosphate monoesters could not be achieved, although both solution controls and the converse reaction between polymer-bound alcohols and soluble phosphates proceeded readily (Blackburn *et al.*, 1969). In view of the mechanism of the solution reaction proposed by Weimann and Khorana (1962), the cited results suggest the impossibility of forming a polymer-bound cyclic trimetaphosphate intermediate (requiring three sites of attachment).
3. Burlitch and Winterton (1975), working with a macroporous resin with a low loading, prepared a linear trimetalide that could not be prepared by solution synthesis due to a cyclization side reaction.
4. Martin *et al.* (1978) were able to prepare stable resin-bound analogues of certain mixed carbonic–carboxylic anhydrides which are known to decompose rapidly in solution.
5. Mazur and Jayalekshmy (1979) found that the ubiquitous dimerization reaction of benzyne could be completely suppressed by attachment to microporous or macroporous resins at reasonable loadings. Delayed trapping experiments showed that the room temperature lifetime of polymer-supported benzyne was more than a minute.

In practice, the balance between site isolation and interaction depends in part on judicious selection of reaction conditions but, more significantly, on the specific chemistry that is to be undertaken on the polymeric support. Crowley and Rapoport (1976) have suggested that some of the examples of site isolation that have been cited (e.g., review by Leznoff, 1974) merely reflect the purification advantage of the solid-phase method, that is, the possibility of selectively obtaining desired products on the basis of their attachment to the support. While this is a useful insight, it does not explain all of the results presented in this section. We concur with the analysis of Mazur and Jayalekshmy (1979), who clearly pointed out the *kinetic* explanation for the variety of results that suggest effective site isolation. Thus, a desired reaction will predominate if it is rapid relative to a competing intersite side reaction. The term "pseudo-dilution" was recommended to express this idea. In contrast, cases of extensive intersite reactions were accounted for by invoking thermodynamically favored equilibration of reversible processes over a lengthy time frame. It is important to point out that while the physical separation of functional sites in the polymeric matrix is critical for certain chemical applications, it is not a requirement for the practical utility of solid-phase peptide synthesis.

2. Reaction Rates

Solid-phase peptide synthesis requires rapid chemical reactions for the repetitive deprotection/coupling cycles. This is to ensure that even in the most unfavorable cases, reactions may be conducted for at least 10 half-lives on a realistic laboratory time scale. Kinetic studies of heterogeneous reactions are technically difficult on both the experimental and theoretical levels. Several general questions have been considered.

a. Uniformity of Rates. In the normal plan of solid-phase couplings (Fig. 2), synthetic units S_τ are invariably added in excess over the amount of initially anchored S_j, with the goal of thus driving the reactions to completion. From this point of view, the critical factor becomes the concentration of reactant remaining in solution at the point where the couplings approach completion, and not the numerical value of excesses used. It would be important to know whether the same (true or apparent) second-order rate constant determined for the initial stages of coupling was applicable to the latter stages as well. The weight of recent evidence (Merrifield, 1973c; Rudinger and Buetzer, 1975) provides an affirmative answer to this question for most cases of practical interest. Rates were determined both with coupling agent or polymer-bound reactant in excess. Some earlier studies, which reached different conclusions, may well have been hampered by technical problems, including insufficient sensitivity of measurements, side reactions leading to depleted concentrations of coupling agents, and misinterpretations of the kinetic data. Limited or slow diffusion through the polymer does not appear to present difficulties, with the equilibrium being reached in the order of seconds.

The common practice of solid-phase synthesis has been to use the same conditions of reactant concentration and time for the different repetitive deprotection and coupling steps. This is the simplest and most pragmatic approach in view both of the automation of the process and of the experimental challenge of rapidly monitoring each step for absolute completion of the reaction. However, it is clearly of interest to know over what range the rate constants vary with differing polymer-bound peptides or incoming activated amino acids or peptide segments. Ideally, the chemistry of the reactions used ought to be such that the fastest and slowest rate constants are within one order of magnitude of one another.

Relative rates of coupling of amino acids have been determined either directly (Merrifield, 1973c) or by a competition technique (Ragnarsson *et al.*, 1971, 1974). The available data indicate that for a given set of conditions, the range of reactivities is at most 20-fold, and more commonly 2- to 5-fold. Thus, general semiquantitative guidelines may be formulated (Section IV) which allow the use of "blind" (probably excessive) conditions, with a strong confidence as to their effectiveness. "Real-time" monitoring (with feedback) of the rates and yields is, in principle, the only unequivocal way to optimize the reaction conditions.

b. Relative Rates in Homogeneous and Heterogeneous Phases. Solid-phase reactions generally proceed at apparent rates two- to threefold slower than the corresponding solution models. From a practical point of view, this difference is entirely acceptable. When the data are corrected for partition effects and resin swellings (Rudinger and Buetzer, 1975; see pp. 17 and 19), the "true" rate constants may for certain cases, in fact, be identical in the homogeneous and heterogeneous phases.

As a diagnostically useful corollary, comparisons of solid-phase and solution rates may often be of value in assessing proposed polymeric supports and reaction solvents. For example, two different commercial batches of chloromethyl-polystyrene resin with nominally identical specifications gave a 20-fold discrepancy in rates with the same model reaction (Rudinger and Buetzer, 1975). The "slower" batch proved to be entirely unsatisfactory for solid-phase peptide synthesis.

3. *Elaboration of the Polymer-Bound Peptide*

The data and conclusions presented sofar in this section, dealing with reaction rates, polymer mobility and solvation, and accessibility of sites, were in general derived for simple models, and it is natural to ask to what extent they apply to an ever-lengthening protected peptide chain being assembled on the solid support.

a. Selection of Sites. Occasionally, special procedures that aim to restrict the solid-phase assembly of a peptide to the physically or chemically most accessible sites have been explored. As examples, (1) limited acidolytic cleavage of an acetoxymethyl-resin was suggested to provide "sterically selected" halomethyl sites, which were then quantitatively esterified and extended in the usual way (Tilak, 1968; Merrifield, 1969, p. 238); (2) a low-swelling solvent was advocated for the initial functionalization (Frank *et al.*, 1977a) or anchoring (Monahan and Gilon, 1973) steps; and (3) "kinetic" site selection was attempted by performing one of the early coupling steps with excess polymer-bound component, followed by termination of unreacted sites (Yamashiro and Li, 1973a). No clear evidence is available regarding the advantages of these interesting concepts. It would be important to establish the possible extent of equilibration of a dynamic polymer structure, i.e., whether the same or different sites are preferentially accessible at different stages of the synthetic process.

Taking a cue from successful applications to affinity chromatography (Cuatrecasas and Anfinsen, 1971), Sparrow (1976) examined the use of a long spacer chain between the peptide and the solid phase. The rationale was to minimize possible steric interference from the backbone of the polymeric support. A threefold increase in the overall yield of a 19-peptide was realized, but the improvement may not necessarily be due solely to the spacer (see p. 58).

b. Loading and Capacity. The selection of an initial loading level will determine the maximum possible yield of a solid-phase peptide synthesis. It has generally been assumed that too high a level of growing chains dispersed throughout the polymeric matrix might result in decreased efficiencies of individual chemical steps—through steric crowding, intersite side reactions, decreased diffusion of reagents, or changes in the solvation properties of the system. In support of this intuitively plausible viewpoint, examples are available from several laboratories of syntheses at high loadings (0.8–2.0 mmol/gm polystyrene resin) that have failed after 10 to 20 cycles. Lower loadings (0.1–0.3 mmol/gm) have been commonly used.

For lack of more detailed information, a protected n-peptide may reasonably be considered as a flexible random coil with a radius of gyration r_G given by $\beta\ (n/6)^{1/2}$, where* β is 11–14 Å (Tanford, 1961, pp. 150–170). Since the average intersite distance in a typical swollen resin at a loading of 0.2 mmol/gm is 34 Å (see p. 24), a peptide of length 12 amino acid residues ($r_G \sim 17$ Å) may be readily accommodated without substantial clashing. For this case, the concomitant weight increase is approximately 0.4 gm peptide/gm starting resin, and it can be assumed that the physical–chemical characteristics of the original support matrix will still predominate. With higher loadings and greater peptide chain lengths, solvation of the protected polypeptide portion itself may lead to changes in the overall swelling of the peptide–resin. Another effect will be the overlapping of reaction spaces ($2r_G >$ average intersite distance). However, this latter phenomenon should not raise difficulties, since long random coils are one to two orders of magnitude less tightly packed than typical globular proteins of the same molecular weight.

We feel that current knowledge is consistent with the working recommendation that the loading of a support for solid-phase peptide synthesis should

* For completely unrestricted rotations and bond angles (classical random walk), $\beta = l_{av} =$ 3.6 Å, the distance between successive amino acid residues in a peptide bond. For typical synthetic organic polymers, $\beta \simeq 3\ l_{av}$ (Tanford, 1961, pp. 159, 403). Data for poly-γ-benzyl-L-glutamate in chloroacetic acid (Tanford, 1961, pp. 408, 519) lead to $\beta = 13.5$ Å.

be inversely proportional to the length n of the desired target peptide in such a way that at the end of a synthesis, the final weight increase due to the protected peptide is less than the original weight of the resin support. In the successful solid-phase assembly of the ribonuclease A sequence (Gutte and Merrifield, 1971), each gram of polystyrene resin eventually gave 0.7 gm of protected protein (0.02 mmol peptide/gm peptide–resin). In that case, the progressive premature acidolysis of the anchoring benzyl ester linkage was a fortunate circumstance, because at the initial loading of 0.21 mmol/gm polystyrene, 4.0 gm of protected protein would have been produced per gram resin. Weight increases higher than the one recommended in this paragraph are quite common in industrial applications (e.g., Colescott *et al.*, 1975). The result on β-lipoprotein (Yamashiro and Li, 1978), where 0.7 gm resin at 0.34 mmol/gm gave 4.3 gm peptide–resin, is a most striking example of a successful synthesis of a large peptide at high loading. Further investigation is clearly required to elucidate the ultimate capacities of microporous gel supports in solid-phase synthesis.

c. Influence of Peptide Chain Length. The work of Gutte and Merrifield (1971) on ribonuclease A showed that the stepwise coupling yields on long peptide chains (>100 amino acid residues) were equivalent to those found at early stages of the synthesis and did not require extensions of the normal reaction time. Furthermore, the structural integrity of the peptide-containing polymeric particles was maintained, and there were no appreciable changes in the swelling properties. These conclusions are supported by results from many other laboratories on a gamut of synthetic applications. In the course of a number of excellent and well-characterized preparations of peptides of moderate size (10–40 amino acid residues) by the solid-phase method, there were no indications of any decreases in yields or efficiencies throughout the various syntheses. In those instances where decreases have been reported, obvious flaws in design and execution were apparent.

4. Yields and Purity of Product

In a solid-phase peptide synthesis, it is desirable but not essential to optimize yields at those steps that occur only once, specifically the initial anchoring to and the final cleavage from the support. Inefficiencies of these reactions will lead primarily to decreased yield rather than decreased purity. In contrast, an exceptionally high standard of fidelity must be applied to those chemical steps that are carried out repetitively in the course of assembling the polymer-supported peptide.

a. Sources of Cumulative Error. It is axiomatic that the starting synthetic units must be both chemically and optically pure, because any preexisting impurities may become incorporated into the final product. The converse statement is not true, as synthesis from analytically pure starting units will not necessarily lead to a homogeneous product.

In general, an understanding of the nature and extent of by-product formation must be accompanied by considerations both of analytical criteria for their detection and of the possibilities of purification strategies for their removal. For example, should racemization occur during coupling, the microheterogeneous mixture of diastereomers will be expected to present insurmountable difficulties in any subsequent purification. Fortunately, the available evidence (p. 128) indicates that this is not a serious problem in stepwise solid-phase synthesis. Furthermore, there are several impressive examples in the literature where by-products racemized at one particular residue could be resolved by appropriate chromatographic or partititon procedures (e.g., Sieber *et al.*, 1970; Izumiya *et al.*, 1971; Jorgensen *et al.*, 1972; Khosla *et al.*, 1973; Yamamoto *et al.*, 1977; Kent *et al.*, 1978; Meyers *et al.*, 1978).

Incomplete deprotection or coupling can, in principle, be due to either temporary or permanent causes. In the former case, which is primarily the consequence of inadequate reaction conditions, deletion peptides may arise; in other words, once the temporary cause is removed, the chain will start to grow again, but will lack the intervening residues. On the basis of the extremely sensitive quantitative preview method of Edman degradation (Section IV,D,4), deletions have often been found to average less than 0.3% per step. Furthermore, it has been demonstrated that even closely related deletion peptides can sometimes be effectively separated. Yamashiro and Blake (1978) showed that most of the single neutral amino acid-deletion peptides of the 31-residue peptide β-endorphin elute at positions different from the desired product upon partition chromatography. Permanent irreversible termination of growing chains can arise from chemical side reactions or by intentional blocking of unreacted sites after each coupling step. This is a less serious problem because terminated peptides may in principle be removed from the desired product at the end of a solid-phase synthesis through the application of a "functional affinity" purification strategy (Section IV,E,2, pp. 163–166).

Premature losses of peptide chains during the synthesis primarily affect the yield rather than the purity of the product. However, at least two low-level side reactions have been recognized to result from further transformations of the resin-bound functional groups created via premature cleavage. Hydroxymethyl groups can, at later stages of the synthesis, serve as sites for the initiation of peptide chains, which will then lack residues at the COOH-terminal end of the desired peptide (Beyerman and in't Veld, 1969; Khosla *et al.*, 1972c), and resin-bound trifluoroacetoxymethyl groups will lead to cumulative irreversible chain termination (Kent *et al.*, 1979).

b. Specific Additional Sources of Error. All of the potential difficulties listed so far share the common feature that the indicated structural errors can arise only in association with the *same* deprotection/coupling cycle at which a given synthetic unit is added. Even when a residue has been safely incorporated among the polymer-supported peptide chains, it may still succumb to damage at subsequent synthetic steps, especially during the final cleavage and removal of permanent protecting groups. Such irreversible modifications or rearrangements generally predominate at specific positions in the peptide structure, and do not usually affect the amino acid sequence and composition. In some instances, the differences between the desired and the defective products may be too subtle to permit their separation during the purification process; alternatively, the nature of the modification may actually facilitate a straightforward purification. The extent of this second category of side reaction is strongly dependent on the nature of compatible sets of protecting groups and the corresponding cleavage conditions chosen for the synthesis.

c. Overall Homogeneity of Product. We define the homogeneity H (in percent) of a product prepared by the stepwise solid-phase method as:

$$H = 100 \prod_{\substack{\tau=1 \\ \tau \neq j}}^{n} \left(1 - \frac{x_\tau}{100}\right) - \sum_{\tau=1}^{n} y_\tau \simeq 100 - [(n-1)\bar{x} + m\bar{y}]$$

where x_τ represents (in percent) the amounts of errors from all sources that arise during the cycle at which unit S_τ is added [(a), preceding 2 pp.]; and y_τ represents the amounts of errors at S_τ from all other sources [(b), above], occurring only at m specific residues out of the total n. Furthermore, $\bar{x}$ and $\bar{y}$ are the appropriate averages which must be low for this formula to hold. In practice, this criterion is usually met, so that peptide chains with two defective sites can safely be excluded from consideration.

In line with the discussion of this section, the equation for homogeneity reveals the requirement to minimize the repetitive error $\bar{x}$, which is relatively constant from step to step. The effect of this factor on the homogeneity of the final product is in direct proportion to the chain length. Also, the above equation shows that the homogeneity of the product depends directly on the number of troublesome residues (i.e., those known or shown to give side reactions), as well as on the numerical extent of these side reactions.

D. Practical Ramifications

The first stages in the undertaking of a solid-phase peptide synthesis are the selection of a suitable polymeric support and the design of the chemical strategy of protecting group combinations and coupling methods. In this section, we overview the subsequent laboratory procedures and manipulations required to carry out these plans. Well-organized and detailed discussions of this topic have been presented by Meienhofer (1973, pp. 155–178) and Stewart and Young (1969).

1. Purity of Materials

a. Solid Supports. The preparation of (underivatized) polymeric resins with acceptable specifications for solid-phase synthesis (Table IV) is generally beyond the scope of most academic chemical or biochemical laboratories. Commercial resins may well contain residual occluded noncovalently bound material arising from the polymerization protocol used, which could conceivably lead to chemical or physical difficulties later on. For example, Stewart (1976, p. 281) mentioned that as much as 20% of the weight of certain commercial batches of polystyrene resin could be extracted with hot trifluoroacetic acid; the loss presumably represents trapped linear polystyrene, residual monomers, cross-linking agents, catalysts, and additives (Patterson, 1971). Therefore, we strongly recommend thorough washing and extraction of resins prior to further synthetic reactions. A recent modification (Mitchell *et al.*, 1978b) of earlier procedures commonly used in our laboratory is as follows: Poly(styrene–co-1% divinylbenzene) resin beads (200 gm) are stirred at 70°C for 30 min at a time with 2 liter lots (twice each) of benzene, methanol, dimethylformamide, dioxane–2 *N* aqueous NaOH (1:1 v/v), dioxane–water (1:1, v/v), dioxane–2 *N* aqueous HCl (1:1 v/v), and dioxane–water (1:1, v/v). Solvent is aspirated off through a filter stick between treatments. The resin is then rinsed on a fritted glass funnel with 4 liters each of hot methanol, benzene, cold methanol, and dichloromethane, followed by drying under vacuum; the final weight of washed resin is typically 80–95% of the starting weight.

Uniformity of particle size is also important, especially with regard to elimination of "fines" which tend to clog the pores of fritted glass filters. The standard polystyrene resin floats in a dense solvent such as dichloromethane and sinks in a lighter solvent such as methanol, whereas the Kel-F–polystyrene graft pellicular resin (Table IV, number 16) sinks in all common solvents. The fine beads can be withdrawn at the bottom of a separatory funnel, or they can be decanted, depending on the density of the solvent.

Physical characterization of the solid support before use for peptide synthesis is recommended, as far as it is possible. Techniques, ranked in order of convenience to set up, include microscopic examination, swelling measurements, infrared spectra, model kinetics, and syntheses of model peptides combined with a sensitive analytical system.

b. Solvents and Reagents. There is no clear consensus or evidence on the purity grade of solvents required for solid-phase synthesis. In view of the repetitive application of large volumes throughout the course of the procedure, one might expect a concentration effect if trace impurities were present that would react to modify a portion of the desired peptide chains. As a striking example, blocking of amino groups was inferred from perchloric acid titrations of peptides synthesized with certain commercial lots of CH_2Cl_2 (Brunfeldt and Christensen, 1972). Meienhofer (1973, pp. 163–164) and Stewart and Young (1969, pp. 30–32) summarize routine recommended practical purification procedures; the book by Perrin *et al.* (1966) is also often worth consulting. As a final cautionary note, a variety of common reagent mixtures possess limited chemical stability, and thus ought always to be freshly prepared.

c. Amino Acid Derivatives. The need for ultrapure starting materials for subsequent incorporation into the growing polymer-supported peptide chain has already been stressed (p. 32). Derivatives should be purified by preparative chromatography or recrystallization until they show only one component at a high loading by a sensitive analytical technique (usually thin-layer chromatography in two solvent systems). Spectral and physical characterizations are also recommended, but these techniques may not always have the necessary limits of sensitivity.

It often happens that acetic acid is used in the course of the purification procedures for protected amino acids, and a small amount could be tenaciously retained in material that passes all of the usual criteria of homogeneity. Coupling of such a contaminated derivative in a solid-phase synthesis employing dicyclohexylcarbodiimide activation would be accompanied by termination through acetylation. Acetic acid can be detected by nuclear magnetic resonance (nmr) spectroscopy or by spraying thin-layer plates with bromcresol green. Even traces can be removed by repeated azeotropic drying with benzene.

As a practical point, N^{α}-protected amino acid derivatives are often obtained as their cyclohexylammonium or dicyclohexylammonium salts (Klieger *et al.*, 1961), for reasons of easier crystallizability as well as increased storage lifetime. Unwitting use of the salts in a coupling reaction will result in the synthesis of a deletion peptide (see Doscher, 1977, p. 593). Hence, just prior to coupling, it is essential to convert the salts back to the free acid, which is then recovered by quantitative extraction into ether or ethyl acetate. Acidification has been achieved in an aqueous (titration to pH 2–3) or two-phase system through the use of (1) a slight excess over the calculated amount of 0.2–1.0 *N* sulfuric acid or hydrochloric acid (Zervas *et al.*, 1963b; Juillerat and Bargetzi, 1976; Mitchell *et al.*, 1978b); (2) excess aqueous sodium citrate buffer, pH 3.5 (Feinberg and Merrifield, 1972; based on Schwyzer *et al.*,

1959); or (3) excess potassium bisulfate buffer, pH 2 (Spangenberg *et al.*, 1971). The latter method has been specifically advocated both to avoid carryover of acid into the organic phase, and to maintain relatively mild neutralization conditions. Another approach utilizes (4) Dowex-50 cation-exchange resin in the H^+-form to effect the neutralization (Otsuka *et al.*, 1966).

The most sensitive way to assess the optical purity of amino acid derivatives is by removal of all protecting groups and conversion of the free amino acid to a (potentially) diastereomeric dipeptide mixture which may be analyzed by ion-exchange chromatography according to Manning and Moore (1968). Details and modifications are given by Feinberg and Merrifield (1972) and Mitchell *et al.* (1978a).

2. Mechanics

a. Reaction Conditions. Solid-phase reactions are generally performed batchwise. However, supports that do not expand (or swell) on a macroscopic level, including several pellicular resins (Tables III and IV), offer the possibility of continuous operation in a column system (Bayer *et al.*, 1970b; Scott *et al.*, 1971). Recently, the continuous-flow mode of operation has been adapted to the standard swollen polystyrene beads, using the pumps and columns of commercial high-pressure liquid chromatography systems (Erickson and Prystowsky, 1979).

Heterogeneous reactions require complete wetting of all solid-phase particles with the appropriate reactant solution. A packed bed of dried polystyrene beads will swell to 95% of its final volume within seconds after the addition of dichloromethane. In a given bead, diffusion is rapid, and the equilibration of solute concentrations follows on the same time scale (see p. 29). The nature or extent of agitation of batchwise heterogeneous reactions does not appear to be a crucial factor (Erickson and Merrifield, 1976, p. 362). Thus, inversion at 4 cycles/min was found to be adequate, although the usual rate is 10–20 cycles/min. Mixing has also been achieved by shaking, stirring, centrifugal flow, bubbling through with nitrogen, and ultrasonic vibration.

b. Washing. Between successive chemical steps of solid-phase peptide synthesis, it is clearly necessary to exchange excess reactants and soluble by-products with fresh solvent by interspersing a series of washing steps. The effective dilution per washing step is governed by the external wash volume V_e, the holdup volume V_h (function of swelling in given solvent), and the distribution coefficient K (see p. 19) of the solute. Extending the formula of Erickson and Merrifield (1976, p. 409), the washout efficiency E_w after N washings is $(V_e/KV_h)^N$. For example, with the typically used V_e of 20 ml/gm of 1% cross-linked polystyrene ($V_h \sim 4$ ml/gm), a single dichloromethane wash gives a fivefold dilution of a solute with $K = 1$; five washes suffice to give a 3000-fold dilution, or 0.03% carryover. For timed chemical steps such as protecting-group removal, a common procedure has been to add the reagent in two or three doses in order to avoid the dilution of the reagent by the solvent retained from the previous wash. A short prewash serves to establish the approximate desired concentration of reagent introduced in the next addition.

Factors relevant to devising a practical washing scheme include the size of the reaction vessel, which limits V_e; the cost of wash solvents, which favors lowering $N \times V_e$; and the measured washout efficiency, which can be increased if necessary by increasing V_e and N, or decreasing V_h. The very last mentioned option was exploited by Corley *et al.* (1972), who alternated washes between solvents that effectively swell (dichloromethane) and shrink (alcohol) the resin. The overall effect was likened to rinsing and squeezing out a wet sponge.

As pointed out previously, equilibration of solute concentrations is very rapid. After solvation of the beads for about 1 min, solvents may be removed either by vacuum or under positive nitrogen pressure. The solid phase is retained on a sintered glass filter. Another way to drain off all the liquid in excess of the holdup volume of the polymeric beads involves use of a specially designed centrifugal reactor (Birr and Lochinger, 1971); this mode of solvent separation has been found to be especially valuable in conjunction with the fragile 0.5% cross-linked polystyrene gelatinous beads (Table IV, number 3) which cannot be handled conveniently in ordinary apparatus.

c. Resin Adhesion. The commonly used polystyrene or polyamide beads, when wet, have some tendency to adhere or clump onto ordinary glass surfaces. This problem does not occur in Teflon, polyethylene, or silanized glass reaction vessels. In our laboratory, glass reaction vessels are cleaned in chromic–sulfuric acid solution (2 hr, 80°C); washed with water and alcohol; dried; silanized with 10% dichlorodimethylsilane in toluene (1 hr, room temperature); washed immediately with methanol; and oven dried.

3. Apparatus

A large variety of designs of reaction vessels, with accompanying methods of agitation, and solvent transfer and removal, have been described and reviewed (Merrified, 1969, pp. 266–268; Stewart and Young, 1969, pp. 65–70; Meienhofer, 1973, pp. 157–163; Erickson and Merrifield, 1976, pp. 432–435; Birr, 1978, pp. 72–80).

a. Manually Operated Systems. The reaction vessel and mechanical shaker initially described by Merrifield (1963) and Merrifield and Corigliano

(1968) are still in general use for manual laboratory work. However, a simple modification of the vessel in which the top is closed with a Teflon-lined screw cap (Gisin and Merrifield, 1972a; Fig. 3) is finding increased favor. Mixing in the former vessel, which has an open side arm and a drying tube, is by rocking through a 100° arc; as a result small amounts of resin can adhere to the walls and work their way out of the reaction solution. Mixing in the closed vessel is through a 180° arc; provided the vessel is over half full, all surfaces and all resin beads will be wet throughout the synthetic reactions.

Figure 3. Reaction vessel and shaker for manually operated solid-phase peptide synthesis in common usage in the authors' laboratory. Photographed by Mark W. Riemen.

A number of other variations in design of manually operated equipment have been described (Kusch, 1966; Khosla *et al.*, 1967; Wieland and Birr, 1967; Gut and Rudinger, 1968; Tritsch and Grahl-Nielsen, 1969; Corley *et al.*, 1972; Park and Regoli, 1972; Burton *et al.*, 1975; Kirby and Warme, 1978).

b. Automation. Part of the original motivation for the exploration of the solid-phase idea was the belief that it would lend itself to automation and, therefore, to the development of a simplified and accelerated technique for the synthesis of polypeptides (Merrifield, 1963). This goal was first realized by Merrifield and Stewart (1965). A simple apparatus was constructed (Merrifield *et al.*, 1966) which provided mechanization of the basic manipulations previously carried out manually: (a) selection of reagent solutions or wash solvents from a set of reservoirs; (b) metering them to a specified volume; (c) delivery into the reaction vessel; (d) mixing for a specified time; and (e) solvent drainage upon completion of chemical reactions or washing steps. The machine consisted of three parts: (1) a reaction vessel; (2) a liquid-handling system (plumbing) for storage, metering, and transfer of solvents into and out of the vessel; and (3) a programmer to control the sequence of events. The glass reaction vessel was fitted with a sintered-glass filter plate through which solvents and reagents were introduced and removed. Reagents were selected by circular multiport valves and were transferred by a metering piston pump. Mixing was performed by repeated inversion of the closed vessel. Liquids were removed from the resin support by vacuum filtration through the fritted disk. The sequence of these events was controlled by a stepping-drum programmer, which activated switches that operated the individual components of the apparatus.

Since that time, several other instruments have been constructed in a variety of academic and commercial laboratories. They are all designed to perform the same chemical reactions, but they differ in construction in many ways. Much more sophisticated control systems have replaced the original stepping drum to provide greater capacity, flexibility, and ease of programming. These include a timed sequence (Loffet and Close, 1968; Scott *et al.*, 1971), teletype (Brunfeldt *et al.*, 1969a), or other punched tapes (Beckman Instruments; Tometsko *et al.*, 1971), and most recently, microprocessors (Vega Engineering) and computers (Villemoes *et al.*, 1976, 1978). Most of the reaction vessels have been glass, but sometimes they are silanized or replaced by Teflon to prevent the resin from adhering to the surface (p. 38). The mechanics of resin agitation and solvent removal have been by the assortment of techniques referred to earlier (pp. 37–38).

Cross-contamination of solvents and reagents must be strictly avoided if quantitative reactions and correct sequences are to be achieved. It is therefore important to avoid a common manifold in which solvents and vapors can be trapped or carried back into reservoirs containing other reagents. The accurately machined rotary valves avoid that problem. However, in most other synthesizers, the liquid handling valves have been replaced by solenoid-operated valves that control the pressurization of individual reagent bottles for the liquid transfer steps. These configurations require a manifold and, therefore, special rinsing provisions to mitigate the risk of cross-contamination.

The instruments described thus far are mechanized but not truly automated. Most of the functions of the machines are regulated by fail-safe monitoring devices that stop the synthesis in case of instrument failure at the transfer, mixing, and filtration steps. However, the feature required for complete automation is a feedback control based on the monitoring of the chemical reactions during the synthesis (Section IV,D). Some way is needed for the instrument to determine whether the last reaction should be repeated, a termination step should be added, or the next chemical step of the synthesis should be started. Efforts along these lines have been underway for some time and three such systems have been described.

Birr (1975) has developed a continuous flow photometric system for monitoring both the coupling reaction and the deprotection reaction. The instrument automatically records the decrease in concentration of α,α-dimethyl-3,5-dimethyloxybenzyloxycarbonyl (Ddz) amino acid reagents as they couple to the growing resin-bound peptide chain. In the next step, the increase in uv absorbance of the Ddz group is recorded as it is released from the resin during the acidolytic deprotection. Such a system is advantageous because data are obtained immediately and no additional chemical reactions are required. The disadvantage lies in the relative insensitivity of the end-point. In the authors' laboratory (Merrifield, 1973c) a similar system has been found to be of great value in measuring rate constants, but less so in establishing the completion of the reactions. In principle the data can readily be incorporated into a feedback loop, but in practice that has not yet been accomplished.

A monitoring system based on picrate titration (Gisin, 1972; Table XI, number 6) has been developed by Hodges and Merrifield (1975a). The precision in following the deprotection step is approximately the same as for the spectrophotometric method, but for quantitation of the extent of the coupling reaction, the picrate procedure is much better, because it is based on the determination of the very low levels of the unreacted amino component (p. 156). The additional chemical reaction required, and the discontinuous nature of the measurement, represent drawbacks of this procedure. When this monitoring system was added to a peptide synthesizer, an automatic feedback was provided which could make one of two decisions: to proceed with the next step or to stop the synthesis if the last step was incomplete. The current version is not able to direct the repetition of the previous step or the insertion of a termination step.

The most advanced automation is that developed by Brunfeldt and collaborators (Villemoes *et al.*, 1976; 1978), which is based on the electrometric titration of amino groups by perchloric acid (Brunfeldt *et al.*, 1969b, 1972a). Through the use of a computer, this system has the capacity to evaluate automatically the data and make the decision to proceed to the next step, to repeat the last step, or to stop the synthesis.

While the mechanical and electronic capability for automated solid-phase peptide synthesis is at hand, it must be realized that the limiting factor has been and continues to be the chemistry of the process. Side reactions have been identified at various stages of the methodology, as documented in subsequent sections of this chapter. Many of these have been overcome, but others remain. Until all of the assorted chemical difficulties are under control, automation cannot reach its full potential.

4. *Analytical Control for Product Quality*

The homogeneity of the crude cleaved product of a solid-phase synthesis is likely to be less than 100%, for reasons already discussed (pp. 32–34). Therefore, fractionation procedures will generally be required, as is the case with any organic synthesis. This is often the most time-consuming aspect of the overall preparation of synthetic peptides. The wealth of techniques for peptide separation, characterization, and structure determination (Bailey, 1967; Hirs, 1967; Needleman, 1970, 1975; Niederwieser and Pataki, 1971; Hirs and Timasheff, 1972, 1977) ought to be marshalled at these stages, both to monitor the progress of the purification scheme and to assess the actual homogeneity of the final purified product.

Barring a technical blunder, it is generally safe to assume that the bulk of the peptide chains assembled by solid-phase synthesis will contain the appropriate amino acid residues in the desired order. Sequential chemical or enzymatic degradation may, nonetheless, reveal covalent modifications of the polypeptide backbone or of the side chains. Spectroscopic and chromatographic tests may also be applied to this issue. Finally, when the target molecule is available from natural sources, a comparison with the synthetic material by all available criteria is essential.

III. THE ANCHOR TO THE POLYMERIC SUPPORT

The physical characteristics and chemical structures of a number of polymeric supports for solid-phase peptide synthesis have been compiled in Table IV (pp. 21–23). In this section, we present and discuss the chemistry involved in (1) functionalization of these supports, (2) attachment of the initial synthetic unit, and (3) cleavage of the completed peptide from the support into solution. The stability of the anchoring linkage to the intervening repetitive steps and the resilience of the desired peptide to the final cleavage conditions are factors of critical concern to the overall success of a synthetic design. These problems have been considered in general terms in Section II,A,2 and 4 (pp. 10–11).

A. Functionalization

Table VI lists some of the methods that have been developed to incorporate a large number of functional groups into polymeric supports. Most of these functionalized supports have actually been used for solid-phase peptide

(*Text continues on p. 56*)

Table VI. Selected Methods for Functionalization of Polymeric Supports[a] (footnotes on p. 55)

No.	Structure[b]	Synthesis	Analytical	Remarks	References
Copolymerization with functionalized monomer					
1.	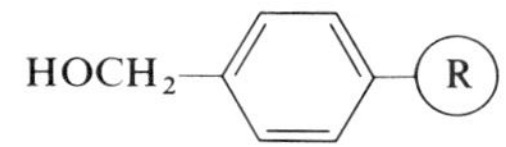	*p*-Hydroxymethylstyrene + styrene + divinylbenzene to give popcorn polymer		Copolymerization said to give better product than derivatization of preformed popcorn polymers (No. 49)	Letsinger *et al.* (1964)
2.	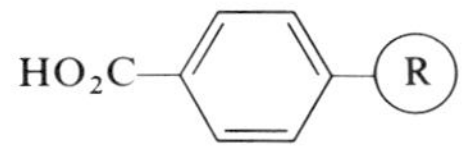	*p*-Carboxystyrene + styrene + divinylbenzene to give popcorn polymer	0.1 *N* NaOH–EtOH (1:1), reflux, and back-titrate	See above. Excess carboxy sites were blocked as methyl esters after conversion to acid chloride	Letsinger *et al.* (1964)
3.	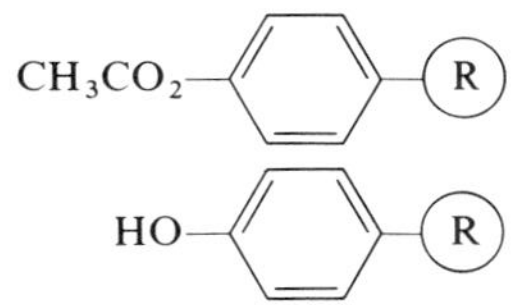	(i) *p*-Acetoxystyrene + styrene + divinylbenzene to give cross-linked gels (ii) $NH_2NH_2 \cdot H_2O$–dioxane (1:9), 25°C, 4 hr	ir of acetoxy (1760 cm^{-1})	Reactivity ratios of monomers determined to be near unity; therefore copolymerization presumably random in nature	Arshady *et al.* (1974); see Hudson *et al.* (1979); Rivaille *et al.* (1979) for applications
4.	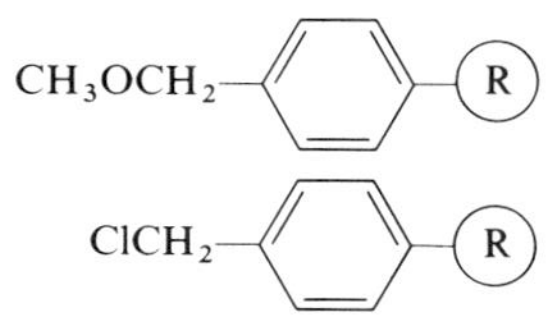	(i) *p*-Methoxymethylstyrene + styrene + divinylbenzene to give cross-linked gels (ii) BCl_3–CCl_4, 0°C	See No. 7	Peptide syntheses on these chloromethyl supports comparable to results with No. 7	Arshady *et al.* (1976)
5.	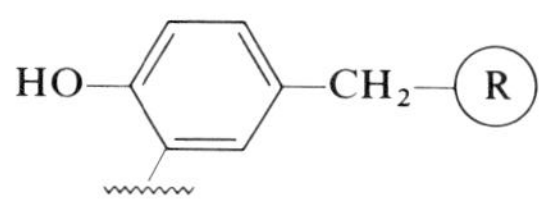	Phenol–formaldehyde copolymer (amorphous)		Related supports have been reviewed	Inukai *et al.* (1968); see also Erickson and Merrifield (1976, pp. 275–277)

Table VI. (*Continued*) (footnotes on p. 55)

No.	Structure[b]	Synthesis	Analytical	Remarks	References
6.	$BocNHCH_2CH_2CONH(CH_2)_6NH$—Ⓡ $H_2NCH_2CH_2CONH(CH_2)_6NH$—Ⓡ	(i) Boc-βAla-N'-acryloyl-hexamethylenediamine and other components to give cross-linked polydimethylacrylamide gel of Table IV, No. 5 (ii) 1 N HCl–HOAc, 30 min	Hydrolysis and amino acid analysis of βAla	An additional Leu or Nle can be coupled as "spacer-internal reference," compare to No. 32	Atherton *et al.* (1975)
Direct derivatization of a preformed polymer[c]					
7.	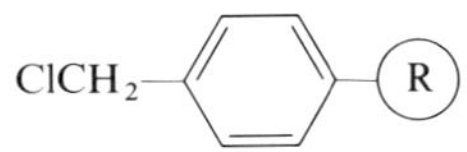	$ClCH_2OCH_3$–CH_2Cl_2 + $ZnCl_2$ 40°C (variable ratios and reaction times). This procedure seems most readily controlled; see, however, footnote *d*	Quaternization in pyridine, 100°C, 2 hr, and determination of Cl^- by Volhard titration	The original, most versatile and widely used functional grouping; undesirable cross-linking possible at high substitutions under Friedel–Crafts alkylation conditions	Merrifield (1963, 1964b); Feinberg and Merrifield (1974); Stewart and Young (1969, pp. 27, 55)
8.	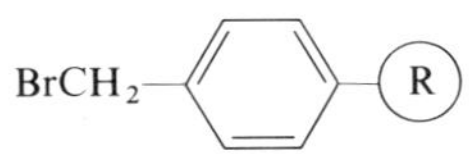	$BrCH_2OCH_3$ (neat) + $SnBr_4$, 0°C, 30 min	Volhard titration, as above	More reactive than chloromethyl (No. 7, above); compare to No. 30	Yajima *et al.* (1970)
9.	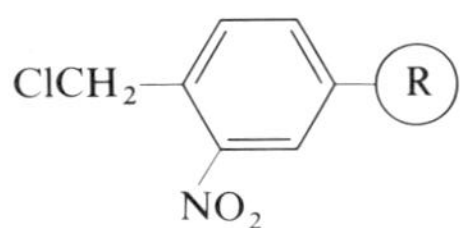	HNO_3 (90%), 0°C after preparation of No. 7	1.03 Nitro groups/aromatic ring by Dumas analysis	Historical interest; also new applications	Merrifield (1963); Laursen (1971); Rich and Gurwara (1973)
10.	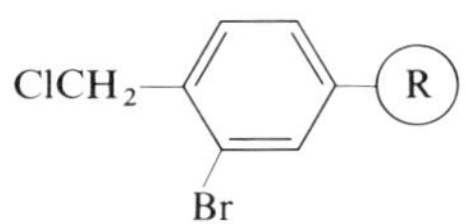	Br_2–CCl_4 + catalytic I_2 after preparation of No. 7	0.97 Br atoms/aromatic ring by elemental analysis	Aromatic Br group "inert" under conditions of peptide synthesis	Merrifield (1963); Li *et al.* (1977)

11.	 (also considerable *meta*?)	*n*BuLi–TMEDA–hexane, 60°C, overnight	Hydrolysis and titration of released base	Milder syntheses available, see No. 42	Fyles and Leznoff (1976)
12.		Br_2–CCl_4 + $Tl(OAc)_3$ catalyst	Elemental analysis	Some conflict in literature on efficacy of various methods	Camps *et al.* (1971); Burlitch and Winterton (1975); Farrall and Fréchet (1976)
13.	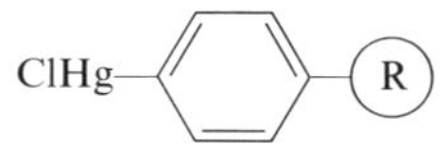	(i) HgO–CF_3COOH–CH_2Cl_2 (ii) $(CH_3)_4NCl$–CH_3OH		Exclusively aromatic mercuration	Burlitch and Winterton (1975)
14.	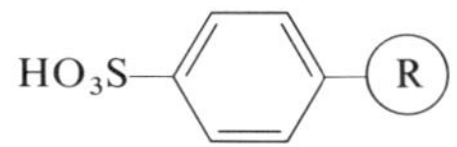	$ClSO_3H$–CH_2Cl_2	Titration with base	See Nos. 45 and 46 for further transformations	Reviewed by Patterson (1971)
15.	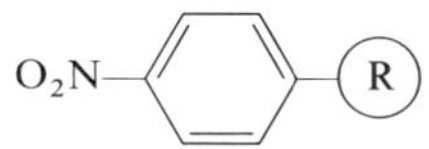 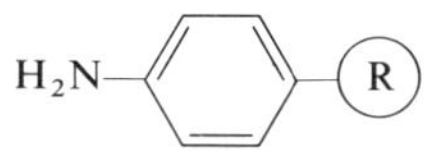	(i) HNO_3 (90%) (ii) $SnCl_2 \cdot 2H_2O$–HCl–DMF, 100°C	Elemental analysis, N	Aromatic amine useful for further transformations.	Dowling and Stark (1969); Patterson (1971)
16.	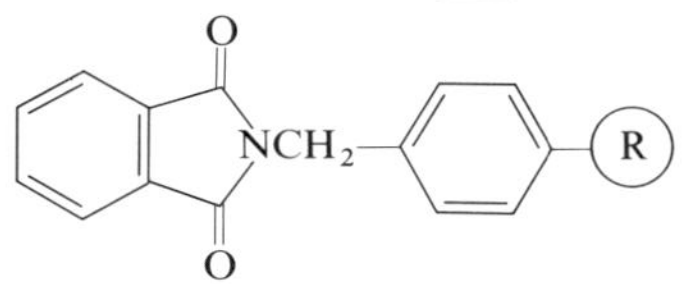 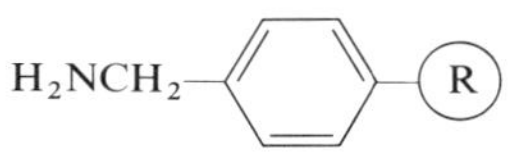	(i) *N*-hydroxymethyl-phthalimide–CF_3COOH–CH_2Cl_2 + 0.1 *M* CF_3SO_3H (ii) 5% NH_2NH_2–EtOH, reflux 16 hr *or* 27% CH_3NH_2–H_2O–CH_2Cl_2, 20°C, 48 hr	(i) Calibrated intensity ratio at 1720:1601 cm^{-1} in ir, or analysis for N (ii) Picrate titration, among others (Table XI)	Preferred method (compare to No. 31) for preparation of aminomethyl resins	Mitchell *et al.* (1976b, 1978b); Skorna *et al.* (1978)

Table VI. (*Continued*) (footnotes on p. 55)

No.	Structure[b]	Synthesis	Analytical	Remarks	References
17.	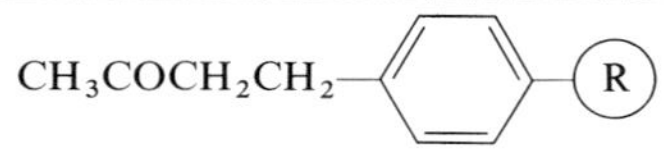	methyl vinyl ketone–HF, 25°C, 30 min		See Nos. 39 and 40 for further transformations	Wang and Merrifield (1969b); alternate route: Wolters *et al.* (1974)
18.	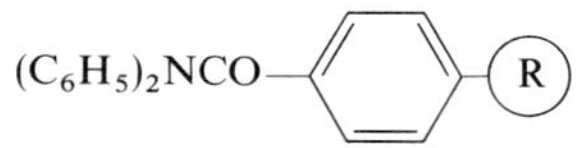	Diphenylcarbamoyl chloride–nitrobenzene + $AlCl_3$		See No. 47 for further transformations	Letsinger *et al.* (1964)
19.	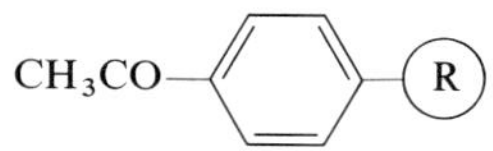	Acetyl chloride–nitrobenzene + $AlCl_3$		Same as above	Letsinger *et al.* (1964)
20.	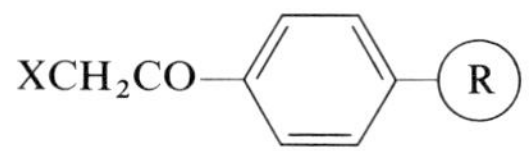 (X = Cl, Br)	Haloacetyl chloride–CH_2Cl_2 + $AlCl_3$	Volhard titration	Polymer-bound phenacyl bromide more reactive than chloride	Weygand (1968); Mizoguchi *et al.* (1970)
21.	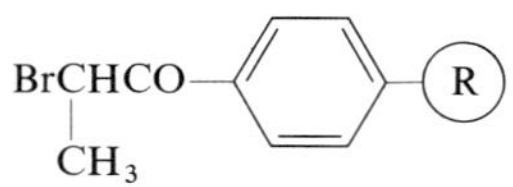	2-Bromopropionyl chloride–CH_2Cl_2 + $AlCl_3$	Elemental analysis, Br	Compare to No. 20, above	Wang (1976)
22.	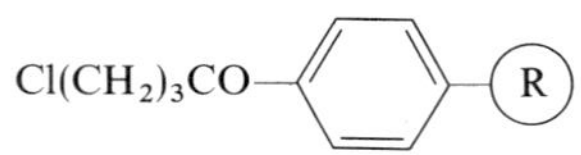 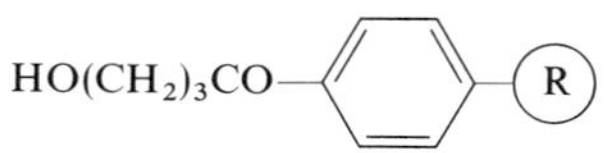	(i) 4-Chlorobutanoyl chloride–nitrobenzene + $AlCl_3$ (ii) HOAc–Et_3N–EtOH reflux (iii) Aqueous NaOH in dioxane–EtOH (1:1), one phase		Compare to Nos. 29, 35 and 58	Tilak and Hollinden (1968b)

23.	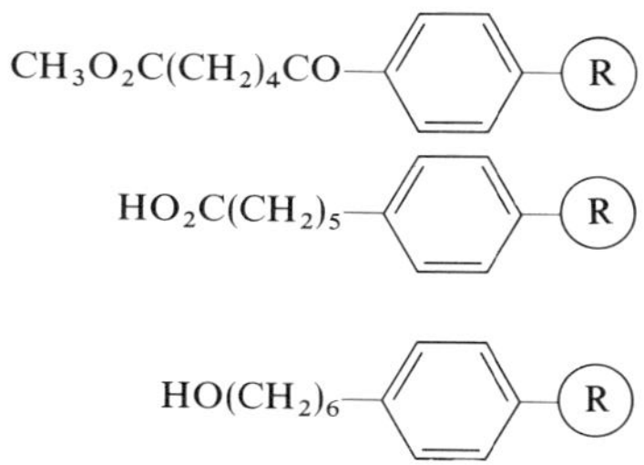	(i) 5-Methoxycarbonyl pentanoyl chloride + $AlCl_3$ (ii) Wolff–Kishner reduction; $NH_2NH_2 \cdot H_2O$–KOH in hot polyethylene glycol (iii) $LiAlH_4$–ether		Compare to Nos. 29 and 58	Bayer *et al.* (1971)
24.	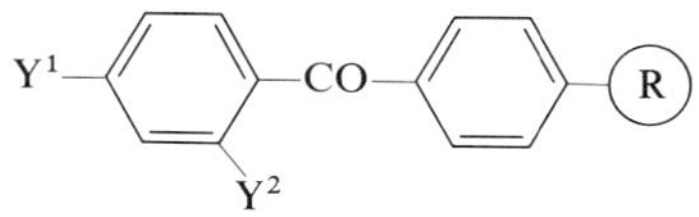 (Y^1, Y^2 = H, Cl, CH_3, OCH_3)	Substituted benzoyl chloride–nitrobenzene + $AlCl_3$	ir of ketone (1660 cm^{-1})	See Nos. 51–55 for further transformations	Southard *et al.* (1971c); Pietta *et al.* (1974); Losse and Meisegeier (1974); Frank *et al.* (1977a)
25.	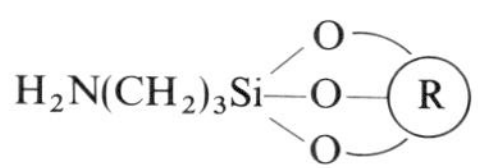	3-Aminopropyl tris-(ethoxy)silane in acetone, 45°C, 24 hr, with glass (surface SiOH sites)	Trinitrobenzene–sulphonic acid reaction	Aminopropyl glass available commercially	Robinson *et al.* (1971)
26.	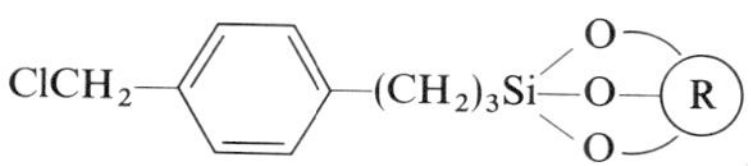	3-(*p*-Chloromethylphenyl)-propyltrichlorosilane in benzene with glass		Compare to No. 7	Parr *et al.* (1974)
27.	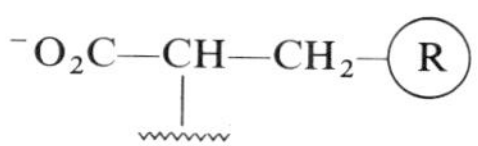	Partial saponification of polydimethylacrylamide support with excess dimethylamine		Support described Table IV, No. 7	Köster and Heidmann (1976)

Table VI. (*Continued*) (footnotes on p. 55)

No.	Structure[b]	Synthesis	Analytical	Remarks	References
Functional transformations					
28.	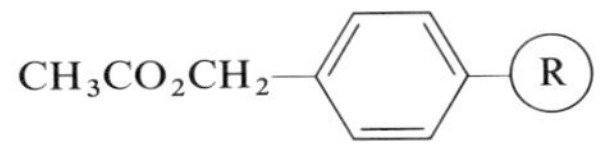	Chloromethyl resin (No. 7) + 1.0 *M* potassium acetate–2-methoxyethanol, 130°C, 24 hr	ir, and Volhard titration on released Cl^-	Proceeds directly to hydroxymethyl resin (No. 29) when reaction is performed 70 hr (acetate cleaved by transesterification)	Stewart and Young (1969, p. 26); Erickson and Merrifield (1973b)
29.	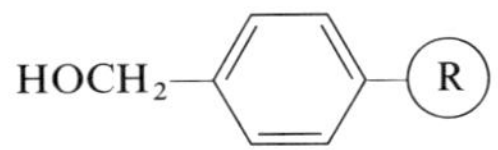	Acetoxymethyl resin (No. 28) + (a) see "remark," No. 28 (easiest method) *or* (b) 0.5 *N* NaOH–dioxane (1:2), 25°C, 48 hr *or* (c) Et_2NH reflux, 12 hr *or* (d) $LiAlH_4$–ether, 25°C, 4 hr *or* (e) anhydrous hydrazine–DMF (1:9), 25°C, 76 hr	Monitor synthetic reactions for completeness (compare to No. 28)	Compare to No. 58	Bodanszky and Sheehan (1966); Stewart and Young (1969, pp. 27, 33); Gisin and Merrifield (1972b); Wang (1975)
30.	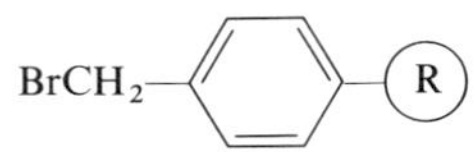	Acetoxymethyl resin (No. 28) + HBr–HOAc		Compare to No. 8	Tilak (1968); Merrifield (1969, p. 238)
31.	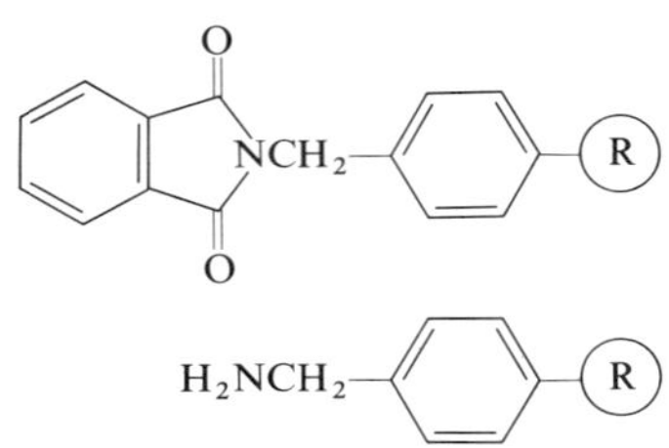	Chloromethyl resin (No. 7) + potassium phthalimide–DMF, 120°C, 5 hr; converted to aminomethyl resin as before (No. 16)	See No. 16	Direct reaction of chloromethyl resin with NH_3 is slow; undesirable cross-linking possible	Weinshenker and Shen (1972); Sparrow (1976); Mitchell *et al.* (1976a); see also Rich and Gurwara (1975a)

No.	Structure	Preparation	Characterization	Comments	Reference
32.	 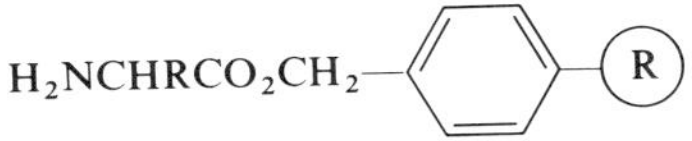	Sections III,B,1,a (esterification) and IV,A (deprotection)	Amino acid analysis	Anchoring handles may be extended off free α-amino groups	
33.		Chloromethyl resin (No. 7) + *p*-hydroxyphenylacetic acid + Et_3N–EtOH, reflux	Hydrolysis and uv of released phenol	Compare to Nos. 3 and 5; phenolic groups were acetylated	Blake and Li (1971)
34.	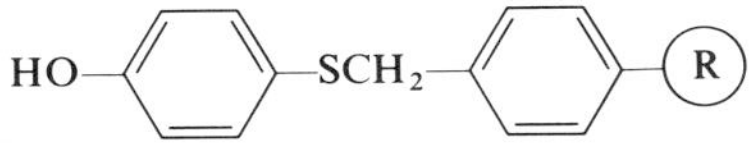	Chloromethyl resin (No. 7) + *p*-mercaptophenol–KOH–DMF reflux		Compare to Nos. 3, 5, and 33	Marshall and Liener (1970); Flanigan and Marshall (1970)
35.	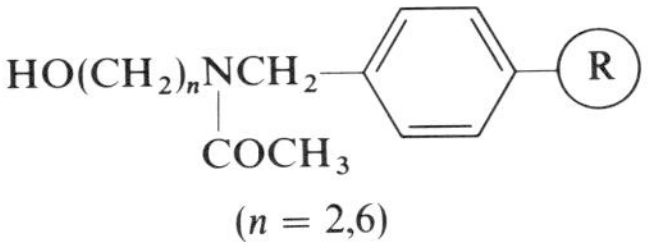	Chloromethyl resin (No. 7) + hydroxyalkylamine in dioxane, reflux, or DMF, 50°C, respectively, followed by acetylation and saponification	ir	Compare to Nos. 22 and 23; undesirable cross-linking possible during amine alkylation step	Tilak and Hollinden (1968b)
36.	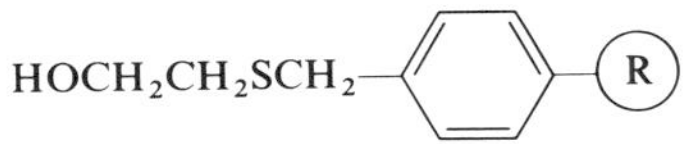 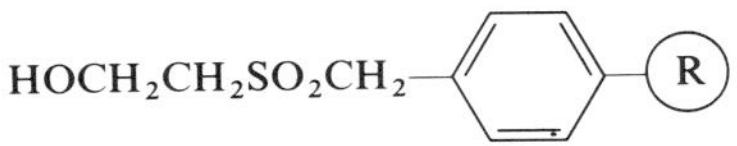	(i) Chloromethyl resin (No. 7) + β-mercaptoethanol + Na in liq. NH_3, then DMF (ii) *m*-Chloroperbenzoic acid–CH_2Cl_2, short spontaneous reflux	Analysis for sulfur, and residual Cl^-	Synthetic reactions apparently quantitative	Tesser *et al.* (1976)
37.	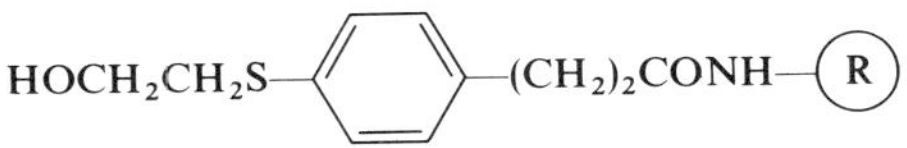	Amino resin (No. 6) + pentachlorophenyl ester of *p*-(2-hydroxyethylthio)phenylpropionic acid		Compare to Nos. 36 and 59	Gait and Sheppard (1976)

Table VI. (*Continued*) (footnotes on p. 55)

No.	Structure[b]	Synthesis	Analytical	Remarks	References
38.	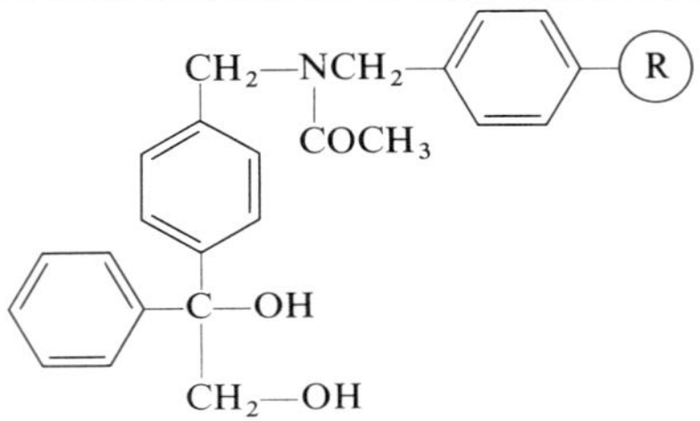	Chloromethyl resin (No. 7) + intermediate from 6-step synthesis (overall 3%); two further synthetic steps (acetylation and saponification) performed on derivatized resin		Elimination of tertiary alcohol promoted by TFA [compare to Eq. (5) on p. 88].	Wieland *et al.* (1972a); Birr *et al.* (1973)
39.	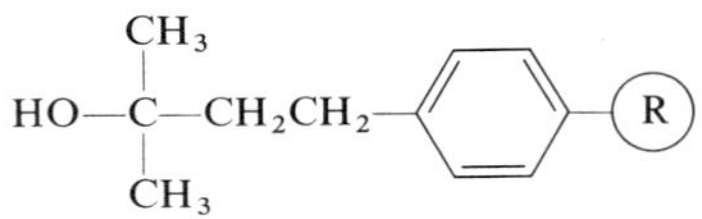	Keto resin (No. 17) + CH_3MgBr–ether	ir	Resin-bound tertiary alcohol	Wang and Merrifield (1969a, 1972)
40.	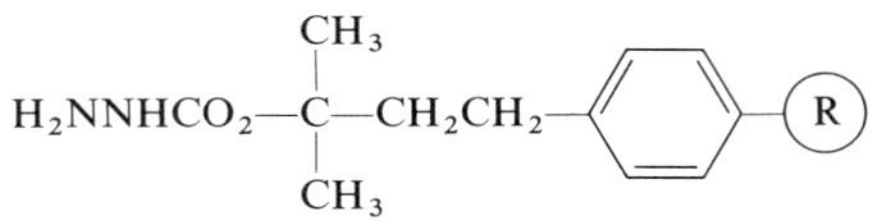	*tert*-Alcohol resin (No. 39) + (i) C_6H_5OCOCl in pyridine–CH_2Cl_2; (ii) $NH_2NH_2 \cdot H_2O$–DMF (1:1), 25°C, 25 hr	Nitrogen analysis, and ir	Compare to Nos. 57 and 60	Same as for No. 17
41.	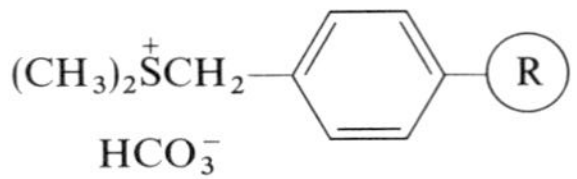	Chloromethyl resin (No. 7) + dimethylsulfide in CH_2Cl_2–CH_3OH–H_2O (1:1:1), 25°C, 4 days; converted with 1 *N* $KHCO_3$	Titration of released ionically and covalently bound Cl^-	Residual methylthio-methyl sites formed after heating were not detrimental	Dorman and Love (1969)
42.	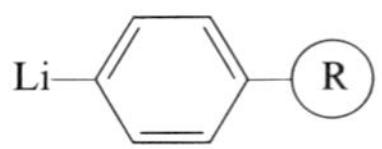	Bromo or mercurochloro resins (Nos. 12 or 13) + *n*BuLi in toluene, 25°C		Compare to No. 11	Same as for No. 12
43.	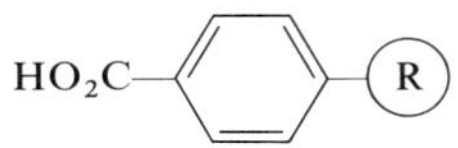	Lithio resin (No. 42) + carbon dioxide	See No. 2	Most convenient route to carboxyl resin	Farrall and Fréchet (1976); Scott *et al.* (1977)

44.	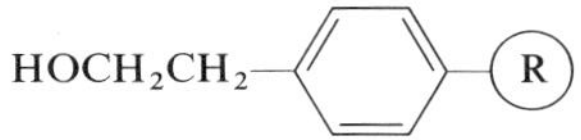	Lithio resin (No. 42) + ethylene oxide in toluene or THF	*N*-analysis on 3,5-dinitrobenzoate	Indistinguishable by ir from copolymerization product	Camps *et al.* (1971); Farrall and Fréchet (1976)
45.	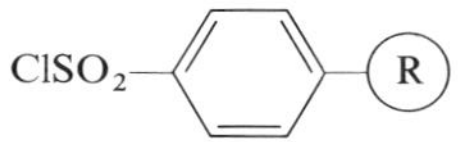	Sulfonic acid resin (No. 14) + $SOCl_2$–DMF		Sulfonyl chloride sites also formed with $ClSO_3H$, see No. 46	Dahlmans (1973)
46.	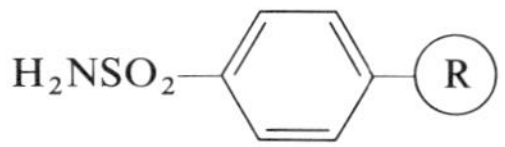	Sulfonic acid resin (No. 14) + (i) $ClSO_3H$; (ii) aqueous NH_4OH		During synthesis: (i) some sulfonic anhydride sites formed; (ii) opened to give 7% recovered sulfonic acid sites	Kenner *et al.* (1971); R. B. Merrifield, unpublished results (1961)
47.	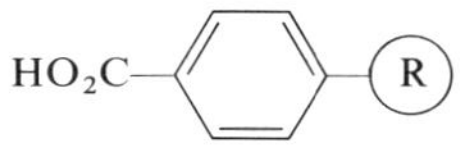	Diphenylcarbamoyl resin (No. 18) + HOAc–H_2SO_4–H_2O (7:5:3), 135°C, 32 hr *or* acetyl resin (No. 19) + hypobromite oxidation	See No. 2	Historical interest; compare to No. 2	Letsinger *et al.* (1964); Fréchet and Haque (1975)
48.	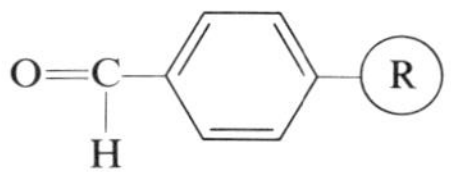	Chloromethyl resin (No. 7) + DMSO–$NaHCO_3$, 155°C, 6 hr	N-analysis on oxime	Yield of DMSO oxidation = 70%	Fréchet and Schuerch (1971)
49.	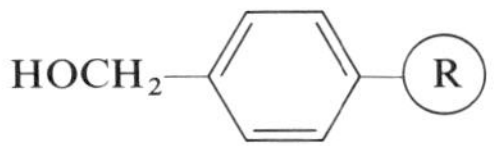	Carboxyl resin (Nos. 2, 43, 47 or 51) + $LiAlH_4$–ether		Compare to Nos. 1 and 29	Letsinger *et al.* (1964)
50.	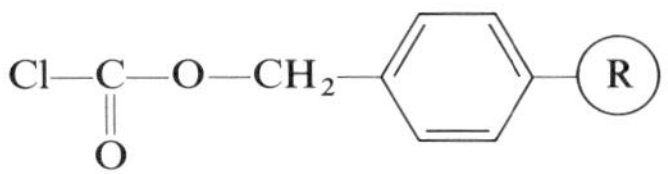	Hydroxymethyl resin (No. 29 or 49) + phosgene–benzene, 25°C, 4 hr	Volhard titration, as for No. 7	Excess chloroformyl sites were hydrolyzed, or reacted with dimethylamine	Letsinger *et al.* (1964); Felix and Merrifield (1970)
51.	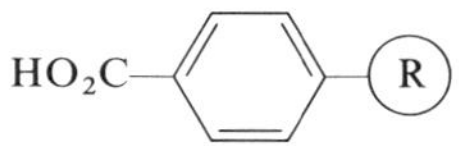	*o*-Cl-Benzoyl resin (No. 24) + potassium *tert*-butoxide in H_2O–dimethoxyethane, reflux, 10 hr	Release of chlorophenyl groups	Standard back-titration with base was criticized. Carboxyl groups were readily esterified with diazomethane	Harrison *et al.* (1975)

Table VI. *(Continued)*

No.	Structure[b]	Synthesis	Analytical	Remarks	References
52.	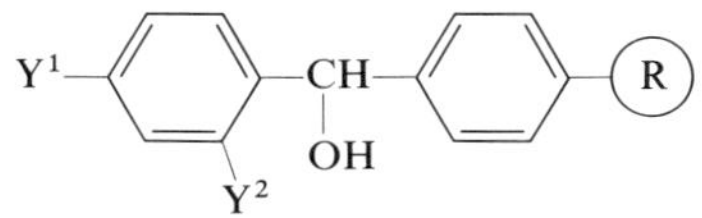	Benzoyl resin (No. 24) + $NaBH_4$–diglyme, 60°C, 4 hr *or* $LiAlH_4$–THF, 20°C, 2 hr	ir	Can be converted to phenyl carbonate; by analogy to Nos. 40 and 50	Same as for No. 24
53.	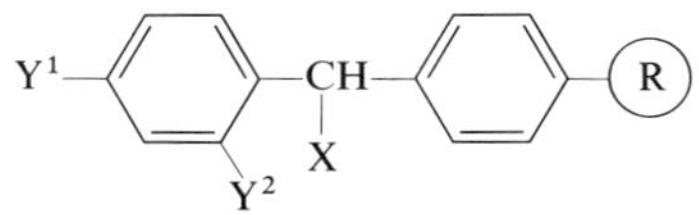 (X = Br, Cl)	Carbinol resin (No. 52), suspended in CH_2Cl_2 + dry hydrogen halide gas, 25°C, 1 hr	Volhard titration, as for No. 7	This synthesis gives quantitative yields	Southard *et al.* (1971c); Pietta *et al.* (1974)
54.	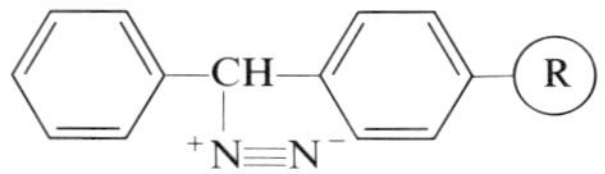	Benzoyl resin (No. 24) + (i) $NH_2NH_2 \cdot H_2O$–*n*BuOH, reflux, 24 hr (ii) CH_3CO_3H–TMG–trace I_2–CH_2Cl_2, 0°C	N_2 evolution after treatment with organic acid	Support stable in dark, 0°C, 6 months	Chapman and Walker (1975)
55.	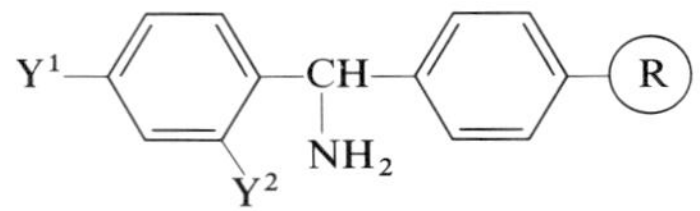	Bromobenzhydryl resin (No. 53) in CH_2Cl_2 + NH_3 gas, 0°C *or* (preferred) Leuckart reaction on benzoyl resin (No. 24): (i) ammonium formate–HCOOH–nitrobenzene, 170°C, 5 hr (ii) ~3 *N* HCl in aqueous EtOH, reflux, 2 hr	Dorman titration, among others (Table XI); nitrogen content after diazotization is direct measure of amount of contaminating secondary amine	An earlier recommended synthetic route via metal hydride reduction of oxime intermediate was shown to give substantial secondary benzhydrylamine by-product	Pietta and Marshall (1970); Rivier *et al.* (1973); Pietta *et al.* (1974); Orlowski *et al.* (1976)

56. 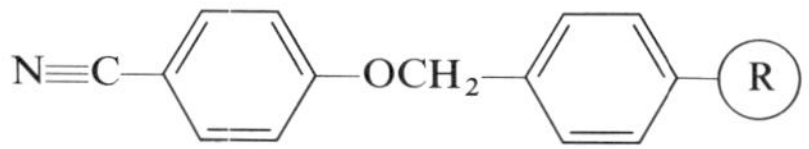 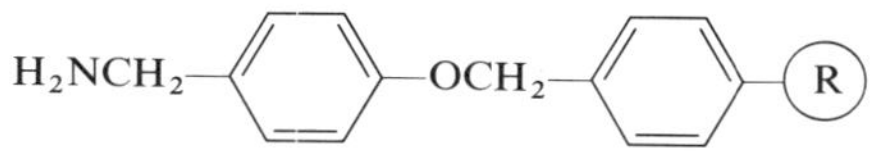	(i) Choromethyl resin (No. 7) + cyanophenol–NaOMe (1:1) in diglyme, 60°C, 2 hr (ii) $LiAlH_4$–ether + stream of dry ammonia *or* multistep route through *p*-alkoxybenzyl resin (No. 58)	No cyano band in ir; see also No. 55 above	Danger of overreduction of nitrile to alcohol. Excess aminomethyl sites were acetylated in presence of trace dimethylaminopyridine	Pietta and Brenna (1975)
57. 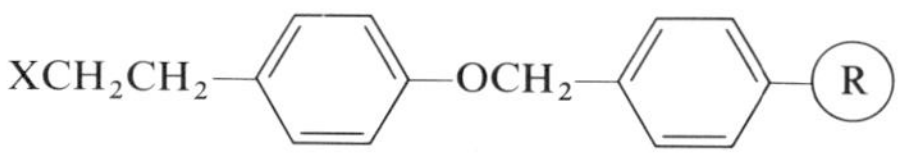 (i) $X = CH_3CO$, (ii) $X = H_2NNHCO_2{-}C(CH_3)_2{-}$	(i) Chloromethyl resin (No. 7) + 4-(*p*-hydroxyphenyl)-2-butanone–NaOMe (1:1) in DMF, 85°C, 24 hr (ii) corresponding sequence of steps as Nos. 17, 39, and 40	ir and elemental analysis	Compare to No. 40; see structure **35** on p. 97	Wang (1975)
58. 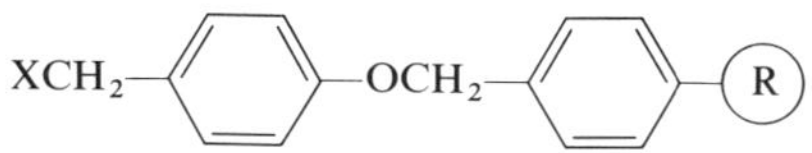 (i) $X = CH_3O_2C$ (ii) $X = HO$	(i) Chloromethyl resin (No. 7) + methyl 4-hydroxybenzoate–NaOMe (1:1) in DMA, 80°C, 24 hr (ii) $LiAlH_4$–ether, 25°C, 6 hr, *or* direct reaction with 4-hydroxybenzyl alcohol, conditions of first step (substantially simpler)	ir, and OCH_3 determination, see also No. 29	Compare to No. 29; see structure **16** on p. 76	Wang (1973)

Table VI. (*Continued*)

No.	Structure[b]	Synthesis	Analytical	Remarks	References
59.		Amino resin (No. 6) + 2,4,5-trichlorophenyl ester of *p*-hydroxymethylphenoxyacetic acid in DMF		Compare to Nos. 58 and 61; strategy ought to be applicable to all supports with amino sites (Nos. 15, 16, 25, 31, 32)	Atherton *et al.* (1978a)
60.	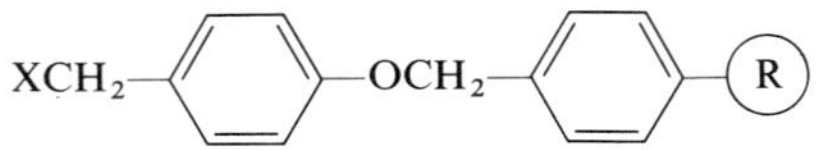 $X = H_2NNHCO_2$—	*p*-Alkoxybenzyl alcohol resin (No. 58) + same synthetic sequence as No. 40		Compare to Nos. 40, 57, and 58	Wang (1973)
61.	 $X = CH_3CO_2$, Br, HO, Protected-$NHCHRCO_2$	Aminomethyl resin (No. 16 or 31) coupled via DCC in CH_2Cl_2 to appropriate *p*-substituted phenylacetic acid derivative		Coupling of protected amino acyl-4-(oxymethyl)phenylacetic acid derivatives represents preferred anchoring strategy (pp. 64–66)	Mitchell *et al.* (1976a, 1978b); Sparrow (1976)
62.		(i) Chloromethyl resin (No. 7) + 30% *n*-hexylamine in CH_2Cl_2, 4°C, 7 days (ii) Symmetrical anhydride of *p*-chloromethylphenylacetic acid		Compare to No. 7; there is a risk of cross-linking in first step of synthetic sequence	Blake and Li (1976)
63.	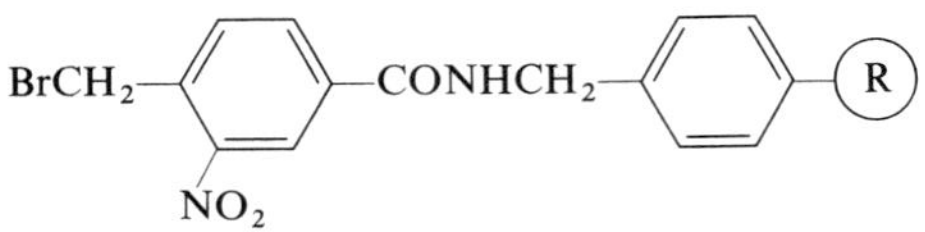	Aminomethyl resin (No. 16 or 31) coupled via DCC in CH_2Cl_2 to 3-nitro-4-bromomethylbenzoic acid		Compare to No. 9	Rich and Gurwara (1975a)

64. 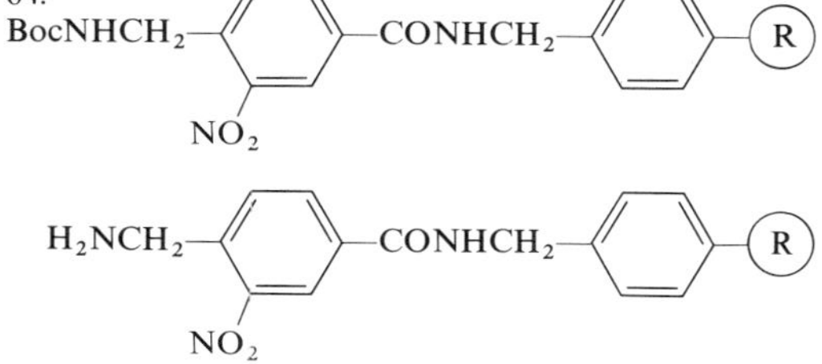	(i) Aminomethyl resin (Nos. 16 or 31) coupled via DCC in CH_2Cl_2 to 3-nitro-4-(Boc-aminomethyl)benzoic acid (ii) Removal of Boc-group		Compare to No. 63	Rich and Gurwara (1975b)
65. 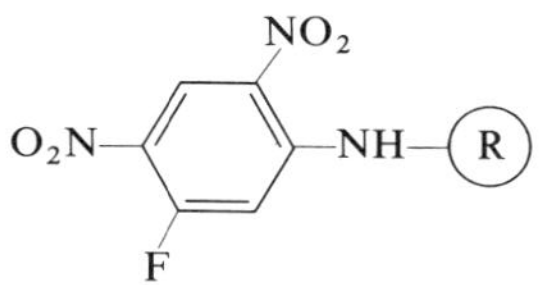	Resin with amino groups (No. 32 was used) + 1,5-difluoro-2,4-dinitrobenzene (large excess) + Et_3N in CH_2Cl_2	Test disappearance free amino groups (ninhydrin)	Resin-bound Sanger reagent; see Fig. 9 on p. 98	Glass *et al.* (1972)

[a] Where several methods are available for preparation of the same functionalized polymer, we have tried to indicate whenever possible which method gives the most reliable chemistry.

[b] (R) represents the remainder of the polymeric support.

[c] The products of electrophilic aromatic substitution reactions have been drawn as the *para* isomer, although usually there is no clear experimental evidence on this point, and a mixture is likely in some cases. Inductive effects preclude any substantial amounts of *meta* isomer, while (large) substituents would not be expected *ortho* to the polymeric backbone on steric grounds.

[d] Chloromethyl methyl ether and (especially) its normal contaminant, bis(chloromethyl) ether have both been designated as potent carcinogens by the Occupational Safety and Health Administration, U.S. Department of Labor, *Federal Register* **38**, 10929 (1973); **39**, 3756 (1974).

Sparrow (1975) has suggested use of the higher boiling, probably also carcinogenic, chloromethyl *ethyl* ether to effect chloromethylation (solvent was hexane; catalyst was boron trifluoride etherate). Chloromethylated polystyrene resins are available commercially from Bio-Rad, Lab Systems, Pierce and other suppliers; these may not necessarily be of comparable quality.

synthesis, while others have been included because of related applications or for their potential as intermediates for anchoring via "handles." At the primary level, the various methods can be categorized as involving either copolymerization with a functionalized monomer (numbers 1–6) or direct derivatization of a preformed polymer (numbers 7–27). Furthermore, desired functional groups may be created by high-yield transformations (numbers 28–65) of the primary derivatized sites. Analytical methods to determine the extent of functionalization are also provided (Table VI, column 3); this information is needed to plan the subsequent attachment of the initial synthetic unit.

The copolymerization route has not been widely applied, since functionalized monomers are occasionally difficult to synthesize in pure form, and few laboratories are equipped or have the experience to carry out the actual polymerization. The substitution level and the extent of cross-linking in the product polymer can be controlled by the composition of the starting polymerization mixture. When the reactivity ratio of the monomers is near unity, the functional groups are expected to become randomly distributed throughout the linear structure of the polymer; in this way, the hypothetical "local clustering" of sites which might occur in the course of direct derivatization could be avoided. On the other hand, it is conceivable that some of the functional groups will be found in sterically unfavorable positions of the three-dimensional polymeric matrix, for example adjacent to cross-links, and thus will subsequently react more slowly.

The direct derivatization approach has been the bulwark of most solid-phase methodology. More specifically, aromatic substitutions (numbers 9–15), and Friedel–Crafts alkylations (numbers 7, 8, 16, and 17) or acylations (numbers 18–24) have served as the principal chemical reactions for the initial introduction of functional groups into the polystyrene-based resins most commonly used in solid-phase peptide synthesis. A major danger attendant to many of these methods lies in the potential for creation of new cross-links in the polymeric matrix. The formation of methylene or sulfone bridges between aromatic rings during chloromethylation (number 7) or sulfonation (number 14) of polystyrene resins is well documented, and is even exploited in the industrial preparation of ion-exchange resins (Patterson, 1971). Chloromethylation of linear (soluble) polystyrene in the absence of a cosolvent yielded a gelatinous insoluble product, presumably due to cross-linking (Green and Garson, 1969). The problem is expected to be accentuated by longer reaction times, more active Friedel–Crafts catalysts, more reactive functional groups (e.g., bromomethyl > chloromethyl; Table VI, number 8 vs. number 7), and higher overall levels of functionalization.

Additional cross-linking can also arise in the course of further transformations. Rich and Gurwara (1975a) found marked changes in the swelling characteristics of a 1% cross-linked polystyrene resin upon conversion of chloromethyl sites to aminomethyl sites by treatment with ammonia under certain conditions. Tjoeng *et al.* (1977) showed that even in the best cases, the titration value of the produced amino groups amounted to only 53% of the original 0.75 mmol/gm substitution of chloromethyl groups, although displacement of chloride was quantitative in accordance with the results of Rich and Gurwara (1975a). An unambiguous route to the desired functionalized support that circumvents the danger of cross-linking involves initial alkylation of a chloromethyl resin with potassium phthalimide (Table VI, number 31), or even better, direct Friedel–Crafts amidoalkylation of underivatized polystyrene with *N*-hydroxymethylphthalimide (Table VI, number 16). The phthaloyl protecting group prevents further alkylation (with cross-linking) of the amino function. The synthetic methods of Table VI that include alkylation steps with amines (numbers 35, 55 [first route], and 62) or thiols (numbers 34 and 36; thioethers might well be converted to sulfonium salts, with ion-exchange properties) ought to be reevaluated for the possibility of cross-linking side reactions.

B. Chemistry of the Anchoring Linkage

Most of the protecting groups that have found application to classical methods of peptide synthesis can in principle be adapted to serve as the basis for an anchor to the polymeric support. There are three increasingly sophisticated types of anchoring strategies.

1. In the simplest and most common case, the polymeric support is converted, by a short and reliable sequence of chemical steps, into a suitable analogue of a reagent used to introduce the appropriate protecting group. Then, the method of attachment of the initial synthetic unit to the solid phase follows the same chemistry as the solution derivatization. For example (see Fig. 1), a benzyl ester linkage can be created by reaction of a protected amino acid salt with a resin-bound benzyl chloride. The nature of the initial functionalization will dictate the yield of the anchoring step and the stability of the ensuing anchoring linkage; the various commensurate goals do not necessarily dovetail.

2. A more flexible approach involves the use of a polymeric support containing a "handle". In the general context required here, a handle is defined as a bifunctional spacer or extender that incorporates the features of the desired protecting group and serves to link the initial synthetic unit of the peptide chain to a previously functionalized solid support in two discrete steps. In this variation, the handle is first attached to the support (for examples, see Table VI, numbers 25, 26, 33–38, 58–65), and the linkage to the synthetic unit is formed second.

3. The most versatile and advantageous anchoring strategy again exploits a handle; but in this variation, the bond to the initial synthetic unit is created first, and its reaction with the functionalized support is carried out second. This allows the fully protected soluble intermediate to be well purified and characterized before attachment to the solid support. Quantitative anchoring can be assured by forming an amide linkage between a carboxyl-containing handle and an amino support, or vice versa. The earliest efforts along these lines from this laboratory have been reviewed (Merrifield, 1969; p. 239; see also Fig. 5 in Erickson and Merrifield, 1976); more recent applications are those of Wieland *et al.* (1970), Atherton *et al.* (1975, 1977), and Mitchell *et al.* (1976a, 1978b).

One of the potential advantages of handle linkages, as applied in variations (2) and (3), has been thought to be a result of the increased distance between the growing peptide chains and the polymeric backbone (see p. 31). Alternatively, the beneficial effects of these approaches may have a chemical basis, in that it is possible to "fine-tune" accurately the stability of the anchoring bond to make it more or less labile as required by the synthetic strategy. The relative merits of these points of view are still uncertain.

Because the preferred variation (3) often requires substantial chemical synthesis and purification prior to the commencement of the solid-phase steps, the easier variations (1) and (2) have usually been applied. The main objection to these latter, chemically more ambiguous, approaches lies in the difficulty of obtaining quantitative yields. Excess free functional sites can occasionally be safely ignored; in other cases, they must be effectively blocked lest they interfere with subsequent chemical steps of the solid-phase synthesis. On the other hand, where it is possible to replace completely the polymer-bound functional groups with the starting synthetic unit, it is important to control the final level of substitution at the initial functionalization step. Side reactions occurring during any of the multiple interconversion and derivatization steps carried out on the resin might result in a heterogenous starting point for the peptide synthesis.

The converse of the anchoring step is the cleavage reaction performed after the desired protected peptide has been assembled on the polymeric support. Depending on the precise cleavage conditions as well as the chemical structure of the anchoring linkage, the originally attached functionality may be released in its free form or as some other useful derivative. For example, the benzyl ester of Fig. 1 can be cleaved by base, methanol, or hydrazine to give, respectively, the peptide acid, methyl ester, or hydrazide.

The final cleavage reaction will not necessarily proceed in quantitative yield. There are three possible general reasons for this. (1) The originally functionalized support may have contained isomeric structures, one of which gives rise to an entirely stable linkage [for a pertinent example, see Orlowski *et al.* (1976), summarized in Table VI, number 55]. (2) The kinetics of cleavage may be such that the reaction is purposely not carried to completion in order to avoid competing, harmful side reactions that might occur upon continued treatment of the cleaved product under the reaction conditions. This problem can sometimes be circumvented by multiple, short treatments followed by rapid quenching of the cleavage reagent. (3) Rearrangement or modification of the anchoring linkage might occur either during the repetitive steps of the synthesis or during the cleavage, so a fixed portion of peptide chains are not removed even upon more extended, drastic treatment. This class of difficulties has sometimes been attributed to steric effects; it may often be mitigated by the use of scavengers throughout the synthetic operations.

We have earlier recommended (pp. 31–32) an optimal desirable range of initial loading for solid-phase peptide synthesis commensurate with the size of the final desired peptide. There are two important caveats to be added with regard to such calculations.

1. With many practical combinations of temporary protecting groups and anchoring linkages, the insufficient chemical selectivity of the deprotection conditions results in a relatively steady per step level of premature loss of chains from the support. Therefore, a correction (which may be substantial for long target sequences) must be made to predict the level of loading at the end of the synthesis.

2. Within the first one or two deprotection/coupling cycles, a unique possibility often arises for significant losses or modifications (termination) of chains through side reactions, such as diketopiperazine formation. The true amount of growing peptide chains should therefore be determined experimentally (by amino acid analysis or titration) at the tri- or tetrapeptide stage.

1. *Substituted Benzyl Esters*

By far the most common anchoring linkage used in solid-phase peptide synthesis has been a substituted benzyl ester of the COOH-terminal group. The lability of the ester bond to acidic reagents can be increased or decreased, respectively, by the introduction of electron-donating or -withdrawing substituents on the aromatic ring. At the same time, the susceptibility of the benzyl ester to nucleophilic reagents is affected in the opposite direction, although not over as wide a range.

a. Synthesis. Benzyl esters have been prepared by (1) O-alkylation with benzyl halide-containing or related supports (Table VI, numbers 4, 7–10, 26, 30, 41, 61–63) using carboxylate salts and (2) acylation of benzyl alcohol supports (Table VI, numbers 29, 58, and 59) with suitable COOH-activated derivatives. The handle method of introducing a substituted benzyl ester linkage is exemplified in Fig. 5, later in this section; this approach avoids essentially all of the known difficulties (see below) associated with techniques relying on direct esterification to a functionalized support as the mode of anchoring.

(i) Alkylation. The original and still widely used method of attachment (Merrifield, 1963, 1964b) involves reaction of chloromethyl-polystyrene resin (1 equiv Cl) with the N^{α}-protected amino acid (1 equiv) and triethylamine (0.9 equiv) in ethyl acetate or ethanol at reflux for 24–65 hr. Most amino acids can be introduced by this procedure and there is no detectable racemization. The yields are 35 to 50% (reviewed by Merrifield, 1969, pp. 234–237). Side reactions are observed with methionine, which can undergo S-alkylation by chloromethyl groups (Sieber and Iselin, 1968b) and with asparagine, which can undergo alcoholysis or deamidation. These are circumvented, respectively, by esterifying methionine sulfoxide derivatives, or in the case of asparagine by changing to dimethylformamide (DMF) as the solvent, at 25°C for 24 hr (Marglin, 1971). Unreacted chloromethyl sites have generally been ignored, but have occasionally been acetylated (Merrifield, 1963). The recent discovery by Schou *et al.* (1976) and Christensen and Schou (1978) of an N-alkylation side reaction involving excess chloromethyl groups (Fig. 4) justifies the need for methods that give quantitative conversion.

An additional drawback of the standard esterification method lies in the creation of quaternary ammonium sites, which confer permanent ion-exchange characteristics to the resin.* The extent of this reaction has been shown to vary with the nature of the solvent, but more useful ways to avoid this problem have relied on the uses of bases other than tertiary amines to prepare the carboxylate salt. Thus, Loffet (1971) obtained reasonable yields (50–70%; no racemization) upon overnight reaction of the chloromethyl resin (1 equiv Cl) with a thoroughly dried *tert*-butyloxycarbonyl-amino acid tetramethylammonium salt (**1**) (1 equiv) in dioxane at 80°C. The procedure has been modified and found to give excellent results in the laboratory of Yamashiro and Li (1973a, 1976). Taking a cue from the work of Marglin (1971), they used DMF as the solvent, at 25°C. For sluggishly reacting substrates, potassium iodide (15 m*M*) was added as a catalyst, and the reaction temperature was raised to 55°C for several hours. Suzuki and Endo (1977) recently reported on the use of the nonnucleophilic tertiary amine 1,5-diazabicyclo[4.3.0]non-5-ene (DBN) (**2**) as a base for the attach-

* An important commercial preparative route to anion-exchange resins involves the treatment of chloromethyl-polystyrene with tertiary amines (e.g., Pepper *et al.*, 1953; see also citations to patent literature quoted therein).

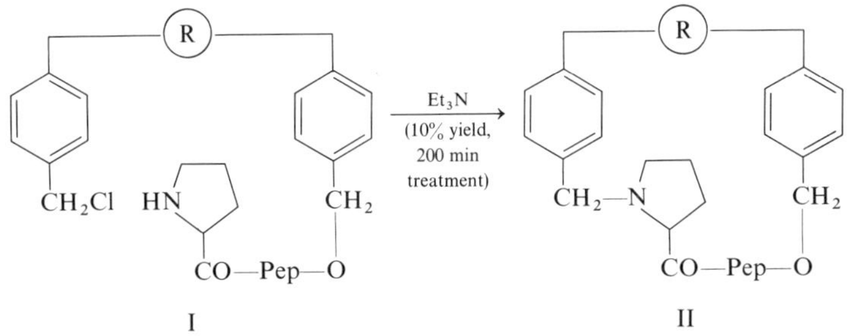

Figure 4. N-Alkylation of proline residues by excess resin-bound chloromethyl groups during solid-phase peptide synthesis, after Schou *et al.* (1976). The by-product II was detected by an increase in perchloric acid titration values, and remained covalently bound to the support after acidolytic cleavage of the benzyl ester linkage.

ment of protected amino acids, and di- and tripeptides. Reactions were in dimethylformamide, at 25°C, for 24 hr; yields were good to excellent (70–100%, based on the amino acid; chloromethyl groups generally in excess) and the absence of racemization was demonstrated by two sensitive chromatographic tests.

(1) (2)

The application of alkali metal salts for the esterification is another obvious way to circumvent the quaternization side reactions. One of the best and most general techniques to prepare resin-bound benzyl esters, for both high yields and convenience, involves the use of cesium salts (Gisin, 1973; see also Wang *et al.*, 1977). A protected amino acid or peptide, in alcohol, is titrated to neutrality with aqueous $CsHCO_3$ or Cs_2CO_3 (the latter reagent is more economical). After repeated evaporation to dryness from DMF, the salt is reacted, without further purification, with a chloromethyl resin in DMF at 50°C, for 16 hr. In this way, essentially all of the chloromethyl sites on the support could be substituted by use of only a slight excess of the cesium salt. Somewhat better reaction rates were obtained by substituting dimethylacetamide or *N*-methylpyrrolidinone for DMF (B. F. Gisin, unpublished observations, 1976).

Potassium salts of *tert*-butyloxycarbonylamino acids formed *in situ* with potassium *tert*-butoxide (0.9 equiv) have been reacted with chloromethyl-resins to yield amino acyl-resins at useful loadings; dimethyl sulfoxide was employed as the reaction solvent, at 80°C for 30 min (Monahan and Gilon, 1973). However, these conditions may lead to the creation of aldehyde sites

(Fréchet and Schuerch, 1971; compare to Table VI, number 48). In another variation, the dried potassium salts prepared from KOH have been used as their 1:1 complex with the crown ether 1,4,7,10,13,16-hexaoxacyclooctadecane (18-crown-6). Quantitative esterification was achieved in DMF at 50°C, for 16 hr (Roeske and Gesellchen, 1976).

Resin-bound substituted benzyl bromides are more reactive than the chlorides, and have been used to achieve anchorings in difficult cases, or when quantitative substitution was desired. By the same token, however, bromomethyl sites are expected to be considerably more prone to alkylation and cross-linking side reactions of the sort already described for chloromethyl sites. Yajima *et al.* (1970; Yajima and Kiso, 1974) were able to attach the dicyclohexylammonium (more effective than triethylammonium) salts of protected peptides (up to length of 4 amino acid residues) to a bromomethyl polystyrene support (Table VI, number 8). Reactions were performed in DMF, at 25°C, for 48 hr, and yields were good (50–70%, based on peptide, since bromomethyl sites were in substantial excess). Rich and Gurwara (1975a) obtained quantitative substitution of a support with an *o*-nitrobenzyl bromide handle (Table VI, number 63) by reverting to the original conditions of Merrifield (1963). Diisopropylethylamine was substituted for triethylamine to minimize quaternization* and a threefold excess each of the *tert*-butyloxycarbonylamino acid and tertiary amine was refluxed in ethyl acetate for 48 hr. For this special case, use of tetramethylammonium salts (**1**) caused some decomposition. There is no clear consensus on the merits of esterifying these supports with alkali metal carboxylate salts (compare Wang *et al.*, 1977, with Tjoeng *et al.*, 1977).

Two intriguing esterification methods, which rely on an intermediate alteration of chloromethyl sites to a more reactive species, seem worthy of mention. Reaction with dimethyl sulfide (Table VI, number 41), followed by ion-exchange binding of a carboxylic acid, gives support **3**; heating of the dry resin *in vacuo* at 80°C for 4 hr resulted in excellent yields (80–95%) of covalently bound resin–ester, as well as dimethyl sulfide (Dorman and Love, 1969; Fontana *et al.*, 1971). An alternate pathway of the reaction is the production of a small amount of the corresponding methyl ester (removable into solution by filtration), thus leaving residual methylthiomethyl sites on the resin; however, these sites do not seem to interfere with the subsequent steps of solid-phase synthesis. Alternatively, a resin-bound α-alkoxyimmonium chloride (**4**) has been invoked as resulting from heating chloromethyl resin with DMF at 70°C for 4 hr. Additional heating under the same conditions with a dicyclohexylammonium salt led to formation of an ester linkage

* Diisopropylethylamine is not an effective base for the standard esterification with chloromethyl-resins.

(yield of reported example was 25%), but without creation of resin-bound tertiary amino sites (Buis *et al.*, 1976). Recently, an effective esterification method (>90% yields) was reported using potassium fluoride as a base in DMF, at 50°C for 24–50 hr (Horiki *et al.*, 1978). The intermediate **4** was again suggested.

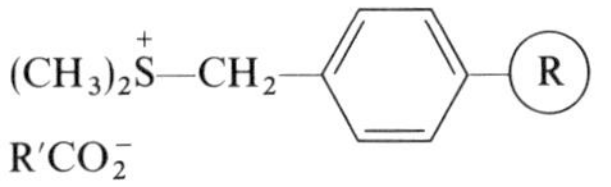

(**3**)

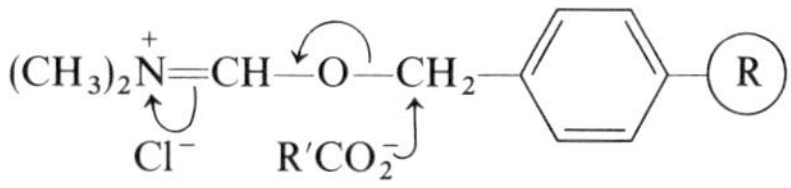

(**4**)

(ii) Acylation. Esterification to resin-bound benzyl alcohols (Table VI, numbers 3, 29, 49, 58, 59, 61) was introduced by Bodanszky and Sheehan (1966) who found *N,N'*-carbonyldiimidazole (4 equiv) in CH_2Cl_2 at 25°C for 24–48 hr to be an effective agent for the condensation of a *tert*-butyloxycarbonylamino acid (4 equiv) and hydroxymethyl-polystyrene (1 equiv OH) (Table VI, number 29). For example, this approach was for a long time (see however, Fig. 5) the only reliable way to directly anchor a free COOH-terminal methionine residue. On the basis of data from several other laboratories (e.g., Stewart and Young, 1969, pp. 32–33; Beyerman and in't Veld, 1969), it seems that most amino acids can be introduced with the aid of *N,N'*-carbonyldiimidazole in 40–80% yields. The excess hydroxymethyl sites must be blocked to prevent initiation later in a solid-phase synthesis of new chains which will be missing amino acids from the COOH-terminal portion of the desired sequence. Erickson and Merrifield (1973b) quantitatively blocked all the sites with acetic anhydride–pyridine–benzene (1:2:2), whereas Wang (1973, 1975) found a mixture of benzoyl chloride (0.3 *M*) and pyridine (0.4 *M*) in CH_2Cl_2 to be optimal for this purpose (half-time of benzoylation = 40 sec at 25°C).

Useful loadings of substituted benzyl alcohol supports have been obtained with the aid of excess *N,N*-dimethylformamide dineopentyl acetal [Schreiber, 1967, based on Brechbühler *et al.*, 1965; see Eq. (1)]; this method was the only one of several tried that allowed esterification of N^α-protected asparagine to the related 2-hydroxyethylsulfonylmethyl support (Table VI, number 36; see also p. 89) without dehydration or other side reactions (Tesser *et al.*, 1976). Oxidation–reduction condensation (compare to Section IV,C,4)

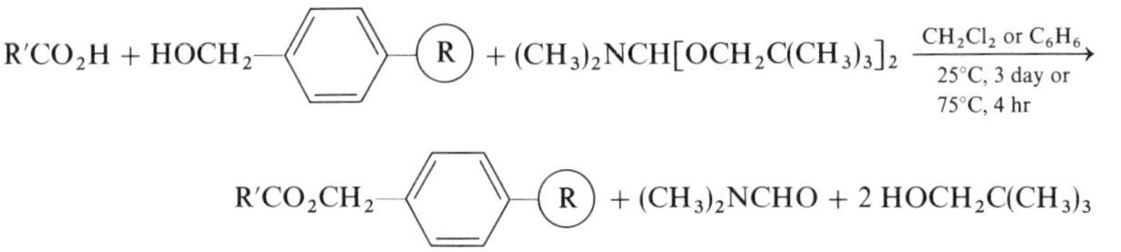

(1)

ROUTE A

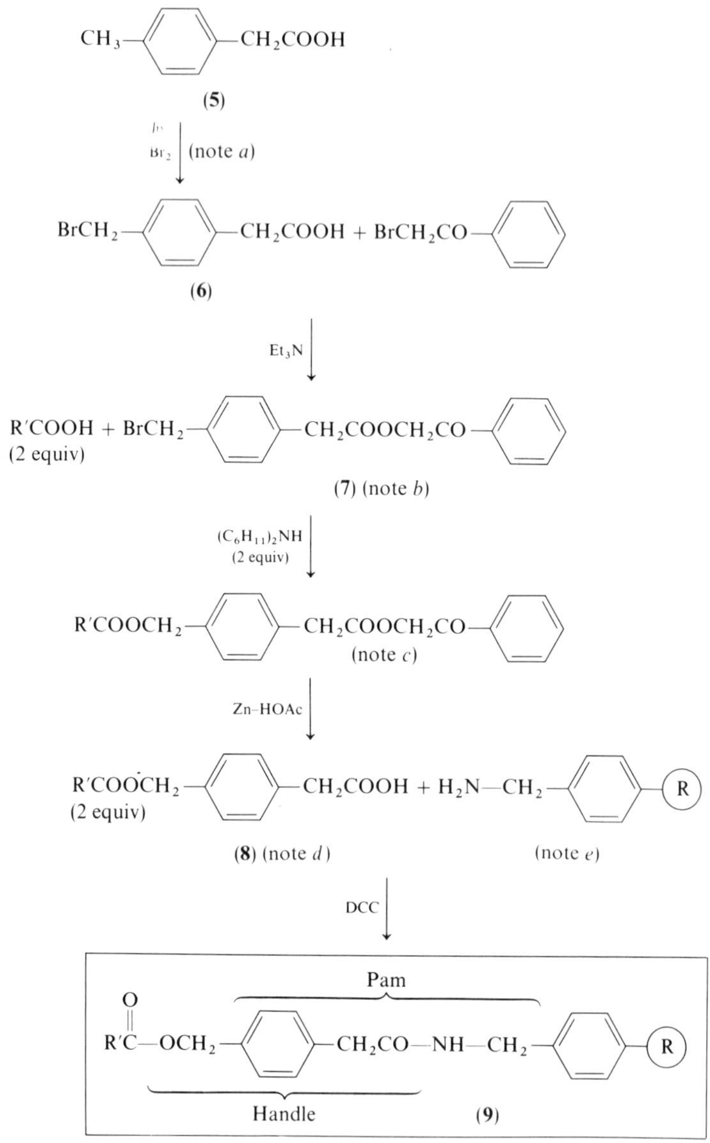

ROUTE B

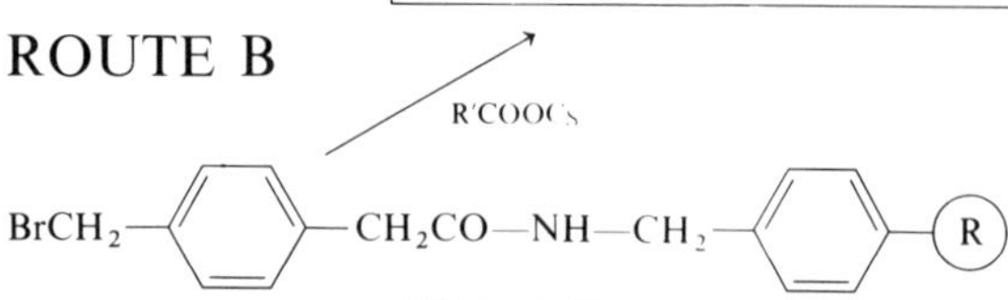

(10) (note f)

mediated by 2,2′-dipyridyl disulfide and triphenylphosphine has also been applied for the esterification (Matsueda *et al.*, 1972); faster rates and better yields were obtained with tri-*n*-butylphosphine (Ueki and Ikeda, 1974). Essentially quantitative substitution can be achieved when couplings are carried out with the use of dicyclohexylcarbodiimide (DCC) in the presence of 4-dimethylaminopyridine (equimolar to DCC and protected amino acid) as a catalyst (Wang, 1975; Chang *et al.*, 1976). Acylation with symmetrical anhydrides has been achieved under similar conditions (Atherton *et al.*, 1978a), and recently, active esters have been successfully applied in the presence of imidazole (1.3 equiv) as a catalyst (Bodanszky and Fagan, 1977).

(iii) Via Handles. The handle strategy for the anchoring of preformed benzyl esters is exemplified in Fig. 5 (Route A). Photobromination of 4-tolyacetic acid (**5**) gave 4-(bromomethyl)phenylacetic acid (**6**) in good yield and purity. Since under the conditions required for direct esterification of **6** self-condensation products were also formed, an extra protection step with bromoacetophenone was incorporated into the synthetic sequence; the phenacyl group was found to be selectively removable by reduction with zinc in acetic acid. Coupling of the resulting 4-(acyloxymethyl)phenylacetic acid derivatives (**8**) to aminomethyl-polystyrene mediated by DCC was readily carried to completion, giving a support **9** with a *p*-(carbamoylmethyl)-benzyl ester linkage to the resin. Alternatively (Route B), the same linkage **9** was accessible by alkylation of the bromomethyl support **10** (compare to Table VI, number 61) by the cesium salt method. Since the use of Route B does not preclude cross-linking or alkylating side reactions, nor ensure a quantitative loading, Route A is preferred for rigorous applications. The supports **9** were originally developed to achieve increased stability to acid for the ester anchoring bond, in comparison with the *p*-alkylbenzyl ester. Formally, the new supports differ by the insertion of a phenylacetamido-methyl group between the methylene group and phenyl ring of the benzyl ester (see Fig. 5). The new resins have thus been described in the literature

---

Figure 5. Anchoring of a carboxylic acid to an amino group of a solid support through a 4-(oxymethyl)phenylacetyl handle, taken from Mitchell *et al.* (1976a, 1978b). In the latter cited work, R′COOH were Boc-LVal-OH, Boc-LLys(Z)-OH, Boc-LAsp(OBzl)-OH, Boc-LSer(Bzl)-OH and Boc-LMet-OH.

[a] Br_2 (excess) in CCl_4 at reflux, with ordinary 150 watt light bulb as photosource—yield of **6**: 50–60% from benzene–hexane.

[b] Yield of **7** : 40% from CH_2Cl_2–petroleum ether.

[c] Used directly for next reaction after simple extractive workup.

[d] Yield of **8**, based on **7**: ~50%, isolated as cyclohexylammonium salt.

[e] See Table VI, number 16.

[f] Compare to Table VI, number 61.

by the convenient term "Pam". Note that this terminology does not fit the definition of a handle as just presented; for that purpose, designation as a 4-(oxymethyl)phenylacetyl derivative is appropriate.

Mitchell *et al.* (1978b) demonstrated that solid-phase syntheses of model peptides anchored on Pam supports derived from Route A, with resins that had originally been functionalized by direct amidoalkylation (Table VI, number 16), gave products which were of improved quality by comparison with control syntheses using alternate anchoring linkages (e.g., *p*-alkylbenzyl ester), strategies (e.g., Route B), and supports (e.g., Table VI, number 31). It seems likely that the same methods will also be effective on other supports that can be unambiguously functionalized with amino sites (Table VI, numbers 6, 25; see also 32 as representative of supports with aliphatic α- or ω-amino acids as spacers). The preformed benzyl ester strategy has also been

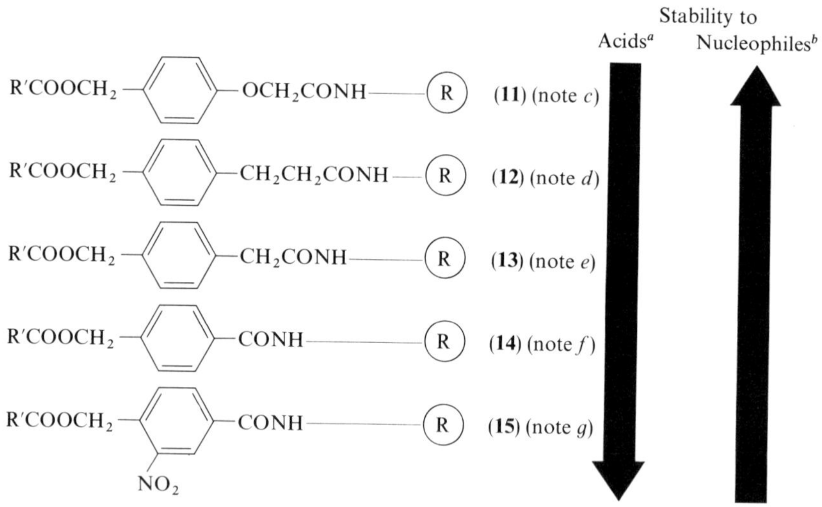

Figure 6. Benzyl ester handles attached to amino group-containing supports (e.g. Table VI, numbers 6, 16, 25, 31, and 32). Synthetic strategies discussed in text; the preformed benzyl esters have so far been used to prepare **12** and **13**.

[a] In order of increased stability to acids.

[b] Qualitative rating; from top to bottom increasing susceptibility to nucleophiles such as hydrazine.

[c] From Atherton *et al.* (1978a), based on Wang (1973); cleaved by TFA, 25°C (see Table VI, number 59).

[d] From Atherton *et al.* (1975); cleaved by HBr, 25°C.

[e] From Mitchell *et al.* (1976a, 1978b) and Sparrow (1976); cleaved by HF, 0°C (see Table VI, number 61, and Fig. 5).

[f] From R. S. Hodges and A. R. Mitchell, unpublished results, 1975; ester linkage stable to HF treatment; see also Atherton and Sheppard (1977), without details.

[g] From Rich and Gurwara (1975a); compare to Wang (1976) (see Table VI, number 63).

applied with formation of a 4-(oxymethyl)phenylpropionyl bridge (**12**) to an amino group-containing polydimethylacrylamide resin (Table IV, number 5 and Table VI, number 6; Atherton *et al*., 1975). The structures of a homologous series of benzyl ester handles attached to amino group-containing supports are shown in Fig. 6.

b. Acidolysis. *(i) Premature Cleavage.* Many practical schemes of solid-phase peptide synthesis utilize temporary protecting groups which are removed under acidic conditions. Therefore, we consider first the extent to which a substituted benzyl ester anchoring linkage is cleaved per deprotection step. This information may be calculated indirectly from the substitution levels determined by amino acid analysis on peptide–resins after a substantial number of real or simulated synthetic cycles (Gutte and Merrifield, 1971; Ueki *et al*., 1977), or more accurately, by direct recovery and quantitation (amino acid analysis, colorimetric reaction, uv chromophore, or radioactivity) of the cleaved peptide (Lin *et al*., 1972; Barany and Merrifield, 1973; Mitchell *et al*., 1976a: Blake and Li, 1976). A variety of results are compiled in Table VII.

To a very good first approximation, the stability of the anchoring ester linkage is determined by the COOH-terminal amino acid. For example, the cleavage rates in TFA–CH_2Cl_2 (1:1) determined on four different COOH-terminal valine-peptides attached to the same resin were within 30% of one another. The absolute rates with TFA (trifluoroacetic acid) under several conditions varied over a threefold range with different COOH-terminal amino acids, in the order Pro > Gly > Val > Phe; glutamic acid anchored through its side chain was at the labile end of this scale. The cleavage of a solution model proceeded at a rate about 1.5–4-fold faster than the corresponding polymer-bound benzyl esters (Kent and Merrifield, 1978; compare to p. 30).

Often, the premature losses of peptide chains from the solid support at the rates shown in Table VII are such as to compromise the chances of success in attempts at the synthesis of long target molecules. These losses affect not only the overall yield, but also indirectly affect the purity of the product. For example, premature cleavage during removal of N^{α}-Boc groups with TFA leads to trifluoroacetoxymethyl groups which can trifluoroacetylate peptidyl α-amino groups and thus cause irreversible chain termination (Kent *et al*., 1979).

The problem described in the previous paragraph can be partially overcome by changing the details of the acidolytic deprotection conditions. Thus, Burton *et al*. (1977) found (Table VII, last column) that in HCl–dioxane, the rates of the losses were decreased, and that steric factors involving the

Table VII. Rates of Cleavage of Benzyl Ester Linkages[a]

COOH-terminal amino acid[b]	$[k_{rel}^{TFA}]$[c]	k (hr^{-1}) with acid, 25°C[d] TFA–CH_2Cl_2 (1:1)	TFA–CH_2Cl_2 (1:4)	HCl(5.8 *N*)–dioxane[e]
Proline	1.4	4.8[e]	2.7[f]	1.1
Glycine	*1.0*	3.4 ± 0.2[e,g]	—	2.0
Lysine (2Cz)	0.6	2.0[e]	—	0.6
Valine	0.7	2.4 ± 0.4[e,h,i]	1.3[j]	0.4 ± 0.1
Phenylalanine	0.5	1.6[k]	—	—
Glu(α-Bzl)[l]	1.3	—	2–3[l]	—

[a] The resins were poly(styrene-co-1% divinylbenzene).

[b] Most of the data in this table were compiled on tetrapeptide resins, unless indicated otherwise. Results with amino acid-resins are quite comparable.

[c] Rates are relative to $k_{Gly} = 1$. The shown data on TFA–CH_2Cl_2 mixtures at 25°C have been combined (entirely consistent) with data of Mitchell *et al.* (1978b) on cleavage of amino acyl 4-(oxymethyl)phenylacetamidomethyl resins (**9**) in refluxing TFA. The rate of the same reaction using a solution model was 4-fold faster.

[d] The rate constant k reported is an arithmetic average of all compiled data; the error (±) represents the true scatter.

[e] Burton *et al.* (1977).

[f] Barany and Merrifield (1973), [^{3}H]Gly-Gln-Lys(Z)-Ser(Bzl)-Pro-Ⓡ.

[g] Ueki *et al.* (1977), H-Gly-Ⓡ.

[h] Mitchell *et al.* (1976a).

[i] Gutte and Merrifield (1971), RNase A(1-124)–Ⓡ. The commonly cited loss of 1.4% of chains/deprotection cycle has been recalculated (to 1.1%), since it was subsequently shown (ref. *j*) that 30% of the chains were already lost at the level of the heptapeptide–Ⓡ.

[j] Lin *et al.* (1972).

[k] Blake and Li (1976).

[l] Anchored through γ-carboxyl group; Krumdieck and Baugh (1969).

COOH-terminal amino acid appeared to play a more significant role (five-fold range in comparison of glycine and valine; the ratio, with these two amino acids was 1.4 with TFA deprotection). Ueki *et al.* (1977) introduced a new α-amino protecting group, the dimethylphosphinothioyl function (see structure **53** on p. 108), which could be removed by an acidolytic mechanism with triphenylphosphine dihydrochloride in CH_2Cl_2. The latter reagent did not cause substantial cleavage of the standard *p*-alkylbenzyl ester linkage; on a per cycle basis, the improvement over the usual combination of *tert*-butyloxycarbonyl α-amino group protection with TFA–CH_2Cl_2 (1:1) for deprotection may be estimated as being about 10-fold.

Premature losses of peptide chains can also be minimized by modulating the stability of the anchoring linkage with electron-withdrawing substituents on the aromatic ring. If acidolysis has also been selected as the mode of final cleavage from the support, some care must be exercised in applying this strategy, but practical compromises have been found. The *p*-alkyl-*o*-bromobenzyl, *p*-(*N*-hexyl)(carbamoylmethyl)benzyl, and *p*-(carbamoylmethyl)benzyl esters (derived from Table VI, numbers 10, 62, and 61) have been shown to be, respectively, 20-, 50-, and 100-fold more stable to TFA–CH_2Cl_2 (1:1) than the standard *p*-alkylbenzyl ester originally used in solid-phase synthesis (rate data, in order given, from Li *et al.*, 1977; Blake and Li, 1976; Mitchell *et al.*, 1976a). Among the various linkages with increased acid stability that have been proposed, it appears that the *p*-(carbamoylmethyl)benzyl esters (Pam resins, Fig. 5) have the best combination of features.

(ii) Final Cleavage. Protected peptides bound to a solid phase through a benzyl ester linkage are converted to the free peptide acids by treatment with a number of strong inorganic or organic acids, under conditions as compiled in Table VIII. Depending on the actual choices made in the original synthetic design, many or all of the permanent side-chain protecting groups are removed at the same time. To minimize modifications of arginine, cysteine, histidine, methionine, tyrosine, and tryptophan residues in the desired peptide chains through alkylation reactions with the generated carbocations, it has been a common and essential practice to run the acidolytic cleavages in the presence of excess amounts of suitable scavengers, which compete for the alkylating species. The chemical structures of several such scavengers are shown in Table IX; they often mimic the susceptible residues that they are intended to protect. A variety of acid-catalyzed intramolecular side reactions proceed despite the presence of scavengers; cases are also known where the scavenger itself reacts with amino acid residues in the peptide chain to produce undesired structures.

There are several factors, not necessarily representing commensurate goals, to be considered in selecting an appropriate combination among the cleavage conditions and scavengers of Tables VIII and IX. Listed in order of importance, these are (1) suppression of chemical side reactions; (2) attainment of a reasonable overall yield of crude product; (3) removal of excess acid as a means of quenching the cleavage reaction; (4) ease of workup of the crude cleavage mixture, including removal of excess or alkylated scavengers; and (5) convenience and reproducibility of the reaction conditions.

A combination of liquid anhydrous hydrogen fluoride (HF) and anisole (9:1 v/v) at 0°C for 30 min (Sakakibara *et al.*, 1967; Scotchler *et al.*, 1970; Feinberg and Merrifield, 1975) represents a useful compromise that is applicable to a wide variety of synthetic problems. Cleavage yields are high

Table VIII. Reaction Conditions for the Acidolytic Cleavage of Protected Peptide Benzyl Esters[a]

No.	Conditions	Remarks	References
1.	TFA, 25°C, 18 hr *or* reflux, 60 min	Not widely used, compare to Table VII for relative rates	Merrifield (1967)
2.	32% HBr in HOAc–TFA (1:1), 25°C, 10–100 min	Convenient for analytical work. Serine and threonine hydroxyls may be acetylated under these conditions	Gisin *et al.* (1978) Acetylation: Guttmann and Boissonnas (1958)
3.	Dry HBr (Br_2-free) gaseous stream through anhydrous TFA, 25°C, 90 min	Generally useful. Shorter reaction times have sometimes been used to limit side reactions. No trifluoroacetylation (compare to No. 2)	Merrifield (1964b) Apparatus: Stewart and Young (1969, pp. 40–41)
4.	Anhydrous HF, 0°C, 30–60 min	Widely useful and often preferred over No. 3. Requires special plastic equipment and handling procedures due to toxicity of reagent	Lenard and Robinson (1967); Experimental details and review: A. B. Robinson, in Stewart and Young (1969, pp. 41–44); Sakakibara (1971) Safety: Sharts and Sheppard (1974, pp. 221–223)
5.	HF–pyridine (9:1), 0°C	The pyridine diluent changes acidity function H_0 of medium from −10.8 (No. 4) to −8.4	Sugano *et al.* (1978)
6.	Pyridinium polyhydrogen fluoride, 25°C, 60 min	Reagent is a stable liquid reservoir for HF (70% is dissolved)	Matsuura *et al.* (1976)
7.	CF_3SO_3H–TFA–CH_2Cl_2 (1:10:10), 25°C, 30 min	Convenient for analytical work	Yajima *et al.* (1974); Mitchell *et al.* (1978b)
8.	CH_3SO_3H–TFA (1:1), 25°C, 30 min	Milder than No. 7	Yajima *et al.* (1975)
9.	1.3 *M* $B(O_2CCF_3)_3$–TFA, 25°C, 30 min	Reagent readily prepared from BBr_3 and TFA (1:3), and commercially available	Pless and Bauer (1973); Bauer and Pless (1975); compare to results of Felix (1974) with BBr_3

[a] Scavengers used in conjunction with these reagents are dealt with in Table IX.

(80–100%) even with difficult COOH-terminal amino acids (e.g., phenylalanine, see Table VII), and a relatively acid-stable *p*-(carbamoylmethyl)-benzyl ester linkage **9** or **13** (Mitchell *et al.*, 1978b). Reactions are conducted in a special all fluorocarbon apparatus (see Table VIII, number 4 for references). The HF is condensed at −78°C into a vessel containing the peptide–resin and anisole. After immersion of the vessel into an ice–water bath (0°C) for the nominally indicated time, excess HF is removed by cautious evaporation at reduced pressure. Because of heat-exchange problems, such a procedure introduces some variability in the precise temperature and time of exposure to acid, which in certain select cases may be critical. Once the last traces of HF have been pumped off, the bulk of the anisole and its degradation products can be separated by trituration of the dried and cleaved resin (still in the cleavage vessel) with ethyl acetate or ether. The liberated peptide may then be extracted by dilute aqueous acetic acid. Occasionally glacial acetic acid or even anhydrous trifluoroacetic acid has been necessary, although there are examples where difficulties have been encountered when the latter reagent was applied for long peptides. A possible N → O acyl shift at serine and threonine residues (p. 210) has been reported to occur slowly in HF and to be readily and quantitatively reversed by treatment with aqueous bicarbonate at pH 7.5, 25°C for several hours (Sakakibara *et al.*, 1962; Lenard and Hess, 1964). Hence, the bicarbonate treatment has occasionally been applied as a precautionary measure.

When cleavage reactions are performed by any of the remaining techniques listed in Table VIII, the mixture that includes the liberated crude peptide is separated from residual resin by filtration. Further washes are frequently performed using TFA, which is an excellent solvent for free peptides but does not react with serine or threonine hydroxyl groups under the strongly acid conditions that still exist in the filtrate (Guttmann and Boissonnas, 1958). Excess volatile acids are removed *in vacuo*; there is still a danger of further acid-catalyzed side reactions, or hydrolysis if any water is present, at this point of the workup. The nonvolatile acids (Table VIII, numbers 7–9) are then either neutralized (buffered) or converted by ion-exchange into a weak species. Additional separation of scavengers and salts can be achieved by precipitation of the peptide products, for example with dry ether, or by a preliminary gel filtration chromatographic step.

(iii) Side Reactions. A number of chemical side reactions have been well documented to occur under the strong acidolytic cleavage conditions described in Table VIII. They can often be minimized, reversed, or circumvented by modifications in experimental technique, or more promisingly, by changes in the protection strategy; details and references are to be found in Section V,B in conjunction with the discussion of each affected amino

Table IX. Scavengers for Use in Acidolytic Cleavage Reactions[a]

Class/structures	Purpose	Remarks and side reactions	References
Phenols and ethers 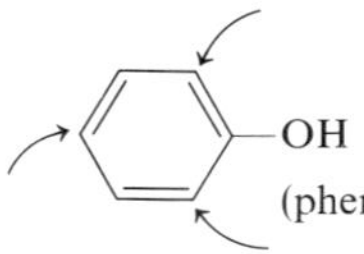 (phenol)	Trapping of carbonium ions; phenols also readily react with oxidizing agents including bromine		Weygand and Steglich (1959); Sakakibara and Shimonishi (1965); Lenard and Robinson (1967)
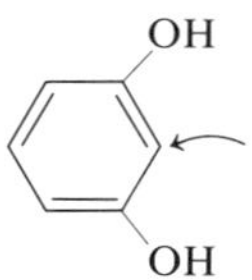 (resorcinol)			
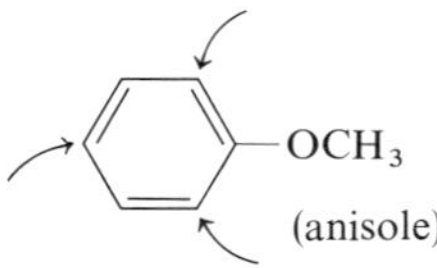 (anisole)		Anisole can suffer demethylation under acid conditions in $B(O_2CCF_3)_3$; can be acylated by glutamyl residues (pp. 196–197) in HF; or transfer its methyl group to methionine (p. 231) in presence of CH_3SO_3H. Anisole is a poor scavenger for *tert*-butyl carbonium ions	Bauer and Pless (1975); Feinberg and Merrifield (1975); Irie *et al.* (1977); Lundt *et al.* (1978)
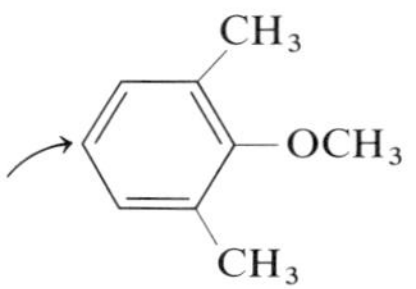 (2,6-dimethylanisole)			Yamashiro and Li (1978)

Thioethers			
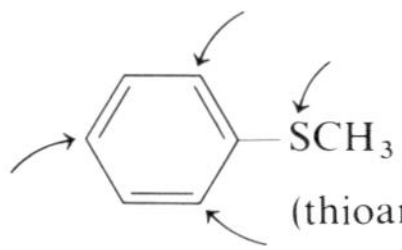 (thioanisole)		Thioanisole preferable to anisole in conjunction with use of $B(O_2CCF_3)_3$; it is also a better scavenger against C-alkylation of tyrosine residues (p. 216). However, the adduct of thioanisole with *tert*-butyl cations can transfer the alkyl group from sulfur to amino acid acceptors.	Bauer and Pless (1975); Kiso *et al.* (1978); Lundt *et al.* (1978)
$CH_3{-}S{-}CH_3$ (dimethyl sulfide) $CH_3{-}S{-}C_2H_5$ (ethyl methyl sulfide)	Compete with internal methionine residues for alkylating species, thus preventing sulfonium salt formation		Guttmann and Boissonnas (1958, 1959); Merrifield (1964c); Kenner and Seely (1972)
$H_2NCH(CO_2H)CH_2CH_2SCH_3$ (methionine)		Added methionine removed by gel filtration	Marglin and Merrifield (1966); Veber *et al.* (1969)

Table IX. *(Continued)*

Class/structures	Purpose	Remarks and side reactions	References
Thiols	Prevention of acid-catalyzed oxidation of tryptophan		
$HOCH_2CH_2SH$ (β-mercaptoethanol)		Polymerization of β-mercaptoethanol in acid ($t_{1/2} \sim 11$ hr) was observed; consequently water formed, and reducing power lowered. Furthermore, in TFA, hydroxyl group is rapidly esterified. It has been stated that 1,2-dithioethane was the best of 9 scavengers tested against *tert*-butyl carbonium ions.	Marshall (1968); Blake and Li (1968)
$HSCH_2CH_2SH$ (1,2-dithioethane)			Sharp *et al.* (1973); Chino *et al.* (1978); Lundt *et al.* (1978)
$HSCH_2(CHOH)_2CH_2SH$ (dithiothreitol: DTT)	DTT is an especially good reducing agent because disulfide-exchange equilibrium shifted by favorable intramolecular cyclization		Discovery of DTT: Cleland (1964). Applications: Li and Yamashiro (1970); Draper (1973)
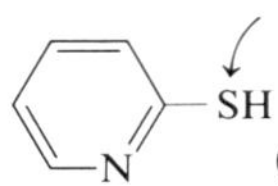 (2-mercaptopyridine)			Matsueda *et al.* (1973)

Indoles			
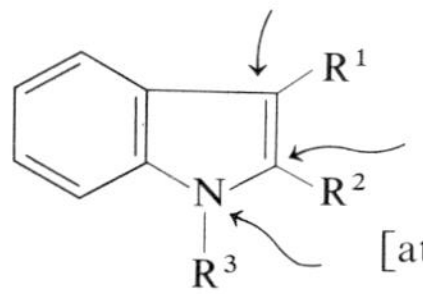 [attack at R = H]	Compete with internal tryptophan residues for alkylating and oxidizing species; has been suggested to be effective scavenger for potentially harmful aldehydes	Compare to Fig. 9 on p. 111 Acid solutions of indole turn dark purple upon standing; an acid-catalyzed dimerization of indoles has been discovered which might lead to complications with tryptophan-containing peptides (see **206** on p. 221)	Wünsch *et al.* (1967); Wünsch and Spangenberg (1972); Stewart and Matsueda (1972); Stewart *et al.* (1976); Omori *et al.* (1976); see Bauer and Pless (1975) for comparative study with other scavengers
$R^1,R^2,R^3 = H$ (indole) $R^1,R^3 = H$, $R^2 = CH_3$ (2-methylindole) $R^1 = CH_3$; $R^2,R^3 = H$ (3-methylindole) $R^1 = CH_2CH(NH_2)CO_2H$; $R^2,R^3 = H$ (tryptophan)			
Phosphorus compounds			
$(C_2H_5O)_2P(=O)H$ (diethyl phosphite)	Advantageous solvent which reacts with water	Could also react with serine and threonine in peptides to give phosphates	Anderson and McGregor (1957); Guttmann and Boissonnas (1958, 1959); see also Iselin (1961); Moore (1963)
$(CH_3O)_3P$ (trimethyl phosphite)	Hydrogen halide acceptor		Young *et al.* (1956)
$(C_6H_5)_3P$ (triphenylphosphine)	Anti-oxidant		Ueki *et al.* (1977); Suzuki *et al.* (1978)

[a] Scavengers are typically used in 10 to 50-fold excess over sensitive residues. Arrows indicate sites of attack by electrophiles.

acid. Thus, there is convincing evidence for (1) succinimide formation and $\alpha \rightarrow \beta$ rearrangement, particularly at aspartyl-glycine, aspartyl-serine, and aspartyl-histidine sequences; (2) acylium ion formation at glutamyl residues leading to pyrrolidone or acylated aromatic products; (3) limited peptide bond cleavage adjacent to methionine residues; (4) S-alkylation of methionine in spite of (or in one case due to) the presence of scavengers; (5) intramolecular alkylation of tyrosine, not prevented by scavengers; and (6) deamidination of the guanido group of arginine, giving rise to ornithine. Furthermore, there is a general danger of partial and irreversible destruction of a number of amino acid side chains, including especially those of tryptophan, tyrosine, cysteine, methionine, asparagine, and glutamine. The acid media may conceivably promote intramolecular lactone formation between nearby acidic and hydroxylic or sulfhydryl residues (Lenard and Hess, 1964) or facilitate oxidative polymerization of peptides containing multiple cysteine residues (p. 246). Finally, it has been found that a number of native, biologically active materials cannot be fully reconstituted following exposure to strong acids. This question of protein stability in HF was first explored by Katz (1954; further results reviewed by Sakakibara, 1971, pp. 61–63). A recent careful study by Aimoto and Shimonishi (1978) under simulated cleavage conditions is also valuable.

The introduction of HF into the arsenal of techniques for peptide synthesis (Sakakibara *et al.*, 1965; Lenard and Robinson, 1967; Sakakibara, 1971) has had far-reaching implications. One school of thought holds that, when used with proper precautions, HF is a reagent of great value and generality, while other workers feel that the HF-method leads to numerous ill-defined, uncontrollable side reactions which therefore compromise its usefulness. According to this latter point of view, the standard HF cleavage conditions must eventually be replaced by more suitable reagents and protocols. At this time, the question of the ultimate practical value of HF is not fully resolved.

(iv) Anchors for Milder Protection Schemes. The *p*-alkoxybenzyl ester linkage **16** (derived from functionalized support number 58 in Table VI;

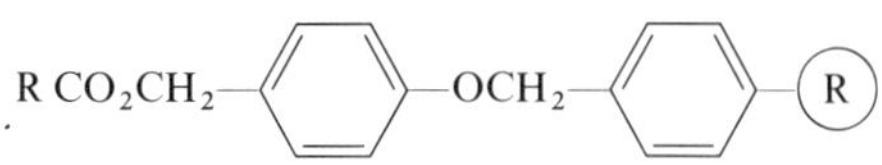

(16)

compare to structure **11** in Fig. 6) developed by Wang (1973, 1975) contrasts with all the others discussed in this section in that it can be cleaved by acid under relatively moderate conditions. When used in conjunction with the 2-(*p*-biphenylyl)propyl(2)oxycarbonyl (Bpoc) group for α-amino protection and 0.5% TFA–CH_2Cl_2 as the deprotecting agent, only about 0.2% of the growing peptide chains were lost from the support per step by premature acidolysis (S. Mojsov, unpublished results, 1977). Final cleavage from the support can be achieved by TFA–CH_2Cl_2 (1:1) at 25°C; for a simple model, the half-time was 5 min (Wang, 1975) and the yield was quantitative.

The anchors **11** and **16** have been routinely applied to the preparation of medium size (up to 10 amino acid residues) protected and free peptides in generally high yields (75–95% after cleavage and crystallization or precipitation) (Wang and Kulesha, 1975; Atherton *et al.*, 1978a; Chang and Meienhofer, 1978). The cleavage yields for longer peptides have been somewhat lower (40–75%) (e.g., Camble and Petter, 1976, for a 47-peptide; S. Mojsov and R. B. Merrifield, unpublished observations, 1978, for a 29-peptide), and further treatments, even with anhydrous HF, did not release additional peptide material from the cleaved resin. It has been suggested that during the course of the synthesis, and especially at the acidolytic cleavage step, the *p*-alkoxybenzyl moiety of the support itself combines irreversibly with released *tert*-butyl or other carbonium ions. This contention is supported by the significant improvements in yields that occurred upon conducting the cleavages in the presence of suitable scavengers.

c. Nucleophilic Cleavage. During the repetitive steps of solid-phase peptide synthesis as commonly practiced, the substituted benzyl ester linkage is not exposed to high concentrations of strong or moderate nucleophiles for any extended periods of time. Nonetheless, in view of the results that follow, the possibilities for low levels of premature cleavage from the support by this mechanism require some attention. Reaction solvents, such as dimethylformamide, or reagents, such as tertiary amines for the neutralization step, ought to be rigorously freed of primary or secondary amines. Also, only hindered alcohols are recommended as wash solvents. In two recent syntheses where the base-labile fluorenylmethyloxycarbonyl (Fmoc) group (structure **59** on p. 113) was used for α-amino protection, only slight if any loss of growing chains from *p*-alkoxybenzyl ester linkages **11** or **16** was observed upon deprotection with piperidine in CH_2Cl_2 or DMF mixtures (Atherton *et al.*, 1978a; Chang and Meienhofer, 1978).

Substituted benzyl ester linkages can be cleaved by alkali, ammonia or other amines, hydrazine, or alkoxides, to yield respectively the peptide acid, amides, hydrazide, or esters. The relative reactivity of a homologous series of esters is qualitatively indicated in Fig. 6. Since a number of side-chain protecting groups that are commonly used in peptide synthesis, including benzyl ethers and *tert*-butyl- or *p*-toluenesulfonyl-based derivatives, are resistant to the action of nucleophilic reagents, this method of cleavage can be of value in the preparation of protected peptide segments for special cases. On the other hand such strongly basic conditions entail a risk of a number of well-documented side reactions including (1) partial racemization at the COOH-terminal residue; (2) $\alpha \rightarrow \omega$ rearrangement at aspartyl and glutamyl residues; (3) succinimide and glutarimide formation in peptides containing COOH-terminal asparaginyl and glutaminyl esters; (4) deamidation at the side chains of asparagine and glutamine; (5) β-elimination of protected derivatives of serine, threonine, and cysteine, leading to the corresponding dehydro residues; and (6) cleavage and scrambling of disulfide bonds.

The first tetrapeptide prepared by the solid-phase method, Leu-Ala-Gly-Val, was removed from a nitrobenzyl-resin (Fig. 1) in 54% yield by saponification with 0.2 *N* NaOH in ethanol–water (9:1), at 25°C for 60 min (Merrifield, 1963). An alkaline mixture consisting of dioxane–methanol–4 *N* NaOH (30:9:1), which was carefully designed to achieve detachment of peptide acids from a β-hydroxyethylsulfonylmethyl support (Table VI, number 36; see also p. 89) was also applied to the standard *p*-alkylbenzyl ester support with comparable cleavage yields (70–80%, 25°C, 3 min, quenched with acetic acid; by-products formed upon longer reaction) (Buis *et al.*, 1976).

The ammonolytic cleavage of a peptide from a solid support was first reported by Bodanszky and Sheehan (1964, 1966). Typical conditions (Manning, 1968; Stewart and Young, 1969, p. 44) involve treatment of a peptide–resin at 25°C for 24–72 hr with a saturated solution (prepared at 0°C) of NH_3 in methanol, ethanol, or alcohol–dimethylformamide mixtures; yields of peptide amides are usually 60–80%. For some sterically unfavorable cases, it seems likely that a significant portion of the actual cleavage is a base-catalyzed transesterification reaction in the alcoholic solvent, with further ammonolysis occurring in solution (Bodanszky and Sheehan, 1966; Beyerman *et al.*, 1968b). The application of benzhydrylamine supports (pp. 93–95), which are cleaved in acid, represents a generally superior way of preparing the desired peptide amides.

Hydrazinolysis of the benzyl ester linkage has occasionally been useful for the preparation of protected peptide hydrazides suitable for further segment condensation by the azide method (reviewed by Meienhofer, Vol. 1, Chapter 4). In an impressive recent example, Chang *et al.* (1976) employed a mixture of anhydrous hydrazine–dimethylformamide (1:9) at 25°C for 16 hr to achieve quantitative cleavage (half-time of reaction was 45 min, by ir) of a pentapeptide; the residual resin containing regenerated hydroxymethyl sites was reesterified with the COOH-terminal *tert*-butyloxycarbonylamino acid, and the entire solid-phase synthetic procedure was repeated five times with the same yields in each run. An additional example, by Ohno *et al.*, (1971), is cited in Table I. Caution is required when α-amino or certain side-chain protecting groups that are labile to the action of hydrazine are used (e.g., N^{α}- or N^{ε}-trifluoroacetyl, formyl, *o*-nitrophenylsulfenyl, or phthaloyl derivatives, N^{i}-formyltryptophan, N^{im}-dinitrophenylhistidine, ω-benzyl or phenacyl esters of aspartate and glutamate and possibly the amides, asparagine and glutamine). For these reasons, solid-phase strategies involving alkoxycarbonyl hydrazide supports (pp. 96–97) are of greater generality.

The tertiary amine-catalyzed alcoholysis of the benzyl ester link was first noted and applied by Beyerman *et al.* (1968b) and by Lombardo and Piasio (cited by, with experimental details, Stewart and Young, 1969, p. 45). Other catalysts include quaternary anion-exchange resins (Pereira *et al.*, 1969) and potassium cyanide (Moore and McMaster, 1978). The rates and yields are dependent on steric features of the alcohol (methanol > ethanol > benzyl alcohol > isopropanol > *tert*-butanol) and the COOH-terminal amino acid (Gly > Ala > Phe > Val, Pro).

An ingenious application of the alcoholysis reaction, which produces protected peptides with a free carboxyl group (see Table I for a typical result), involves initial self-catalyzed transesterification with excess dimethylaminoethanol (mixture with DMF, 1:1, 25°C, 70 hr) (Barton *et al.*, 1973; Savoie and Barton, 1974). For sluggishly reacting substrates, further catalysis was provided by thallous 2-(dimethylamino)ethoxide (0.1–0.25 equiv). Based on an earlier observation by Greben *et al.* (1968), the resulting dimethylaminoethyl ester (**17**) was found to be readily hydrolyzable in aqueous DMF, occasionally with sodium bicarbonate or imidazole as basic catalysts.

$$R'CO_2CH_2CH_2N(CH_3)_2$$

(**17**)

It has been reported that under the transesterification conditions, dimethylformamide was partially decomposed, and peptide dimethylamides have been found (Birr, 1978, p. 67). Therefore, another variation was developed in which methanolysis served to release the peptide from the support, and transesterification with dimethylaminoethanol followed by hydrolysis proceeded in solution (Trudelle and Birr, 1975; compare also to Larsson *et al.*, 1976). Both racemization of the COOH-terminal residue, and $\alpha \rightarrow \beta$ aspartyl rearrangement occur to some extent. From a practical point of view, the levels of these side reactions may be acceptably low when certain recommended experimental precautions are followed.

d. Reductive Safety-Catch Cleavage. While a functionalized support containing an 2-oxymethyleneanthraquinone moiety has not yet been developed for solid-phase peptide synthesis, the solution chemistry of the corresponding "Maq" esters **18** elucidated by Kemp and Reczek (1977) contains a number of interesting features [Eq. (2)].

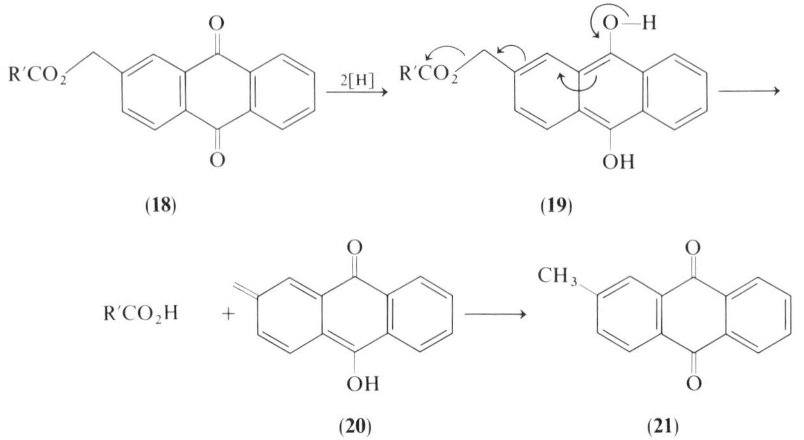

(18) (19) (20) (21) (2)

Due to the electron-withdrawing nature of the *meta*- and *para*-"benzophenone" substituents in the parent quinone structure, Maq esters **18** are surprisingly stable to acid. For example, in HBr–HOAc, they are about 250-fold more slowly cleaved than are simple benzyl esters. On the other hand, the hydroquinone form **19**, which is easily obtained by a variety of specific reductive procedures, undergoes ready fission of the alkyl–oxygen bond, releasing the free acid. The by-product xylylidene derivative **20** rapidly tautomerizes to quinone **21** without participating in undesirable side reactions with nucleophiles.

e. Catalytic Hydrogenolysis. The recent discovery by Mazur and collaborators (Schlatter *et al.*, 1977; Jones, 1977) of conditions under which peptides can be successfully removed from the resin by hydrogenation represents a welcome and exciting addition to the repertoire of methods available for solid-phase peptide synthesis, especially in view of the specificity, mildness, and widespread usefulness of this technique in solution synthesis. A peptide–resin is equilibrated with a solution of palladium acetate (0.2 *M*) in dimethylformamide, and the reaction typically performed with hydrogen at 60 psi, 40°C, 24 hr; the active catalyst is palladium black deposited *in situ*. After filtration and washing, the procedure can be repeated with a fresh palladium acetate solution to obtain additional crude cleaved material. The crude yield of an N^{α}-Boc-pentapeptide was 88%; on the other hand, removal of an 18-peptide proceeded in low yield (R. H. Mazur, personal communication, 1977). In a recent variation, Khan and Sivanandaiah (1978b) applied a catalytic transfer hydrogenation technique using cyclohexene as the reducing agent to obtain N^{α}-Boc-bradykinin in 60% yield from the peptide–resin.

Since *tert*-butyl-based derivatives are unaffected by these conditions, hydrogenolytic cleavage has considerable potential for the preparation of protected peptide segments suitable for further condensation. The technique should also avoid many of the acid-catalyzed side reactions that can occur during acidolytic cleavages. Progress in this direction has already been briefly reported (Atherton *et al.*, 1979a). By analogy to experience obtained in solution, one might predict that special care will be required to handle peptides containing cysteine and methionine residues.

f. Photolysis. Rich and Gurwara (1975a) found that irradiation of an *o*-nitrobenzyl ester linkage (derived from functionalized support number 63 in Table VI; see structure **15** in Fig. 6) released protected peptide acids, generally with COOH-terminal glycine, in good yields (50–80%). Photolysis was performed with resins suspended in anhydrous methanol or ethanol, under nitrogen, with light at wavelengths centered at 3500 Å (40% $CuSO_4$ solution acted to filter out all wavelengths <3200 Å), at 25°C for 18–24 hr (compare also with Wang, 1976; Tjoeng *et al.*, 1977). Under these conditions, *tert*-butyl-, benzyl-, and *p*-toluenesulfonyl-based protecting groups are stable, as are the phenolic and indole side chains of tyrosine and tryptophan residues. However, N^{im}-dinitrophenylhistidine residues undergo photodegradation, and if this protected derivative is included in the synthetic scheme, a preliminary thiolytic removal step prior to photolysis would be well advised.

The reaction mechanism of the photolytic cleavage presumably involves an internal oxidation–reduction [Barltrop *et al.*, 1966; see Eq. (3)]. The product, nitrosobenzaldehyde **22**, is prone to further transformation in

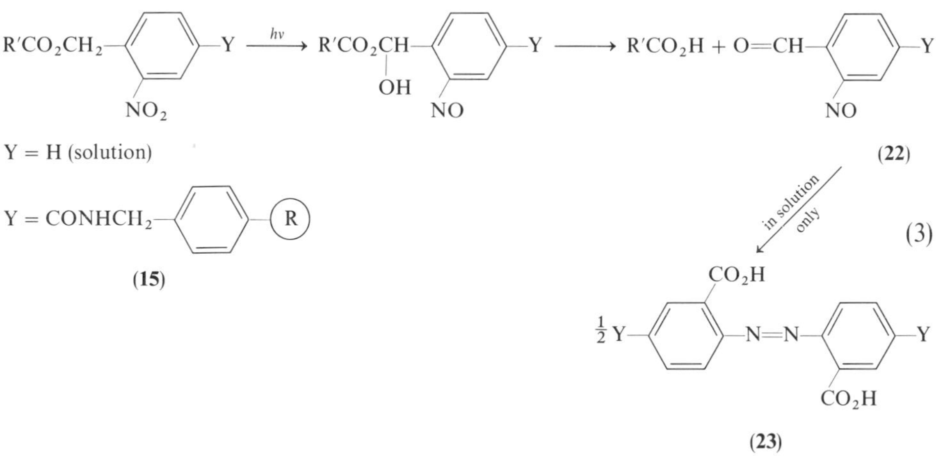

(3)

solution to the diazobenzene **23**, which acts as an efficient internal light filter and prevents further cleavage (for example, yield with $R' = C_6H_5$, $Y = H$, was only 17%). Evidently, the effective site separation on the solid support (compare to p. 27ff) is such as to minimize this deleterious side reaction, and there is no need to add any aldehyde scavengers. The initial functionalized *o*-nitrobenzyl bromide support (Table VI, number 63) can suffer some based-catalyzed side reactions during esterification (see p. 62). Pronounced darkening of the resin is often noted; the resulting chromophores might be expected to serve as quenching agents that reduce yields in the subsequent photolytic cleavage. The problem does not arise when the already described optimal esterification conditions developed by Rich and Gurwara (1975a) are used.

g. Side Reactions. Peptide chains anchored as a substituted benzyl ester may be prematurely lost from the solid support through gradual acidolysis (p. 67ff). On the other hand, the benzyl alcohol moiety of the ester linkage is a relatively poor leaving group in S_N2 reactions, and under the common mild conditions of solid-phase synthesis, the anchor is almost entirely resistant to the action of nucleophilic reagents (p. 77). In fact, the only significant nucleophiles throughout the synthesis will be the α-amino terminal groups of the growing peptide chains, and only for the periods of time before they are acylated in the coupling steps.

There is one important exception to the generalization of the previous paragraph. At the level of the dipeptide–benzyl ester, the intermediates are uniquely juxtaposed to permit formation of six-membered diketopiperazine rings by intramolecular aminolytic cyclization. This side reaction (see Fig. 7 for a specific illustration) was first noted by Lunkenheimer and Zahn (1970) and its scope further defined by Brunfeldt *et al.* (1972a), Gisin and Merrifield (1972a), Khosla *et al.* (1972c), and Rothe and Mazánek (1974). The quantitative level of the cleavage is strongly dependent on steric and chiral features of the original dipeptide structure, since in the transition state a peptide bond is forced into the usually less favored cis-conformation. The order, for the COOH-terminal amino acid residue, is *N*-methylamino acid > Pro > Gly > Val. The same overall ordering is probably also valid for the NH_2-terminal residue, although here the kinetic (unhindered nucleophile) and thermodynamic (conformation of transition state) steric requirements counteract, and the effects are less pronounced. The rate of the reaction was markedly increased upon catalysis by weak carboxylic acids (pK_a 3–5), and a probable mechanism has been proposed (Gisin and Merrifield, 1972a; Fig. 7). Both the trifluoroacetate or hydrochloride salts of the dipeptide–resin, as well as the free amine in the presence of a tertiary amine, were relatively stable. Hence, it could be concluded that diketopiperazine release will occur

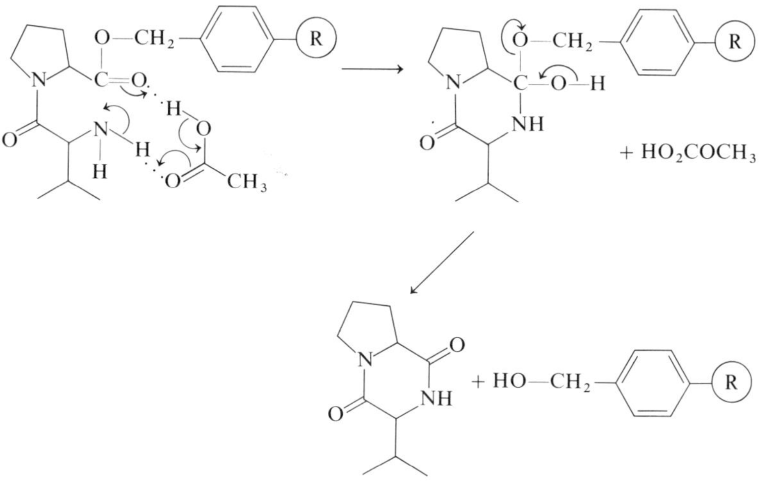

Figure 7. Postulated mechanism of acetic acid-catalyzed diketopiperazine formation from H-Val-Pro-resin, redrawn from Gisin and Merrifield (1972a).

primarily during the coupling step of the standard scheme of solid-phase peptide synthesis, and not during the deprotection or neutralization steps (Gisin and Merrifield, 1972a). However, other workers (Khosla *et al.*, 1972c, Rothe and Mazánek, 1974) reported specialized conditions whereby significant diketopiperazine formation was promoted by base catalysis during the neutralization step.

Free hydroxymethyl groups, which might later serve as initiation sites for new peptide chains, are generated concomitant with the release of the diketopiperazine from the support, and it is mainly for this reason that the reaction must be avoided. In the practical course of a solid-phase peptide synthesis, employing any of the 20 commonly occurring L-amino acids, the diketopiperazine side reaction can be safely ignored (level $\leqslant 5\%$) unless either of the first two residues is glycine or proline. In these latter cases, the reaction can still be effectively suppressed by following any of a variety of precautionary procedures: (1) Avoidance of direct exposure of the dipeptide–resin to the carboxyl group of an N^{α}-protected amino acid (Gisin and Merrifield, 1972a). This can be done either by the use of preactivated acylating species, or in the case of dicyclohexylcarbodiimide coupling, by "reverse addition", i.e., prior addition of the condensing agent; (2) *in situ* neutralization of the hydrochloride salt of the dipeptide–resin with the *N*-methylmorpholine salt of the N^{α}-protected amino acid, in the presence of the condensing agent (Suzuki *et al.*, 1975); or (3) bypassing the dipeptide–resin stage either by direct anchoring of a protected tripeptide to the support, or by coupling a protected dipeptide to an amino acid-resin (Rothe *et al.*, 1976; Chang and Meienhofer, 1978).

2. Benzhydryl Esters

Benzhydryl ester linkages, which are substantially more labile in acid than simple benzyl esters, were explored by Southard *et al.* (1969, 1971a) for use in milder schemes of solid-phase peptide synthesis. They have been prepared by (1) reaction of the dicyclohexylammonium salt of an N^{α}-protected amino acid with a bromobenzhydryl-support (Table VI, number 53), presumably quantitative in CH_2Cl_2 at 25°C for 96 hr (Southard *et al.*, 1971a); (2) *N,N'*-carbonyldiimidazole-mediated coupling of the acid to a secondary alcohol support (Table VI, number 52) in about 35% yield (Losse and Meisegeier, 1974); and (3) reaction of the free acid with a diazo support (Table VI, number 54) in chloroform at 25°C for 60 min (>85% by nitrogen evolution) (Chapman and Walker, 1975; compare to Aboderrin *et al.*, 1965).

For these supports, the premature acidolytic loss of chains per deprotective cycle amounted to 9% in conjunction with the N^{α}-(2-benzoyl-1-methylvinyl) (Bmv) protecting group (structure **55** on p. 109) (Southard *et al.*, 1971a), and 0.4% with the 2-(*p*-biphenylyl)propyl(2)oxycarbonyl (Bpoc) group (Losse and Meisegeier, 1974). High-yield cleavage of the completed peptide from the resin was achieved with TFA–$CHCl_3$ (1:19) at 25°C for 30 min (Southard *et al.*, 1971a).

3. Phenacyl Esters

Phenacyl ester linkages were introduced for solid-phase peptide synthesis by Weygand (1968) and Mizoguchi *et al.* (1969, 1970) based on their properties as reversible carboxyl protecting groups, which were discovered and developed by Sheehan and Daves (1964) and Stelakatos *et al.* (1966). Phenacyl esters are exceptionally stable to anhydrous acids, including liquid HF and refluxing TFA, but are one to two orders of magnitude more susceptible to nucleophiles than simple benzyl esters. Cleavage can also be achieved, in solution, by reductive methods such as catalytic hydrogenolysis (Stelakatos *et al.*, 1966; Taylor-Papadimitriou *et al.*, 1967), use of zinc in glacial or aqueous acetic acid (Hendrickson and Kandall, 1970), or photolysis in protic solvents (Sheehan and Umezawa, 1973). However, partial reduction of the phenacyl carbonyl has been reported as a serious side reaction in at least one case (Taylor-Papadimitriou *et al.*, 1967), since the resultant ethylphenyl esters resist further facile reduction. Hence, the applicability of the listed reductive methods for solid-phase synthesis still requires further assessment.

a. Synthesis. Bromoacetyl or 2-bromopropionyl supports (Table VI, numbers 20 and 21) are readily esterified in essentially quantitative yields by reaction with the cesium salts of protected amino acids and short peptides (Wang, 1976; Birr *et al.*, 1977; compare to pp. 60–63 for general discussion of esterification methods and see Mizoguchi *et al.*, 1970, for tabulations of results with triethylammonium salts).

b. Nucleophilic Cleavage. The general considerations (Section III,B,1,c) for nucleophilic cleavage of benzyl esters also apply to phenacyl esters, although the rates are much faster in the latter case. As a consequence, the final cleavage from the support can be shortened, and the accompanying side reactions lessened. On the other hand, the risk of deleterious side reactions due to extraneous nucleophiles will be greater during the synthesis with phenacyl esters.

Several reagents have been recommended for cleavage of phenacyl ester linkages (Weygand, 1968; Mizoguchi *et al.*, 1969, 1970), including (1) 1.0 *M* sodium thiophenoxide in DMF, 25°C, 5–24 hr; (2) 0.5 *N* NaOH in dioxane–H_2O (2:1), 25°C, 6–24 hr; (3) saturated ammonia in methanol, 25°C, 20 hr; (4) methanol containing 1.5 *M* triethylamine; and (5) hydrazine hydrate, 10–50 equiv in methanol or DMF, 25°C, 0.5–3 hr (see also Maki *et al.*, 1975; Wang, 1976; better yields are sometimes obtained with anhydrous hydrazine). Furthermore, (6) benzyltrimethylammonium hydroxide in methanol–dioxane (1:3) has been routinely applied (Birr *et al.*, 1977) for saponification of small protected peptide segments. A comparative study of several methods to remove protected peptide acids from phenacyl supports motivated the development of a reagent with advantageous properties, namely (7) 0.1 *M* potassium cyanide as its 1:1 complex with the crown ether, dicyclohexyl-18-crown-6, in dry DMF or *N*-methylpyrrolidinone, 25°C, 2 × 12 hr (Tam *et al.*, 1977). Under these conditions, the N^i-formyl and N^{im}-dinitrophenyl protecting groups for tryptophan and histidine, respectively, were cleaved, but the β-benzyl ester of aspartic acid was recovered essentially unchanged.

c. Photolysis. The mechanism of phenacyl ester photolysis appears to involve radical scission of the alkyl–oxygen bond (Sheehan and Umezawa, 1973). The reaction solvent must provide protons both for the resultant carboxylic acid, and for the acetophenone by-product; for example, acetaldehyde is formed when ethanol is the solvent. Alternatively, a hydrogen radical donor such as cumene may be employed in an inert solvent, for example benzene. The photoreactivity follows the sequence α-methylphenacyl > *p*-methoxyphenacyl > unsubstituted phenacyl.

Based on the above findings, Wang (1976) studied an α-methylphenacyl ester-anchoring linkage (derived from functionalized support number 21 in Table VI) for solid-phase peptide synthesis. A protected tetrapeptide with COOH-terminal glycine was assembled in the usual way and then removed from the resin in 92% crude yield by irradiation at wavelengths above 3500 Å (uranium glass filter), at 20°C for 72 hr (half-time approximately 5 hr). The suspending solvent was dimethylformamide and argon was swept through the system to ensure anaerobic conditions. The results were considered superior to those obtained with an *o*-nitrobenzyl ester linkage (p. 81). Photolytic cleavages of α-methylphenacyl ester-anchored peptides containing COOH-terminal amino acids other than glycine (e.g., Ala, Leu, Thr(Bzl), Ile) proceeded at rates two- to fivefold slower.

d. Side Reactions. Amino acid phenacyl esters (**24**) generated during the very first deprotection and neutralization cycle of a solid-phase peptide synthesis can undergo intramolecular attack on the phenacyl carbonyl by the free α-amino group, followed by dehydration to give the structure **25** [Birr *et al.*, 1977; Tam *et al.*, 1977; see Eq. (4)]. Often, as much as 30% of

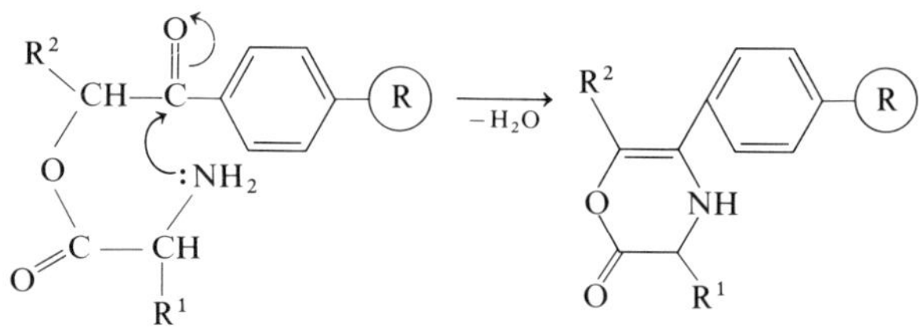

(4)

(24) **(25)**

R^1 = H, other side chains
R^2 = H, CH_3

the COOH-terminal amino acid will be lost by this irreversibly terminating side reaction, which is accompanied by a bright yellow coloration of the support. No certain way to avoid the formation of **25** has yet been found; the extent of cyclization will be determined by the competition with the bimolecular coupling step involving the resin-bound amino group and the next protected amino acid or peptide.

Diketopiperazine formation and loss of loaded sites from the support (compare to Section III,B,1,g on pp. 82–84) has been observed at the dipeptide–phenacyl ester level (Birr *et al.*, 1977). The overall problem is expected to be more serious than for benzyl ester linkages because phenacyl esters are more susceptible to aminolysis.

Both of the side reactions discussed here can be tolerated, but they can also be completely circumvented by directly coupling a protected tripeptide to a bromoacetyl or bromopropionyl support (Birr *et al.*, 1977).

4. Substituted Phenyl Esters

Phenyl esters derived from a variety of functionalized supports (Table VI, numbers 3, 5, 33, and 34) have been successfully applied as anchoring linkages for solid-phase peptide synthesis (details in the suitable references of Table VI). Residual phenolic sites ought to be blocked, for example by acetylation. The phenyl esters are reasonably stable to acids but are readily cleaved by bases or nucleophiles, either directly or after safety-catch activation of a phenyl ring substituent (e.g., hydrogen peroxide or *m*-chloroperbenzoic acid oxidation of a *p*-alkylthiophenyl ester, based on Table VI, number 34, to a *p*-alkylsulfonylphenyl ester; D. Marshall and Liener, 1970; Flanigan and G. Marshall, 1970; see Table I for an example). It seems possible that a portion of the actual cleavage from the support may be via oxazolone formation; such a mechanism would predict partial racemization of the COOH-terminal residue. The mild nucleophilic cleavage of (unactivated) phenyl esters by peroxide ions in partially aqueous media at pH 10.5, 20°C, 10–30 min, in the presence of dimethyl sulfide as a scavenger (Kenner and Seely, 1972), which does not give substantial racemization, has been applied to the synthesis of a protected octapeptide (D. Hudson, cited by Kenner, 1977, pp. 446 and 455; see also Hudson *et al.*, 1979).

5. Alkyl Esters

n-Alkyl ester linkages are entirely stable to most strong anhydrous acids (e.g., Table VIII) at 25°C. Therefore, cleavage is generally performed under basic (usually hydrolytic) conditions. The most promising approach along these lines is incorporation of a good electron-withdrawing group as a β-ethyl ester substituent, either directly or in the course of safety-catch activation. Subsequently, nucleophilic cleavage of the ester by a β-elimination mechanism can proceed under the influence of a relatively mild anhydrous base.

In one instance, a *tert*-alkyl ester linkage was examined for solid-phase peptide synthesis (Wang and Merrifield, 1972; based on functionalized support number 39 in Table VI). The anchor is readily cleaved by TFA–CH_2Cl_2 (1:1), at 25°C for 120 min, and is stable to strong nucleophiles. Unfortunately, the initial esterification step proved to be difficult, and the best conditions found (activation of 1 equiv N^{α}-phthaloylamino acid with 2 equiv *p*-toluenesulfonyl chloride in dry pyridine; based on Brewster and Ciotti, 1955) were shown to be accompanied by partial racemization. Hence, the *p*-alkoxybenzyl ester linkage (**16** on p. 76) is preferred for applications requiring cleavage from the support with anhydrous acids of moderate strength.

a. Synthesis. N^{α}-protected amino acids are esterified to primary alcohol supports (Table VI, numbers 22, 23, 35–37, and 44) by the usual methods (pp. 63 and 65); specific procedures may generally be found in the same publications describing functionalization of the support, but it would be surprising if more recently developed improved procedures were not equally applicable. Buis *et al.* (1076) recommended blocking residual free hydroxyl-groups by formylation with acetic anhydride–formic acid–pyridine–benzene (8:3:0.1:10), 25°C, 20 hr. Alternatively, isopropenyl formate (10-fold excess) in CH_2Cl_2 may be applied for the same purpose (van Melick and Wolters, 1972).

b. Safety-Catch Activation. Guided by the analogy to the carboxyl protecting groups of Hardy *et al.* (1968) and Miller and Stirling (1968), several authors have investigated the oxidation of β-sulfide ethyl esters (based on functionalized supports numbers 36 and 37 of Table VI). Conditions include hydrogen peroxide in acetone with ammonium molybdate as catalyst (Marshall and Flanigan, cited by Marshall and Merrifield, 1971, pp. 130–131) or *N*-chlorosuccinimide (0.2 *N*) in 0.2 *M* phosphate buffer (pH 7.5)–dioxane (1:1) (Gait and Sheppard, 1976). Evidently, these methods are limited to peptides not containing cysteine, methionine, tyrosine, or typtophan residues. Activation of these anchors by S-methylation (Amaral *et al.*, 1966; compare to Kunz, 1976b) does not appear to have been tried in the solid-phase mode.

A safety-catch linkage that is activated under acidic conditions has been developed by Wieland *et al.* (1970) and adaptated to solid-phase peptide synthesis (Table VI, functionalized support number 38). In the solution model [Eq. (5)], dehydration at the tertiary hydroxyl group of the primary

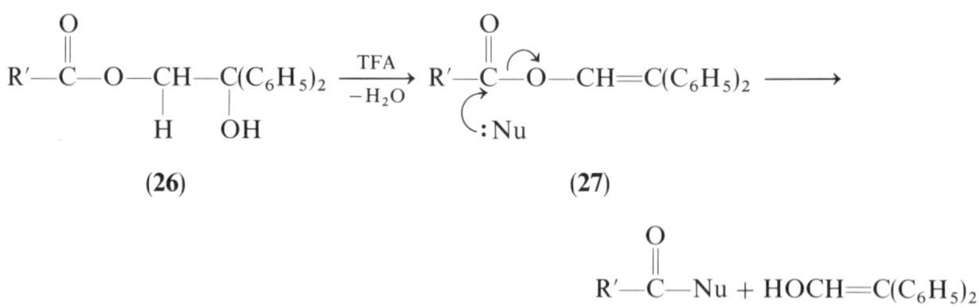

(26) (27) (5)

glycol ester **26**, using anhydrous trifluoroacetic acid, at 20°C for 60 min, gives the enol ester **27**, which in turn is readily cleaved by nucleophiles. This clever scheme may be limited because the acidic activation conditions preclude use of the anchor in conjunction with a variety of useful protecting groups.

c. Nucleophilic Cleavage. Ordinary *n*-alkyl esters have essentially the same spectrum of lability to bases and nucleophiles as do benzyl esters (Section III,B,1,c); hence, we restrict this discussion to the β-activated *n*-alkyl esters. The β-sulfonylethyl esters **28** obtained by safety-catch activation (see above) or through direct anchoring and peptide chain extension on the solid phase (starting with the oxidized form of functionalized support number 36, Table VI) have been shown to be quite resistant both to tertiary amines, as used repetitively in the neutralization steps, and to the strong nucleophile sodium thiophenoxide, as used to remove phenacyl esters. Release of a protected peptide acid from these supports can be achieved

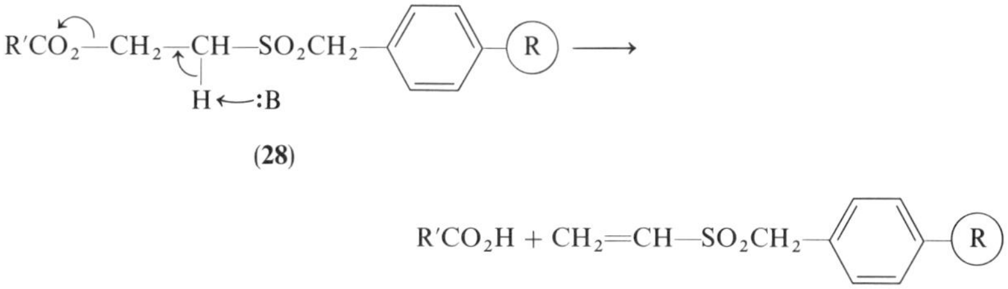

(28)

(6)

[Eq. (6)] by (1) methoxide present in a mixture consisting of dioxane–methanol–4 *N* NaOH (30:9:1), at 25°C, for 3 min followed by quenching with acetic acid (Tesser *et al*., 1976; Buis *et al*., 1976) or (2) 0.2 *N* NaOH in dioxane–H_2O (1:1); at 25°C, for 10 min (Gait and Sheppard, 1976; sofar applied only to oligonucleotides). The available evidence suggests that methods (1) and (2) release peptides attached through the standard *p*-alkylbenzyl ester linkage at comparable rates and yields.

The β-sulfonylethyl esters are also cleaved by (3) anhydrous tetramethylammonium fluoride (slurry 80 mg/ml) in dimethylformamide, at 40°C, for 90 min (Teutsch, 1976). Method (3) has the advantage that exposure to methanol during the base step is unnecessary, hence ruling out the low levels of transesterification (1–2% methyl esters) that have been observed by method (1).

The β-trimethylsilylethyl esters studied by Sieber (1977) might be usefully adapted to solid-phase peptide synthesis. Cleavage by a β-elimination reaction involving anhydrous tetraethyl- or tetrabutylammonium fluoride in dimethylformamide or dimethyl sulfoxide, at 25°C for 5–60 min, may be additionally favored by the high bond energy of the formed silicon–fluorine bond (Eaborn, 1960). Sieber (1977) also discussed side reactions encountered during treatment of protected peptides under these conditions. Thus, N^{α}-benzyloxycarbonyl groups gave rise to hydantoins, benzyl esters were appreciably cleaved, disulfides rapidly disproportionated, COOH-terminal asparagine and glutamine were converted to the imide derivatives, and $\alpha \rightarrow \beta$ rearrangement at protected aspartic acid residues was noted. The rapid cleavage of the β-trimethylsilylethyl esters induced by anhydrous trifluoroacetic acid at 25°C may be a more reliable means of deprotection than fluoridolysis (compare to Carpino *et al.*, 1978).

Ugi *et al.* (1977) briefly described a series of β-haloester anchoring linkages for solid-phase peptide synthesis (e.g., **29**). A carboxyl component was

(29)

reacted in CH_2Cl_2 with dichloroacetaldehyde and an isocyanide support (Skorna *et al.*, 1978) by the Passerini reaction (compare to Eckert, 1977), and cleavage from the support was through use of a "supernucleophile", cobalt(I) phthalocyanine anion (Schrauzer and Deutsch, 1969). Further information regarding the stability and applicability of this interesting anchor **29** is awaited.

6. Sulfonamides

A resin-bound analogue of the *p*-toluenesulfonyl amino protecting group (prepared quantitatively by reaction of an amino acid *tert*-butyl ester with functionalized support number 45 in Table VI) has been explored for application to an N → C scheme of peptide synthesis (Dahlmans, 1973) but does not appear to offer any particular advantages due to the relative severity of the cleavage conditions (PH_4I in TFA). The potential of this type of support for side-chain anchoring of arginine and histidine residues, which can be deprotected under milder acidic conditions, remains to be explored.

Acyl sulfonamides **30** were among the first carboxyl anchors used (Kenner *et al.*, 1971) to demonstrate the safety-catch principle; a specific application is included in Table I. The linkage was established, in about 25% yield, by treating a sulfonamide support (Table VI, number 46) with an N^{α}-protected amino acid 2,4,5-trichlorophenyl ester (2 equiv) and triethylamine (1 equiv) in DMF, at 25°C for 48 hr. The residual free sulfonamide sites evidently did not react in subsequent active ester coupling steps, for which the weaker base 2,6-lutidine was used in place of triethylamine.

Acyl sulfonamides are stable to strong anhydrous acids including HBr–HOAc at 25°C, as well as to aqueous alkali. The latter property is presumably due to prior basic attack to ionize the acidic NH group ($pK_a \sim 2.5$). Safety-catch activation by N-methylation with diazomethane in ether–acetone (solvent mixture chosen to suppress competitive O-methylation) gives the labile species **31** which is readily cleaved by alkali (1 equiv 0.5 *N* NaOH) or other nucleophiles [Eq. (7)]. The basic conditions may still be strong enough to promote certain side reactions of the type already cited (p. 78); partial

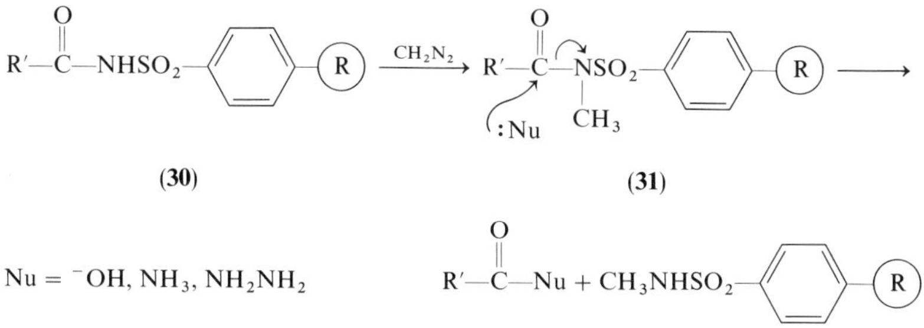

(30) (31)

(7)

racemization of the COOH-terminal residue and $\alpha \rightarrow \beta$ rearrangement of an aspartyl-glycine sequence have in fact been observed.

7. *Anchoring through Peptide Bonds*

Peptide structures are readily elaborated and entirely stable throughout the normal repetitive steps of a solid-phase synthesis. However, from the litany of peptide degradative procedures (reviewed by Hirs, 1967; Spande *et al.*, 1970; Needleman, 1975) there are at least three types of reactions that are sufficiently discriminatory to be applicable to selective cleavage from the support. In these cases, only the peptide bond adjacent to a specific amino acid residue is broken, while the desired peptide structure is isolated intact.

a. α-Cleavage at Dehydroalanine Residues. Gross *et al.* (1973, 1975) developed an effective solid-phase route to protected peptide amides [Eq. (8)], based on the known lability of enamines to hydrolysis (reviewed in Spande *et al.*, 1970, pp. 102–114). The method was also demonstrated in

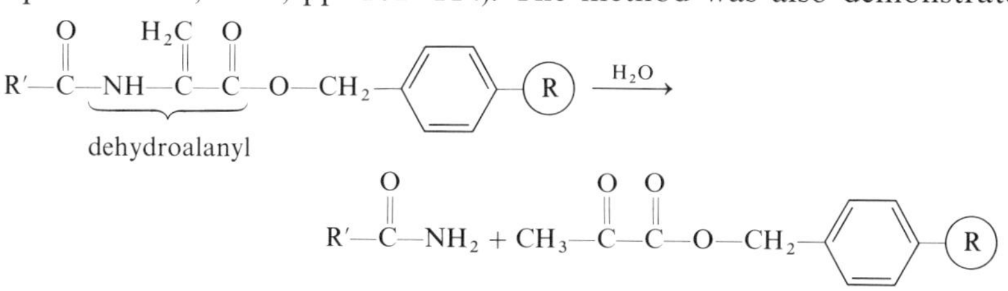

(8)

the case of side-chain anchoring of aspartic acid or glutamic acid residues, which are recovered as their ω-amides.*

* The solid-phase synthesis of free and protected peptides containing COOH-terminal asparagine can lead to an assortment of difficulties (Section V,B,5), both during anchoring through the carboxyl group and final cleavage (especially under basic conditions). Side-chain anchoring of aspartic acid to functionalized supports that produce the β-carboxamide upon cleavage appears to be an attractive alternative strategy.

Dehydroalanine residues must be introduced indirectly because the free amino acid readily decomposes with loss of ammonia. In a preferred scheme, a protected amino acyl-*O*-*p*-toluenesulfonylserine dipeptide is treated with 0.1 *N* NaOH in aqueous alcohol (Photaki, 1963) to give a protected dipeptide with COOH-terminal dehydroalanine, which is then esterified as a benzyl ester to the support in the usual way (Section III,B,1,a). There is a danger of partial racemization of the NH_2-terminal amino acid of the initial dipeptide (which will eventually be the COOH-terminal residue of the peptide amide) during the β-elimination step, even under the given conditions that were especially recommended to minimize this problem. Evidently, the naked dehydroalanine residue remains unaffected by many of the standard steps of the repetitive deprotection/coupling cycles, including treatments with anhydrous trifluoroacetic acid and tertiary amines; an oxytocin analogue (9 residues) was safely assembled applying this linkage (Rosamond and Ferger, 1976). On the other hand, it might seem prudent to avoid the use of thiols or strong nucleophiles, which could add across the enamine double bond. In another approach (Gross and Matsuura, 1975), an *S*-methylcysteine residue incorporated by the usual methods becomes activated by a two-step safety-catch procedure, involving oxidation to the sulfone and β-elimination, all while the peptide remains on the solid support (see Rich and Tam, 1975, for details of comparable chemistry carried out in solution).

The hydrolytic cleavage procedure [Eq. (8)] worked out by Gross *et al.* (1973) involves treatment with water (1 equiv per dehydroalanine residue) in 1.0 *N* HCl–HOAc, at 50°C for 30 min. Yields are high (60–90%, after purification) and the *tert*-butyloxycarbonyl protecting group is removed at the same time.

b. Electrophilic Attack. The cleavage of the peptide bond at the carboxyl end of a methionine residue after specific S-alkylation with cyanogen bromide (Gross and Witkop, 1962; Gross, 1967) has been modified for removal of a peptide from a polymeric support (Hancock and Marshall, 1975). As described, the most widely applied C → N scheme of stepwise solid-phase synthesis is used to introduce, in order, two glycine (presumed to decrease steric hindrance) and one methionine residue, followed by the desired sequence. Cleavage proceeds in good yield (55–98%) using cyanogen bromide (50 equiv) in propionic acid, at 25°C for 20–40 hr; the *tert*-butyloxycarbonyl protecting group is also lost under these conditions. The cleaved peptide includes a COOH-terminal homoserine (or homoserine lactone) residue which can, however, be removed enzymatically (p. 93) with acceptable specificity in some cases.

A more flexible adaptation of cyanogen bromide cleavage reactions to solid-phase methodology would result in retention of the homoserine moiety on the support, so that the desired peptide may be obtained directly in solution. Three, as yet untested, proposals follow: (1) Methionine could be anchored to the support through its amino group, with further peptide synthesis conducted in the N → C direction. To achieve chain extension in the usual C → N direction, a conceivable approach involves (2) coupling of the ε-amino group of lysine to the free α-carboxyl group of an N^{α}-methionyl-resin. Further synthesis gives a peptide containing a side-chain anchored COOH-terminal lysine. Another optional strategy exploits the fact that (3) peptide bond cleavage occurs at the amino end of a cyanylated *S*-methyl-cysteinyl residue (Gross and Morell, 1974; Gross and Matsuura, 1975); the reaction conditions are more strenuous than for cleavage at methionine, and there is a competing β-elimination pathway. Incidentally, incorporation of internal methionine residues of a desired peptide sequence prepared by any of the schemes discussed in this section ought to be readily possible through the use of methionine sulfoxide derivatives; the remaining 19 naturally occurring amino acids are probably compatible without any additional precautions.

c. Enzymatic Hydrolysis. The idea of using proteolytic enzymes to achieve cleavage from polymeric supports was advanced and tested by Anfinsen (1968) and by Blecher and Pfaender (1973). While this approach is almost ideal from the point of view of mildness of the reaction conditions, it is presently limited by the need for a hydrophilic medium, and by the insufficient specificity of action of most endopeptidases. The major enzyme with useful properties in this regard is trypsin, which cleaves peptide bonds at the carboxyl side of arginine, lysine, and *S*-aminoethylcysteine residues. Internal lysine residues will not be cleaved if the ε-amino side chain is blocked by any of a variety of reversibly removable protecting groups (Section V,B,2,a).

A peptide may occasionally be cleaved from a solid support still containing amino acids superfluous to the desired sequence. Manageable examples include COOH-terminal homoserine from cyanogen bromide cleavage (p. 92) and COOH-terminal lysine or arginine from enzymatic trypsin cleavage or from an ε-amino side-chain anchored peptide synthesis. The specificities of carboxypeptidase A (reviewed by Pétra, 1970) and carboxypeptidase B (reviewed by Folk, 1970) are perfectly matched to solve these problems, since the former enzyme catalyzes the hydrolysis of the COOH-terminal peptide bonds of many amino acid residues but specifically not those of the basic residues, while the latter enzyme removes the basic residues but no others.

8. Carboxamides

The synthesis of peptide amides following attachment of the COOH-terminal residue through its carboxyl group to certain suitable amino group-containing functionalized supports (benzhydrylamine resins) was pioneered by Pietta and Marshall (1970). Acidolytic or photolytic cleavage occurs on the other side of the ensuing carboxamide linkage, resulting in transfer of an equivalent of ammonia to the peptide portion. This approach is more general than ammonolysis of benzyl or phenacyl ester linkages (Sections III,B,1,c and B,3,b) but there is not enough information to make a comparison with the much less widely used dehydroalanyl method [Eq. (8) on p. 91]. Side chain anchoring of the ω-carboxyl group of aspartic acid or glutamic acid residues to facilitate the synthesis of peptides containing COOH-terminal asparagine or glutamine has been demonstrated recently (Li *et al.*, 1976; Wang *et al.*, 1979; M. Shimizu, personal communication, 1975; compare to footnote on p. 91). Peptides with a COOH-terminal *N*-methylamide group can be prepared in an exactly analogous manner using an *N*-methylbenzhydrylamine resin (Rivier *et al.*, 1977). As with the substituted benzyl ester linkages (Section III,B,1), the lability of the anchoring bond of the carboxamides described in this section can be modulated by introduction of electron-donating or -withdrawing substituents on one of the aromatic rings.

a. Synthesis. Carboxamides are prepared by coupling an activated carboxyl component to an amino group-containing functionalized support (Table VI, numbers 55, 56, and 64; since these supports are often stored as their hydrochloride salts, a preliminary neutralization step with a tertiary amine solution will be necessary). Dicyclohexylcarbodiimide is often used as the condensing agent, and since the conditions are essentially the same as in the later repetitive cycles of a solid-phase synthesis, the anchoring step can be incorporated into an automated scheme in the same reaction vessel. Yields are high (70–100%) but any unreacted sites are covered by acetylation, for example with 0.2 *M* acetic anhydride–CH_2Cl_2 in the presence of catalytic amounts of dimethylaminopyridine, at 25°C for 30 min (Pietta and Brenna, 1975). In an interesting application, a protected 14-peptide (1 equiv) prepared in solution was loaded onto a benzhydrylamine–polystyrene resin (1.8 equiv amino sites) in 40% yield, with the aid of dicyclohexylcarbodiimide (1 equiv = 20 m*M*) and 1-hydroxybenzotriazole (2 equiv) in dimethylformamide (Hagenmaier, 1975). Cleavage with HF (see below) produced the free peptide amide.

b. Acidolysis. Carboxamides are considerably more stable to anhydrous acids than are the analogous esters (compare this Section to Sections III,B,1,b and III,B,2). The premature loss in TFA–CH_2Cl_2 (1:1) of peptide chains anchored through a benzhydrylamide linkage was shown (Burton *et al.*, 1977) to proceed at a rate 3.8-fold slower than the corresponding value determined with the standard *p*-alkylbenzyl ester linkage (compare to

Table VII). Under standard cleavage conditions (anhydrous HF–anisole (9:1), 0°C, 30–60 min) peptides containing COOH-terminal glycinamide or prolinamide are obtained in useful yields (50–90%); but removal of peptides with COOH-terminal phenylalaninamide is difficult (10–30%). An approximate order of lability is His(Tos) > Gly > Glu(OBzl), Thr(Bzl) > Pro, Met > Val, Leu > Phe (Stewart *et al.*, 1976). Aspartic and glutamic acids, anchored through their side chains, are near the stable end of the scale. To circumvent the poor cleavage yields obtained with certain peptides, the somewhat more acid-labile linkages derived from *p*-methylbenzhydrylamine (Stewart *et al.*, 1976) or *p*-methoxybenzhydrylamine (Pietta and Marshall, 1970; Orlowski *et al.*, 1976) supports (Table VI, number 55 with $Y^1 = CH_3$ or CH_3O) have been recommended; these linkages will probably be unacceptable for lengthy syntheses due to excessive premature losses of chains from the solid support.

The *p*-alkoxybenzylamide linkage (derived from support number 56 in Table VI; compare to structure **16**) is substantially more acid-labile than the others discussed in this section and is therefore suitable for the preparation of fully protected peptide amides (Pietta and Brenna, 1975). Premature acidolysis was 0.7% per hr in 1% TFA–CH_2Cl_2, and final cleavage using TFA–CH_2Cl_2 (1:1) at 25°C for 30 min gave crystalline amino acid and peptide amides in acceptable yields (35–80%).

c. Photolysis. By analogy with their earlier results (see p. 81), Rich and Gurwara (1975b) photolyzed *o*-nitrobenzylamide linkages (derived from Table VI, support number 64); a protected decapeptide amide was obtained in 65% yield after chromatography and crystallization. This particular peptide contained COOH-terminal glycine, but a model experiment with Boc-Val-resin suggested that steric hindrance by the COOH-terminal residue might not be an important effect in controlling cleavage yields. This is in contrast to results with α-methylphenacyl esters (pp. 85–86).

A series of light-sensitive nonpeptide amides of the general structures **32** and **33** have been developed by Amit and Patchornik (1973; Amit *et al.*, 1976a,b). Depending on the amide structure, photocleavage provided the

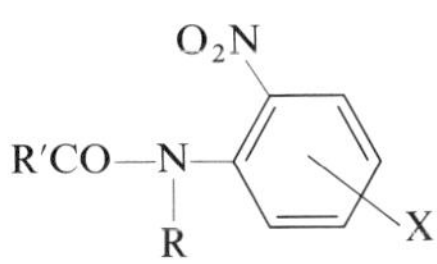

(32)

$R = CH_3$, n-C_4H_9, C_6H_{11}, $C_6H_5CH_2$, C_6H_5

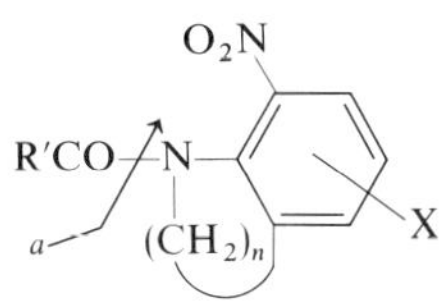

(33)

$X = H, CH_3O, Br$
$n = 2, 3$

free acid, or photosolvolysis—in appropriate solutions including water, alcohols, or ammonia as nucleophilic components—gave the acid, ester, or amide. Several of the derivatives are labile to visible light and require special handling. Based on the work just described, 1-peptidyl-5-bromo-7-nitroindolines have been assessed as carboxyl-protected derivatives of peptides (Goissis *et al.*, 1977). Construction of the secondary amide bond (*a* in formula **33**) was difficult due to the electron-withdrawing substituents; hence acylation of indoline was performed prior to modification of the aromatic ring. In spite of the various complications, further research along these lines may lead to a new practical photolysable linkage for solid-phase peptide synthesis.

9. Substituted Hydrazides

Protected peptide hydrazides (reviewed by Meienhofer, Vol. 1, Chapter 4) are useful precursors for peptide synthesis in that (1) reaction with nitrous acid gives the azides suitable for segment condensation and (2) oxidation with *N*-bromosuccinimide or other agents gives diimides, which react with nucleophiles such as water, ammonia, or amino acid or peptide esters to provide respectively the free peptide acid, peptide amide, or an extended peptide. The latter class of applications was explored for a safety-catch scheme of solid-phase synthesis by Wieland *et al.* (1970) with moderate success. An elegant aspect of this work was the anchoring strategy, in which a protected amino acid was attached first to a *p*-hydrazinobenzoic acid

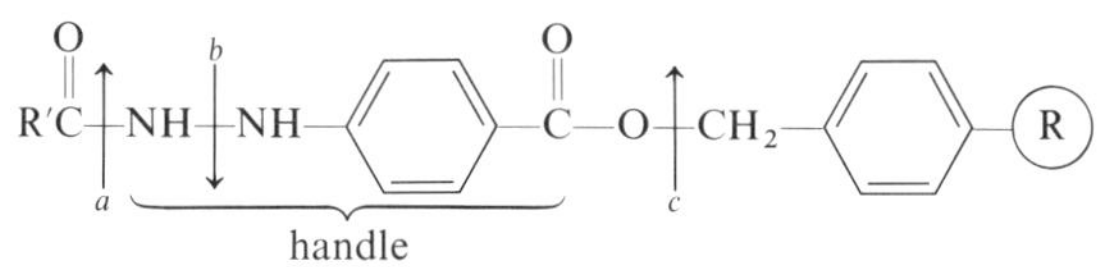

(34)

handle (bond a in formula **34**) prior to esterification to the support (creating bond c). Oxidation (bond b) followed by nucleophilic attack (bond a) was a conceivable option for cleavage reactions. However, the completed peptides were best removed as their *p*-carboxyphenylhydrazides by acidolysis (bond c). Limitations of the oxidative approach have already been discussed (p. 88).

General routes to protected peptide hydrazides have been developed by Wang and Merrifield (1969b; Wang, 1973, 1975; see Table I for an illustration). Alkoxycarbonylhydrazide supports (Table VI, numbers 40, 57, and 60) can be acylated with an N^{α}-2-(*p*-biphenylyl)propyl(2)oxycarbonyl-amino acid using standard dicyclohexylcarbodiimide mediated coupling. Smooth acidolytic cleavage of the ensuing linkage proceeds upon treatment with TFA–CH_2Cl_2 (1:1) at 25°C for 30 min. Taking into account both the ease of functionalization and a need to minimize premature acidolytic loss of chains from the support, the 3-(*p*-benzyloxyphenyl)-1,1-dimethylpropyl-oxycarbonylhydrazide linkage (**35**) appears to have the best combination of features (Wang, 1975; and personal communication, 1978). This approach avoids the potential side reactions encountered in methods employing the

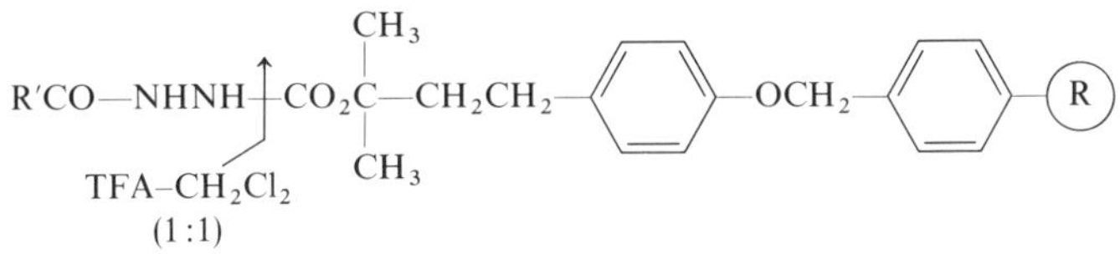

(**35**)

direct hydrazinolysis of benzyl and phenacyl ester linkages to the solid support (pp. 78–79 and 85).

10. Substituted Urethanes

Urethane linkages have been applied for solid-phase synthesis according to several strategies: (1) Attachment of the α-amino group of a protected amino acid or peptide in conjunction with an N → C coupling scheme; (2) N^{ω}-side-chain anchoring of ornithine or lysine residues; and (3) polymer-bound protection of α-amino groups for dual anchoring approaches (see Section II,B and Table I for examples, references, and general discussion). It seems that significant improvements and expansions on the available methodology described below ought to be possible, using the ample precedents from solution chemistry as a guide.

a. Synthesis. The amino group-containing components can be reacted with supports bearing a chloroformyl or phenyl carbonate group (Table VI, number 50; see also remark about number 52) in DMF or $CHCl_3$ at room temperature in the presence of triethylamine. Yields vary depending on the substrate and reaction time from 20 to >90%, but the excess starting amine can be readily recovered. Excess functional sites on the resin should be derivatized by treatment with, for example, dimethylamine (2.0 *M*) in chloroform at 25°C for 3 hr (Felix and Merrifield, 1970).

b. Acidolysis. Urethanes are one to two orders of magnitude more sensitive to acids than the corresponding esters (Ben-Ishai and Berger, 1952). This fact should be considered before embarking on a multi-step synthesis using acid-labile, temporary protecting groups (compare to Section III,B,1,b,i on pp. 67–69). Cleavage of a completed peptide attached through a polymer-bound *p*-alkylbenzyloxycarbonyl linkage proceeds

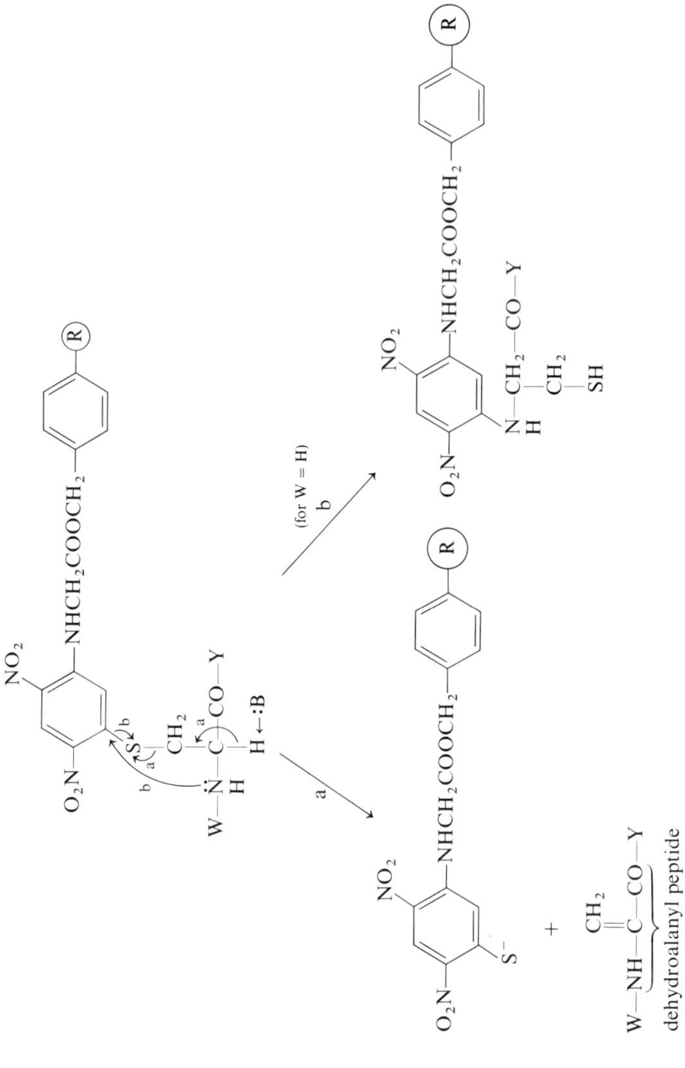

Figure 8. Base-catalyzed side reactions of *S*-dinitrophenylene bridged cysteine peptides: (a) β-elimination; (b) S → N migration, redrawn after Glass *et al.* (1974).

readily and in high yields upon treatment with strong acids (Table VIII) including HBr in acetic acid (Letsinger *et al.*, 1964; Sklyarov and Shashkova, 1969); a polymeric benzhydryloxycarbonyl linkage used in a dual anchoring strategy can be cleaved with TFA–CH_2Cl_2 (1:2), at 25°C for 20 min (Frank *et al.*, 1977a).

11. Dinitrophenylene Linkages

The susceptibility of the acid-stable *S*-, *O*-, and N^{im}-dinitrophenyl derivatives of cysteine, tyrosine, and histidine to thiolysis (Shaltiel, 1967) has been exploited by Glass *et al.* (1972, 1974) in the development of schemes for side-chain anchoring and bidirectional solid-phase peptide synthesis (see Table I for examples).

a. Synthesis. The appropriate functionalized fluorodinitrophenylene support (Table VI, number 65) can be reacted with the free side-chain component (10 equiv), at 25°C for 24 hr, in dimethylformamide with triethylamine (10 equiv) as the base. The reported yields for this anchoring step were good (70–80%, based on glycine as an internal standard); it is not clear from the data whether unreacted fluoro sites remained on the support, but no additional blocking step was employed.

b. Nucleophilic Cleavage. To avoid base-catalyzed side reactions (Fig. 8) of *S*-dinitrophenylene linkages, Glass *et al.* (1974) found it necessary to avoid the use of triethylamine during neutralization steps of the repetitive deprotection/coupling cycles. As a substitute, neutralization with a mixture of *N*-methylmorpholine (1.0 *N*) and acetic acid (0.5 *N*) in dimethylformamide, referred to as "buffered DMF", was found to permit subsequent acylation. Due to the presence of acetic acid from the "buffered DMF," the usual dicyclohexylcarbodiimide activation procedure for coupling is not applicable, and protected amino acids or peptide segments have to be introduced via their active esters or anhydrides.

Cleavage of a protected nonapeptide from the support (70% crude yield) was achieved by treatment with β-mercaptoethanol–"buffered DMF" (1:9), at 25°C for 2 × 24 hr. An N^{im}-dinitrophenylene linkage has been quantitatively cleaved using a mixture of β-mercaptoethanol (0.67 *M*) and triethylamine (17 m*M*) in dimethylformamide, at 25°C for 9 hr (Glass *et al.*, 1972).

c. Side Reactions. Two base-catalyzed side reactions of *S*-dinitrophenylene linkages have been mentioned (Fig. 8), both of which serve to lower overall yields without apparently affecting the homogeneity of the product. A β-elimination reaction leads to loss of chains from the support, while an S → N migration—occurring only when the α-amino group of the anchored cysteine is free—leads to irreversible retention of chains because the N^{α}-dinitrophenylene linkage is resistant to thiolysis.

IV. THE REPETITIVE DEPROTECTION/COUPLING CYCLE

Once an initial synthetic unit has been covalently anchored to a polymeric support by the chemical procedures described in the previous section, the major concern of a solid-phase synthesis becomes the systematic elaboration of an everlengthening peptide chain. This section deals with the scope and limitations of methods for removal of temporary protecting groups, and the following coupling of incoming protected amino acids or peptide segments, as discussed in general terms in Sections II,A,3 and A,5. It is essential that these repetitive steps proceed in high yields and with minimal side reactions (see Table II, p. 18, as well as Section II,C,4,a, pp. 32–33). Problems warranting special attention include (1) the potential for racemization, (2) premature irreversible termination of growing peptide chains, (3) monitoring to evaluate the extent of completion of reactions, and (4) intentional blocking of unreacted chains.

A. Temporary Protecting Groups*

There are several stringent requirements which must be satisfied by a temporary protecting group for peptide synthesis (see Fig. 2 on p. 8, groups B and C). In order to survive during the coupling steps, the protecting group must be adequately resistant to (1) mild acids (stable to α-carboxyl group), (2) mild bases or nucleophiles (stable to α-amino group), and (3) acylating agents (not attacked by the incoming activated protected amino acid or peptide segment). Additionally, (4) the optical integrity of the incoming species must be preserved upon coupling, and (5) removal of the protecting group must be rapid, quantitative, and free of side reactions. Three optional features of protecting groups are (6) straightforward introduction into amino acids or peptides (working time, cost of reagents, ease of purification), (7) a "solubilizing" effect on the protected intermediate (this feature is considerably more pertinent to solution than to solid-phase strategies), and (8) an *orthogonal* mode of deprotection (complementary to that used to remove any other protecting groups; see pp. 11–12).

Assembly of the peptide chain in the C → N direction is preferred for a variety of practical chemical considerations. Hence, in these cases, the temporary group satisfying the criteria just outlined will be needed to protect the α-amino function. Amino group protection and the preparations of derivatives has been reviewed (Barton, 1973; Carpino, 1973); the general treatises listed in the opening sentence of this chapter also contain valuable

* Throughout this section on temporary protecting groups, structural formulas for protecting groups will include the functional group being protected (usually the α-amino) in boldface.

information on these topics. In several instances (see Table I), assembly has been in the N → C direction or even bidirectional. Here temporary protection will be required for the α-carboxyl group (reviewed by Haslam, 1973) or its hydrazide (reviewed by Meienhofer, Vol. 1, Chapter 4).

Protection of the last residue (either NH_2-terminal or COOH-terminal) of a peptide sequence offers special opportunities in a synthetic design since, depending on the final objective, a protecting group different from the temporary group removed repetitively throughout the synthesis can be used. As the most trivial illustration, the α-amino protecting group of the NH_2-terminal residue can be of the same level of stability as the permanent protecting groups [e.g., Weitzel *et al.*, 1973, coupled N^{α},N^{ω},N^{δ}-tris(benzyloxycarbonyl)arginine]. Also, some naturally occurring peptides are blocked at one or both termini as, for example, *N*-formyl, *N*-acetyl, and amino-terminal pyroglutamyl or carboxyl-terminal amide derivatives. The blocking group can be part of the last synthetic unit coupled, or it can be introduced in an additional chemical step on the solid phase after removal of the last temporary group. In all the cited cases, the blocked derivative will survive strong acidolytic conditions for cleavage of the peptide from the support. Finally, a different α-amino protecting group might be preferred for solid-phase preparation of protected peptide segments, in the case where an otherwise valuable temporary group would not survive the chosen cleavage conditions. Alternatively, certain protecting groups not suited for general repetitive use might still lend themselves to special applications (e.g., *N*-trityl group, see pp. 107–108).

Chemical reactions that are sufficiently mild and specific to be applied to the repetitive removal of the temporary protecting groups include acidolysis, basic or nucleophilic cleavage, and reduction (including thiolysis). Procedures such as oxidative cleavage, photolysis, electrolysis, metal reduction, and catalytic hydrogenolysis, while valuable for solution applications, are either less general or more difficult technically when attempted for repetitive usage at each step of solid-phase schemes.

In making a choice of deprotective conditions for a particular application to solid-phase peptide synthesis, factors that must be considered include reaction rates swelling of the support by the medium, stability of the anchoring linkage, and convenience of preparation and stability of the reagent. The use of carboxylic acids as cosolvents or reagents leaves open the risk of physical or chemical "carryover," as is further discussed in Section IV,B (pp. 120–121).

1. *tert-Alkyl Urethanes Removable by Acidolysis*

The *tert*-butyloxycarbonyl (Boc) group (**36**) (Carpino, 1957; McKay and Albertson, 1957; Anderson and McGregor, 1957) is by far the most widely used functionality for α-amino protection in solid-phase peptide synthesis. The urethane structure effectively shields the activated *tert*-butyloxycarbonylamino acid from racemization, since oxazolones do not form. This remarkable attribute has been rationalized as being due either to the decreased nucleophilicity of the urethane carbonyl, or to the lowered acidity of the urethane proton relative to the carbonyl or proton of carboxamides (discussed by Young, 1967). *tert*-Butyloxycarbonyl derivatives, while slightly thermolabile, are very resistant to alkali, nucleophiles, and catalytic hydrogenolysis. Their importance for peptide synthesis stems from their smooth deprotection by acids of moderate strength.

The *tert*-amyloxycarbonyl (Aoc) (**37**) (Sakakibara *et al.*, 1965) and adamantyloxycarbonyl (Adoc) (**38**) (Haas *et al.*, 1966; Moroder *et al.*, 1976) groups have closely comparable properties, and appear to offer no intrinsic advantages over the Boc group. However, in selected instances, these latter derivatives are preferred because of superior solubility characteristics and greater ease of preparation (the chloroformates are more stable). The 1-methylcyclobutyloxycarbonyl (Mcb) group (**39**) is about 20-fold more stable to acid than the Boc group, and has been advocated for protection of the NH_2-terminus of peptide segments because of its enhanced stability during purification procedures (Brady *et al.*, 1977).

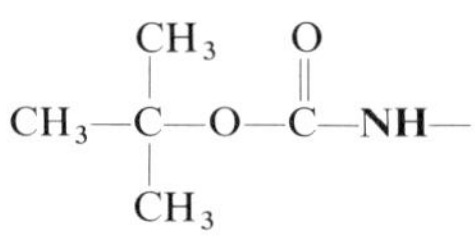

(**36**) Boc

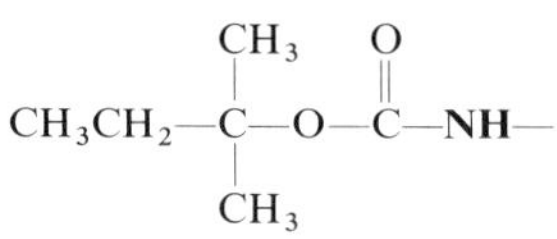

(**37**) Aoc

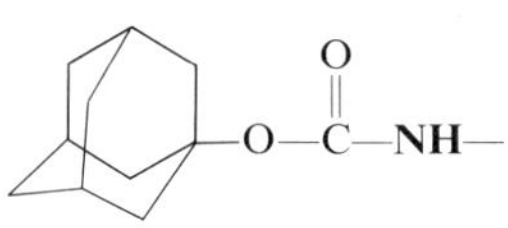

(**38**) Adoc

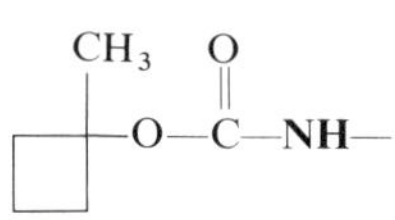

(**39**) Mcb

a. Deprotection. The *tert*-butyloxycarbonyl group has been removed by a variety of inorganic (protic or Lewis) and organic (including carboxylic) acids, in several (usually anhydrous) solvent mixtures. Among the combinations that have been found useful for solid-phase synthesis are (1) 1.0 *N* HCl–HOAc, 25°C for 30 min (Merrifield, 1964a,b); (2) 4.0 *N* HCl–dioxane, 25°C for 30 min (Stewart and Woolley, 1965; Stewart and Young, 1969, pp. 18, 30–31; warning: reagent accumulates peroxides upon standing); (3) 0.1 *N* HCl–formic acid, 25°C for 30 min (Ohno *et al.*, 1972; see also Yonezawa *et al.*,

1978; tryptophan residues formylated *in situ* but reversed with hydrazine upon completion of synthesis); (4) neat anhydrous trifluoroacetic acid (TFA), 15 min (Kusch, 1966; Takashima *et al.*, 1968; based on Kappeler and Schwyzer, 1960); (5) TFA–CH_2Cl_2 in ratios (v/v) from 1:4 to 1:1, 25°C, 20–30 min (e.g., Gutte and Merrifield, 1971); (6) β-mercaptoethanesulfonic acid–acetic acid (1:4), 25°C for 30 min (Loffet and Dremier, 1971; note that the reagent serves as internal scavenger); (7) 0.01 *M* methanesulfonic acid and 0.1 *M* TFA in CH_2Cl_2–*m*-xylene (1:1), 25°C for 30 min (Erickson and Wang, 1975); (8) 2 *M* methanesulfonic or *p*-toluenesulfonic acid in dioxane, 25°C for 30 min (Suzuki *et al.*, 1978); (9) 0.05 *M* (1 equiv in examples) trimethylsilyl perchlorate in benzene, CH_2Cl_2, or benzene–acetonitrile (1:1), 25°C for 5 min (Vorbrüggen and Krolikiewicz, 1975; the perchlorate salt is formed, proton donor unclear; not yet applied to solid-phase peptide synthesis); and (10) boron trifluoride etherate (final 0.7 *M*)–acetic acid–formic acid (1:6:5 v/v), 25°C for 30 min (Buis *et al.*, 1976; based on Hiskey *et al.*, 1971; scavengers to protect tryptophan residues were not required with this reagent). With the commonly used polystyrene resins, reagents (2), (5), (7), and (8) provide excellent swelling of the support, whereas (1), (3), (4), (6), and (10) do not, Scavengers (Table IX, on p. 72ff) are frequently added to the deprotection mixtures; the pertinence of this precaution is suggested by nmr observations of Brady *et al.* (1977) who showed that the by-product after treatment of *tert*-butyloxycarbonylamino acids with trifluoroacetic acid is not gaseous isobutylene, but rather *tert*-butyl trifluoroacetate, a soluble source of carbonium ions (see also Lundt *et al.*, 1978).

The deprotective reactions listed above are usually conducted under pseudo first-order conditions, and run a minimum of 10 half-lives, as meameasured on a sluggish substrate (e.g., Boc-isoleucine). Applicable semiquantitative data are available only in a few cases (Merrifield, 1964b; Karlsson *et al.*, 1970b). Incomplete deprotection has been persuasively demonstrated only on rare occasions [e.g., Loffet, 1973; using reagent (6) on NH_2-terminal Boc-His(Bzl)], and even so, has not necessarily been reproducible (e.g., compare original findings of Chou *et al.*, 1971, with reexamination from the same, Shapira *et al.*, 1972, or other laboratories: Stewart and Matsueda, 1972; C. H. Li, cited by Meienhofer, 1973, p. 98). In general, caution is advisable when dealing with marginal deprotective conditions (e.g., Reid, 1976) and/or unusual amino acid residues. For example, TFA–CH_2Cl_2 (3:7) effected practically no deprotection of a Boc-N^{π}-methylhistidine residue, and only neat trifluoroacetic acid sufficed to remove the Boc group (M. Rosenblatt, personal communication, 1976).

b. Potential Side Reaction. A potential *N-tert*-butylation side reaction [Eq. (9)] was looked for but not found (<0.05%) when N^{α}-*tert*-butyloxycarbonyl protection was used in conjunction with deprotection in TFA–CH_2Cl_2 (1:1) under solid-phase conditions (Mitchell and Merrifield, 1976).

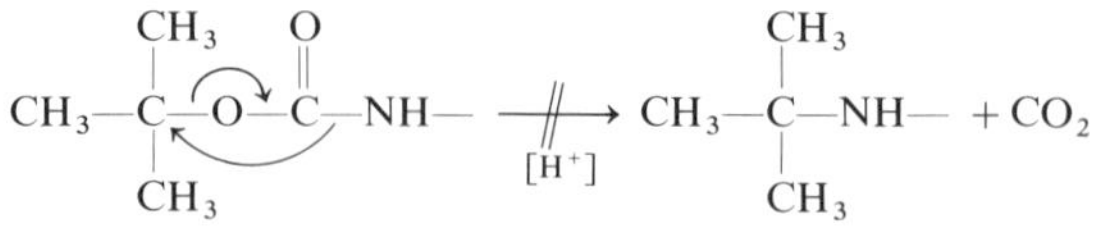

(9)

Such a side reaction would be undesirable because it would introduce a permanent hindered secondary amine into the peptide chain. In contrast to the results with the Boc group, removal of the benzyloxycarbonyl function from α- or ε-amino groups gave between 0.6 and 3.3% of the *N*-benzyl derivatives, with solutions of TFA or CH_3SO_3H. However, in stronger acids such as HBr or HF, no N-alkylation was detectable. The mechanistic evidence suggests that N-alkylation is an intramolecular process that is insensitive to scavengers.

2. *Substituted Arylisopropyl Urethanes Removable by Acidolysis*

Several substituted 2-arylisopropyl urethanes have been suggested for use in mild protection schemes for solid-phase peptide synthesis that are based on graduated acid lability. These include the 2-(*p*-biphenylyl)propyl(2)oxycarbonyl (Bpoc) (R = *p*-biphenyl, **40**) (Sieber and Iselin, 1968a,b; Wang and Merrifield, 1969a; Feinberg and Merrifield, 1972); 2-(*p*-phenylazophenylyl)-propyl(2)oxycarbonyl (Azoc) (R = *p*-phenylazophenyl, **41**) (Tun-Kyi and Schwyzer, 1976; not yet applied to solid-phase synthesis but recommended by the authors on the basis of the intense yellow color); α,α-dimethyl-3,5-dimethyloxybenzyloxycarbonyl (Ddz) (R = 3,5-dimethoxyphenyl, **42**) (Birr *et al.*, 1972); and 2-phenylpropyl(2)oxycarbonyl (Poc) (R = phenyl, **43**) (Sandberg and Ragnarsson, 1974, 1975) derivatives.

The highly acid-labile derivatives described in this section are near the natural boundary of tolerable stability imposed by the α-carboxyl group. Thus, the free acids slowly decompose in solution or upon storage at room temperature. For this reason, the N^{α}-protected amino acids are often advantageously stored in the cold as their cyclohexylammonium or dicyclohexylammonium salts (see p. 36).

In a standard mixture (Sieber and Iselin, 1968a) of acetic acid–water (4:1) at 25°C, the relative rates of cleavage are Boc:Poc:Ddz:Bpoc = 1:700:1400:3000; the Azoc group has not been tested, but is said to be closely comparable to Bpoc. For solid-phase peptide synthesis, solutions of trifluoroacetic acid ranging from 0.5% to 5% in CH_2Cl_2 at 25°C for 15 to 30 min have been advocated for quantitative removal of the groups **40**–**43**.

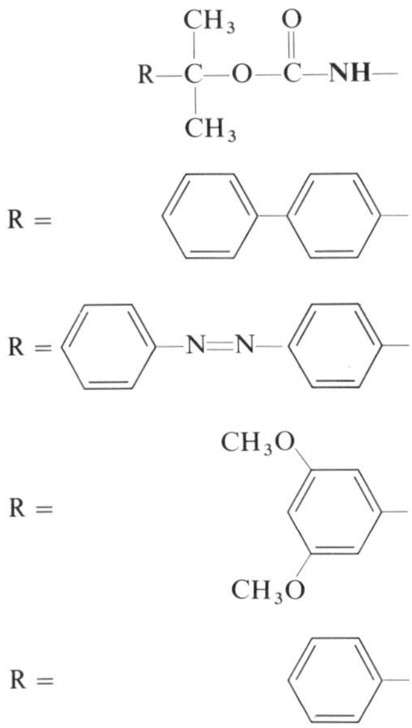

(40) Bpoc

(41) Azoc

(42) Ddz

(43) Poc

The relative rates of solution cleavage in 0.5% TFA–CH_2Cl_2 are Boc:Ddz:Bpoc = 1:2.3 × 10^4:1.3 × 10^5 (Matsueda and Stewart, 1975).

The Bpoc group (**40**) is also cleaved by pyridinium salts of strong inorganic acids (Klostermeyer and Schwertner, 1973); the rate is too slow to serve as a feasible general means of deprotection, but sufficiently fast to interfere with certain titration methods (e.g., Dorman, 1969). The Ddz group (**42**) can be removed not only by acidolysis but also by photolysis at wavelengths above 2800 Å. This latter mode of deprotection is not normally applied during the repetitive steps of solid-phase synthesis. All of the groups **40**–**43** resist alkali and catalytic hydrogenolysis.

Completion of deprotection reactions can be judged by quantitation of the ultraviolet absorbing arylisoprene by-products released into solution. Similarly, the rate of uptake of the protecting group chromophore out of solution into the resin-bound growing peptide chain can be usefully applied to monitor the coupling steps. The methoxy singlets in the nmr spectrum of Ddz derivatives **42** may also be exploited for analytical purposes. The steric bulk of the aryl substituent of these protecting groups does not appear to impair the efficiency of coupling. To prove this point, Ragnarsson *et al.* (1974) performed competition experiments which showed that the extent of incorporation of amino acids was the same regardless of whether the N^α-protecting group was Boc, Poc, Bpoc, or Moz (4-methyloxybenzyloxycarbonyl).

The arylisopropyl urethane protecting groups **40**–**43** have not been evaluated for low levels of N-alkylation side reactions under the acidolytic deprotection conditions of solid-phase peptide synthesis [compare to Eq. (9)]. The legitimacy of such a concern is underscored by the model results of Sheradsky *et al.* (1971) with the Bpoc derivative (**44**) of *O*-(5-nitro-2-pyridyl)hydroxylamine [Eq. (10)]. N-Alkylation has also been observed

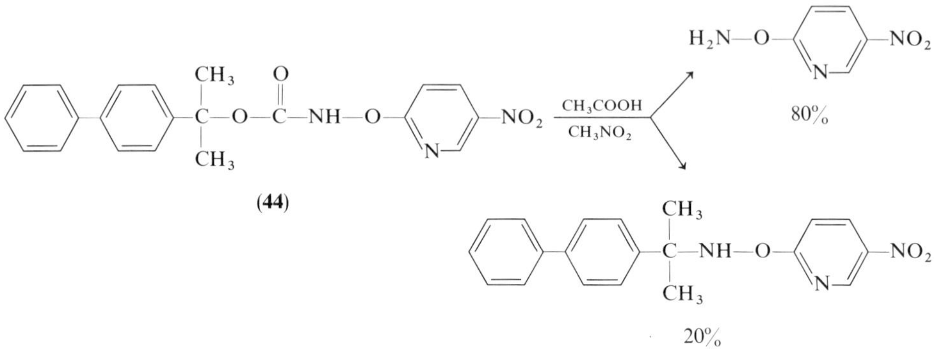

(10)

upon cleavage of dicyclopropylmethyloxycarbonylaniline (**45**) with hexafluoro-2-propanol ($pK_a = 9.3$), presumably due to the stability of the

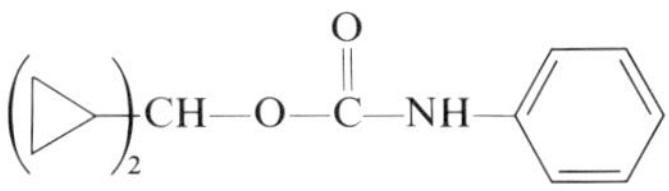

(**45**)

dicyclopropylmethyl carbocation (Carpino, 1973).

3. *Other Amino-Protecting Groups Removable by Acidolysis*

The benzyloxycarbonyl (Z) group (**46**) (Bergmann and Zervas, 1932), which is removed by strong acids such as HBr in acetic acid (Ben-Ishai and Berger, 1952), has a place of the first order in the history of peptide synthesis. It was the first group examined for solid-phase synthesis (see Fig. 1). Given

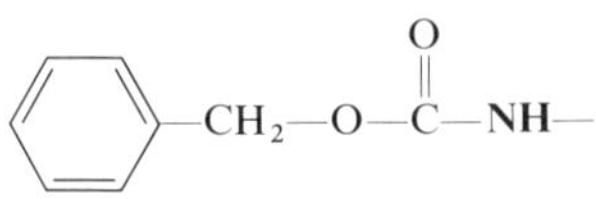

(**46**) Z

the overall status of current methodology, N^{α}-protection with the benzyloxycarbonyl group will probably be restricted to the NH_2-terminal residue of a peptide. The widely applied protection scheme for solution synthesis, relying on the hydrogenolyzable N^{α}-benzyloxycarbonyl group and acidolyzable side-chain *tert*-butyl derivatives (Schwyzer and Sieber, 1963) has not been applicable for solid-phase synthesis.

The furfuryloxycarbonyl (Foc) (**47**) (Jeschkeit *et al.*, 1966; Losse and Neubert, 1970) and *p*-methoxybenzyloxycarbonyl (Moz) (**48**) (McKay and Albertson, 1957; Weygand and Ragnarsson, 1966; Ohno *et al.*, 1971) groups

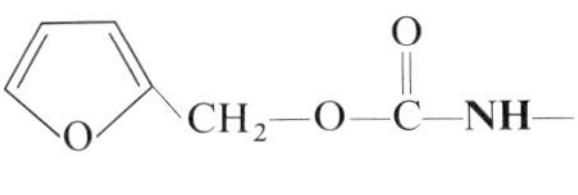

(**47**) Foc

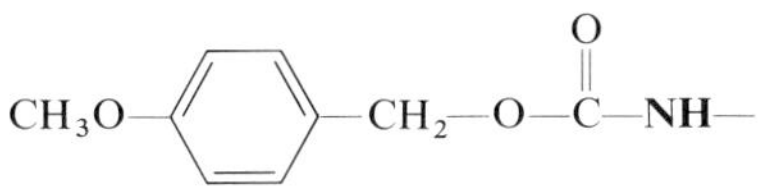

(**48**) Moz

are primary urethanes that are deprotected under the same acidolytic conditions and at essentially the same rates as the *tert*-butyloxycarbonyl group (Section IV,A,1; see Losse and Neubert, 1968b for comparative rate data). Hence, they have occasionally been used in stepwise solid-phase synthesis. The ultraviolet absorbing chromophore of **48** facilitates spectrophotometric monitoring of deprotection and coupling steps.

The triphenylmethyl (trityl, Trt) group (**49**) (Helferich *et al.*, 1925; Zervas and Theodoropoulos, 1956) has only been applied once, for the stepwise solid-phase synthesis of a model tetrapeptide (Losse and Neubert, 1968a).

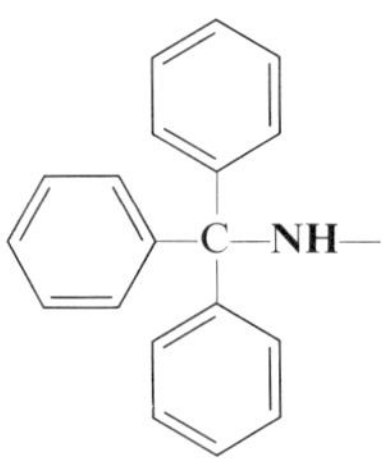

(**49**) Trt

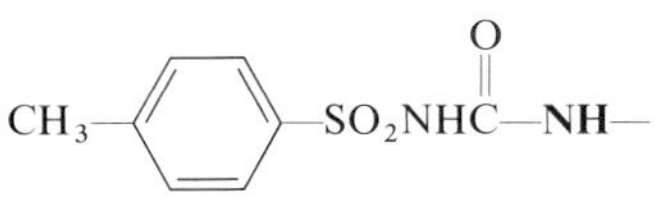

(**50**) Tac

In general, this means of protection is not recommended for the repetitive cycles, especially in view of steric hindrance during some coupling steps. A case in point is the low incorporation of Trt-Gly-OH observed in a solid-phase competitive coupling experiment (Ragnarrson *et al.*, 1974; cf. p. 105ff). On the other hand, a renaissance of specialized applications may be expected which exploit NH_2-terminal trityl-protected peptide segments prepared by solid-phase synthesis. Hence, the N^{α}-trityl group is very resistant to alkali and hydrazine, and is surprisingly stable to anhydrous acids (same order of magnitude as the Boc group; Losse and Neubert, 1968b). In the standard aqueous acid mixture (acetic acid–water = 4:1), however, the relative rates of cleavage are Boc:Bpoc:Trt = 3×10^{-3}:1:7. Even greater kinetic selectivity (about 350-fold) in the presence of Bpoc groups is achieved by use of 90% trifluoroethanol as a solvent, as "pH 4" on the glass electrode of an autotitrator (Riniker *et al.*, 1975).

The *p*-toluenesulfonylaminocarbonyl (Tac) group (**50**) (Künzi and Studer, 1975) does not appear acceptable for repetitive use because of hydantoin formation upon activation of N^{α}-Tac amino acids. However, **50** may turn out to be of value for temporary protection of the NH_2-terminal residue. The sulfonyl carbamate could be introduced into resin-bound peptides by reaction with *p*-toluenesulfonyl isocyanate and would survive cleavage of the protected peptide by moderately strong anhydrous acids, including TFA and HBr in acetic acid. A side reaction to give the ureido derivative through loss of the *p*-toluenesulfonyl moiety takes place in HF. Selective deprotection of the Tac group occurs under essentially neutral conditions in refluxing aqueous alcohols.

The well-known acid sensitivity of the phosphorus–nitrogen bond, first explored for peptide chemistry in the Zervas laboratory (Cosmatos *et al.*, 1961), has recently been followed up by two groups (Kenner *et al.*, 1976; Ueki and Ikeda, 1976; Ueki *et al.*, 1977). Model compounds of **51**–**53** all exhibited a lability to acid only slightly greater than that of the *tert*-

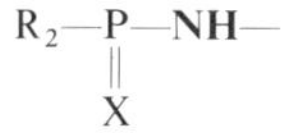

(**51**)	R = C_6H_5, X = O	Dpp
(**52**)	R = C_6H_5, X = S	Ppt
(**53**)	R = CH_3, X = S	Mpt

butyloxycarbonyl group, and were resistant to alkali, nucleophiles, and catalytic hydrogenolysis. In addition, coupling reactions with this mode of N^{α}-protection proceeded without noticeable racemization. Of the groups listed, the dimethylphosphinothioyl (Mpt) derivatives (**53**) were selected for further study, since deprotection and coupling reactions of the bulkier derivatives **51** and **52** could not always be brought to completion in reasonable periods of time. A new reagent was developed to mediate the acidolytic cleavage; thus, stock solutions of triphenylphosphine dihydrochloride (van den Akker and Jellinek, 1967) were prepared by saturating a solution of triphenylphosphine (1 equiv) in CH_2Cl_2 with hydrogen chloride (2 equivs taken up). Removal of the dimethylphosphinothioyl (Mpt) group was achieved with 0.2 *N* HCl–0.2 *M* triphenylphosphine–CH_2Cl_2, at 25°C for 2 × 30 min; the excess triphenylphosphine was believed to suffice as a scavenger against oxidizing agents. The procedure was applied to the solid-phase synthesis of Mpt-$(Trp)_3$-OMe; an interesting feature was the comparative stability of the standard *p*-alkylbenzyl ester anchoring linkage (p. 68).

Two highly acid-sensitive α-amino protecting groups that have been tested for mild schemes of solid-phase peptide synthesis are the 2-hydroxy-1-naphthylidene group (**54**) (Khosla *et al.*, 1972a; based on Sheehan and Grenda, 1962; compare also to Esko *et al.*, 1968) and the 2-benzoyl-1-methylvinyl (Bmv) group (**55**) (Southard *et al.*, 1969, 1971a,b; based on Dane *et al.*, 1962). The Schiff's base **54** is hydrolyzed in 0.5 *N* HCl in

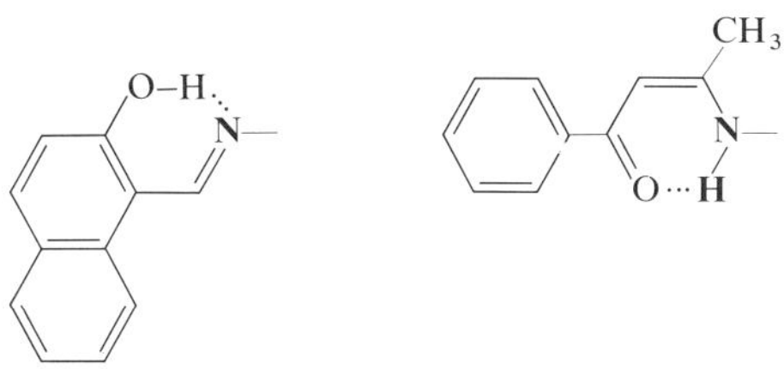

(**54**) Bmv (**55**)

dioxane–water (3:1), 25°C for 90 min (monitored by discharge of yellow color); alternatively, it can be rapidly cleaved by a primary amine (e.g., 0.4 *M* benzylamine–CH_2Cl_2, at 25°C for 30 min). Transamination of the protecting group onto an attacking α-amino group nucleophile during the coupling step, similar to side reactions described in solution (Sheehan and Grenda, 1962), might limit this protection technique, but the question has not been examined in solid-phase synthesis. 2-Benzoyl-1-methylvinylamino acids are stored as their dicyclohexylammonium salts, and have been coupled as such in the presence of *p*-toluenesulfonic acid for *in situ* neutralization. Deprotection is achieved with 0.4 *N* HCl in tetrahydrofuran–water (15:1) at 25°C for 30 min; 1.0 *N* *p*-toluenesulfonic acid monohydrate in tetrahydrofuran has also been used. An attempted loading of 2-benzoyl-1-methylvinylglycine onto a hydrazide resin (functionalized support number 40 in Table VI) resulted in substantial levels of a side reaction, evidently involving hydrazone formation of the 2-benzoyl-1-methylvinyl carbonyl (Wolters *et al*, 1974). Although protecting groups **54** and **55** are interesting for their chemical novelty, their application to peptide synthesis has been largely superseded by the highly acid-labile urethane groups **40**–**43** discussed previously.

4. *Arylsulfenyl Groups Removable by Acidolysis and Nucleophilic Cleavage*

The *o*-nitrophenylsulfenyl (Nps) group (**56**) (Zervas *et al.*, 1963b; reviewed by Photaki, 1976, pp. 149–162) was applied with considerable success in some early syntheses by the solid-phase method (Kessler and Iselin, 1966, 1 cycle; Najjar and Merrifield, 1966, 18 cycles; Sklyarov and Shashkova, 1969, 5 cycles). From extensive experiences in solution syntheses, a variety of advantages as well as drawbacks to the use of **56** and related protecting

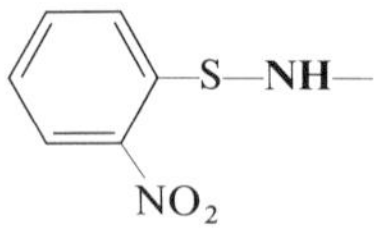

(**56**) Nps

groups are known. Recently, the development of new cleavage reagents, and of a new coupling method involving N^{α}-Nps amino acid carboxyanhydrides (see Table X, p. 146), suggests that the full potential of arylsulfenylamino protecting groups for solid-phase peptide synthesis has not yet been realized.

N^{α}-Protected Nps-amino acids can be activated and introduced into peptides without racemization. The NH_2-terminal Nps group of peptides is relatively stable to alkali and hydrazine [e.g., Halstrøm *et al.* (1976) describe hydrazinolysis of a benzyl ester linkage to the resin in the presence of an Nps group]. Removal is effected by acidolysis or by certain nucleophiles (Fig. 9). This sensitivity imposes some limitations on compatible conditions for peptide synthesis; for example, the Nps group is not entirely stable under the conditions of couplings with dicyclohexylcarbodiimide and 1-hydroxybenzotriazole (Geiger *et al.*, 1973), nor to treatments with pyridine hydrochloride (Dorman, 1969; Klostermeyer and Schwertner, 1973), nor to perchloric acid titrations (Christensen *et al.*, 1976; 0.6% lost per cycle). Furthermore, Nps-amino acids with a free carboxyl group decompose autolytically at 25°C. They have been reported to be stable at 0°C for 24 hr in the dry state (Juillerat and Bargetzi, 1976). Long-term storage is in the form of dicyclohexylammonium salts (p. 36).

Acidolytic removal of the Nps group during solid-phase peptide synthesis has been achieved with (1) 0.1 *N* HCl in chloroform–acetic acid (10:1), 25°C for 10 min (Najjar and Merrifield, 1966), (2) 0.2 *N* HCl–dioxane, 25°C for 30 min (Sklyarov and Shashkova, 1969), and (3) 0.24 *N* HCl in acetic acid–mercaptoacetic acid–chloroform (7:2:20) 25°C for 30 min (Christensen *et al.*, 1976). In principle, only two equivalents of hydrogen halide are required (1 equiv when methanol is solvent, see Poduška, 1967), but in a solid-phase approach, excesses are used. Although acidolysis is the fastest and easiest way to remove the Nps group, the kinetic selectivity with respect to *tert*-butyl based protecting groups is generally not great enough for that combination of protecting groups and cleavage conditions to be feasibly applied. The main disadvantage of methods relying on hydrogen halide cleavage is the

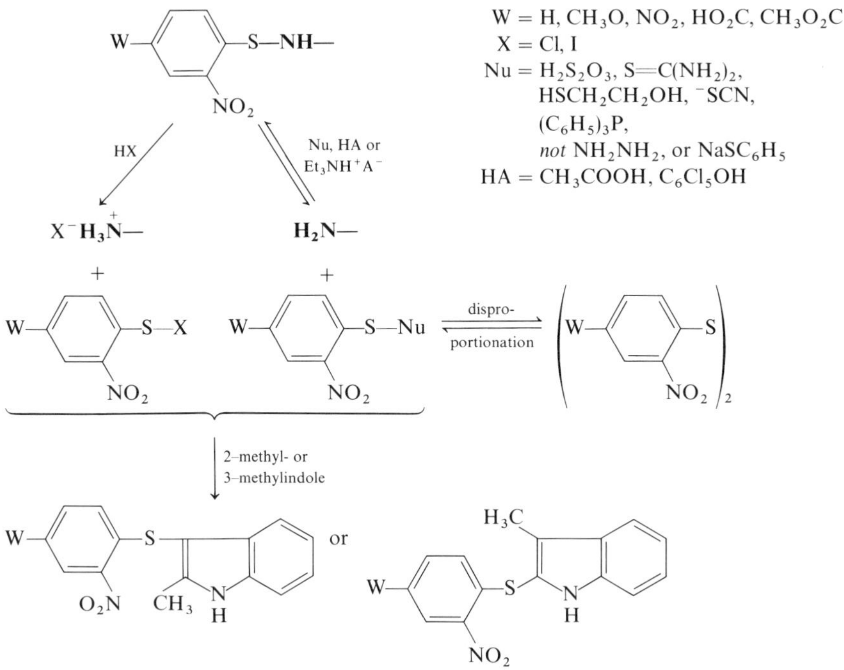

Figure 9. Removal of the *o*-nitrophenylsulfenyl (Nps) and related amino protecting groups. References are given in text of Section IV,A,4.

ready addition reaction of the product sulfenyl halide to the 2-position of the indole nucleus of tryptophan residues (Fontana *et al.*, 1966; Wünsch *et al.*, 1967; Scoffone *et al.*, 1968). Scavengers such as 2-methyl- or 3-methylindole help somewhat (Fig. 9).

The alternative way to remove the *o*-nitrophenylsulfenyl group, which avoids the tryptophan alkylation problem and also permits selective deprotection in the presence of *tert*-butyl based groups, involves the use of any of a variety of nucleophilic reagents (Fig. 9). The reaction is prominently catalyzed by weak acids or their salts. It appears to be an equilibrium that can usually be driven to completion by scavengers, or by the intense insolubility of the by-products obtained either directly or by disproportionation. This last consideration, insolubility of by-products, could complicate applications to solid-phase synthesis, for example by clogging of filters. The relative reactivity of the Nps group can be judged by the fact that clean and selective removal of phenacyl esters from NH_2-terminal Nps-peptides is possible (70–90% isolated yields) using sodium thiophenoxide in dimethylformamide (L. Zervas, cited by Photaki, 1976, p. 155).

In an extensive study, Kessler and Iselin (1966) concluded that 1.0 *M* thioacetamide in acetic acid–methanol (9:41), at 25°C for 30 min, was the most suitable reagent for removal of the Nps group under nearly neutral conditions for solid-phase synthesis. However, under these identical conditions, Ragnarsson *et al.* (1970a) were unable to drive the deprotection to completion. While these authors attributed the results to poor swelling of the resin in the cleavage medium, their limited data are also consistent with the rapid establishment of a 9:1 equilibrium mixture of free and protected amine. A substantial improvement is provided by the work of Juillerat and Bargetzi (1976), who recommended use of methyl 3-nitro-4-mercaptobenzoate (**57**) in CH_2Cl_2 (5 to 10 equiv = 0.5 *M*, 25°C, 4 hr). In addition, 2 m*M* dicyclohexylammonium acetate was present as a catalyst. The unsymmetrical disulfide between the Nps moiety **56** and thiol **57** has good solubility. A 30-peptide was assembled by this method, but no details were provided. Very recently, Tun-Kyi (1978) described the solid-phase synthesis of a pentapeptide from Nps-amino acids, using 1 *M* 2-mercaptopyridine (**58**)–1 *M* acetic acid–CH_2Cl_2, at 25°C for 30 min, for deprotection. Other reagents for the removal of the Nps group include dithiothreitol at pH 8 (Polzhofer and Ney, 1971), ammonium thiocyanate in CH_2Cl_2–acetic acid in the presence of methylindole scavengers (Wünsch and Spangenberg, 1972), and triphenylphosphine with pentachlorophenol in methanol (Matsueda *et al.*, 1975).

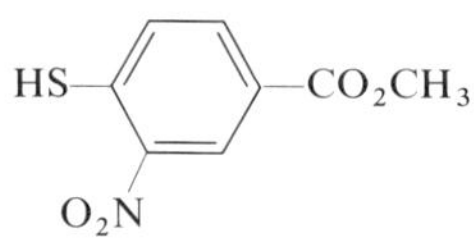

(**57**)

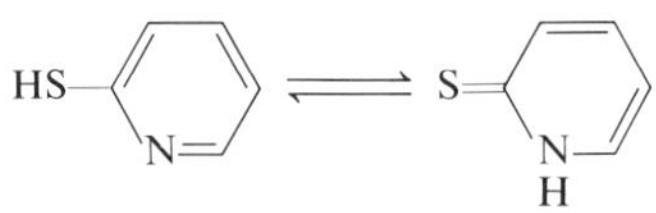

(**58**)

5. Urethanes Removable under Basic Conditions

Several reasonably acid-stable urethane amino protecting groups are available that are cleaved by a β- or a 1,6-elimination mechanism, either directly or after safety-catch activation [e.g., Le Corre *et al.* (1978) and references cited therein; see Eq. (6) on p. 89 for a prototype β-elimination and Eq. (12) on p. 114 for a 1,6-elimination]. The vinylic by-products can possibly react with the released amino group, and hence nucleophilic scavengers are generally required and are incorporated into the cleavage mixtures.

To date, only the 9-fluorenylmethyloxycarbonyl (Fmoc) group (**59**) (Carpino and Han, 1972) has been applied as the temporary protecting group for stepwise solid-phase peptide synthesis. The cleavage conditions [Eq. (11)], which are orthogonal to those for *tert*-butyl-based side-chain protecting groups, are (1) piperidine–CH_2Cl_2 (1:1), 25°C for 5 + 25 min

(Chang and Meienhofer, 1978) or (2) piperidine–dimethylformamide (1:4), 25°C for 3 + 7 min (Atherton *et al.*, 1978a,b). This chemistry is especially

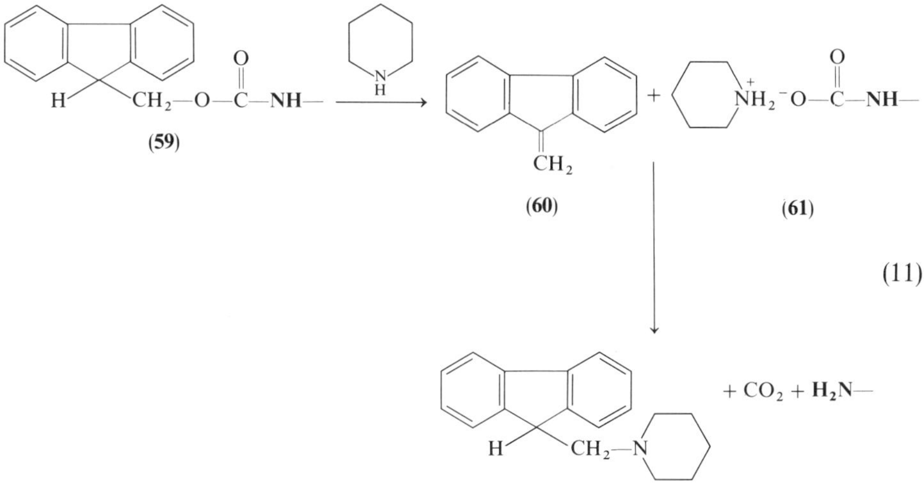

(59) **(60)** **(61)**

(11)

suitable to the solid-phase approach because the unfavorable solubility properties of Fmoc-peptides and by-products from the cleavage are by-passed in this mode. Furthermore, the dibenzofulvene intermediate **60** is trapped as an adduct with excess piperidine, and additional steps to decompose the presumed carbamate intermediate **61** do not appear to be required. The Fmoc moiety **59**, while extraordinarily acid stable, is slowly cleaved by solutions of tertiary amines [e.g., half-life in diisopropylethylamine–dimethylformamide (1:1) ~ 10 hr]. It is removed more readily than the methylsulfonylethyloxycarbonyl (Msc) group (**62**) (Wolters *et al.*, 1974; Tesser and Balvert-Geers, 1975; compare also to Kunz, 1976b). It remains

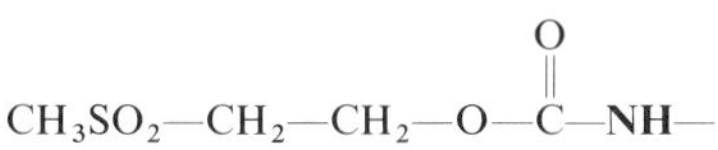

(**62**) Msc

to be seen whether the quantitative succinimide formation at COOH-terminal asparagine observed (König and Volk, 1977) upon very brief basic treatment to remove the Msc group will be observed under the milder conditions for removal of the Fmoc group. Incidentally, anchoring of an N^{α}-Fmoc-amino acid to a benzyl alcohol support may not necessarily be straightforward [compare Atherton *et al.* (1978a, 1979a), who pointed out the difficulties in the formation of ester bonds in the presence of base-labile groups with Chang and Meienhofer (1978), who reported no problem in anchoring an Fmoc-amino acid].

The 5-benzisoxazolylmethyleneoxycarbonyl (Bic) group (**63**) (Kemp and Hoyng, 1975) probably does not meet the ordinary standards for repetitive application in solid-phase synthesis, but might be suitable for some applications based on the unusual and elegant method for its removal [Eq. (12)].

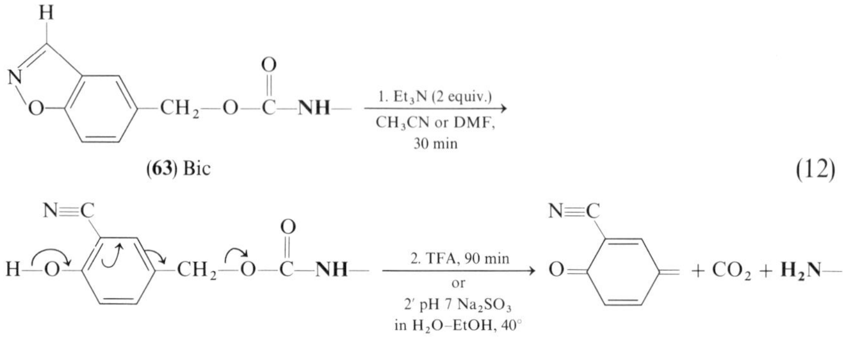

(**63**) Bic

(12)

At the present time, the same prognosis seems appropriate for a range of relatively acid-stable β-halourethanes (e.g., **64**–**68**) that are cleaved by (1) intramolecular mechanisms in suitable solvents (Carpino *et al.*, 1970),

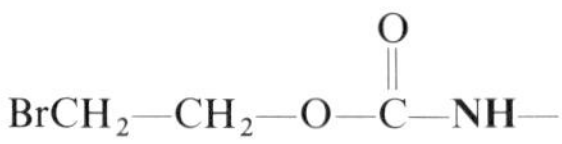

(**64**)

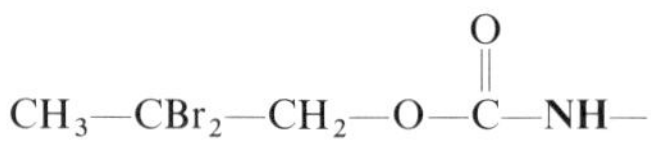

(**65**)

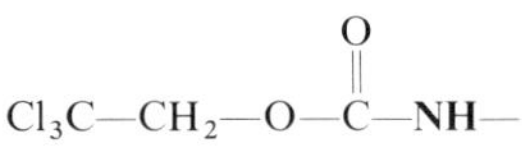

(**66**)

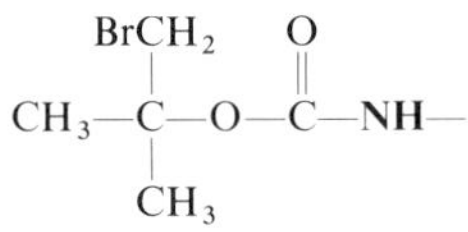

(**67**)

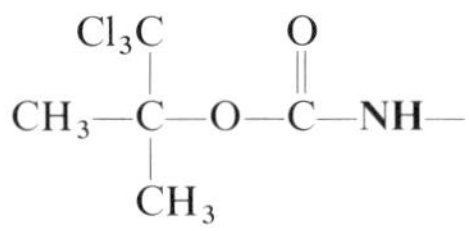

(**68**)

(2) "supernucleophiles" (Eckert *et al.*, 1975; see also Ugi *et al.*, 1976; Eckert *et al.*, 1978), (3) treatment with tertiary phosphines followed by β-elimination catalyzed by mild alkali in the presence of scavengers (Kunz, 1976a,c), as well as (4) various reductive techniques (e.g., Windholz and Johnston, 1967). Side reactions encountered for several of these groups include (1) intramolecular N-alkylation (e.g., Ohnishi *et al.*, 1972), (2) lability to base, and (3) partial reduction to a no longer cleavable species (e.g., Semmelhack and Heinsohn, 1974).

6. Dithiasuccinoyl Group Removable by Reduction

The 1,2,4-dithiazolidine-3,5-dione heterocycle (reviewed by Zumach and Kühle, 1970) has been adapted as the basis of a new amino protecting group for peptide synthesis (Barany and Merifield, 1977). These disulfide-containing amine derivatives **69**, termed dithiasuccinoyl (Dts) amines by analogy with their carbocyclic analogues, are deprotected by treatment with thiols (Fig. 10) and other reducing agents. The reductions are driven to completion by loss of two equivalents of gaseous carbonyl sulfide. Dts compounds are resistant to strong acids, including HBr–acetic acid, at 25°C, and 6 *N* hydrochloric acid, at 110°C, and to weak bases such as aqueous sodium bicarbonate and the α-amino groups of peptides. The Dts-heterocycle also shows acceptable stability under photolytic conditions. Cleavage at the Dts carbonyl occurs with strong bases, yielding the free amine with aqueous sodium hydroxide and the urea derivative with benzylamine. A possible side reaction of the thiolytic deprotection step, in which competing nucleophilic attack of a thiol at the Dts carbonyl would give some of the undesired N^{α}-thiourethane terminated derivative in addition to the free amine, does not appear to occur.

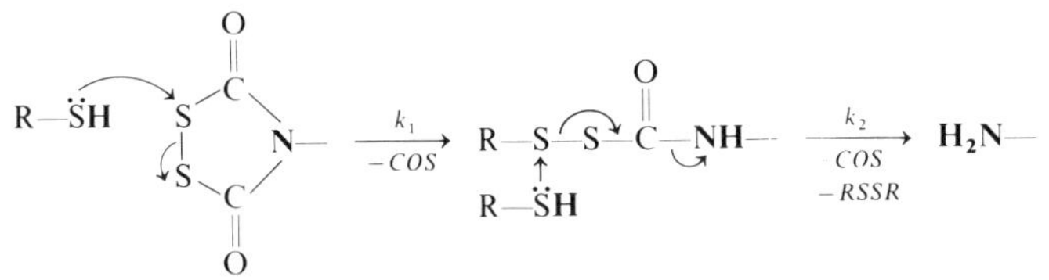

(**69**) Dts (**70**)

R	$\kappa = k_2/k_1$	k_{rel}^{env}
$HOCH_2CH_2$—	0.1–5	1
$HSCH_2CHOHCHOHCH_2$—	>20	9–12
$CH_3O_2CCH_2CH_2$—	0.5–2	1.0–1.4
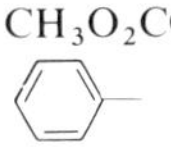	~ 10^5	0.7–1.6

Figure 10. Mechanism of thiolytic removal of the dithiasuccinoyl (Dts) amino protecting group (Barany and Merrifield, 1979). The quantity k^{env}, given here relative to the results for β-mercaptoethanol, is given by $k_1k_2/(k_1 + k_2)$. It can be shown to provide a fairly accurate, although still conservative, "envelope" approximation to the rate law of the complex consecutive pseudo first-order reactions. Thus, for $\geqslant 75\%$ completion of the reaction, product obtained is always $\geqslant 100\% [1 - \exp(-k^{env}t)]$.

A kinetic investigation of the thiolytic removal of the Dts group (Barany and Merrifield, 1979) was facilitated by the direct determination of Dts compounds **69** on a standard automated amino acid analyzer. Integration constants obtained ranged from 10 to 60% of those for the corresponding free amino acids; apparently the active reducing agent is hydrindantin, and the released amine reacts *in situ* with ninhydrin to give a purple color. Furthermore, the open-chain carbamoyl disulfide intermediates **70** were chromatographically separable, with integration constants near unity.

The reaction mechanism of the reductive deprotection (Fig. 10) involves two consecutive pseudo first-order reactions with rate constants k_1 and k_2. The ratio of rates, $\kappa = k_2/k_1$, was found to vary from essentially 0 to $\geqslant 500$ under a wide variety of conditions of thiol, solvent, and base. Thiolytic removal of the Dts group in anhydrous solvents was markedly catalyzed by tertiary amines, such as pyridine and triethylamine; for example, the half-time for reductive deprotection of Dts-glycine in 0.1 *M* β-mercaptoethanol in neat pyridine was 2.9 min, with $\kappa = 0.12$. Open-chain carbamoyl disulfide intermediates were never observed when deprotection was mediated by thiophenol or dithiothreitol; in the latter case, the second step of the reaction mechanism is expected to be very fast because it is kinetically favored by an intramolecular cyclization to the oxidized disulfide. In aqueous solutions, the rates of the two steps were measured independently, giving a calculated value $\kappa \sim 375$. Using β-mercaptoethanol (20 m*M*, half-time of 55 sec at pH 8), log k_1 was shown to vary linearly with pH from pH 5.5–9.0; the rate of reductive deprotection was 1.8×10^4-fold faster than that of background hydrolysis. Data with other thiols in aqueous solutions fit a Brønsted correlation, and confirmed the idea that the reactive species of the mechanism is the thiol anion.

Quantitative conversion of Dts-amino acids to the parent amino acids could be achieved at 25°C with 0.1 *N* sodium hydroxide (within 30 min), 0.1 *M* triphenylphosphine in dioxane–H_2O (9:1) (within 100 min), 2 m*M* tri-*n*-butylphosphine in dioxane–H_2O (9:1) or water (within 15 sec), 0.01 *M* alcoholic sodium borohydride (within 3 min), as well as with thiols. Reaction rates with various Dts-amino acids lie within one order of magnitude of each other (Barany and Merrifield, 1979). Optimal conditions for the application of Dts-amino acids to peptide synthesis remain to be worked out, but pilot studies reveal racemization-free coupling as well as quantitative deprotection as judged by the absence of diastereomeric or deletion peptides in the sensitive analytical system developed by Kent *et al.* (1978).

7. Miscellaneous Amino Protecting Groups

The structures and methods for removal of several amino protecting groups that have been only rarely if at all used for solid-phase synthesis are given here for completeness. None of them is really suitable for repetitive use, but applications can be envisaged for blocking of the NH_2-terminal residue prior to cleavage of a protected peptide segment (see p. 101). Also, some of these groups have been applied to side-chain protection of lysine and ornithine (Section V,B,2).

The phthaloyl (Pht) group (**71**) (King and Kidd, 1949; Sheehan and Frank, 1949; applied in anchoring step of a solid-phase synthesis by Wang and Merrifield, 1972; see p. 87) and the 4,5-diphenyl-4-oxazoline-2-one function

(**72**) (Sheehan and Guziec, 1973) are, along with the dithiasuccinoyl group (**69**) (pp. 114–116), the only known examples of reversibly removable functionalities in which both protons of the amino group are replaced. The groups **71** and **72** are both resistant to vigorous acidolytic conditions; the

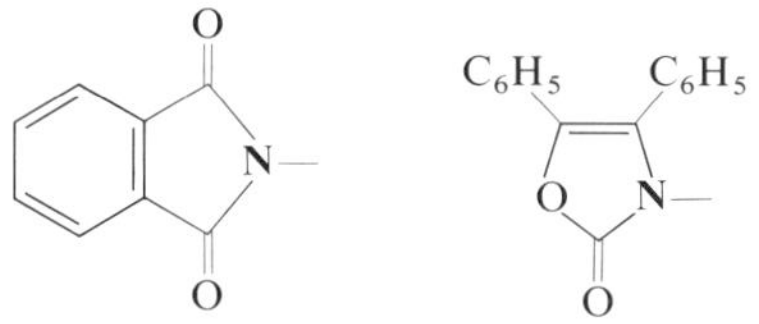

(**71**) Pht (**72**)

preferred means of deprotection are, respectively, hydrazinolysis (Ing and Manske, 1926; Schwyzer *et al.*, 1963a; see also Wolfe and Hasan, 1970, for alternative hydrolytic route) and catalytic hydrogenolysis. An oxidative procedure for the opening of heterocycle **72** is also available, and the same approach is used to deprotect the very acid-stable (ethylthio)carbonyl group (**73**) (Kollonitsch *et al.*, 1956a,b). The limitations of oxidations have been covered earlier (p. 88).

The piperidino-oxycarbonyl group (**74**) (Stevenson and Young, 1969) is of interest because it is relatively resistant to moderately strong anhydrous

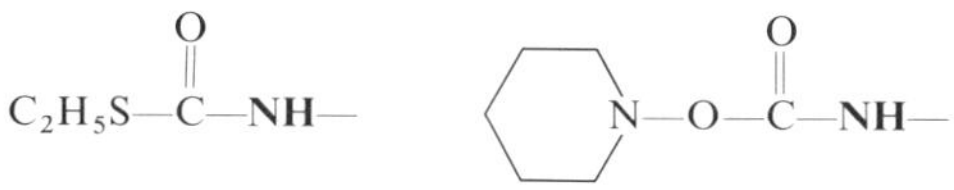

(**73**) (**74**)

acids, including cold HBr in acetic acid, but is smoothly removed by catalytic hydrogenolysis, electrolysis, and by reduction with zinc or sodium dithionite in aqueous acetic acid.

N-Acyl-type groups are generally not used for temporary protection of amino acids because oxazolones, leading to racemization, readily form upon activation in the coupling step. The acid-stable trifluoroacetyl (Tfa) group (**75a**) (Weygand and Csendes, 1952) is removed by 1 *M* aqueous piperidine (Goldberger and Anfinsen, 1962), or by alcoholic sodium borohydride at 60°C (Weygand and Frauendorfer, 1970). The chloroacetyl group (**75b**) (Holley and Holley, 1952; Masaki *et al.*, 1968) has now been found to be removable under mild conditions with 2-aminothiophenol (Glass *et al.*, 1979). An arginyl residue has been applied as a trypsinolysable protecting group (Meyers and Glass, 1975). The relatively acid-stable *p*-toluenesulfonyl (Tos) group (**76**) (Fischer and Bergmann, 1913) was used to cover the NH_2-terminal residue of a solid-phase peptide product (Meienhofer *et al.*, 1970b);

removal subsequent to cleavage from the support was performed with sodium in liquid ammonia (du Vigneaud and Behrens, 1937).

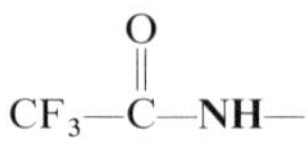

(**75a**) Tfa

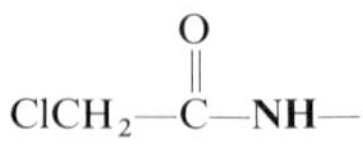

(**75b**)

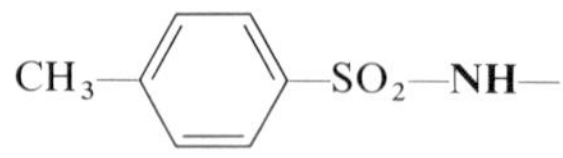

(**76**) Tos

8. Carboxyl or Hydrazide Protecting Groups

Relatively little work has been done on solid-phase elongation in the N → C direction. In the original researches of Letsinger *et al.* (1964), amino acid ethyl esters were coupled, and then deprotected by saponification. Mukaiyama's group (Matsueda *et al.* 1973; Maruyama *et al.*, 1976) has developed an approach involving preparation in solution of peptide segments with the COOH-terminal carboxyl groups blocked as *tert*-butyl esters (**77**) (Roeske, 1959, 1963; Anderson and Callahan, 1960). After coupling (Section IV,C), the free carboxyl group is liberated by treatment with TFA–CH_2Cl_2 (1:1) containing 5% β-mercaptoethanol, at 25°C for 30 min. In another approach, amino acyl *tert*-butyloxycarbonyl hydrazides (**78**) (Schwyzer, 1959; Boissonnas *et al.*, 1960) were coupled to a resin-bound azide (Felix and

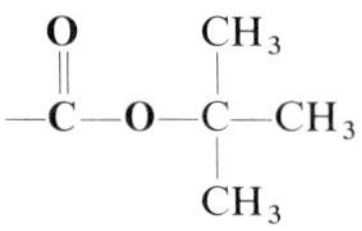

(**77**) O*t*Bu

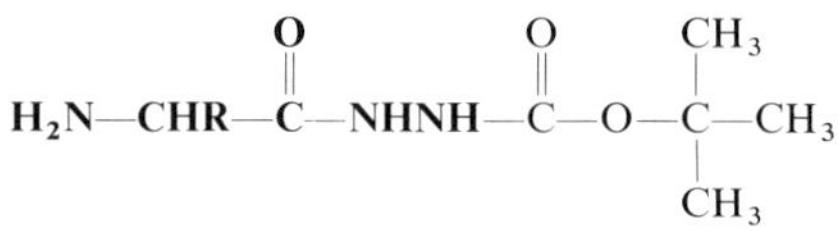

(**78**)

Merrifield, 1970). The *tert*-butyloxycarbonyl group was removed with 4 *N* HCl–dioxane, at 25°C for 30 min, and the cycle was repeated after diazotization of the released hydrazide.

B. Neutralization

Usually, when the N^{α}-protecting group is removed by an acidolytic mechanism (Section IV,A,1–4), the released α-amino group is obtained in the form of its salt. Since a protonated amine is a poor nucleophile and will not react readily with acylating agents, an additional step, termed "neutralization" is required. This is usually achieved by exposure of the peptide–resin salt to an excess (2.5% to 10%, v/v) of a suitable tertiary amine in a solvent such as CH_2Cl_2, $CHCl_3$, or DMF, in which the produced tertiary ammonium salt will have a reasonable solubility. The equilibrium exchange to give the free α-amino peptide–resin is rapidly established (possibly diffusion-controlled) so that brief treatments (e.g., 2 × 2 min) are more than adequate. This solid-phase mode of carrying out the neutralization step provides decided advantages over corresponding solution procedures. Thus, the precise stoichiometry of neutralizations in solution is difficult to control, and the excess base, as well as the equivalent of formed tertiary ammonium salt, can in many cases catalyze racemization (reviewed by Kemp, Vol. 1, Chapter 7) during segment condensation reactions and might promote low levels of other side reactions as well. However, in a solid-phase synthesis, all salts and excess base can be removed from the reaction medium prior to the coupling step, by simple washing.

In the original work on solid-phase synthesis (Merrifield, 1963), triethylamine was used as the base; in aprotic solvents, this reagent is several thousandfold more basic than α-amino groups of peptides (Williams and Young, 1972). Recently, the comparably strong, but hindered diisopropylethylamine (Hünig and Kiessel, 1958; first applied to solid-phase peptide synthesis by Wang and Merrifield, 1969b) has been widely preferred, with the aim of minimizing the possibility of quaternary ammonium site formation with halomethyl or halophenacyl supports. It should be noted that some of the newer supports functionalized by primary amidoalkylation cannot give rise to quaternary sites (e.g., Table VI, number 16; see also p. 66).

Triethylamine reacts with CH_2Cl_2 to form chloromethyltriethylammonium chloride (Fankhauser *et al.*, 1973); the rate of ionic chloride generation in a 10% (v/v) solution is 0.5 μmol ml^{-1} hr^{-1} (unpublished results of our laboratory, cited by Erickson and Merrifield, 1976, p. 378). The side reaction is two to three orders of magnitude slower using diisopropylethylamine (Hodges and Merrifield, 1975a). In view of these results, stock solutions of tertiary amines in CH_2Cl_2 or $CHCl_3$ should be freshly prepared (i.e., daily). Careful purification of the bulk tertiary amine to remove contaminating primary and secondary amines, for example by steps including distillation from ninhydrin, acetic anhydride, or α-naphthyl isocyanate, seem desirable.

Concerns have also been noted about reprotonation of the α-amino group during the course of washings with bulk solvents (e.g., chloroform contaminated with traces of HCl as reported by Wieland *et al.*, 1971a). Incomplete couplings are a possible consequence of such problems. For this reason, the entire sequence of steps comprising neutralization and coupling are commonly repeated with fresh reagents (so-called "double-coupling" protocol). Alternatively, Yamashiro and Li (1974) introduced the practice of adding a small amount of a tertiary amine solution towards the end of symmetrical anhydride coupling reactions.

The standard neutralization conditions are evidently excessive and milder procedures may be worth considering, especially when somewhat base-labile anchoring linkages and side-chain protecting groups are used in the overall synthetic plan. Thus, good results have been obtained using *N*-methylmorpholine (Li *et al.*, 1977; Blake and Li, 1978), which has a basicity closely comparable to that of the α-amino group of peptides (Williams and Young, 1972). The equilibrium to the free α-amino peptide–resin is driven by mass action. A mixture of *N*-methylmorpholine (1.0 *N*) and acetic acid (0.5 *N*) in dimethylformamide, referred to as "buffered DMF" because the absorbance of a bromocresol green indicator remained constant upon addition of glycine ethyl ester hydrochloride from 0 to 0.16 *M*, has also been shown to be effective for the neutralization step (Glass *et al.*, 1974; see p. 99). In this case, the released α-amino group may exist partially as its acetate. Therefore, incoming protected amino acids or peptide segments must be preactivated either as active esters or anhydrides in order to minimize acetylation that would occur if an activating agent such as DCC were used for the coupling reaction.

Neutralization can also be performed *in situ*, in combination with the activation and coupling step. This approach requires that the peptide–resin be obtained as its hydrohalide salt, either directly upon deprotection or after exchange with, for example, the appropriate pyridine hydrohalide salt (Dorman, 1969). Dicyclohexylcarbodiimide (3 equiv) has been added with the *N*-methylmorpholine salt of a *tert*-butyloxycarbonylamino acid (3 equiv) in CH_2Cl_2 (Suzuki *et al.*, 1975, 1976a), or with a Boc-amino acid (1.5 equiv) plus the Boc-amino acid benzyltrimethylammonium salt (1.5 equiv) in dioxane (Andreatta and Rink, 1973; only applied to liquid-phase procedure). Other salts of Boc-amino acids have not been considered for solid-phase procedures and a systematic study seems warranted. However, it must be remembered that dissolved hydrohalide salts retained from neutralization *in situ* might have adverse effects, including the promotion of racemization.

Carboxylic acids are frequently employed as reagents or cosolvents in the α-amino group deprotection step. Any residual acid retained through physical or chemical "carryover" could be activated (directly or by exchange) during the subsequent coupling step, thus leading to acylation and irreversible termination of growing peptide chains (compare to results in solution due to Fletcher *et al.*, 1973). Retention of carboxylic acids in solid-phase synthesis has been observed despite a neutralization step with a tertiary amine, and has been attributed to reaction vessel design or materials (Brunfeldt *et al.*, 1972a; Yamashiro and Li, 1974). However, none of the bases for neutralization that have been cited earlier is of sufficient strength to release carboxylic acids ionically bound to quaternary ammonium sites, nor will these bases react with acids that are covalently bound to the support as esters (Kent *et al.*, 1979; see following paragraph). Acetylation of peptide chains has been shown by radioactive incorporation and mass spectrometry (Brunfeldt *et al.*, 1972a,b); and the occurrence of trifluoroacetylation (estimated at about 1% per step when dicyclohexylcarbodiimide activation was used) was deduced on the basis of chromatography or electrophoresis and end-group determination before and after base treatment, and of fluorine incorporation by elemental analysis (Ondetti *et al.*, 1971; Bush *et al.*, 1972; Noble *et al.*, 1977). Formylation during active ester couplings in dimethylformamide (which might be contaminated with trace residual formic acid) has also been alluded to (Atherton *et al.*, 1975; see also Birr, 1978, p. 41). Improved yields and homogeneity of products, attributable to lower levels of these acylation side reactions, were found in some cases upon switching to different coupling methods. These include the use of symmetrical anhydrides (Noble *et al.*, 1977) and dicyclohexylcarbodiimide with 1-hydroxybenzotriazole (Fletcher *et al.*, 1973; demonstrated in solution).

An additional chemical mechanism of trifluoroacetylation in solid-phase peptide synthesis has now been elucidated and shown to be independent of the coupling step (Kent *et al.*, 1979); this mechanism may account for the quantitatively significant portion of some of the observations referred to in the preceding paragraph. Trifluoroacetoxymethyl groups are generated from pre-existing hydroxymethyl sites during treatment with TFA–CH_2Cl_2 in the deprotection step, or they are formed directly by premature acidolysis of the benzyl ester anchoring linkage (see p. 67). An intersite trifluoroacetyl transfer from hydroxyl to amine occurs during the subsequent neutralization step, once the nucleophilic α-amino group is liberated from its salt by treatment with the tertiary amine. This mechanism predicts that the undesirable termination of growing peptide chains by trifluoroacetylation will be increased by (1) lengthening of the time period for neutralization; (2) the presence of unblocked hydroxymethyl sites on the functionalized support; (3) the occurrence of proline at the amino-terminus of a growing chain. The level of the side reaction was 1–2% per cycle when the standard *p*-alkylbenzyl ester linkage was employed, but it could be eliminated ($<0.02\%$ per cycle) by switching to the more acid-stable *p*-(carbamoylmethyl)benzyl ester linkage (Pam-resin, see pp. 64–66 and 69). The chemically unambiguous route (Fig. 5) for the synthesis of these latter supports without the creation of extraneous functionalities was emphasized as a key reason for the successful prevention of trifluoroacetylation according to the described mechanism.

During the brief time that the α-amino group exists in its free form subsequent to neutralization, but prior to coupling, the opportunity exists for a variety of intramolecular aminolysis side reactions. These may lead to the loss or termination of growing polymer-supported peptide chains. The same dangers are conceivable, in principle, when those α-amino protecting groups that give the free α-amine directly are employed (Section IV,A,4–6). These side reactions are often found to be catalyzed by weak carboxylic acids (including the next incoming N^{α}-protected amino acid) or tertiary amines; the major examples are (1) diketopiperazine formation from a dipeptide–resin (pp. 82–83); (2) dihydro-oxazinone formation from an amino acid phenacyl ester resin (p. 86); (3) pyroglutamyl formation at NH_2-terminal glutamine residues (p. 205), or (4) potential diazacycloheptane formation at NH_2-terminal protected arginine residues (p. 173). Also, (5) low-level intermolecular aminolytic cleavage of the anchoring benzyl ester linkage has occasionally been reported (p. 27); and (6) intermolecular trifluoroacetylation occurs as described in the preceding paragraph.

C. Formation of the Peptide Bond

The mechanism of peptide bond formation involves the attack of a nucleophilic α-amino group upon the activated carboxyl function of a protected amino acid or peptide segment. The activated species can be preformed in a separate chemical step, as in the use of anhydrides or active esters; alternatively, a suitable condensing agent such as dicyclohexylcarbodiimide can be employed to mediate coupling via *in situ* activation. It is generally recognized that no practical coupling method proceeds through activation of the free amino group (Bodanszky *et al.*, 1976, p. 109).

The degree of activation of the carboxyl group will control the rate of the desired coupling reaction, but can also influence the levels of side reactions.

One of the major potential side reactions that must be avoided is racemization. The quantitative level of racemization, which proceeds through a 5(4H)-oxazolone mechanism (Bergmann and Zervas, 1928), is influenced by a variety of factors, the most important of which is whether N^{α}-protected amino acids or peptide segments are activated (reviewed by Goodman and Glaser, 1970; Kemp, Vol. 1, Chapter 7). In the former case, the chemical structure of the α-amino protecting group is highly relevant to this issue. In the common practice of stepwise solid-phase peptide synthesis applying the groups of Section IV,A, racemization is not a serious concern. For polymer-supported segment condensation (Section II,B,2), the extent of racemization will depend on the rate at which the activated intermediate couples, relative to the overall rate of the competing pathway comprising rearrangement to, and racemization and coupling of the 5(4H)-oxazolone. Thus, the negligible racemization observed upon coupling small soluble peptides in model studies may in some cases have reflected a kinetic effect, and the same results might not necessarily hold with larger, slowly coupling peptides. One strategy to side-step the question of racemization is to restrict the choice of carboxyl component segments to those containing COOH-terminal glycine or proline residues (although under some conditions, in solution, racemization of the penultimate amino acid residue has been observed; e.g., Weygand *et al.*, 1966a; Dzieduszycka *et al.*, 1973). At the present time, a few techniques for segment condensation are available which give acceptably low levels of racemization.

A second serious potential side reaction attributable to overactivation is peptide backbone acylation. Such a process, which would lead to branching of peptide chains or amino acid insertion, has been observed under drastic solution conditions (Brenner, 1967). The weight of the available evidence (see Merrifield *et al.*, 1974) suggests that even under exaggerated conditions, backbone acylation does not occur in solid-phase synthesis.

Based on this discussion, it becomes possible to consider the relative merits of N → C, C → N, and bidirectional strategies for peptide chain elongation in solid-phase synthesis (Section II,B). Stepwise condensation in the C → N direction (Table I, number 1a) involves activation of the α-carboxyl group of an N^{α}-protected amino acid, and coupling to a polymer-supported peptide chain. This mode has three important advantages: (1) the racemization issue can be practically ignored; (2) temporary α-amino protecting groups are available with excellent characteristics (Section IV,A); and (3) any rearranged by-products of the activated species are soluble and are not incorporated into the growing chain. In contrast, when stepwise condensation under the best conditions is carried out in the N → C direction,* (1) some racemization can be expected at each step, because the activated carboxyl group is contributed by the peptide moiety and (2) rearranged by-products of the activated species are retained on the support, giving rise to terminated peptides in the final cleaved product. This latter factor [number (2)] is also a drawback when polymer-supported segment condensation is carried out in the N → C direction, but not in the C → N direction. However, from the standpoint of racemization, the possible directional strategies of segment condensation (Table I, 2a and 2b) are equivalent because a peptide carboxyl group must be activated in any case.

Most of the chemical methods used to achieve couplings in solution have been tested for solid-phase peptide synthesis, but only a few have been generally adopted. In the following, we discuss the activation of protected amino acids and peptide segments and consider reaction mechanisms, rates, appropriate solvents, side reactions, and racemization.

1. Carbodiimides

The use of carbodiimides for peptide bond formation (Sheehan and Hess, 1955) has been widely adopted in solid-phase methodology. Commonly (e.g., Merrifield, 1963, 1964b; Marshall and Merrifield, 1965) an N^{α}- and side-chain protected amino acid (three molar equivalents over initial loading of support) is dissolved in slightly more than the minimal volume of CH_2Cl_2

* Evidently, the indicated potential difficulties have not deterred Nature from carrying out stepwise solid-phase synthesis in the N → C direction on ribosome supports with transfer RNA activation. The specificity of binding sites on enzymes allows couplings to proceed without racemization.

required to fully swell, and subsequently readily suspend, the resin. Occasionally, addition of 10 to 25% (v/v) dimethylformamide is also necessary to solubilize certain derivatives. If the derivative is available as the dicyclohexylammonium salt, a preliminary neutralization and extraction to obtain the free acid is necessary (p. 36–37). The deprotected and neutralized peptide–resin is preequilibrated with the amino acid solution for 5 min, and dicyclohexylcarbodiimide (DCC, equimolar to carboxyl component) is then added. The reaction is allowed to proceed, at 25°C for 30 to 120 min. The major by-product, dicyclohexylurea, is only sparingly soluble in solvents of low dielectric constant. The solubility in dimethylformamide, ethanol, and acetic acid is respectively 4, 10, and 17 mg/ml (Merrifield, 1963). The urea is removed in the course of thorough washing of the peptide–resin following the coupling reaction, and the last traces are entirely solubilized by the acidic solvents of the subsequent deprotection step. Diisopropylcarbodiimide, which leads to a more soluble urea, has recently been preferred in the solid-phase synthesis of a somatostatin analogue (Sarantakis *et al.*, 1976).

There are two fundamental variations of the standard dicyclohexylcarbodiimide coupling protocol just described. As already stated (p. 121), a variety of intramolecular aminolysis side reactions are catalyzed by weak carboxylic acids, such as the next incoming N^{α}-protected amino acid. At those critical points where such side reactions are anticipated, "reverse addition" has been found very helpful (Gisin and Merrifield, 1972a). Specifically, dicyclohexylcarbodiimide is added to the peptide–resin prior to the carboxyl component. The "adsorption coupling" method (Esko and Karlsson, 1970; Elliot *et al.*, 1972; Ragnarsson *et al.*, 1974) is based on the observation (H. Klostermeyer, personal communication, 1968) that when an α-amino peptide–resin is briefly equilibrated with a carboxylic acid in a solvent of low dielectric constant such as CH_2Cl_2, or even better, CCl_4 or benzene, one equivalent of the carboxylic acid is retained as a resin-bound ion-pair despite extensive washing with these same solvents.* When DCC in CH_2Cl_2 is now added (1.5–3 equiv over amino sites on resin), incorporation of the adsorbed amino acid often proceeds in excellent yields (94–99%). More polar solvents, such as dimethylformamide, are not suitable for this procedure because they tend to dissociate the resin-bound complex. Strikingly, when a *tert*-butyloxycarbonyl dipeptide was coupled by the adsorption method in CH_2Cl_2, no racemization was detectable (Esko and Karlsson, 1970). By contrast, racemization by the standard procedure in CH_2Cl_2 was 2%, and in DMF it was 18% (Barton *et al.*, 1973). These lines of evidence suggest that the activated species is not free in solution for any appreciable

* The excess carboxyl component can be recovered by filtration, thus permitting a high efficiency of synthesis (see footnote on pp. 10–11).

amount of time. The noted adsorption effects are of theoretical importance, but have not been extensively applied to practical situations.

a. Mechanism. The mechanism and stoichiometry of carbodiimide-mediated activation and coupling is complex and quite sensitive to reaction conditions, including the order of addition and structures of reactants, and is still not fully understood (reviewed, over the years, by Khorana, 1953; Schröder and Lübke, 1965, pp. 108–111; Rich and Singh, Vol. 1, Chapter 5). Moreover, the "adsorption coupling" results suggest that mechanisms in solid-phase synthesis may not necessarily follow those in solution; for example, the 2:1 carboxylic acid: carbodiimide stoichiometry found to be optimal under some solution conditions (DeTar and Silverstein, 1966a; DeTar *et al.*, 1966) can not hold in the case of a resin-bound ion-pair salt. The principal pathways and side reactions are summarized in Fig. 11.

The initial adduct between a carboxylic acid **79** and dicyclohexylcarbodiimide (**80**) is believed to be the *O*-acylisourea **81**. Isolable model compounds for the adduct **81**, which show the expected reactivity, include compounds **82** (Hegarty and Bruice, 1970) and **83** (Hegarty *et al.*, 1977). The *O*-acylisourea

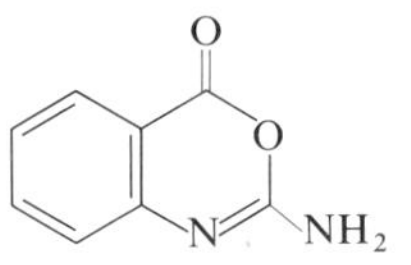

(82)

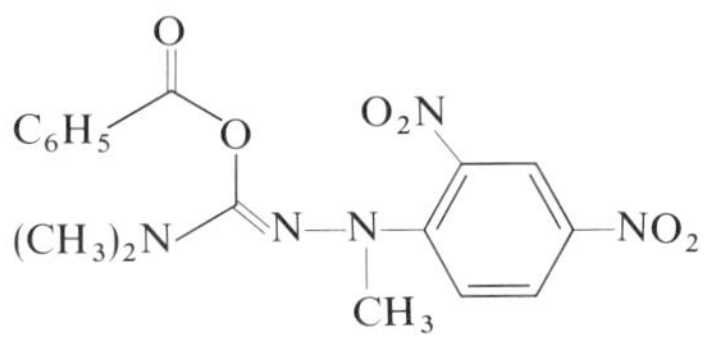

(83)

81 can undergo further reactions according to a variety of pathways. Intramolecular rearrangement produces the *N*-acylurea **84** (pathway a) or 5(4H)-oxazolone **85** (pathway b). Addition of a second equivalent of the carboxyl component **79** gives symmetrical anhydride **86** (pathway c), and interception with a nucleophilic hydroxylamine compound **87** gives the active ester **88** (pathway d). Finally, the isourea can undergo nucleophilic attack by the amino component **89** to give the desired peptide **90** directly (pathway e). An equivalent of dicylcohexylurea (**91**) is formed in all the reactions of isourea **81** (pathways b–e). In addition, all of the intermediates **85**, **86**, and **88** react further with the amine nucleophile **89** to produce a peptide bond.

Several techniques have been developed to assess the extent to which the carbodiimide reaction proceeds through the intermediacy of the symmetrical anhydride **86** (Rebek and Feitler, 1973, 1974; Rebek *et al.*, 1975). A mixture of three resin-bound amine nucleophiles was equilibrated with an excess of labeled Z-Gly-OH (three equivalents over total amino groups), followed

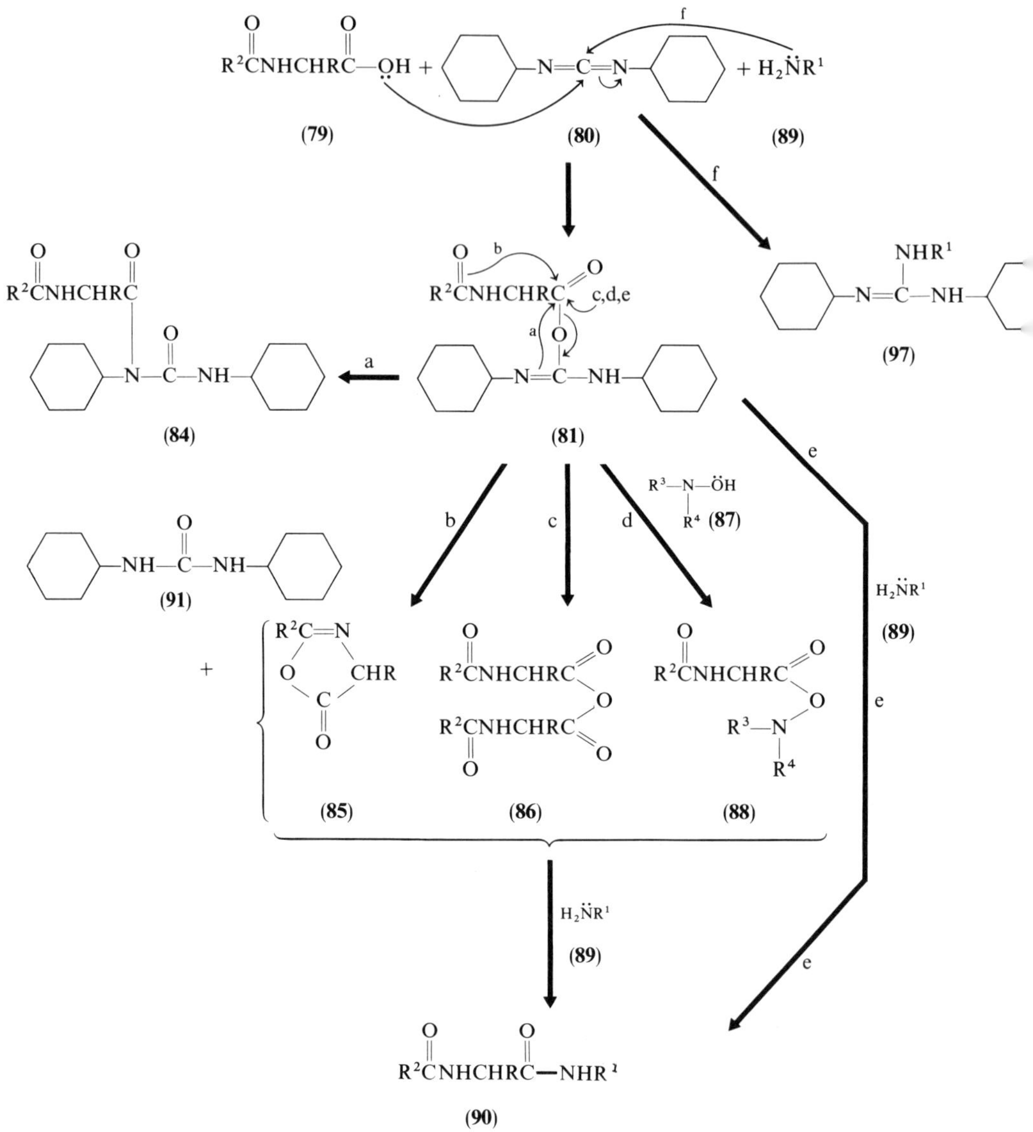

Figure 11. Pathways of peptide bond formation, and of several side reactions, mediated by dicyclohexylcarbodiimide. Detailed explanation in text of Section IV,C,1,a and b.

by a limiting quantity (0.3 equiv) of DCC or labeled $(Z\text{-}Gly)_2O$. A tritium label was used in the DCC experiment, and carbon-14 was used for the symmetrical anhydride experiment. From the fact that the $^3H/^{14}C$ ratios as well as the yield distributions among all three possible dipeptide products were identical within experimental error, these authors concluded that acylation in solid-phase peptide synthesis proceeds through a common intermediate, namely the symmetrical anhydride. This result was in contrast to findings from a similarly designed experiment in solution, where half of the pathway was believed to proceed directly via the *O*-acylisourea. However, it is worth commenting that the indicated experimental conditions do not duplicate standard solid-phase protocols. In another approach, the effect of order of addition was studied in solution by looking for kinetic resolution in the reaction of an achiral carboxylic acid and a racemic amine with an optically active carbodiimide. The product amide was racemic when the carbodiimide was added last as the limiting reagent, which is consistent with the intermediacy of an achiral symmetrical anhydride. Rotation in the product amide was observed under solution conditions mimicking "reverse addition" and "adsorption coupling" protocols. Finally, a double-resin trapping method, used with success in discerning soluble intermediates in intermolecular reaction mechanisms, was not applied to the issue of carbodiimide activation in solid-phase synthesis, although it seems ideally suited for this purpose.

b. Side Reactions. The standard coupling method utilizing dicyclohexylcarbodiimide is suited for the rapid, high-yield, and racemization-free *stepwise* incorporation of essentially all amino acids. The major exception involves the occurrence of a dehydration side reaction when asparagine or glutamine derivatives with an unprotected carboxamide side chain are activated; hence, other methods are required to introduce these two residues (Section V,B,5).

(i) N-Acylureas. The formation of *N*-acylureas (**84**) is a solvent-dependent, base-catalyzed process that is acceptably slow in CH_2Cl_2 or CH_3CN but excessive in DMF (Khorana, 1955; Sheehan *et al.*, 1956; Merrifield and Woolley, 1956; DeTar and Silverstein, 1966a,b). Thus, quantitative incorporation of Boc-proline in a solid-phase synthesis using DCC in DMF required eight equivalents of reagents because of *N*-acylurea formation (Merrifield, 1964b). Several unsymmetrical carbodiimides tested by Ito *et al.* (1977) were shown to lead to lower levels of *N*-acylureas. This advantage would be negated if the carbodiimides were to disproportionate.

The cyclohexylureido derivative **92** was observed as a significant by-product during the DCC-mediated coupling of nonanoic acid (**93**) to

2,2-dimethylaziridine (**94**) (Fenwick, 1973). Evidently [Eq. (13)] rearrangement to the *N*-acylurea **95** was followed by facile cleavage by the nucleophile **94** to account for the anomalous urea **92** and the soluble cyclohexylamide

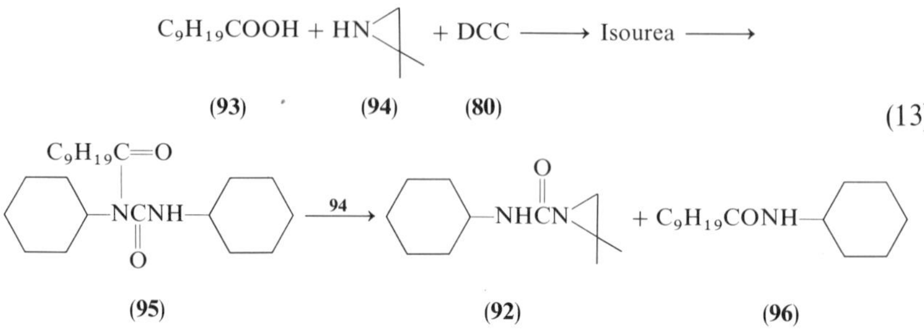

(13)

96. Were a similar side reaction to occur under solid-phase conditions, even at low levels, irreversibly terminated peptides would result. This possibility may be ruled out (<0.2%) by the experiments using DCC labeled with ^{14}C at the 1-carbon (Merrifield *et al.*, 1977a; see following page).

In summary, the side reaction leading to *N*-acylurea formation merely serves to deplete the activated species of a DCC-mediated coupling reaction, but does not appear to lead to undesirable modifications of the resin-bound growing peptide chains.

(ii) Racemization. 5(4H)-Oxazolones (**85**) are the racemizing intermediate in the DCC-mediated activation of acyl amino acids or peptides, but are not formed in appreciable amounts when N^{α}-urethane or N^{α},N^{α}-diacyl protected derivatives are coupled (Section IV,A). Thus, a solid-phase study using N^{α}-*tert*-butyloxycarbonylamino acids revealed no detectable racemization (<0.02% per step; Kent *et al.*, 1978); while a very low, but real, amount of racemate was quantitated (0.03%) by an isotope-dilution assay on the product of a solid-phase dipeptide synthesis (Rebek and Feitler, 1975). It is possible, however, that the racemization observed in the latter case may have been due to the conditions of cleavage by base-catalyzed transesterification rather than to the coupling. Earlier, less sensitive methods for the evaluation of the extent of racemization have been reviewed (Erickson and Merrifield, 1976, pp. 412–415).

Substantial racemization due to α-proton abstraction does occur in the course of stepwise coupling of certain N^{α},N^{im}-derivatives of histidine (see pp. 186–189 for documentation and methods to suppress this side reaction).

(iii) Potential Guanidine Formation. The reaction of dicyclohexylcarbodiimide with amines **89** to give a guanidine structure **97** (Fig. 11, pathway f) (DeTar *et al.*, 1966) has been shown not to occur (<0.2%) under

the conditions of solid-phase peptide synthesis (Merrifield *et al.*, 1977a). In the latter study, incorporation of [1-^{14}C]DCC was used as a sensitive probe of the potential side reaction. However, the guanidine **98** was readily formed in high yield by reaction of DCC with a *protonated* valine-resin. The product **98** was characterized by rapid triethylamine-catalyzed cleavage from the support to give imidazolone **99** [Eq. (14); compare to solution results of Muramatsu *et al.*, 1963]. The guanidine adduct of di- and tripeptide–resins

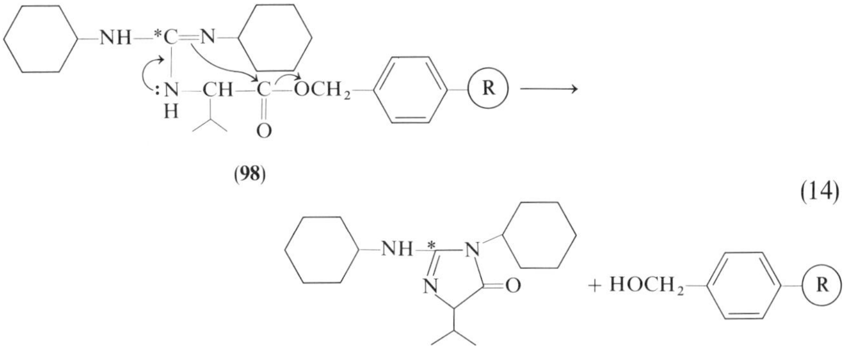

(98) (14) (99)

took up an equivalent of picric acid during monitoring, but was not acylated under standard coupling conditions. Acylation was, however, possible in the presence of a quaternary ammonium hydroxide.

(iv) Miscellaneous. The dimerization of dicyclohexylcarbodiimide to give 1,3-dicyclohexyl-2,4-bis(cyclohexylimino)-1,3-diazetidine (**100**) [Eq. (15)] has been noted under some conditions of peptide bond formation in solution (Jakubke and Klessen, 1977). The reaction is mechanistically similar to the amidine formation discussed above. The dimer **100** itself will not be detri-

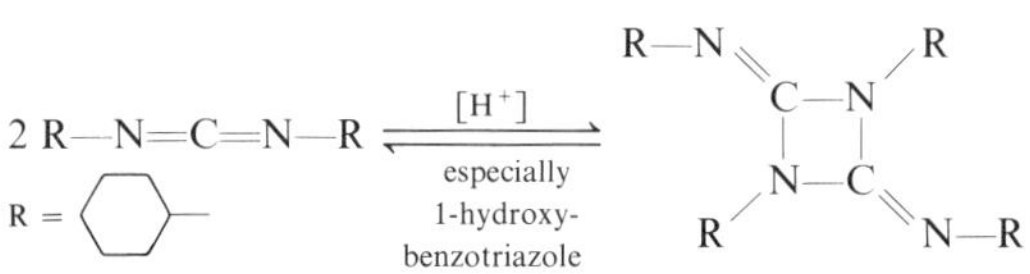

(80) (100) (15)

mental to the progress of a synthesis, except that it reduces the effective concentration of DCC.

Dicyclohexylcarbodiimide is itself a weak base that is qualitatively of lesser strength than pyridine (DeTar and Silverstein, 1966b). Nonetheless, the excess reagent could in principle serve to catalyze any of a variety of the above-mentioned side reactions.

c. Rates and Yields of Stepwise Synthesis. Under solid-phase conditions, the life-time in CH_2Cl_2 of the effective acylating species arising with DCC-activation is on the order of hours. This was shown by preincubation and delayed trapping experiments (Tometsko, 1973; Dorman, 1974). Slow decomposition ($t_{1/2} = 3–8$ hrs) will give the *N*-acylurea, or possibly a urethane-acylated dipeptide derived from rearrangement of a symmetrical anhydride [Eq. (17) on p. 139], although neither of these by-products was identified in the cited experiments.

The rates of DCC-mediated couplings in solid-phase synthesis are primarily dependent on the structure of the activated amino acid, somewhat less so on that of the attacking α-amino peptide–resin nucleophile, and hardly at all on the nature of the urethane α-amino protecting group. Where a direct quantitative comparison is possible (Merrifield, 1973c), second-order rate constants are about 20- to 30-fold less than those for preformed symmetrical anhydrides (p. 138). Thus, if the latter species is the one actually responsible for acylation, the rate-determining step of the overall process may be the initial DCC-activation. The second-order rate constants—determined by a continuous-flow monitoring technique—on a synthesis of Leu-Ala-Gly-Val-resin, using N^{α}-2-(*p*-biphenylyl)propyl(2) oxycarbonyl protection and DCC coupling, ranged from 0.2 to 0.5 liter mol^{-1} sec^{-1}; thus, for a typical synthesis in which the nominal concentration of resin-bound amine is 0.02 *M* and the incoming amino acid and DCC are both 0.06 *M*, the time for 99% completion of the reaction can be calculated as ranging from 4 to 9 min. This is in good accord with the semiquantitative data of Esko *et al.* (1968) on the synthesis of an isomeric model tetrapeptide using N^{α}-*tert*-butyloxycarbonyl protection. In solution, the second-order rate constants for the reaction of benzyloxycarbonyl and benzoyl amino acids with dicyclohexylcarbodiimide in acetonitrile were determined to range from 0.3 to 0.8 liter mol^{-1} sec^{-1} based on consumption of carbodiimide judged in the ir spectrum (DeTar *et al.*, 1966). These data, taken with those presented above, are again consistent with the idea that generation of the true acylating species, rather than its aminolysis, is rate determining.

Competition experiments (Ragnarsson *et al.*, 1971, 1974) established that under solid-phase conditions, *tert*-butyloxycarbonylglycine activated with DCC was incorporated onto an alanine- or valine-resin in CH_2Cl_2 at the fastest rate, while most N^{α}-Boc side-chain protected amino acids reacted at roughly equal rates, within 40 to 70% of that of Boc-glycine. The β-substituted amino acids, threonine, valine, and especially isoleucine, as well as a model dipeptide, coupled 5- to 20-fold more slowly than Boc-glycine. From a steric point of view, an -Ile-Ile- sequence should be the most difficult to prepare, but quantitative yields were obtained (Esko and Karlsson, 1970) both by standard (single coupling) and adsorption coupling (treatment was repeated) procedures. Essentially quantitative yields are obtained by the standard procedure in the DCC-activated condensation of di- and tripeptide segments as well (e.g., Sakakibara *et al.*, 1968b; Gisin *et al.*, 1969; see Table I), although larger segments cannot always be coupled to completion (e.g., Omenn and Anfinsen, 1968).

The extremely fast coupling rates just cited have served to justify development of an accelerated protocol, where the standard 120 min single-coupling reaction period was reduced to two 3–5 min couplings (Corley *et al.*, 1972; Eastlake *et al.*, 1976). Also, so long as depletion of the activated species by rearrangements does not occur to a substantial extent over the standard coupling period, much smaller excesses (e.g., 1.5 equiv) of *tert*-butyloxycarbonylamino acids and dicyclohexylcarbodiimide have given acceptable yields (e.g., Merrifield, 1964b; Hodges and Merrifield, 1975a; Upson and Hruby, 1976; Live *et al.*, 1977; see footnote on pp. 10–11).

The overall efficiency of the synthetic cycles, ignoring termination side reactions, can be assessed by measuring the amounts of deletion peptides. Experimentally, this is done either directly by chromatographic determination, or indirectly by sequential "preview" Edman degradation (pp. 156–158). Using these techniques, coupling yields in solid-phase synthesis have been deduced to range from 99.3 to 99.7% per step, when DCC-activation has been used (Erickson and Merrifield, 1973b; Birr and Frank, 1975; van Rietschoten *et al.*, 1975b; Mitchell *et al.*, 1978b). These yields, while excellent, are nevertheless consistently lower than what might be predicted on the basis of the extrapolated rate data. Furthermore, the levels of unreacted amino groups upon completion of the acylation steps can usually be reliably established as <0.2% (e.g., Hodges and Merrifield, 1975a; Tan and Kaiser, 1976). Hence, certain amino groups must be temporarily inaccessible at each step, either to acylating or monitoring reagents. This effect could in principle be due either to physical–chemical properties of the polymeric support or to some chemical side reactions. The new phenylacetamidomethyl (Pam) supports prepared by a handle approach utilizing resins that had been derivatized by direct amidoalkylation (see pp. 64–66, esp. Fig. 5) were applied to a synthesis of a model tetrapeptide in which the levels of deletion peptides were reduced to 0.05% (Mitchell *et al.*, 1978b). Hence, coupling yields greater than 99.9% per step are possible when optimal chemical procedures are used.

d. Diimides in the Presence of Additives. The scope of the coupling reactions mediated by dicyclohexylcarbodiimide has been remarkably expanded by the use of certain additives that react rapidly with the initial *O*-acylisourea adduct **81** to give active esters **88**, which are still excellent acylating agents (Fig. 11, pathway d). Furthermore, intermolecular formation of **88** competes effectively with intramolecular side reactions (Fig. 11, pathways a and b) leading to an *N*-acylurea **84** or the racemizing intermediate 5(4H)-oxazolones **85**.

(i) N-Hydroxysuccinimide. The originally proposed additive, *N*-hydroxysuccinimide (HOSu) (**101**) (Weygand *et al.*, 1966b) was soon tested with good results for segment condensation in solid-phase synthesis (Weygand and Ragnarsson, 1966; Omenn and Anfinsen, 1968; Ragnarsson *et al.*, 1975). In the procedures used, most of the polymer-bound amino sites were suc-

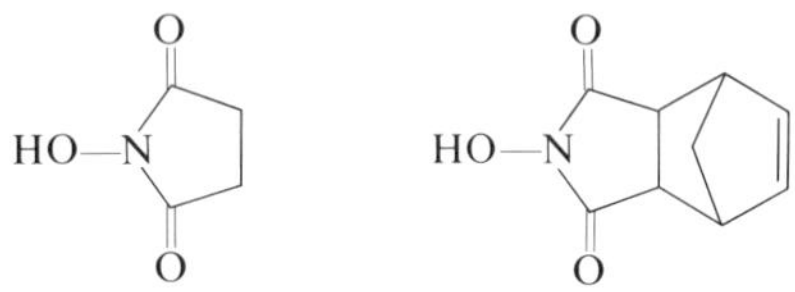

(**101**) HOSu (**102**) HONB

cessfully acylated; racemization in CH_2Cl_2 was negligible ($\leqslant 1\%$) but considerable in dimethylformamide ($\sim 9\%$). The most extensive application of the DCC–*N*-hydroxysuccinimide coupling method was in the segment condensation solid-phase assembly of a trypsin inhibitor sequence (Yajima and Kiso, 1974; see Table I). Four protected segments, ranging from 9 to 19 residues and all containing COOH-terminal glycine, were coupled in average yields of $\sim 80\%$, using molar ratios of peptide–resin:carboxyl component: HOSu:DCC of about 1:2:2:4. Reactions were carried out in DMF, at 25°C for 4 days, followed by acetylation steps to block any unreacted sites. This synthesis is noteworthy for the good incorporations of significantly sized segments. Since the presence of free amino groups on the resin at the end of each coupling step was not measured, it is unclear whether the reported yields reflect kinetic inaccessibility of certain polymer-supported sites, or a competing terminating side reaction.

Dicyclohexylcarbodiimide (**80**) and *N*-hydroxysuccinimide (**101**) react in a molar ratio of 1:2 to give compound **103** and dicyclohexylurea (**91**) (Gross and Bilk, 1968). The active ester (**103**) can acylate amino groups, resulting in

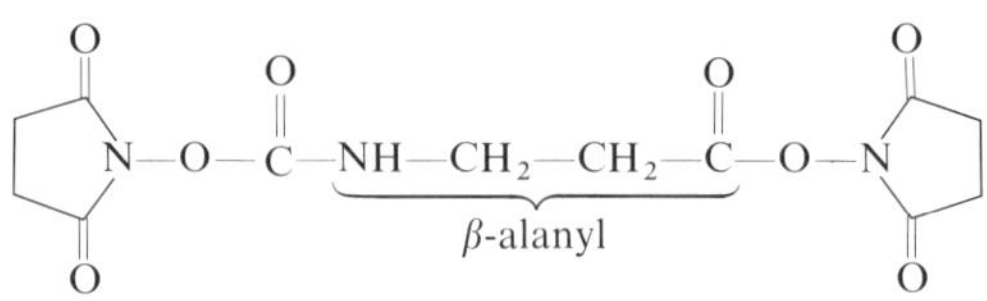

(**103**)

the creation of terminated peptides containing β-alanine. In a solid-phase coupling of Boc-His(Bzl)-OH to H-Glu(OBzl)-resin (Windridge and

Jorgensen, 1971b) mediated by DCC–HOSu, no racemization was observed but the overall yield was only 70%. Upon acid hydrolysis, 30% of β-alanine was quantitated, suggesting that **103** had indeed been formed and incorporated into the peptide chain. A corresponding side reaction does not appear to occur with the use of the structurally analogous but more rigid *N*-hydroxy-5-norbornene-2,3-dicarboximide (**102**) (Fujino *et al.*, 1974); this latter reagent has, however, not been tested in solid-phase peptide synthesis.

(ii) 1-Hydroxybenzotriazole. The systematic researches of König and Geiger (1970a,b,c) have led to the recommendation of 1-hydroxybenzotriazole (HOBt) (**104**) as a most efficacious *in situ* additive for DCC-mediated couplings in peptide synthesis. The active species in solution is presumably the

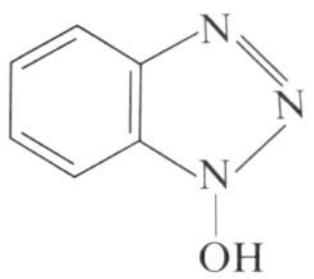

(**104**) HOBt

1-hydroxybenzotriazole ester, which is aminolyzed at a rate about 10^3-fold faster than *N*-hydroxysuccinimide and similar active esters, perhaps because of a concerted bifunctional catalytic mechanism (compare König and Geiger, 1972a; Horiki, 1977a,b; McCarthy *et al.*, 1977). The net overall rate is of the same order of magnitude as that for activation and coupling with DCC alone. In contrast to several other additives that have been tested, HOBt does not react with DCC to give active acylating agents that could terminate peptide chains. Since HOBt has acidic ($pK_a = 4.3$) as well as nucleophilic characteristics, some caution is required with the choice of α-amino and side chain protecting groups. In particular, the N^{α}-*o*-nitrophenylsulfenyl group and the N^{im}-tosyl group for histidine are known to be incompatible (see pp. 110 and 180, respectively); the N^{α}-2-(*p*-biphenylyl)propyl(2)oxycarbonyl group (Section IV,A,2) is marginally stable (0.3% cleaved per simulated cycle with 0.03 *M* HOBt; S. Mojsov, personal communication, 1978). Furthermore, HOBt (**104**) is a particularly good catalyst for the dimerization side reaction of DCC [Eq. (15) on p. 129].

In cases where coupling with DCC alone in solid-phase synthesis has given inadequate yields and unacceptable by-products, couplings with DCC–HOBt improved matters. This has been shown both for the racemization-free introduction of Boc-His(Bzl)-OH (Windridge and Jorgensen, 1971b) and for the dehydration-free introduction of asparagine and glutamine (Hruby *et al.*, 1973; Blake and Li, 1975b). In some laboratories, 1-hydroxybenzotriazole is added as a matter of routine to all DCC-mediated couplings (e.g., Moore *et al.*, 1977). The ratios of peptide–resin: carboxyl component:HOBt:DCC have ranged from 1:x:1:x to 1:x:2x:x, where x varies from 1.5 to 5. Usually, DCC is introduced last after a brief pre-equilibration period involving the other three reactants, as in the standard coupling protocol; sometimes "reverse addition," in which the N^{α}-protected amino acid is added last, is used instead. Alternatively, the 1-hydroxybenzotriazole ester has been preformed by incubation of the N^{α}-protected amino acid with DCC–HOBt at 0°C for 10 to 60 min, followed by filtration to remove precipitated dicyclohexylurea.

Some DMF is always present in the milieu to solubilize the 1-hydroxybenzotriazole (HOBt), and often DMF is the exclusive solvent. When sufficient excesses are employed, reactions are generally complete in 60 to 120 min.

(iii) Polymer-Supported Segment Condensation with DCC–HOBt. The most successful recent results with polymer-supported segment condensation have all relied on the combination of DCC and HOBt. The weight of the literature evidence suggests that the DCC–HOBt procedure gives maximal reaction rates and yields, with markedly reduced amounts of racemization (e.g., Table VI in Kemp, Vol. 1, Chapter 7). Almost all studies have utilized DMF as a solvent, and long coupling periods have been employed (25°C, 1–5 days).

In the impressive synthesis of glucagon by the Protein Synthesis Group of Shanghai (1975; see Table I), four protected peptide segments ranging in size from penta- to nonapeptides were prepared in solution and condensed on a solid support. The division of the glucagon sequence into protected segments was closely comparable to that used in the successful all-solution synthesis of Wünsch (1967). In the Chinese synthesis, peptide chain growth was in the C → N direction. Coupling yields on the solid support, as assessed by residual amino sites, were >95% per step, using peptide–resin:carboxyl component: HOBt:DCC in molar ratios near 1:3:1:3; fortunately, the protected segments were sufficiently soluble in DMF that net initial concentrations of 0.05 to 0.1 *M* were possible during these steps. The coupling yields at three steps of the synthesis of an insulin A-chain analogue (Weber and Andre, 1975), where HOBt was present in 6 equiv, ranged from 70 to 90% but could not be raised by repeated coupling. It is unclear what effect the unusually high initial loading of the support (1.2 mmol/gm) had on these latter results. The success of a given coupling step seems to be dependent on the size and sequence of both the carboxyl and amino components, irrespective of whether solution or solid-phase methods are used (e.g., Camble and Petter, 1976).

The comparative study of Di Bello *et al.*, (1977) on the coupling yields of

Z-Phe-Arg-OH to a resin-bound 12-peptide established the clear superiority of the DCC–HOBt technique in this particular model system. The yield was 96% with DCC–HOBt, and ranged from 10–40% with DCC–HOSu, DCC alone, Woodward's reagents, or EEDQ.

Two recent results by polymer-supported segment condensation, both of which relied on the DCC–HOBt coupling method, suggest the current frontiers of this approach (Fig. 12). The syntheses are also noteworthy in that all peptide bond-forming steps were carried out by the solid phase method. Since specific structural features of the target sequences facilitated the details of the chosen chemical strategies, the described procedures are not entirely general.

Gisin *et al.* (1977) prepared four segments covering the entire linear sequence proposed for alamethicin by the stepwise C → N solid-phase method. The peptide chain was then extended in the N → C direction (Fig. 12A), with anchoring each time at the COOH-terminus. At all stages of the synthesis, an effectively protected peptide acid was obtainable by simple HF cleavage of the anchoring benzyl ester linkage. This was possible because the NH_2-terminal aminoisobutyryl residue of the peptide was permanently protected as the desired acetyl derivative, and because there were no residues requiring side-chain protection. After each HF cleavage, the peptide segments were purified and then coupled onto the next of four peptide–resins. Even a 17-peptide could be coupled in excellent yield (69%, after cleavage and purification) by DCC–HOBt.

Birr *et al.* (1977) outlined a solid-phase strategy towards the 22-residue mast-cell degranulating peptide already synthesized by their group using solution methodology. Four protected segments (I–IV, from the NH_2- to COOH-terminus) were prepared by stepwise solid-phase synthesis (Fig. 12B) while bound to a support with a phenacyl ester linkage. Segments I and III were cleaved from the support, purified, and coupled to resin-bound segments II and IV respectively. The protected segment corresponding to the NH_2-terminal half was obtained in 53% yield after cleavage and purification, and was quantitatively incorporated onto the resin-bound segment from the COOH-terminus to provide the fully protected completed peptide on the support in a reported yield of 32%. The appropriate protection strategy relied on the acid-labile, but base-stable, N^{α}-Ddz temporary group and base-stable permanent groups. There were also no base-sensitive residues or sequences in the target structure (cf. to p. 78). The soluble carboxyl components in the coupling steps were always selected to be limiting; and based on their uptake into the polymer-supported peptide chains, the nominal coupling yields with DCC–HOBt were judged to range from 73% to essentially quantitative at the various steps. Unreacted amino sites were blocked with 3-nitrophthalic anhydride (Table XII, number 5, on p. 162).

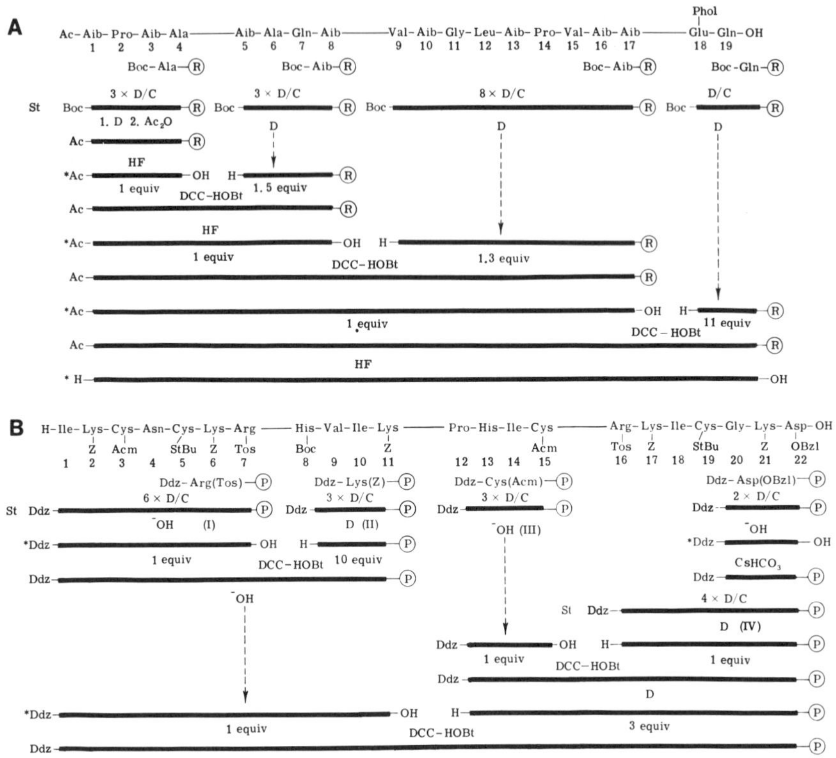

Figure 12. Polymer-supported segment condensation syntheses, with couplings mediated by dicyclohexylcarbodiimide and 1-hydroxybenzotriazole. Ⓡ and Ⓟ are polystyrene resins incorporating, respectively, the standard benzyl ester or a phenacyl ester anchoring linkage to the support. *St* means peptide sequences on these lines were assembled by stepwise solid-phase synthesis; *D* indicates one deprotective step; *D/C* denotes one deprotection/coupling cycle by standard methods; an asterisk means that a soluble intermediate was purified.

(A) An alamethicin-like peptide (Gisin *et al.*, 1977). Aib = aminoisobutyric acid, which for steric reasons is extremely difficult to couple with or to. This was the principal reason why a segment condensation approach was selected. Phol = L-phenylalaninol, attached as an amide to the γ-carboxyl group of the glutamate residue at position-18.

(B) Mast-cell-degranulating peptide (Birr *et al.*, 1977). See text for description of synthesis. As an additional interesting feature, note that the COOH-terminal tripeptide was cleaved from the support, purified, and reattached. This approach was followed to circumvent formation of **25** [Eq. (4) on p. 86] and diketopiperazine release from the phenacyl ester linkage. Also, segment III was obtained in only 20% yield after cleavage from the support, and hence, for further applications, was obtained in solution (60%) by condensation of two dipeptides.

2. Symmetrical Anhydrides

The preparation of symmetrical anhydrides (**86**) (Wieland *et al.*, 1950; Schüssler and Zahn, 1962) of N^{α}-protected amino acids and their use [Eq. (16)] as the activated species for the couplings in stepwise solid-phase peptide synthesis has been recognized in several laboratories (e.g., Wieland *et al.*, 1971c; Hagenmaier and Frank, 1972; Yamashiro and Li, 1974; Atherton *et al.*, 1975, 1977; Yamashiro *et al.*, 1976; Chang and Meienhofer, 1978) as a worthwhile alternative to the standard carbodiimide-mediated methods. The suggestion that symmetrical anhydrides actually represent a significant fraction of the true acylating species when DCC-activation is carried out with dicyclohexylcarbodiimide (pp. 125–127) has undoubtedly contributed to this trend. All amino acids can be introduced through their symmetrical anhydrides, but other methods are preferred for N^{ω}-blocked arginine (Section V,B,1) and asparagine or glutamine with unprotected side-chain carboxamide groups (Section V,B,5).

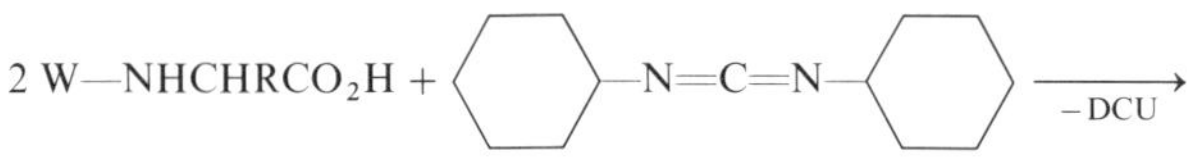

(**80**)

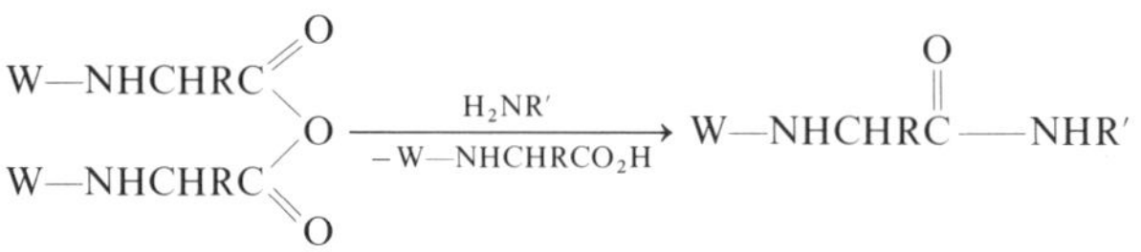

(16)

(**86**) (**90**)

W = Boc, Bpoc, Fmoc, etc.

Analytically pure symmetrical anhydrides have been prepared by reaction of sodium salts of *tert*-butyloxycarbonylamino acids with phosgene in tetrahydrofuran at −40°C (Wieland *et al.*, 1973). A simple preparative procedure exploiting water-soluble carbodiimides (Chen *et al.*, 1978) also appears to have merit. In current practice, the symmetrical anhydrides are usually preformed *in situ* in CH_2Cl_2 solution immediately before required for the coupling reaction (see preceding paragraph for references). The protected amino acid and DCC, in a molar ratio of 2:1, are reacted at 0°C for 15 min, at the end of which time the precipitated dicyclohexylurea is removed by filtration. The anhydride solution can be used as such, or it can be taken to dryness and redissolved in DMF.

a. Rates and Yields. The concentrations of the symmetrical anhydrides used during the couplings have been 0.1 *M* or higher, and usually tests for unreacted amino groups have been negative within 10 min. In some protocols, a tertiary amine was added, in slight molar excess over the theoretical yield of released carboxylic acid, for the final stages of the coupling; the level of deletion peptides could thus be reduced five- to eightfold (Yamashiro *et al.*, 1976). Overall coupling yields can be estimated at 99.6 to 99.8% per step, which compares favorably with the best results by the standard method with dicyclohexylcarbodiimide. The importance of maintaining adequate swelling of the peptide-containing polymeric support during the coupling step has also been stressed, and addition of trifluoroethanol (20% v/v) has been reported to be helpful in selected cases (Yamashiro *et al.*, 1976).

The second-order rate constants for symmetrical anhydride couplings (Merrifield, 1973c) were found to range from 4 liter mol^{-1} sec^{-1}, for the reaction of Bpoc-alanine with alanine-resin, to 0.2 liter mol^{-1} sec^{-1} for Bpoc-isoleucine to phenylalanine-resin. These rates were essentially the same as those for the mixed anhydride couplings of Bpoc amino acylethyl carbonates (see Table VIII, p. 367, in Erickson and Merrifield, 1976), but this latter set of derivatives is not as attractive for stepwise solid-phase peptide synthesis because terminated derivatives are possible from "wrong-way" opening of the anhydride. Making the same assumptions as in an earlier calculation (p. 130), the time for 99% completion of the symmetrical anhydride reaction is estimated as 15–30 sec for couplings of most amino acids, and 3–8 min for the sterically difficult cases. The significant exception to this generalization was observed in the coupling of N^{α},N^{ω}-disubstituted arginine derivatives (Section V,B,1). The net rate was some 200-fold slower than that of most symmetrical anhydride couplings, and 15- to 30-fold slower than corresponding couplings mediated by dicyclohexylcarbodiimide. To explain these results, the initial anhydride was assumed to rapidly react intramolecularly to form an internal δ-lactam (Fig. 16). The latter species will still be a moderately active acylating agent, but the time for 99% completion of the reaction was increased to 40–80 min.

The symmetrical anhydride method has been criticized for being wasteful in that one equivalent of the protected amino acid to be incorporated into the peptide chain is released as a by-product of the acylation [Eq. (16)]. However, the starting material can be recovered in 80 to 90% yields by treatment of the filtrate with aqueous sodium bicarbonate, followed by acidification and extraction (Flor *et al.*, 1973).

b. Side Reactions. Intramolecular rearrangement of symmetrical anhydrides **86** gives bis-protected dipeptidyl derivatives **105** [Eq. (17); Weygand and Reiher, 1955; Tadema *et al.*, 1964; DeTar *et al.*, 1966]. This side reaction, which occurs preferentially with sterically unhindered amino acid residues such as glycine, is first of all liable to reduce the concentration of the active acylating species. Second, the carboxyl group of the dipeptide **105** can be reactivated by anhydride exchange, leading to incorporation of

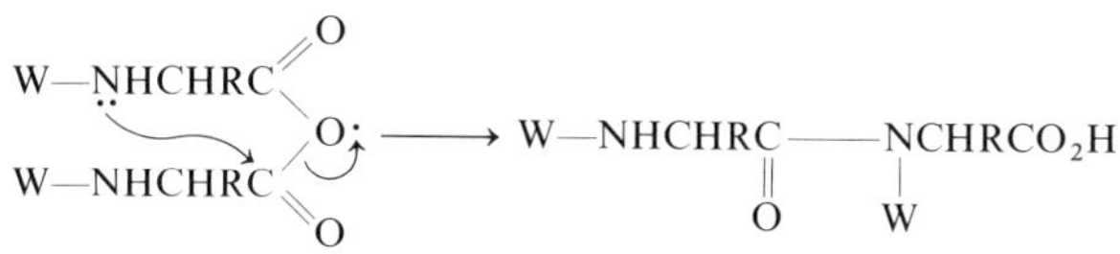

(17)

(86) **(105)**

an additional equivalent of the amino acid residue into the peptide chain. The relevance of these two problems to solid-phase synthesis was studied by Merrifield *et al.* (1974), who showed that when Leu-Ala-Gly-Val was synthesized by the mixed anhydride method, the crude product mixture included 5–8% Leu-Ala-Gly-Gly-Val. The half-life of the activated species from either an equimolar mixture of Bpoc-glycine and DCC, or the preformed symmetrical anhydride of Bpoc-glycine, was found to be approximately 2 hr at 25°C, as measured by delayed trapping with valine-resin. In spite of the substantial amount of rearranged diglycyl derivative formed during this time and recovered in solution, only a trace ($\leqslant$0.3%) was in an activated form which reacted with valine-resin to give Gly-Gly-Val-resin. However, when triethylamine hydrochloride was added to the preformed symmetrical anhydride to simulate the conditions of the mixed anhydride method, activation of the diglycyl derivative occurred and Gly-Gly-Val comprised 7 to 15% of the peptide product mixture. These data suggest that urethane acylation and net incorporation of an additional amino acid residue need not be a concern under the standard conditions of solid-phase peptide synthesis, using either the standard DCC or the symmetrical anhydride method. Even the mixed anhydride method is acceptable if careful control of temperature is observed. This conclusion was confirmed by Edman degradation of a peptide synthesized by the solid-phase method with symmetrical anhydride activation (Atherton *et al.*, 1977). Extra glycine residues that might have been inserted by this mechanism could not be detected.

3. Active Esters

Following the early success of Bodanszky and Sheehan (1964), a substantial number of active esters (reviewed by Bodanszky, Vol. 1, Chapter 3) have been tested for use in solid-phase peptide synthesis. The rates for coupling active esters are two to four orders of magntiude slower than those for dicyclohexylcarbodiimide- or anhydride-mediated reactions; hence, the use of active esters has generally been limited to special applications. Most important among these are (1) incorporation of side-chain unprotected asparagine and glutamine without dehydration to the cyano derivatives, and (2) low racemization during coupling of small protected peptide segments as their preformed active esters. The use of active esters formed *in situ* by reaction of dicyclohexylcarbodiimide and suitable additives containing hydroxyl groups has already been covered (pp. 131–136).

tert-Butyloxycarbonylamino acids can be incorporated into peptide chains through their *p*-nitrophenyl esters (Bodanszky and du Vigneaud, 1959); the reaction solvent is generally dimethylformamide. The actual concentration of the ester, rather than the nominal number of equivalents over resin-bound amino groups, will determine the coupling rate. With concentrations ranging from 0.1 to 0.4 *M* or more (nearly saturated), the half-times of the pseudo first-order couplings with most amine nucleophiles at 25°C can be roughly stated as 0.5 to 1.5 hr, which means that the reactions usually need to be carried out for 8 to 18 hr (e.g., data of Karlsson *et al.*, 1970a; Kovács and Penke, 1973; Yamashiro and Li, 1974; Upson and Hruby, 1976). Sterically restricted active ester couplings, however, can take much longer. Following observations in solution due to König and Geiger (1973), the practice of carrying out solid-phase couplings with *p*-nitrophenyl esters in the presence of an equivalent amount of 1-hydroxybenzotriazole (**104**) has permitted reduction of the reaction period to 3.5–5 hr (Atherton *et al.*, 1975; Upson and Hruby, 1976, Khan and Sivanandaiah, 1978a).

Stepwise solid-phase synthesis has been carried out with other active esters, including phthalimido esters (Nefkens and Tesser, 1961; Merrifield, 1965), 2,4,5-trichlorophenyl esters (Pless and Boissonnas, 1963; Kenner *et al.*, 1971), pentachlorophenyl esters (J. Kovacs *et al.*, 1966; Visser *et al.*, 1970a; Johnson, 1973), pentafluorophenyl esters (K. Kovács and Penke, 1973), *o*-nitrophenyl esters (Bodanszky and Funk, 1973), and 5-chloro-8-hydroxyquinoline esters (Jakubke and Baumert, 1974b). Less widely used bifunctional catalysts include 1,2,4-triazole (Beyerman *et al.*, 1968a; but compare to Ragnarsson *et al.*, 1970b), imidazole (Li and Yamashiro, 1970), and 3-hydroxy-4-oxo-3,4-dihydroquinazoline (**106**) (Wissmann *et al.*, 1973).

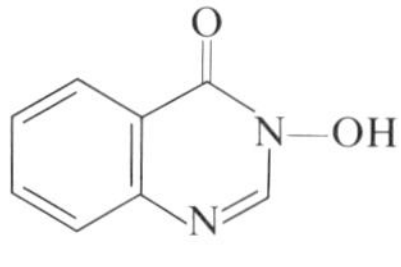

(106)

Toluene has been advocated to be a superior solvent (Khan and Sivanandaiah, 1976; see also Bodanszky and Fagan, 1977, but compare to Merrifield, 1963). The effective solution chemistry recently reported regarding the activating agent 4,6-diphenylthieno[3,4-*d*]dioxol-2-one 5,5-dioxide (**107**) [Hollitzer *et al.*, 1976; Eq. (18)] might also warrant examination for stepwise solid-phase synthesis. The reaction can be carried out in one step starting with both carboxyl and amino components, or the activated enol ester **108** can be isolated and crystallized beforehand.

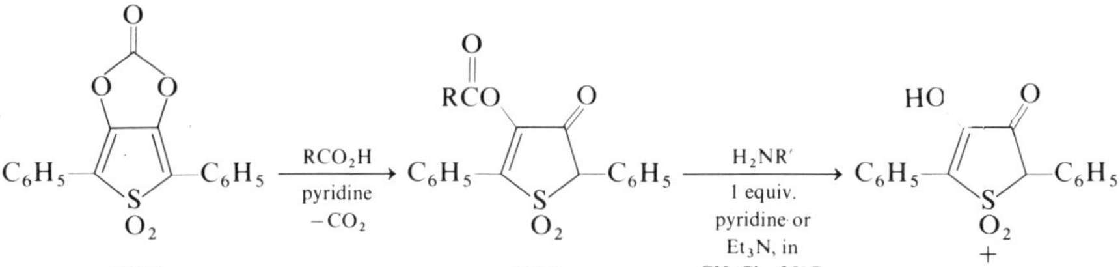

(107) **(108)**

(18)

The esters of peptides with *N*-hydroxysuccinimide (**101**) are relatively stable, isolable compounds which can be purified and subsequently used as acylating agents. This approach was first illustrated in a solid-phase system by Weygand and Ragnarsson (1966) who were able to achieve quantitative incorporation, with negligible racemization, of the dipeptide Moz-LLeu-LPhe-OSu (0.1 *M* in CH_2Cl_2, 25°C for 24 hr). Sklyarov and Shashkova (1969, see Table I) first assembled the linear sequence of a gramicidin by solid-phase synthesis employing side-chain anchoring through a lysine residue. The methyl ester protecting the COOH-terminal glycine residue was removed by saponification, and the polymer-bound peptide was converted to its *N*-hydroxysuccinimide ester. After removal of the N^{α}-*tert*-butyloxycarbonyl (Boc) protecting group, the desired intramolecular cyclization was effected by heating the peptide–resin at 80°C for 6 hr in 2% pyridine–dioxane. Neubert and Jakubke (1975) worked out a solid-phase synthesis of a heptapeptide through three successive couplings (0.08 *M* carboxyl component in CH_2Cl_2, 25°C for 22 hr) of Boc-dipeptidyl-5-chloro-8-hydroxyquinoline esters (**109**), each of which in turn was prepared in solution by the "backing-off" procedure (Goodman and Stueben, 1959).

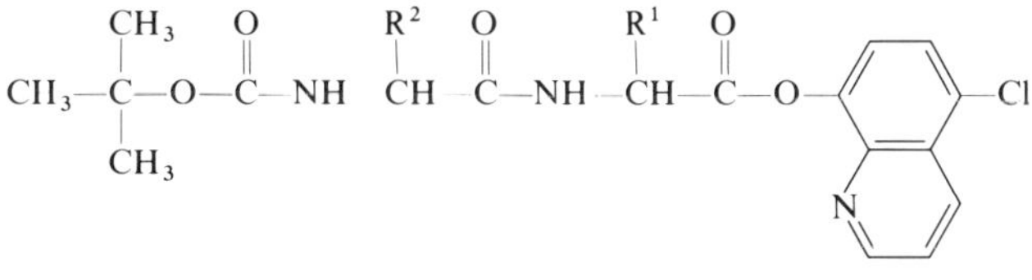

(109)

It was concluded that for the special case under consideration, the segment condensation approach had no advantages over the standard stepwise procedure using DCC-activation and coupling.

4. Oxidation–Reduction Condensation

The novel oxidation–reduction method (Fig. 13) for peptide bond formation, developed by Mukaiyama *et al.* (1968, 1970; reviewed in this volume, Chapter 5) has permitted a variety of important solid-phase applications. For stepwise synthesis, equimolar amounts of an N^{α}-protected amino acid, 2,2′-dipyridyl disulfide (**110**) and a triarylphosphine (**111**) are dissolved in CH_2Cl_2 or DMF, all in amounts representing 3 to 10 equivalents over the initial loading of the support. The reactants are added to the peptide–resin in the given order, and the reaction is allowed to proceed at 25°C for 1 to 6 hr (e.g., Matsueda *et al.*, 1973; Birr *et al.*, 1975).

A concerted mechanism has been proposed (Mukaiyama *et al.*, 1970) for this reaction, involving **112**, but a 2-pyridyl thioester (**113**) or even a symmetrical anhydride generated in a rate-limiting step, could also be considered as possibilities for the true acylating species. The discovery that disulfide **110** is a suitable oxidant for the reaction was a significant breakthrough since the product 2-mercaptopyridine (**58**) tautomerizes to the thione form (Fig. 13) and therefore does not require scavenging with heavy metal salts or vinyl compounds, which had characterized some earlier formulations

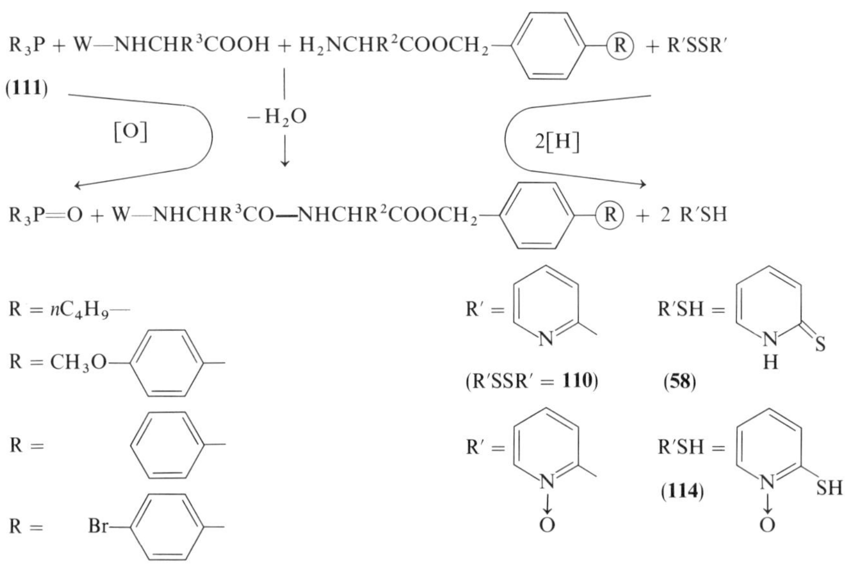

Figure 13. Schematic course of oxidation–reduction condensation in solid-phase peptide synthesis. The reducing phosphine R_3P and the disulfide R′SSR′ that serves as the net oxidizing agent are ranked in the order of reactivity. Thus, combinations of the compounds at the tops of the lists result in the most rapid couplings. The most reactive combinations also have the corresponding greatest tendency to lead to racemization in the carboxyl component.

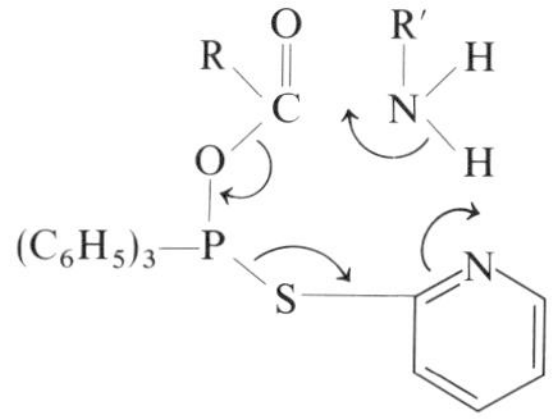

(112)

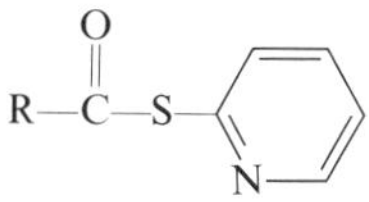

(113)

of the method. Both the pyridinethione and triarylphosphine are readily soluble in organic solvents and are readily removed from the solid phase by washing.

The available kinetic data from our laboratory indicate that oxidation–reduction couplings with these reagents proceed at a rate about 10-fold slower than those mediated by dicyclohexylcarbodiimide. Birr *et al.* (1975) used the more reactive trianisylphosphine (Fig. 13) in combination with 2,2′-dipyridyl disulfide (**110**) to mediate the stepwise synthesis of a decapeptide. The recorded pattern by continuous flow ultraviolet absorbance monitoring suggested two overlapping phases of the mechanism, namely uptake of starting materials and release of products. Hence, quantitative analysis of the data was not possible; but qualitatively, a half-time of the order of 5 min seemed to be indicated. The peptide product was also evaluated by Edman degradation-preview analysis (p. 156ff), and was found to 95.7% pure. This corresponds to an average efficiency of 99.5% per deprotection/coupling cycle; however, the yields at each step were not necessarily the same.

In contrast to the standard DCC procedure, side-chain unprotected asparagine and glutamine derivatives can apparently be safely incorporated into peptides by the oxidation–reduction method without the danger of dehydration. The phosphine component of the coupling milieu will remove the *o*-nitrophenylsulfenyl- or dithiasuccinoyl α-amino protecting groups (Section IV,A,4 and 6), so these are incompatible with the methodology. Nonetheless, this latter fact can occasionally be used to advantage (e.g., Mukaiyama *et al.*, 1968).

The oxidation–reduction method has been used to achieve segment condensations on solid supports. Yields have been excellent and racemization has been low. An especially noteworthy feature is the possibility for peptide chain elongation in the N → C direction (e.g., Matsueda *et al.*, 1973, 1975; Maruyama *et al.*, 1976; see Table I). Typically, the α-carboxyl group of the polymer-supported peptide anchored through its α-amino group (Section III,B,10) is deprotected, and three equivalents of the incoming protected peptide segment with a free amino group are added. Couplings are mediated with 1.0 *M* each of 2,2′-dipyridyl disulfide, 2-mercaptopyridine, and triphenylphosphine (added last) in CH_2Cl_2, at −15°C for 6 hr, and overnight at 25°C. Peptide segments containing COOH-terminal glycine or proline are coupled immediately at 25°C and the mercaptopyridine is omitted. Progress of the reactions can be monitored by quantitation of the soluble amino component, and the half-times have been reckoned as ranging from 1 to 2 hr. Segments of up to 5 amino acid residues have been tested. Dimethylformamide, which is a better solvent than CH_2Cl_2 for protected peptides, was not recommended for oxidation–reduction couplings.

The conditions developed represent a compromise between coupling yields and extent of racemization. In a recent study, Matsueda *et al.* (1977) documented the structural effects of the oxidation–reduction components on the solid-phase synthesis of the model tetrapeptide Leu-Ile-Asn-Leu from the suitable dipeptidyl precursors. The observed trend, incorporating some earlier results (Mukaiyama *et al.*, 1970; Ueki and Shishikura, 1973; Ueki and Ikeda, 1974) observed in solution, is summarized in Fig. 13. The amount of D-*allo*-isoleucine found in the hydrolysate of the peptide product using the standard oxidation–reduction segment condensation protocol outlined above ranged from 5 to 8%, irrespective of which end of the chain was anchored to the solid-phase. As a control, the D-*allo*-isoleucine found from a stepwise oxidation–reduction synthesis was 1.4%, which is probably close to the background racemization occurring during acid hydrolysis. In the key experiment, racemization was only 0.6% above this background when activation and coupling was performed with equimolar 2,2′-dithiodipyridine-1,1′-dioxide and triphenylphosphine, in the presence of two equivalents of 2-mercaptopyridine *N*-oxide (**114**), first at 0°C for 2 hr, followed by 25°C for 22 hr.

5. Other Coupling Methods and Reagents

The major techniques for peptide bond formation in solid-phase synthesis have already been covered in some detail. Table X lists additional preformed activated derivatives of the carboxyl component and additional *in situ* activating reagents. Some of these have been extensively tested and found to be of limited generality because of slow coupling rates, low yields, deleterious side reactions, or appreciable racemization, but may nevertheless serve for some special applications. Several promising new methods for which the scope and limitations are still under investigation have also been included in Table X.

The novel four component synthesis (reviewed by Ugi, 1962, 1971, and in this volume, Chapter 4) has been elaborated for (1) creation of two new peptide bonds and simultaneous *de novo* formation of the internal amino acid residue, and (2) condensation of peptide segments in a first-order

Table X. Miscellaneous Coupling Methods and Reagents for Solid-Phase Peptide Synthesis (footnotes on p. 148)

No.	Name/structure	Conditions	Remarks	References
Preformed activated species of carboxyl component[a]				
1.	Azide $R—C(=O)—N_3$	(i) Diazotization of resin-bound hydrazide (p. 118): *n*-butyl nitrite–HCl (5 equiv each) in THF, −30°C, 1 hr (ii) Neutralize; 10% Et_3N–THF (iii) Couple amino acyl–Boc-hydrazide (0.04 *M*) in THF, −30°C, 2 hr; 0°C, 4 hr; 25°C, 16 hr	Tested for stepwise $N \rightarrow C$ scheme (Table I), with urethane anchor (p. 97); terminating side reaction involving amide formation	Felix and Merrifield (1970), based on protocol of Honzl and Rudinger (1961)
		or Segment condensation: DMF solution of azide, start −20°C, final 5°C or 25°C, total several days	Yields in segment condensation variable (60–90%); rates are slow, but method is nearly racemization-free especially on solid-phase where all excess bases can be removed by filtration and washing	Segment condensation: Visser and Kerling (1970b); Weber and André (1975); general review, Meienhofer Vol 1, Chapter 4
2.	Mixed anhydride $R—C(=O)—O—C(=O)—OR'$ ($R' = C_2H_5$, *sec*-C_4H_9)	(i) Activation: $RCO_2H:R'OCOCl:Et_3N$ = 1:0.9:1; 0°C, 30 min in CH_2Cl_2 or toluene (ii) Coupling: same solution; 25°C, 2 hr; after filtering off $Et_3N\cdot HCl$	Rapid reaction rates, comparable to symmetrical anhydrides. "Wrong-way" opening of mixed anhydride might occur to small extent at each step, giving rise to terminated peptides. Urethane acylation and coupling of dipeptidyl unit has been observed (p. 139)	Letsinger *et al.* (1964); Tilak and Hollinden (1968b); Jakubke and Baumert (1974a); Merrifield *et al.* (1974); general review, Meienhofer Vol. 1, Chapter 6

Table X. *(Continued)* (footnotes on p. 148)

No.	Name/structure	Conditions	Remarks	References
3.	*N*-Carboxyanhydride and related 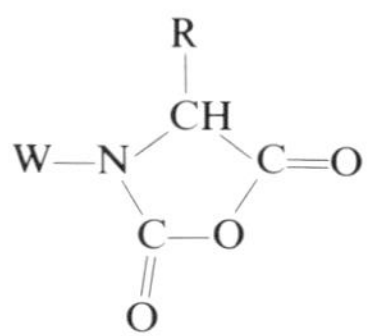 W = H (NCA)	(a) Polymerization: NCA (0.16 *M* = 8-fold excess over resin-bound amino initiator sites), in DMF at 25°C, 24 hr	Multiple cycles of addition because amino group regenerated after acylation. Probably no racemization[b]	Ingwall *et al.* (1968); based on Katchalski and Sela (1958)
		(b) Stepwise: NCA (0.03 *M*) in $CHCl_3$ + resin-bound *N*-trimethylsilylamine, 25°C, 90 min; followed by 1% HCl–EtOH and trimethylsilylation preparatory to repeat coupling cycle	Only one cycle of addition, because carbamate intermediate is trapped as its trimethylsilyl derivative, which is subsequently deprotected in dilute acid	Maher *et al.* (1972); based on Kricheldorf and Greber (1971)
	W = 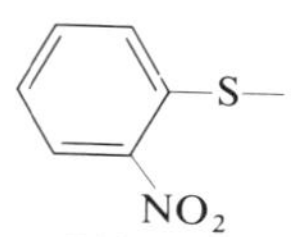 (Nps—NCA)	(c) Stepwise: Nps–NCA (0.25 *M*)–CH_2Cl_2, 25°C, 4 hr	Base-catalyzed racemization hazard during preparation of derivatives. Not applicable to proline residues. See Section IV,A,4, pp. 109–112 on Nps protecting groups	Halstrøm *et al.* (1976); Christensen *et al.* (1976); based on Kricheldorf and Fehrle (1974); see also Katakai (1975)

	Activating reagents			
4.	Isoxazolium Salts 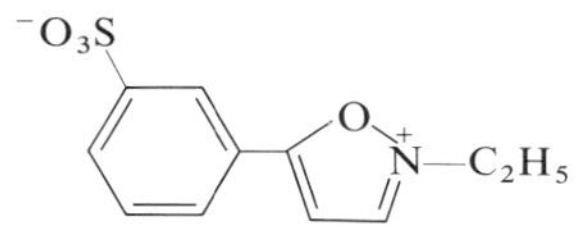 *N*-ethyl-5-phenylisoxazolium-3′-sulfonate (Woodward K or NEPIS) compare to 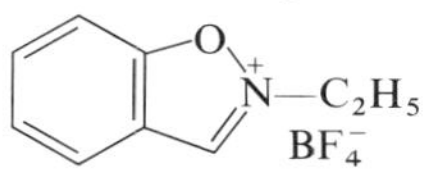 (Kemp) to date only used in solution.	RCO_2H:NEPIS:Et_3N = 1:1:1 in DMF, 25°C, 4 to 48 hr	Reaction mechanism proceeds through enol ester, which can rearrange to give a yield-diminishing imide side reaction. In solid-phase synthesis, yields are variable (20–95%) both in stepwise and segment condensation strategies; racemization is also dependent on conditions	Woodward *et al.* (1961); reviewed by Woodman (1974); compare to Kemp and Chien (1967). For solid-phase: Khosla *et al.* (1968, 1974); Jakubke and Baumert (1974a); Omenn and Anfinsen (1968)
5.	1-Alkoxycarbonyl-2-alkoxy-1,2-dihydroquinoline 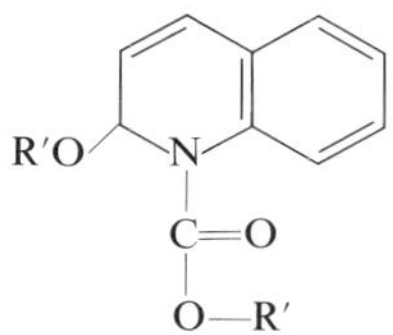 R′ = C_2H_5 (EEDQ) R′ = *sec*-C_4H_9 (IIDQ)	Reagent equimolar to RCO_2H, in DMF or DMF–CH_2Cl_2, 25°C, 24 hr or longer	Active intermediate is the mixed anhydride (compare to No. 2; quinoline is the by-product) which is generated slowly but reacts rapidly. Nitrile formation was not observed during stepwise incorporation of side-chain unprotected Asn and Gln. Racemization in segment condensation is 3–8% in model systems	Belleau and Malek (1968); Sipos and Gaston (1971); Yajima and Kawatani (1971); Barton *et al.* (1973); Kiso *et al.* (1973); C. T. Wang *et al.* (1974); Protein Synthesis Group (1975)

Table X. *(Continued)*

No.	Name/structure	Conditions	Remarks	References
6.	*o*-Phenylene chlorophosphite 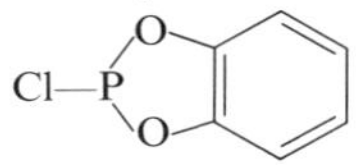 (PCP)	RCO_2H:PCP:Et_3N = 1:1:1, in diethyl phosphite, activate at <25°C, couple at 100°C	Used to prepare nylon-like oligomers	Rothe *et al.* (1966), based on Anderson and Young (1952)
7.	Diethyl phosphorocyanidate 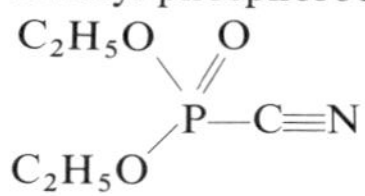 (DEPC)	RCO_2H:DEPC:Et_3N = 1:1:1, in DMF, 25°C, 2 hr (stepwise) to 48 hr (segment)	Reactions rapid, provided DMF, not CH_2Cl_2, is solvent. Racemization only slight (1% by Izumiya test on solid-phase; 5% by Young test in solution)	Yamada *et al.* (1975); see also Yamada *et al.* (1973); Shioiri *et al.* (1976)

[a] In structures for numbers 1 and 2, RCO_2H represents the carboxyl component that is activated. In structure for number 3, the original amino acid is $H_2NCHRCO_2H$.

[b] Stepwise controlled addition of NCA's is recognized as a standard for racemization-free coupling in solution (Hirschmann *et al.*, 1967; see also Manning and Moore, 1968). Disadvantages include marginal stability characteristics, and possible "wrong-way" opening to give ureido derivatives. The sulfur analogues 1,3-thiazolidine-2,5-diones (NTA's) are more stable than NCA's but racemization occurs during coupling (Dewey *et al.*, 1971). Hence, the latter derivatives would not be recommended for solid-phase synthesis.

(concentration independent) intramolecular rearrangement. This latter mode could potentially be profitably adapted to solid-phase synthesis. Theoretical advantages have also been discussed for the "amine-capture" method of peptide synthesis, so far tested only in solution (Kemp and Vellaccio, 1975).

D. Monitoring

Table XI lists some of the available techniques for following the course of individual chemical steps of the deprotection/coupling cycles of solid-phase peptide synthesis. Several of these provide information about the overall substitution level of growing peptide chains anchored to the polymeric support, while others are designed to rationally evaluate endpoints of reactions. Methods that are irreversible or destructive need to be carried out on aliquots of the peptide–resin, but nondestructive methods can be applied to the entire batch of peptide–resin as well. The ideal monitoring method should be rapid, accurate, sensitive, and easily automated; it should not interfere with the course of the synthesis. These characteristics are covered below and in Table XI; interesting critical discussions are also available elsewhere (Hirt *et al.*, 1973; Hodges and Merrifield, 1975a; Brunfeldt, 1977; Birr, 1978).

1. Experimental Approaches

"Real-time" monitoring is achieved by following reactions as they occur in the entire heterogeneous system of the peptide–resin and soluble components. In practice, such monitoring has only been possible by continuous-flow methods relying on either spectrophotometric quantitation of the uptake of soluble reactants, or the release of soluble products (Table XI; number 1B, see also number 3). An alternative approach to monitoring involves interruption of the reaction, filtration, thorough washing of the peptide–resin, and finally analysis of the recovered soluble components (numbers 1A and 2). In this case, the speed of the method depends on the washing time, as well as the adjunct analytical system. The continuous and discontinuous batchwise methods just described are advantageous because they do not alter the fundamental schedule of chemical steps of the synthesis. In contrast, monitoring of resin-bound components requires additional chemical reactions, since to date there are no accurate spectroscopic probes for this purpose (see, however, application of radioactivity, number 2). Nondestructive analytical methods carried out on the entire batch of peptide–resin (numbers 4–6) add between 30 min and 2 hr to each synthetic cycle. Destructive analyses on aliquots of the peptide–resin (numbers 7–12) can take from 10 min to 24 hr depending on the chemistry involved. Of these, only the rapid, qualitative techniques (numbers 7–10) are suited for routine use to detect completion of coupling reactions and to judge whether or not

Table XI. Selected Methods for Monitoring a Solid-Phase Peptide Synthesis

No.	Reagent/conditions	Measure	Remarks	References
		MONITORING ON TOTAL BATCH		
		Analysis of Soluble Reactants or Products (Continuous or Discontinuous)		
1.	W—NHCHRCO—X uptake of W upon coupling release of X upon coupling release of W upon deprotection	(A) *Discontinuous mode*: Thorough washing of peptide–resin; measure uv	Discontinuous best for accurate quantitation	(A) Bodanszky and Sheehan (1966); Najjar and Merrifield (1966); Wang and Merrifield (1969a); Yamashiro *et al.* (1972); Wolters *et al.* (1974)
		(B) *Continuous mode*: Use circulating flow system	Continuous mode, "real-time" Need to follow at isosbestic point if both W and X absorb in uv	(B) Gut and Rudinger (1968); Merrifield (1973c); Birr (1975, 1978); Birr *et al.* (1975); Gut (1975); Rudinger and Buetzer (1975)
	X = *p*-nitrophenyl (starting material) (product) X = 5-chloro-8-quinoline X = 2-thiopyridine W = *o*-nitrophenylsulfenyl W = Bpoc W = Ddz	 uv: 270 nm ($\varepsilon = 10^4$) uv: 314 nm ($\varepsilon = 1.1 \times 10^4$) uv: 330 nm uv: 410 nm uv: 384 nm ($\varepsilon = 4 \times 10^3$) uv: 254 nm ($\varepsilon = 2.2 \times 10^4$) uv: 280 nm ($\varepsilon = 2.3 \times 10^3$) and 224 nm ($\varepsilon = 7.2 \times 10^3$)	These methods are also suited in principle to ir, or other spectroscopic probes, depending on the chromophores of W and X	
2.	[^{14}C]Boc-NHCHRCO$_2$H	Liquid scintillation counting, with suitable corrections for quenching of organic solvent or polymer (combustion prior to counting will circumvent the quenching problem)	Determination of radioactivity on peptide–resin is check for complete deprotection, provided *tert*-butylation side reactions do not take place. This method is expensive and impractical for routine use.	Krumdieck and Baugh (1969)

3.	Unreacted protected peptide segment in polymer-supported segment condensation by oxidation–reduction coupling	Small aliquot of circulating solution withdrawn and quantitated in void volume of high pressure liquid chromatogram. Analysis time = 15 min, including elution of 1-nitronaphthalene internal standard	This discontinuous assay, plotted as a function of time, gives the kinetic profile. Stated accuracy = ± 0.3–1.4%	Maruyama *et al.* (1976)
		MONITORING ON TOTAL BATCH (occasionally ALIQUOT)		
		Analysis of Resin-Bound Reactants or Products		
4.	Pyridine hydrochloride (0.3 *M*) in $CH_2Cl_2 \rightarrow$ hydrochloride salt of resin-bound amino groups, as well as N^{im}-protected histidine residues	Elute Cl^- with Et_3N–CH_2Cl_2; determine chloride in filtrates potentiometrically with $AgNO_3$. Time = 20 min washings and 10–20 min for titration	Pyridine hydrochloride is a sufficiently strong acid to partially remove N^{α}-Nps, N^{α}-Bpoc and N^{im}-Tos groups. Spurious chloride from Et_3N–$CHCl_3$ solutions, or the presence of quaternary ammonium sites on the resin, will lead to unjustifiably high backgrounds. Sensitivity = 2 μmol	Dorman (1969); see also Klostermeyer and Schwertner (1973); Beyerman *et al.* (1974)
5.	Perchloric acid (0.04 *N*) in acetic acid $\rightarrow$ titration of α-amino groups and other basic centers as in No. 4, above	Potentiometric titration with glass and calomel electrodes inserted directly into reaction vessel. Time = 2 hr, including washes and duplicate determinations In aliquot mode, minimum 250 mg peptide–resin required A fully automated, computer-controlled mode has been developed	Acid sensitive groups may be partially lost in perchloric acid; acetic acid cosolvent can give acetylation of α-amino groups. Stated accuracy = ± 2 μmol	Brunfeldt *et al.* (1969b; 1972a); Schou *et al.* (1975); Schou (1976); Villemoes *et al.* (1976)

Table XI. *(Continued)*

No.	Reagent/conditions	Measure	Remarks	References
6.	Picric acid (0.1 *M*) in CH_2Cl_2 or imidazolium picrate (0.1 *M*) in DMF–CH_2Cl_2 (1:9) yields picrate salt of resin-bound amino groups 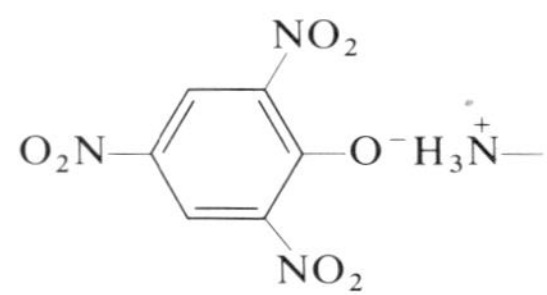	Elute picrate with diisopropylethylamine in CH_2Cl_2 (direct neutralization) or with pyridine hydrochloride (0.1 *M*) in CH_2Cl_2 (ion-exchange); collect washes, dilute with 95% ethanol, and measure uv: 358 nm ($\varepsilon = 1.45 \times 10^4$) Time = 45 min The method has been fully automated	Advantages and limitations comparable to Nos. 4 and 5. Imidazolium picrate preferred because it allows monitoring N^{im}-protected histidine peptides, and does not cleave the N^{α}-Bpoc group. Rising background with increased chain length on peptide–resin largely eliminated by thorough washing. Stated precision = ±0.2%; Sensitivity = 0.2–1.0 μmol	Gisin (1972); Hancock *et al.* (1975); Hodges and Merrifield (1975a)
		MONITORING ON ALIQUOTS Analysis of Resin-Bound Reactants or Products		
7.	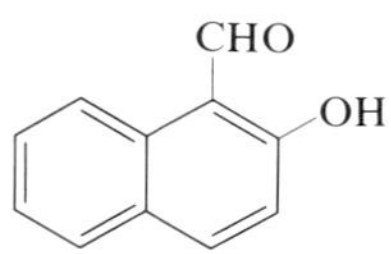 2-Hydroxy-1-naphthaldehyde (50-fold excess ~0.2–1 *M*), 25°C in ethanol = 12 hr in dioxane = 1 hr in CH_2Cl_2 = 2 hr gives Schiff's base **54** with α-amino groups (10 mg peptide–resin samples)	Release with benzylamine (0.4 *M*) in CH_2Cl_2, 30 min Measure uv: 420 nm ($\varepsilon = 10^4$)	Relatively time-consuming method, but furnishes accurate and reproducible results. Cannot be performed batchwise because regenerated NH_2-terminal residue partially racemizes. Sensitivity ~ 0.5 μmol. Not applicable to monitoring of amino-terminal proline	Esko *et al.* (1968); Esko and Karlsson (1970); see also Corley *et al.* (1972); Yamashiro *et al.* (1972); Eastlake *et al.* (1976)

8.	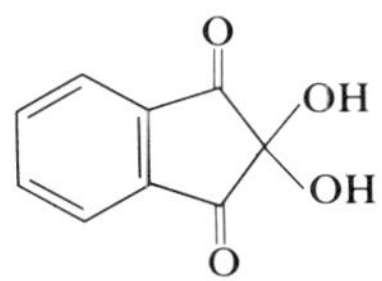 Ninhydrin reagent, 100°C, 5 min (10 to 20 mg peptide–resin samples)	Qualitatively assess blue color of beads and solution. Slight positive with 5 μmol amino groups/gm resin = 99% coupling with initial load of 0.5 mmol/gm	Occasionally gives false negative results. False positives can result from reaction of ninhydrin with by-products of coupling reaction. Amino-terminal proline or Asp(OBzl) give brown or reddish-brown color, difficult to distinguish from background color of resin	Kaiser *et al.* (1970); Hirt *et al.* (1973); Tregear (1975)
9.	Fluorescamine. See p. 161 for structures and further references. Aliquot of peptide–resin (5 to 10 mg) collected in Pasteur pipette with glass wool plug, washed, neutralized, washed, and finally soaked with a few drops Et_3N–$CHCl_3$ (1:9). React with 1 ml of reagent (3 mg/ml), 25°C, 10 min, then thoroughly wash and filter beads	Qualitatively assess fluorescence, compared to blank, when viewed under long wave uv light (366 nm). Test still positive with 0.6 μmol amino group/gm resin = 99.9% coupling with initial load of 0.5 mmol/gm resin	10-fold more sensitive than No. 8. False positives are possible with spurious fluorophors; false negatives are possible when fluorescence is quenched for chemical (lactonization of initial adduct at elevated temperature or in acid) or physical (as with nitrated polystyrene resins) reasons. Amino-terminal proline gives a nonfluorophoric product	Felix and Jimenez (1973); A. M. Felix, personal communication, 1976; see also Smith (1976); Stewart (1976, p. 277)
10.	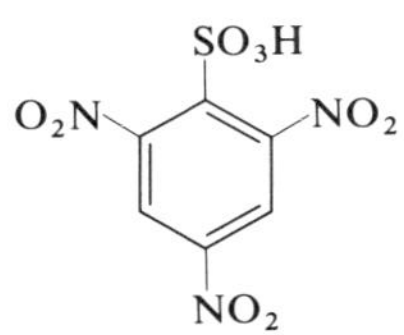 2,4,6-Trinitrobenzenesulfonic acid yields trinitrophenylated resin-bound amine. Reaction time = 10 min (1 mg peptide–resin sample)	Qualitative microscopic examination of beads: orange-red color with 3 μmol amino groups/gm resin	Sensitivity comparable to No. 8, but has advantage in that here, supernatant solution is colorless	Hancock and Battersby (1976)

Table XI. *(Continued)*

No.	Reagent/conditions	Measure	Remarks	References
11.	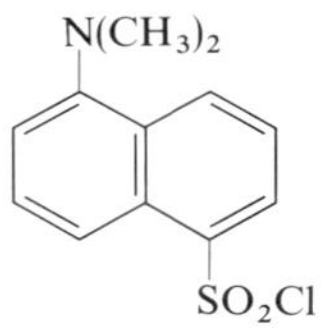 5-Dimethylamino-1-naphthalenesulfonyl chloride (Dns-Cl) yields fluorescent Dns-peptide-resin. (25 mg peptide-resin)	Hydrolysis of Dns-peptide-resin (6 *N* HCl, 110°C, 24 hr), followed by thin layer chromatography of Dns-amino acids from end-group *or* (for quantitation) hydrazinolysis to remove Dns-peptide and direct fluorescence measurement (exitation 350 nm; emission 517 nm)	End-group methodologies are sensitive, but time-consuming. They are expected to be particularly useful to follow polymer-supported segment condensation	Zhukova *et al.* (1970); Garden and Tometsko (1972); based on Gray (1967)
12.	Acid hydrolysis: (a) 12 *N* HCl-acetic acid-phenol (2:1:1) (v/v), 110°C, 24 hr *or* (b) 12 *N* HCl–propionic acid (1:1) (v/v), 130°C, 2 hr *or* (c) methanesulfonic acid–propionic acid–H_2O (5:12:3) (v/v), 110°C, 22 hr (5 to 10 mg samples of peptide–resin)	Workup and quantitative amino acid analysis. (a) filter, wash with 1 *N* HCl, extract with $CHCl_3$ to remove phenol degradation products, evaporate aqueous phase, take up in buffer, and Millipore filter. (b) direct rotary evaporation of filtrate (c) neutralization of hydrolysate filtrate with 3.5 *N* NaOH and dilution to volume with H_2O	Hydrolysis mixture must swell peptide–resins for quantitative recoveries. (a) is most accurate and reproducible; (b) is quicker but not recommended with labile amino acids: (c) is especially valuable with labile amino acids. See footnote[†] on p. 169 for further remarks and references.	(a) Gutte and Merrifield (1971); (b) Westall *et al.* (1972); (c) Hodges and Merrifield (1975b) Problems with earlier techniques discussed by Stewart and Young (1969, p. 53); Erickson and Merrifield (1976, p. 311)

to continue with the next step of the synthesis. A further qualitative or semi-quantitative monitoring approach involves release of peptides from the solid support by acidolysis, hydrazinolysis, or alcoholysis at intermediate stages of the assembly, followed by product analysis in solution by chemical (end-groups and sequence) and chromatographic techniques. The details of these methods, which provide useful checkpoints for lengthy or difficult synthetic endeavors, are beyond the scope of this section.

2. *Quantitation*

The substitution level of growing polymer-supported peptide chains is needed to calculate necessary excesses of soluble reactants, and to establish the extents of termination, branching, and premature cleavage side reactions. The experimental determination of resin-bound components involves a new series of heterogeneous chemical monitoring reactions, which must satisfy the same criteria for quantitativeness as the solid-phase deprotection and coupling reactions they are designed to control. For most applications, substitution levels need to be known only to an absolute accuracy of $\pm 10\%$, which is well within the range of a variety of quantitative methods. Discrete batchwise determinations of soluble products (number 1A), or of acidic titrants (numbers 4–6) applied to measure free resin-bound amino groups after deprotection, are particularly well suited for quantitative comparative studies. In contrast, continuous-flow measurements of these components give relative and not absolute concentrations because of partition effects (p. 19). Quantitations by the assorted irreversible or destructive methods (numbers 7, 11, 12) are subject to sampling errors arising from variable drying and weighing of the resin. The use of internal standards can sometimes help in this regard. Furthermore, in order to make a valid comparison with the initial loading of the support, substitution levels experimentally determined on aliquots of a peptide–resin must be corrected in the obvious way for the weight gain due to the peptide component.

3. *Evaluation of Reaction Endpoints*

Continuous-flow monitoring of the incorporation of a soluble reactant into growing polymer-supported peptide chains will give the necessary raw data to determine second-order rate constants for coupling reactions. In principle, effective endpoints can be calculated by extrapolation of measured rate constants, assuming that these are uniform through $>99.5\%$ completion of reactions (p. 29). However, these methods are inherently insensitive for direct judgments of endpoints. To illustrate this point, the absorbance of an activated incoming amino acid originally in two fold excess over the resin-bound α-amino groups will drop from 2.0 units to 1.05 and 1.00 units if the coupling yields are 95% and 100%, respectively. The difference of 5% between the false and true endpoints is near the precision limit of most spectroscopic techniques. Monitoring of the appearance of a soluble product of a deprotection or coupling reaction, where the measured change ranges from 0 to 0.95 or 1.00 units, is of insufficient precision for the same reasons. Direct measurement of resin-bound products, such as α-amino groups appearing after deprotection or new residues incorporated by coupling, is even more difficult and has a similar margin of uncertainty.

Methods designed to measure the amount of unreacted resin-bound peptide chains are inherently more sensitive for the evaluation of endpoints of coupling reactions. For example, the titer of free α-amino groups will vary over a fivefold range, 0.05 vs. 0.01 units, when 95% and 99% yields, respectively, have been achieved in the coupling steps. Relatively large errors in the quantitative analysis are reflected by only a small error in the calculated extent of coupling. The sensitivity limits of various suitable quantitative and qualitative tests are given in Table XI (numbers 4–10). In a number of instances, repeated couplings have resulted in constant but high background levels that are well above the stated sensitivity limits. This may be due either to nonspecific binding or irreversible reaction of the titrant or analytical probe with the peptide moiety, or to inefficient washing procedures.

4. Analysis of Deletion Peptides

Deletion peptides arise in stepwise solid-phase synthesis when a deprotection or coupling reaction fails go to completion for a temporary reason, either physical or chemical. The levels of the deletion peptides Leu-Gly-Val and Leu-Ala-Val formed upon synthesis of the model tetrapeptide Leu-Ala-Gly-Val can be determined directly, following cleavage from the support, by ion-exchange chromatography on standard amino acid analyzers (Dorman *et al.*, 1971; Merrifield *et al.*, 1974). The average level of incomplete reactions in model peptides can be amplified by selecting sequences with repeating amino acids, preferably charged (Erickson and Merrifield, 1973b; Yamashiro *et al.*, 1976). For example, incorporation of 10 residues of Boc-Lys(Z)-OH onto a valine-resin at an average overall efficiency of 99.4% per deprotection/coupling cycle gave 5.0 mol% of Lys_9Val, easily resolved from the main product $Lys_{10}Val$ by chromatography on carboxymethylcellulose employing a sodium chloride gradient (Erickson and Merrifield, 1973b).

The methods described in the above paragraph are not general, and also require cleavage of the peptide from the support prior to analysis. An elegant sensitive method, termed "preview Edman degradation," has recently been developed which circumvents these objections (Niall *et al.*, 1972; Fankhauser *et al.*, 1974; Birr and Frank, 1975; Tregear, 1975; Tregear *et al.*, 1977; for applications, see additionally Birr *et al.*, 1975; van Rietschoten *et al.*, 1975b;

Atherton *et al.*, 1977; Rosenblatt *et al.*, 1977a). Chemically, the method is based on serial reactions of peptides with phenylisothiocyanate, followed by acid-catalyzed cleavage and loss of the NH_2-terminal residue as the thiazolidinone, later converted to the phenylthiohydantoin for identification purposes (Edman, 1950, 1970; applied to solid-phase systems by Laursen, 1971). In earlier work, a single cycle of the Edman chemistry was applied as an end-group method to monitor solid-phase peptide syntheses (Weygand and Obermeier, 1968; Niall *et al.*, 1972; Geising and Hörnle, 1973).

The principles of the preview technique are illustrated in the hypothetical case of Fig. 14. Suppose that the desired product A–B–C–D–E–F–Ⓡ comprises 90% of the peptide–resins, that deletions of A, B, and C have occurred at 0.4%, 0.4%, and 0.2%, respectively, and that due to a significant error at the step to incorporate D, 9% of the chains lack this latter residue. After one cycle of Edman degradation, the residue B is quantitated by sensitive analytical techniques at 0.4% of the expected A, even though B should only be "viewed" in the second cycle. Thus, the 0.4% "preview" of B reflects that amount of deletion of A. The second cycle of Edman degradation in this example shows the amplification effect inherent in the method, because now 0.8% preview of C is noted. In fact, as the third cycle shows, it is usually technically difficult to ascribe low levels of deletion to specific positions in the chain. At this stage of the analysis, it is justified to conclude an average deletion of 0.3% per synthetic cycle. The precision in determining average deletion is improved with an increased number of Edman steps. The fourth degradation of the example illustrates the dramatic jump in preview reflecting the failure in the corresponding stage of the synthesis. Because this level is large, the specific location of the deletion can be identified. The level of the last previewed residue will indicate the total amount of deletion peptides

	Edman cycle →		1	2	3	4
Correct	A–B–C–D–E–F–Ⓡ	90 %	A	B	C	D
des (A)	B–C–D–E–F–Ⓡ	0.4%	**B**	**C**	**D**	**E**
des (B)	A–C–D–E–F–Ⓡ	0.4%	A	**C**	**D**	**E**
des (C)	A–B–D–E–F–Ⓡ	0.2%	A	B	**D**	**E**
des (D)	A–B–C–E–F–Ⓡ	9.0%	A	B	C	**E**
	Percentage "preview" with respect to expected residue		0.4%	0.8%	1.0%	10%

Figure 14. Amplification effect in analysis of deletion peptides from solid-phase synthesis by sequential degradation, modified from Niall *et al.* (1972) and Tregear *et al.* (1977). Previewed residue is in boldface.

contaminating the desired peptide. The preview method provides no information on modified or terminated peptides.

The preview Edman degradation method has been carried out on peptide–resin aliquots as small as 5 mg; the sensitivity is limited mainly by the adjunct analytical technique for quantitation of the released amino acid derivatives. An experimental complication in this latter regard is the need to have available side-chain-protected standards. The other side of the issue is the inadequate stability of some protecting groups to the acidic conditions of the Edman degradation, or partial destruction of thermolabile derivatives when gas chromatographic analysis is applied. Due to incomplete yields of the Edman cycles, there will usually be a significant carryover (2–10%) or "postview" of the main amino acid residue identified in the prior step; for example, detection of B in the third cycle of Fig. 14. If two residues of the same amino acid occur consecutively in the sequence, carryover will obscure preview. However, since the effects of deletions are cumulative, the results will be noted at later cycles of sequential degradation.

At the present time, there is no true information on the minimum levels of deletion peptides accurately quantifiable by preview. Double cleavage, which has been observed to a very minor extent in sequence studies on natural structures, would account for artifactual preview. Furthermore, the Edman procedure has been found to lead to low levels of "background" phenylthiohydantoins due to random cleavage of the peptide chain.

5. *Radiolabeled Peptides*

The incorporation of one or two radiolabeled amino acid residues into peptides prepared by solid-phase synthesis substantially facilitates monitoring of purification steps once the peptide is cleaved from the resin. Generally, N^{α}-protected ^{3}H- or ^{14}C-labeled amino acids with functionally inert side chains are selected for this purpose. For maximally efficient utilization of the label, coupling should be carried out with an excess of resin-bound α-amino groups, followed by a repeated coupling using cold reagents (e.g., Barany and Merrifield, 1973). In these cases, the specific activity of the peptide must be determined after the fact of the incorporation, for example by liquid scintillation counting and amino acid analysis of a hydrolysate. Should a predetermined specific activity be required, the more profligate common coupling methods employing excess acylating agent must of course be used (see footnotes on pp. 10–11 and 124).

In planning the synthesis of a medium-sized peptide or small protein in our laboratory, we frequently find it convenient to incorporate a ^{3}H-labeled amino acid near the COOH-terminus, and a ^{14}C-labeled amino acid near the NH_2-terminus. The goal of these studies is purification of products to a constant ratio of $^{3}H/^{14}C$ specific activities. This ratio will be abnormally high in terminated peptides, and abnormally low in peptides initiated at late stages of the synthesis.

E. Systematic Elimination of Deletion and Terminated Peptides

Several general techniques have been proposed for solid-phase peptide synthesis which aim to minimize the formation of deletion and terminated peptide by-products (Section II,C,4,a) at the level of the repetitive deprotection/coupling cycles. Other methods concede the formation of by-products but are designed to facilitate their resolution from the principal desired product once the peptide mixture has been cleaved from the support. It should be noted that these techniques, as described below, do not bear on the systematic elimination of modified or rearranged peptide structures (Section II,C,4,b).

1. Blocking of Unreacted Amino Groups to Suppress Formation of Deletion Peptides

Incomplete removal of an α-amino protecting group, W, from a peptide–resin W–A–Ⓡ, will lead to a mixture of protected peptide–resins W–A–Ⓡ and W–B–A–Ⓡ after coupling of an incoming activated N^{α}-protected amino acid, W–B. Deprotection and incorporation of residue C in the normal way will give resin-bound deletion peptide C–A–Ⓡ along with the resin-bound desired peptide C–B–A–Ⓡ. Another plausible pathway for production of the same deletion peptide, which could occur when α-amino deprotection is complete, involves incomplete acylation of the free α-amino groups of peptide–resin A–Ⓡ by activated W–B. Table XII lists a number of acylating reagents that have been devised to react with residual free α-amino groups to form stable adducts and thus suppress the production of deletion peptides by the latter mechanism. According to this reasoning, a potential family of deletion peptides will be replaced by a new family of terminated peptides, which might be easier to separate from the desired product because of differences in charge or solubility conferred by the blocking group.

The utility of the reagents listed in Table XII has invariably been demonstrated in model systems where couplings were prevented from going to completion by application of less than one equivalent of each incoming activated amino acid. The presence of unreacted α-amino groups, and their subsequent quantitative acylation by the intentional terminating reagents, could be shown by any of the variety of monitoring techniques listed in Table XI. Furthermore, in several cases (Table XII, numbers 4–6) the blocked peptides were easily separated to give the analytically pure desired model

Table XII. Reagents for Intentional Irreversible Termination of Peptide Chains in Solid-Phase Synthesis

No.	Reagent	Product	Conditions	Remarks	References
1.	$CH_3—\overset{O}{\overset{\parallel}{C}}O\overset{O}{\overset{\parallel}{C}}—CH_3$ (acetic anhydride)	$CH_3—\overset{O}{\overset{\parallel}{C}}—NH—$	Reagent (~1 *M*) + triethylamine (~0.3 *M*) in DMF or CH_2Cl_2, 20 min to 2 hr *or* reagent–pyridine (1:1), 25°C, 2 hr	Original termination method	Merrifield (1963); Blake and Li (1968); Jorgensen *et al.* (1969); Bayer *et al.* (1970a); Mitchell *et al.* (1978b)
2.	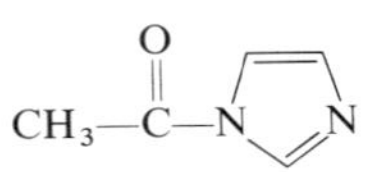 (*N*-acetylimidazole)	$CH_3—\overset{O}{\overset{\parallel}{C}}—NH—$	Reagent (10 equiv over original amino groups) in CH_2Cl_2, 25°C, 60 min	More effective than No. 1 in model system	Markley and Dorman (1970)
3.	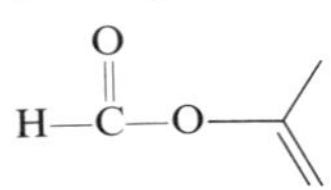 (isopropenyl formate)	$H—\overset{O}{\overset{\parallel}{C}}—NH—$	Reagent (10% v/v) in CH_2Cl_2, 25°C, 60 min	Since leaving group is acetone, this reagent is mild enough to use in presence of N^{α}-Bpoc groups	van Melick and Wolters (1972); Wolters *et al.* (1974); compare to Tesser *et al.* (1976)

4.	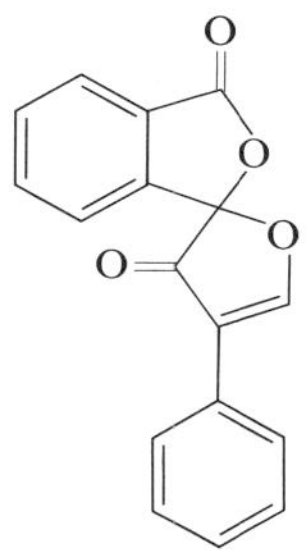 (fluorescamine)	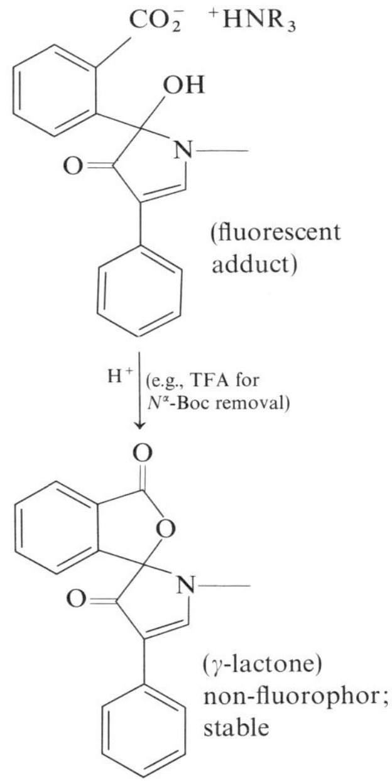	Reagent (1 equiv over original amino groups ~0.3 *M*) + diisopropylethylamine (1 equiv) in CH_2Cl_2, 25°C, 15 min	More effective than No. 1 in model system; terminated spirolactones readily separable by extraction into ethyl acetate. The fluorescamine termination procedure should not be applied with amino-terminal proline, because the product in this case contains a free carboxyl group	Felix *et al.* (1975); see also Moore *et al.* (1977); based on Weigele *et al.* (1972)

Table XII. *(Continued)*

No.	Reagent	Product	Conditions	Remarks	References
5.	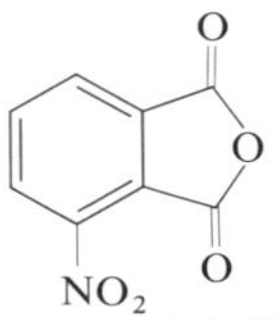 (3-nitrophthalic anhydride)	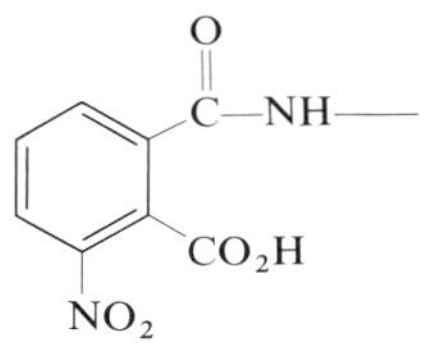	Reagent (0.1 *M*) in pyridine, 25°C, 10 min	The presence of the 2-carboxyl group (pK_a ~ 2.3) does not appear to interfere with subsequent DCC-mediated or symmetrical anhydride couplings; it facilitates separation of terminated by-products by ion-exchange chromatography. The blocked product is reported to be stable to acidolytic cleavage conditions of solid-phase peptide synthesis, but to be quantitatively cleaved in the acid step of the Edman degradation (TFA, 40°C, 45 min), and by 20% aqueous citric acid (60°C, 45 min).	Wieland *et al.* (1969b, 1971b); Penke and Birr (1974); Birr (1978, p. 61); Birr and Penke, personal communication, 1979
6.	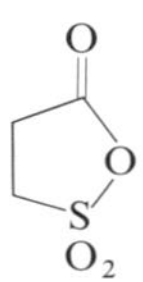 (3-sulfopropionic acid anhydride)	$HO_3S-CH_2CH_2-C(=O)-NH-$	Reagent (0.7 *M*) + *N*-ethylmorpholine (1.4 *M*) in DMF-tetrahydrofuran (1:1), 25°C, 1 hr	Useful acidic handle for purification, as No. 5 above. Acylation of peptide bonds may occur as side reaction of the blocking step. The structurally closely related sulfobenzoic anhydride has also been mentioned.	Wissmann and Geiger (1970); compare to Wieland *et al.* (1969b) and Birr *et al.* (1973)

peptide in one fractionation step. For these reasons, a number of laboratories have adopted protocols for intentional termination into the routine schedule of the repetitive cycles, while in other instances, termination has been used only when a need was indicated by monitoring. A case can also be made for eliminating intentional termination altogether when optimal synthetic methodology is used (p. 131). The rationale for these approaches is intrinsically circular: If the residual α-amino groups are accessible both to the reagents of Table XII, and later to new activated N^{α}-protected amino acids, it is natural to expect that the deletion in question can be prevented merely by multiple couplings. The possibility of peptide backbone acylation with the more reactive of these reagents, especially when in large excess, has been raised but not proven. The resulting diacylimides would be predicted to be cleavable under mildly basic conditions, to give either the regenerated peptide or smaller fragments.

2. Affinity Purification to Remove Terminated Peptides

As already described, intentionally terminated by-products can be separated from the desired peptide at the conclusion of a solid-phase synthesis on the basis of properties of the blocking group. A reciprocal strategy (Fig. 15; Krieger *et al.*, 1976) involves functional selection of all polymer-supported peptide chains that are still growing by chemical reaction with

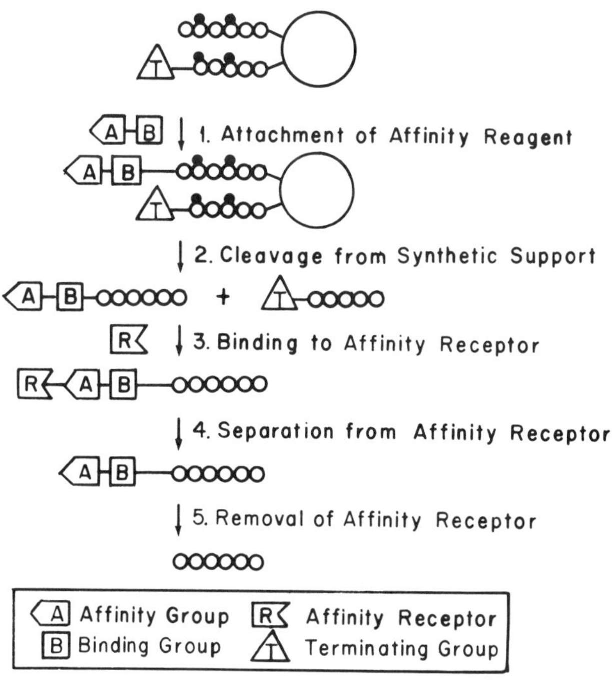

Figure 15. A general strategy for affinity purification of solid-phase peptides, taken from Krieger *et al.* (1976). ●, protecting group; ○, amino acid residues; ◯, resin support.

Table XIII. Reagents for Intentional Reversible Modification of Peptide Chains that Facilitate Purification Steps

No.	Affinity-blocked product	Introduction	Affinity receptor	Removal	Remarks	References
1.	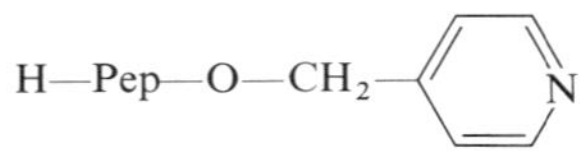 (4-picolyl ester)	C^{α}-protecting group for stepwise synthesis in solution	Cation-exchange column	H_2/Pd	Picolyl ester stable to strong acids	Camble *et al.* (1968)
2.	H—Cys—Met—Pep—O—Ⓡ	Protected methionine and cysteine coupled by standard stepwise solid-phase synthesis	Organomercurial column; elute with thiols *or* sulfhydryl column which binds by disulfide exchange	CNBr	Use of acid stable Met(O) and Cys(Acm) for internal residues should increase generality of this method	Krieger *et al.* (1976); Lindeberg *et al.* (1978) Compare to Seliger *et al.* (1978) for related application in solid-phase polynucleotide synthesis
3.	H—Lys—Pep—O—Ⓡ	Protected lysine coupled in usual way (see No. 2)	Cation-exchange column	Edman degradation	Requires differential protection for internal Lys residues	Suzuki *et al.* (1976a)
4.	H_2N—CH(CH_2SO_3H)—CO—Pep—OH (Cysteic acid)	Solution coupling of Boc-cysteic acid *N*-hydroxysuccinimide ester	Readily separable by countercurrent distribution	Edman degradation	Hydrophilic products possess enhanced water solubility	Hubbuch *et al.* (1977)

5.	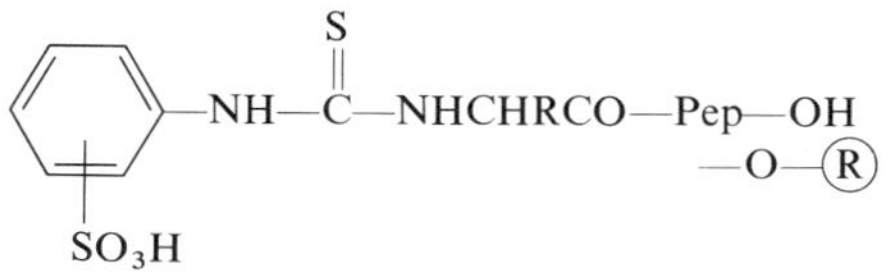 (*m* or *p*)	Proposed use: couple one extra amino acid, then react with *m*- or *p*-sulfophenylisothiocyanate	Mild anion-exchange column	Edman degradation	Probably will cyclize during HF cleavage from the support	Birr *et al.* (1970); Dwulet and Gurd (1976)
6.	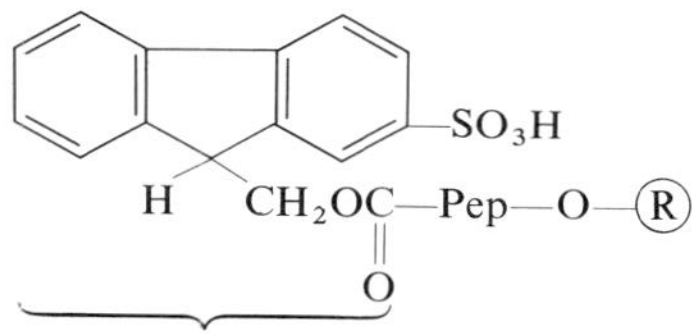 [9-(2-sulfo)fluorenylmethyloxycarbonyl (Sulmoc)]	Sulfmoc-Cl (2 equiv) + diisopropylethylamine (3 equiv) in CH_2Cl_2, 25°C, 2 hr	Mild or strong anion-exchange column	Anhydrous morpholine or piperidine; dilute aqueous base	Sulfmoc group stable to HF, HBr, and pyridine	Merrifield and Bach (1978)
7.	$(C_6H_5)_3\overset{+}{P}—CH_2CH_2—O—\overset{O}{\overset{\|}{C}}—Pep—OH$	Reaction with 2-(triphenylphosphonio)ethoxycarbonyl chloride		0.1 *N* NaOH, 25°C, 1 min	Potentially useful along lines of No. 6	Kunz (1976a)

a specially chosen blocking group, termed an "affinity reagent." In this way, the desired chains are distinguished from all by-products that have been irreversibly terminated either by choice, as outlined earlier to suppress deletion sequences, or terminated due to chemical side reactions. An affinity reagent A–B contains an affinity group A and a reversible bonding group B. After cleavage from the support, the affinity-labeled peptides are separated from terminated peptides which lack the affinity reagent by selective bonding to an affinity receptor R. The affinity-labeled peptides are then dissociated from the affinity receptor, and finally the affinity reagent is removed to give the desired peptide. Deletion peptides are not removed by this technique. Note that the affinity blocking group, introduced while the peptide is still anchored to the solid support, must be largely stable to the conditions for cleavage of the peptide into solution.

Several reagents compatible with the affinity purification concept are listed in Table XIII. Affinity groups A have been designed to facilitate separations on the basis of charge and polarity (numbers 1, 3–7) or specific reactions of thiols (number 2). The reversible bonding groups B have been removed by catalytic hydrogenolysis (number 1), specific peptide bond cleavage with cyanogen bromide (number 2), or the Edman method (numbers 3–5), or by β-elimination promoted by mild base (numbers 6–7). Of the affinity reagents shown, numbers 2, 3, and 6 have actually been applied with benefit to the purification of peptides synthesized by the solid phase method. In addition, the others, particularly 4 and 7, are potentially useful for the same purpose.

V. PROTECTED AMINO ACIDS

There are 20 genetically coded amino acid residues, and quite a few more that are found in naturally occurring products, such as peptide antibiotics, or are of interest for analogues in structure–function studies. In this section, we discuss the important amino acid residues with regard to their incorporation into peptide structures through use of the solid-phase method. Thus, given a peptide of arbitrary amino acid sequence, it should be possible to design rationally several viable approaches towards its synthesis, and to envisage the chemical difficulties that might arise.

A. Chemistry of the Protection Scheme

Selection of the anchor to the polymeric support, and of the temporary protecting group removed at each repetitive deprotection/coupling cycle, has already been discussed (Sections III,B and IV,A, respectively). Permanent side-chain protection is normally required for lysine, aspartic acid, glutamic acid, and cysteine, strongly preferred for arginine, histidine, serine, threonine, and tyrosine, and optional for asparagine, glutamine, methionine, and tryptophan. Those residues in the intermediate category will, if unprotected, undergo variable levels of acylation at each coupling step, with the production of branched peptides. Some of the residues in the optional category have been temporarily protected for the anchoring or coupling step, since once they have been incorporated into the polymer-supported peptide chain, no further modification of their side chains is ordinarily expected. The chosen permanent protecting groups must be entirely stable during all deprotection/coupling cycles, yet must be smoothly removable at the end of the synthesis (Section II,A,5).

In the common practice of solid-phase peptide synthesis, most of the side-chain functionalities are protected as ethers, esters, and urethanes of substituted benzyl alcohols or of *tert*-butanol. Thus, they can all be deprotected in a single treatment, generally with acid. The ease of acidolysis is a function of the linkage that is broken, and of the stability of the ensuing carbonium ion. For a homologous series of derivatives a qualitative order, from most labile to most stable, is urethane ($N^{\alpha} > N^{\varepsilon} > N^{im} > N^{g}$) $\sim$ *O*-carbonate > ester $\sim$ ether > > thioether > amide.

The properly chosen benzyl derivatives are compatible with all of the temporary protecting groups listed in Section IV,A, and the permanent groups are removed by strong acids (Table VIII on p. 70) or occasionally by catalytic hydrogenolysis in solution. Usually, benzyl side-chain protection is applied in tandem with the N^{α}-*tert*-butyloxycarbonyl group, and a *p*-alkyl or *p*-(carbamoylmethyl)benzyl ester anchoring linkage, so that deprotection is carried out concurrently with acidolytic cleavage of the peptide from the support. Alternatively, the peptide can be cleaved from the support while retaining the side-chain benzyl protecting groups when an orthogonal mode such as nucleophilic attack or photolysis is applied. Suitable anchoring is provided by the usual substituted benzyl esters, by substituted *o*-nitrobenzyl or phenacyl esters, or by *n*-alkyl esters after safety-catch activation (see Section III,B for details and limitations of these various approaches). Benzyl protected peptide acids or hydrazides of moderate size are also accessible when protection schemes based on graduated acid-lability are employed. Appropriate N^{α}-protecting groups are the 2-(*p*-biphenylyl)propyl(2)oxycarbonyl (Bpoc), α,α-dimethyl-3,5-dimethyloxy-benzyloxy-carbonyl (Ddz), 2-phenylpropyl(2)oxycarbonyl (Poc), or 2-benzoyl-1-methylvinyl (Bmv) functions (Sections IV,A,2 and 3), while compatible anchoring linkages to the resin include *p*-alkoxybenzyl or benzhydryl esters or a 3-(*p*-benzyloxyphenyl)-1,1-dimethylpropyloxycarbonyl hydrazide (pp. 76, 84, and 97, respectively).

The *tert*-butyl-based derivatives resist catalytic hydrogenolysis, alkali, and nucleophiles, but are deprotected by acids of moderate strength [e.g., TFA–CH_2Cl_2 (1:1) at 25°C]. Hence, they lend themselves to mild schemes of solid-phase peptide synthesis involving highly acid-sensitive or orthogonally removable temporary protecting groups, including the Bpoc, Ddz, Nps, Fmoc, or Dts functions. Anchoring linkages such as the substituted phenacyl, *p*-alkoxybenzyl, and *o*-nitrobenzyl esters are among those that will be compatible. Several of the obvious combinations have not yet been implemented, but this is an important direction in current research on methodological improvements, since it represents a way around several harmful side reactions promoted by strong acids.

A handful of other functionalities round out the list of useful permanent protecting groups that are removed by acidolysis. Most notable among these is the *p*-toluenesulfonyl (Tos) group (reviewed by Rudinger, 1973), which has been advantageously applied to the protection of arginine and histidine.

Many of the amino acids can be protected in an alternate way by groups that are stable to acid but removable by base, catalytic hydrogenolysis, metal reduction, thiolysis, electrolysis, or other means. In fact, the only protecting groups known for methionine and tryptophan are resistant to acidolysis, as are derivatives such as N^{ε}-trifluoroacetyl-lysine, N^{im}-dinitrophenylhistidine, aspartic acid β-phenacyl ester, and *S*-acetamidomethylcysteine. These are preferred for a variety of solid-phase applications. The acid-stable protecting groups can be removed while the peptide is still anchored to the support. Alternatively, the peptide can be first released into solution by an acidolytic mechanism, which generally serves to remove most of the other side-chain protecting groups. The partially protected peptide thus obtained can be purified, and as an added benefit, can serve as an analogue of the desired peptide obtained after a final deprotection step.

Some of the orthogonal modes cited in the previous paragraph for the removal in solution of acid-stable protecting groups were originally demonstrated on simple model peptides. Attempts to extend the methodologies to more complicated peptides, where unforeseen steric effects can intervene or where several residues need to be deprotected simultaneously, have occasionally resulted in poor yields. Specific recommendations on these types of strategies therefore need to be considered on a case to case basis (Section V,B, which follows). An as yet unexplored protection scheme using the common N^{α}-*tert*-butyloxycarbonyl temporary group, and thiolyzable permanent side-chain groups along with a thiolyzable anchoring linkage, would allow the critical final cleavage and deprotection steps of a solid-phase synthesis to be performed under entirely neutral conditions.

B. Chemistry of the Individual Amino Acid Residues

We describe here for each amino acid: (1) suitable side-chain protected derivatives;* (2) chemical side reactions and other considerations for initial anchoring or peptide chain elongation with that residue in solid phase synthesis; (3) side reactions occurring at the already incorporated residue in the course of the repetitive deprotection/coupling cycles, or under the conditions for the final cleavage from the support; and (4) where pertinent, analytical techniques that permit specific detection or quantitation of the appropriate residue,† and/or allow verification of the levels of any side reactions.

1. Arginine

a. Protection. The strongly basic δ-guanidino group ($pK_a = 12.5$) of arginine has been protected by the N^{ω}-nitro (NO_2) (**115**) (Bergmann *et al.*, 1934) and the N^{ω}-*p*-toluenesulfonyl (Tos) (**116**) (Ramachandran and Li, 1962) groups in most solid-phase syntheses. Other means of arginine side-

* Throughout the course of this section, structural formulas for protecting groups will include the functional group that is being protected in boldface. An *asterisk* next to the formula number indicates that the diagrammed side-chain protected derivative is currently commercially available with suitable N^{α}-protection, generally the *tert*-butyloxycarbonyl (Boc) group (**36**). Occasionally, the *tert*-amyloxycarbonyl (Aoc, **37**) derivative is made available because of superior solubility characteristics. A good number of N^{α}-2-(*p*-biphenylyl)propyl(**2**)oxycarbonyl (Bpoc, **40**) derivatives, and a somewhat smaller set of N^{α}-*o*-nitrophenylsulfenyl (Nps, **56**), N^{α}-α,α-dimethyl-3,5-dimethyloxybenzyloxycarbonyl (Ddz, **42**), and N^{α}-9-fluorenylmethyloxycarbonyl (Fmoc, **59**) derivatives are also available, both with benzyl-based or *tert*-butyl-based side-chain protection. While most N^{α}-benzyloxycarbonyl (Z, **46**) derivatives are available, these are usually not directly applicable to solid-phase procedures for reasons already explained (pp. 7 and 107).

† A satisfactory amino acid analysis, within the $\pm 3\%$ confidence limitation imposed by modern instrumentation, is a necessary but not sufficient criterion for the purity of a synthetic peptide (Moore, 1972). Optimal conditions for hydrolysis of peptide–resins (see Table XI, no. 12, p. 154) and for peptides (Moore and Stein, 1963) have been reviewed. It has been pointed out (Uy and Wold, 1977) that hundreds of naturally occurring amino acids are either destroyed by hydrolysis or overlooked because of overlap with other peaks on analyzer chromatograms; this remark clearly carries over to side-chain protected or structurally modified amino acids encountered from the manipulations of peptide synthesis.

In this section, by "standard amino acid analyzers," we refer to the methodologies in use in our laboratory, employing Beckman Model 120 and 121 machines. Sulfonated polystyrene cation-exchange resins of the AA-15 or PA-35 type are packed into a long (0.9 × 54 cm) and short (0.9 × 15 cm) column. The sodium acetate buffers, operating conditions, and the order of elution of the amino acids are as described in the classic paper of Spackman, Moore, and Stein (1958). Subtle differences in order of elution and resolution occur by changing resin types, and buffer compositions and pH. Additional chromatographic protocols and allied considerations have been thoroughly described (Bailey, 1967, pp. 62–109; Benson and Patterson, 1971; Blackburn, 1978).

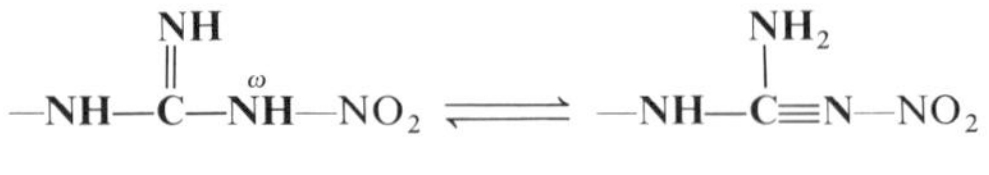

*(**115**) NO_2

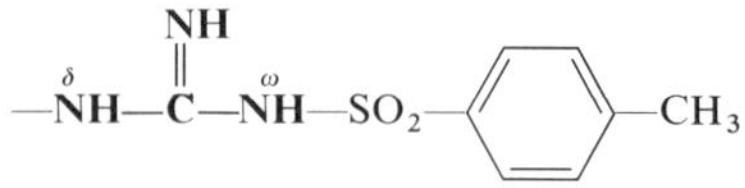

*(**116**) Tos

chain protection (reviewed by Schröder and Lübke, 1965, pp. 167–175) known from solution chemistry include protonation (**117**), and N^{ω}- (**118**) or N^{δ},N^{ω} di-substitution (**119**) as urethanes, and several of these have been tried in solid-phase synthesis. Finally, following a suggestion due to Fruton

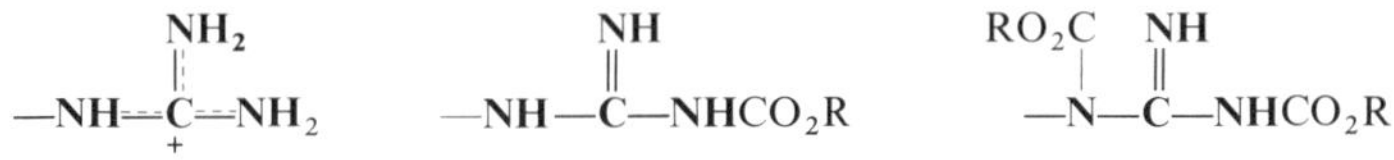

(**117**) (**118**) (**119**)

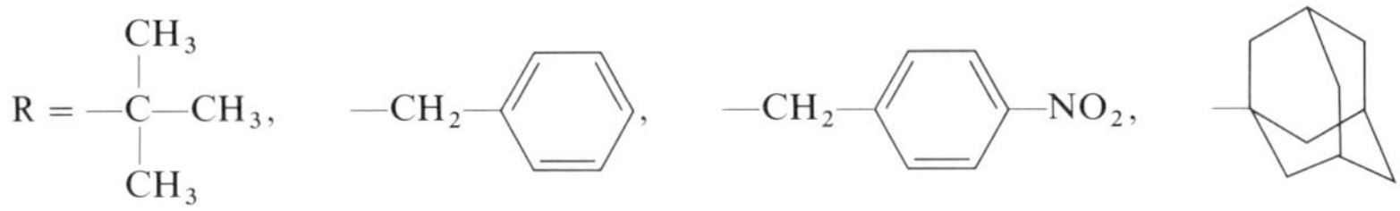

Boc Z 4Nz Adoc

(1949), an N^{δ}-protected ornithine residue (Section V,B,2) can be incorporated at the position intended for arginine; and following deprotection, the δ-amino group can be converted to an δ-guanido group by reaction with *O*-methylisourea [Eq. (19)], *S*-methylisothiourea, or 1-guanyl-3,5-dimethyl-pyrazole (Barrass and Elmore, 1957; Bannard *et al.*, 1958). This chemical

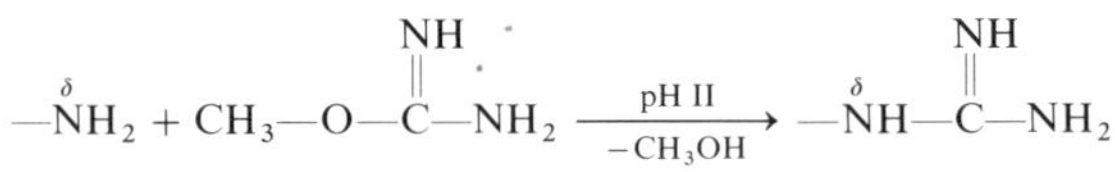 (19)

strategy† has been applied to complex arginine and homoarginine peptides synthesized in solution (e.g., Bodanszky *et al.*, 1964; Borin *et al.*, 1972) and on a solid-phase (Lindeberg *et al.*, 1976; Cosand and Merrifield, 1977; Granier *et al.*, 1978).

For many solid-phase applications, N^{ω}-tosylarginine has been preferred. The side-chain protecting group is readily removed simultaneously with

† The work of Cosand and Merrifield (1977) on analogues of the neurotoxic 18-peptide, apamin, illustrated the concept of an *internal structural control.* Thus, the observation that [$Orn^{13,14}$] apamin prepared by solid-phase synthesis was biologically inactive could be interpreted to mean that one or both arginine residues at positions 13 and 14 of the natural molecule play an important role in the neurotoxic action, or else that the supposed analogue contained an incorrect sequence or a structural modification due to an unrecognized difficulty in the synthesis. The latter possibility was ruled out by the demonstration that synthetic [Har^4] apamin obtained by guanidination of synthetic [$Orn^{13,14}$] apamin possessed full biological activity equivalent both to native apamin as well as to [Har^4] apamin derived by guanidination of the native material.

substituted benzyl-based groups during cleavage of the peptide from the support in liquid HF at 0°C for 30 min in the presence of anisole (Mazur and Plume, 1968). The tosyl group is resistant to HBr–TFA, nucleophiles including alkali, ammonia, and hydrazine, and to catalytic hydrogenolysis. The earlier literature contains numerous examples and experimental details (e.g., Stewart and Young, 1969, pp. 46–48) for final detosylation with sodium in liquid ammonia (du Vigneaud and Behrens, 1937). This latter procedure has recently been shunned for most complex peptides due to the greater convenience and reliability of the HF treatment, although specific cases exist where degradative losses were much higher in HF (e.g., Tzougraki *et al.*, 1978). It should be noted that even under the most controlled conditions, sodium in liquid ammonia has been observed to promote considerable cleavage at some peptide bonds, degradation of certain amino acid side chains, and racemization. Detosylation by strong acid reagents other than HF (see Table VIII) often requires forcing conditions. The standard conditions using CH_3SO_3H gave only partial reaction, and either FSO_3H at 25°C or CF_3SO_3H at 40°C was required to achieve complete acidolysis (Yajima *et al.*, 1974, 1975). Treatment of Aoc-Arg(Tos)-OH with pyridinium polyhydrogen fluoride at 25°C for 60 min gave free arginine in 95% yield. The deprotection with boron tris(trifluoroacetate) in TFA appears to go to completion, although it may be slow (compare Pless and Bauer, 1973, with Nishimura and Fujino, 1976). In this regard, the somewhat more acid-sensitive *p*-methoxybenzenesulfonyl (Mbs) (**120**) (Nishimura and Fujino, 1976) or mesitylene-2-sulfonyl (Mts) (**121**) (Yajima *et al.*, 1978) groups, both so far only tested in solution, might replace the tosyl group for guanido protection.

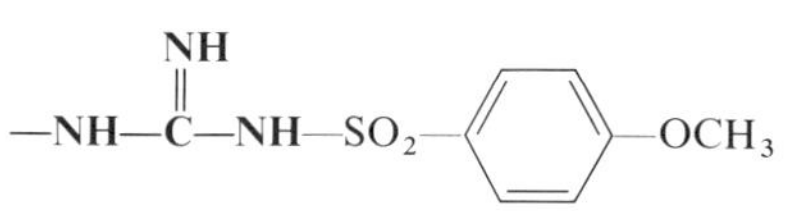

(**120**) Mbs

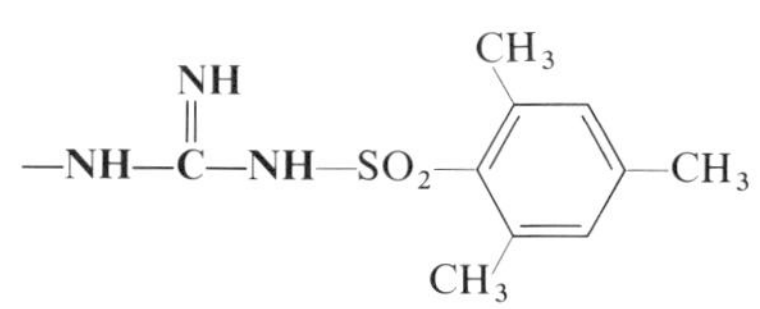

(**121**) Mts

The nitro protecting group **115** is also compatible with a protection scheme where the final cleavage is carried out in liquid HF in the presence of anisole or a similar scavenger as a nitronium acceptor. The nitration and denitration of arginine is an equilibrium reaction catalyzed by HF (Lenard, 1967). Among other acids, CH_3SO_3H, CF_3SO_3H, and pyridinium polyhydrogen fluoride effect only partial removal; BBr_3 gives partial cleavage and deamidination; and boron tris(trifluoroacetate) and FSO_3H lead to quantitative deprotection (Pless and Bauer, 1973; Felix, 1974; Yajima *et al.*, 1974, 1975). A nitroarginine residue in peptides will be stable when cleavage from the support is carried out with $TFA–CH_2Cl_2$ (1:1), for example in a protection scheme using 2-(*p*-biphenyl)propyl(2)oxycarbonyl α-amino group protection and *p*-alkoxybenzyl ester anchoring (p. 167), and will also be resistant to the stronger combination of HBr–TFA. The nitroarginine-containing peptide can then be deprotected in solution by catalytic hydrogenolysis (Bergmann *et al.*, 1934; for examples and details, see Merrifield, 1964b; Stewart and Young, 1969, pp. 45–46). Catalysts used have included freshly prepared palladium black (Wieland, 1912) or 5% palladium oxide on barium sulfate (Kuhn and Haas, 1955); solvents are generally methanol–aqueous acetic acid mixtures, and reaction conditions typically range from atmospheric pressure to 50 psi for 24 hr or more. For complex peptides, catalytic hydrogenolysis is often slow; and stable, partially reduced intermediates may be formed to various extents (see p. 174).

The denitration by catalytic hydrogenolysis according to the common conditions has occasionally been successful for sulfur-containing peptides. A more general method recently introduced by Meienhofer and Kuromizu (1974; see also Felix *et al.*, 1978) in which catalytic hydrogenolysis is carried out in liquid ammonia as a solvent, appears to have considerable merit for removal of the nitro group from cysteine- or methionine-containing peptides. The N^g-nitro group may also be removed by reduction with zinc (Pless and Guttmann, 1967; Turán *et al.*, 1975) or titanium(III) trichloride (Freidinger *et al.*, 1978), or by electrolysis (Scopes *et al.*, 1965). The susceptibility of nitroarginine to side reactions in the course of lengthy exposure to nucleophilic reagents (see p. 174) should be considered if ammonolytic or hydrazinolytic cleavage of the peptide is an aspect of the synthetic plan.

Several N^ω-(**118**) or N^δ,N^ω-bis (**119**) urethane derivatives of arginine have been synthesized since the introduction of this class of compounds into peptide synthesis by Zervas *et al.* (1959b); di-substitution is required to suppress fully the basicity of the δ-guanidino group. The merits of this interesting form of protection are not fully in focus; the obvious advantages are counterbalanced by considerations of ease of synthesis, potential steric hindrance during coupling, selective deblocking of an α-amino group, and solubility of the derivative. The side-chain urethane is removable by acidolysis at a slightly slower rate than the corresponding N^α-urethane. The bisadamantyloxycarbonyl derivative (Jäger and Geiger, 1970) is perhaps the most promising, being compatible with a scheme relying primarily on *tert*-butyl based groups for side-chain protection. A solid-phase coupling of Z-Arg(Adoc)$_2$-OH succeeded well to give an NH_2-terminal arginine peptide (Syrier and Beyerman, 1975), and the same starting material was successfully used by the Protein Synthesis Group (1975) in the solution synthesis of segments later coupled on the solid phase. For stepwise incorporation of Arg(Adoc)$_2$ into the interior of peptide chains, a selectively removable α-amino protecting group is required.

Recently, Camble and Petter (1976) have defined conditions under which peptide segments prepared in solution without covalent protection of the guanido side chain of arginine could be successfully handled in the course of a solid-phase synthesis. The relatively weak base, *N*-methylmorpholine, was added only during coupling and acetylation steps, and in strictly limited amounts (1 to 2 equiv relative to available α-amino groups). Hence, the guanido group remained in the protonated form **117** and was not acylated through the course of four additional coupling steps. In general, protection through protonation cannot be recommended for extended stepwise syntheses, because of side reactions that invariably will arise.

b. Anchoring and Coupling. The esterification of COOH-terminal N^{α},N^{ω}-protected arginine to chloromethyl supports (pp. 60–63) has often been achieved satisfactorily (e.g., Merrifield, 1964b). Satisfactory loading of Boc-Arg(Tos)-OH by the tetramethylammonium salt procedure required harsher conditions than usual (Yamashiro and Li, 1976).

As a practical remark, it should be mentioned that N^{α}-Boc- or N^{α}-Aoc-tosylarginine or nitroarginine are poorly soluble in CH_2Cl_2; DMF (10–25% v/v) must often be added to obtain adequate concentrations for the activation and coupling steps. The DCC method gives excellent incorporations into resin-bound peptide chains, and the symmetrical anhydride method has also often been successfully applied. The quantitative rate data (Merrifield, 1973c; discussed on p. 138) suggest that in the latter case, the true acylating species may be the δ-lactam, so the DCC method would seem to be preferred for the introduction of arginine.

c. Side Reactions. Figure 16 shows two intramolecular reactions of protected arginine derivatives that are well documented in solution, but the relevance of these to solid-phase conditions is unclear. The formation (pathway a) of a 1,3-diazacycloheptane (**122**) (Paul *et al.*, 1961; Jäger and Geiger, 1970; Künzi *et al.*, 1974), which occurs even with N^{ω},N^{δ}-bis-protected derivatives, would be a source of concern following deprotection of an NH_2-terminal arginine residue. Similar reactions are known to be catalyzed by weak acids and bases; should this process ever arise to give terminated peptides in solid-phase synthesis, one might predict that it could be suppressed by the same types of precautions used to minimize diketopiperazine formation (p. 83).

The formation (pathway b) of a δ-lactam (**123**) (Bodanszky and Sheehan, 1960; Paul *et al.*, 1961; Arold and Reissmann, 1968) occurs whenever N^{α},N^{ω}-disubstituted derivatives are activated for coupling, but is avoided when the N^{δ}-position is blocked. The δ-lactam has been invoked (see above)

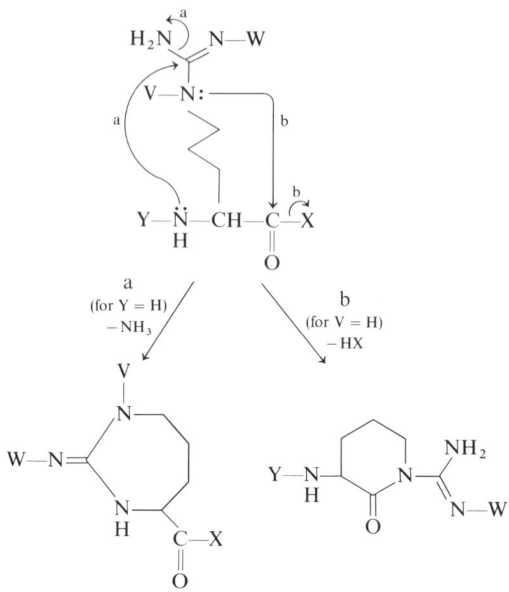

(122) **(123)**

Figure 16. Potential intramolecular side reactions of protected arginine derivatives during solid-phase peptide synthesis: (a) 1,3-diazacycloheptane (**122**); (b) δ-lactam (**123**).

to explain the anomalous kinetics of coupling of arginine derivatives by the symmetrical anhydride method. The lactam **123** is also capable of transferring the protected guanyl moiety onto nucleophilic amines, thus resulting in their premature termination.

Ornithine peptides occasionally arise in the course of manipulations involving arginine peptides. In syntheses where arginine has been protected only by protonation, the appearance of ornithine in amino acid hydrolysates is diagnostic for guanidino group acylation and branching (Camble and Petter, 1976; Photaki and Yiotakis, 1976). Ornithine by-products have also been observed after ammonolysis (Künzi *et al.*, 1974; up to 50%) or after HBr–TFA or HF treatment (Yamashiro *et al.*, 1972, up to 20%) of peptides containing N^g-nitroarginine, but not the same peptides containing N^g-tosylarginine. The latter side reaction may be sequence dependent, however, and may thus be only indirectly related to the cleavage conditions. The phenomenon was noted in one model peptide containing histidine residues adjacent to the nitroarginine in question, but not in a different model. The catalytic hydrogenolysis or alternative reductive cleavages of $Arg(NO_2)$ in peptides may occassionally stop at nitroso or aminoarginine (Gros *et al.*, 1961; Iselin, 1966; Turán *et al.*, 1975).

d. Analytical. The presence of nitroarginine in peptides can be assessed by monitoring the ultraviolet absorbance at 271 nm ($\varepsilon = 1.7 \times 10^4$; Riniker and Schwyzer, 1961), and as a peak eluting near methionine on standard automated amino acid analyzers. Hydrolysis with 6 *N* HCl or similar reagents at 110°C gives a mixture of free Arg, Orn, and starting Arg(NO_2) which quantitatively accounts for all the material (Merrifield, 1964b; Stewart and Young, 1969, pp. 53–54; Künzi *et al.*, 1974); while under these same conditions, tosylarginine is entirely converted to free arginine. The quantitative Sakaguchi reaction (Weber, 1930; see Stewart and Young, 1969, p. 56) facilitates monitoring of the deprotection reactions of Arg(NO_2) and Arg(Tos), as well as the localization of arginine-containing peptides upon chromatographic fractionation. As already stated, arginine residues in peptides sometimes give rise to ornithine-containing by-products. If the peptide also contains lysine, amino acid analysis of a hydrolysate under standard short column conditions will usually not resolve traces of Orn from an excess of Lys, but this separation is readily achieved under modified conditions on the long column.

2. Lysine and Ornithine

a. Protection. Many of the amino protecting groups described in Section IV,A have been used on the δ- or ε-amino groups ($pK_a = 10.5$) of ornithine and lysine. The various *tert*-alkyl urethanes (Section IV,A,1) are compatible with protection schemes involving highly acid-labile (Section IV,A,2 and 3) or orthogonally removable (Section IV,A,4–6) α-amino protecting groups. There are also a number of relatively acid-stable protecting groups (not all of which have been tried in solid-phase synthesis) which may be suitably applied in the preparation of fully or selectively protected peptide segments.* These include the 9-fluorenylmethyloxycarbonyl (Fmoc) (**59**), methylsulfonylethyloxycarbonyl (Msc) (**62**), β,β,β-trichloroethyloxycarbonyl (**66**) (Yajima *et al.*, 1971), dithiasuccinoyl (Dts) (**69**), phthaloyl (Pht) (**71**), 4,5-diphenyl-4-oxazoline-2-one (**72**), piperidino-oxycarbonyl (**74**), trifluoroacetyl (Tfa) (**75a**), chloroacetyl (**75b**), and *p*-toluenesulfonyl (Tos) (**76**) functions, the properties and deprotection methods of which have already been discussed. Among these, the trifluoroacetyl group (**75a**) is considered especially useful (Anfinsen *et al.*, 1967; Ontjes and Anfinsen, 1969a; also Cosand and Merrifield, 1977, further applied as explained in footnote on p. 170). The trifluoroacetyl derivatives are sufficiently stable to anhydrous acids so that the amino side chain remains fully blocked both during the synthesis (negligible branching: Ontjes and Anfinsen, 1970) and after

* Peptides containing internal lysine residues incorporated as such side-chain protected derivatives are resistant to the action of trypsin (see p. 93).

cleavage of the peptide from the support in liquid HF at 0°C. Removal occurs in 1 *M* aqueous piperidine, pH ~ 12, at 0°C for 2 hr. To the above listing may be added the *p*-nitrobenzyloxycarbonyl group (4Nz) (**124**) (Channing *et al.*, 1951; Carpenter and Gish, 1952) which is substantially more difficult to remove in acids than the benzyloxycarbonyl group (Noda *et al.*, 1970). The 4Nz group is readily removed by catalytic hydrogenolysis with the production of *p*-toluidine as a by-product (see Ohno *et al.*, 1971, for a solid-phase

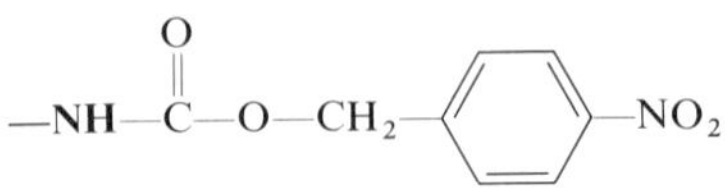

(**124**) 4Nz

application). Alternatively, stannous chloride in HBr–HOAc has been used as the deprotective reagent (Yaron and Schlossman, 1968); the rate is only several-fold faster than in the absence of the $SnCl_2$. Very recently, sodium dithionite or titanium (III) trichloride in aqueous solutions at neutral pH have been used to effect reductive cleavage (Guibé-Jampel and Wakselman, 1979). The probable mechanism of these latter reactions was suggested to involve initial reduction of **124** to a labile *p*-aminobenzyloxycarbonyl derivative [compare to Eq. (12) on p. 114].

The most common protection scheme for stepwise solid-phase synthesis utilizes the N^{α}-*tert*-butyloxycarbonyl group and relies on the graduated acid lability of the side-chain protecting groups and anchoring linkage. Initially (Merrifield, 1964c), the benzyloxycarbonyl (Z) function was applied for the protection of the ε-amino group of lysine, but it soon became evident that a steady loss of side-chain protecting groups occurs concomitant to α-amino deprotection (Yaron and Schlossman, 1968). As a result, peptide chain "branches" could grow off the liberated side chains. The rate ratio for acidolysis of the N^{α}-*tert*-butyloxycarbonyl (Boc) and N^{ε}-benzyloxycarbonyl (Z) groups has been established in the order of 10^4 (Erickson and Wang, 1975); thus, 0.1% of the N^{α}-Boc group would remain at a point when 0.1% of the N^{ε}-Z group had already been lost.

One of the earlier approaches to solve the branching problem was to switch to groups, such as the Tfa (**75a**) or 4Nz (**124**) functions, that are much more acid-stable and only removable in separate steps in solution under orthogonal conditions (see above paragraph). Alternatively, groups that are chemically unrelated to the benzyloxycarbonyl function have been considered in the hope that they would fortuitously fall in the proper stability range to survive repeated treatments with the acidolytic temporary deprotective reagent, yet be cleanly removable at a final cleavage step by liquid HF. The N^{ε}-2,4-dimethyl-3-pentyloxycarbonyl group (also called diisopropylmethyloxycarbonyl, Dmc) (**125**) (Sakakibara *et al.*, 1970), which is removed with

HF at 0°C for 3 hr, or 20°C for 1 hr, may be somewhat too stable for sensitive applications (used in solid-phase synthesis by Izumiya *et al.*, 1972; Schaich *et al.*, 1973). The recently developed *p*-tolylmethylsulfonyl (Pms) group

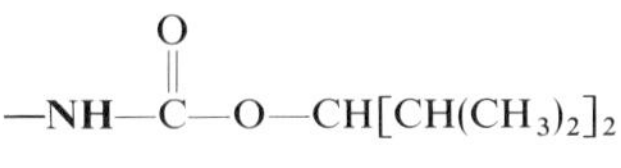

*(**125**) Dipmoc

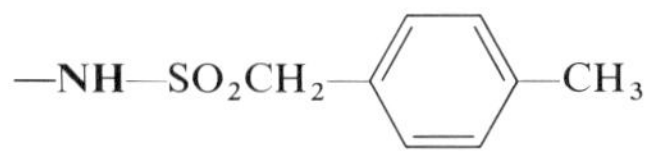

(**126a**) Pms

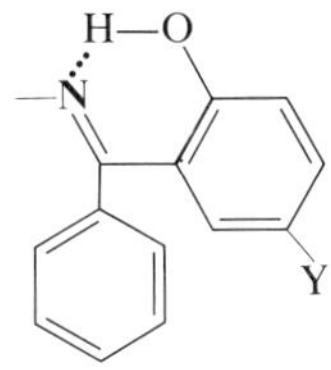

(**126b**) Y = Cl, CH_3

(Fukuda *et al.*, 1978; compare to Milne and Peng, 1957) is cleaved surprisingly readily in HF at −20°C for 60 min. Although the Pms group was said to be completely stable to TFA at 20°C for 24 hr, possible low-level premature side-chain deprotection cannot be ruled out in the absence of more discriminating experimental data. The 2-hydroxydiarylmethylene (Schiff's base) derivatives **126b** (Abdipranoto *et al.*, 1977; compare to structure **54** on p. 109 may also be of interest, because of their stability once incorporated into polystyrene-based peptide-resins. This stability was attributed to the inability of water needed for hydrolysis to penetrate the hydrophobic beads. The group **126b** was essentially resistant to TFA–CH_2Cl_2, HCl-dioxane, and triethylamine–methanol, and was cleaved with boron tris(trifluoroacetate).

The most productive line of research, which has led to the practical elimination of branching, has involved systematic "fine-tuning" of the stability of the benzyloxycarbonyl group by the introduction of electron-withdrawing substituents by analogy to the pioneering approach of Bláha and Rudinger (1965). The apparent first-order rates of deprotection in TFA–CH_2Cl_2 (1:1) of seven chlorinated benzyloxycarbonyl derivatives of lysine were measured and found to vary over a 1000-fold range (Erickson and Merrifield, 1973b). One *meta*- or two *ortho*-halo substituents gave the maximum quantitative difference, which was near or beyond the limit to be of value for protection schemes based on graduated acid lability; a single *para*-halo substituent did not make a sufficient difference, since the rate was only lowered 3-fold. The best practical compromise has involved a halo atom in the *ortho*-position, as in **127** (Yamashiro and Li, 1973a; Erickson and Merrifield, 1973b). These N^{ε}-*ortho*-halo-benzyloxycarbonyl derivatives

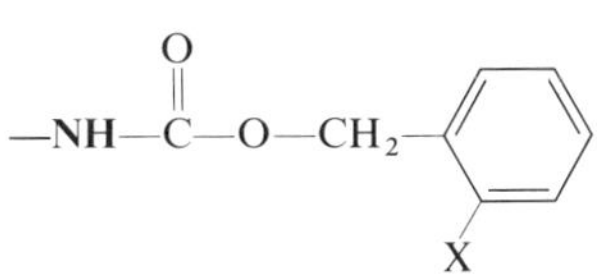

*(**127**) X = Cl, Br

are 60-fold more stable than the N^{ε}-benzyloxycarbonyl group, yet are fully cleaved in HF at 0°C for 30 min. It should be added that N^{ε}-2,4-dichlorobenzyloxycarbonyllysine possesses the same acid-lability as **127** but has the beneficial attribute of crystallinity of its N^{α}-*tert*-butyloxycarbonyl derivative, and that the N^{ε}-*m*-chlorobenzyloxycarbonyl group, which is deprotected in HF at 0°C for 60 min, is now advocated for lysine protection in long peptides (B. W. Erickson, personal communication, 1978).

The *tert*-alkyl urethanes as well as the substituted benzyloxycarbonyl groups resist the actions of nucleophiles, thus allowing these modes of side-chain protection to be applied when protected peptide segments are released from the support with alkali, ammonia, or hydrazine.

b. Anchoring and Coupling. Handling of N^{α},N^{ε}-protected lysine derivatives in solid-phase synthesis presents no particular pitfalls. Coupling yields are 99.4–99.8% per step (Erickson and Merrifield, 1973b; Yamashiro *et al.*, 1976).

Two examples of peptides that have been anchored to a solid support through a urethane linkage (Section III,B,10) of the δ- or ε-amino group of ornithine or lysine are given in Table I (numbers 4a and 4c). Clearly, permanent protection will be required for the α-carboxyl group of the synthetic unit that becomes anchored.

c. Analytical. The merits of various N^{ε}-protected derivatives have been most sensitively elucidated on model peptides containing oligolysine sequences (Yaron and Schlossman, 1968; Grahl-Nielsen and Tritsch, 1969; Ontjes and Anfinsen, 1970; Erickson and Merrifield, 1973b; Yamashiro *et al.*, 1976). Separation on the basis of charge has been achieved by ion-exchange chromatography on carboxymethylcellulose employing salt gradients and peaks have been quantitated by color development with ninhydrin (see Yaron *et al.*, 1972, for correction factor of color constant due to chain length). The earlier peaks containing less than the required number of ionizable groups represented deletion peptides, while peaks eluting beyond the major component were families of branched peptides. Identification of small amounts of branching in an ordinary sequence not possessing the "amplification" effect of the oligolysine models may be a formidable problem.

Erickson and Merrifield (1973b) observed that 30 mol% of the peptide products from a synthesis of $(Lys)_{10}$ Val with N^{α}-*tert*-butyloxycarbonyl and N^{ε}-benzyloxycarbonyl protection were branched. From the distribution of the branched peptides, the apparent first-order rate constant for removal of the benzyloxycarbonyl group from a resin-bound Lys(Z) residue by TFA–CH_2Cl_2 (1:1) could be indirectly determined. This value was about half that determined in solution for deprotection of H-Lys(Z)-OH in the same medium, thus bearing out the earlier generalization (p. 30) about relative rates in the heterogeneous and homogeneous phase.

3. Histidine

a. Protection. The weakly basic imidazole group† ($pK_a = 6.0$) of histidine has been preferably protected for solid-phase synthesis by the N^{im}-*p*-toluenesulfonyl (Tos) (**128**), (Fujii and Sakakibara, 1970, 1974), N^{τ}-2,4-dinitrophenyl (Dnp) (**129**) (Shaltiel, 1967; Chillemi and Merrifield, 1969; Shaltiel and Fridkin, 1970; Stewart *et al.*, 1972) and various N^{im}-urethane groups (**130**) (Ontjes and Anfinsen, 1969a, 1970; Yamashiro *et al.*,

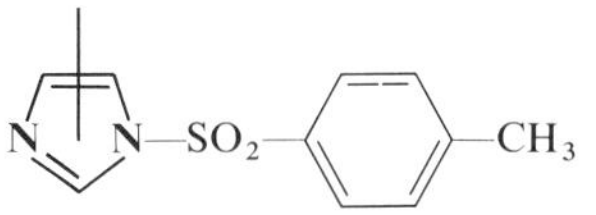

*(**128**) Tos

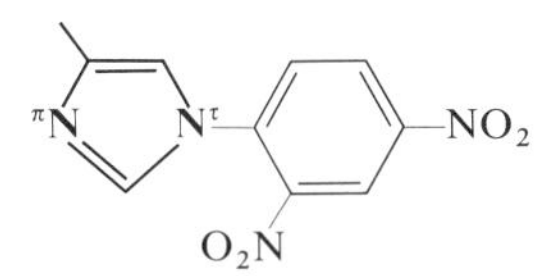

*(**129**) Dnp

(**130**) R as in **118** (* for Boc)

1972). In the past, the imidazole group has been left unprotected (**131**) (e.g. Loffet, 1967; Gutte and Merrifield, 1971) or has been blocked as the N^{im}-benzyl (Bzl) derivative (**132**) (du Vigneaud and Behrens, 1937), but currently there are good reasons not to recommend these two approaches.

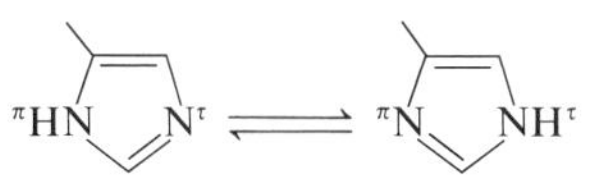

*(**131**)

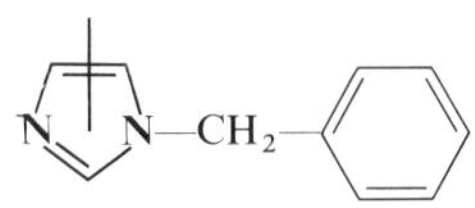

*(**132**) Bzl

Several other interesting means of histidine side-chain protection have been proposed but not extensively tested for solid-phase synthesis; prominent among these are fluorinated acetals [Weygand *et al.*, 1967; Seltzman and Chapman, 1974; see Eq. (20) later in this section and accompanying discussion for structures and details], the N^{im}-diphenyl-4-pyridylmethyl group (**133**)

† The N^{π} and N^{τ}-atoms of the imidazole ring of histidine are defined as shown in formula **131**. There are both nmr spectral (Mathews and Rapoport, 1973) and chemical (Henkart, 1971; Bell and Jones, 1974) criteria to ascertain the position of N^{im}-substituents, although these have been applied only infrequently (see **129** and **134**). Where there is an ambiguity, the imidazole ring has been drawn as in **128**. On steric grounds, substitution is expected at N^{τ}, but N^{π}-derivatives can be synthesized by reaction of an acylating agent with N^{τ}-tritylhistidine, followed by selective removal of the trityl group with mild aqueous acid (Jones and Ramage, 1978).

(Coyle and Young, 1976), and the N^{π}-phenacyl (Pac) group (**134**) (Jones and Ramage, 1978).

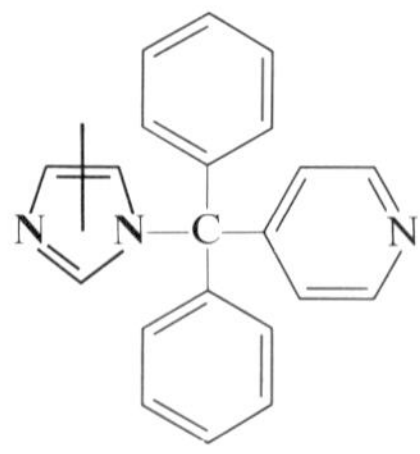

(**133**)

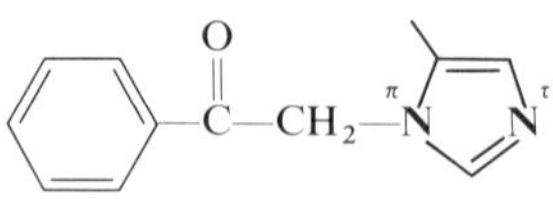

(**134**) Pac

As will be seen, several of the groups listed in the above paragraph provide permanent protection for histidine throughout the synthesis, others are gradually lost in subsequent deprotection/coupling cycles due to exposure to various reagents, and still others only provide temporary protection for the actual coupling step at which histidine is being incorporated into the peptide chain. The consequences of a free imidazole side chain are covered later (pp. 184–186).

The N^{im}-tosyl group (**128**) has achieved wide-spread use because it was reported to confer essentially permanent protection on histidine, yet to be readily removed at the end of a synthesis by liquid HF, at 0°C (Fujii and Sakakibara, 1970). N^{im}-Tosylhistidine is substantially more labile to a host of reagents than N^{ω}-tosylarginine, which in turn is more labile than N^{α}-tosylamino acids. In fact, Boc-His(Tos)-OH is of such marginal stability that it must be stored as its dicyclohexylammonium salt. Incorporation of N^{im}-tosylhistidine into peptide chains imposes certain limitations on the range of compatible chemical reagents which will not cause loss of side-chain protection during the repetitive deprotection/coupling cycles. (1) The removal of N^{α}-Boc groups should be carried out with TFA–CH_2Cl_2 mixtures since substantial partial detosylation (~10%/hr, 25°C) is promoted by 4 *N* HCl–dioxane (Stewart *et al.*, 1972; Fujii and Sakakibara, 1974). Although it has been found that H-His(Tos)-OH is 100-fold more stable in TFA–CH_2Cl_2 (1:1) than common benzyl-based ester and ether derivatives (Erickson and Merrifield, 1973a), indirect results obtained both in solution (Schwartz and Katsoyannis, 1973) and on the solid phase (Hodges and Merrifield, 1975a) are consistent with appreciable detosylation of histidine in peptides after several cycles of exposure to TFA. (2) The N^{im}-tosyl group is quantitatively removed by pyridine hydrochloride (10 equiv in DMF, 25°C for 2 hr; Beyerman *et al.*, 1974), a reagent that is used in several monitoring procedures (e.g., Dorman, 1969). (3) The N^{im}-tosyl group is also lost upon exposure to 1-hydroxybenzotriazole (**104**) when this reagent is used as an additive for

DCC-mediated couplings (Fujii and Sakakibara, 1974; compare to p. 133). (4) In model studies in solution, low levels of aminolytic transfer of the N^{im}-tosyl group onto an α-amino group have been found (Fujii and Sakakibara, 1974); it is problematic whether this side reaction would occur under solid-phase conditions. As a preparative application, N^{im}-tosyl groups are lost under conditions of aminolytic removal of peptides from a solid support, using DMF or methanol as solvents (Stewart *et al.*, 1972; Fujii and Sakakibara, 1974; Chang *et al.*, 1976). It should be noted that the incompatible conditions represented in (1) and (3) are often indispensable in dealing with asparagine and glutamine residues (see specifically pp. 206 and 201). The properties (2) and (4) can be exploited preparatively for mild removal of the N^{im}-tosyl group. Additional removal methods, mentioned for completeness, include catalytic hydrogenolysis and sodium in liquid ammonia.

The N^{im}-dinitrophenyl group (**129**) reduces the pK_a of the imidazole ring to 3.0 (Henkart, 1971). It is very resistant to strong acids—including hot 6 *N* HCl, TFA at 25°C, and liquid HF—but is readily removed by nucleophiles including alkali above pH 12, ammonia in DMF, hydrazine, unsolvated cyanide ion, and thiols such as β-mercaptoethanol and thiophenol (Shaltiel and Fridkin, 1970; Stewart *et al.*, 1972). There is no evidence that the sensitivity of a His(Dnp) residue to nucleophiles leads to complications during the repetitive deprotection/coupling cycles, but this possibility should be kept in mind (for example, it would seem to preclude the use of thiols as scavengers for *tert*-butyl carbocations released during the deprotection of N^{α}-*tert*-butyloxycarbonyl groups). At the conclusion of a solid-phase synthesis, the peptide can be released into solution, for example by acid, with the dinitrophenyl (Dnp) group remaining intact for subsequent orthogonal removal (e.g., Chillemi and Merrifield, 1969; see also Goren and Fridkin, 1978, for a kinetic scheme of solution deprotection, and a review of earlier literature). In an alternative and usually preferred approach the Dnp group can be cleaved while the peptide remains anchored to the solid support (e.g., Lin *et al.*, 1972; Stewart *et al.*, 1972); this approach becomes imperative if the mode of cleavage will involve photolysis, unsolvated fluoride ion, and possibly catalytic hydrogenolysis (Rich and Gurwara, 1975a; Teutsch, 1976; see Sections III,B,1,f; 5,c; and 1,e, respectively) because the His(Dnp) residue will be destroyed under those conditions. An especially reliable and selective deprotection method involves treatment of the peptide–resin with 0.1 *M* thiophenol in DMF, at 25°C for 30 min, as followed by spectrophotometric monitoring for 2,4-dinitrophenyl phenyl sulfide ($\lambda_{max} = 337$ nm, $\varepsilon = 10^4$).

In strong and moderate anhydrous acids, any given N^{im}-urethane **130** is slightly more stable than the corresponding N^{α}- or N^{ε}-urethanes (Sections IV,A,1 and 3; and V,B,2,a), but the N^{im}-urethane group can be selectively removed by hot weak acids, and by nucleophiles sometimes as mild as water and methanol. The N^{α},N^{im}-bis-*tert*-butyloxycarbonyl derivative provides good temporary protection for histidine during the coupling step (Yamashiro *et al.*, 1972). Despite the fact that the free imidazole side chain is liberated at the first N^{α}-*tert*-butyloxycarbonyl deprotection step, this synthetic strategy has met with good success (Yamashiro and Li, 1973a; Li *et al.*, 1975). The N^{im}-*tert*-butyloxycarbonyl group is cleaved not only by TFA, but also by hydrazine hydrate or ammonia in methanol (Schnabel *et al.*, 1971). One might expect that the N^{im}-*tert*-butyloxycarbonyl group would be largely retained in a synthesis based on the N^{α}-2-(*p*-biphenylyl)propyl(2)oxycarbonyl group, but this is not so (Yamashiro *et al.*, 1972). Perhaps, the N^{im}-adamantyloxycarbonyl group (Haas *et al.*, 1966) may be better in this regard; indeed, the Protein Synthesis Group (1975), following Wünsch *et al.* (1968), applied N^{α},N^{im}-bis(Adoc)-histidine to protect the NH_2-terminal residue in the polymer-supported segment condensation synthesis of glucagon, and quantitative stepwise incorporation of this derivative in a solid-phase model study has also been reported (Schaich *et al.*, 1973). A greater measure of permanent protection is provided by the N^{im}-benzyloxycarbonyl group (Patchornik *et al.*, 1957; Sakiyama *et al.*, 1958; Inouye and Otsuka, 1962) which has been successfully applied in the solid-phase syntheses of several large peptides (Ontjes and Anfinsen, 1969a, 1970; Aoyagi *et al.*, 1972; Blake and Li, 1978). This group is readily removed in HBr–acetic acid and HF, and the rate of premature acidolysis in TFA–CH_2Cl_2 is about $\frac{1}{5}$ that of N^{ε}-benzyloxycarbonyl (Yamashiro and Blake, 1978). The main concern with the N^{im}-benzyloxycarbonyl group is eventual intramolecular transfer resulting in termination of α-amino groups (see p. 185).

The N^{im}-benzyl group (**132**) is stable to strong acids, including HF, and is removed by sodium in liquid ammonia or by catalytic hydrogenolysis. The former conditions give rise to side reactions as noted earlier (p. 171), whereas the latter reaction is frequently slow or incomplete (e.g. Woolley, 1966). It may also be accompanied by overreduction, as in the conversion of phenylalanine to β-cyclohexylalanine (Windridge and Jorgensen, 1971b). The N^{im}-1-alkoxycarbonylamino-2,2,2-trifluoroethyl groups **135** and **136** (Weygand *et al.*, 1967) are removed by urethane cleavage in acid, followed by spontaneous hydrolysis of the resultant amino-2,2,2-trifluoroethyl derivative to 2,2,2-trifluoroacetaldehyde hydrate (fluoral) and ammonia [Eq. (20)].

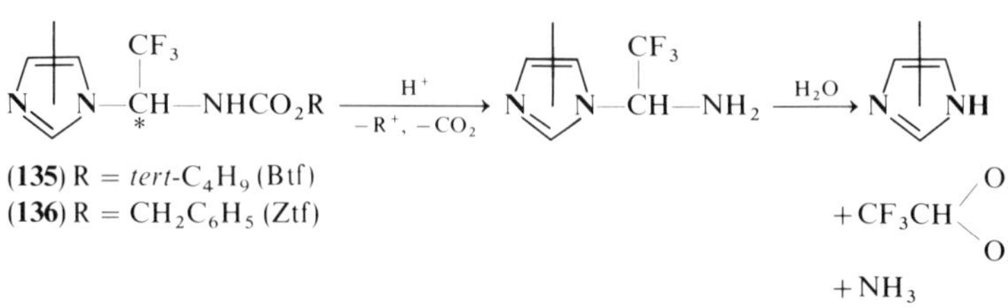

(**135**) R = *tert*-C_4H_9 (Btf)
(**136**) R = $CH_2C_6H_5$ (Ztf)

(20)

Disadvantages of this novel approach include difficulties in synthesizing the appropriate histidine derivatives and in the fact that these contain a second optical center (*). Furthermore, preliminary results suggest that yields of solid-phase couplings are marginal, presumably due to the bulk of the protecting group (R. B. Merrifield, unpublished results, 1968; Schaich *et al.*, 1973; Schou *et al.*, 1975). The N^{im}-1,1,1,3,3,3-hexafluoro-2-(*p*-chlorophenoxymethoxy)propyl group (**137**) (Seltzman and Chapman, 1974), which is removed by 2.4 *N* HCl in acetic acid–H_2O (1:4), at 25°C for 30 min, is optically inactive. Two acid-stable imidazole protecting groups are the

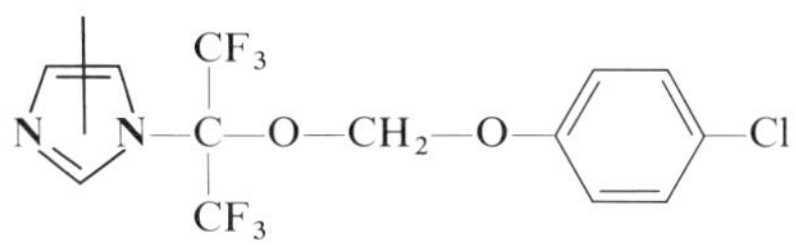

(**137**)

N^{im}-diphenyl-4-pyridylmethyl function (**133**). removable by catalytic hydrogenolysis, zinc in acetic acid, and electrolysis, and the N^{π}-phenacyl group (**134**), removable by zinc in acetic acid.

b. Anchoring and Coupling. The esterification of N^{α},N^{im}-protected histidine to chloromethyl supports (pp. 60–63) is complicated by the possibility of alkylation side reactions. For example, Stewart *et al.*, (1967; see also Stewart and Young, 1969, pp. 8–9) found that although Boc-His(Bzl)-OH was well incorporated by the standard procedure, as judged by weight gain, little benzylhistidine could be recovered upon acid hydrolysis. The desired Boc-His(Bzl)-resin was obtained by acylation of a hydroxymethyl support. The use of a hydroxymethyl-resin appears to be required for anchoring of Boc-His(Tos)-OH as well, due to a base-catalyzed detosylation under standard conditions (Stewart, 1976, p. 274). On the other hand, Boc-His(Dnp)-OH could be esterified without difficulty by the usual procedure involving reaction of the triethylammonium salt with a chloromethyl support (Chillemi and Merrifield, 1969), evidently because the 2,4-dinitrophenyl function (in contrast to the N^{im}-benzyl group) sufficiently depresses the basicity of the imidazole ring.

Side-chain anchoring of histidine residues via a resin-bound dinitrophenylene linkage (Glass *et al.*, 1972) has been detailed earlier (Section III,B,11).

Regarding the activation and coupling of various N^{α},N^{im}-derivatives of histidine in stepwise solid-phase synthesis, it is important to recognize that in these cases, racemization-free incorporation can never be automatically assumed (pp. 186–189). The weight of the available evidence suggests that racemization is minimal with N^{im}-tosyl (**128**), N^{im}-dinitrophenyl (**129**), and

N^{im}-urethane (**130**) derivatives, but unacceptably high with an unprotected imidazole (**131**) or N^{im}-benzyl (**132**) group.

Some practical remarks about several protected derivatives of histidine follow. Boc-His-OH, with a free imidazole side chain, is nearly insoluble in CH_2Cl_2; this derivative must therefore be coupled in neat DMF or at best in DMF–CH_2Cl_2 (1:1). Coupling rates and yields in this case have occasionally been poor (see below), possibly due to a combination of chemical and solubility reasons. The solubility of Boc-His(Bzl)-OH is slightly better, but only those N^{α},N^{im}-derivatives in which the basicity of the imidazole ring is entirely suppressed, e.g., N^{im}-tosyl (**128**), N^{im}-dinitrophenyl (**129**), and N^{im}-urethanes (**130**), are readily soluble in CH_2Cl_2. It has been the experience of our laboratory and others that stocks of dry, solid Boc-His(Dnp)-OH undergo decomposition upon standing as evidenced by heterogeneity on thin layer chromatography; recrystallization from hot isopropanol on the same day as the intended coupling seems to be adequate to avoid introducing impurities into the resin-bound peptide. Several comparative studies on the solid-phase coupling of many histidine derivatives under standardized conditions have been published (Stewart *et al*., 1972; Schaich *et al*., 1973; Schou *et al*., 1975).

Histidine analogs including C^{β}-(2-pyridyl)alanine, 4-fluorohistidine, homohistidine, 4-imidazolylglycine, N^{π}-methylhistidine, and N^{τ}-methylhistidine have been incorporated into peptides by solid-phase procedures (Jorgensen *et al*., 1972; Dunn *et al*., 1974; van Batenburg *et al*., 1976a,b, 1977). The racemization question was generally not investigated in these studies, although in certain cases sizeable levels were unavoidable.

c. Side Reactions. *(i) Modifications of Free Imidazole Ring.* The coupling of N^{α}-protected side-chain unprotected histidine has been observed to proceed at slower than normal rates (B. Floyd, cited by Erickson and Merrifield, 1976, p. 329); in some cases, the yields were found to be unacceptably low (e.g., Ontjes and Anfinsen, 1970; Yamashiro *et al*., 1972). The intramolecular cyclization of the initial activated species **138**, to give the azolide **139** [Eq. (21); Sheehan *et al*., 1959; Veber, 1975], could help

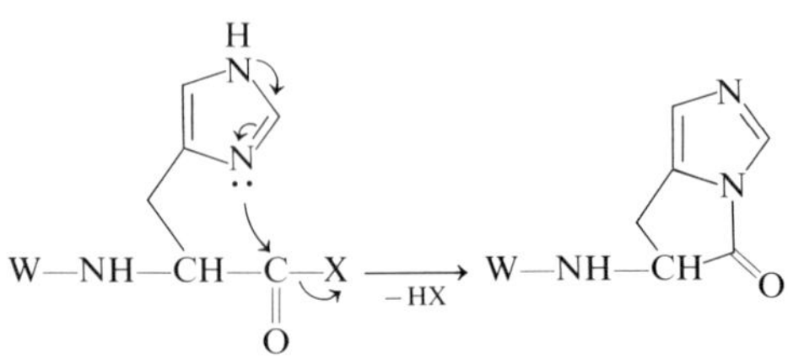

(21)

(138) **(139)**

account for these results. The activated amide **139** will still react with the peptidyl amino group, but at a slower rate than **138**.

A free imidazole ring may be present in histidine residues of the polymer-supported peptide chains either because histidine was introduced with the side chain unprotected, or else because the N^{im}-protecting group is prematurely lost by acidolysis during the N^{α}-deprotection step (see p. 180ff) During the steps to couple subsequent protected amino acids, which are added in excess to drive the acylation of the α-amino group to completion, substantial N^{im}-acylation (25–70%) also takes place (B. Gutte and R. B. Merrifield, unpublished results, 1968; Yamashiro *et al.*, 1972). Usually, there is no chance for a branched peptide chain to grow, because the imidazole group is quantitatively deacylated in the TFA–CH_2Cl_2 (1:1) applied in the next deprotection step to remove an N^{α}-*tert*-butyloxycarbonyl group. Deacylation is probably not significantly promoted by milder acids used to remove N^{α}-2-(biphenylyl)propyl(2)oxycarbonyl and related groups (Section IV,A,2); thus, Yamashiro *et al.* (1972) found that after four coupling cycles subsequent to incorporation of Bpoc-His(Boc)-OH, the titer of free resin-bound amino groups rose by 35%. These data suggested partial side-chain deprotection followed by branching; but no branched peptides were detected by chromatography. It is quite likely that deacylation will occur readily in conjunction with the removal of base-labile or thiolysable α-amino protecting groups (Section IV,A,4–6), although quantitative data are not available. The occurrence of N^{im}-acylation means that N^{α}-protected amino acids are required in excess during coupling steps; this process might explain the observation of a deletion peptide found specifically in a synthesis where the imidazole of histidine was free (Stewart *et al.*, 1972). In the sequence reported, Boc-nitroarginine was poorly incorporated in the fourth coupling cycle after introduction of His^6 of angiotensin.

Another potential problem with N^{im}-acylation is a low-level intermolecular transfer of the acyl group to the α-amino group of another peptide chain, leading eventually to amino acid insertion. Such a side reaction is consistent with the reactivity of acyl imidazoles, but it is unlikely to be important in most synthetic schemes because the relative rates and sequence of steps minimize the time that unprotonated free α-amino groups and acylated imidazole side chains will be present simultaneously. N^{im}-Urethanes are not as sensitive to nucleophiles as N^{im}-acyl compounds, but with a sufficiently acid-stable urethane, the risk of an intermolecular $N^{im} \rightarrow N^{\alpha}$ shift accompanied by chain termination needs to be considered.

Dicyclohexylcarbodiimide reacts with a free imidazole group to give the guanidine addition product **140** (Rink and Riniker, 1974). The side reaction is mechanistically similar to several discussed earlier (pp. 128–129). Of particular interest is the marked extent of catalysis by weak acids such as

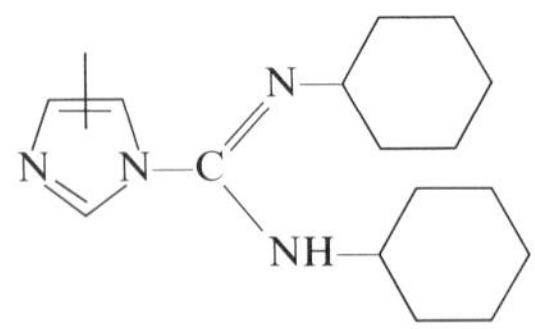

(140) $\pi:\tau = 1:3$

1-hydroxybenzotriazole (**104**) and pyridine hydrochloride. The reaction was found to be best reversed by heating with methanol at reflux, 2 hr. Alternatively, cleavage has been stated to be complete under the conditions for N^{α}-Boc deprotection with TFA (Ivanov *et al.*, 1976).

(ii) Racemization in Stepwise Synthesis. The fact that N^{α}-urethane protection is not sufficient to prevent racemization of certain histidine derivatives in stepwise solid-phase synthesis was first discovered by Windridge and Jorgensen (1971a,b), and has been further studied both on the solid phase (Lin *et al.*, 1972, Khosla *et al.*, 1972b) as well as in solution (Syrier and Beyerman, 1974; Beyerman *et al.*, 1974; Veber, 1975; Jones and Ramage, 1978). A likely mechanism for this unique (contrast to p. 128) racemization involves reversible intramolecular abstraction of the α-proton of the initial activated species **138** by the basic N^{π} of the imidazole ring [Eq. (22); compare to Eq. (21)]. This mechanism predicts that

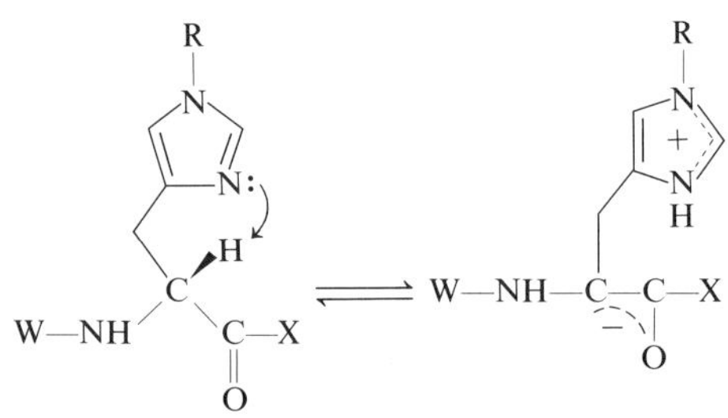 (22)

(138)

racemization will be suppressed by (1) N^{im}-protection with electron-withdrawing groups which lower the basicity of the imidazole ring, specifically by the N^{im}-tosyl (**128**), N^{im}-dinitrophenyl (**129**), N^{im}-urethane (**130**), and N^{im}-1-benzyloxycarbonylamino-2,2,2,-trifluoroethyl (Ztf) (**136**) groups and (2) specific protection of the N^{π} position, as in the phenacyl derivative **134**. Both of these predictions have been generally borne out experimentally (Table XIV). In addition, (3) the beneficial effects of 1-hydroxybenzotriazole (p. 133) have been shown; the difference (Table XIV, compare numbers 6 and 10) is too great to be simply due to a kinetic effect. It is advised that every synthetic peptide containing histidine be checked for the optical integrity of this residue (Section V,B,3,d, as follows).

d. Analytical. The monitoring by titration of a solid-phase peptide synthesis once a histidine residue has been incorporated is fraught with difficulties, especially at the stage of analysis where low values of free α-amino groups must be measured (Schaich *et al.*, 1973; Hodges and Merrifield, 1975a; Schou *et al.*, 1975; compare to Table XI). N^{im}-Tosylhistidine is incompatible, for one reason or another, with procedures based on pyridine hydrochloride (Dorman, 1969), picric acid (Gisin, 1972), or perchloric acid (Brunfeldt *et al.*, 1972a). In the most accurate and sensitive modification of prior methodology, Hodges and Merrifield (1975a) were able to nearly suppress the titration of N^{im}-dinitrophenylhistidine residues by the use of 0.1 *M* imidazolium picrate in DMF–CH_2Cl_2 (1:9). The extent of the couplings reactions could thus be estimated to $\pm 1\%$.

The reactions of an N^{im}-dinitrophenylhistidine residue in peptides can be monitored spectrophotometrically (Shaltiel and Fridkin, 1970). His(Dnp) survives hydrolysis in 6 *N* HCl at 110°C and can be quantitated as a peak between phenylalanine and lysine on standard amino acid analysers (Stewart and Young, 1969, p. 54; Henkart, 1971). In contrast, N^{im}-tosyl and N^{im}-urethane groups are lost upon acid hydrolysis and free histidine must then be quantitated. The quantitative Pauly reaction (see Stewart and Young, 1969, pp. 56–57) is primarily specific for a free, not blocked, imidazole group, and thus facilitates detection and analysis of histidine-containing peptides.

The question of histidine racemization can be addressed at three levels of sensitivity. Enzymatic digestion of the peptide is the most widely applicable but least sensitive method, while the chromatographic separation of diastereomers of a model peptide is the most precise method, but is of only limited scope (Windridge and Jorgensen, 1971a). Acid hydrolysis of a complex histidine peptide followed by specific determination of DHis chromatographically (Lin *et al.*, 1972; following Manning and Moore, 1968) or enzymatically (Jorgensen *et al.*, 1970; Syrier and Beyerman, 1974) is readily feasible, but the accuracy of the results is no better than ± 0.5–1% because of the need to correct for background racemization ($\sim 3\%$) during the hydrolysis. Incidentally, Khosla *et al.* (1973) reevaluated their widely cited earlier results (Khosla *et al.*, 1972b), in which nontrivial amounts of racemization were found even with N^{im}-tosyl protection, and concluded that the observed inconsistencies were due to racemization during their hydrolysis step.

Table XIV. Racemization of Histidine in Synthetic Peptides

No.	Derivative	Coupling method[a]	Solvent	DHis (%)	Analytical method[b]	References[c]
1.	Boc-His-OH	DCC(SPS)	DMF	5	M	A
2.	Boc-His-OH	DCC(SPS)	DMF–CH_2Cl_2	2.3	H	D
3.	Boc-His-OH	MA(Sol)	DMF	<1	H	E
4.	Boc-PyAla-OH[d]	DCC + HOBt (SPS)	CH_2Cl_2	~50[d]	H	C
5.	Boc-His(Bzl)-OH	DCC(Sol)	CH_2Cl_2	1.8	M	A
6.	Boc-His(Bzl)-OH	DCC(SPS)	CH_2Cl_2	11	M	A
7.	Boc-His(Bzl)-OH	DCC(SPS)	CH_2Cl_2	12	E,H	A
8.	Boc-His(Bzl)-OH	DCC(SPS)	DMF	25	M	A
9.	Boc-His(Bzl)-OH	DCC + HOSu (Sol and SPS)	CH_2Cl_2	<0.1	M	A
10.	Boc-His(Bzl)-OH	DCC + HOBt (SPS)	CH_2Cl_2	<0.3	M	B
11.	Boc-His(Bzl)-OH	NEPIS(Sol)	DMF	24	M	A
12.	Boc-His(Bzl)-OH	NEPIS(SPS)	DMF	36	M	A
13.	Boc-His(Bzl)-OH	EEDQ(SPS)	CH_2Cl_2	10	M	A
14.	Boc-His(Bzl)-OH	MA(Sol)	DMF	7	H	E
15.	Boc-His(Bzl)-OH	CDI(SPS)	CH_2Cl_2	0.3	M	A
16.	Boc-His(Bzl)-ONp	AE(SPS)	DMF	4.3	M	A
17.	Boc-His(Ztf)-OH	MA(Sol)	DMF	4	H	E
18.	Boc-His(Ztf)-OH	DCC(SPS)	CH_2Cl_2	0.5	M	A
19.	Boc-His(Tos)-OH	DCC(SPS)	CH_2Cl_2	0.5	H	D
20.	Boc-His(Tos)-OH	MA(Sol)	DMF	<1	H	E
21.	Boc-His(Dnp)-OH	DCC(SPS)	CH_2Cl_2	1	M	A
22.	Boc-His(Dnp)-OH	DCC(SPS)	CH_2Cl_2	0.6	H	D
23.	Boc-His(Dnp)-OH	MA(Sol)	DMF	6	H	E

24.	Boc-His(Boc)-OH	MA(Sol)	DMF	<1	H	E
25.	Z-His(Z)-OH	MA(Sol)	DMF	<1	H	E
26.	Z-His(τPac)-OH	DCC(Sol)	DMF	35	H	F
27.	Z-His(πPac)-OH	DCC(Sol)	DMF	<2	H	F
28.	Boc-His(πPac)-OH	DCC(SPS)	DMF–CH_2Cl_2	2	H	G

[a] All couplings performed at 25°C. The yields are not necessarily quantitative, for reasons discussed elsewhere. Abbreviations: Sol. = in solution; SPS = solid-phase conditions; MA = mixed anhydride; NEPIS = *N*-ethyl-5-phenylisoxazolium-3′-sulfonate (Table X, number 4); EEDQ = 1-ethoxycarbonyl-2-ethoxy-1,2-dihydroquinoline (Table X, number 5); CDI = *N*,*N*′-carbonyldiimidazole; AE = active ester.

[b] Analytical method (see p. 187):
M = chromatographic resolution of diastereomeric model dipeptides (±0.1%)
H = analysis of acid hydrolysate (±1%)
E = enzymatic digest (±5%)

[c] References:
A = Windridge and Jorgensen (1971a);
B = Windridge and Jorgensen (1971b);
C = Jorgensen *et al.* (1972);
D = Lin *et al.* (1972);
E = Beyerman *et al.* (1974);
F = Jones and Ramage (1978);
G = Fletcher *et al.* (1979).

[d] PyAla = C^{β}-(2-pyridyl)-L-alanine. Note that ~50% of the D-isomer indicates that racemization was essentially complete.

4. Aspartic and Glutamic Acid

a. Protection. The acidic β- and γ-carboxyl groups (p$K_a \simeq 4$) of aspartic and glutamic acid have been protected as substituted benzyl esters (**141**–**144**) (Guttmann and Boissonnas, 1958; Ledger and Stewart, 1965; Prestidge *et al.*, 1976; Suzuki *et al.*, 1976b, 1977; Yamashiro, 1977), phenacyl (OPac) esters

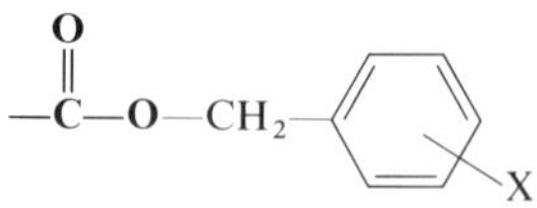

(**141**) X = *p*-OCH_3 Mob
*(**142**) X = H Bzl
(**143a**) X = *p*-Br
(**143b**) X = *p*-Cl
(**144**) X = *p*-NO_2 ONb

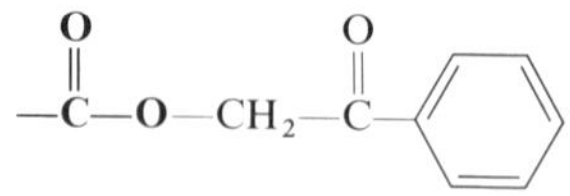

*(**145**) OPac

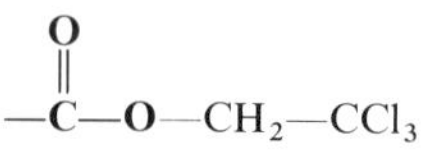

(**146**) OTce

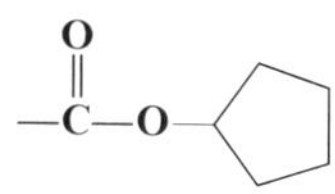

(**147**)

(**145**) (Sheehan and Daves, 1964; Yang and Merrifield, 1976; Suzuki *et al.*, 1976b, 1977), cyclopentyl esters (**147**) (Blake, 1979), and *tert*-butyl esters (O*t*Bu) (**77**) (Roeske, 1959, 1963; Anderson and Callahan, 1960; Schwyzer and Kappeler, 1961; Schwyzer and Dietrich, 1961). Much of the known chemistry of carboxyl protection (reviewed by Haslam, 1973) has also been channeled toward the development of anchoring linkages for solid-phase synthesis (Sections III,B,1–6).

The choice of carboxyl protecting groups will vary depending on the synthetic strategy, including considerations of ways to minimize certain serious side reactions (described on pp. 193–198). In the most common solid-phase protection scheme of N^{α}-*tert*-butyloxycarbonyl group removal with TFA and permanent protecting group removal in HBr–TFA or HF, the unsubstituted benzyl esters **142** provide the proper balance of rates of acidolysis. Based on quantitative data (Erickson and Merrifield, 1973a), only 0.01% of the ω-benzyl ester groups are estimated to be lost at each N^{α}-deprotection step. Although this margin of safety is more than adequate for even large-sized target peptides, two research groups have recommended the *p*-halobenzyl derivatives **143** for solid-phase peptide synthesis (Yamashiro *et al.*, 1973; Prestidge *et al.*, 1976; Yamashiro, 1977). The reported extent of enhanced stability to acid ranged from 1.5- to 5-fold, but removal was quantitative in liquid HF at 0°C within 10 min. Finally, the cyclopentyl esters (**147**) are 14-fold more stable in TFA than **142**, but are again readily cleaved in HF at 0°C (Blake, 1979).

The *p*-nitrobenzyl (ONb) (**144**) and phenacyl (OPac) (**145**) esters are both essentially stable to the strong acidolytic final cleavage conditions. Both groups are removed later, in solution, by reductive methods such as catalytic hydrogenolysis or zinc in aqueous acetic acid (e.g., Taylor-Papadimitriou *et al.*, 1967; J. D. Young *et al.*, 1967; Hendrickson and Kendall, 1970; Suzuki *et al.*, 1976b, 1977). The properties of the acid-stable but zinc-labile β,β,β-trichloroethyl esters (**146**) (Woodward *et al.*, 1966; Marinier *et al.*, 1973) are of comparative interest. The listed reductive methods occasionally do not go to completion, especially on larger substrates, and serious side reactions have also been reported in certain cases. A reliable and quantitative method for removal of the phenacyl protecting group **145**, which can be carried out while the peptide is still covalently attached to the solid support, involves treatment with 1 *M* sodium thiophenoxide in dimethylformamide at 25°C for 8 hr (Yang and Merrifield, 1976; based on Sheehan and Daves, 1964; $t_{1/2}$ in solution was 11 min).

The primary alkyl ester side-chain protecting groups **141**–**146** are susceptible to nucleophilic reagents in the same way as are the common substituted benzyl ester linkages to the solid support (Section III,B,1,c). Side-chain deprotection will occur in alkali, partial conversion to asparagine and glutamine residues will ensue upon aminolysis, the unnatural side-chain hydrazides will form upon exposure to hydrazine, and transesterification will take place with alcohols in the presence of tertiary amines. Hence, selective cleavage of protected peptide segments incorporating aspartic or glutamic acid residues is impractical by conventional strategies. Limited selectivity is possible when the anchor to the resin is a phenacyl ester, with cleavage by potassium cyanide–crown ether complex (Tam *et al.*, 1977; see p. 85), or with an activated *n*-alkyl ester cleaved by fluoridolysis (p. 89); however, these procedures introduce new difficulties. Preservation of side-chain ester protection can be achieved, however, when a photolysable anchor is employed (pp. 81–82 and 85–86).

When highly acid-labile or orthogonally removable α-amino protecting groups are used, the *tert*-butyl esters (**77**) may be selected for side-chain protection. These are readily removable in trifluoroacetic acid solutions, and as an additional advantage, resist the action of nucleophiles. For completeness, *p*-methoxybenzyl esters (**141**) (Weygand and Hunger, 1962) are mentioned. These derivatives, which can be conveniently prepared through the diazo compound (Offord *et al.*, 1976) possess a lability to acids close to that of *tert*-butyl esters.

b. Anchoring. The creation of an ester linkage to the solid support involving the α-carboxyl group of protected derivatives of aspartic or glutamic acid is possible through a wide range of methods. The use of

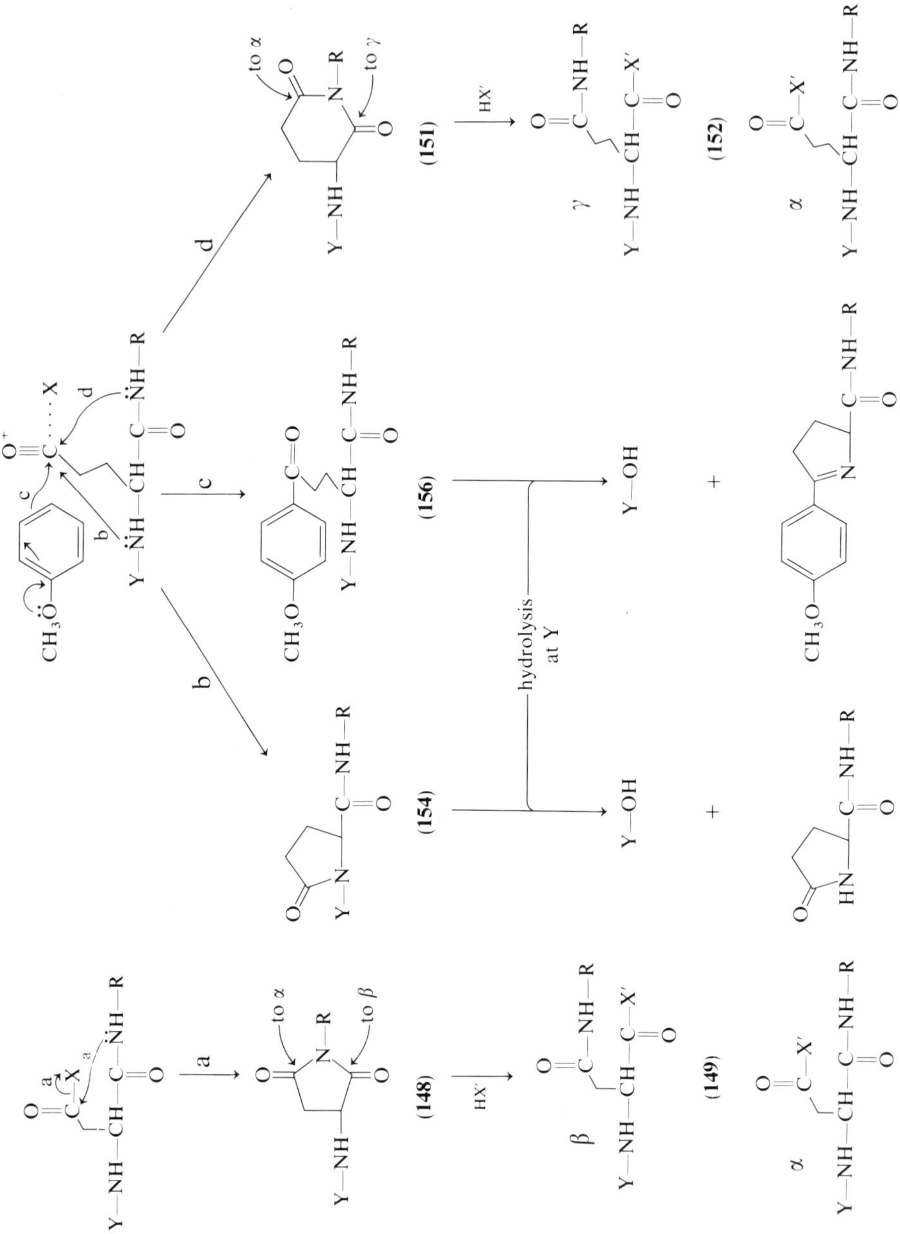

alcoholic solvents during this step should be avoided because of the possibility of transesterification of the ω-carboxyl protecting group (Bayer *et al.*, 1968; Marglin, 1971). Side-chain anchoring of glutamyl residues to the support through the ω-carboxyl group has been demonstrated by Krumdieck and Baugh (1969), Meienhofer *et al.* (1970a), and Grzonka *et al.* (1974) (see Table VII, note 1 on p. 68).

c. Side Reactions. Serious side reactions have been observed at aspartic and glutamic acid residues under certain conditions of the final cleavage and removal of permanent protecting groups in solid-phase synthesis. The principal pathways are schematized in Fig. 17; it is interesting to note that the difference of a single methylene group distinguishing aspartate and glutamate is accompanied by a strikingly different spectrum of side reactions. Fortunately, methods are available to minimize or circumvent most of these problems.

(i) Intramolecular Acylation Leading to Imides. The $\alpha \rightarrow \beta$ rearrangement at aspartyl residues (pathway a) proceeds through the isolable succinimide derivative **148** and is sequence dependent. The reaction was originally demonstrated in solution under the base-catalyzed conditions of alkaline saponification of methyl or benzyl esters (Battersby and Robinson, 1955; see also Iselin and Schwyzer, 1962). Cleavage of model peptides containing an aspartyl-alanine sequence by saponification of a benzyl ester linkage to the solid support produced an equilibrium mixture containing predominantly the undesired β-isomer **149** (75–85%; Marshall and Merrifield, 1965). In contrast, acidolytic cleavage with HBr–TFA (Table VIII, number 3) gave exclusively the α-isomer **150**.

The acid-catalyzed intramolecular acylation is most severe at the sterically favorable aspartyl-glycine or aspartyl-serine sequences (Merrifield, 1967; Ondetti *et al.*, 1968; S. S. Wang *et al.*, 1974; Yang and Merrifield, 1976; Blake, 1979), and has also been noted in solution with an aspartyl-histidine sequence, where additional intramolecular catalysis could be a factor (Baba *et al.*, 1973). The reaction is promoted by stronger acids (e.g., anhydrous HF, Table VIII, number 4), longer reaction times, and higher temperatures. The accumulated data were consistent with the conclusion that for substantial

---

Figure 17. Schematic pathways of, and possible products arising from side reactions of protected or free derivatives of aspartic acid, asparagine, glutamic acid, and glutamine, as further discussed in the text of Sections V,B,4,c and 5,c (X,X′ = OH, NH_2, OBzl, O*t*Bu; Y = H, Boc, peptidyl, etc.; R = H or remainder of peptide). (a) $\alpha \rightarrow \beta$ rearrangement through succinimide intermediate **148**; (b) pyrrolidone **154** formation; (c) γ-acylation of anisole; (d) $\alpha \rightarrow \gamma$ rearrangement through glutarimide intermediate **151**.

levels of the side reaction to be observed, both the α-amino and α-carboxyl groups of the dipeptide unit must be blocked, and the β-carboxyl should preferably be blocked as an ester. The succinimide intermediate **148** is the only by-product formed directly, and the equilibrium mixture of α- and β-peptides is established after treatment with 1% aqueous ammonium bicarbonate or triethylamine. For example, HF cleavage of Boc-Glu(OBzl)-Asp(OBzl)-Gly-Thr(Bzl)-resin gave 99% of the imide and a 3:7 α:β ratio after hydrolysis. Acid-catalyzed imide formation from this model peptide did not occur to any appreciable extent when the β-carboxyl group was free.

The base-catalyzed succinimide formation followed by α → β rearrangement is substantially more facile than the acid-catalyzed one, as suggested by the results cited earlier on the aspartyl-alanine sequence [see Bodanszkey and Kwei (1978), for a systematic study of the sequence dependence, determined in solution on dipeptide models]. The danger of this acylation side reaction, together with the susceptibility of esters to nucleophiles, renders those solid-phase procedures relying on basic reagents to cleave the anchoring linkage unsatisfactory for most aspartic acid-containing peptides.

Several results observed in solution suggest the scope of the discussed side reaction. The model compound Boc-Asp(OBzl)-Asn-Tyr(Bzl)-OMe was converted to the imide with a half-time of 1 hr by 0.03 *M* triethylamine (1 equiv) in DMF; but surprisingly, the rate of the side reaction was suppressed by two orders of magnitude with the addition of 1-hydroxybenzotriazole (**104**) (1 equiv) to the triethylamine (2 equiv)–DMF solution (Bodanszky *et al.*, 1978; see Martinez and Bodanszky, 1978, for a study of the effects of other additives). These workers also showed that the β-phenacyl ester of an aspartyl-valine protected model peptide underwent base-catalyzed imide formation at a rate 40-fold faster than that measured for the corresponding β-benzyl ester (Bodansky and Martinez, 1978). Quantitative succinimide formation from Boc-Asp(OBzl)-Gly-OMe has been noted within 30 min upon mild treatment with unsolvated cyanide ion (J. P. Tam, unpublished results, 1978; compare to p. 85). Even the normally very stable aspartic acid β-*tert*-butyl ester residue, as incorporated into several peptide segments, may lose *tert*-butanol, presumably via imide formation, when subjected to alkali (Schwyzer *et al.*, 1963b), hydrazine (Smithers, 1971), or unsolvated fluoride ion (Sieber, 1977; see p. 89). However, no α → β rearrangement of Z-Asp(O*t*Bu)-Gly-OC_6H_5 was observed after a sequential 3-step treatment consisting of: (1) selective peroxide cleavage of the phenyl ester; (2) catalytic hydrogenolysis of the benzyloxycarbonyl group; and (3) acidolysis of *tert*-butyl ester in TFA (Kenner and Seely, 1972; compare to p. 87).

It has generally been assumed that intramolecular acylation is not promoted to any appreciable extent by any of the reagents used during the repetitive deprotection/coupling cycles used to elaborate the polymer-supported growing peptide chain during solid-phase synthesis. Experimental support for this idea is readily provided in the cases of small peptides containing vulnerable sequences, because as will be shown, subtle variations in the protection strategy and cleavage conditions are reflected in dramatic differences in the levels of isolated undesirable β-peptides. It now appears that low-level imide formation also occurs at the neutralization step (Section IV,B) of the synthesis, and for a long peptide, the amount of by-product accumulated through many cycles could be comparable to the amount formed in a single acid cleavage step (Blake, 1979). The level of the side reaction was estimated at 0.7% per step when an aspartyl-glycine sequence protected by a β-benzyl ester was exposed to 5% diisopropylethylamine–CH_2Cl_2 and was even higher when the base was triethylamine. On the other hand, the side reaction was negligible with an aspartyl-leucine sequence. As a general precautionary measure, the use of the milder base, *N*-methylmorpholine, has been recommended, (Blake and Li, 1978).

Based on the mechanistic information just presented, several strategies have been proposed and developed to circumvent imide formation and $\alpha \rightarrow \beta$ rearrangement. Jorgensen *et al.* (1971a) were interested in preparing an angiotensin-II analog with NH_2-terminal aspartyl-glycine. The des(Asp)-peptide was prepared by standard solid-phase methodology and coupled with N^{α}-benzyloxycarbonyl β-benzyl aspartate *p*-nitrophenyl ester in solution. The desired peptide was obtained after catalytic hydrogenolysis to remove the benzyl-based protecting groups (see Yang and Merrifield, 1976, note 17, for an example of an aspartyl-glycine-containing peptide which gave appreciable amounts of the imide upon catalytic hydrogenolysis). A more general approach (S. S. Wang *et al.*, 1974) utilized *tert*-butyl esters for side-chain protection, along with the N^{α}-2-(*p*-biphenylyl)propyl(2)oxycarbonyl group and *p*-alkoxybenzyl ester anchor. Succinimide formation did not occur during cleavage from the support and side-chain carboxyl deprotection in TFA–CH_2Cl_2 (1:1). Furthermore, additional protecting groups could be safely removed in liquid HF, because the peptide with a free β-carboxyl group was not expected to be prone to cyclization. In a procedure based on the same principle, but retaining the common solid-phase combination of N^{α}-*tert*-butyloxycarbonyl protection and a *p*-alkylbenzyl ester anchoring linkage to the support, aspartic acid was introduced as its β-phenacyl ester (**145**) (Yang and Merrifield, 1976). The side-chain protecting group was removed prior to HF cleavage, and imide formation occurring during the latter step was reduced as much as 40-fold by comparison to the control synthesis using β-benzyl ester (**142**) protection. In another sequence examined (Glu-Asp-Ser-Leu), imide formation was 25% with the β-benzyl ester, and 9.5% with the free β-carboxyl group previously protected as the phenacyl ester. Bearing in mind the results of Bodanszky and Martinez (1978) as well as Blake (1979) cited earlier, it is entirely possible that much of the residual imide formed when β-phenacyl protection was used may have been due to a base-catalyzed reaction during the synthesis. Suzuki *et al.* (1976b) have reported that with model dipeptides, the acid-stable phenacyl ester could be removed in solution (see below) *subsequent* to HF cleavage from the support. The same strategy was also advocated (see also Suzuki and Endo, 1978) in the application of β-*p*-nitrobenzyl esters (**144**). However, since Nitecki *et al.* (1971) showed that use of these latter esters provided even less protection against imide formation than did the ordinary β-benzyl esters (**142**), the interesting strategic proposal of Suzuki *et al.* (1976b) cannot be accurately evaluated at the present time.

As already mentioned, a good way to minimize low-level base-catalyzed imide formation during solid-phase synthesis is to control the tertiary amine used for neutralization. A different approach involves a change to the β-cyclopentyl ester (**147**) (Blake, 1979). The rates of rearrangement of an aspartyl-glycine model sequence protected respectively by **147** and **142** (unsubstituted β-benzyl ester) were comparable in acid, but differed by more than an order of magnitude in base due to the steric hindrance of the secondary alkyl ester. Hence, **147** may be advantageous for syntheses of long peptides.

The $\alpha \rightarrow \gamma$ rearrangement of glutamyl peptides (Fig. 17, pathway d) occurs readily in alkali (Battersby and Robinson, 1955; Lieflander, 1960; see also Sondheimer and Holley, 1954). The intermediate glutarimides **151** are considerably less stable than the succinimides **148**, and open in a way to give almost exclusively the undesired γ-peptide **152**. In the most common modes of solid-phase synthesis, where cleavage and deprotection reactions are conducted in acid, formation of either glutarimides or γ-peptides has never been observed, but other side reactions prevail (Feinberg and Merrifield, 1975).

(ii) γ-Acylation at Glutamic Acid. In a very strong acid such as liquid HF (Table VIII, number 4) the free γ-carboxyl side chain of glutamic acid becomes protonated and reversibly loses water to form an acylium ion (Olah *et al.*, 1970). Intramolecular γ-acylation of the glutamyl nitrogen (Fig. 17, pathway b) then gives the five-membered pyrrolidone ring **154**; as stated above, the alternate pathway to give the six-membered glutarimide **151** does not appear to occur. Formation of **154** in neat HF at 25°C for 2 hr occurred in 50–70% yield, and the by-product was further characterized by the facile, mild basic hydrolysis of the diacylamine to give the predicted mixture of fragments, all prepared by alternate, unambiguous routes (Feinberg and Merrifield, 1975). When HF cleavage is performed in the presence of anisole as a scavenger (Table IX), the acylium ion is trapped intermolecularly (Fig. 17, pathway c). The anisylated adduct **156** (Sano and Kawanishi, 1975; Feinberg and

Merrifield, 1975) was observed in 50–70% yield in HF–anisole (9:1) at 25°C for 2 hr, while at the same time, the amount of pyrrolidone **154** was 10%. With NH_2-terminal acylated glutamic acid, a cyclic Schiff base **157** was noted.

The γ-acylation side reactions are suppressed in three different ways.

1. Careful control of the time and temperature of the HF–anisole cleavage mixture was studied by Feinberg and Merrifield (1975). Reaction at 0°C for 30 min was considered to be a workable compromise under which conditions the desired intact unmodified model glutamyl peptide represented 99% of the product mixture, and the side-chain protecting groups were fully removed.
2. Weaker acids have been used, including HBr–TFA (Table VIII, number 3; tested by Feinberg and Merrifield, 1975) or HF with its acidity attenuated by pyridine as an inert diluent (Table VIII, number 5; Sugano *et al.*, 1978). The acids are still of sufficient strength to deblock side-chain benzyl-based derivatives. γ-Acylation is not expected when a protection scheme utilizing *tert*-butyl-based derivative and no acid stronger than trifluoroacetic is employed.
3. When acid-stable protecting groups, specifically *p*-nitrobenzyl (**144**) and phenacyl (**145**) esters, are used, γ-acylation cannot take place during the HF–anisole step (Suzuki *et al.*, 1977). Removal is via orthogonal routes already mentioned.

It is of interest to contrast the mechanistic requirements of the acid-catalyzed side reactions of aspartyl and glutamyl residues (Fig. 17). In HF, succinimides readily form from susceptible sequences, while the homologous glutarimides have not been detected. Acylation of anisole by the β-carboxyl of an aspartyl residue does not appear to occur, but the corresponding γ-acylation involving glutamyl residues is significant under conditions already outlined. The γ-acylation proceeds with a free side-chain carboxyl group, whereas succinimide formation occurs more readily with a β-ester.

(iii) Miscellaneous. Whereas low level reactions of the benzyl ester linkage to the solid support with nucleophiles merely lead to lower yields due to loss of chains, the same reactions of side-chain benzyl ester protecting groups give rise to undesired modified peptides. As a case in point, Southard (1971) characterized a peptide containing an Asp (OMe) residue, which was obtained from a solid-phase synthesis of a pentapeptide containing an Asp(OBzl) residue. The side reaction was attributed to the proximity of a triethylamine treatment to a methanol wash during the synthetic cycle. It could be reversed by saponification for 1 hr at pH 13. Switching to a hindered alcohol wash solvent ought also to suppress the observed transesterification.

As described later (pp. 205–206), pyroglutamyl formation is a significant side reaction occurring at NH_2-terminal glutamine in peptides. Indirect evidence for the occurrence of this process with β-benzyl protected glutamic acid (e.g., Polzhofer and Ney, 1970a; Bonora *et al.*, 1974; Birr and Pipkorn, 1979) has generally been open to alternative explanations. However, Rosenblatt *et al.* (1977b) have now given convincing proof, based on chromatographic isolation, end-group analysis, Edman degradation sequencing, and enzymatic digestion with pyroglutamyl aminopeptidase, that the postulated side reaction took place with Glu(OBzl) under solid-phase conditions.

It has long been known that the peptide bond on the carboxyl side of an aspartic acid residue is some 100-fold more labile to hot dilute *aqueous* acid than most peptide bonds (reviewed by Schultz, 1967). The dipeptide sequence aspartyl-proline is especially susceptible, even at ambient temperature (Piszkiewicz *et al.*, 1970). Only 0.3 to 0.5% peptide bond hydrolysis was found with Boc-Asp(OBzl)-Pro-OBzl following HF–anisole treatment at 0°C and aqueous work up (Suzuki *et al.*, 1976b).

d. Analytical. On simple model systems, α- and β-aspartyl peptides **150** and **149** are often chromatographically or electrophoretically separable and quantifiable (Marshall and Merrifield, 1965; Kenner and Seely, 1972; Yang and Merrifield, 1976), although for more complex systems, their separation has been essentially impossible. The most straightforward way to check for $\alpha \rightarrow \beta$ rearrangement is by leucine aminopeptidase digestion, since the enzyme will not cleave a β-peptide bond (Khairallah *et al.*, 1963). A chemical criterion (Marshall and Merrifield, 1965) involves esterification, lithium aluminum hydride reduction, acid hydrolysis, and amino acid analysis. Formation of homoserine and α-aminobutyrolactone is diagnostic of an α-peptide bond, while aspartic acid in a β-linkage gives rise to β-homoserine and β-aminobutyrolactone. A subtractive method to test for ω-linkages (Miller *et al.*, 1977) involves esterification with *O*-pivaloylhydroxylamine followed by a Lossen rearrangement at pH 8.5. Internal α-linkages remain intact while the side chain is modified, but COOH-terminal aspartic or glutamic acid in a β- or γ-linkage are degraded.

An nmr method to distinguish α- and β-isomers has also been described (Baba *et al.*, 1973). The succinimide intermediates are readily recognized on the basis of the charge difference with respect to the desired peptide (e.g., Ondetti *et al.*, 1968; S. S. Wang *et al.*, 1974; Yang and Merrifield, 1976).

The anisylated by-product **156** derived from glutamyl peptides is best quantitated by its ultraviolet absorption at 276 nm ($\varepsilon = 1.7 \times 10^4$; Feinberg and Merrifield, 1975). Recovery of glutamic acid in acid hydrolysates of the peptides is lowered in direct accordance with the level of the side reaction (Feinberg and Merrifield, 1975; Sano and Kawanishi, 1975). When peptide–resins are hydrolyzed with 12 *N* HCl–dioxane (1:1), one of the early reagents for this purpose (Marshall and Merrifield, 1965), recovery of glutamic acid is low due to the formation of γ-esters with the alcoholic breakdown products of dioxane (Stewart and Young, 1969, p. 53; Polzhofer and Ney, 1970b).

5. Asparagine and Glutamine

a. Protection. The neutral β- and γ-carboxamide groups of asparagine and glutamine have usually been left unprotected (**158**) for solid-phase synthesis (e.g., Marglin and Merrifield, 1966), although this approach has forced modifications of the usual coupling methods whenever these two residues have been introduced. An alternative, early strategy (e.g., Inukai *et al.*, 1968) to create ω-carboxamide groups involves aminolysis of β- or γ-esters of aspartic or glutamic acid following routine incorporation of these latter derivatives (Section V,B,4) at the desired position by solid-phase procedures; this approach is infrequently tried due to the risks of $\alpha \rightarrow \omega$ rearrangement during aminolysis (see, however, Birr, 1978, p. 67). Finally, the development of a variety of acidolyzable amide dervatives, including 9-xanthenylamides (**159**) (Dickman and Westcott, 1954; Shimonishi *et al.*, 1962; Dorman *et al.*, 1972), substituted mono- and bisbenzylamides (**160**–**162**) (Weygand *et al.*, 1968b,c; Pietta *et al.*, 1968, 1971, 1976), and substituted benzhydrylamides (**163**–**165**) (Sakakibara *et al.*, 1967; König and Geiger, 1970d; Hruby *et al.*, 1972) has made viable approaches in which asparagine and glutamine can be incorporated by standard coupling procedures (e.g., Wissmann and Geiger, 1970; Hruby *et al.*, 1972; Birr, 1973; Stewart *et al.*, 1976; Atherton *et al.*, 1977).

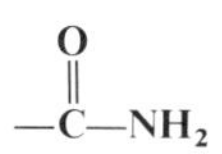

*(**158**)

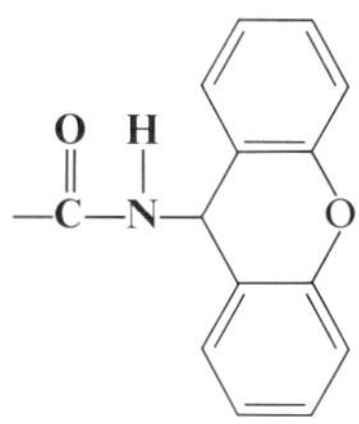

*(**159**) Xan

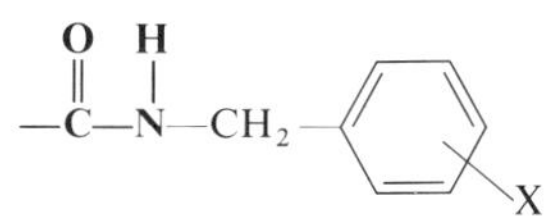

(**160**) X = *p*-OCH_3 Mob
(**161**) X = 2,4-$(OCH_3)_2$ Dmob

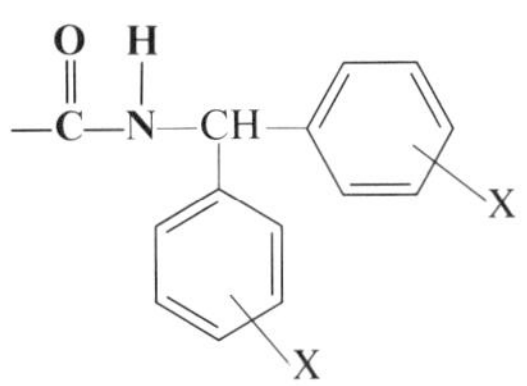

(**163**) X = *p*-OCH_3 Ddm
(**164**) X = *p*-CH_3
(**165**) X = H Dpm

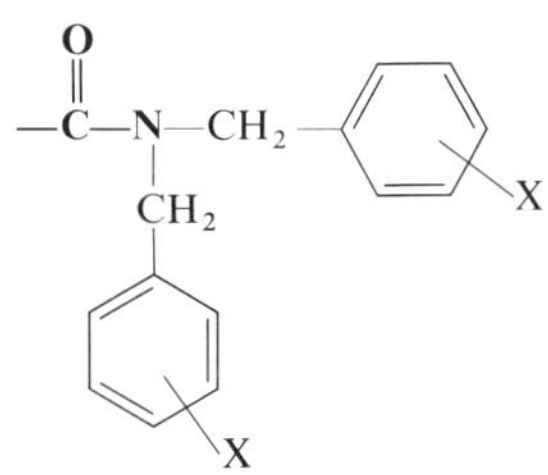

(**162**) X = 2,4-$(OCH_3)_2$ Dmob

In contrast to the corresponding ester derivatives, all of the amides (**159**–**165**) are resistant to catalytic hydrogenolysis. For glutamine derivatives, where intramolecular processes do not intervene, the amides are also resistant to the actions of nucleophilic reagents as alkali, alkoxides, and ammonia.

The relative ease of acidolysis follows the series: 9-xanthenyl (Xan) (**159**) > 4,4′-dimethoxydiphenylmethyl (Ddm) (**163**) > diphenylmethyl (Dpm) (**165**) ~ 2,4-dimethoxybenzyl (Dmob) (**161**) > *p*-methoxybenzyl (Mob) (**160**). Selective removal of the N^{α}-*tert*-butyloxycarbonyl temporary protecting group in the presence of N^{ω}-xanthenyl (**159**) or N^{ω}-4,4′-dimethoxy diphenylmethyl (**163**) functions is not possible. The "fine-tuning" of carboxamide protecting groups for protection schemes involving graduated acid lability offers three possibilities.

1. When the N^{α}-*tert*-butyloxycarbonyl group is used, essentially permanent carboxamide protection seems to be best provided by the N^{ω}-diphenylmethyl (**165**) or N^{ω}-2,4-dimethoxybenzyl (**161**) groups, which are sufficiently stable to repetitive treatments with TFA, yet are removable by HF at 0°C for 60 min (Sakakibara *et al.*, 1967; Pietta *et al.*, 1976). The N^{ω}-*p*-methoxybenzyl (**160**) function is definitely too stable.
2. Also, when the N^{α}-2-(*p*-biphenylyl)propyl(2)oxycarbonyl group is used, the 9-xanthenyl (**159**) and 4,4′-dimethoxydiphenylmethyl (**163**) groups are lost after several cycles of repetitive deprotection. Nonetheless, the temporary carboxamide protection provided during the initial anchoring or coupling step is probably all that is required, since often no significant side reactions are anticipated once the next amino acid residue has been incorporated.
3. When the N^{α}-*tert*-butyloxycarbonyl and related highly acid-sensitive groups are used, the 4,4′-dimethoxydiphenylmethyl (**163**) function is a good choice for the preparation of protected peptide segments. Removal requires TFA–anisole (9:1) at 25°C for 2 to 3 hr (König and Geiger, 1970d), which represents conditions somewhat more vigorous than those required to remove most *tert*-butyl-based side-chain protecting groups. Removal of the 9-xanthenyl (**159**) function is rapid in 3 *N* HCl–dioxane (Dorman *et al.*, 1972).

b. Anchoring and Coupling. Esterification of N^{α}-protected glutamine and especially asparagine to supports containing chloromethyl or hydroxymethyl groups by the most common methods (Section III,B,l,a) has been found to be accompanied by a variety of difficulties (Marglin, 1971; Dorman *et al.*, 1972; Tesser *et al.*, 1976) including dehydration, deamidation, and alcoholysis. Acceptable results are obtained whenever the ester linkage is prepared by O-alkylation reactions at moderate temperatures where the solvent is dimethylformamide, as is the case in the procedures of Marglin (1971), Gisin (1973), and Yamashiro and Li (1974), among others. Esterification to a dimethylsulfonium support **3** (Dorman and Love, 1969) also seems to work well. Esterification by O-acylation is best mediated by *N*,*N*-dimethylformamide dineopentyl acetal (Schreiber, 1967), or with the use of oxidation–reduction condensation (Matsueda *et al.*, 1972).

A novel approach for the synthesis of peptides containing COOH-terminal asparagine or glutamine that circumvents the usual problems is based on the initial coupling of N^{α},C^{α}-protected aspartic or glutamic acid through the free ω-carboxyl group to a substituted benzhydrylamine support (pp. 93–95; see also footnote on p. 91). The ω-carboxamide group, protected throughout the synthesis via the benzhydrylamide anchor, is released in a final acidolytic cleavage step. Successful applications of this idea have been recently reported (Li *et al.*, 1976; Wang *et al.*, 1979).

When derivatives of asparagine or glutamine with an unprotected carboxamide side chain are coupled in solid-phase synthesis using dicyclohexylcarbodiimide as the condensing agent, a major dehydration side reaction is normally found (see pp. 202–203). However, Corley *et al.* (1972) reported that dehydration did not occur if coupling times were kept short. Similarly, there are reports that the uses of symmetrical anhydrides (Hagenmaier and Frank, 1972; Hemmasi and Bayer, 1974) or of the coupling reagent 1-ethoxycarbonyl-2-ethoxy-1,2-dihydroquinoline (EEDQ, Table X, number 5; C. T. Wang *et al.*, 1974) avoid the side reaction. It is more probable that in each of these cases dehydration did occur but was largely reversed during the subsequent HF cleavage and aqueous work up (Mojsov *et al.*, 1979).

Preformed active esters (Bodanszky and du Vigneaud, 1959; see p. 140) have been widely used for the safe introduction of asparagine and glutamine residues, although coupling rates are often slow and the reactions may not readily reach completion (e.g., Tan and Kaiser, 1976). Oxidation–reduction condensation of these residues also appears to proceed without dehydration (see p. 143). The best available general method for the solid-phase coupling of N^{α}-protected asparagine and glutamine involves the use of DCC–HOBt (König and Geiger, 1970a; see p. 133). As a practical point, most of the various asparagine and glutamine derivatives required for the approaches cited in these paragraphs are poorly soluble in CH_2Cl_2, and coupling reactions are generally carried out in CH_2Cl_2–DMF mixtures or even neat DMF.

A different trend for incorporating asparagine and glutamine without risk of a dehydration side reaction has involved the use of side-chain carboxamide protected derivatives (pp. 199–200). The xanthenylamides (**159**) are often poorly soluble; the more soluble benzhydrylamides (**163**–**165**) possess a substantial bulk so that even couplings by the standard dicyclohexylcarbodiimide method occasionally proceed slowly.

Subsequent to the safe introduction of each glutamine residue to a growing resin-bound peptide chain, there exists at the next deprotection step a risk of intramolecular cyclization to a peptide **155** containing amino-terminal pyroglutamic acid (**166**). Prudent precautions to minimize this sequence-dependent, terminating side reaction are detailed later (pp. 205–206). Alternatively, when proglutamyl peptides are desired intentionally, the amino acid **166** can be activated and coupled to resin-bound peptides directly (Fluoret, 1970; Stewart *et al.*, 1972) because no further acylation of

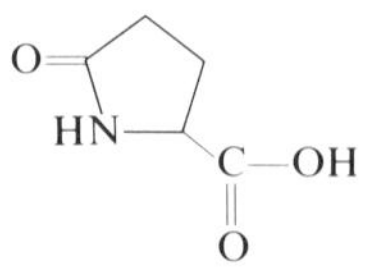

(166)

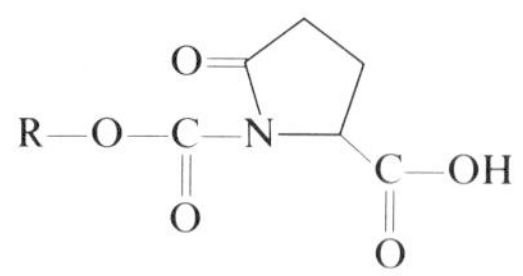

(167) R = $C_6H_5CH_2$, *tert*-C_4H_9

the secondary amide nitrogen will occur under standard coupling conditions. Alkoxycarbonyl pyroglutamic acids have also been used (Rivaille and Milhaud, 1971; Sievertsson *et al.*, 1972) primarily because of their superior solubility characteristics. A third approach to pyroglutamyl peptides has involved intentional cyclization of an NH_2-terminal glutamine residue, promoted in solution by aqueous acetic acid at reflux for 10 min, or at 40°C overnight (e.g., Matsuo *et al.*, 1971; Carraway and Leeman, 1974).

c. Side Reactions. The carboxamide side chains of asparagine and glutamine generally do not require protection, in so far as intermolecular acylation followed by branching do not occur. Nonetheless, these functions are susceptible to a variety of side reactions which require special attention.

(i) Dehydration of Carboxamide to Nitrile. The activation and coupling of N^{α}-protected asparagine, and to a lesser extent glutamine, with a good condensing agent such as dicyclohexylcarbodiimide, has long been known to be accompanied by the formation and incorporation into peptide chains of substantial amounts of the corresponding dehydration products (Gish *et al.*, 1956; Ressler, 1956; Katsoyannis *et al.*, 1958; Ressler and Ratzkin, 1961). Several lines of mechanistic investigation have suggested that, for asparagine, the key intermediate of the dehydration is the isoimide **168** [Eq. (23)] (Stammer, 1961; Liberek, 1962; Kashelikar and Ressler, 1964; Paul and Kende, 1964). In the initial step, dicyclohexylcarbodiimide could react with

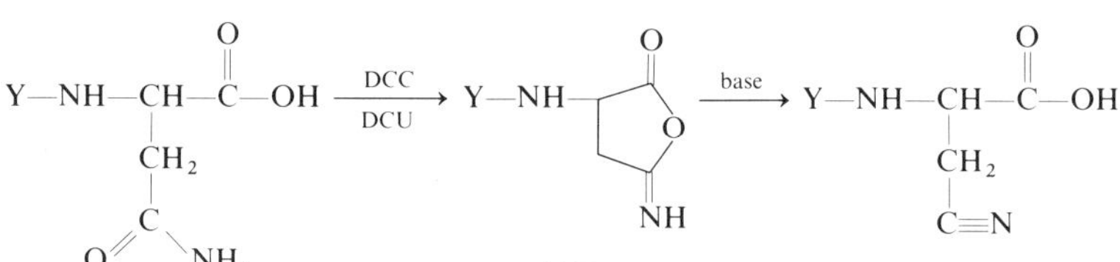
(23)

(168)

either the α-carboxyl or β-carbonyl groups; in any event, dehydration requires the assistance of the free asparagine carboxyl group. An important practical corollary, which has been well verified experimentally, is that once asparagine or glutamine has been incorporated intact into the peptide chain, there is no further risk of side-chain dehydration upon additional repeated exposure to DCC during subsequent coupling steps.

A variety of coupling methods have been evaluated to determine the extent of the dehydration side reactions, and the conclusions have already been summarized (p. 201). Recently, it was discovered that β-cyanoalanyl-glycine was nearly quantitatively rehydrated in the standard workup of the cleavage reaction with HF (Mojsov *et al.*, 1979); this finding has implications that facilitate a reinterpretation of earlier literature data. Treatment of the model dipeptide with TFA–CH_2Cl_2 (1:1) at 25°C did not cause significant rehydration. Therefore, further solid-phase model studies were designed involving the coupling of Boc- or Bpoc-asparagine to a glycine *p*-alkoxybenzyl ester resin. The extent of dehydration accompanying different coupling methods was evaluated in this reliable system, using a sensitive chromatographic technique for quantitations of results. *N,N′*-Dicyclohexylcarbodiimide alone, and preformed symmetrical anhydrides, each gave dehydration levels in the range of 30–50%. Although the nitrile form could be rehydrated in HF, this further treatment does not alleviate the overall problem with asparagine because other, stable by-products were also formed. For example, when either of the described coupling methods was used, 10–15% of a product tentatively identified as β-aspartamidinoglycine was also present. On the other hand, when DCC–HOBt coupling was used, the desired asparaginyl-glycine dipeptide was the principal product, and the amount of nitrile detected was less than 1%. The results of Mojsov *et al.* (1979) with symmetrical anhydrides contradicted the conclusion of Hemmasi and Bayer (1974) that this method would prevent dehydration by insuring that no excess DCC would be present due to the reaction stoichiometry.

Dehydration does not occur during the coupling of pure *p*-nitrophenyl esters of N^α-protected asparagine and glutamine, although a nitrile does form during the dicyclohexylcarbodiimide-mediated coupling of *p*-nitrophenol in the course of preparing the starting esters. The nitriles are removable during recrystallization of the desired product (Bodanszky and du Vigneaud, 1959). With the more reactive pentafluoro and pentachlorophenyl active esters, intramolecular displacement of the phenol by nucleophilic attack of the amide group oxygen gives primarily the isoimide **168** precursor of the dehydration product. Some succinimide **148** or glutarimide **151** by-products are formed as well, through the attack of the amide group nitrogen [Eq. (24); Kisfaludy *et al.*, 1975].

As already noted, nitrile formation does not occur when the ω-carboxamide group is blocked during the activation and coupling step.

(ii) Intramolecular Acylation Leading to Imides. The cyclization of asparagine and glutamine peptides and esters to give the succinimide (**148**) and glutarimide (**151**) intermediates (Fig. 17 on p.192) is readily promoted by base (Sondheimer and Holley, 1954; Battersby and Robinson, 1955).

There are two kinds of mechanisms that may be considered. The first is as drawn in Fig. 17 (pathway a, X = NH_2) and involves acylation of an internal peptide bond, with loss of ammonia from the carboxamide side chain. Subsequent hydrolytic openings of the ring will lead to a mixture of α- and β-peptides containing aspartic acid. This sequence of events is expected to depend on the steric nature of the adjacent amino acid residue, in the same way as for $\alpha \rightarrow \beta$ rearrangement of aspartyl peptides discussed earlier (p. 193). The tendency for imide formation from asparagine peptides is less than that for β-esters of aspartyl peptides, but greater than that for aspartyl peptides with a free β-carboxyl group. For the sterically most favored asparaginyl-glycine sequence, imide formation is not expected to be significant under even strong acid cleavage conditions of solid-phase synthesis. However, at mildly alkaline pH, asparaginyl-glycine sequences readily rearrange, and this is the basis for a specific mild chemical method of peptide bond cleavage (Bornstein and Balian, 1977).

Another route to imides is shown in Eq. (24). The mechanism is closely related to that of dehydration [Eq. (23) on p. 202], and in fact, under certain conditions, both processes take place. In the present context, the imide forms by nucleophilic attack of the side-chain carboxamide nitrogen on the α-car-

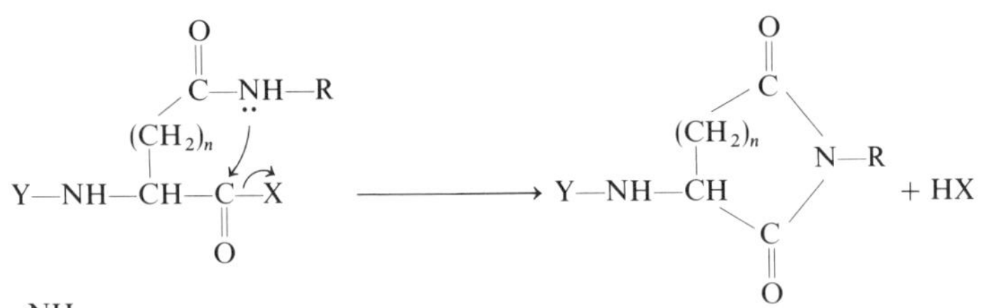

(24)

(**169**) X = NH_2

(**170**) X = NH-R′

(**171**) X = ON 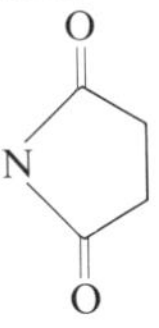

(**172**) X = OC_6H_5

(**173**) X = O-*t*-C_4H_9

(**174**) X = OCH_2— 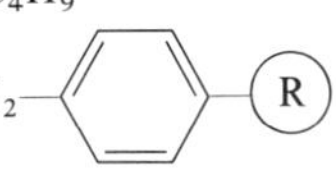

(**175**) X = $OCH_2CH_2SO_2$— 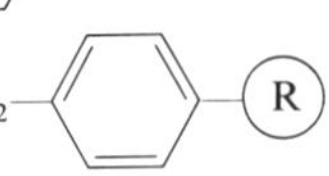

n = 1, 2

R = H; carboxamide protecting group

Y = peptidyl; amino protecting group

(**148**) R = H, *n* = 1

(**151**) R = H, *n* = 2

(**176**) R = protecting group, *n* = 1

not observed for

R = protecting group, *n* = 2

bonyl group, with concomitant cleavage of a peptide or ester bond. The level of the side reaction through this mechanism depends on the group X. For an internal asparagine or glutamine residue, as in **170**, peptide bond cleavage is not likely. On the other hand, the reaction is quite facile under certain essentially neutral conditions with active esters such as **171** and **172** (Meyers *et al.*, 1969; Zahn and Fölsche, 1969; Kisfaludy *et al.*, 1975). Succinimide formation at COOH-terminal asparagine is quantitative in the presence of one equivalent of alkali to remove an N^{α}-methylsulfonylethyloxycarbonyl group (**62**), even when a *tert*-butyl ester (**173**) is used to protect the carboxyl group (Roeske, 1963; König and Volk, 1977). Tetraethylammonium fluoride in anhydrous media was also sufficiently strong as a base to promote the same side reaction (Sieber, 1977). In a solid-phase application, Tesser *et al.* (1976) were unable to prepare peptides containing COOH-terminal asparagine using an alkali-labile β-ethylsulfonyl ester linkage to the support (**175**). The N^{ω}-4,4′-dimethoxydiphenylmethyl (**163**) and N^{ω}-2,4-dimethoxybenzyl (**161**) functions have been reported to protect against alkali-catalyzed glutarimide formation, but not against aspartimide formation (König and Geiger, 1970d; Pietta *et al.*, 1971). Nonetheless, Tesser *et al.* (1976) were able to detach successfully Nps-Asn (Ddm)-OH from the β-ethylsulfonyl ester-resin through the use of base (see p. 89); but they also remarked on the limited applicability of this approach due to the steady loss of the 4,4′-dimethoxydiphenylmethyl group during repetitive deprotections in the course of a synthesis. To get around these difficulties, this group (Wolters *et al.*, 1974) prepared an insulin A chain octapeptide with COOH-terminal asparagine by an alternate route involving coupling of the COOH-terminal protected dipeptide in solution to a protected hexapeptide hydrazide product of a solid-phase synthesis.

(iii) Pyrrolidone Formation at NH_2-Terminal Glutamine. The intramolecular aminolytic cyclization of NH_2-terminal glutamine in peptides (Fig. 17, pathway b, Y = H, Z = NH_2) gives irreversibly terminated pyroglutamyl peptides (**155**). This side reaction (reviewed by Blömback, 1967) is notorious as a nuisance in the sequential degradation of peptides (Smyth *et al.*, 1962). Pyroglutamyl formation has been unequivocally shown to occur at levels from 5 to 30% under certain conditions of both solution (e.g., Schnabel *et al.*, 1967) and solid-phase peptide synthesis (Barany and Merrifield, 1973; Tregear, 1975) by direct isolation and analysis of terminated peptides; occurrence of the side reaction can also be inferred from some earlier data (e.g., Takashima *et al.*, 1968).

The systematic studies of Dorman *et al.* (1972) and especially of Beyerman *et al.* (1973) clarified the conditions under which pyroglutamyl formation occurs. Their results are closely comparable to those of Gisin and Merrifield

(1972a) on diketopiperazines (p. 82ff) since both side reactions have been found to be catalyzed by weak carboxylic acids. Furthermore, in both cases, the appropriate peptide–resins are essentially stable as salts of a strong acid, or as the free base even in the presence excess triethylamine. In both cases, the side reaction depends on the sequence of the susceptible peptide, and on the rate of coupling of the incoming activated protected amino acid.

Based on the available evidence, two precautions seem well-advised whenever the deprotection of, and coupling to a glutamine residue arises in the course of a solid-phase synthesis.

1. The N^α-Boc group is best removed with 4 *N* HCl–dioxane (see p. 102). When the alternative choices of 1 *N* HCl–HOAc, or TFA–CH_2Cl_2 (1:1) are selected as deprotective agents, the NH_2-terminal deblocked glutamine residue becomes exposed to weak carboxylic acid catalysts, especially after several washing steps which serve to dilute the stronger acids. Incidentally, one might predict that for this very reason, pyroglutamyl formation could be a significant problem when the N^α-Bpoc and related highly acid-sensitive groups are removed by dilute acids. The extent and avoidance of the side reaction under those conditions remain open questions.
2. A "reverse-coupling" protocol (see pp. 83 and 124), as the next protected amino acid is introduced, will also minimize the intramolecular cyclization of NH_2-terminal glutamine (e.g., Syrier and Beyerman, 1975).

Pyroglutamyl formation is probably prevented by protection of the side-chain carboxamide (Pietta *et al.*, 1968, 1971) although no data are available addressing the question at high levels of sensitivity. Peptides with NH_2-terminal Gln(Ddm) can be directly cyclized in solution to pyroglutamyl peptides by heating in TFA–anisole (König and Geiger, 1972b). It seems almost certain that under those conditions, loss of the N^ω–4,4′-dimethoxydiphenylmethyl group precedes cyclization.

(iv) Deamidation. Side-chain carboxamide groups of internal asparagine and glutamine residues are entirely hydrolyzed in 2 *N* aqueous HCl, at 100°C within 2 hr (reviewed by Wilcox, 1967). On the other hand, the carboxamide group appears to be stable in a strong anhydrous acid such as HF (Sakakibara, 1971). A Lewis acid of comparable strength, BBr_3, did cause low (5–10%) levels of deamidation (Felix, 1974). Basic cleavage reagents introduce an especially grave risk of partial deamidation.

Robinson's group systematically studied the deamidation in aqueous solutions of a large series of model peptides synthesized by the solid-phase method (summarized by Robinson, 1974). Deamidation was favored by increased pH, increased ionic strength, and increased temperature (McKerrow and Robinson, 1971; Scotchler and Robinson, 1974). In phosphate buffer at pH 7.5, ionic strength 0.2, at 37°C, the half-times for this sequence-dependent process ranged fron 6 to 500 days for peptides containing asparagine, and from 96 to 3,400 days for peptides containing glutamine. In several cases, extrapolations of these relatively slow rates to zero time suggested that up to 20% deamidation of asparagine residues had occurred through the cumulative course of synthesis, HF-cleavage, workup, and purification steps (Robinson and Tedro, 1973).

The partial loss of side-chain carboxamide groups, and their conversion to ω-hydrazides in solution, has been reported in a special instance under the conditions for hydrazinolysis of α-carboxylic esters (MacLaren *et al.*, 1958).

(v) A Side Reaction Associated with Carboxamide Protection. Because amides are not as easily cleaved as esters under acidolytic conditions (p. 167), it follows that for the same strength acid to deblock both a carboxamide and a carboxyl group, protection of the former must be based on a functionality that gives rise to a much more stable carbonium ion. Specifically, the xanthenyl (Xan) and 4,4′-dimethoxydiphenylmethyl (Ddm) carbonium ions (compare to structures **159** and **163**) are substantially more stable than *tert*-butyl carbocations; consequently, the former are more prone to alkylate tryptophan, cysteine, methionine, histidine, and tyrosine residues. To illustrate this point, Southard (1971) treated Bmv-Gln(Xan)-Trp-Leu-Met-Asn-Thr(*t*Bu)-resin with 0.4 *N* HCl in tetrahydrofuran to remove the N^{α}-2-benzoyl-1-methylvinyl group **55**, and showed that the xanthenyl group was lost from glutamine and transferred irreversibly to the adjacent tryptophan, presumably at the 2-position of the indole ring.

d. Analytical. Information about the integrity of asparagine and glutamine residues cannot be obtained by amino acid analysis of a 6 *N* HCl acid hydrolysate, since under these conditions, both the parent carboxamide amino acids and their dehydration products are converted to aspartic and glutamic acid. After enzymatic hydrolysis, the various standards and conceivable modified products elute on the standard amino acid analyzer in the following order: β-cyanoalanine, aspartic acid, asparagine + glutamine, α-amino-γ-cyanobutyric acid, glutamic acid (Spackman *et al.*, 1958; Hemmasi and Bayer, 1974; Mojsov *et al.*, 1979). With modified buffers based on lithium citrate, asparagine and glutamine are resolved from one another and from glutamic acid, which elutes in between (Benson *et al.*, 1967). A physical criterion for dehydration is the infrared spectral band near 2250 cm^{-1}; the limit of detection of this technique is probably 3–5% (e.g, C. T. Wang *et al.*, 1974; Wolters *et al.*, 1974). Identification of nitriles has also been achieved by mass spectrometry (Hemmasi and Bayer, 1974) and by a chemical method involving conversion to the appropriate diaminocarboxylic acid by sodium in liquid ammonia with methanol (Ressler and Kashelikar, 1966).

Detection of succinimides, glutarimides, and $\alpha \rightarrow \omega$ rearranged products is by methods already covered (p.198). Pyroglutamyl peptides are best identified by isolation, amino acid analysis, and the absence of a free NH_2-terminus.

The fluorescence of the xanthenyl (Xan) group (**159**) facilitates monitoring of its fate but will also interfere with the fluorescamine test for completion of coupling (Stewart, 1976, p. 277).

6. *Serine and Threonine*

a. Protection. The neutral, weakly nucleophilic primary and secondary β-hydroxyl groups respectively of serine and threonine have been most widely protected as substituted benzyl ethers (**177**–**179**) (Okawa, 1956; Mizoguchi *et al.*, 1968; Hruby and Ehler, 1970; S. S. Wang *et al.*, 1974; Sugano and Miyoshi, 1976; Yamashiro, 1977) or *tert*-butyl ethers (**180**) (Beyerman and Bontekoe, 1962; Callahan *et al.*, 1963). In addition, infrequent use has been

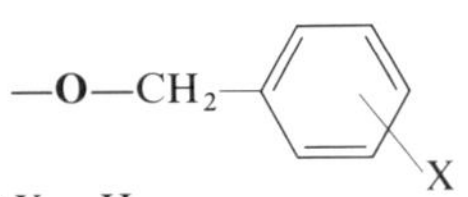

*(**177**) X = H
(**178**) X = *p*-Br, Cl
(**179**) X = *o*-Cl

Bzl

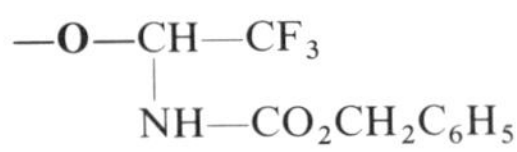

(**181**) Ztf

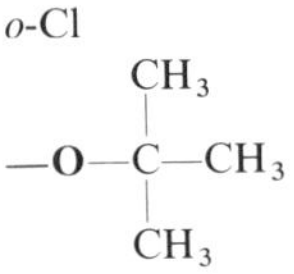

*(**180**) *t*Bu

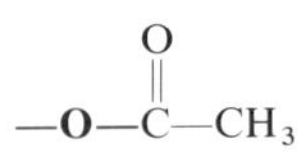

(**182**) Ac

made of the 1-benzyloxcarbonylamino-2,2,2-trifluoroethyl (Ztf) group (**181**) (Weygand *et al.*, 1968a., applied by Polzhofer, 1969; Pietta *et al.*, 1973) and the acetyl group (**182**) (Sakami and Toennies, 1942; applied by Tilak and Hollinden, 1968a). During the early days of solid-phase synthesis, the unavailability of a suitable β-hydroxy-protected derivative of threonine forced investigators to couple this amino acid without side-chain protection (e.g., Marglin and Merrifield, 1966; see also Yajima and Kiso, 1974). However, this approach leaves open the serious danger of O-acylation and branching of peptide chains (Stewart *et al.*, 1967). The branching side reaction has also been noted with unprotected serine, even when only active ester couplings are employed (Garner *et al.*, 1971). The branched chains can be cleaved by alkali at the end of a synthesis, but an intramolecular O→N acyl shift during the first neutralization step after introduction of side-chain unprotected serine or threonine could lead to a peptide containing an additional amino acid inserted into the backbone. Hydroxyl group protection has been reviewed (Reese, 1973).

Unsubstituted β-benzyl ethers (**177**) are lost to the extent of 0.01% per each cycle of acidolytic N^{α}-*tert*-butyloxycarbonyl removal; this value is the same as for ω-benzyl esters (Erickson and Merrifield, 1973a; compare to p. 190). This level of stability is normally more than adequate, but an additional 5- and 50-fold increase in stability is provided respectively through *p*-halo (**178**) (Yamashiro, 1977) and *o*-chloro (**179**) (Erickson and Merrifield, 1974) substitution of the benzyl ring. All benzyl ethers are smoothly cleaved in liquid HF at 0°C; the unsubstituted benzyl ethers (**177**) are also cleaved by HBr–TFA. Benzyl ethers (**177–179**) are stable in the presence of alkali and other nucleophiles; they are readily cleaved by catalytic hydrogenolysis.

The *tert*-butyl ethers (**180**) are cleaved with trifluoroacetic acid, and are generally useful for milder synthetic schemes based on acidolysis of *tert*-butyl derivatives. The Ztf group (**181**) is labile under the standard acidolytic or hydrogenolytic conditions for removal of the benzyloxycarbonyl group (**46**) [compare to Eq. (20) on p. 182]. The acetyl group (**182**) is removable by saponification. However, as soon as the free α-amino group of an NH_2-terminal *O*-acetylserine or *O*-acetylthreonine residue becomes liberated during the neutralization step, a facile O→N acyl shift is to be expected (Guttmann and Boissonnas, 1958). The subsequent peptide chains will therefore be terminated by N-acetylation, but may become branched through subsequent coupling at the exposed hydroxyl side chain.

O-Methylserine is an interesting amino acid analogue of serine and threonine for structure-function studies, because its side chain can serve only as a hydrogen bond acceptor and not a donor. The *O*-methyl ether function is stable to the acid conditions used in solid-phase peptide synthesis, including HF at 0°C for at least 4 hr (Hodges and Merrifield, 1974, 1975b).

b. Coupling. Because of its β-substituent, the coupling of protected threonine proceeds at a rate which is an order of magnitude lower than that for the coupling of sterically unhindered amino acids (see p. 130). Hence, either longer than usual reaction times or, preferably, explicit monitoring for unreacted amino sites, is advisable whenever threonine is to be incorporated into a peptide chain (see van Batenburg *et al.*, 1976a, for a recent example). Due to slower coupling rates, competing termination side reactions of the NH_2-terminal residue of the resin-bound peptide might proceed to a greater than usual extent, with the net apparent result of incomplete coupling.

c. Side Reactions. In the presence of a strong acid such as HBr, the hydroxyls of serine and threonine are partially acylated in acetic acid but not in trifluoracetic acid (Guttmann and Boissonnas, 1958; see Table VIII).

An acetyl group (**182**) is removable by saponification. However, the alkaline conditions could also lead to β-elimination to give dehydro residues (compare to p. 92) and, furthermore, alkali has been reported to promote substantial amounts of peptide bond cleavage at both the amino and carboxyl sides of serine residues containing a free hydroxyl group (Harris and Fruton, 1951; Zahn and Schnabel, 1957; Guttmann and Boissonnas, 1958).

Serine and threonine peptides undergo a slow intramolecular N→O acyl shift in acid that is rapidly reversed in base (reviewed by Schröder and Lübke, 1965, pp. 213–214; Iwai and Anda, 1967). Thus, intramolecular alcoholysis gives a 2-hydroxy-1,3-oxazolidine intermediate (**183**), and the equilibrium is shifted by protonation of the liberated α-amino group [Eq. (25)]. This side

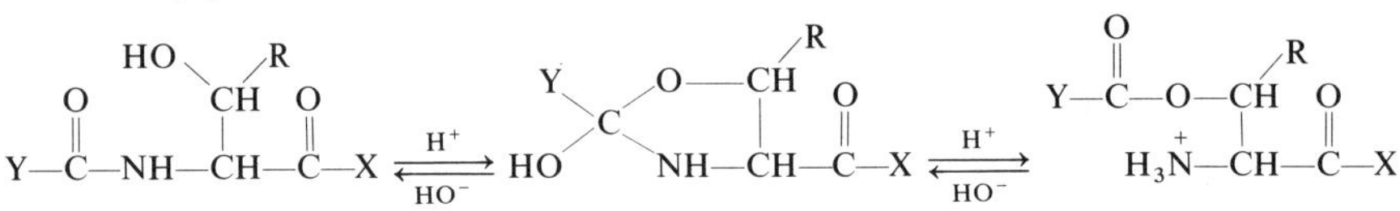

N-acyl (**183**) *O*-acyl

R = H, CH_3 (25)

reaction has not been reported to occur to any serious extent under the usual solid-phase cleavage conditions, and the simple precaution of bicarbonate treatment after an HF cleavage of a peptide ought to ensure that the isolated peptide will be exclusively in the thermodynamically favored amide form (see p. 71 for details). However, Fujino *et al.*, (1978) have presented evidence implying a significant ($\sim 50\%$) level of N → O acyl shifts promoted by CH_3SO_3H–anisole (10:1), at 25°C for 1 hr (Table VIII, number 8); again 0.5 *M* aqueous NH_4OH at 0°C for 30 min reversed the side reaction.

The aspartyl-serine partial sequence is especially prone to $\alpha \rightarrow \beta$ rearrangement, as already discussed (p. 193).

b. Analytical. Serine and threonine are partially deaminated during acid hydrolysis with 6 *N* HCl at 110°C; a suitable correction for, respectively, 10% and 5% destruction in 24 hr is commonly applied (Hirs *et al.*, 1954). The methyl ether of *O*-methylserine is substantially hydrolyzed in 6 *N* HCl (64% in 24 hr) to give serine.

The recovery of *O*-methylserine, a readily quantifiable and separable peak on amino acid analyzer chromatograms, is excellent ($\geqslant 97\%$) when complete peptide hydrolysis is carried out at 110°C with 3 *N* *p*-toluenesulfonic acid containing 0.2% indole, or with 4 *N* methanesulfonic acid, or for peptide–resins with methanesulfonic acid–propionic acid–water (5:12:3) (Hodges and Merrifield, 1974, 1975b).

Peptides that have undergone an N → O acyl shift at serine or threonine residues carry an extra positive charge and are readily distinguished by electrophoresis (Sakakibara *et al.*, 1962). The level of this side reaction can be estimated by difference through the use of sodium nitrite as a deamination reagent, followed by hydrolysis and amino acid analysis (Fujino *et al.*, 1978).

7. Tyrosine

a. Protection. The weakly acidic phenolic hydroxyl group $pK_a = 10$) of tyrosine has been left unprotected (**184**) (e.g., Anfinsen *et al.*, 1967) during solid-phase synthesis, or has been blocked as a substituted benzyl ether (**185**–**188**) (Wünsch *et al.*, 1958; Yamashiro and Li, 1972, 1973a; Erickson and Merrifield, 1973a; Amit *et al.*, 1977; Lemaire *et al.*, 1977); a secondary alkyl ether (**189**) (Engelhard and Merrifield, 1978); a *tert*-butyl ether (**190**) (Beyerman and Bontekoe, 1962; Callahan *et al.*, 1963; Schröder, 1963); or a 2,4-dinitrophenyl ether (**191**) (Fridkin *et al.*, 1977). Furthermore, the substituted benzyloxycarbonyl groups (**192** and **193**) (Overell and Petrow, 1955; Semkin *et al.*, 1968; Erickson and Merrifield, 1973a; Yamashiro and Li, 1973b), and the ethoxycarbonyl (**194**) (Geiger *et al.*, 1968), and carbamoyl (**195**) (Jäger *et al.*, 1968) groups are available for tyrosine protection. Each strategy for the introduction of tyrosine residues suffers fron certain disadvantages, but workable compromises are available in most cases.

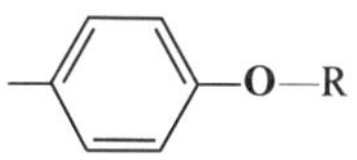

*(**184**) R = H

(**189**) R = 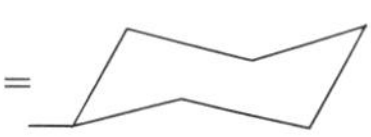 cHex

*(**190**) R = 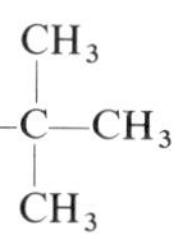 tBu

(**191**) R = 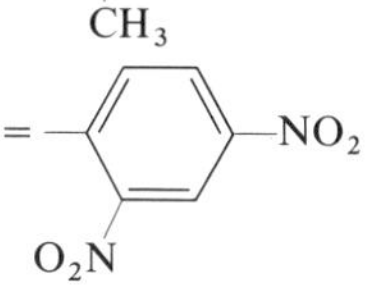 Dnp

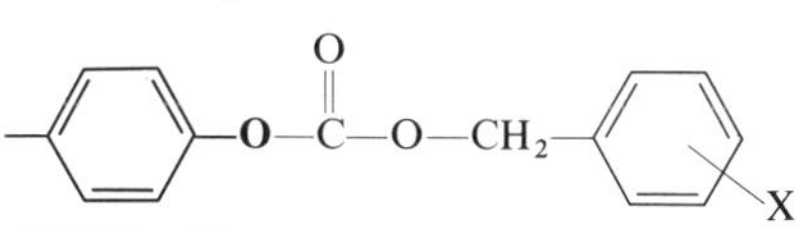

(**192**) X = H
*(**193**) X = *o*-Br

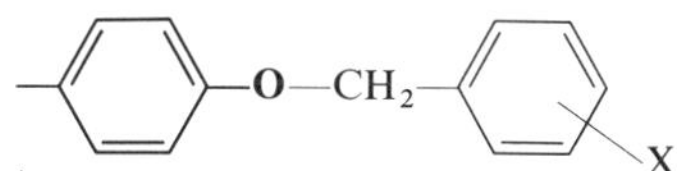

*(**185**) X = H Bzl
(**186**) X = *m*-Br
*(**187**) X = 2.6-Cl_2
(**188**) X = *o*-NO_2

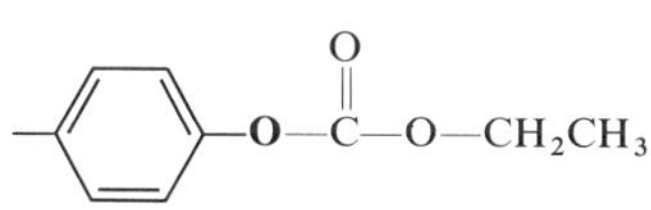

(**194**) Ec

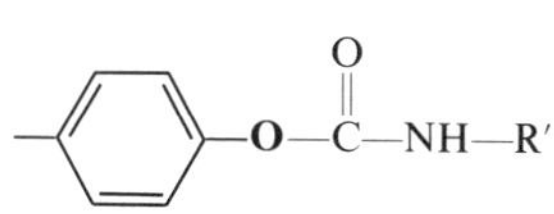

(**195**) R′ = H, *sec*-C_4H_9, C_6H_5, $CH(C_6H_5)_2$

The introduction of tyrosine in the unprotected form (**184**) appears to be an acceptable strategy when this residue is near the NH_2-terminal end of a readily purifiable, small to medium-sized peptide (e.g., Meienhofer *et al.*,

1970b; Hruby *et al.*, 1977; Live *et al.*, 1977; Fischman *et al.*, 1978). It is helpful to use small excesses of relatively mild acylating agents in subsequent coupling steps. A obvious advantage of this approach is elimination of the 3-alkylation side reaction (see p. 215ff). On the debit side, the free phenolic hydroxyl is inevitably subject to O-acylation (for solution precedents, see Paul, 1963; Ramachandran and Li, 1963) and subsequent branching. Because these products are activated esters, the *O*-acyl group might also be transferred in subsequent cycles to free α-amino groups of growing chains, resulting finally in peptides that contain one or several inserted amino acid residues.

Unsubstituted *O*-benzyltyrosine, which was widely used in the earlier literature (e.g., Marshall and Merrifield, 1965), can no longer be recommended for two reasons: (1) The rate of premature acidolytic loss of side-chain protection upon exposure to TFA–CH_2Cl_2 (1:1) to remove the N^{α}-*tert*-butyloxycarbonyl group is too great, estimated at 0.8% per each deprotection step (Erickson and Merrifield, 1973a); and (2) final acidolytic removal of the *O*-benzyl group (**185**) is accompanied by intramolecular 3-alkylation (p. 215ff). To circumvent the first of these problems, the *m*-bromo-(**186**) and 2,6-dichlorobenzyl (**187**) derivatives have been proposed. These are respectively 50- and 5000-fold more stable to acid, yet fully cleaved within 10 min by liquid HF at 0°C (Yamashiro and Li, 1972, 1973a; Erickson and Merrifield, 1973a). However, these changes do not serve to eliminate entirely the alkylation side reaction. Towards this latter goal, Engelhard and Merrifield (1978) reasoned that a sterically hindered ether might have improved features and synthesized model compounds of four secondary alkyl ethers of tyrosine. The cyclohexyl derivative **189** was recommended as providing the proper balance of rates of acidolysis, being 125-fold more stable to TFA–CH_2Cl_2 (1:1) than the unsubstituted benzyl ether **185**, yet smoothly deblocked by HF–anisole at 0°C within 30 min. The primary *O*-alkyl ether of *O*-methyltyrosine is stable to HBr–acetic acid at 25°C, and peptides containing this amino acid as a tyrosine analog have been made by solid-phase synthesis (Stewart and Woolley, 1967; Chaturvedi *et al.*, 1970).

The *tert*-butyl ethers (**190**) are compatible with milder synthetic schemes in which trifluoroacetic acid is the reagent for final acidolytic deprotection; and they are also desirable because 3-*tert*-butylation of the phenol ring does not appear to occur. The acid-stable *O*-*o*-nitrobenzy ether (**188**) is removable by the orthogonal mode of photolysis at wavelengths centered at 3500 Å [Amit *et al.*, 1977; compare to Eq. (3) on p. 81]. Photolysis of *O*-*o*-nitrobenzyltyrosine in methanol gave a 63% yield of free tyrosine; the yield was quantitative in the presence of semicarbazide hydrochloride, which was the most effective of three aldehyde scavengers tested. It follows that 3-alkylation is avoided by photolytic cleavage. Another orthogonally removable phenolic hydroxyl protecting group is the 2,4-dinitrophenyl function (**191**) (Fridkin *et al.*, 1977), which is stable in TFA and HF, but cleaved within 1 hr by β-mercaptoethanol–dimethylformamide (1:1) or by 0.05 *M* β-mercaptoethanol in pH 9.0 aqueous buffer. The rate of thiolysis of the *O*-dinitrophenyl(Dnp) group is an order of magnitude faster than that of the N^{im}-Dnp (**129**) group (p. 181). The *O*-Dnp group is also removed (60–70% in 24 hr at 25°C) by aminolysis with 0.1 *M* glycine ethyl ester in dimethylformamide, or by hydrolysis at pH 11; these susceptibilities might preclude certain solid-phase applications.

The *O*-benzyloxycarbonyl group (**192**) is of some merit because its acidolytic removal proceeds without intramolecular 3-alkylation, but it is still much too prone to premature acidolysis, as already discussed for the *O*-benzyl group (**185**), to be recommended for solid-phase synthesis (Erickson and Merrifield, 1973a). The *O*-*o*-bromobenzyloxycarbonyl group (**193**) (Yamashiro and Li, 1973b) is suitably "fine-tuned" for protection schemes based on graduated acid lability in which the N^{α}-*tert*-butyloxycarbonyl group is removed by TFA, and removal of permanent groups is effected in HF. The *O*-ethoxycarbonyl group (**194**) (applied in solid-phase synthesis by Wissmann and Geiger, 1970) is stable to acid, but removed by alkaline hydrolysis, ammonolysis, and hydrazinolysis.

All of the alkyl ethers (**185**–**190**) resist the actions of nucleophiles; therefore tyrosine can be protected in these ways when final cleavage of protected peptide segments from the support is carried out by alkali, ammonia, or hydrazine. In contrast, all of the acyl derivatives (**192**–**195**) are labile to the same nucleophilic reagents; this lability is of serious consequence even at the level of the repetitive deprotection/coupling cycles of a solid-phase synthesis. As a cautionary example, intermolecular aminolytic conversion of H-Tyr(Z)-OH in 10% diisopropylethylamine–CH_2Cl_2 to a mixture of Z-Tyr(Z)-OH and free tyrosine was substantial in 2 hr at 25°C (25% starting material remaining; Erickson and Merrifield, 1973a). Such a side reaction in solid-phase synthesis will lead to a termination of one resin-bound peptide chain; the further fate of the second peptide chain with a liberated free phenolic hydroxyl group has already been discussed (p. 212). In addition, mercaptans that are often added to the deprotective media as scavengers for *tert*-butyl carbonium ions could conceivably also be of sufficient nucleophilic reactivity to prematurely deblock the side chain of an acyl-protected tyrosine residue.

Biological studies on peptides, including radioimmunoassay, metabolic tracing, and binding measurements, often require radiolabeled material of high specific activity. An ideal way to do this is presented if the molecule contains a tyrosine residue, because the phenolic ring is readily iodinated with ^{125}I or ^{131}I [Eq. (26); reviewed by Glazer, 1976, pp. 57–63]. This

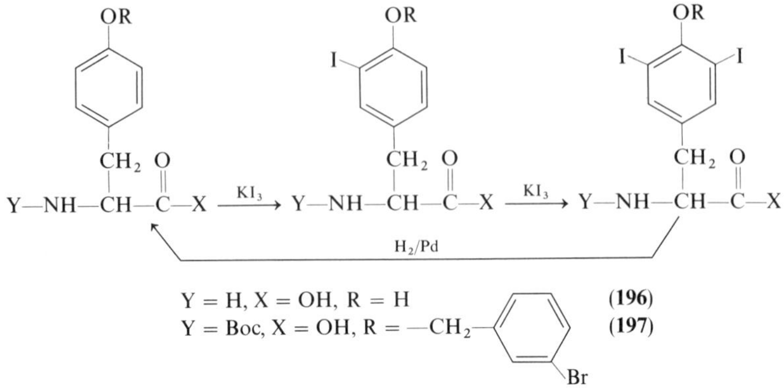

(26)

(**196**)
(**197**)

powerful technique still suffers from certain ambiguities and a new approach has recently been proposed by Lemaire *et al.* (1977). These workers introduced 3,5-diiodotyrosine (**196**) (pK_a = 6.4) into peptides by total synthesis, through its N^α-Boc-*O*-(*m*-bromobenzyl) derivative (**197**). The choice of the *m*-bromobenzyl protecting group (**186**) is safe in this case because both *o*-positions of the phenolic ring are already iodinated, and the 3-alkylation side reaction is impossible. The 3,5-diiodotyrosine-containing peptide* can be purified, and catalytic reduction with tritium was proposed to provide a labeled peptide containing tritiated tyrosine of high specific activity. Very recently, the successful application of this idea was reported (Houghten and Li, 1978).

b. Side Reactions. The characteristic chemically reactive sites of the phenolic nucleus of tyrosine are the *o*-positions, which are readily attacked by electrophiles and oxidizing agents [compare to Eq. (26)]. This sensitivity imposes limitations on some common procedures of peptide synthesis. For example, oxidative methods for cleaving peptides from a support (Section III,B,5,b and III,B,9) cannot be carried out on tyrosine-containing peptides. The use of reagents such as mercury salts intended to deblock cysteine residues (Section V,B,11,a and c) may lead upon prolonged treatment to irreversibly mercurated tyrosine residues (B. J. Williams and G. T. Young, personal communication, 1978). Precautions are also necessary in conjunction with standard acidolytic deprotection. The "scrubbing" of HBr gas

* The 3,5-diiodotyrosine-containing peptide is an analogue of the desired peptide, and catalytic reduction with hydrogen to regenerate an underivatized tyrosine residue in the parent compound could serve as the chemical basis for an internal structural control study (compare to footnote on p. 170).

by bubbling through resorcinol solutions is necessary to prevent bromination of tyrosine by contaminating bromine (Marglin and Merrifield, 1966).

Recently, it was shown (Nakaie *et al.*, 1977) that when the TFA–$CHCl_3$ solution used to remove the N^{α}-*tert*-butyloxycarbonyl group was more than a day old, 5 to 8% of the peptide product purified at the conclusion of the solid-phase synthesis contained a residue of 3-chlorotyrosine in place of tyrosine. This side reaction did not occur when freshly prepared trifluoroacetic acid solutions, preferably in CH_2Cl_2, were used.

The major difficulty with *O*-alkyl ether protection of tyrosine is the occurrence of an acid-catalyzed intramolecular 3-alkylation, also referred to as an O → C rearrangement [Eq. (27)]. The by-product **198** (Y = H, X = OH, R = $CH_2C_6H_5$) was first observed in 6 *N* HCl hydrolysates of *O*-benzyltyrosine (Iselin, 1962), and later found in amounts of 15 to 35% upon treat-

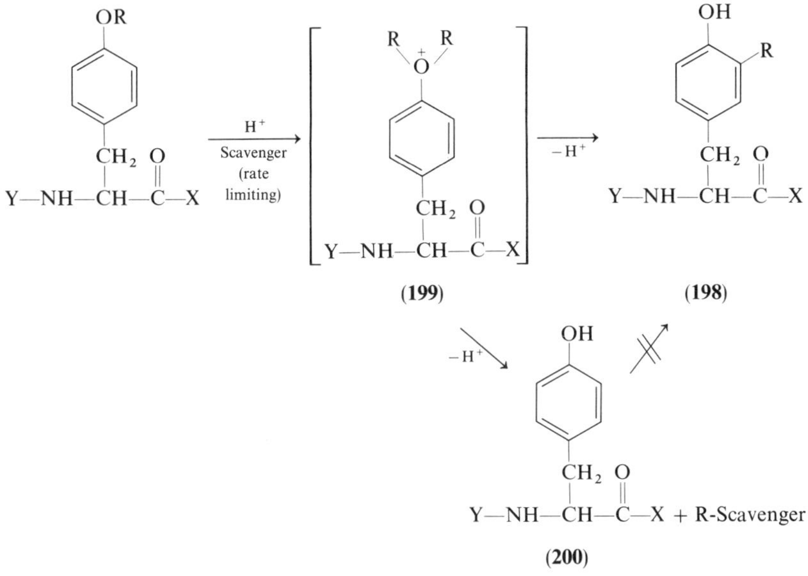

(27)

ment with either TFA–CH_2Cl_2 (1:1) or HF–anisole (9:1) (Erickson and Merrifield, 1973a). The latter authors showed that the levels of by-product formation were not reduced by added scavengers (Table IX) and postulated a common intermediate **199** formed in a rate-limiting step. This intermediate collapses intramolecularly to give **198**, or else dissociates to the deprotected species **200**; the ratio of **198** to **200** formed is constant throughout the course of the deprotection. In the absence of scavengers, additional amounts of **198** were observed, implying that 3-alkylation may also proceed by an intermolecular pathway. Not too surprisingly, appreciable levels of 3-benzyltyrosine have now been found upon examination of the peptide products from solid-phase syntheses in which *O*-benzyltyrosine was used, and where the final cleavage and deprotection was carried out in HF–anisole (e.g., Galardy *et al.*, 1976; Nakaie *et al.*, 1977).

When other strong acid–scavenger combinations (Tables VIII and IX) are tried, recoveries of free tyrosine from *O*-benzyltyrosine are invariably low. The yields, in all cases in the presence of anisole, were 66% in HF–pyridine (9:1), 68% in pyridinium polyhydrogen fluoride, 11% in CF_3SO_3H–CH_2Cl_2, 30% in CH_3SO_3H–TFA, and 81% in BBr_3–CH_2Cl_2 (see Table VIII for references). In a new comparative study on the extent of rearrangement under a variety of acidolytic conditions, the amount of 3-benzyltyrosine formed ranged from 40% with no scavenger to 13–20% in the presence of anisole to 1–4% in the presence of thioanisole (Kiso *et al.*, 1978). Although not identified as such, irreversible modification of tyrosine by C-alkylation was evidently significant in HBr–TFA but not in HBr–HOAc (Trudelle and Spach, 1972). The latter reagent is not particularly general, because the hydroxyls of serine, threonine and tyrosine are partially acetylated. Nonetheless, Bodanszky *et al.* (1978) recently stressed the special scavenging action of acetic acid and proposed two new reagents designed to minimize benzyl migration. They recommended use of TFA–HOAc (7:3) for repetitive removal of the N^{α}-*tert*-butyloxycarbonyl group from a protected peptide and showed that only 1% 3-benzyltyrosine formed when final deprotection was carried out by HBr in phenol–*p*-cresol (1:1). The latter medium warrants further investigation as a cleavage reagent for solid-phase synthesis.

The 3-alkylation side reaction does not occur when the *O*-benzyl group is removed under non-acidic conditions, such as catalytic hydrogenolysis subsequent to release of the peptide from the support. In acid, and in the presence of anisole, no (< 0.2%) 3-benzyltyrosine was formed upon deprotection of *O*-benzyloxycarbonyltyrosine, presumably because the additional carbon dioxide moiety efficiently separates the incipient benzylic carbocation from the phenolic ring (Erickson and Merrifield, 1973a; see also Yamashiro and Li, 1973b).

Because reservations have been expressed about the desirability of acyl protection for tyrosine (see p. 213), a different trend to try to eliminate 3-alkylation has been to change the nature of the *O*-alkyl ether (Erickson and Merrifield, 1973a; Engelhard and Merrifield, 1978; Yamashiro and Blake, 1978). With this goal in mind, the *O*-2,6-dichlorobenzyl (**187**) protecting group has been shown to reduce the side reaction during HF treatment from 20–70% to 2–9%, whereas the *O*-cyclohexyl group (**189**) reduced it to 0.3% or less. Thus, **189** is the best group yet available. In conjunction with strategies relying on TFA for the final deprotection, the *O*-*tert*-butyl group can be readily removed from tyrosine without any intramolecular alkylation.

c. Analytical. Recovery of tyrosine from acid hydrolysates of peptides is good, provided small amounts of antioxidants are added to the 6 *N* HCl [e.g., 0.2% phenol (w/v), 0.1% mercaptoacetic acid (v/v); Sanger and Thompson, 1963; Salnikow *et al.*, 1973]. The various 3-alkyltyrosine rearrangement by-products are all stable to hydrolysis, and can be quantitated on standard automated amino acid analyzers (Erickson and Merrifield, 1973a; Engelhard and Merrifield, 1978). Peptides containing tyrosine absorb in the ultraviolet ($\lambda_{max} = 278$ nm, $\varepsilon = 1.1 \times 10^3$); the position and shape of the peak at different pH values is a characteristic criterion for the integrity of this residue. Detection upon chromatographic fractionation is also facilitated by the quantitative Folin–Lowry method (Lowry *et al.*, 1951; see Stewart and Young, 1969, p. 57). Partial chemical cleavage of the peptide bond at the carboxyl end of a tyrosine residue through oxidation with *N*-bromosuccinimide in mildly acidic aqueous media is occasionally of analytical value (reviewed by Spande *et al.*, 1970, pp. 187–191).

8. *Phenylalanine*

As has been stated earlier (Schröder and Lübke, 1965, p. 140), the synthesis of peptides containing phenylalanine presents no special problems, and phenylalanine is a standard amino acid to test new protecting groups and coupling methods. The weak ultraviolet absorption ($\lambda_{max} = 259$ nm, $\varepsilon = 187$) is occasionally analytically useful.

The only difficulty known to arise with phenylalanine residues occurs when catalytic hydrogenolysis is employed to remove N^{im}-benzyl or N^{g}-nitro groups from histidine and arginine residues (Visser *et al.*, 1970a; Schafer *et al.*, 1971; Windridge and Jorgensen, 1971b). Thus, Windridge and Jorgensen (1971b) found that 8% of β-cyclohexylalanine had formed at a time when 2% His(Bzl) remained [H_2/Pd(C), 30 psi, 10% aqueous acetic acid, 48 hr]. Khosla *et al.* (1976) described a routine screening for this side reaction by amino acid analysis; no β-cyclohexylalanine was formed under their hydrogenolysis conditions [H_2/Pd black or 5% Pd($BaSo_4$), 30 psi, methanol–acetic acid–H_2O (5:1:1), 48 hr].

9. *Tryptophan.*

a. Protection. The weakly nucleophilic β-indole group of tryptophan has been left unprotected (**201**) (Marshall, 1968) for solid-phase synthesis, or has been protected by the *N*-formyl (For) group (**202**) (Previero *et al.*, 1967; Ohno *et al.*, 1972, 1973; Yamashiro and Li, 1973c). Recently, the acid-labile N^{i}-benzyloxycarbonyl group has been proposed (Chorev and Klausner, 1976), but further information will be required before its place in peptide chemistry can be evaluated. Indeed, the relative merits of the principal methods, utilizing **201** or **202**, remain to be sorted out. Difficulties are associated with each approach, but they can often be kept at a controllable level.

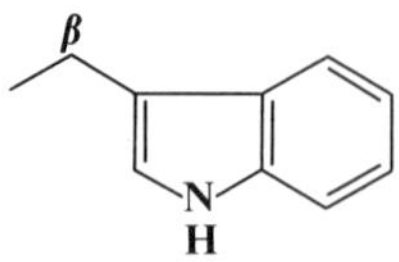

*(**201**)

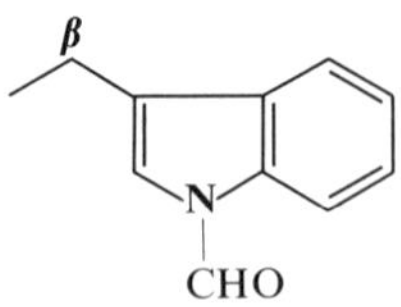

*(**202**) For

The extreme susceptibility of tryptophan to oxidative degradation by atmospheric oxygen in acidic media (p. 220) has necessitated modifications of the standard solid-phase synthetic protocols (see Stewart and Young, 1969, pp. 48–49). In their outstanding synthesis of ACTH (39 amino acid residues), Yamashiro and Li (1973a) incorporated N^{α}-Boc-tryptophan and carried out eight more deprotection/coupling cycles [total time of exposure to TFA–CH_2Cl_2 (1:1) = 2 hr] to extend the polymer-supported peptide chain to its NH_2-terminus in an automated Beckman peptide synthesizer operated under nitrogen. The net yield of purified product after HF–anisole cleavage and three chromatographic steps was 7%. The discovery of a significant N^{i}-*tert*-butylation side reaction with the unprotected indole ring during acidolytic removal of the N^{α}-*tert*-butyloxycarbonyl group (p. 222), establishes the need for more stringent safety precautions, and such an approach has been followed by most authors. The addition of scavengers (Table IX, p. 72 note references) to the common acid reagents (Section IV,A,1,a, starting on p. 102, numbers 1, 2, and 5) has been generally beneficial. The amount of *tert*-butylated tryptophans formed from *tert*-butyloxycarbonyltryptophan at 25°C for 50 min ranged from 14% in neat TFA to 1.1% in TFA–1,2-dithioethane–dimethyl sulfide (19:20:1) as the most effective of several combinations tested (Sakakibara, 1977, p. 440; Chino *et al.*, 1978). Application of β-mercaptoethanesulfonic acid (Section IV,A,1,a; number 6) combines scavenger and acidolytic features in a single reagent. Scavengers are not required when reagent mixtures including formic acid are used (Section IV,A,1,a; number 3 and 10), but these results are due to *in situ* formylation as explained below.

Other approaches to incorporate tryptophan with an unprotected indole side chain have depended on changing the α-amino protecting group to one removable under milder acidic conditions, or under neutral or basic conditions. A synthesis where four cycles of N^{α}-Bpoc deprotection were carried out in the presence of trytophan with TFA–β-mercaptoethanol–CH_2Cl_2 (3:2:195) gave good results (Wang and Merrifield, 1969a), and the new method involving removal of the dimethylphosphinothioyl (Mpt) function (**53**) with 0.2 *N* HCl–0.2 *M* triphenylphosphine–CH_2Cl_2 seems promising (Ueki *et al.*, 1977; see p. 108). Nucleophilic cleavage of the N^{α}-*o*-nitrophenylsulfenyl group (pp. 109–112) was demonstrated to be compatible with syntheses of tryptophan peptides by Marshall (1968). Recently the base-labile N^{α}-fluorenylmethyloxycarbonyl (Fmoc) group (**59**) [Eq. (11) on p. 113], was successfully used in the stepwise solid-phase synthesis of the 14-peptide, dihydrosomatostation, containing tryptophan at position 8 (Chang and Meienhofer, 1978).

All of the studies designed to minimize destruction of tryptophan during the repetitive deprotection/coupling cycles are contingent on the assumption that final cleavage of the peptide from the solid support can be safely carried out. Destruction of trytophan in model peptides is negligible in HF at 0°C when reaction times are 1 hr or less, and scavengers such as anisole or 1,2-dithioethane are present; these conditions are definitely preferable to HBr in TFA (Guttmann and Boissonnas, 1959; Marshall, 1968; Sakakibara *et al.*, 1968a; Sakakibara, 1977, p. 440; Chino *et al.*, 1978). Successful cleavage of tryptophan peptides from more acid-labile linkages such as the *p*-alkoxybenzyl ester (**16**) will require effective scavenging for *tert*-butyl and other carbocations. The abundance of cleavage methods involving nucleophilic attack (Sections III,B,1,c;3,b; and 5,c) are useful for tryptophan peptides because of the stability of the β-indole group to basic reagents.

Tryptophan in peptides and proteins becomes partially formylated to **202** in HCl–formic acid, and the reaction is reversed in midly basic aqueous solution at pH $\geqslant 9$ (Previero *et al.*, 1967). Formic acid-containing media for N^{α}-*tert*-butyloxycarbonyl removal (Ohno *et al.*, 1972; Buis *et al.*, 1976) have been used, without scavengers, in solid-phase synthesis; peptides of good quality containing intact tryptophan were isolated following cleavages mediated, respectively, by hydrazine in dimethylformamide and sodium hydroxide in dioxane–methanol. These encouraging data have been systematized with a series of studies in which N^{i}-formyltryptophan was incorporated as such into peptides (Ohno *et al.*, 1973; Yamashiro and Li, 1973c, 1974; Lemaire *et al.*, 1976). It was shown that whereas 1 *N* HCl–HOAc promotes considerable oxidative degradation of tryptophan, this undesired pathway is entirely suppressed by the N^{i}-formyl group. Further exposure to the same acidic reagent led to gradual loss of the formyl group, whereupon tryptophan was degraded as before.

The N^{i}-formyl group is obviously expected to confer resistance to N^{i}-*tert*-butylation upon N^{α}-*tert*-butyloxycarbonyl removal; hence scavengers are unnecessary [see, however, Rees and Offord (1976), who argue the opposing point of view that indole formylation makes no difference]. It is of interest that Li *et al.* (1975) reported overall yields of 25% and 10%, respectively, from syntheses of the 18-peptide β-MSH in which tryptophan was or was not protected. Also, Birr (1978, p. 53) has stated that symmetrical anhydrides formed from Boc-Trp(For)-OH were pure, whereas side reactions occurred when Boc-Trp-OH with the free indole was thus converted.

The N^i-formyl group is stable to the usual acidic and basic conditions of the repetitive deprotection/coupling cycles and also remains intact upon cleavage in HF. A variety of orthogonal removal conditions have been proposed, including (1) hydrazine hydrate–DMF (1:10), 25°C for 48 hr; (2) 0.1 *N* aqueous piperidine or triethylamine, 0°C for 10 min, quenched by lyophilization; (3) 1.0 *M* aqueous ammonium bicarbonate, pH 9, 25°C for 24 hr, quenched by lyophilization; (4) liquid ammonia, preferably containing 100 equiv hydroxylamine hydrochloride for 2 hr; and (5) aqueous solution, pH 11.5, 25°C for 3 min, quenched by acidification to pH 4.5.

There are several limitations associated with the methods for N^i-deformylation. Transfer of the formyl group from the indole nitrogen to the NH_2-terminus of a peptide is a major side reaction, quantitated at ~50% and ~20% in 0.01 *M* aqueous ammonium bicarbonate, pH 9, and in liquid ammonia, respectively (Yamashiro and Li, 1973c). The modified conditions (3) and (4) described in the previous paragraph lowered transformylation to under 10%. Another source of difficulties is the slow reaction rates observed when mild deprotection conditions are applied to large formyltryptophan-containing peptides. On the other hand, the N^i-formyl group is cleanly removed by the same bases and nucleophiles, discussed earlier, that cleave benzyl, phenacyl, and activated *n*-alkyl ester linkages to the solid support (Sections III,B,1,c;3,b; and 5,c). These latter conditions can often lead to side reactions with other susceptible amino acid residues, although not with tryptophan.

Blake and Li (1975a) have described several structural analogues of tryptophan and their incorporation into peptides by solid-phase methodology.

b. Side Reactions. The indole ring of tryptophan in peptides is sensitive to oxidation and alkylation (reviewed by Fontana and Toniolo, 1976). Tryptophan residues are particularly labile in acidic media, where atmospheric oxygen is sufficiently potent to initiate the oxidation sequence schematized in Eq. (28); each of the intermediates can be isolated under suitable

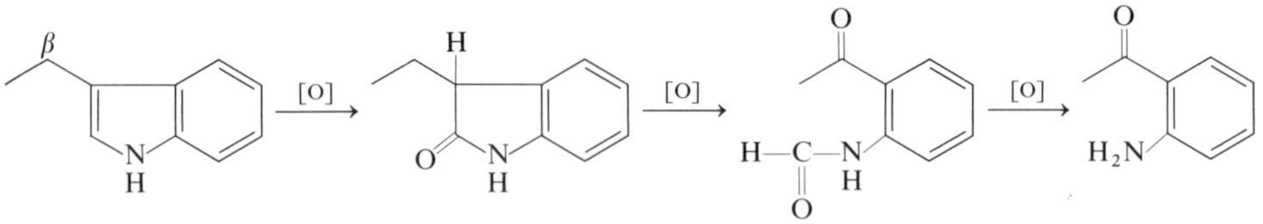

(201) indolyl — (203) oxindolyl — (204) formylkynureninyl — (205) kynureninyl

(28)

conditions. The *β*-oxindolyl derivative (**203**) is formed partially in HBr–HOAc (Theodoropoulos and Fruton, 1962) and in good yields with *N*-bromosuccinimide in mildly acidic aqueous solution (Patchornik *et al.*, 1960); further reaction leads to peptide bond cleavage at the carboxyl end of the tryptophanyl residue (Spande *et al.*, 1970, pp. 158–161; Fontana and Toniolo. 1976, pp. 340–344). Quantitative and specific conversion of **201** to **203** can be achieved with a mixture of dimethyl sulfoxide and concentrated aqueous hydrochloric acid (Savige and Fontana, 1977a). Further oxidation to open the indole heterocycle leads to formylkynurenine (**204**), which can finally lose carbon monoxide to give the acid-stable kynurenine (**205**). The limitations and precautions with procedures of peptide synthesis already mentioned for tyrosine (p. 214) apply with even greater urgency to tryptophan.

Electrophiles add to the indole nucleus primarily at the 2-position (Fig. 9). This exceedingly facile, quantitative and irreversible reaction is the reason that N^{α}-*o*-nitrophenylsulfenyl (**56**) protecting group removal with hydrogen halides (pp. 110–111) is incompatible with synthesis of tryptophan peptides.* Another instance of 2-alkylation was reported by Southard (1971; see p. 207) and should be a source of concern whenever side-chain blocked derivatives, for example of asparagine, glutamine, and cysteine, are used that give rise to very stable carbonium ions upon acidolytic deprotection.

Recently, Omori *et al.* (1976) reported that in acidic media such as HF and TFA, the model compound, N^{α}-acetyltryptophan methyl ester, was significantly transformed; the structure of the Ehrlich-positive product was assigned on the basis of nmr and mass spectrometry to the dimer **206**. The

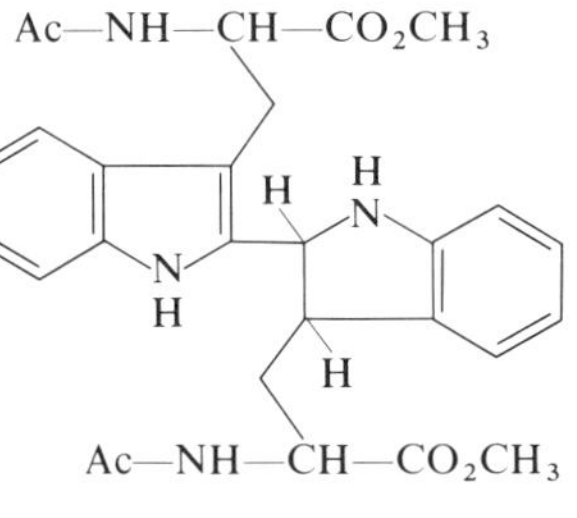

(**206**)

yield of **206** from a 15% (w/v) solution of the starting indole in TFA, at 25°C overnight was 40% (28% yield of crystalline material). These data suggest

* Tryptophan peptides substituted in the 2-position with an *o*-nitrophenylsulfenyl group are regenerated to the unmodified indole ring by catalytic hydrogenolysis (Marche *et al.*, 1975); the methionine thioether is not split under these conditions. This novel chemical observation might be applied for specific radiolabeling with tritium gas, for internal structural control studies (see last paragraph of Section V,B,7,a and accompanying footnote, p. 214, and as the basis for reversible protection of tryptophan in peptide synthesis.

that the addition of indoles as scavengers to TFA solutions used to remove the N^{α}-*tert*-butyloxycarbonyl group might be counter-productive, since the scavenger indole could in principle react with an indole group of a tryptophan residue, leading to irreversible modification of the resin-bound peptide. The formation of by-products similar to **206** might also account for the aggregation and inactivation of the enzyme lysozyme (six Trp) upon treatment with HF at 0°C for 60 min or longer (Aimoto and Shimonishi, 1975).

When the N^{α}-*tert*-butyloxycarbonyl group is removed acidolytically from tryptophan-containing peptides in the absence of carbonium ion scavengers, significant levels of *tert*-butylated by-products are observed (Alakhov *et al.*, 1970; Wünsch *et al.*, 1972, 1977; Sakakibara, 1977, pp. 439–440; Chino *et al.*, 1978; Ogawa *et al.*, 1978; Jaeger *et al.*, 1978a,b; Löw *et al.*, 1978a,b). The side reaction was discovered by mass spectrometry, and structural assignments were made by nmr. The principal site of *tert*-butylation is the indole nitrogen (up to 50%). The preliminary indications are that the N^{i}-alkylation is an equilibrium in HF which can, therefore, be entirely suppressed or reversed by the addition of 1,2-dithioethane as a scavenger. Protection of the indole nitrogen with the formyl function (**202**) is expected to and in fact does prevent the side reaction at this site (Löw and Kisfaludy, 1979; Masui *et al.*, 1979). Alternative sites of *tert*-butylation of the indole ring have been identified in low amounts. The C-alkylated by-products may be formed by migration of an initial N^{i}-adduct, and do not revert to the free indole species under anhydrous acid conditions.

Hydrogenolytic removal of an N^{α}-benzyloxycarbonyl group from a peptide containing N^{i}-formyltryptophan was accompanied by production of substantial levels of the corresponding derivative with a 2,3-dihydroindole side chain (Löw and Kisfaludy, 1979). Under the more vigorous conditions needed to hydrogenolyze nitroarginine, the indole nucleus (unprotected, in these cases) has been reported to undergo transformation resulting in the appearance of octahydrotryptophan-containing peptides (Keil *et al.*, 1962; Bajusz *et al.*, 1973).

Photolytic destruction of tryptophan is a potential hazard which seems not to occur upon irradiation at wavelengths above 3500 Å (Rich and Gurwara, 1975a; see p. 81).

c. Analytical. Hydrolysis of peptides with 6 *N* HCl at 110°C leads to oxidative destruction of tryptophan by a complex mechanism that is facilitated by compounds such as cystine, serine, or carbohydrates. There are three practical procedures to obtain tryptophan for quantitative amino acid analysis: (1) hydrolysis with 3 *N* *p*-toluenesulfonic acid in the presence of 0.2% 3-(2-aminoethyl)indole (Liu and Chang, 1971), 3 *N* β-mercaptoethanesulfonic acid (Penke *et al.*, 1974), or, possibly best, 4 N methanesulfonic acid (Simpson *et al.*, 1976). Recovery of tryptophan by these methods, in the absence of carbohydrates, is better than 90%. If β-oxindolyl-alanine (**203**) is present, it will also be stable to the hydrolytic conditions, and chromatographically distinguishable (Savige and Fontana, 1977a); (2) base hydrolysis with 4.2 N NaOH in the presence of starch as an antioxidant (Hugli and Moore, 1972); and (3) enzymatic hydrolysis (detailed by Stewart and Young, 1969, pp. 54–55). The first two of these methods will reconvert formyltrytophan to tryptophan. The *tert*-butylated tryptophan by-products previously mentioned elute as a broad peak after tryptophan on the amino acid analyzer (Sakakibara, 1977, p. 440; Chino *et al.*, 1978).

The characteristic and strong ultraviolet absorption spectrum of tryptophan ($\lambda_{max} = 279$ nm, $\varepsilon = 5.2 \times 10^3$) is particularly diagnostic for the structural integrity of this residue. Upon oxidative degradation, the height and shape of the spectral pattern changes (initial increase in extinction near 282 nm), as documented by Ohno *et al.*, (1972). Reactions of formyltryptophan can be conveniently followed at the absorption maximum around 300 nm (Previero *et al.*, 1967; Ohno *et al.*, 1972). The Ehlich reagent, *p*-dimethylaminobenzaldehyde, gives a colored product with many compounds containing the β-indole nucleus (Spies and Chambers, 1950), although not with formyltryptophan (Ohno *et al.*, 1973). Chemical modification of deblocked tryptophan in a synthetic peptide with sulfenyl chlorides (following Scoffone *et al.*, 1968, compare to Fig. 9) is a particularly specific test for the absence of oxidation by-products (Ohno *et al.*, 1972). Cleavage of the peptide bond at the carboxyl end of a tryptophan residue is now best carried out with 2-(2-nitrophenylsulfenyl)-3-methyl-3′-bromoindolenine (BNPS-skatole, reviewed by Fontana and Scoffone, 1972) or a mixture of dimethyl sulfoxide and aqueous concentrated hydrogen bromide (Savige and Fontana, 1977c). A more detailed treatment of these and other topics in the characterization of tryptophan peptides is provided by Fontana and Toniolo (1976).

10. Methionine

a. Protection. The weakly nucleophilic γ-methyl thioether group of methionine has been left unprotected (**207**) in solid-phase synthesis (e.g., Merrifield, 1964c), or been protected as the sulfoxide (**208**) (Iselin, 1961; Hofmann *et al.*, 1965b; applied by Gutte and Merrifield, 1971; Yamashiro and Li, 1973a).

Chemical transformations of methionine derivatives are summarized in Fig. 18. Under different conditions of peptide synthesis, purification, or analytical characterization, one or more of these pathways predominate; it is important to be aware of all the possible by-products and interrelationships of the different derivatives.

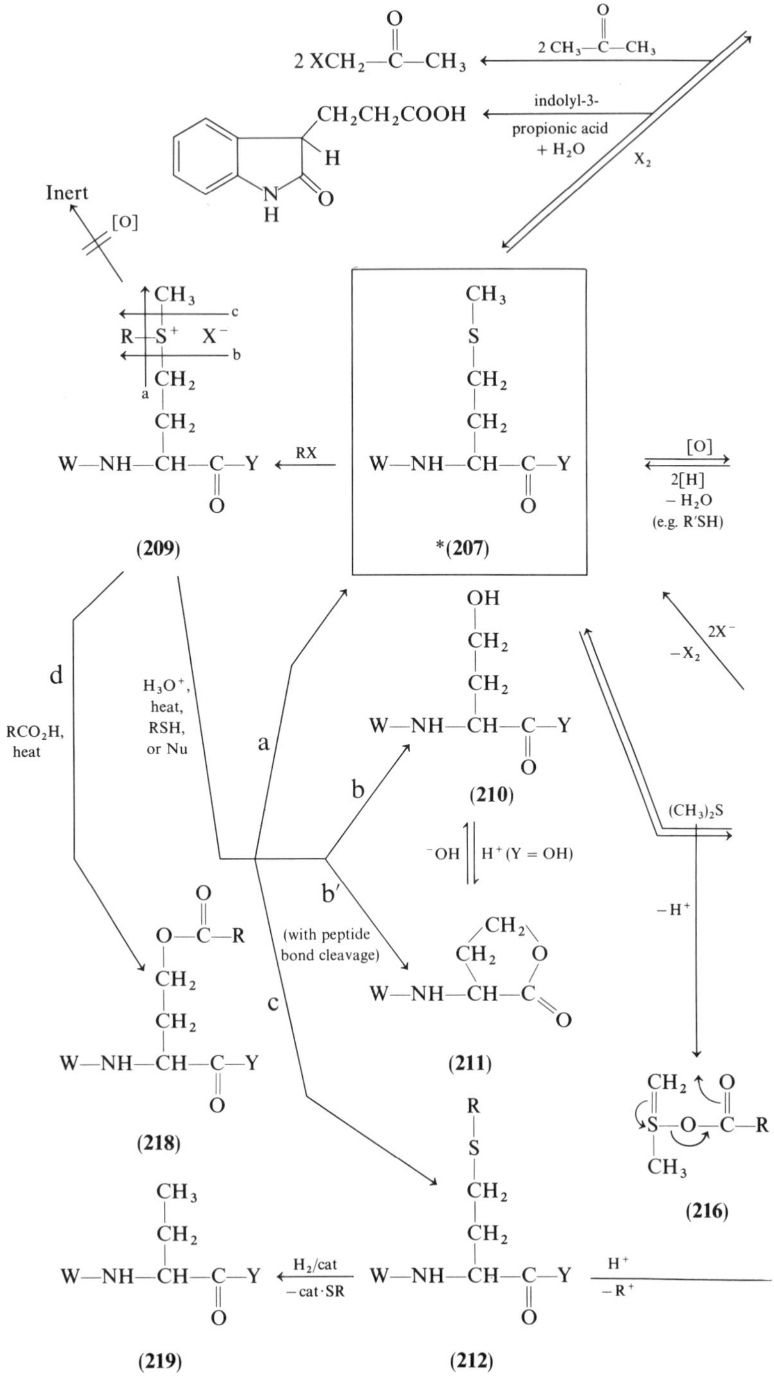
2 CH3—C—CH3
2 XCH2—C—CH3
indolyl-3-
propionic acid
+ H2O
CH2CH2COOH
X2
Inert
[O]
RX
[O]
2[H]
– H2O
(e.g. R′SH)
(209)
*(207)
(210)
(211)
(212)
(216)
(218)
(219)
d
RCO2H,
heat
H3O+,
heat,
RSH,
or Nu
a
b
b′
(with peptide
bond cleavage)
c
2X−
−X2
(CH3)2S
−H+
H2/cat
−cat·SR
H+
−R+

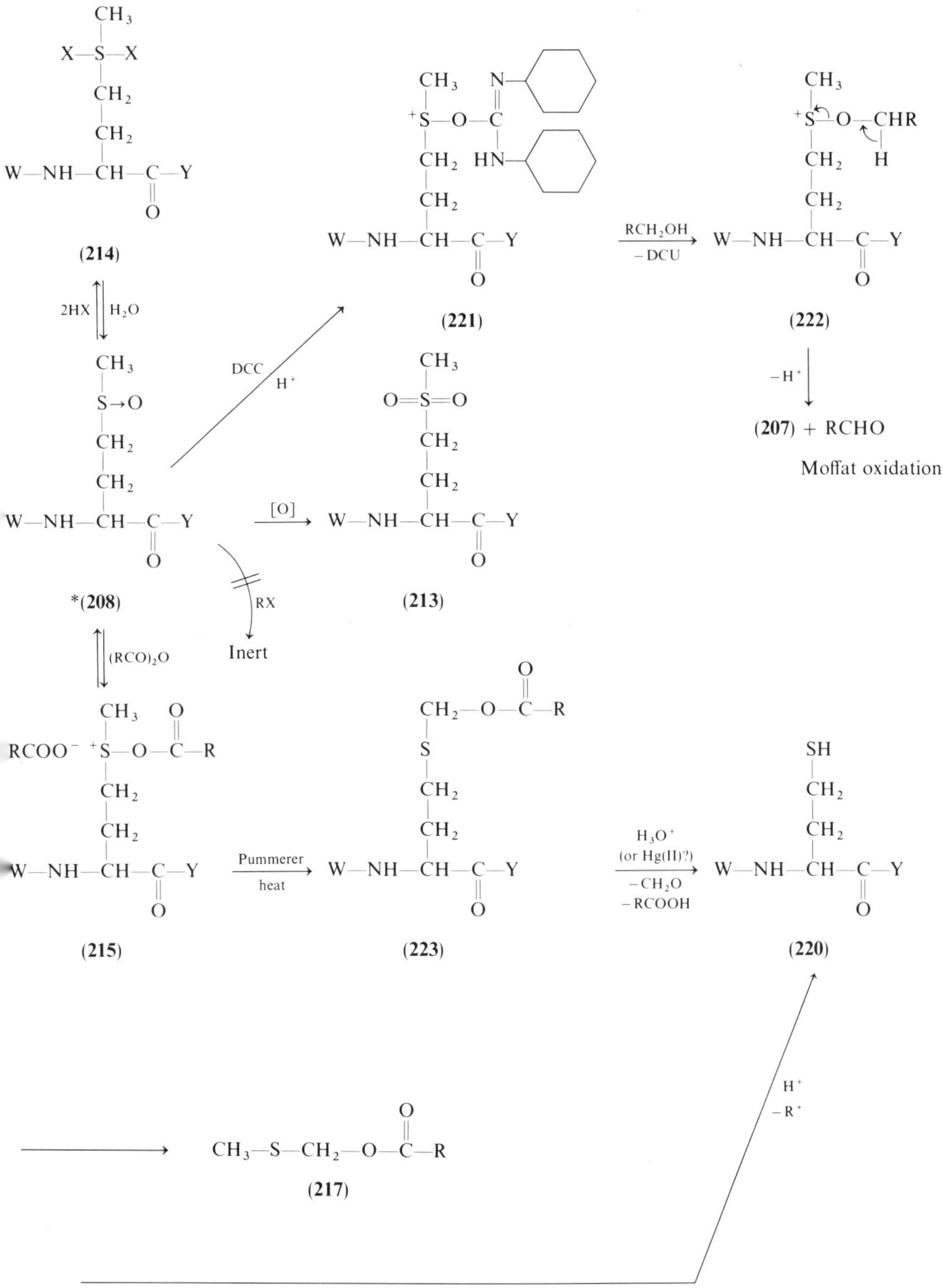

Figure 18. Transformations of methionine derivatives. Detailed explanation in text of Section V,B,10,a and c.

Unprotected methionine (**207**) is labile to spontaneous oxidation to the sulfoxide (**208**) by exposure to atmospheric oxygen (Hofmann *et al.*, 1965a). Norris *et al.*, (1971) showed that no sulfoxide was formed from an *N*-acetyl-methionyl-resin after 32 simulated coupling cycles in an automated instrument operated under nitrogen; however, a protected methione-containing hexa-peptide was oxidized to the extent of 10% after standing in glacial acetic acid solution for 96 hr.

The major side reaction of unprotected methionine (see p. 230ff) is alkylation with carbocations under acidic conditions to give the relatively stable sulfonium salts **209**. This process can take place either during the repetitive deprotection steps or in the final acidolytic cleavage, and is sometimes but not always suppressed by the addition of scavengers. Sulfonium salts serve in some sense as a protecting group, since they are entirely effective in preventing further oxidation (Neumann *et al.*, 1962; see Stirling, 1977, for a general review). Decomposition of the salts back to sulfides is promoted by aqueous acid, by nucleophiles or thiols at alkaline pH, and by heat (Hofmann elimination); the three possible products (Toennies and Kolb, 1945; Gundlach *et al.*, 1959) are residues of regenerated methionine (**207**) (Fig. 18, pathway a), of homoserine (**210**) (pathway b), and of S-alkylated homocysteine (**212**) (pathway c). In addition, intramolecular cyclization to homoserine lactone (**211**) (pathway b′) is accompanied by cleavage at the peptide bond containing the carboxyl group of the original methionine residue.

Pyrolytic acid-catalyzed fragmentation of sulfonium salts generally proceeds in a direction so that the predominant sulfide is the one with the two least electron-donating substitutents. For example, after heating the *S*-carboxymethylmethionine sulfonium salt at 100°C for 1 hr at pH 6.5, homoserine (**210**) and carboxymethyl methyl sulfide were the almost exclusive products (Gundlach *et al.*, 1959; see Ramirez *et al.*, 1973, for similar results with the *S*-methylmethionine sulfonium salt). Hydrolysis in strong acid gave a complex mixture of products. In order to direct the fragmentation along the desired pathway giving predominantly regenerated methionine, thiolysis procedures have been developed (Naider and Bohak, 1972; Jones *et al.*, 1976). In these cases, the probable mechanism involves S_N2 attack of a thiol, so the rate and product distribution depend on the susceptibility to nucleophiles of the group to be dealkylated. Conditions used were rather drastic; demethylation of *S*-methylmethionine sulfonium salt by dithiothreitol at pH 10.5 and 37°C proceeded with a half-life of 3.2 hr. Low levels of homoserine and/or *S*-alkylated homocysteine by-products derived from the attacking thiol cannot be ruled out from the available data.

The *S*-*tert*-butylmethionine sulfonium salts (**209**; R = *tert*-C_4H_9) are substantial by-products whenever unprotected methionine is used in peptide synthesis along with *tert*-butyl-based protecting groups removable by acidolysis. *tert*-Butyl sulfonium salts are known to revert rapidly to the parent sulfides upon brief boiling in aqueous solution or upon heating in the solid-state, with unimolecular elimination of *iso*-butene (Hughes and Ingold, 1933; Lavine *et al.*, 1954). The optimal conditions to recover methionine peptides from the *tert*-butyl sulfonium form have been explored by Noble *et al.* (1976), and their conclusions are summarized in Fig. 19. The best yields of the methionine peptides were found (Routes A and E) when either the free peptide or the peptide–resin had been exposed to heating steps; conversely, the sulfonium peptide was the almost exclusive product when the work-up was carried out rapidly in the cold (Route C). The side reaction was markedly less (compare Routes D and B) when the N^{α}-*tert*-butyloxycarbonyl group was removed with trifluoroacetic acid while the peptide was still on the resin; Noble *et al.* (1976) recommend this step as a routine precaution. Alkylation was significant ($>50\%$) even when the Boc-peptide resin was treated with HF in the presence of anisole as a scavenger. These data have been confirmed and extended in our laboratory (M. W. Riemen and R. B. Merrifield, unpublished results, 1978). Reversion of *tert*-butylsulfonium methionine to methionine was found to occur spontaneously and in high yield simply upon standing several days in neutral aqueous solution at 25°C. A small amount (2–10%) of homoserine (**210**) or homoserine lactone (**211**) (Fig. 18, pathways b or b′) was always found under several conditions (see also Jendrek *et al.*,

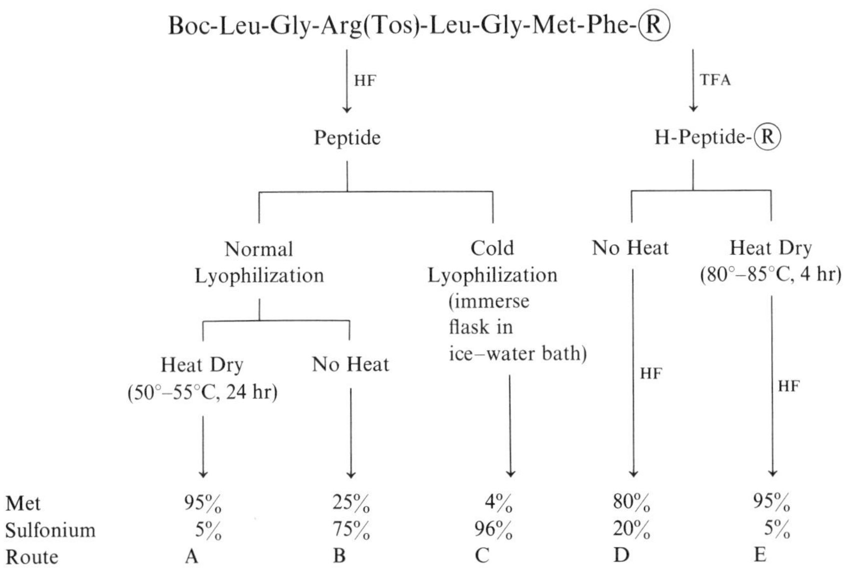

Figure 19. Regeneration of a methionine residue from the *tert*-butylated sulfonium salt by-product; taken from Noble *et al.* (1976). Copyright (1976) by the American Chemical Society. Reprinted by permission of the copyright owner.

1967, for a relevant model reaction). Wrong-way decomposition (Fig. 18, pathway c) to *tert*-butylhomocysteine (**212**; R = *tert*-C_4H_9) could not be definitively ruled out, but an upper limit of $\leqslant 2\%$ was established. Thiolysis of the sulfonium salt is contraindicated, because with these reagents, S_N2 attack led to loss of *tert*-butyl methyl sulfide and a high yield of homoserine.

Alkylation side reactions are essentially avoided by the use of methionine sulfoxide (**208**) (Iselin, 1961; Neumann *et al.*, 1962; see Oae, 1977, for general review). An experimental complication with the use of sulfoxides stems from the second optically active center in the amino acid. Methionine sulfoxide is stable to most strong acids, such as HF (Gutte and Merrifield, 1971) and CH_3SO_3H (Irie *et al.*, 1977), and is also stable to base and nucleophiles, as for example in the conditions for aminolytic cleavage from the support (Norris *et al.*, 1971). Possible overoxidation to the sulfone (**213**), premature reversion to the thioether (**207**) in acid, or reactions with dicyclohexylcarbodiimide or acylating agents are all potential drawbacks that have not generally been quantitated under conditions of peptide synthesis (pp. 231–232).

Several methods are available for the reduction of methionine sulfoxide to methionine in peptides. Reaction rates are often slow or conditions are too harsh: (1) Aqueous mercaptoacetic acid, 1–10% (v/v), at 37°–50°C for 24 hr or more (Iselin, 1961; Hofmann *et al.*, 1966; Fujino *et al.*, 1978). Even repeated applications, separated by ion-exchange chromatography, have not necessarily given complete reduction (Hofmann *et al.*, 1966), and mercaptoacetylation of α-amino groups (3–45%, depending on pH) has been noted as a side reaction under these conditions (Kessler and Iselin, 1966; Houghten and Li, 1977, 1979); (2) Aqueous *N*-methylmercaptoacetamide, 10% (v/v), 37°C for 12 hr has been advocated as the thiolytic condition of choice (Houghten and Li, 1977, 1979); (3) Dithrothreitol, 1 *M* at pH 9 (Polzhofer and Ney, 1971; Irie *et al.*, 1977) or 0.1 *M* in 0.1 *N* acetic acid (Yamashiro and Li, 1973a) at 50°C for 24–48 hr; (4) In strong aqueous or anhydrous hydrogen halides, equilibration of the sulfoxide with the dihalogen sulfide species **214** takes place (leading references in Savige and Fontana, 1977a). The reaction is driven to methionine **207** by a halogen acceptor (Fig. 18, top left), e.g., acetone, 10 min, 25°C, subsequent to treatment with HBr–HOAc (Iselin, 1961), or indolyl-3-propionic acid (100 equiv) added to 12 *N* HCl–HOAc (1:2), for 10 min at 25°C (Savige and Fontana, 1977b), or 12 *N* HCl–dimethyl sulfide (9:1), for 5 min at 25°C (Savige and Fontana, 1977b): (5) The sulfoxide oxygen readily reacts with strong acylating agents to yield acylsulfonium salts **215**, which are in turn quantitatively reducible with iodide (Allenmark, 1966). This one-step method has been adopted both for iodometric titration of resin-bound methionine sulfoxide, and as a means of preparative reduction to methionine;

conditions are sodium iodide (1.3 *M*) and acetyl chloride (0.9 *M*) in DMF, 25°C for 10 min (Norris *et al.*, 1971). A related technique that gives excellent yields, but has not as yet been applied to peptide synthesis, involves treatment of a sulfoxide (0.3 *M*) in CH_2Cl_2 with trifluoroacetic anhydride (1.2 equiv) and dimethyl sulfide (1.5 equiv) for a few seconds at 0°C (Tanikaga *et al.*, 1977). In a plausible mechanism, the initial acylsulfonium salt **215** exchanges with dimethyl sulfide, giving the reduced sulfide and the species **216**. Pummerer rearrangement then gives the stable α-acyloxysulfide (**217**; $R = CF_3$).

The problems associated with methionine protection can be entirely avoided (e.g., Kenner, 1977) by incorporating isosteric norleucine residues at the requisite positions. Working with *Staphylococcus aureus* nuclease, Anfinsen's group has justified this type of approach in two ways: (1) The enzyme in which norleucine had replaced methionine biosynthetically, isolated from a methionine auxotroph, was fully active (Anfinsen and Corley, 1969); and (2) a 42-residue fragment containing this substitution at two positions was prepared by solid-phase synthesis, and full activity was realized upon reconstitution with a natural fragment (Ontjes and Anfinsen, 1969b). In general, the synthesis of analogues containing norleucine will provide straightforward insights into the role of methionine residues in situations where chemical modification of methionine by alkylation or oxidation gives ambiguous results (e.g., Rocchi *et al.*, 1969; Rosenblatt *et al.*, 1976).

b. Anchoring. The esterification of COOH-terminal N^{α}-protected methionine to chloromethyl supports (p. 60ff) is complicated by a significant *S*-alkylation side reaction (Sieber and Iselin, 1968b) that occurs when the original procedure of Merrifield (1963, 1964b) is applied. In this case, an *S*-*p*-alkylbenzylhomocysteine residue (**212**; $R = -CH_2-C_6H_4-$Ⓡ) with a free α-carboxyl group is bound to the resin. The standard esterification procedure is successful with N^{α}-protected methionine sulfoxide (Marglin, 1971; Norris *et al.*, 1971); the latter authors formed Boc-methionine-resin by reduction of the resin-bound sulfoxide with acetyl chloride–sodium iodide in DMF (see above). It is puzzling that, at the end of an attempted solid-phase synthesis of an 11-peptide starting with Boc-Met(O)-resin, hardly any sulfoxide remained on the resin, as quantitated by iodometric titration (Syrier and Beyerman, 1975).

Several groups have been able to esterify Boc-methionine cesium salt in dimethylformamide or dimethylacetamide to a chloromethyl resin without apparent significant *S*-alkylation (Coy *et al.*, 1976; Showell *et al.*, 1976; following Gisin, 1973; compare to Monahan and Gilon, 1973). Coupling of Boc-methionine to hydroxy-containing supports mediated by *N*,*N*′-carbonyldiimidazole also seems satisfactory (Stewart and Young, 1969, pp. 9, 33).

The application of cyanogen bromide for the cleavage of peptides anchored to the polymeric support through a methionine residue (Gross and Witkop, 1962; Hancock and Marshall, 1975) has been discussed earlier (p. 92).

c. Side Reactions. The scope and limitations of synthetic manipulations and chemical modifications applied to peptides containing methionine are defined by the susceptibility of this residue to alkylation and oxidation. For example, iodination of tyrosine [Eq. (26)] is accompanied by conversion of methionine (**207**) to the sulfoxide (**208**), undoubtedly via hydrolysis of the intermediate **214** (Fig. 18; see Rosenblatt *et al.*, 1976, for a pertinent illustration). Overoxidation to methionine sulfone (**213**) is an inevitable further side reaction; whereas the sulfoxide can often be reduced back to methionine by a variety of reagents (pp. 228–229), regeneration of methionine from the sulfone is not possible.

Acidolytic removal of the benzyloxycarbonyl group from methionine peptides leads to partial formation of *S*-benzylhomocysteine (**212**; R = $CH_2C_6H_5$) (Dekker and Fruton, 1948; Albertson and McKay, 1953) through the sometimes isolable *S*-benzylmethionine sulfonium salt (**209**) (Gawron and Draus, 1958). Scavengers (Table IX) suppress but do not necessarily eliminate this intermolecular side reaction (Guttmann and Boissonnas, 1958; Iselin, 1961). In spite of the known resistance of sulfoxides to S-alkylation, Iselin (1961) still observed 10 to 15% *S*-benzylhomocysteine, after treatment of Z-Met(O)-OH with HBr–HOAc. It seems plausible that the by-product arose from reaction of the benzyl carbocation with the dibromosulfonium salt **214** or even reduced methionine (**207**) formed under the conditions, rather than with methionine sulfoxide (**208**).

The formation of *S*-*tert*-butylmethionine salts (**209**; R = *tert*-C_4H_9) upon acidolytic removal of *tert*-butyl-based protecting groups has been documented in several laboratories (Veber *et al.*, 1969; Sieber *et al.*, 1970; Noble *et al.*, 1976; Raschig and Schneider, 1976; M. W. Riemen and R. B. Merrifield, unpublished results, 1978). Carbonium ion scavengers, such as anisole and water, or reducing agents such as thiols seem to be only partially effective in suppressing the side reaction (Noble *et al.*, 1976; Raschig and Schneider, 1976); but competing thioethers, such as free methionine in very large excess, have been used with some success (Marglin and Merrifield, 1966; Veber *et al.*, 1969). There is no clear correlation relating the level of *tert*-butylation to the acid strength. In anhydrous HF in the absence of scavengers, the side reaction was essentially quantitative, even with only one equivalent of the *tert*-butyl carbocation derived from the N^{α}-*tert*-butyloxycarbonyl group (Noble *et al.*, 1976; see Fig. 19, Route C); and in trifluoroacetic acid containing *tert*-butanol, the sulfonium salt was the predominant product (Sieber *et al.*,

1970). A claim that by-product formation in solution was much less upon Boc group removal in 4 *N* HCl–dioxane, as opposed to in trifluoroacetic acid (Raschig and Schneider, 1976), may be worth following up, or may simply be accounted for by the insolubility of the target peptide in the former medium. Methods for essentially complete reversal of *tert*-butylation have already been covered (p. 227). It might be pointed out that positively charged sulfonium salts could bind N^{α}-protected amino acids by ion-exchange during the repetitive coupling steps of solid-phase peptide synthesis. Decomposition of these salts at elevated temperatures, in the same way as in the esterification method of Dorman and Love (1969) (see structure **3** on p. 63), could give a peptide branched at homoserine (**218** in Fig. 18).

During the cleavage reactions of methionine peptides in CH_3SO_3H–anisole. Irie *et al.* (1977) not only confirmed the relative ineffectiveness of anisole in scavenging against S-alkylation by benzyl or *tert*-butyl carbocations, but also observed the major formation of *S*-methylmethionine sulfonium salt (**209**; $R = CH_3$). The methyl group was shown to be derived from anisole through a "facilated" acidolytic cleavage of the aromatic ether, giving phenol as the by-product. When phenetole was the scavenger, the *S*-ethylmethionine sulfonium salt (**209**; $R = C_2H_5$) was formed by an analogous process. An unanswered question from the work of Irie *et al.* (1977) concerns the extent of these side reactions in TFA, or in HF and other strong acids (Table VIII). A trivial way to avoid the reported *S*-methylation would be to use phenol or resorcinol as scavengers in place of anisole as scavengers, but it is not clear how effective these compounds would be in suppressing side reactions of tryptophan (Section V,B,9,b) or other labile amino acids. Irie *et al.* (1977) showed that S-methylation was entirely prevented by incorporation of methionine into peptides as the sulfoxide.

Catalytic hydrogenolysis of protecting groups from methionine-containing peptides may fail because of poisoning of the catalyst (White, 1934; Dekker *et al.*, 1949); but under modified conditions, including more catalyst and longer reaction times, reasonable yields have often been achieved. A frequent difficulty is partial desulfurization of methionine (Guttmann and Boissonnas, 1958), giving rise to contaminating peptides containing α-amino-*n*-butyric acid (**219**) (see Finn and Hofmann, 1976, p. 173). Catalytic hydrogenolysis in liquid ammonia is a promising new method that may circumvent these problems (Meienhofer and Kuromizu, 1974; Felix *et al.*, 1978). Deblocking with sodium in liquid ammonia (e.g., Hofmann *et al.*, 1957) can be accompanied by demethylation of methionine (**207**) to homocysteine (**220**) (Stekol, 1941; Dekker *et al.*, 1949). Using the model peptide methionyl-glycine, Lenard *et al.* (1964) showed 60% cleavage of the peptide bond after 12 hr in HF at 30°C and quantitative cleavage in 36 hr. Cleavage was shown to be prevented by oxidation to the sulfone, and it has been assumed that methionine sulfoxide also provides protection against this side reaction (see Spande *et al.*, 1970, p. 213, for a proposed mechanism).

Application of methionine sulfoxide seems to prevent most of the side reactions of methionine discussed so far, but this approach has known or potential drawbacks as well (p. 228). Gutte and Merrifield (1971) analyzed synthetic ribonuclease A that had been assembled using methionine sulfoxide and subsequently converted to the *S*-sulfonate, reduced with β-mercaptoethanol, and reoxidized to form the disulfide bonds. The methionine content was 80% of theory, with the remainder recovered as the sulfone. The unsuccessful experiences of Syrier and Beyerman (1975) with Boc-Met(O)-resin have already been alluded to (p. 229).

In general, one can never exclude the risk that, under conditions of peptide synthesis, some methionine sulfoxide reverts prematurely to methionine by the speculative pathways outlined in Fig. 18. These include intermolecular oxidation of a nearby indole ring of tryptophan (see Savige and Fontana, 1977b, p. 458, for a possible precedent) or reaction with dicyclohexylcarbodiimide under acid-catalyzed conditions to give the potent oxidizing adduct **221** (Pfitzner and Moffat, 1965). Sulfoxides also react readily with acylating agents to give acylsulfonium salts (**215**). It would be of interest to know to what extent **215** forms under coupling conditions of solid-phase peptide synthesis, and what its fate is under acidolytic deprotection conditions. The well-documented conditions of the Pummerer rearrangement (Horner and Kaiser, 1959; Johnson *et al.*, 1967), which results in net α-acylation and reduction of the sulfoxide to yield **223**, are probably too harsh to be relevant, although they do provide an experimentally verifiable route to a homocysteine residue (**220**) (Oae *et al.*, 1963).

d. Analytical. The quantitation of methionine and its derivatives upon hydrolysis and amino acid analysis is a complex problem, because of the various possible reductive and oxidative interconversions (Fig. 18). The recoverable standards of interest elute from the analyzer column in the following order: homocysteic acid, methionine sulfoxide, aspartic acid, methionine sulfone, threonine, serine, homoserine, glutamic acid, alanine, α-amino-*n*-butyric acid, valine, methionine, phenylalanine, homocystine, histidine, *S*-methylmethionine sulfonium salt, ammonia, homoserine lactone, and arginine (Spackman *et al.*, 1958; Gundlach *et al.*, 1959; Jones *et al.*, 1976; Irie *et al.*, 1977).

Ray and Koshland (1962) found that methionine sulfoxide largely reverted to methionine during acid hydrolysis of proteins, whereas Floyd *et al.* (1963) repeated the experiment on the pure sulfoxide and discerned a mixture of methionine (50–60%), *S*-methylmethionine sulfonium salt (10%), homocystine (10%), and homocysteic acid (10–20%). The availability of a suitable acceptor of positive halogen, such as tryptophan in the former case, has been invoked to rationalize this discrepancy (Savige and Fontana, 1977b). Hydrolysis in 6 *N* HCl containing *β*-mercaptoethanol (1/2000 v/v) (Keutmann and Potts, 1969) or in 3 *N* mercaptoethanesulfonic acid (Penke *et al.*, 1974) has been reported to give good recoveries of methionine, concomitant with reduction of any methionine sulfoxide.

Methionine sulfoxide in the presence of methionine can be accurately quantitated by (1) base hydrolysis (Neumann *et al.*, 1962; Ray and Koshland, 1962), (2) exhaustive enzymatic digestion (Hofmann *et al.*, 1966; Gutte and Merrifield, 1971), and (3) hydrolysis with 3 *N* *p*-toluenesulfonic acid in the absence of scavengers against tryptophan destruction (Savige and Fontana, 1977b). The latter conditions were also found to lead to excellent recoveries of *S*-methylmethionine sulfonium salt (Jones *et al.*, 1976). By comparison, *S*-carboxymethylmethionine sulfonium salt has been shown to be moderately stable in hot 6 *N* HCl, but hydrolysis under the more vigorous standard conditions for 22 hr gave a mixture of methionine (20%), homoserine and its lactone (10%), and *S*-carboxymethylhomocysteine (50%) (Fig. 18, pathways a, b, b′, and c), along with trace homocysteic acid (Gundlach *et al.*, 1959). Amino acid analyses on hydrolysates of peptides modified by *tert*-butylation of methionine were not diagnostic of this side reaction, since free methionine was quantitatively obtained (Noble *et al.*, 1976).

An indirect method for the quantitation of methionine, methionine sulfoxide, and methionine sulfonium salts was developed by Neumann *et al.* (1962), relying on the stability of sulfoxides to alkylation and of sulfonium salts to oxidation. Optimal conditions for performic acid oxidation, with quantitative formation of methionine sulfone from methionine or its sulfoxide, have been described by Moore (1963). Performic acid oxidation, followed by hydrolysis of a peptide containing *tert*-butylmethionine sulfonium salt, gave the composition: homoserine lactone (10%), methionine sulfoxide (10%), methionine sulfone (20%), homoserine (10%), methionine (10%), and unaccounted (40%) (Noble *et al.*, 1976).

Additional evidence for sulfonium salts of methionine peptides is provided by nmr (Jones *et al.*, 1976; Noble *et al.*, 1976; Irie *et al.*, 1977) and by their behavior on ion-exchange chromatography or electrophoresis (Sieber *et al.*, 1970; Noble *et al.*, 1976; Irie *et al.*, 1977). The cleavage of methionine peptides with cyanogen bromide is of frequent analytical value (Gross and Witkop, 1962; reviewed by Gross, 1967; by Spande *et al.*, 1970, pp. 205–210).

11. Cysteine

a. Protection. The neutral, strongly nucleophilic *β*-thiol group of cysteine has been protected in most solid-phase syntheses as substituted benzyl thioethers (**224**–**227**) (du Vigneaud *et al.*, 1930; Akabori *et al.*, 1964; Erickson and Merrifield, 1973a, 1974; Yamashiro *et al.*, 1973; Cosand and Merrifield, 1977), triphenylmethyl thioether (trityl, Trt) (**228**) (Amiard *et al.*, 1956; Zervas and Theodoropoulos, 1956) *tert*-butyl thioether (*t*Bu) (**229**) (Beyerman and Bontekoe, 1962; Callahan *et al.*, 1963; Fujino and Nishimura, 1976; Pastuszak and Chimiak, 1977; Felix *et al.*, 1978; Nishimura *et al.*, 1978), the acetamidomethyl (Acm) derivative (**230**) (Veber *et al.*, 1972b; applied by Ontjes *et al.*, 1973; Trudelle and Birr, 1975; van Rietschoten *et al.*, 1977; Sandberg and Ragnarsson, 1978), or as mixed disulfides (**231**) (Inukai *et al.*, 1967; Weber and Hartter, 1970; Wünsch and Spangenberg, 1971; van Rietschoten *et al.*, 1975b, 1977; Birr *et al.*, 1977). Other cysteine derivatives

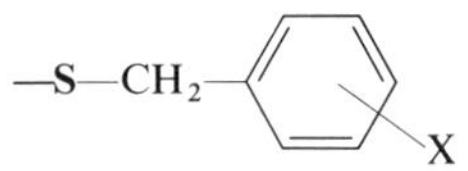

*(**224**) X = H Bzl
*(**225**) X = *p*-CH_3 Dmb
*(**226**) X = 3,4-$(CH_3)_2$
*(**227**) X = *p*-OCH_3 Mob

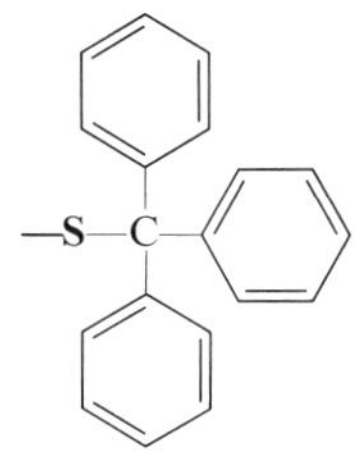

*(**228**) Trt

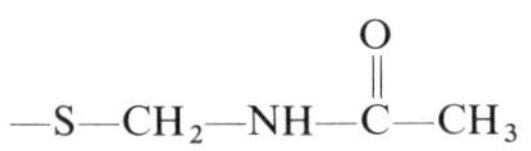

*(**230**) Acm

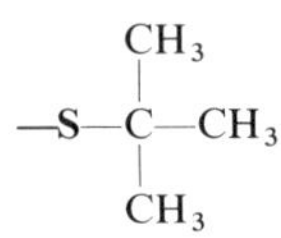

*(**229**) *t*Bu

—S—S—R′

(**231**) R′ = CH_3, C_2H_5, *n*-C_3H_7, *sec*-C_3H_7, **tert*-C_4H_9

that have been briefly examined and might be of value for special applications include the substituted benzhydrylthioethers (**232** and **233**) (Zervas and Photaki, 1962; Hanson and Law, 1965; Photaki *et al.*, 1970), mono- and dithioacetals (**234**–**236**) (Holland and Cohen, 1958; Brownlee *et al.*, 1964; Hiskey and Sparrow, 1970), and the *S*-acetyl (**237**), *S*-benzoyl (**238**) (Zervas *et al.*, 1963a), *S*-alkoxycarbonyl (**239**) (Berger *et al.*, 1956; Zervas *et al.*, 1963a; Photaki, 1970; Schnabel *et al.*, 1971), and *S*-ethylcarbamoyl (**240**) (Guttmann, 1966; Storey *et al.*, 1972) groups. Thiol protection and the related problem of rational formation of the disulfide bond has been covered in depth in several outstanding reviews (Hiskey *et al.*, 1973; Friedman, 1973; pp. 241–251; Photaki, 1976; pp. 115–147). The crowning achievement in this field was the solution synthesis of insulin by Sieber *et al.* (1974, 1977), through the selective formation of the three disulfide bonds. The adaptation of this sophisticated chemistry to solid-phase systems is an exciting area for future investigation (pp. 241–243).

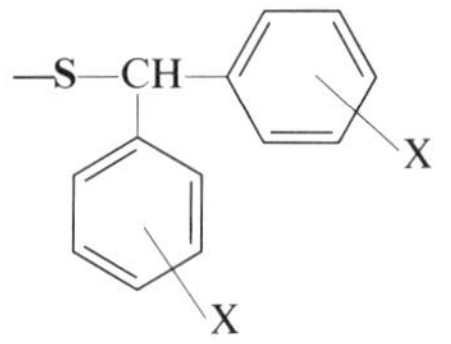

(**232**) X = *p*-OCH_3 Ddm
(**233**) X = H Dpm

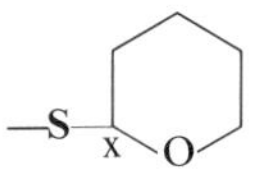

second optical center (x)

(**234**) Thp

$—S—CH_2—O—CH_2—CH(CH_3)_2$ *no* second optical center

(**235**) iBom

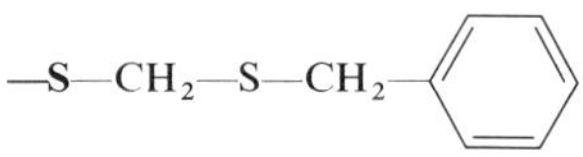

(**236**) Btm

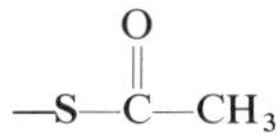

(**237**) Ac

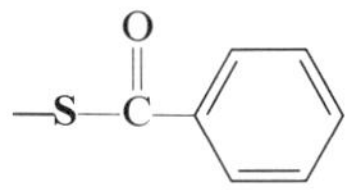

(**238**) Bz

$—S—CO_2R'$

(**239**) R′ = Bzl, Mob, *t*Bu

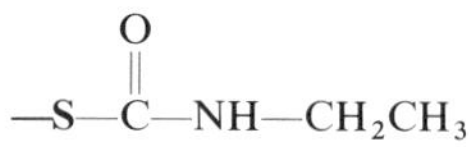

(**240**) Ec

(**241**)

The choice of the thiol protecting group will vary, depending on the synthetic strategy. In the past (e.g., Marglin and Merrifield, 1966; Gutte and Merrifield, 1971), the *S*-benzyl group (**224**) was the one most commonly used for solid-phase synthesis. Since this thioether is stable to strong nucleophiles, to catalytic hydrogenolysis, and to HBr–HOAc, any one of these reagents can be used to cleave a benzyl ester anchoring linkage to the support and yield an *S*-benzylcysteine-containing peptide in solution. For a long time, the only available method to remove the *S*-benzyl group was sodium in liquid ammonia (Sifferd and du Vigneaud, 1935), but HF at 20°C for 30 min (Sakakibara *et al.*, 1967) is now preferred. Since both these conditions are unnecessarily harsh toward complex peptides, one recent trend has involved "fine-tuning" to develop a substituted *S*-benzyl group quantitatively removable under standard acidolytic cleavage conditions of HF at 0°C and yet stable to repeated removal of the N^{α}-*tert*-butyloxycarbonyl group with TFA. The *S*-*p*-methoxybenzyl group (**227**) (Akabori *et al.*, 1964) is only two- to fivefold more stable in TFA–CH_2Cl_2 (1:1) at 25°C than the N^{α}-benzyloxycarbonyl group (Erickson and Merrifield, 1973a, 1974; Yamashiro *et al.*, 1973; compare to p. 176) and, thus, is expected to be suitable only for syntheses of relatively short cysteine peptides (e.g., Hammerstrom *et al.*, 1970; Live *et al.*, 1977; van Rietschoten *et al.*, 1977; see however the successful efforts of Tan and Kaiser, 1976, where at least 10% of the resin-bound chains containing *S*-*p*-methoxybenzylcysteine residues ought to have been prematurely deprotected according to the earlier kinetic data). The recommended cysteine protecting groups for the common solid-phase scheme involving acidolytic final deprotection are the *S*-*p*-methylbenzyl (**225**) (Erickson and Merrifield, 1973a, 1974; applied by Cosand and Merrifield, 1977; Live *et al.*, 1977) and the *S*-3,4-dimethylbenzyl (**226**) (Yamashiro *et al.*, 1973; also applied by Hruby *et al.*, 1977) groups. These are removed by HF–anisole at 0°C within 60 min and 10 min, respectively.

Several cysteine derivatives, in addition to the substituted benzyl thioethers discussed in the above paragraph, can be deprotected by anhydrous acids under conditions that are practical for peptide syntheses. The relative ease of acidolysis follows the series 4,4′-dimethyloxybenzhydryl (**232**) > tetrahydropyranyl (**234**) ~ isobutyloxymethyl (**235**) ~ trityl (**228**) > benzhydryl (**233**) > benzylthiomethyl (**236**) ~ benzyl (**224**). The first four of these do not appear to be entirely stable to the conditions of repetitive acidolytic deprotection of N^{α}-Boc-peptide-resins with HCl–HOAc (Hammerström *et al.*, 1970), but only the 4,4′-dimethoxybenzhydryl function (**232**) (Hanson and Law, 1965; Photaki *et al.*, 1970) is sufficiently labile to be removed with trifluoroacetic acid at a rate comparable to those for cleavage of *tert*-butyl-based urethanes, esters, and ethers. In a careful study, Photaki *et al.* (1970) defined conditions for the optimal acidolytic cleavage of diphenylmethyl and trityl thioethers and showed that the stable carbonium ions derived in these reactions readily recombine with the liberated thiol groups. Detritylation in TFA was found to be almost completely suppressed by triphenylcarbinol (5 equiv). The equilibrium of *S*-trityl (**228**) and free cysteine residues in TFA was established within 30 min; and could be shifted predominantly in the direction of the cleaved products by the addition of phenol as a scavenger. The best yields of free sulfhydryl groups seem to be obtained with a trifluoroacetic acid–thiophenol mixture (Büllesbach *et al.*, 1979). The *S*-diphenylmethyl group (**233**) and the *S*-trityl group (**228**) can both be removed by HF–anisole, at 0°C within 30 min (Sakakibara *et al.*, 1967; Zahn *et al.*, 1968).

The thiol esters **237** and **238**, thiocarbonates **239**, and thiocarbamate **240** are of interest because of their lability to nucleophiles and basic conditions. In addition, the thiocarbonates **239** are cleavable under acidolytic conditions similar to but slightly more severe than those used for breaking the oxygen–alkyl bond of the analogous esters and urethanes. The *S*-acetyl (**237**) and *S*-benzoyl (**238**) groups were found to be practically resistant to refluxing TFA and HBr–HOAc, but were both removed without significant β-elimination or racemization, by orthogonal cleavage with 0.1 *N* sodium methoxide in methanol at 25°C in 5 to 10 min (Zervas *et al.*, 1963a; compare to Hiskey *et al.*, 1970). Methanolysis also removed the *S*-benzyloxycarbonyl group, at a somewhat slower rate. The major drawback to the *S*-acyl protecting groups **237– 39**, which precludes their application in stepwise solid-phase synthesis, is the facile intramolecular S → N acyl shift that takes places as soon as the free α-amino group of a cysteinyl peptide–resin is liberated (documented by Hammerström *et al.*, 1970; citing solution precedents of Zervas *et al.*, 1965; Guttmann, 1966; Hiskey *et al.*, 1966b). With *S*-acetyl and *S*-benzyloxycarbonylcysteine, chain termination was essentially quantitative even with a modified coupling protocol, although with *S*-benzoylcysteine, the putative side reaction was relatively minor. It was speculated that the deblocked thiol groups after an S → N shift could react with residual chloromethyl sites on the support (Hammerström *et al.*, 1970). The *S*-acyl groups **237–239** could still be profitably applied to protect *internal* cysteine residues of peptide segments for polymer-supported condensation. In a plausible strategy, removal of *S*-acyl functions would occur simultaneously with cleavage of the anchoring linkage by transesterification, while leaving any *S*-thioethers and *S*-thioacetals intact.

The relatively acid-stable *S*-ethylcarbamoyl (Ec) group (**240**) is related to the protecting groups just described; it was specificaly developed to avoid the S → N acyl shift (Guttmann, 1966). Hence, **240** is suitable for stepwise solid-phase synthesis (Hammerström *et al.*, 1970; Rosamond and Ferger, 1976). Mild deprotection has been achieved with liquid ammonia at −20°C for 4 hr, or with 1 *N* hydrazine–methanol, at 25°C for 2 hr. By comparision, the *S*-(*N*-methoxymethyl)carbamoyl group (**241**) (Tschesche and Jering, 1973), known from chemical modification studies in protein chemistry, is removable in aqueous buffers at pH 9.5, at 25°C in 1 hr. The possibilities for partial premature deblocking of an *S*-ethylcarbamoylcysteine residue in a resin-bound peptide upon exposure to excess triethylamine during neutralization, or by intermolecular aminolysis during repetitive coupling steps, remain open questions.

A further, productive approach in thiol protection lies in the development of neutral deblocking methods. Mixed disulfides **231**, also known as *S*-alkylmercapto- or *S*-alkylsulfenylcysteine derivatives, are an attractive choice because of the variety of mild and specific methods for their reductive or oxidative cleavage (reviewed by Friedman, 1973; pp. 199–229; Field, 1977), generally carried out in solution after removal of a peptide from the solid support (see, however, Inukai *et al.*, 1968; discussed on p. 243). Reduction of disulfides can be carried out (1) as an equilibrium process driven with a large excess of a thiol such as β-mercaptoethanol, or the much more nucleophilic thiophenol, in aqueous solution at pH 7.5 to 9; (2) with a small excess of dithiothreitol, where the equilibrium is driven by intramolecular cyclization of the reagent (Cleland, 1964); or (3) with essentially one equivalent of tri-*n*-butylphosphine at pH **8** (reviewed by Rüegg and Rudinger, 1977). Disulfides can also be cleaved by oxidative sulfitolysis (Bailey and Cole, 1959); the resulting *S*-sulfonates are readily reduced by these same methods to the desired thiols (see p. 241 for further discussion).

In assessing the suitability of various *S*-alkylsulfenylcysteine derivatives for applications to solid-phase peptide synthesis, the tendency of mixed disulfides to disproportionate in very strong acid as well as in the presence of relatively weak bases must be considered (compare to p. 242). Disulfide exchange has been found to be minimal with the sterically most crowded *tert*-butylmercapto derivative (Weber *et al.*, 1970; Wünsch and Spangenberg, 1971). The *tert*-butylmercapto derivative was found to be quite labile in HF (van Rietschoten *et al.*, 1975b) and for this reason, the *S*-ethylmercapto derivative (Inukai *et al.*, 1967) has been preferred for syntheses of the 18-peptide apamin and analogues which contain four half-cystine residues (van Rietschoten *et al.*, 1977; Granier *et al.*, 1978). However, difficulties in solid-phase syntheses with the *S*-ethylmercapto derivative have been reported as well (e.g., Seidel *et al.*, 1976). The available literature data suggest that the mixed disulfides used to protect cysteine are surprisingly stable upon exposure to nucleophilic reagents, such as ammonia or alkali, which are of sufficient strength to detach peptides from phenolic or phenacyl resins (Inukai *et al.*, 1968; Birr *et al.*, 1977).

Thiols can also be deblocked by electrophilic reagents such as heavy metal salts, sulfenyl chlorides, iodine, and thiocyanogen. Those methods giving disulfides directly, and corresponding references, are covered later (p. 241). The ease of these reactions largely follows the stability of the resultant carbonium ion according to the series tetrahydropyranyl (**234**) > trityl (**228**) ~ isobutyloxymethyl (**235**) > *tert*-butyl (**229**) ~ acetamidomethyl (**230**) ~ benzhydryl (**233**) ~ benzylthiomethyl (**236**). All of these groups can also be cleaved by anhydrous or aqueous acids, but the latter reactions are inconveniently slow because of the difficulty in protonation of the sulfur atom of thioethers, thioacetals, and especially thiazolidines. Unlike acidolysis, cleavage with heavy metal salts is essentially irreversible. Regeneration of thiols from mercaptides is effected by hydrogen sulfide, a large excess of mercaptans, or, in the case of silver salts, with hydrogen chloride in $CHCl_3$ or DMF (e.g., Zervas and Photaki, 1962; Veber *et al.*, 1969, 1972b). Both the heavy metal cleavage, and the thiol regeneration conditions, will lead to scrambling of pre-existing disulfides (Hiskey and Harpold, 1968; Kamber, 1971a).

The *S*-tetrahydropyranyl group (**234**) has been rapidly cleaved with 0.5 *N* aqueous silver nitrate, at 0°C (Holland and Cohen, 1958); the *S*-trityl group (**228**) has been removed with silver nitrate (1 equiv) in pyridine–methanol at 0°C (Zervas and Photaki, 1962) or with mercury (II) acetate (1.1 equiv) in neutral solvents at 25°C or higher for a few hr (Zervas and Photaki, 1962;

Zervas *et al.*, 1965; Hiskey *et al.*, 1966a; Kamber, 1971a). The latter group of authors showed that *S*-detritylation could be carried out selectively in the presence of the *S*-diphenylmethyl (**233**) or *S*-benzoyl (**238**) groups. It has been stated (Hiskey *et al.*, 1973, p. 261) that mercury(II) acetate is generally more effective than silver nitrate. Mercury(II) acetate has also been used to remove the *S*-acetamidomethyl group (**230**) (Veber *et al.*, 1972b; several equiv at pH 4, 25°C for 4 hr), and the same conditions have also been found to be applicable to cleavage of *tert*-butyl (**229**) thioethers (Felix *et al.*, 1978; compare to Nishimura *et al.*, 1978). The *S*-*p*-methoxybenzyl (**227**) and *S*-benzylthiomethyl (**236**) groups, but not the *S*-benzyl (**224**) group, have been cleaved with mercury(II) acetate (several equiv) in TFA (0°C for 30 min) or 80% aqueous formic acid (25°C for 20 min) respectively (Nishimura *et al.*, 1978; Brownlee *et al.*, 1964). Finally, the base-labile *S*-ethylcarbamoyl group (**240**) has been shown to be stable to *p*-chloromercuribenzoate at pH 4.6, but to be readily cleaved by this reagent at pH 7.0 or higher (Storey *et al.*, 1972); in this case, the metal could be serving merely as a catalyst for "facilitated" hydrolysis.

A number of S-protecting groups are cleaved by sulfenyl chlorides, with the concomitant formation of disulfides [Kamber, 1973; Fontana, 1975; Hiskey *et al.*, 1975; Pastuszak and Chimiak, 1977; see Eq. (29)]. These

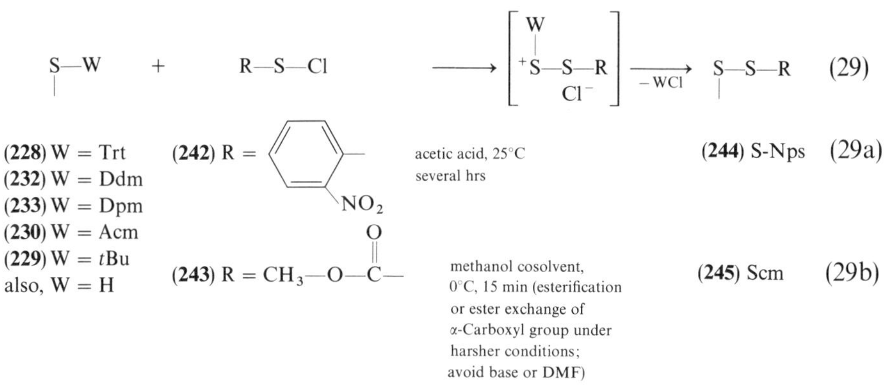

reactions are incompatible with tryptophan-containing peptides (see Fig. 9 on p. 111, and also p. 221). The *S*-*o*-nitrophenylsulfenyl derivatives (**244**) can be reduced by the usual methods (p. 237); the *S*-carbomethoxysulfenyl (Scm) derivatives (**245**) are most advantageously applied to the directed formation of unsymmetrical disulfide bonds [Brois *et al.*, 1970; see Eq. (32) on p. 243].

It should be mentioned that prior to the development of the deprotection methods mentioned in the previous two paragraphs, the only known way to cleave the *S*-*tert*-butyl group was by treatment with HF–anisole at 20°C for

1 hr (Sakakibara *et al.*, 1967); this fact emphasizes the extraordinary stability of this group to conditions of peptide synthesis. The *S*-acetamidomethyl group was reported to be stable to strong nucleophiles such as hydrazine and to the strong acid, HF (Veber *et al.*, 1972b). Nonetheless, at the end of a solid-phase synthesis of the 18-peptide apamin, in which four Cys(Acm) residues were used, and cleavage was in HF–anisole at 0°C for 1 hr, 20% of the thiol groups were found to have been deblocked (van Rietschoten *et al.*, 1977).

b. Anchoring and Coupling. The esterification of N^{α}-urethane, *S*-alkyl-protected derivatives of cysteine to chloromethyl supports by standard methodology (pp. 60–63) seems to proceed without complications (e.g., Mizoguchi and Woolley, 1967; Rivier *et al.*, 1975b) despite the theoretical possibility of an S-alkylation side reaction as noted for methionine (p. 229).

Side-chain anchoring of cysteine residues via a resin-bound dinitrophenylene linkage (Glass *et al.*, 1974) has been detailed earlier (pp. 98–99), including a discussion of ways to minimize base-catalyzed side reactions (Fig. 8). A resin-bound *S*-alkylcarbamoyl linkage (compare to structure **240**), cleaved by 1 *N* aqueous sodium hydroxide (3 equiv) in methanol, has now been reported (Smith *et al.*, 1979) for side-chain anchoring.

Hammerström *et al.* (1970) investigated the coupling of 10 different *S*-protected derivatives of N^{α}-Boc-cysteine using standard dicyclohexylcarbodiimide activation. Excellent incorporation was obtained in all cases, including that with the sterically hindered *S*-trityl function (**208**). In contrast, Wolters *et al.* (1974) remarked on the failure to obtain quantitative incorporation of Bmv-Cys(Trt)-OH which in retrospect might have been due to a competing side reaction involving residual *tert*-alkoxyhydrazine sites on the functionalized support used.

Acylation of resin-bound α-amino groups with bis-N^{α}-Boc-cystine was described by Lunkenheimer and Zahn (1970) as part of a strategy to prepare symmetrical cystine peptides. Recently, Kullmann and Gutte (1978) were able to couple the asymmetrical cystine peptide

Bpoc-Asn-Tyr(Dcb)-Cys-Asn-OBzl   Boc-Cys-Gly-OH

(the two Cys residues joined by a disulfide bridge)

to a hexapeptide resin in 75% yield using the DCC–HOSu method.

c. Formation of Disulfide Bonds.* Cysteine peptides prepared by solid-phase synthesis are usually first cleaved from the resin support, and the

* Garsky *et al.* (1976) reported a cyclic analog of somatostatin prepared by a combination of solid-phase and solution synthesis, in which the disulfide bond was replaced by an amide bond formed by DCC–HOBt mediated coupling of the α-carboxyl of COOH-terminal aspartate with the γ-amino group of NH_2-terminal γ-aminobutyric acid. Leading references to other nonsulfur containing disulfide analogs, all of which have been prepared entirely by solution methods, are found in Smith *et al.* (1978).

thiol protecting group is removed either simultaneously or later by an orthogonal mode in solution. As a penultimate step in the synthetic scheme, conversions of the deblocked species to *S*-sulfonates (**246**) by oxidative sulfitolysis [Swan, 1957; Bailey and Cole, 1959; Cole, 1967; Chan, 1968; see Eq. (30)] has frequently proven to be a valuable maneuver which facilitates purification. This technique was of key importance in all of the syntheses

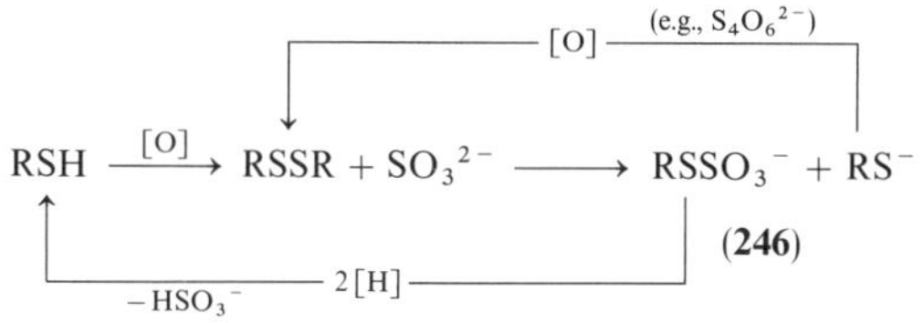

(30)

of insulin by classical methods (Meienhofer *et al.*, 1963; Katsoyannis *et al.*, 1964; Kung *et al.*, 1965), and has also been applied for the manipulation of peptides prepared by the solid-phase approach (e.g., Marglin *et al.*, 1966; Okuda and Zahn 1969; Gutte and Merrifield, 1971). The *S*-sulfonates (**246**) ($pK_a = 1 - 2$) are somewhat more stable to conditions of peptide chemistry than are mixed disulfides (Hiskey *et al.*, 1973, p. 297); they even survive the various acidic and basic conditions of the Edman degradation (Brandenburg *et al.*, 1975). They are labile in warm aqueous acid and are reduced back to thiols by the same reagents that reduce mixed disulfides (p. 237).

Formation of intramolecular disulfide bonds from sulfhydryl group-containing peptides is generally accomplished in dilute solutions by oxidation with molecular oxygen (e.g., White, 1961; Anfinsen and Haber, 1961; for typical applications to products of solid-phase synthesis, see Gutte and Merrifield, 1971; van Rietschoten *et al.*, 1975b), potassium ferricyanide (Hope *et al.*, 1962), or 1,2-diiodoethane (Weygand and Zumach, 1962). Furthermore, oxidative methods exist for the direct conversion of S-protected half-cystine residues to the disulfides. Iodine (1 equiv), in methanol, at 25°C for 1 hr, transformed *S*-trityl and *S*-acetamidomethyl functions in this way without apparent side reactions that modify tyrosine or methionine, and with no interchange of preexisting disulfides (Kamber and Rittel, 1968; Kamber 1971a,b; Marbach and Rudinger, 1974; Sieber *et al.*, 1977). In trifluoroethanol, the oxidation was specific for *S*-trityl in the presence of *S*-acetamidomethyl (Sieber *et al.*, 1976). All of the methods just cited are intrinsically chemically ambiguous; any chances for non-randomness in the pairings of half-cystine residues depend on the molecular dynamics of peptide or protein folding. Furthermore, there is always a risk of dimerization and polymerization during oxidation.

Much research has gone into the development of controlled chemical methods for the synthesis of unsymmetrical disulfides (reviewed by Hiskey *et al.*, 1973). The foremost methods involve thiolysis of sulfenylthiocyanates

(**247**) [Hiskey and Tucker, 1962; Hiskey *et al.*, 1967; Hiskey and Ward, 1970; Eq. (31)] or of *S*-carbomethoxysulfenyl (Scm) derivatives (**245**) [Brois *et al.*, 1970; Kamber, 1973; Hiskey *et al.*, 1975; Eq. (32)]. The synthesis of the derivatives **245** is given in Eq. (29b); they are stable to trifluoroacetic acid but labile to alkali and hydrazine.

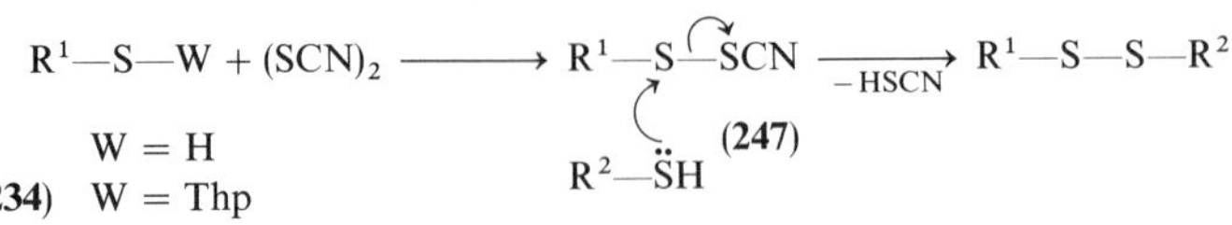
(31)

W = H
(**234**) W = Thp
(**235**) W = iBom
(**228**) W = Trt
(**233**) W = Dpm

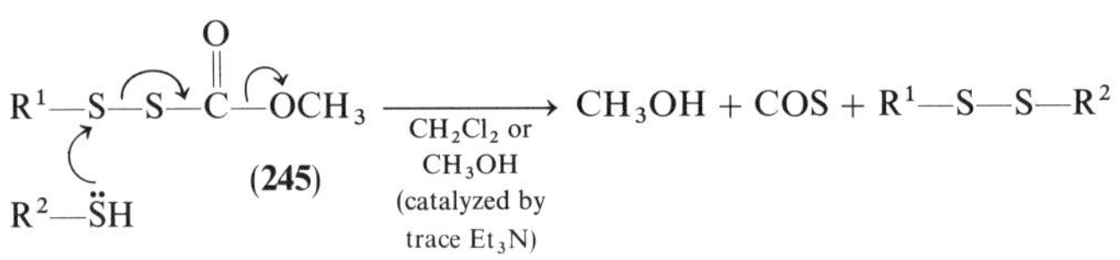
(32)

A prerequisite for the utility of these procedures is that, in subsequent chemical steps, no disulfide exchange leading to disproportionated or polymerized by-products take place (Sanger, 1953; Jensen, 1959; Zervas *et al.*, 1959a). In strong anhydrous acids, including trifluoroacetic acid (Kamber, 1971a, p. 409; Sieber *et al.*, 1978, p. 359), HBr–TFA (Lunkenheimer and Zahn, 1970), and HF (Kullmann and Gutte, 1978), or with tertiary amines used for neutralization (Lunkenheimer and Zahn, 1970), disproportionation has not been detected. In fact, elongation of peptides already containing a disulfide bond (as originally explored by Zervas *et al.*, 1959a; Rydon and Serrão, 1964; Zahn and Schmidt, 1970) appears to be possible under safe conditions (Kamber *et al.*, 1976). In contrast, exposure of disulfide-containing peptides to aqueous media at pH 9 or above, especially in the presence of catalytic amounts of thiols, or to strong nucleophiles such as methoxide, hydrazine, and unsolvated fluoride or cyanide ions, will lead to scrambling, cleavage, or modifications of the disulfides, and hence these techniques cannot be used once the half-cystine residues have been specifically paired (Kamber, 1971a; Sieber, 1977; Sieber *et al.*, 1978; reviewed by Friedman, 1973, pp. 199–229).

With this background, the small number of studies on the preparation of disulfides of peptides while anchored to a solid phase can be evaluated in perspective, and new directions in research can be envisaged (e.g., adaptation of Eqs. (29b) and (32) to solid-phase synthesis). In principle, an S-protected cysteine residue in the resin-bound peptide can be deblocked by an orthogonal cleavage method. Advantages of the solid-phase mode include (1) circumvention of solubility problems which occasionally lead to poor yields in solution and (2) the possibility of recycling reagents to force to completion those cleavages that are equilibrium reactions. Furthermore, (3) intramolecular oxidations to disulfides are favored by the pseudo–dilution provided by the polymeric support (p. 28).

Bondi *et al.* (1968) prepared a Trt-Cys(Trt)-Cys(Trt)-Ala-resin, cleaved the *S*-trityl groups with silver nitrate in pyridine, and obtained an 80% yield of H-Cys-Cys-Ala-OH after oxidation with 1,2-diiodoethane and cleavage from the support. Of comparative interest is the observation that S-detritylation of a resin-bound peptide with mercury(II) acetate, followed by hydrogen sulfide, gave problems because of the irreversible adsorption of mercury(II) sulfide to the resin (Kuhlmann and Gutte, 1978). Inukai *et al.* (1968) used thiophenol to deblock the *S*-ethylmercapto protecting groups of linear protected oxytocin anchored to a phenolic resin and formed the disulfide by air oxidation. After ammonolytic cleavage from the support and purification, the yield of oxytocin was 25%. Lunkenheimer and Zahn (1970) defined conditions for both reduction and intermolecular reoxidation of resin-bound cystine residues.

The most challenging application of the chemistry described in this section lies in the controlled formation of two or more cystine bridges in the same molecule. This task requires several classes of cysteine protecting groups which must be selectively removable under conditions which do not promote disulfide exchange. The desired disulfide bonds are then formed in turn either by oxidative combination of two specific cysteine residues that were originally blocked by the same protecting group, or else by the directed attack of the free thiol of one residue on the suitably activated derivative of the other residue [Eqs. (31) and (32)]. The pioneering work of Hiskey, and the dramatic all-solution total synthesis of insulin by Sieber *et al.* (1974, 1977) has already been referred to. Solid-phase synthetic methodology can also contribute to this area (Birr and Pipkorn, 1979). A fully protected insulin A-chain was assembled and then released from the support by aminolysis of the *β*-phenacyl ester (through side chain of COOH-terminal aspartic acid) linkage. Under these conditions, the *p*-methoxybenzyl (**227**), acetamidomethyl (**230**), and *tert*-butyl disulfides (**231**) employed to protect the four half-cystine residues were stable. Selective deprotection and disulfide formation in solution, including a recombination step with a natural insulin B-chain, resulted in a good yield of active semisynthetic hormone.

d. Side Reactions. Protection of the *β*-thiol of cysteine for use in peptide synthesis has long been recognized as imperative, due to the high degree of nucleophilicity of bivalent sulfur as well as to the ready tendency for oxidation to disulfides. However, the point that S-protected cysteine is still vulnerable to alkylation and oxidation side reactions, analogous to those observed at methionine (Section V,B,10,c), does not appear to have been adequately stressed.* For example, upon counter-current distribution of the crude product—obtained by cleavage in HBr–TFA–anisole—from a solid-phase synthesis of an octapeptide containing *S*-benzylcysteine at position 3, two by-products were isolated, with amino acid compositions differing from the major, desired component (Barany and Merrifield, 1973). The hydrophilic by-product ($\sim 15\%$), originally assigned as a deletion peptide because it lacked *S*-benzylcysteine, could also have been either a peptide containing a sulfonium salt derived in an alkylation reaction or a dehydroalanine peptide formed by β-elimination (see following paragraph). The hydrophobic by-product ($\sim 6\%$) lacked cysteine, cystine, cysteic acid, or *S*-benzylcysteine, but contained a new amino acid peak eluting near ammonia on a standard analyzer. In contrast to the original interpretation, it now appears likely that this new peak was the sulfoxide or sulfone of *S*-benzylcysteine (compare to **208** and **213** in Fig. 18). By-products modified at cysteine ($\sim 25\%$ of total yield) have also been noted in oxytocin syntheses in which *p*-methylbenzyl (**225**) or *p*-methoxybenzyl (**227**) protecting groups were removed in HF–anisole at 0°C for 60 min (Live *et al.*, 1977).

The base-catalyzed β-elimination of *S*-alkyl, *S*-acyl, sulfone, and sulfonium salt derivatives of cysteine to yield the corresponding dehydroalanine derivatives is particularly facile (Patchornik and Sokolovsky, 1964; reviewed by Spande *et al.*, 1970, pp. 102–107) and often serves to limit applicable manipulations of peptide synthesis. For example, ordinary tertiary amines were too strong to effect neutralization (Section IV,B) of peptide–resins containing an *S*-dinitrophenylene anchoring linkage (Glass *et al.*, 1974; see Fig. 8), and "buffered DMF" was required for this step (see p. 99 for details). In a definitive study, Hiskey *et al.* (1970) quantitated the amount of β-elimination by determining pyruvic acid in 6 *N* HCl hydrolysates and found 2 to 5% of this side reaction during saponification of a C^{α}-alkyl ester off an *S*-alkylcysteine peptide. Methanolysis to remove the *S*-benzoyl group (**238**) gave only 0.2 to 0.4% β-elimination in methanol, but a change to a more polar solvent such as dimethylacetamide, frequently required to solubilize the peptide substrate, resulted in 15 to 25% of this side reaction. The disulfide bond of cystine in peptides and proteins has also been observed to undergo β-elimination in strong alkali (Bohak, 1964).

Partial racemization of *S*-benzylcysteine upon hydrazinolysis (MacLaren *et al.*, 1958) or upon base treatment of N^{α}-urethanyl active esters (Bodanszky

* An early example from a solution synthesis, employing benzylthiomethyl thioether protection (**236**), is due to Katsoyannis (1961), but was disputed by Brownlee *et al.* (1964). In the former work, removal of the N^{α}-Z group with HBr–HOAc was accompanied by the formation of three ninhydrin positive impurities; these were largely absent when deblocking was carried out in the presence of a mixture of diethyl phosphite and methyl ethyl sulfide (see Table IX). The obvious implication is that carbocation scavengers suppressed alkylation of the thiol protecting group.

and Bodanszky, 1967) is a related problem. A β-elimination-readdition mechanism may be in effect, and direct abstraction of the acidic α-proton has also been favored (Kovacs *et al.*, 1968). Racemization of cysteine introduced as the *S*-acetamidomethyl derivative was not observed (Veber *et al.*, 1972b), and this result is probably general for most other protecting groups.

The base-catalyzed intramolecular S → N shift of acyl (**237** and **238**), alkoxycarbonyl (**239**), or 2,4-dinitrophenyl groups (reviewed by Friedman, 1973, pp. 183–190) occurs with cysteine residues under even milder conditions than the O → N acyl shift at serine and threonine [Eq. (25) on p. 210]. The limitations imposed by this side reaction on the applicability of the appropriate thiol protecting groups (Section V,B,11,a) and anchoring linkages (Fig. 8) have already been covered. The possibility of the inverse N → S migration promoted by strong acid has been raised by Sakakibara (1971, p. 81). While glutathione was recovered unchanged after treatment with HF at 0°C for 60 min, these conditions may have given rise to 10% *S*-glycylcysteine upon deprotection of benzyloxycarbonylglycyl-*S*-*p*-methoxybenzylcysteine. Sakakibara (1971, p. 81) cautioned that regeneration of the peptide to the desired *N*-acyl form by treatment with base could be accompanied by partial racemization of both cysteine and the residue on the amino side of the peptide bond; such racemization has been precedented (Hirotsu *et al.*, 1967a) upon opening of thiazolines **248**, which are the dehydrated form of the intermediate **249** [compare to structure **183** in Eq. (25)]. Thiazolines **248** form directly in high yields upon treatment of cysteine peptides containing a free sulfhydryl group, an *S*-trityl, or an *S*-benzoyl function with HCl in dry methanol or chloroform (Stoffel and Craig, 1961; Hirotsu *et al.*, 1967b, 1970), but not at all with trifluoroacetic acid (Photaki *et al.*, 1970).

(248) **(249)**

As already stated (p. 235ff) the choice of a cysteine protecting group is related to the selection of final deblocking conditions. Naturally, there will be a wide variability of results depending on the actual target peptide or protein. Each protection scheme suffers certain drawbacks, but there is enough flexibility to expect at least one approach to be manageable. The, as yet untested, strategy of relying on a moderate reagent, for example, TFA–phenol to remove cysteine side-chain protecting groups such as the *S*-4,4′-dimethoxydiphenylmethyl (**232**) or *S*-isobutoxymethyl (**235**) functions might be accompanied by side reactions of these reactive carbocations with susceptible residues (cf. p. 207). The application of HF, even with scavengers, has been observed to lead to low yields, incomplete regeneration of activity, and/or partially intractable polymer films with various complex synthetic or natural materials containing several free sulfhydryl groups per polypeptide chain (discussion to Ivanov *et al.*, 1976; Rocchi *et al.*, 1976; Aimoto and Shimonishi, 1977; van Rietschoten *et al.*, 1977; Sandberg and Ragnarsson, 1978; W. L. Cosand and R. B. Merrifield, unpublished results, 1975). In a number of instances, improvements were obtained by switching to relatively acid-stable cysteine protecting groups, such as the *S*-acetamidomethyl (**230**) and *S*-alkylmercapto (**231**) functions, which are removable under orthogonal conditions. Electrophilic removal with heavy metals has occasionally failed due to solubility problems (e.g., Hiskey and Smith, 1968), and oxidative removal with iodine has been noted to give rise to cysteic acid containing by-products (R. A. Jessop and G. T. Young, personal communication, 1978). Overexposure to these reagents could also lead to modifications of tyrosine, tryptophan, and methionine residues.

Treatment of cysteine, cystine, and *S*-benzylcysteine peptides with sodium in liquid ammonia may lead to 3 to 7% desulfurization per mole of half-cystine (Katsoyannis, 1966; Zahn *et al.*, 1967; J. M. Manning *et al.*, 1972). The last-mentioned authors showed that the ensuing alanine was largely racemic. The remarks made earlier for methionine (p. 231) apply equally to the catalytic hydrogenolysis of cysteine peptides.

e. Analytical. The accurate quantitation of cysteine, cystine, and their derivatives in peptides by hydrolysis and amino acid analysis is a complex problem (Moore and Stein, 1963; Blackburn, 1968, pp. 127–131). These amino acids are destroyed upon alkaline hydrolysis, and recoveries from hydrolysates in 6 *N* HCl at 110°C depend on the trace levels of oxidizing agents, including molecular oxygen or tyrosine, in the acid medium. The *S*-benzylcysteine derivative is stable to acid hydrolysis and elutes after phenylalanine on standard amino acid analyzers (Stewart and Young, 1969). Under these conditions, destruction of *S*-acetamidomethylcysteine is variable, but this amino acid can be determined after enzymatic hydrolysis. It coelutes either with aspartic acid in the standard sodium citrate buffer system of analysis or with threonine in the modified lithium citrate system (Veber *et al.*, 1972b). Cysteine is not found as half-cystine when acid hydrolysis is carried out in the presence of β-mercaptoethanol (Keutmann and Potts, 1969); the reduction products coelute with proline but have a different ratio of absorbances, at 440 and 570 nm, of color developed with ninhydrin. Both cysteine and cystine residues in peptides are converted in excellent yield (94%) to the stable cysteic acid by oxidation with performic acid (Moore, 1963); tryptophan is destroyed under these conditions and methionine is converted to the sulfone. Cysteic acid elutes in the void volume ahead of acidic amino acids on standard analyzers. Cysteine with a free sulfhydryl group is best determined (Crestfield *et al.*, 1963) after alkylation with iodoacetate to effect conversion to the acid-stable *S*-carboxymethyl-cysteine, which elutes before methionine sulfoxide (Spackman *et al.*, 1958). A thorough summary of the procedures for the quantitative assay of sulfhydryl groups and disulfide bonds in proteins should be consulted for further information (Friedman, 1973, pp. 275–310).

A particularly elegant way to establish the pairing of half-cystine residues in peptides involves mapping by the diagonal electrophoresis method (Brown and Hartley, 1966), but this important analytical control has yet to be applied to test synthetic products. The recent spectacular studies of Sieber *et al.* (1978) established that, in the case of insulin, the mispaired disulfide analogues synthesized by an unambiguous chemical route were resolved from insulin itself by simple thin layer chromatography.

12. Proline and N-Methylamino Acids

The imino acid proline has several characteristic stereo-electronic features that are reflected in ways relevant to peptide synthesis and attendant procedures (Schröder and Lübke, 1965, pp. 146–148).

1. The pK_a of an α-imino group in proline-containing peptides is 1 to 1.5 pH units higher than that of α-amino groups of peptides. Hence, *N*-alkylation of proline residues, but not α-amino acid residues, by excess resin-bound chloromethyl groups can be a problem under some conditions of solid-phase peptide synthesis (Schou *et al.*, 1976, Fig. 4); also, proline residues are especially susceptible to trifluoroacetylation by the intersite mechanism described earlier (Kent *et al.*, 1979; see Section IV,B, p. 121). The higher pK_a also means that the application of bases weaker than the standard tertiary alkyl amines in the neutralization step of solid-phase peptide synthesis (Section IV,B) may not necessarily lead to complete deprotonation of α-imino groups; however, no significant problems with the incorporation of proline or subsequent amino acids were evident when *N*-methylmorpholine ($pK_a \sim 8$) was the base for neutralization (Blake and Li, 1973; Li *et al.*, 1977).

2. In general, acylation of the α-imino group of proline occurs in excellent yields and at suitable rates during solid-phase synthesis, in contrast to the usual difficulty in acylation of secondary amino groups (see p. 249). In solution, coupling of a dipeptide onto proline *tert*-butyl ester has been found to give amplified levels of racemization in the middle amino acid of the resultant tripeptides (König and Geiger, 1970b); mixed anhydride acylation of proline peptides has led under some conditions to nearly quantitative yields of the "wrong-way" opened terminated products (e.g., de Leer and Beyerman, 1973; reviewed by Meienhofer, Vol. 1, Chapter 6). Both of these latter results suggest that rates of coupling onto α-imino groups are nevertheless slower than rates with α-amino groups.

3. The five-membered pyrrolidine ring has an important effect on the local conformation of the polypeptide linkage, with the consequence that certain chemical or enzymatic cleavages of peptide bonds become unusually facile, while others are suppressed to a dramatic extent. For example, in the Edman degradation the peptide bond of a phenylthiocarbamoyl-prolyl-histidine sequence is cleaved at a rate 10^3-fold slower than cleavage of a phenylthiocarbamoyl-histidine-proline sequence, and these are the two extremes among all dipeptide sequences tested (Tarr, 1977). The special tendency of proline-containing dipeptide-resins to form diketopiperazines with concomitant loss of chains is of considerable practical significance (p. 83). The prolyl-glycine bond is sensitive to strong alkali, and the aspartyl-proline bond is labile in acid.

4. For a combination of plausible reasons, including the absence of an enolizable amide bond, and the likely difficulty of forming a contiguous five-membered ring as required in an oxazolone intermediate, activation of COOH-terminal proline has been considered a safe strategy for segment condensation approaches aimed at minimizing racemization (p. 122). This has often been borne out experimentally. As a drawback, peptide segments with COOH-terminal proline, or even N^{α}-protected proline, are especially prone to form the *N*-acylurea (**84**) upon dicyclohexylcarbodiimide-activation in a polar solvent such as dimethylformamide (p. 127).

5. The tertiary amide bond of proline peptides is an attractive site for selective cleavage by strongly basic nucleophiles, because unlike the usual secondary amide bonds, the former cannot ionize as a first step and thus repel further attack (Spande *et al.*, 1970, pp. 130–143). The reductive cleavage of acylproline bonds upon deprotection by sodium in liquid ammonia (Hofmann and Yajima, 1961) was a special nuisance during synthetic studies on insulin (Marglin and Merrifield, 1966; see also Marglin, 1972).

6. Among various monitoring procedures proposed for solid-phase peptide synthesis (Table XI), the Schiff's base titration and fluorescamine tests are negative with proline peptides, while the ninhydrin test is of marginal utility. On amino acid analysis, color development with ninhydrin gives a low absorbance maximum at 440 nm (Spackman *et al.*, 1958). Detection of proline with fluorescamine is possible after an additional *N*-chlorosuccinimide oxidation step (Felix and Terkelsen, 1973).

Two proline analogs that have been successfully incorporated in solid-phase synthesis are 3,4-dehydroproline (**250**) (Felix *et al.*, 1973, 1977) and thiazolidine-4-carboxylic acid (**251**) (Felix *et al.*, 1973; Rosamond and Ferger,

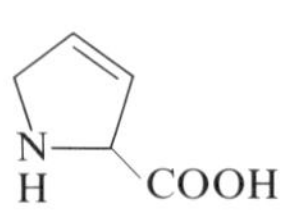

(**250**) Δ^3 Pro

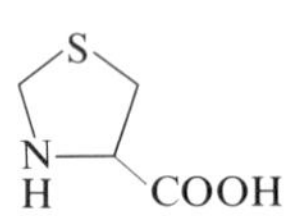

(**251**) Thz

1976). The derivative **250** is of interest because the double bond can be nearly quantitatively reduced by catalytic hydrogenation, deuteration, or tritiation to regenerate proline in a fully assembled and purified peptide (Felix *et al.*, 1977; compare to footnotes on pp. 214 and 221). Incorporation of **251**, which is the condensation product of cysteine and formaldehyde, requires special consideration because of its lability to iodine, mercury(II), aqueous alkali, and sodium in liquid ammonia. However, the thiazolidine moiety is stable to acid of moderate strength, and an oxytocin analog could be prepared via Boc-Cys(Ec)-Tyr-Ile-Gln-Asn-Cys(Ec)-Thz-Leu-Gly-dehydroalanyl-resin (Rosamond and Ferger, 1976).

N-Methylamino acid residues have been incorporated at interior positions of peptides by solid-phase synthesis in several laboratories (Andreatta and Scheraga, 1971; Jorgensen *et al.*, 1971b; Peña *et al.*, 1974; Khosla *et al.*, 1976; Rivier *et al.*, 1976). Among the various considerations listed for proline, numbers (1), (3), (5), and (6) essentially hold for *N*-methylamino acids as well. Since acylation of a secondary amino group compares unfavorably with that of a primary amino group (Schröder and Lübke, 1965, pp. 143–144), longer reaction times (e.g., 16 hr) are necessary for dicyclohexylcarbodiimide-mediated couplings onto *N*-methylamino acids (Andreatta and Scheraga, 1971; Peña *et al.*, 1974). Good results using Woodward's reagent K (Table X, number 4) for condensation have also been reported (Khosla *et al.*, 1976). Diketopiperazine formation, cyclization off the resin, and acylation of the liberated hydroxymethyl sites has been invoked to explain the isolation of the NH_2-terminal hexapeptide from an attempted solid-phase synthesis of an octapeptide with COOH-terminal *N*-methylphenylalanine (Khosla *et al.*, 1972c). Solution model studies have revealed *N*-methylamino acids to be prone to racemization at comparable levels under the same conditions as optically active α-amino acids, in contrast to the racemization-resistant property of proline (McDermott and Benoiton, 1973a,b). *N*-Methylamino acids are quantifiable on amino acid analyzers, but the ninhydrin color constants, depending on the side chain, are only 1 to 15% of those for α-amino acids (Spackman *et al.*, 1958; and all references cited previously in the paragraph).

13. Glycine, Alanine, Valine, Leucine, Isoleucine, and Norleucine

Incorporation of any of the amino acid residues with aliphatic side chains into peptides by standard solid-phase methodology is entirely straightforward. Differences in chemical reactivity are governed by steric factors. Because of its position on the least hindered end of this scale, glycine residues are more prone to certain side reactions involving cyclic transition states, including (1) diketopiperazine formation (pp. 82–83); (2) formation of **25** from amino acid phenacyl ester linkages [Eq. (4) on p. 86]; (3) urethane acylation and amino acid insertion in mixed as well as symmetrical anhydride couplings [Eq. (17) on p. 139]; and (4) succinimide formation at aspartyl-glycine and asparaginyl-glycine sequences (p. 193ff and p. 204ff). Ways to minimize, eliminate, or circumvent all these problems have already been discussed. Glycine, which is optically inactive, is an advantageous COOH-terminal residue for segment condensation strategies. Protected peptide segments anchored through COOH-terminal glycine are cleaved from polymeric supports by nucleophiles (Section III,B,1,c) and by photolysis (Section III,B,3,c) at rates two- to fivefold faster than those for peptides with other COOH-terminal residues.

Relative rates in the coupling reactions of solid-phase synthesis have been given earlier (Sections IV,C,1,c and 2,a); the sterically difficult Ile–Ile bond was formed 20-fold slower than Ala–Ala, but still in satisfactory yield. Racemization of isoleucine at the α-carbon yields its diastereomer, D-*allo*-isoleucine, which elutes between methionine and leucine on standard amino acid analyzers (Moore *et al.*, 1958). This observation has been adopted for a test for racemization (Bodanszky and Conklin, 1967; see also Matsueda *et al.*, 1977); the sensitivity limit is the 1 to 3% of background racemization occurring during acid hydrolysis. The amino acids with aliphatic side chains are useful diagnostic markers for analytical control of solid-phase peptide synthesis. Incorporation of a radiolabeled residue at a specific selected position of the polypeptide chain is particularly useful (p. 158); considerations of yields and economics make those amino acids requiring only α-amino protection the reasonable choice for this purpose.

VI. CONCLUSIONS AND OUTLOOK

Readers of this volume need not be reminded of the important role that peptides play as structural and functional elements in biochemistry and pharmacology. The applications of peptides in studies aimed at the elucidation of fundamental scientific questions is limited only by the ingenuity of the investigators. Because peptides and, especially, their structural analogues are generally difficult to obtain from natural sources,* much research has been devoted to providing these materials synthetically, in good yields and purity. Solid-phase peptide synthesis is one of the ways to achieve these aims.

The enthusiasm generated at the inception of the solid-phase principle has been qualified in certain respects, but has been justified in the final analysis by a substantial array of accomplishments. There have, of course,

* The recent advances in molecular biology, encapsulated in that much misunderstood word "clone", hold forth great promise for efficiently achieving some of the same goals that have classically occupied peptide chemists. The new technology will likely complement, rather than supersede, our efforts.

been failures, but these have often been due to an uncritical overextension of routine methodologies. Some workers have blamed poor results on ill-defined physical–chemical limitations of the polymeric support itself and have taken this to prove an inherent inadequacy in the overall concept. However, further careful examination, when carried out, has invariably established chemical reasons for the synthetic difficulties.

Chemical problems associated with the repetitive stepwise procedures have been largely overcome by several recent experimental improvements in resins, functionalization procedures, anchoring linkages, and reaction protocols. The deprotection and coupling steps can be made to proceed rapidly and in near quantitative yields, and without significant racemization. Terminated peptide chains can be removed from growing chains at the end of a synthesis by the application of new affinity methods.

The proper choice of a protection scheme, which in turn dictates the severity of chemical conditions that a peptide must survive during the synthesis and at the final cleavage, is of paramount importance in the recovery of intact unmodified products. Practical schemes based on graduated acidolysis for protecting-group removal have been and are continuing to be used with success. A number of side reactions arising under acidolytic conditions have been identified; their level of occurrence has often been markedly attenuated by subtle changes in conditions including temperature and the nature of cleavage reagents and scavengers. The development and application of mild, orthogonal protection schemes for solid-phase peptide synthesis is a top priority of future research.

Another rapidly developing area is the use of polymeric supports for segment condensation. The requisite protected peptides are available by solution synthesis, by solid-phase methodology that incorporates an extra level of selectivity in the protection scheme, or by derivatization of purified fragments obtained by chemical degradation or proteolysis of natural materials. Much progress has been made in obtaining more favorable solubility characteristics in protected segments, and in optimizing coupling yields. Because the intermediates can be isolated, purified, and characterized, the segment condensation approach may prove to be the most advantageous way to prepare small proteins by solid-phase synthesis.

Some additional new and original ideas put forward during the past few years merit further study. These include anchoring of amino acids through their side chains to facilitate both mono- and bidirectional chain extension, the simultaneous application of both a liquid-phase and a solid-phase support for dual anchoring, and the simultaneous application of two insoluble supports for solid-phase synthesis in a single vessel followed by separations based on differences in resin density or size. The concept of pseudo-dilution, or time-dependent isolation of sites, has illuminated discussions of chemical pathways of polymer-supported reactants. Application of the idea of internal structural controls will improve the confidence in conclusions drawn from results on inactive synthetic analogs.

Several large-scale preparations of important biologically active peptides have been achieved by economically practical solid-phase schemes. It has even been possible to recycle the solid support without reductions in the yield and purity of the cleaved products.

The status of the chemistry of solid-phase synthesis and monitoring procedures has lagged behind the technology available for automation and complete feed-back control. Recent developments encourage us to believe that this gap will continue to close. Finally, it is incumbent upon all peptide chemists to apply appropriate fractionation tools to purify their synthetic materials and then to establish critically the identity and homogeneity of the final products by all available analytical and functional criteria. Powerful and discriminating techniques including affinity chromatography, high-pressure liquid chromatography, and preview sequence analysis have now been added to the repertoire of methods suited to these purposes.

It has been our aim in this chapter to evaluate much of the available data on the chemistry of solid-phase peptide synthesis, and to show how this information can be integrated with and related to the body of results obtained by classical solution approaches. We hope that a thorough familiarity with the advantages and shortcomings of the many known techniques will aid the reader in the rational and effective application of solid-phase peptide synthesis.

ACKNOWLEDGMENTS

We thank Drs. Stephen B. H. Kent and Alexander R. Mitchell for critical reading and discussions of this review, and Ms. D. Graefe and Mary Kerins for expert and conscientious assistance with the preparation of the manuscript. We also thank colleagues who communicated their findings in advance of publication.

This work was supported by U.S. Public Health Service Research Fellowship 1 F32 AM 05806-01 to G.B. and Grant AM 01260 to R.B.M.

ADDENDUM

The Sixth American Peptide Symposium was held from June 17–22, 1979 [to be published as "Peptides: Structure and Biological Function" (E. Gross and J. Meienhofer, eds.), Pierce Chemical Co., Rockford, Illinois, 1979]. The following is a selective synopsis of new methodological developments reported at the meeting, presented in parallel with and cross-referenced to this chapter.

The Anchor to the Solid Support (Section III). The advantages of handles for anchoring (see pp 57–58) are becoming widely recognized. An alternate route to Boc-amino acyl-4-

(oxymethyl)-phenylacetic acid derivatives (**8** in Fig. 5) was worked out (Sallay *et al.*). The *p*-(α-phenyl)aminomethylphenoxyacetic acid handle (**252**) was suitable for direct preparation of peptide α-carboxamides in high yields after cleavage with HF (Matsueda and Gaehde, cf to Section III,B,8). "Two multi-detachable" anchoring linkages [**253** and **254**] were described (Tam *et al.*), which contain two orthogonally cleavable ester bonds (*a*, cleaved by strong acids p. 70, and *b*, cleaved by photolysis, pp. 81 and 86, or by nucleophiles, p. 85). These are separated by a "collapsible spacer," the oxymethylphenylacetoxy moiety, which undergoes decarboxylative 1,6-elimination (see arrows in structure **254**) in the presence of unsolvated cyanide ions. Hence, supports **253** and **254** can give rise to free peptides (by HF cleavage), to fully protected peptides (by photolysis) which can be purified and reattached to an aminomethyl support (Pam-linkage, p. 64ff) for further elongation and to fully protected peptide acids (by nucleophilic cleavage with

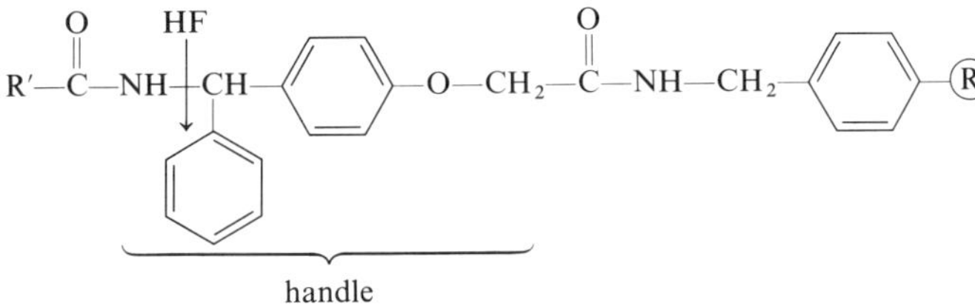

(**252**)

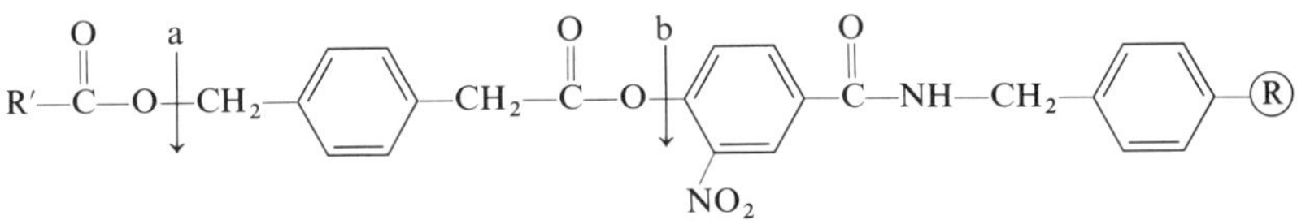

(**253**) Pon

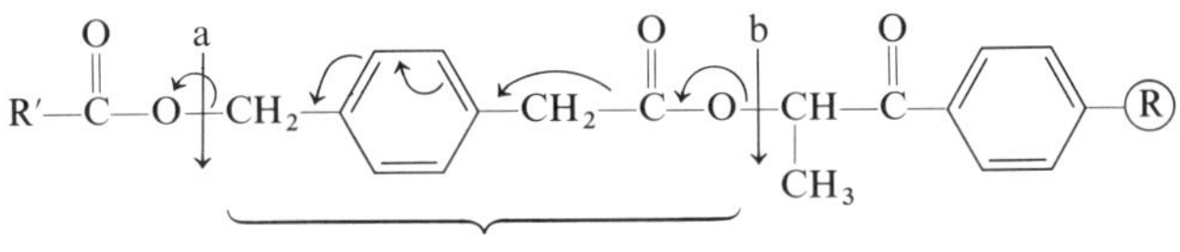

(**254**) Pop

cyanide) for segment condensation. An additional advantage of this versatile approach is that bond *b* is not adjacent to the polypeptide backbone, and cleavage yields are independent of the COOH-terminal amino acid residue.

Temporary Protecting Groups (Section IV,A). The widely accepted view (p. 102) that oxazolones do not form upon activation of N^{α}-urethane-protected amino acids has been challenged (Benoiton and Chen). Nonetheless, the oxazolones that were formed under specialized conditions reacted with amine nucleophiles to give peptides *without* racemization, so long as tertiary amines were absent.

Interest in the use of arylsulfenyl groups (pp. 109–112) for milder protection schemes continues. The Nps group (**56**) was applied to an improved solid-phase synthesis of the HF-labile peptide hormone gastrin (Fries *et al.*). Deprotection was achieved with 10 equiv. of the trifluoroacetate salt of 2-mercaptopyridine (**58**), and final deblocking of six *tert*-butyl side-chain protecting groups and cleavage of the *p*-alkoxybenzyl ester linkage (**16**) to the resin was carried out in TFA:water:1,2-dithioethane (98:1:1). The 3-nitro-2-pyridinesulfenyl group (**255**) was described (Matsueda and Walter); it is stable to TFA but partially cleaved in HCl–dioxane.

Orthogonal removal is with triphenylphosphine + alcohol or with 2-mercaptopyridine *N*-oxide (**114**). A novel oxidation–reduction coupling of these derivatives is also possible [Eq. (33);

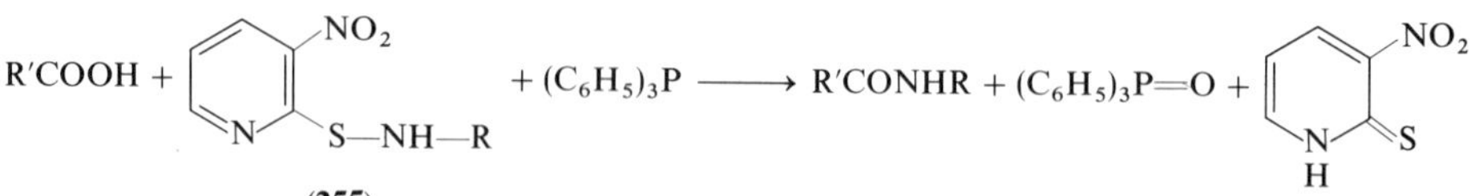

(**255**)

(33)

compare to Section IV,C,4]. Further applications of the 9-fluorenylmethyloxycarbonyl (Fmoc) group (59) (see pp. 112–113) in orthogonal protection and anchoring schemes were discussed (Atherton *et al.*), and the mechanism and kinetics of thiolytic removal of the dithiasuccinoyl (Dts) group (69) (pp. 114–116) were reported (Barany).

Formation of the Peptide Bond (Section IV,C). In one particular 32-peptide example of a solid-phase synthesis, better results were obtained with DCC-coupling than with symmetrical anhydrides (Tregear *et al.*). An α-methylamino acid was satisfactorily incorporated in stepwise solid-phase synthesis via its *N*-carboxyanhydride (Ciarkowski and Marshall; compare to Table X, number 3).

Monitoring (Section IV,D). A new qualitative color test for monitoring completeness of coupling in solid-phase peptide synthesis, employing chloranil, was reported (Christensen) to have a sensitivity comparable to the Kaiser ninhydrin test (Table XI, number 8). The latter test was scaled down to a micro-level (Burton *et al.*), and applied to monitor the synthesis of a 107-residue antibody fragment. Once a negative test was obtained, an extra coupling step and chain termination with *N*-acetylimidazole was performed; the cumulative preview (see Fig. 14) averaged 0.3% per cycle.

Affinity Purification to Remove Terminated Peptides (Section IV,E,2). The procedure with Sulfmoc-Cl (Table XIII, number 6) has been successfully extended to purify some large synthetic fragments (Merrifield *et al.*). Other affinity groups that have been explored (Wilchek and Miron) include the biotinyl-methionine (avidin binding) and dinitrophenyl (specific antibody binding) moieties.

Chemistry of the Individual Amino Acid Residues (Section V,B). The N^{im}-isobutyloxycarbonyl group (Grønvald *et al.*; compare to p. 181–182) was shown to be stable under conditions of N^{α}-Boc removal with TFA–CH_2Cl_2, and couplings with DCC activation proceeded smoothly without significant racemization of histidine. Partial cleavage by 1-hydroxybenzotriazole, and some $N^{im} \rightarrow N^{\alpha}$ transfer were observed. Deprotection was achieved within 2 hr at 50°C by treatment with methanolic triethylamine or pyridine hydrochloride.

An example of imide formation at aspartyl peptides with a free β-carboxyl group (cf p. 193ff) under strongly acidic conditions was reported (Schon and Kisfaludy). Both the acid and base-catalyzed side reactions were reduced when the new β-cyclohexyl esters (Tam *et al.*, reported by Merrifield) were compared to the usual β-benzyl ester protecting groups.

High specific activity tritiation of biologically active peptides containing phenylalanine has now been achieved via synthetically incorporated tri-halogenated phenylalanine (Escher *et al.*; compare to discussion on pp. 213–214.)

Reduction of methionine sulfoxide (cf pp. 228–229) was carried out with iodide in aqueous trifluoroacetic acid (Beyerman *et al.*); side reactions were observed with tryptophan but other sensitive residues such as tyrosine, histidine, and cysteine could be handled satisfactorily. Two side reactions occurring during the use of *S*-acetamidomethylcysteine (**230**) were discussed (Williams and Young).

Outlook. It was encouraging to see these advances in methodology, as well as the widespread application of solid-phase synthesis to the preparation of many biologically important peptides that were reported at the meeting. Continued progress in this field can be confidently expected.

REFERENCES

Abdipranoto, A., Hope, A. P., and Halpern, B. (1977). *Aust. J. Chem.* **30**, 2711–2715.

Aboderrin, A. A., Delpierre, G. R., and Fruton, J. S. (1965). *J. Am. Chem. Soc.* **87**, 5469–5472.

Aimoto, S., and Shimonishi, Y. (1975). *Bull. Chem. Soc. Jpn.* **48**, 3291–3297.

Aimoto, S., and Shimonishi, Y. (978). *Bull. Chem. Soc. Jpn.* **51**, 205–23.

Akabori, S., Sakakibara, S., Shimonishi, Y., and Nobuhara, Y. (1964). *Bull. Chem. Soc. Jpn.* **37**, 433–434.

Alakhov, Y. B., Kiryushkin, A. A., Lipkin, V. M., and Milne, G. W. (1970). *J. Chem. Soc. D* pp. 406–407.

Albertson, N. F., and McKay, F. C. (1953). *J. Am. Chem. Soc.* **75**, 5323–5326.

Albertsson, P. A. (1971). "Partition of Cell Particles and Macromolecules," 2nd ed. Wiley, New York.

Allenmark, S. (1966). *Acta Chem. Scand.* **20**, 910–911.

Altgelt, K. H., and Segal, L., eds. (1971). "Gel Permeation Chromatography." Dekker, New York.

Amaral, M. J., Barrett, G. C., Rydon, H. N., and Willett, J. E. (1966). *J. Chem. Soc. C* pp. 807–813.

Amiard, G., Heymès, R., and Velluz, L. (1956). *Bull. Soc. Chim. Fr.* pp. 698–700.

Amit, B., and Patchornik, A. (1973). *Tetrahedron Lett.* pp. 2205–2208.

Amit, B., Ben-Efraim, D. A., and Patchornik; A. (1976a). *J. Chem. Soc., Perkin Trans.. 1* pp. 57–63.

Amit, B., Ben-Efraim, D. A., and Patchornik, A. (1976b). *J. Am. Chem. Soc.* **98**, 843–844.

Amit, B., Hazum, E., Fridkin, M., and Patchornik, A. (1977). *Int. J. Pept. Protein Res.* **9**, 91–96.

Anderson, G. W., and Callahan, F. (1960). *J. Am. Chem. Soc.* **82**, 3359–3363.

Anderson, G. W., and McGregor, A. C. (1957). *J. Am. Chem. Soc.* **79**, 6180–6183.

Anderson, G. W., and Young, R. W. (1952). *J. Am. Chem. Soc.* **74**, 5307–5309.

Andreatta, R. H., and Rink, H. (1973). *Helv. Chim. Acta* **56**, 1205–1218.

Andreatta, R. H., and Scheraga, H. A. (1971). *J. Med. Chem.* **14**, 489–492.

Anfinsen, C. B. (1968). *Pure Appl. Chem.* **17**, 461–487.

Anfinsen, C. B., and Corley, L. G. (1969). *J. Biol. Chem.* **244**, 5149–5152.

Anfinsen, C. B., and Haber, E., (1961). *J. Biol. Chem.* **236**, 1361–1363.

Anfinsen, C. B., Ontjes, D., Ohno, M., Corley, L., and Eastlake, A. (1967), *Proc. Natl. Acad. Sci. U.S.A.* **58**, 1806–1811.

Aoyagi, H., Yonezawa, H., Takahashi, N., Kato, T., Izumiya, N., and Yang, C. C. (1972). *Biochim. Biophys. Acta* **263**, 823–826.

Arold, H., and Reissmann, S. (1968). *Z. Chem.* **8**, 107.

Arshady, R., Kenner, G. W., and Ledwith, A. (1974). *J. Polym. Sci., Polym. Chem. Ed.* **12**, 2017–2025.

Arshady, R., Kenner, G. W., and Ledwith, A. (1976). *Makromol. Chem.* **177**, 2911–2918.

Arshady, R., Atherton, E., Gait, M. J., Lee, K., and Sheppard, R. C. (1979). *J. Chem. Soc., Chem. Commun.*, pp. 423–425.

Atherton, E., and Sheppard, R. C. (1977). *In* "Peptides: Proceedings of the Fifth American Peptide Symposium" (M. Goodman and J. Meienhofer, eds.), pp. 503–505. Wiley, New York.

Atherton, E., Clive, D. L., and Sheppard, R. C. (1975). *J. Am. Chem. Soc.* **97**, 6584–6585.

Atherton, E., Caviezel, M., Over, H., and Sheppard, R. C. (1977). *J. Chem. Soc., Chem. Commun.* pp. 819–821.

Atherton, E., Fox, H., Harkiss, D., Logan, C. J., Sheppard, R. C., and Williams, B. J. (1978a). *J. Chem. Soc., Chem. Commun.* pp. 537–539.

Atherton, E., Fox, H., Harkiss, D., and Sheppard, R. C. (1978b). *J. Chem. Soc., Chem. Commun.* pp. 539–540.

Atherton, E., Sheppard, R. C., and Williams, B. J. (1979a). "Peptides, 1978" (I. Z. Siemion and G. Kupryszewski, eds.), pp. 207–210 Wroclaw Uni. Press, Wroclaw, Poland.

Atherton, E., Gait, M. J., Sheppard, R. C., and Williams, B. J. (1979b). *Bioorganic Chemistry* **8**, 351–370.

Axén, R., Porath, J., and Ernback, S. (1967). *Nature (London)* **214**, 1302–1304.

Baba, T., Sugiyama, H., and Seto, S. (1973). *Chem. Pharm. Bull.* **21**, 207–209.

Bailey, J. L. (1967). "Techniques in Protein Chemistry," 2nd ed. Elsevier, Amsterdam.

Bailey, J. L., and Cole, R. D. (1959). *J. Biol. Chem.* **234**, 1733–1739.

Bajusz, S., Turán, A., Fauszt, I., and Juhász, A. (1973). *In* "Peptides 1972" (H. Hanson and H. D. Jakubke, eds.), pp. 93–96. North-Holland Publ., Amsterdam.

Bannard, R. A., Casselman, A. A., Cockborn, W. F., and Brown, G. M. (1958). *Can. J. Chem.* **36**, 1541–1549.

Barany, G., and Merrifield, R. B. (1973). *Cold Spring Harbor Symp. Quant. Biol.* **37**, 121–126.

Barany, G., and Merrifield, R. B. (1977). *J. Am. Chem. Soc.* **99**, 7363–7365.

Barany, G., and Merrifield, R. B. (1979). *Anal. Biochem.* **95**, 160–1700.

Barltrop, J. A., Plant, P. J., and Schofield, P. (1966). *Chem. Commun.* pp. 822–823.

Barrass, B. C., and Elmore, D. T. (1957). *J. Chem. Soc.* pp. 3134–3139.

Barton, J. W. (1973). *In* "Protective Groups in Organic Chemistry" (J. F. W. McOmie, ed.), pp. 43–93. Plenum, New York.

Barton, M. A., Lemieux, R. U., and Savoie, J. Y. (1973). *J. Am. Chem. Soc.* **95**, 4501–4506.

Battersby, A., and Robinson, J. C. (1955). *J. Chem. Soc.* pp. 259–269.

Bauer, W., and Pless, J. (1975). *In* "Peptides: Chemistry, Structure and Biology" (R. Walter and J. Meienhofer, eds.), pp. 341–345. Ann Arbor Sci. Publ., Ann Arbor, Michigan.

Bayer, E., Hagenmaier, H., Jung, G., and König, W. (1968). *In* "Peptides 1968" (E. Bricas, ed.), pp. 162–170. North-Holland Publ., Amsterdam.

Bayer, E., Eckstein, H., Hägele, K., König, W., Brüning, W., Hagenmaier, H., and Parr, W. (1970a). *J. Am. Chem. Soc.* **92**, 1735–1738.

Bayer, E., Jung, G., Halàsz, I., and Sebestian, I. (1970b). *Tetrahedron Lett.* pp. 4503–4505.

Bayer, E., Breitmaier, E., Jung, G., and Parr, W. (1971). *Hoppe-Seyler's Z. Physiol. Chem.* **352**, 759–760.

Bell, J. R., and Jones, J. H. (1974). *J. Chem. Soc., Perkin Trans. 1* pp. 2336–2339.

Belleau, B., and Malek, G. (1968). *J. Am. Chem. Soc.* **90**, 1651–1652.

Ben-Ishai, D., and Berger, A. (1952). *J. Org. Chem.* **17**, 1564–1570.

Benson, J. V., and Patterson, J. A. (1971). *In* "New Techniques in Amino Acids, Peptides and Proteins" (A. Niederwieser and G. Pataki, eds.), pp. 1–73. Ann Arbor Sci. Publ., Ann Arbor, Michigan.

Benson, J. V., Gordon, M. J., and Patterson, J. A. (1967). *Anal. Biochem.* **18**, 228–240.

Berger, A., Noguchi, J., and Katchalski, E. (1956). *J. Am. Chem. Soc.* **78**, 4483–4488.

Bergmann, M., and Zervas, L. (1928). *Biochem. Z.* **203**, 280–292.

Bergmann, M., and Zervas, L. (1932). *Ber. Dtsch. Chem. Ges.* **65**, 1192–1201.

Bergmann, M., Zervas, L., and Rinke, H. (1934). *Hoppe-Seyler's Z. Physiol. Chem.* **224**, 40–44.

Beyerman, H. C., and Bontekoe, J. S. (1962). *Recl. Trav. Chim. Pays-Bas* **81**, 699–709.

Beyerman, H. C., and in't Veld, R. A. (1969). *Recl. Trav. Chim. Pays-Bas* **88**, 1019–1026.

Beyerman, H. C., Boers-Boonekamp, C. A., and Maassen van den Brink-Zimmermannova H. (1968a). *Recl. Trav. Chim. Pays-Bas* **87**, 257–273.

Beyerman, H. C., Hindriks, H., and de Leer, E. W. (1968b). *Chem. Commun.* p. 1668.

Beyerman, H. C., de Leer, E. W., and van Vossen, W. (1972). *J. Chem. Soc., Chem. Commun.* pp. 929–930.

Beyerman, H. C., Lie, T. S., and van Veldhuizen, C. J. (1973). *In* "Peptides 1971" (H. Nesvadba, ed.), pp. 162–164. North-Holland Publ., Amsterdam.

Beyerman, H. C., Hirt, J., Kranenburg, P., Syrier, J. L., and van Zon, A. (1974). *Recl. Trav. Chim. Pays-Bas* **93**, 256–257.

Birr, C. (1973). *Justus Liebigs Ann. Chem.* pp. 1652–1662.

Birr, C. (1975). *In* "Peptides, 1974" (Y. Wolman, ed.), pp. 117–122 Wiley, New York.

Birr, C. (1978). "Aspects of the Merrifield Peptide Synthesis" *In* "Reactivity and Structure Concepts in Organic Chemistry" (K. Hafner, C. W. Rees, B. M. Trost, J. M. Lehn, P. von Ragué Schleyer, and R. Zahradnik, eds.), Vol. 8. Springer-Verlag, Berlin and New York.

Birr, C., and Frank, R. (1975) *FEBS Lett.* **55**, 68–71.

Birr, C., and Pipkorn, R. (1979). "Peptides, 1978" (I. Z. Siemion and G. Kupryszewski, eds.) pp. 625–629. Wroclaw Uni. Press, Wroclaw, Poland.

Birr, C., and Lochinger, W. (1971). *Synthesis* pp. 319–321.

Birr, C., Reitel, C., and Wieland, T. (1970). *Angew. Chem., Int. Ed. Engl.* **9**, 731.

Birr, C., Lochinger, W., Stahnke, G., and Lang, P. (1972). *Justus Liebigs Ann. Chem.* **763**, 162–172.

Birr, C., Flor, F., Fleckenstein, P., and Wieland, T. (1973). *In* "Peptides 1971" (H. Nesvadba, ed.), pp. 175–184. North-Holland Publ., Amsterdam.

Birr, C., Ueki, M., and Frank, R. (1975). *In* "Peptides: Chemistry, Structure and Biology" (R. Walter and J. Meienhofer, eds.), pp. 409–417. Ann Arbor Sci. Publ., Ann Arbor. Michigan.

Birr, C., Wengert-Müller, M., and Buku, A. (1977). *In* "Peptides: Proceedings of the Fifth American Peptide Symposium" (M. Goodman and J. Meienhofer, eds.), pp. 510–513, Wiley, New York.

Blackburn, G. M., Brown, M. J., Harris, M. R., and Shire, D. (1969). *J. Chem. Soc., C* pp. 676–683.

Blackburn, S., ed. (1978). "Amino Acid Determination." 2nd ed. Dekker, New York.

Bláha, K., and Rudinger, J. (1965). *Collect. Czech. Chem. Comm.* **30**, 585–598.

Blake, J., (1979). *Int. J. Pept. Protein Res.* **13**, 418–425.

Blake, J., and Li, C. H. (1968). *J. Am. Chem. Soc.* **90**, 5882–5884.

Blake, J., and Li, C. H. (1971). *Int. J. Protein Res.* **3**, 185–189.

Blake, J., and Li, C. H. (1973). *Int. J. Pept. Protein Res.* **5**, 123–125.

Blake, J., and Li, C. H. (1975a). *J. Med. Chem.* **18**, 423–426.

Blake, J., and Li, C. H. (1975b). *Int. J. Pept. Protein Res.* **7**, 495–501.

Blake, J., and Li, C. H. (1976). *J. Chem. Soc., Chem. Commun.* pp. 504–505.

Blake, J., and Li, C. H. (1978). *Int. J. Pept Protein Res.* **11**, 315–322.

Blecher, H., and Pfaender, P. (1973). *Justus Liebigs Ann. Chem.* pp. 1263–1268.

Blömback, B. E. (1967). *In* "Methods in Enzymology" (C. H. W. Hirs, ed.), Vol. 11, pp. 398–411. Academic Press, New York.

Bodanszky, M. (1960). *Ann. N.Y. Acad. Sci.* **88**, 655–664.

Bodanszky, M. (1970). *In* "Peptides: Chemistry and Biochemistry" (B. Weinstein and S. Lande, eds.), pp. 1–14. Dekker, New York.

Bodanszky, M. (1977). *In* "Peptides: Proceedings of the Fifth American Peptide Symposium" (M. Goodman and J. Meienhofer, eds.), pp. 1–13. Wiley, New York.

Bodanszky, M., and Bodanszky, A. (1967). *Chem. Commun.* pp. 591–593.

Bodanszky, M., and Conklin, L. E. (1967). *Chem. Commun.* pp. 773–774.

Bodanszky, M., and du Vigneaud, V. (1959). *J. Am. Chem. Soc.* **81**, 5688–5691.

Bodanszky, M., and Fagan, D. T. (1977). *Int J. Pept. Protein Res.* **10**, 375–379.

Bodanszky, M., and Funk, K. W. (1973). *J. Org. Chem.* **38**, 1296–1300.

Bodanszky, M., and Kwei, J. Z. (1978). *Int. J. Pept. Protein Res.* **12**, 69–74.

Bodanszky, M., and Martinez, J. (1978). *J. Org. Chem.* **43**, 3071–3073.
Bodanszky, M., and Sheehan, J. T. (1960). *Chem. Ind.* (*London*) pp. 1268–1269.
Bodanszky, M., and Sheehan, J. T. (1964). *Chem. Ind.* (*London*) pp. 1423–1424.
Bodanszky, M., and Sheehan, J. T. (1966). *Chem. Ind.* (*London*) pp. 1597–1598.
Bodanszky, M., Ondetti, M. A., Birkhimer, C. A., and Thomas, P. L. (1964). *J. Am. Chem. Soc.* **86**, 4452–4459.
Bodanszky, M., Ondetti, M. A., Levine, S. D., and Williams, N. J. (1967). *J. Am. Chem. Soc.* **89**, 6753–6757.
Bodanszky, M., Klausner, Y. S., and Ondetti, M. A. (1976). "Peptide Synthesis," 2nd ed. Wiley, New York.
Bodanszky, M., Tolle, J. C., Deshmane, S. S., and Bodanszky, A. (1978). *Int. J. Pept. Protein Res.* **12**, 57–68.
Bohak, Z. (1964). *J. Biol. Chem.* **239**, 2878–2887.
Boissonnas, R. A., Guttmann, S., and Jaquenoud, P. A. (1960). *Helv. Chim. Acta* **43**, 1349–1358.
Bondi, E., Fridkin, M., and Patchornik, A. (1968). *Isr. J. Chem.* **6**, 22p.
Bonora, G. M., Toniolo, C., Fontana, A., Di Bello, C., and Scoffone, E. (1974). *Biopolymers* **13**, 157–167.
Borin, G., Toniolo, C., Moroder, L., Marchiori, F., Rocchi, R., and Scoffone, E. (1972). *Int. J. Protein Res.* **4**, 37–45.
Bornstein, P., and Balian, G. (1977). *In* "Methods in Enzymology" (C. H. W. Hirs and S. N. Timasheff, eds.), Vol. 47, pp. 132–145. Academic Press, New York.
Brady, S. F., Hirschmann, R., and Veber, D. F. (1977). *J. Org. Chem.* **42**, 143–146.
Brandenburg, D., Biela, M., Herbertz, L., and Zahn, H. (1975). *Hoppe-Seyler's Z. Physiol. Chem.* **356**, 961–979.
Brechbühler, H., Büchi, H., Hatz, E., Schreiber, J., and Eschenmoser, A. (1965). *Helv. Chim. Acta* **48**, 1746–1771.
Breitenbach, J. W. (1968) *Adv. Macromol. Chem.* **1**, 139–167.
Brenner, M. (1967). *In* "Peptides" (H. C. Beyerman, A. van de Linde, and W. Maassen van den Brink, eds.) pp. 1–7. North-Holland Publ., Amsterdam.
Brewster, J., and Ciotti, C. (1955). *J. Am. Chem. Soc.* **77**, 6214–6215.
Brois, S. J., Pilot, J. F., and Barnum, H. W. (1970). *J. Am. Chem. Soc.* **92**, 7629–7631.
Brown, J. R., and Hartley, B. S. (1966). *Biochem. J.* **101**, 214–228.
Brownlee, P. J., Cox, M. E., Handford, B. O., Marsden, J. C., and Young, G. T. (1964). *J. Chem. Soc.* pp. 3832–3840.
Brunfeldt, K. (1977). *Acta Physica Chemica* **23**, 499–514.
Brunfeldt, K., and Christensen, T. (1972). *FEBS Lett.* **19**, 345–348.
Brunfeldt, K., Halstrøm, J., and Roepstorff, P. (1969a). *Acta Chem. Scand.* **23**, 2830–2838.
Brunfeldt, K., Roepstorff, P., and Thomsen, J. (1969b). *Acta Chem. Scand.* **23**, 2906–2907.
Brunfeldt, K., Christensen, T., and Roepstorff, P. (1972a). *FEBS Lett.* **22**, 238–244.
Brunfeldt, K., Christensen, T., and Roepstorff, P. (1972b). *FEBS Lett.* **25**, 184–188.
Buis, J. T., Tesser, G. I., and Nivard, R. J. (1976). *Tetrahedron* **32**, 2321–2325.
Büllesbach, E. E., Danho, W., Helbig, H., and Zahn, H. (1979). "Peptides, 1978" (I. Z. Siemion and G. Kupryszweski, eds.) pp. 643–646. Wroclaw Uni. Press, Wroclaw, Poland.
Burlitch, J. M., and Winterton, R. C. (1975). *J. Am. Chem. Soc.* **97**, 5605–5606.
Burton, J., Poulsen, K., and Haber, E. (1975). *Biochemistry* **14**, 3892–3898.
Burton, J., Margolies, M. N., and Haber, E. (1977). *In* "Peptides: Proceedings of the Fifth American Peptide Symposium" (M. Goodman and J. Meienhofer, eds.), pp. 525–527. Wiley, New York.
Bush, M. E., Alkan, S. S., Nitecki, D. E., and Goodman, J. W. (1972). *J. Exp. Med.* **136**, 1478–1483.

Callahan, F. M., Anderson, G. W., Paul, R., and Zimmerman, J. E. (1963). *J. Am. Chem. Soc.* **85**, 201–207.
Camble, R., and Petter, N. N. (1976). *In* "Peptides 1976" (A. Loffet, ed.), pp. 299–307. Editions de l'Université de Bruxelles, Belgium.
Camble, R., Garner, R., and Young, G. T. (1968). *Nature (London)* **217**, 247–248.
Camps, F., Castells, J., Ferrando, M. J., and Font, J. (1971). *Tetrahedron Lett.* pp. 1713–1714.
Carpenter, F. H., and Gish, D. T. (1952). *J. Am. Chem. Soc.* **74**, 3818–3821.
Carpino, L.A. (1957). *J. Am. Chem. Soc.* **79**, 4427–4431.
Carpino, L. A. (1973). *Acc. Chem. Res.* **6**, 191–198.
Carpino, L. A., and Han, G. Y. (1972). *J. Org. Chem.* **37**, 3404–3409.
Carpino, L. A., Parameswaran, K. N., Kirley, R. R., Spiewak, J. W., and Schmitz, E. (1970). *J. Org. Chem.* **35**, 3291–3295.
Carpino, L. A., Tsao, J. H., Ringsdorf, H., Fell, E., and Hettrich, G. (1978). *J. Chem. Soc., Chem. Commun.* pp. 358–359.
Carraway, R., and Leeman, S. E. (1975). *J. Biol. Chem.* **250**, 1912–1918.
Chan, W. W. (1968). *Biochemistry* **7**, 4247–4254.
Chang, C. D., and Meienhofer, J. (1978). *Int. J. Pept. Protein Res.* **11**, 246–249.
Chang, J. K., Shimizu, M., and Wang, S. S. (1976). *J. Org. Chem.* **41**, 3255–3258.
Channing, D. M., Turner, P. B., and Young, G. T. (1951). *Nature (London)* **167**, 487–488.
Chapman, P. H., and Walker, D. (1975). *J. Chem. Soc., Chem. Commun.* pp. 690–691.
Chaturvedi, N. C., Park, W. K., Smeby, R. R., and Bumpus, F. M. (1970). *J. Med. Chem.* **13**, 177–181.
Chen, F. M., Kuroda, K., and Benoiton, N. L. (1978). *Synthesis*, pp. 928–929.
Chillemi, F., and Merrifield, R. B. (1969). *Biochemistry* **8**, 4344–4346.
Chino, N., Masui, Y., and Sakakibara, S. (1978). *In* "Peptide Chemistry 1977" (T. Shiba, ed.), pp. 27–32. Protein Res. Found., Osaka, Japan.
Chorev, M., and Klausner, Y. S. (1976). *J. Chem. Soc., Chem. Commun.* p. 596.
Chou, F. C., Chawla, R. K., Kibler, R. F., and Shapira, R. (1971). *J. Am. Chem. Soc.* **93** 267–268.
Christensen, T., and Schou, O. (1978). *Int. J. Pept. Protein Res.* **12**, 121–129.
Christensen, T., Halstrøm, J., Kovacs, K., Villemoes, P., and Brunfeldt, K. (1976). *In* "Peptides 1976" (A. Loffet, ed.), pp. 213–218. Editions de l'Université de Bruxelles, Belgium.
Cleland, W. W. (1964). *Biochemistry* **3**, 480–482.
Cole, R.D. (1967). *In* "Methods in Enzymology" (C. H. W. Hirs, ed.), Vol. 11, pp. 206–208. Academic Press, New York.
Colescott, R. L., Bossinger, C. D., Cook, P. I., Dailey, J. P., Enkoji, T., Flanigan, E., Geever, J. E., Groginsky, C. M., Kaiser, E., Laken, B., Mason, W. A., Olsen, D. B., Reynolds, H. C., and Skibbe, M. O. (1975). *In* "Peptides: Chemistry, Structure and Biology" (R. Walter and J. Meienhofer, eds.), pp. 463–467. Ann Arbor Sci. Publ., Ann Arbor, Michigan.
Collman, J. P., Hegedus, L. S., Cooke, M. P., Norton, J. R., Dolcetti, G., and Marquardt, D. N. (1972). *J. Am. Chem. Soc.* **94**, 1789–1790.
Corley, L., Sachs, D. H., and Anfinsen, C. B. (1972). *Biochem. Biophys. Res. Commun.* **47**, 1353–1359.
Cosand, W. L., and Merrifield, R. B. (1977). *Proc. Natl. Acad. Sci. U.S.A.* **74**, 2771–2775.
Cosmatos, A., Photaki, I., and Zervas, L. (1961). *Chem. Ber.* **94**, 2644–2655.
Coy, D. H., Kastin, A. J., Schally, A. V., Morin, O., Caron, N. G., Labrie, F., Walker, J. M., Fertel, R., Berntson, G. G., and Sandman, C. A. (1976). *Biochem. Biophys. Res. Commun.* **73**, 632–638.
Coyle, S., and Young, G. T. (1976). *J. Chem. Soc., Chem. Commun.* pp. 980–981.
Crestfield, A. M., Moore, S., and Stein, W. H. (1963). *J. Biol. Chem.* **238**, 622–627.

Crowley, J. I., and Rapoport, H. (1976). *Acc. Chem. Res.* **9**, 135–144.
Crowley, J. I., Harvey, T. B., and Rapoport, H. (1973). *J. Macromol. Sci., Chem.* **7**, 1117–1126.
Cuatrecasas, P., and Anfinsen, C. B. (1971). *Annu. Rev. Biochem.* **40**, 259–278.
Dahlmans, J. J. (1973). *In* "Peptides 1972" (H. Hanson and H. D. Jakubke, eds.), pp. 171–172. North-Holland Publ., Amsterdam.
Dane, E., Drees, F., Konrad, P., and Dockner, T. (1962). *Angew. Chem., Int. Ed. Engl.* **1**, 873.
Davies, C. W., and Owen, B. D. (1956). *J. Chem. Soc.* pp. 1676–1680.
Dekker, C. A., and Fruton, J. S. (1948). *J. Biol. Chem.* **173**, 471–477.
Dekker, C. A., Taylor, S. P., and Fruton, J. S. (1949). *J. Biol. Chem.* **180**, 155–173.
de Leer, E. W., and Beyerman, H. C. (1973). *Recl. Trav. Chim. Pays-Bas* **92**, 174–184.
DeTar, D. F., and Silverstein, R. (1966a). *J. Am. Chem. Soc.* **88**, 1013–1019.
DeTar, D. F., and Silverstein, R. (1966b). *J. Am. Chem. Soc.* **88**, 1020–1023.
DeTar, D. F., Silverstein, R., and Rogers, F. F. (1966). *J. Am. Chem. Soc.* **88**, 1024–1030.
Dewey, R. S., Schoenewaldt, E. F., Joshua, H., Paleveda, W. J., Schwam, H., Barkemeyer, H., Arison, B. H., Veber, D. F., Strachan, R. G., Milkowski, J., Denkewalter, R. G., and Hirschmann, R. (1971). *J. Org. Chem.* **36**, 49–59.
Di Bello, C., Marigo, A., Buso, O., and Lucchiari, A. (1977). *Tetrahedron Lett.* pp. 1135–1136.
Dickman, S. R., and Westcott, W. L. (1954). *J. Biol. Chem.* **210**, 481–489.
Dorman, L. C. (1969). *Tetrahedron Lett.* pp. 2319–2321.
Dorman, L. C. (1974). *Biochem. Biophys. Res. Commun.* **60**, 318–325.
Dorman, L. C., and Love, J. (1969). *J. Org. Chem.* **34**, 158–165.
Dorman, L. C., Markley, L. D., and Mapes, D. A. (1971). *Anal. Biochem.* **39**, 492–497.
Dorman, L. C., Nelson, D. A., and Chow, R. C. (1972). *In* "Progress in Peptide Research" (S. Lande, ed.). pp. 65–68. Gordon & Breach, New York.
Doscher, M. S. (1977). *In* "Methods in Enzymology" (C. H. W. Hirs and S. N. Timasheff, eds.), Vol. 47, pp. 578–617. Academic Press, New York.
Dowling, L. M., and Strak, G. R. (1969). *Biochemistry* **8**, 4728–4734.
Draper, M. W., Merrifield, R. B., and Rizack, M. A. (1973). *J. Med. Chem.* **16**, 1326–1330.
Dunn, B. M., Di Bello, C., Kirk, K. L., Cohen, L. A., and Chaiken, I. M. (1974) *J. Biol. Chem.* **249**, 6295–6301.
du Vigneaud, V., and Behrens, O. K. (1937). *J. Biol. Chem.* **117**, 27–36.
du Vigneaud, V., Audrieth, L. F., and Loring, H. S. (1930). *J. Am. Chem. Soc.* **52**, 4500–4504.
Dwulet, F. E., and Gurd, F. R. (1976). *Anal. Biochem.* **76**, 530–538.
Dzieduszycka, M., Smulkowski, M., and Taschner, E. (1973). *In* "Peptides 1972" (H. Hanson and H. D. Jakubke, eds.), pp. 103–107. North-Holland Publ., Amsterdam.
Eaborn, C. (1960). "Organosilicon Compounds," p. 90, Butterworths, London.
Eastlake, A., Curd, J. G., and Schechter, A. N. (1976). *J. Biol. Chem.* **251**, 6426–6430.
Eckert, H. (1977). *Synthesis* pp. 332–333.
Eckert, H., Schrauzer, G., and Ugi, I. (1975). *Tetrahedron* **31**, 1399–1401.
Eckert, H., Listl, M., and Ugi, I. (1978). *Angew. Chem., Int. Ed. Engl.* **17**, 361–362.
Edman, P. (1950). *Acta Chem. Scand.* **4**, 283–293.
Edman, P. (1970). *In* "Protein Sequence Determination" (S. B. Needleman, ed.), pp. 211–256. Springer-Verlag, Berlin and New York.
Elliot, D. F., Moritz, P., and Wade, R. (1972). *J. Chem. Soc., Perkin Trans. 1* pp. 1862–1866.
Engelhard, M., and Merrifield, R. B. (1978). *J. Am. Chem. Soc.* **100**, 3559–3563.
Erickson, B. W., and Merrifield, R. B. (1973a). *J. Am. Chem. Soc.* **95**, 3750–3756.
Erickson, B. W., and Merrifield, R. B. (1973b). *J. Am. Chem. Soc.* **95**, 3757–3763.
Erickson, B. W., and Merrifield, R. B. (1974). *Isr. J. Chem.* **12**, 79–85.
Erickson, B. W., and Merrifield, R. B. (1976). *In* "The Proteins" (H. Neurath and R. L. Hill, eds.), 3rd ed., Vol. 2, pp. 255–527. Academic Press, New York.
Erickson, B. W. and Prystowsky, M. B. (1979). *In* "International Liquid Chromatography

Symposium II, Biological/Biomedical Applications of Liquid Chromatography" (G. L. Hawk, ed.), Dekker, New York, pp. 293–305.
Erickson, B. W., and Wang, C. Y. (1975). *Abstr., 170th Natl. Meet., Am. Chem. Soc., 1975* ORGN 081.
Esko, K., and Karlsson, S. (1970). *Acta Chem. Scand.* **24**, 1415–1422.
Esko, K., Karlsson, S., and Porath, J. (1968). *Acta Chem. Scand.* **22**, 3342–3343.
Fankhauser, P., Fries, P., and Brenner, M. (1973). *Helv. Chim. Acta* **56**, 2516–2520.
Fankhauser, P., Fries, P., Stahala, P., and Brenner, M. (1974). *Helv. Chim. Acta* **57**, 271–277.
Farrall, M. J., and Fréchet, J. M. (1976). *J. Org. Chem.* **41**, 3877–3882.
Farrall, M. J., and Fréchet, J. M. (1978). *J. Am. Chem. Soc.* **100**, 7998–7999.
Feinberg, R. S., and Merrifield, R. B. (1972). *Tetrahedron* **28**, 5865–5871.
Feinberg, R. S., and Merrifield, R. B. (1974). *Tetrahedron* **30**, 3209–3212.
Feinberg, R. S., and Merrifield, R. B. (1975). *J. Am. Chem. Soc.* **97**, 3485–3496.
Felix, A. M. (1974). *J. Org. Chem.* **39**, 1427–1429.
Felix, A. M., and Jimenez, M. H. (1973). *Anal. Biochem.* **52**, 377–381.
Felix, A. M., and Merrifield, R. B. (1970). *J. Am. Chem. Soc.* **92**, 1385–1391.
Felix, A. M., and Terkelsen, G. (1973). *Arch. Biochem. Biophys.* **157**, 177–182.
Felix, A. M., Jimenez, M. H., Vergona, R., and Cohen, M. R. (1973). *Int. J. Pept. Protein Res.* **5**, 201–206.
Felix, A. M., Jimenez, M. H., Vergona, R., and Cohen, M. R. (1975). *Int. J. Pept. Protein Res.* **7**, 11–22.
Felix, A. M., Wang, C. T., Liebman, A. A., Delaney, C. M., Mowles, T., Burghardt, B. A., Charnecki, A. M., and Meienhofer, J. (1977). *Int. J. Pept. Protein Res.* **10**, 299–310.
Felix, A. M., Jimenez, M. H., Mowles, T., and Meienhofer, J. (1978). *Int. J. Pept. Protein Res.* **11**, 329–339.
Fenwick, G. R. (1973). *Chem. Ind. (London)* pp. 636–637.
Field, L. (1977). *In* "Organic Chemistry of Sulfur" (S. Oae, ed.), pp. 303–382. Plenum, New York.
Finn, F. M., and Hofmann, K. (1976). *In* "The Proteins" (H. Neurath and R. L. Hill, eds.), 3rd ed., Vol. 2, pp. 105–253. Academic Press, New York.
Fischer, E., and Bergmann, M. (1913). *Justus Liebigs Ann. Chem.* **398**, 96–124.
Fischman, A. J., Riemen, M. W., and Cowburn, D. (1978). *FEBS Lett.* **94**, 236–240.
Flanigan, E., and Marshall, G. R. (1970). *Tetrahedron Lett.* pp. 2403–2406.
Fletcher, A. R., Jones, J. H., Ramage, W. I., and Stachulski, A. V. (1979). "Peptides, 1978" (I. Z. Siemion and G. Kupryszewski, eds.), pp. 169–171. Wroclaw Uni. Press, Wroclaw, Poland.
Fletcher, G. A., Löw, M., and Young, G. T. (1973). *J. Chem. Soc., Perkin Trans. 1* pp. 1162–1164.
Flor, F., Birr, C., and Wieland, T. (1973). *Justus Liebigs Ann. Chem.* pp. 1601–1605.
Floyd, N. F., Cammaroti, M. S., and Lavine, T. F. (1963). *Arch. Biochem. Biophys.* **102**, 343–345.
Fluoret, G. (1970). *J. Med. Chem.* **13**, 843–845.
Folk, J. E. (1970). *In* "Methods in Enzymology" (G. E. Perlmann and L. Lorand, eds.), Vol. 19, pp. 504–508. Academic Press, New York.
Fontana, A. (1975). *J. Chem. Soc., Chem. Commun.* pp. 976–977.
Fontana, A., and Scoffone, E. (1972). *In* "Methods in Enzymology" (C. H. W. Hirs and S. N. Timasheff, eds.), Vol. 25, pp. 482–494. Academic Press, New York.
Fontana, A., and Toniolo, C. (1976). *In* "Progress in the Chemistry of Organic Natural Products" (W. Herz, H. Grisebach, and G. W. Kirby, eds.), Vol. 33, pp. 309–449. Springer-Verlag, New York.
Fontana, A., Marchiori, F., Moroder, L., and Scoffone, E. (1966). *Tetrahedron Lett.* pp. 2985–2987.
Fontana, A., Veronese, F. M., and Boccú, E. (1971). *Z. Naturforsch., Teil B* **26**, 314–317.

Frank, H., and Hagenmaier, H. (1975). *Experientia* **31**, 131–133.
Frank, H., Meyer, H., and Hagenmaier, H. (1977a). *Chem. Ztg* **101**, 188–193.
Frank, H., Hagenmaier, H., Bayer, E., and Desiderio, D. M. (1977b). *In* "Peptides: Proceedings of the Fifth American Peptide Symposium" (M. Goodman and J. Meienhofer, eds.), pp. 514–517. Wiley, New York.
Fréchet, J. M., and Haque, K. E. (1975). *Macromolecules* **8**, 130–134.
Fréchet, J. M., and Schuerch, C. (1971). *J. Am. Chem. Soc.* **93**, 492–496.
Freidinger, R. M., Hirschmann, R., and Veber, D. F. (1978). *J. Org. Chem.* **43**, 4800–4803.
Fridkin, M., Patchornik, A., and Katchalski, E. (1965). *J. Am. Chem. Soc.* **87**, 4646–4648.
Fridkin, M., Hazum, E., Tauber-Finkelstein, M., and Shaltiel, S. (1977). *Arch. Biochem. Biophys.* **178**, 517–526.
Friedman, M. (1973). "The Chemistry and Biochemistry of the Sulfhydryl Group in Amino Acids, Peptides and Proteins." Pergamon, Oxford.
Fruton, J. S. (1949). *Advan. Protein Chem.* **5**, 1–82 (esp. p. 64).
Fujii, T., and Sakakibara, S. (1970). *Bull. Chem. Soc. Jpn.* **43**, 3954–3956.
Fujii, T., and Sakakibara, S. (1974). *Bull. Chem. Soc. Jpn.* **47**, 3146–3151.
Fujino, M., and Nishimura, O. (1976). *J. Chem. Soc., Chem. Commun.* p. 998.
Fujino, M., Kobayashi, S., Obayashi, M., Fukuda, T., Shinagawa, S., and Nishimura, O. (1974). *Chem. Pharm. Bull.* **22**, 1857–1863.
Fujino, M., Wakimasu, M., Shinagawa, S., Kitada, C., and Yajima, H. (1978). *Chem. Pharm. Bull.* **26**, 539–548.
Fukuda, T., Kitada, C., and Fujino, M. (1978). *J. Chem. Soc., Chem. Commun.* pp. 220–221.
Fyles, T. M., and Leznoff, C. C. (1976). *Can. J. Chem.* **54**, 935–942.
Gait, M. T., and Sheppard, R. C. (1976). *J. Am. Chem. Soc.* **98**, 8514–8516.
Galardy, R. E., Bleich, H. E., Ziegler, P., and Craig, L. C. (1976). *Biochemistry* **15**, 2303–2309.
Garden, J., and Tometsko, A. M. (1972). *Anal. Biochem.* **46**, 216–223.
Garner, R., Schafer, D. J., and Young, G. T. (1971). *In* "Peptides 1969" (E. Scoffone, ed.), pp. 102–108. North-Holland Publ., Amsterdam.
Garsky, V. M., Clark, D. E., and Grant, N. H. (1976). *Biochem. Biophys. Res. Commun.* **73**, 911–916.
Gawron, O., and Draus, F. (1958). *J. Org. Chem.* **23**, 1040–1041.
Geiger, R., Jäger, G., Volk, A., and Siedel, W. (1968). *Chem. Ber.* **101**, 2189–2196.
Geiger, R., Treuth, G., and Burow, F. (1973). *Chem. Ber.* **106**, 2339–2346.
Geising, W., and Hörnle, S. (1973). *In* "Peptides 1971" (H. Nesvadba, ed.), pp. 146–152. North-Holland Publ., Amsterdam.
Gish, D. T., Katsoyannis, P. G., Hess, G. P., and Stedman, R. J. (1956). *J. Am. Chem. Soc.* **78**, 5954.
Gisin, B. F. (1972). *Anal. Chim. Acta* **58**, 248–249.
Gisin, B. F. (1973). *Helv. Chim. Acta* **56**, 1476–1482.
Gisin, B. F., and Merrifield, R. B. (1972a). *J. Am. Chem. Soc.* **94**, 3102–3106.
Gisin, B. F., and Merrifield, R. B. (1972b). *J. Am. Chem. Soc.* **94**, 6165–6170.
Gisin, B. F., Merrifield, R. B., and Tosteson, D. C. (1969). *J. Am. Chem. Soc.* **91**, 2691–2695.
Gisin, B. F., Kobayashi, S., and Hall, J. E. (1977). *Proc. Natl. Acad. Sci. U.S.A.* **74**, 115–119.
Gisin, B. F., Ting-Beall, H. P., Davis, D. G., Grell, E., and Tosteson, D. C. (1978). *Biochim. Biophys. Acta* **509**, 201–217.
Glass, J. D., Schwartz, I. L., and Walter, R. (1972). *J. Am. Chem. Soc.* **94**, 6209–6211.
Glass, J. D., Talansky, A., Grzonka, Z., Schwartz, I. L., and Walter, R. (1974). *J. Am. Chem. Soc.* **96**, 6476–6480.
Glass, J. D., Pelzig, M., and Pande, C. S. (1979). *Int. J. Pept. Protein Res.*, **13**, 28–34.
Glazer, A. N. (1976). *In* "The Proteins" (H. Neurath and R. L. Hill, eds.), 3rd ed., Vol. 2, pp. 1–103. Academic Press, New York.

Goissis, G., Erickson, B. W., and Merrifield, R. B. (1977). *In* "Peptides: Proceedings of the Fifth American Peptide Symposium" (M. Goodman and J. Meienhofer, eds.), pp. 559–561, Wiley, New York.
Goldberger, R. F., and Anfinsen, C. B. (1962). *Biochemistry* **1**, 401–405.
Goodman, M., and Glaser, C. B. (1970). *In* "Peptides: Chemistry and Biochemistry" (B. Weinstein and S. Lande, eds.), pp. 267–335. Dekker, New York.
Goodman, M., and Stueben, K. C. (1959). *J. Am. Chem. Soc.* **81**, 3980–3983.
Goren, H. J., and Fridkin, M. (1978). *Int. J. Pept. Protein Res.* **11**, 1–8.
Grahl-Nielsen, O., and Tritsch, G. L. (1969). *Biochemistry* **8**, 187–192.
Granier, C., Pedroso, E. M., and van Rietschoten, J. (1978). *Eur. J. Biochem.* **82**, 293–299.
Gray, W. R. (1967). *In* "Methods in Enzymology" (C. H. W. Hirs, ed.), Vol. 11, pp. 139–151. Academic Press, New York.
Greben, A. E., Martinov, V. F., and Tutov, M. I. (1968). *Zh. Obshch. Khim.* **38**, 664–665; *Chem. Abstr.* **69**, 97143j (1968).
Green, B., and Garson, L. R. (1969). *J. Chem. Soc., C* pp. 401–405.
Greenstein, J. P., and Winitz, M. (1961). "Chemistry of the Amino Acids," 3 Vol. Wiley, New York.
Gregor, H. P., Hoeschele, G. K., Potenza, J., Tsuk, A. G., Feinland, R., Shida, M., and Teyssié, P. (1965). *J. Am. Chem. Soc.* **87**, 5525–5534.
Gros, C., de Garilhe, M. P., Costopanagiotis, A., and Schwyzer, R. (1961). *Helv. Chim. Acta* **44**, 2042–2048.
Gross, E. (1967). *In* "Methods in Enzymology" (C. H. W. Hirs, ed.), Vol. 11, pp. 238–255. Academic Press, New York.
Gross, E., and Matsuura, S. (1975). *In* "Peptides: Chemistry, Structure, and Biology" (R. Walter and J. Meienhofer, eds.), pp. 351–358. Ann Arbor Sci. Publ., Ann Arbor, Michigan.
Gross, E., and Morell, J. L. (1974). *Biochem. Biophys. Res. Commun.* **59**, 1145–1150.
Gross, E., and Witkop, B. (1962). *J. Biol. Chem.* **237**, 1856–1860.
Gross, E., Noda, K., and Nisula, B. (1973). *Angew. Chem., Int. Ed. Engl.* **12**, 664–665.
Gross, E., Noda, K., and Matsuura, S. (1975). *In* "Peptides, 1974" (Y. Wolman, ed.), pp. 403–413, Wiley, New York.
Gross, H., and Bilk, L. (1968). *Tetrahedron* **24**, 6935–6939.
Grzonka, Z., Glass, J. D., Schwartz, I. L., and Walter, R. (1974). *J. Med. Chem.* **17**, 1294–1298.
Guibé-Jampel, E., and Wakselman, M. (1979). "Peptides, 1978" (I. Z. Siemion and G. Kupryszewski, eds.) pp. 89–91. Wroclaw Uni. Press, Wroclaw, Poland.
Gundlach, H. G., Moore, S., and Stein, W. H. (1959). *J. Biol. Chem.* **234**, 1761–1764.
Gut, V. (1975). *Collect. Czech. Chem. Commun.* **40**, 129–136.
Gut, V., and Rudinger, J. (1968). *In* "Peptides 1968" (E. Bricas, ed.), pp. 185–186. North-Holland Publ., Amsterdam.
Gutte, B., and Merrifield, R. B. (1969). *J. Am. Chem. Soc.* **91**, 501–502.
Gutte, B., and Merrifield, R. B. (1971). *J. Biol. Chem.* **246**, 1922–1941.
Guttmann, S. (1966). *Helv. Chim. Acta* **49**, 83–96.
Guttmann, S., and Boissonnas, R. A. (1958). *Helv. Chim. Acta* **41**, 1852–1867.
Guttmann, S., and Boissonnas, R. A. (1959). *Helv. Chim. Acta* **42**, 1257–1264.
Haas, W. L., Krumkalns, E. V., and Gerzon, K. (1966). *J. Am. Chem. Soc.* **88**, 1988–1992.
Hagenmaier, H. (1975). *Hoppe-Seyler's Z. Physiol. Chem.* **356**, 777–785.
Hagenmaier, H., and Frank, H. (1972). *Hoppe-Seyler's Z. Physiol. Chem.* **353**, 1973–1976.
Haller, W. (1965). *J. Chem. Phys.* **42**, 686–693.
Halstrøm, J., Christensen, T., and Brunfeldt, K. (1976). *Hoppe-Seyler's Z. Physiol. Chem.* **357**, 999–1002.
Hammerström, K., Lunkenheimer, W., and Zahn, H. (1970). *Makromol. Chem.* **133**, 41–51.
Hancock, W. S., and Battersby, J. E. (1976). *Anal. Biochem.* **71**, 260–264.

Hancock, W. S., and Marshall, G. R. (1975). *J. Am. Chem. Soc.* **97**, 7488–7489.
Hancock, W. S., Battersby, J. E., and Harding, D. R. (1975). *Anal. Biochem.* **69**, 497–503.
Hanson, R. W., and Law, H. D. (1965). *J. Chem. Soc.* pp. 7285–7297.
Hardy, P. M., Rydon, H. N., and Thompson, R. C. (1968). *Tetrahedron Lett.* pp. 2525–2526.
Harris, J. I., and Fruton, J. S. (1951). *J. Biol. Chem.* **191**, 143–151.
Harrison, C. R., Hodge, P., Kemp, J., and Perry, G. M. (1975). *Makromol. Chem.* **176**, 267–274.
Harrison, I. T., and Harrison, S. (1967). *J. Am. Chem. Soc.* **89**, 5723–5724.
Haslam, E. (1973). *In* "Protective Groups in Organic Chemistry" (J. F. W. McOmie, ed.), pp. 183–215. Plenum, New York.
Hayatsu, H., and Khorana, H. G. (1967). *J. Am. Chem. Soc.* **89**, 3880–3887.
Hegarty, A. F., and Bruice, T. C. (1970). *J. Am. Chem. Soc.* **92**, 6568–6574.
Hegarty, A. F., McCormack, M. T., Ferguson, G., and Roberts, P. J. (1977). *J. Am. Chem. Soc.* **99**, 2015–2016.
Heitz, W., and Platt, K. L. (1969). *Makromol. Chem.* **127**, 113–140.
Helferich, B., Moog, L., and Jünger, A. (1925). *Ber. Dtsch. Chem. Ges.* **58**, 872–886.
Hemmasi, B., and Bayer, E. (1974). *Hoppe-Seyler's Z. Physiol. Chem.* **355**, 481–489.
Hendrickson, J. B., and Kandall, C. (1970). *Tetrahedron Lett.* pp. 343–344.
Henkart, P. (1971). *J. Biol. Chem.* **246**, 2711–2713.
Hirotsu, Y., Shiba, T., and Kaneko, T. (1967a). *Bull. Chem. Soc. Jpn.* **40**, 2945–2949.
Hirotsu, Y., Shiba, T., and Kaneko, T. (1967b). *Bull. Chem. Soc. Jpn.* **40**, 2950–2954.
Hirotsu, Y., Shiba, T., and Kaneko, T. (1970). *Bull. Chem. Soc. Jpn.* **43**, 1564–1567.
Hirs, C. H. W., ed. (1967). "Methods in Enzymology," Vol. 11. Academic Press, New York.
Hirs, C. H. W., and Timasheff, S. N., eds. (1972). "Methods in Enzymology," Vol. 25. Academic Press, New York.
Hirs, C. H. W., and Timasheff, S. N., eds. (1977). "Methods in Enzymology," Vol. 47. Academic Press, New York.
Hirs, C. H. W., Stein, W. H., and Moore, S. (1954). *J. Biol. Chem.* **211**, 941–950.
Hirschmann, R., Strachan, R. G., Schwam, H., Schoenewaldt, E. F., Joshua, H., Barkemeyer, B., Veber, D. F., Paleveda, W. J., Jacob, T. A., Beesley, T. E., and Denkewalter, R. G. (1967). *J. Org. Chem.* **32**, 3415–3425.
Hirschmann, R., Nutt, R. F., Veber, D. F., Vitali, R. A., Varga, S. L., Jacob, T. A., Holly, F. W., and Denkewalter, R. G. (1969). *J. Am. Chem. Soc.* **91**, 507–508, and preceding four communications.
Hirt, J., de Leer, E. W., and Beyerman, H. C. (1973). *In* "The Chemistry of Polypeptides" (P. G. Katsoyannis, ed.), pp. 363–387. Plenum, New York.
Hjertén, S. (1971). *In* "New Techniques in Amino Acids, Peptides and Proteins" (A. Niederwieser and G. Pataki, eds.), pp. 227–247. Ann Arbor Sci. Publ., Ann Arbor, Michigan.
Hiskey, R. G., and Harpold, M. A. (1968). *J. Org. Chem.* **33**, 559–563.
Hiskey, R. G., and Smith, R. L. (1968). *J. Am. Chem. Soc.* **90**, 2677–2681.
Hiskey, R. G., and Sparrow, J. T. (1970). *J. Org. Chem.* **35**, 215–220.
Hiskey, R. G., and Tucker, W. P. (1962). *J. Am. Chem. Soc.* **84**, 4794–4797.
Hiskey, R. G., and Ward, B. F. (1970). *J. Org. Chem.* **35**, 1118–1121.
Hiskey, R. G., Mizoguchi, T., and Igeta, H. (1966a). *J. Org. Chem.* **31**, 1188–1192.
Hiskey, R. G., Mizoguchi, T., and Inui, T. (1966b). *J. Org. Chem.* **31**, 1192–1195.
Hiskey, R. G., Mizoguchi, T., and Smithwick, E. L. (1967). *J. Org. Chem.* **32**, 97–102.
Hiskey, R. G., Upham, R. A., Beverly, G. M., and Jones, W. C. (1970). *J. Org. Chem.* **35**, 513–515.
Hiskey, R. G., Beacham, L. M., Matl, V. G., Smith, J. N., Williams, E. B., Thomas, A. M., and Wolters, E. T. (1971). *J. Org. Chem.* **36**, 488–490.
Hiskey, R. G., Rao, V. R., and Rhodes, W. G. (1973). *In* "Protective Groups in Organic Chemistry" (J. F. W. McOmie, ed.), pp. 235–308. Plenum, New York.

Hiskey, R. G., Muthukumaraswamy, N., and Vunnam, R. R. (1975). *J. Org. Chem.* **40**, 950–952.
Hodges, R. S., and Merrifield, R. B. (1974). *J. Org. Chem.* **39**, 1870–1872.
Hodges, R. S., and Merrifield, R. B. (1975a). *Anal. Biochem.* **65**, 241–272.
Hodges, R. S., and Merrifield, R. B. (1975b). *J. Biol. Chem.* **250**, 1231–1241.
Hofmann, K., and Yajima, H. (1961). *J. Am. Chem. Soc.* **83**, 2289–2293.
Hofmann, K., Jöhl, A., Furlenmeier, A. E., and Kappeler, H. (1957). *J. Am. Chem. Soc.* **79**, 1636–1641.
Hofmann, K., Haas, W., Smithers, M. J., Wells, R. D., Wolman, Y., Yanaihara, N., and Zanetti, G. (1965a). *J. Am. Chem. Soc.* **87**, 620–631.
Hofmann, K., Haas, W., Smithers, M. J., and Zanetti, G. (1965b). *J. Am. Chem. Soc.* **87**, 631–639.
Hofmann, K., Finn, F. M., Limett, M., Montibeller, J., and Zanetti, G. (1966). *J. Am. Chem. Soc.* **88**, 3633–3639.
Holland, G. F., and Cohen, L. A. (1958). *J. Am. Chem. Soc.* **80**, 3765–3769.
Holley, R. W., and Holley, A. D. (1952). *J. Am. Chem. Soc.* **74**, 3069–3074.
Hollitzer, O., Seewald, S., and Steglich, W. (1976). *Angew. Chem., Int. Ed. Engl.* **15**, 444–445.
Honzl, J., and Rudinger, J. (1961). *Collect. Czech. Chem. Commun.* **26**, 2333–2344.
Hope, D. B., Murti, V. V., and du Vigneaud, V. (1962). *J. Biol. Chem.* **237**, 1563–1566.
Horiki, K. (1977a). *Tetrahedron Lett.* pp. 1897–1900.
Horiki, K. (1977b). *Tetrahedron Lett.* pp. 1901–1904.
Horiki, K., Igano, K., and Inouye, K. (1978). *Chem. Lett.* pp. 165–168.
Horner, L., and Kaiser, P. (1959). *Justus Liebigs Ann. Chem.* **626**, 19–25.
Horvath, C. G., Preiss, B. A., and Lipsky, S. R. (1967). *Anal. Chem.* **39**, 1422–1428.
Houghten, R. A., and Li, C. H. (1977). *In* "Peptides: Proceedings of the Fifth American Peptide Symposium" (M. Goodman and J. Meienhofer, eds.), pp. 458–460. Wiley, New York.
Houghten, R. A., and Li, C. H. (1978). *Int. J. Pept. Protein Res.* **11**, 325–326.
Houghten, R. A., and Li, C. H. (1979). *Anal. Biochem.* **98**, 36–40.
Hruby, V. J., and Ehler, K. W. (1970). *J. Org. Chem.* **35**, 1690.
Hruby, V. J., Muscio, F. A., Brown, W., and Gitu, P. M. (1972). *In* "Chemistry and Biology of Peptides" (J. Meienhofer, ed.), pp. 331–333. Ann Arbor Sci. Publ., Ann Arbor, Michigan.
Hruby, V. J., Muscio F. A., Groginsky, C. M., Gitu, P. M., Saba, D., and Chan, W. Y. (1973). *J. Med. Chem.* **16**, 624–629.
Hruby, V. J., Upson, D. A., and Agarwal, N. S. (1977). *J. Org. Chem.* **42**, 3552–3556.
Hubbuch, A., Dahno, W., and Zahn, H. (1977). *In* "Peptides: Proceedings of the Fifth American Peptide Symposium" (M. Goodman and J. Meienhofer, eds.), pp. 540–542. Wiley, New York.
Hudson, D., Kenner, G. W., Sharpe, R., and Szelke, M. (1979). *Int. J. Pept. Protein Res.* **14**, 177–185.
Hughes, E. D., and Ingold, C. K. (1933). *J. Chem. Soc.* pp. 1571–1576.
Hugli, T. E., and Moore, S. (1972). *J. Biol. Chem.* **247**, 2828–2834.
Hugli, T. E., and Erickson, B. W. (1977). *Proc. Natl. Acad. Sci. U.S.A.* **74**, 1826–1830.
Hünig, S., and Kiessel, M. (1958). *Chem. Ber.* **91**, 380–392.
Ing, H. R., and Manske, R. H. F. (1926). *J. Chem. Soc.* pp. 2348–2351.
Ingwall, R. T., Scheraga, H. A., Lotan, N., Berger, A., and Katchalski, E. (1968). *Biopolymers* **6**, 331–368.
Inman, J. K., and Dintzis, H. M. (1969). *Biochemistry* **8**, 4074–4082.
Inouye, K., and Otsuka, H. (1962). *J. Org. Chem.* **27**, 4236–4243.
Inukai, N., Nakano, K., and Murakami, M. (1967). *Bull. Chem. Soc. Jpn.* **40**, 2913–2918.
Inukai, N., Nakano, K., and Murakami, M. (1968). *Bull. Chem. Soc. Jpn.* **41**, 182–186.

Irie, H., Fujii, N., Ogawa, H., Yajima, H., Fujino, M., and Shinagawa, S. (1977). *Chem. Pharm. Bull.* **25**, 2929–2934.

Iselin, B. (1961). *Helv. Chim. Acta* **44**, 61–78.

Iselin, B. (1962). *Helv. Chim. Acta* **45**, 1510–1515.

Iselin, B. (1966). *In* "Peptides" (L. Zervas, ed.), pp. 27–33. Pergamon Press, Oxford and New York.

Iselin, B., and Schwyzer, R. (1962). *Helv. Chim. Acta* **45**, 1499–1509.

Ito, H., Takamatsu, N., and Ichikizaki, I. (1977). *Chem. Lett.* pp. 539–542.

Ivanov, V. T., Mikhaleva, I. I., Volpina, O. M., Myagkova, M. A., and Deigin, V. I. (1976). *In* "Peptides 1976" (A. Loffet, ed.), pp. 219–231. Editions de l'Université de Bruxelles, Belgium.

Iwai, K., and Ando, T. (1967). *In* "Methods in Enzymology" (C. H. W. Hirs, ed.), Vol. 11, pp. 263–282. Academic Press, New York.

Izumiya, N., Muraoka, M., and Aoyagi, H. (1971). *Bull. Chem. Soc. Jpn.* **44**, 3391–3395.

Izumiya, N., Waki, M., Kato, T., Ohno, M., Aoyagi, H., and Mitsuyasu, N. (1972). *In* "Chemistry and Biology of Peptides" (J. Meienhofer, ed.), pp. 269–279. Ann Arbor Sci. Publ., Ann Arbor, Michigan.

Jaeger, E., Thamm, P., Knof, S., Wünsch, E., Löw, M., and Kisfaludy, L. (1978a). *Hoppe-Seyler's Z. Physiol. Chem.* **359**, 1617–1628.

Jaeger, E., Thamm, P., Knof, S., and Wünsch, E. (1978b). *Hoppe-Seyler's Z. Physiol. Chem.* **359**, 1629–1636.

Jäger, G., and Geiger, R. (1970). *Chem. Ber.* **103**, 1727–1747.

Jäger, G., Geiger, R., and Siedel, W. (1968). *Chem. Ber.* **101**, 2762–2770.

Jakubke, H. D., and Baumert, A. (1974a). *J. Prakt. Chem.* **316**, 67–74.

Jakubke, H. D., and Baumert, A. (1974b). *J. Prakt. Chem.* **316**, 241–248.

Jakubke, H. D., and Klessen, C. (1977). *J. Prakt. Chem.* **319**, 159–162.

Jendrek, J. P., Barker, R. H., and Altschul, A. M. (1967). *Biochim. Biophys. Acta* **136**, 409–411.

Jensen, E. V. (1959). *Science* **130**, 1319–1323.

Jeschkeit, H., Losse, G., and Neubert, K. (1966). *Chem. Ber.* **99**, 2803–2812.

Johnson, B. J. (1973). *J. Pharm. Sci.* **62**, 1019–1021.

Johnson, C. R., Sharp, J. C., and Phillips, W. G. (1967). *Tetrahedron Lett.* pp. 5299–5302.

Jones, D. A. (1977). *Tetrahedron Lett.* pp. 2853–2856.

Jones, J. H., and Ramage, W. I. (1978). *J. Chem. Soc., Chem. Commun.* pp. 472–473.

Jones, W. C., Rothgeb, T. M., and Gurd, F. R. (1976). *J. Biol. Chem.* **251**, 7452–7460.

Jorgensen, E. C., Windridge, G. C., Patton, W., and Lee, T. C. (1969). *J. Med. Chem.* **12**, 733–737.

Jorgensen, E. C., Windridge, G. C., and Lee, T. C. (1970). *J. Med. Chem.* **13**, 352–356.

Jorgensen, E. C., Windridge, G. C., and Lee, T. C. (1971a). *J. Med. Chem.* **14**, 631–633.

Jorgensen, E. C., Rapaka, S. R., Windridge, G. C., and Lee, T. C. (1971b). *J. Med. Chem.* **14**, 904–906.

Jorgensen, E. C., Windridge, G. C., Hsieh, K. H., and Lee, T. C. (1972). *In* "Chemistry and Biology of Peptides" (J. Meienhofer, ed.), pp. 513–520. Ann Arbor Sci. Publ., Ann Arbor, Michigan.

Juillerat, M., and Bargetzi, J. P. (1976). *Helv. Chim. Acta* **59**, 855–866.

Jung, G., Bovermann, G., Göhring, W., and Heusel, G. (1975). *In* "Peptides: Chemistry, Structure and Biology" (R. Walter and J. Meienhofer, eds.), pp. 433–437. Ann Arbor Sci. Publ., Ann Arbor, Michigan.

Kaiser, E., Colescott, R. L., Bossinger, C. D., and Cook, P. I. (1970). *Anal. Biochem.* **34**, 595–598.

Kamber, B. (1971a). *Helv. Chim. Acta* **54**, 398–422.

Kamber, B. (1971b). *Helv. Chim. Acta* **54**, 927–930.

Kamber. B. (1973). *Helv. Chim. Acta* **56**, 1370–1381.

Kamber, B., and Rittel, W. (1968). *Helv. Chim. Acta* **51**, 2061–2064.
Kamber, B., Riniker, B., Sieber, P., and Rittel, W. (1976). *Helv. Chim. Acta* **59**, 2830–2840.
Kappeler, H., and Schwyzer, R. (1960). *Helv. Chim. Acta* **43**, 1453–1459.
Karlsson, S., Lindeberg, G., and Ragnarsson, U. (1970a). *Acta Chem. Scand.* **24**, 337–338.
Karlsson, S., Lindeberg, G., Porath, J., and Ragnarsson, U. (1970b). *Acta Chem. Scand.* **24**, 1010–1014.
Kashelikar, D. V., and Ressler, C. (1964). *J. Am. Chem. Soc.* **86**, 2467–2473.
Katakai, R. (1975). *J. Org. Chem.* **40**, 2697–2702.
Katchaski, E., and Sela, M. (1958). *Adv. Protein Chem.* **13**, 243–492.
Katchalski, E., Silman, I., and Goldman, R. (1971). *Adv. Enzymol. Relat. Areas Mol. Biol.* **34**, 445–536.
Katsoyannis, P. G. (1961). *J. Am. Chem. Soc.* **83**, 4053–4057.
Katsoyannis, P. G. (1966). *Am. J. Med.* **40**, 652–661.
Katsoyannis, P. G., Gish, G. T., Hess, G. P., and du Vigneaud, V. (1958). *J. Am. Chem. Soc.* **80**, 2558–2562.
Katsoyannis, P. G., Fukuda, K., Tometsko, A. M., Suzuki, K., and Tilak, M. A. (1964). *J. Am. Chem. Soc.* **86**, 930–932.
Katz, J. J. (1954). *Arch. Biochem. Biophys.* **51**, 293–305.
Keil, B., Zikàn, J., Rexovà, L., and Šorm, F. (1962). *Collect. Czech. Chem. Commun.* **27**, 1678–1686.
Kemp, D.S., and Chien, S. W. (1967). *J. Am. Chem. Soc.* **89**, 2743–2745.
Kemp, D. S., and Hoyng, C. F. (1975). *Tetrahedron Lett.* pp. 4624–4628.
Kemp, D. S., and Reczek, J. (1977). *Tetrahedron Lett.* pp. 1031–1034.
Kemp, D. S., and Vellaccio, F. (1975). *J. Org. Chem.* **40**, 3003–3004.
Kemp, D. S., Roberts, D., Hoyng, C., Grattan, J., Vellaccio, F., and Reczek, J. (1975). *In* "Peptides: Chemistry, Structure and Biology" (R. Walter and J. Meienhofer, eds.), pp. 295–305. Ann Arbor Sci. Publ., Ann Arbor, Michigan.
Kenner, G. W. (1977). *Proc. R. Soc. London* **353**, 441–457.
Kenner, G. W., and Seely, J. H. (1972). *J. Am. Chem. Soc.* **94**, 3259–3260.
Kenner, G. W., McDermott, J. R., and Sheppard, R. C. (1971). *J. Chem. Soc. D* pp. 636–637.
Kenner, G. W., Moore, G. A., and Ramage, R. (1976). *Tetrahedron Lett.* pp. 3623–3626.
Kent, S. B. H., and Merrifield, R. B. (1978). *Isr. J. Chem.* **17**, 243–247.
Kent, S. B. H., Mitchell, A. R., Barany, G., and Merrifield, R. B. (1978). *Anal. Chem.* **50**, 155–159.
Kent, S. B. H., Mitchell, A. R., Engelhard, M., and Merrifield, R. B. (1979). *Proc. Natl. Acad. Sci. U.S.A.* **76**, 2180–2184.
Kessler, W., and Iselin, B. (1966). *Helv. Chim. Acta* **49**, 1330–1344.
Keutmann, H. T., and Potts, J. T. (1969). *Anal. Biochem.* **29**, 175–185.
Khairallah, P. A., Bumpus, F. M., Page, J. H., and Smeby, R. R. (1963). *Science* **140**, 672–674.
Khan, S. A., and Sivanandaiah, K. M. (1976). *Synthesis* pp. 614–615.
Khan, S. A., and Sivanandaiah, K. M. (1978a). *Int. J. Pept. Protein Res.* **12**, 164–169.
Khan, S. A., and Sivanandaiah, K. M. (1978b). *Synthesis* pp. 750–751.
Khorana, H. G. (1953). *Chem. Rev.* **53**, 145–166.
Khorana, H. G. (1955). *Chem. Ind.* (*London*) pp. 1087–1088.
Khosla, M. C., Smeby, R. R., and Bumpus, F. M. (1967). *Science* **156**, 253–254.
Khosla, M. C., Chaturvedi, N. C., Smeby, R. R., and Bumpus, F. M. (1968). *Biochemistry* **7**, 3417–3421.
Khosla, M. C., Smeby, R. R., and Bumpus, F. M. (1972a). *In* "Progress in Peptide Research" (S. Lande, ed.), pp. 41–47. Gordon & Breach, New York.
Khosla, M. C., Smeby, R. R., and Bumpus. F. M. (1972b). *In* "Chemistry and Biology of Peptides" (J. Meienhofer, ed.), pp. 227–230. Ann Arbor Sci. Publ., Ann Arbor, Michigan.

Khosla, M. C., Smeby R. R. and Bumpus, F. M. (1972c). *J. Am. Chem. Soc.* **94**, 4721–4724.

Khosla, M. C., Hall, M. M., Smeby, R. R., and Bumpus, F. M. (1973). *J. Med. Chem.* **16**, 829–832.

Khosla, M. C., Hall, M. M., Smeby, R. R., and Bumpus, F. M. (1974). *J. Med. Chem.* **17**, 1156–1160.

Khosla, M. C., Mũnoz-Ramírez, H., Hall, M. M., Smeby, R. R., Khairallah, P.A., Bumpus, F. M., and Peach, M. J. (1976). *J. Med. Chem.* **19**, 244–250.

King, F. E., and Kidd, D. A. (1949). *J. Chem. Soc.* pp. 3315–3319.

Kirby, T. W. and Warme, P. K. (1978). *Anal. Biochem.* **85**, 367–376.

Kisfaludy, L., Schön, I., Rényei, M., and Görög, S. (1975). *J. Am. Chem. Soc.* **97**, 5588–5589.

Kiso, Y., Kai, Y., and Yajima, H. (1973). *Chem. Pharm. Bull.* **21**, 2507–2510.

Kiso, Y., Isawa, H., Kitagawa, K., and Akita, T. (1978). *Chem. Pharm, Bull.* **26**, 2562–2564.

Klieger, E., Schröder, E., and Gibian, H. (1961). *Justus Liebigs Ann. Chem.* **640**, 157–167.

Klostermeyer, H., and Schwertner, E. (1973). *Z. Naturforsch. Teil B* **28**, 334–338.

Kollonitsch, J., Gábor, V., and Hajós, A. (1956a). *Chem. Ber.* **83**, 2288–2292.

Kollonitsch, J., Hajós, A., and Gabár, V. (1956b). *Chem. Ber.* **83**, 2293–2301.

König, W., and Geiger, R. (1970a). *Chem. Ber.* **103**, 788–798.

König, W., and Geiger, R. (1970b). *Chem. Ber.* **103**, 2024–2033.

König, W., and Geiger, R. (1970c). *Chem. Ber.* **103**, 2034–2040.

König, W., and Geiger, R. (1970d). *Chem. Ber.* **103**, 2041–2051.

König, W., and Geiger, R. (1972a). *In* "Chemistry and Biology of Peptides" (J. Meienhofer, ed.) pp. 343–350. Ann Arbor Sci. Publ., Ann Arbor, Michigan.

König, W., and Geiger, R. (1972b). *Chem. Ber.* **105**, 2872–2882.

König, W., and Geiger, R. (1973). *Chem. Ber.* **106**, 3626–3635.

König, W., and Volk, A. (1977). *Chem. Ber.* **110**, 1–11.

Köster, H., and Heidmann, W. (1976). *Angew. Chem., Int. Ed. Engl.* **15**, 546–548.

Kovacs, J., Giannotti, R., and Kapoor, A. (1966). *J. Am. Chem. Soc.* **88**, 2282–2292.

Kovacs, J., Mayers, G. L., Johnson, R. H., and Ghatak, U. R. (1968). *Chem. Commun.* pp. 1066–1067.

Kovács, K., and Penke, B. (1973). *In* "Peptides 1972" (H. Hanson and H. D. Jakubke, eds.), pp. 187–188. North-Holland Publ., Amsterdam.

Kraus, M. A., and Patchornik, A. (1974). *J. Polym. Sci., Polym. Symp.* **47**, 11–18.

Kricheldorf, H. R., and Fehrle, M. (1974). *Chem. Ber.* **107**, 3533–3547.

Kricheldorf, H. R., and Greber, G. (1971). *Chem. Ber.* **104**, 3168–3175.

Krieger, D. E., Erickson, B. W., and Merrifield, R. B. (1976). *Proc. Natl. Acad. Sci. U.S.A.* **73**, 3160–3164.

Krumdieck, C. L., and Baugh, C. M. (1969). *Biochemistry* **8**, 1568–1572.

Kuhn, R., and Haas, H. J. (1955). *Angew. Chem.* **67**, 785.

Kullmann, W., and Gutte, B. (1978). *Int. J. Pept. Protein Res.* **12**, 17–26.

Kung, Y. T., Du, Y. C., Huang, W. T., Chen, C. C., Ke, L. T., Hu, S. C., Jiang, R. Q., Chu, S. Q., Niu, C. I., Hsu, J. Z., Chang, W. C., Cheng, L. L., Li, H. S., Wang, Y., Loh, T. P., Chi, A. H., Li, C. H., Shi, P. T., Yieh, Y. H., Tang, K. L., and Hsing, C. Y. (1965). *Sci. Sin.* **14**, 1710–1716.

Kunin, R., Meitzner, E., and Bortnick, N. (1962). *J. Am. Chem. Soc.* **84**, 305–306.

Kunz, H. (1976a). *Chem. Ber.* **109**, 2670–2683.

Kunz, H. (1976b). *Chem. Ber.* **109**, 3693–3706.

Kunz, H. (1976c). *Justus Liebigs Ann. Chem.* pp. 1674–1679.

Künzi, H., and Studer, R. O. (1975). *Helv. Chim. Acta* **58**, 139–146.

Künzi, H., Manneberg, M., and Studer, R. O. (1974). *Helv. Chim. Acta* **57**, 566–572.

Kusch, P. (1966). *Kolloid-Z. Z. Polym.* **208**, 138–142.

Larsson, L. E., Melin, P., and Ragnarsson, U. (1976). *Int. J. Pept. Protein Res.* **8**, 39–44.

Laursen, R. A. (1971). *Eur. J. Biochem.* **20**, 89–102.

Laursen, R. A., ed. (1975). "Solid-Phase Methods in Protein Sequence Analysis." Pierce Chemical Company, Rockford, Illinois.

Lavine, T. F., Floyd, N. F., and Cammaroti, M. S. (1954). *J. Biol. Chem.* **207**, 107–117.

Le Corre, G., Guibé-Jampel, E., and Wakselman, M. (1978). *Tetrahedron* **34**, 3105–3112.

Ledger, R., and Stewart, F. H. (1965). *Aust. J. Chem.* **18**, 1477–1484.

Lemaire, S., Yamashiro, D., and Li, C. H. (1976). *J. Med. Chem.* **19**, 373–376.

Lemaire, S., Yamashiro, D., Behrens, C., and Li, C. H. (1977). *J. Am. Chem. Soc.* **99**, 1577–1580.

Lenard, J. (1967). *J. Org. Chem.* **32**, 250–251.

Lenard, J., and Hess, G. P. (1964). *J. Biol. Chem.* **239**, 3275–3281.

Lenard, J., and Robinson, A. B. (1967). *J. Am. Chem. Soc.* **89**, 181–182.

Lenard, J., Schally, A. V., and Hess, G. P. (1964). *Biochem. Biophys. Res. Commun.* **14**, 498–502.

Lerman, L. S. (1953). *Proc. Natl. Acad. Sci. U.S.A.* **39**, 232–236.

Letsinger, R. L., and Kornet, M. J. (1963). *J. Am. Chem. Soc.* **85**, 3045–3046.

Letsinger, R. L., and Mahadevan, V. (1966). *J. Am. Chem. Soc.* **88**, 5319–5324.

Letsinger, R. L., Kornet, M. J., Mahadevan, V., and Jerina, D. M. (1964). *J. Am. Chem. Soc.* **86**, 5163–5165.

Leznoff, C. C. (1974). *Chem. Soc. Rev.* **3**, 65–85.

Li, C. H., and Yamashiro, D. (1970). *J. Am. Chem. Soc.* **92**, 7608–7609.

Li, C. H., Yamashiro, D., and Lemaire, S. (1975). *Biochemistry* **14**, 953–956.

Li, C. H., Lemaire, S., Yamashiro, D., and Doneen, B. A. (1976). *Biochem. Biophys. Res. Commun.* **71**, 19–25.

Li, C. H., Yamashiro, D., Tseung, L. F., and Loh, H. H. (1977). *J. Med. Chem.* **20**, 325–328.

Liberek, B. (1962). *Bull. Acad. Pol. Sci. Ser. Sci. Chem.* **10**, 227–231.

Lieflander, M. (1960). *Hoppe Seyler's Z. Physiol. Chem.* **320**, 35–57.

Lin, M. C., Gutte, B., Caldi, D. G., Moore, S., and Merrifield, R. B. (1972). *J. Biol. Chem.* **247**, 4768–4774.

Lindeberg, G., Melin, P., and Larsson, L. E. (1976). *Int. J. Pept. Protein Res.* **8**, 193–198.

Lindeberg, G., Tengborn, J., Bennich, H., and Ragnarsson, U. (1978). *J. Chromatography* **156**, 366–369.

Liu, T. Y., and Chang, Y. H. (1971). *J. Biol. Chem.* **246**, 2842–2848.

Live, D. H., Agosta, W. C., and Cowburn, D. (1977). *J. Org. Chem.* **42**, 3556–3561.

Loffet, A. (1967). *Experientia* **23**, 406–407.

Loffet, A. (1971). *Int. J. Protein Res.* **3**, 297–299.

Loffet, A. (1973). *In* "Peptides 1972" (H. Hanson and H. D. Jakubke, eds.), pp. 242–244. North-Holland Publ., Amsterdam.

Loffet, A., and Close, J. (1968). *In* "Peptides 1968" (E. Bricas, ed.), pp. 189–193. North-Holland Publ., Amsterdam.

Loffet, A., and Dremier, C. (1971). *Experientia* **27**, 1003.

Losse, A. (1973). *Tetrahedron* **29**, 1203–1208.

Losse, G., and Meisegeier, B. (1974). *J. Prakt. Chem.* **316**, 629–637.

Losse, G., and Neubert, K. (1968a). *Z. Chem.* **8**, 228–229.

Losse, G., and Neubert, K. (1968b). *Z. Chem.* **8**, 387.

Losse, G., and Neubert, K. (1970). *Tetrahedron Lett.* pp. 1267–1270.

Löw, M., and Kisfaludy, L. (1979). *Hoppe-Seyler's Z. Physiol. Chem.* **360**, 13–18.

Löw, M., Kisfaludy, L., and Sohár, P. (1978b). *Hoppe-Seyler's Z. Physiol. Chem.* **359**, 1643–1651.

Löw, M., Kisfaludy, L., Jaeger, E., Thamm, P., Knof, S., and Wünsch, E. (1978a). *Hoppe-Seyler's Z. Physiol. Chem.* **359**, 1637–1642.

Lowry, O. H., Rosebrough, N. J., Farr, A. L., and Randall, R. J. (1951). *J. Biol. Chem.* **193**, 265–275.

Lundt, B. F., Johansen, N. L., Vølund, A., and Markussen, J. (1978). *Int. J. Pept. Protein Res.* **12**, 258–268.

Lunkenheimer, W., and Zahn, H. (1970). *Justus Liebigs Ann. Chem.* **740**, 1–17.

McCarthy, D. G., Hegarty, A. F., and Hathaway, B. J. (1977). *J. Chem. Soc., Perkin Trans. 2* pp. 224–231.

McDermott, J. R., and Benoiton, N. L. (1973a). *Can. J. Chem.* **51**, 2555–2561.

McDermott, J. R., and Benoiton, N. L. (1973b). *Can. J. Chem.* **51**, 2562–2570.

McKay, F. C., and Albertson, N. F. (1957). *J. Am. Chem. Soc.* **79**, 4686–4690.

McKerrow, J. H., and Robinson, A. B. (1971). *Anal. Biochem.* **42**, 565–568.

MacLaren, J. A., Savige, W. E., and Swan, J. M. (1958). *Aust. J. Chem.* **11**, 345–359.

Maher, J. J., Furey, M. E., and Greenberg, L. J. (1972). *Tetrahedron Lett.* pp. 1581–1584.

Maki, Y., Sugiyama, H., Kikuchi, K., and Seto, S. (1975). *Chem. Pharm. Bull.* **23**, 929–931.

Manning, J. M., and Moore, S. (1968). *J. Biol. Chem.* **243**, 5591–5597.

Manning, J. M., Marglin, A., and Moore, S. (1972). *In* "Progress in Peptide Research" (S. Lande, ed.), pp. 173–183. Gordon & Breach, New York.

Manning, M. (1968). *J. Am. Chem. Soc.* **90**, 1348–1349.

Manning, M. (1976). *In* "Encyclopedia of Polymer Science and Technology" Suppl. Vol. 1, pp. 492–510. Wiley, New York.

Manning, M., Coy, E., and Sawyer, W. H. (1970). *Biochemistry* **9**, 3925–3930.

Marbach, P., and Rudinger, J. (1974). *Helv. Chim. Acta* **57**, 403–414.

Marche, P., Girma, J. P., Morgat, J. L., and Fromageot, P. (1975). *Eur. J. Biochem.* **50**, 375–382.

Marglin, A. (1971). *Tetrahedron Lett.* pp. 3145–3148.

Marglin, A. (1972). *Int. J. Protein Res.* **4**, 47–55.

Marglin, A., and Merrifield, R. B. (1966). *J. Am. Chem. Soc.* **88**, 5051–5052.

Marinier, B., Kim, Y. C., and Navarre, J. M. (1973). *Can. J. Chem.* **51**, 208–214.

Markley, L. D., and Dorman, L. C. (1970). *Tetrahedron Lett.* pp. 1787–1790.

Marnett, L. J., Neckers, D. C., and Schaap, A. P. (1976). *In* "Application of Biomedical Systems in Chemistry, Part II" (J. B. Jones, ed.), pp. 995–1044. Wiley, New York.

Marshall, D. L., and Liener, I. E. (1970). *J. Org. Chem.* **35**, 867–868.

Marshall, G. R. (1968). *In* "Pharmacology of Hormonal Polypeptides and Proteins" (N. Back, L. Martini, and R. Paoletti, eds.), pp. 48–52. Plenum, New York.

Marshall, G. R., and Merrifield, R. B. (1965). *Biochemistry* **4**, 2394–2400.

Marshall, G. R., and Merrifield, R. B. (1971). *In* "Biochemical Aspects of Reactions on Solid Supports" (G. R. Stark, ed.), pp.111–169. Academic Press, New York.

Martin, G. E., Shambhu M. B., Shakhshir, S. R., and Digenis, G. A. (1978). *J. Org. Chem.* **43**, 4571–4574.

Martinez, J., and Bodanszky, M. (1978). *Int. J. Pept. Protein Res.* **12**, 277–283.

Maruyama, H., Matsueda, R., Kitazawa, E., Takahagi, H., and Mukaiyama, T. (1976). *Bull. Chem. Soc. Jpn.* **49**, 2259–2267.

Masaki, M., Kitahara, T., Kurita, H., and Ohta, M. (1968). *J. Am. Chem. Soc.* **90**, 4508–4509.

Masui, T., Chino, N., Sakakibara, S., Tanaka, T., and Kita, H. (1979). *In* "Peptide Chemistry 1978" (N. Izumiya, ed.), pp. 189–194. Protein Res. Found, Osaka, Japan.

Mathews, H. R. and Rapoport, H. (1973). *J. Am. Chem. Soc.* **95**, 2297–2303.

Matsueda, G. R., and Stewart, J. M. (1975). *In* "Peptides: Chemistry, Structure and Biology" (R. Walter and J. Meienhofer, eds.), pp. 333–339. Ann Arbor Sci. Publ. Ann Arbor, Michigan.

Matsueda, R., Kitazawa, E., Maruyama, H., Takahagi, H., and Mukaiyama, T. (1972). *Chem. Lett.* pp. 379–384.

Matsueda, R., Maruyama, H., Kitazawa, E., Takahagi, H., and Mukaiyama, T. (1973). *Bull. Chem. Soc. Jpn.* **46**, 3240–3247.

Matsueda, R., Maruyama, H., Kitazawa, E., Takahagi, H., and Mukaiyama, T. (1975). *J. Am. Chem. Soc.* **97**, 2573–2575.
Matsueda, R., Takahagi, H., and Mukaiyama, T. (1977). *Chem. Lett.* pp. 719–722.
Matsuo, H., Arimura, A., Nair, R. M., and Schally, A. V. (1971). *Biochem. Biophys. Res. Commun.* **45**, 822–827.
Matsuura, S., Niu, C. H., and Cohen, J. S.(1976). *J. Chem. Soc., Chem. Commun.* pp. 451–452.
Mazur, R. H., and Plume, G. (1968). *Experientia* **24**, 661.
Mazur, S. and Jayalekshmy, P. (1979). *J. Am. Chem. Soc.* **101**, 677–683.
Medzihradszky, K., and Rádóczy, J. (1966). *In* "Peptides" (L. Zervas, ed.), pp. 49–54. Pergamon Press, Oxford and New York.
Meienhofer, J. (1973). *In* "Hormonal Proteins and Peptides" (C. H. Li, ed.), Vol. 2, pp. 45–267. Academic Press, New York.
Meienhofer, J., and Kuromizu, K. (1974). *Tetrahedron Lett.* pp. 3259–3262.
Meienhofer, J., and Trzeciak, A. (1971). *Proc. Natl. Acad. Sci. U.S.A.* **68**, 1006–1009.
Meienhofer, J., Schnabel, E., Bremer, H., Brinkhoff, O., Zabel, R., Sroka, W., Klostermeyer, H., Brandenburg, D., Okuda, T., and Zahn, H. (1963). *Z. Naturforsch., Teil B* **18**, 1120–1121.
Meienhofer, J., Jacobs, P. M., Godwin, H. A., and Rosenberg, I. H. (1970a). *J. Org. Chem.* **35**, 4137–4140.
Meienhofer, J., Trzeciak, A., Havran, R. T., and Walter, R. (1970b). *J. Am. Chem. Soc.* **92**, 7199–7202.
Merrifield, R. B. (1962). *Fed. Proc., Fed. Am. Soc. Exp. Biol.* **21**, 412.
Merrifield, R. B. (1963). *J. Am. Chem. Soc.* **85**, 2149–2154.
Merrifield, R. B. (1964a). *J. Am. Chem. Soc.* **86**, 304.
Merrifield, R. B. (1964b). *Biochemistry* **3**, 1385–1390.
Merrifield, R. B. (1964c). *J. Org. Chem.* **29**, 3100–3102.
Merrifield, R. B. (1965). *Endeavour* **24**, 3–7.
Merrifield, R. B. (1967). *Recent Prog. Horm. Res.* **23**, 451–482.
Merrifield, R. B. (1969). *Adv. Enzymol. Relat. Areas Mol. Biol.* **32**, 221–296.
Merrifield, R. B. (1973a). *Harvey Lect.* **67**, 143–173.
Merrifield, R. B. (1973b). *In* "The Chemistry of Polypeptides" (P. G. Katsoyannis, ed.), pp. 336–361.Plenum, New York.
Merrifield, R. B. (1973c). Reported at the International Symposium on Peptide Synthesis, Madison, Wisconsin, May 30, 1973.
Merrifield, R. B., and Bach, A. E. (1978). *J. Org. Chem.* **43**, 4808–4816.
Merrifield, R. B., and Corigliano, M. A. (1968). *Biochem. Prep.* **12**, 98–102.
Merrifield, R. B., and Littau, V. (1968). *In* "Peptides 1968" (E. Bricas, ed.), pp. 179–182. North-Holland Publ., Amsterdam.
Merrifield, R. B., and Stewart, J. M. (1965). *Nature (London)* **207**, 522–523.
Merrifield, R. B., and Woolley, D. W. (1956). *J. Am. Chem. Soc.* **78**, 4646–4649.
Merrifield, R. B., Stewart, J. M., and Jernberg, N. (1966). *Anal. Chem.* **38**, 1905–1914.
Merrifield, R. B., Mitchell, A. R., and Clarke, J. E. (1974). *J. Org. Chem.* **39**, 660–668.
Merrifield, R. B., Gisin, B. F., and Bach, A. E. (1977a). *J. Org. Chem.* **42**, 1291–1295.
Merrifield, R. B. Barany, G., Cosand, W. L., Engelhard, M., and Mojsov, S. (1977b). *In* "Peptides: Proceedings of the Fifth American Peptide Symposium" (M. Goodman and J. Meienhofer, eds.), pp. 488–502. Wiley, New York.
Meyers, C. A., and Glass, J. D. (1975). *Proc. Nat'l Acad. Sci. U.S.A.* **72**, 2193–2196.
Meyers, C. A., Havran, R. T., Schwartz, I. L., and Walter, R. (1969). *Chem. Ind. (London)* pp. 136–137.
Meyers, C. A., Coy, D. H., Huang, W. Y., Schally, A. V., and Redding, T. A. (1978). *Biochemistry* **12**, 2326–2331.

Meyers, W. E., and Royer, G. P. (1977). *J. Am. Chem. Soc.* **99**, 6141–6142.
Millar, J. R., Smith, D. G., Marr, W. E., and Kressman, T. R. (1963). *J. Chem. Soc.* pp. 218–225.
Miller, A. W., and Stirling, C. J. (1968). *J. Chem. Soc. C* pp. 2612–2617.
Miller, M. J., DeBons, F. E., and Loudon, G. M. (1977). *J. Org. Chem.* **42**, 1750–1761.
Milne, H. B., and Peng, C. H. (1957). *J. Am. Chem. Soc.* **79**, 639–644.
Mitchell, A. R., and Merrifield, R. B. (1976). *J. Org. Chem.* **41**, 2015–2019.
Mitchell, A. R., Erickson, B. W., Ryabtsev, M. N., Hodges, R. S., and Merrifield, R. B. (1976a). *J. Am. Chem. Soc.* **98**, 7357–7362.
Mitchell, A. R., Kent, S. B. H., Erickson, B. W., and Merrifield, R. B. (1976b). *Tetrahedron Lett.* pp. 3795–3798.
Mitchell, A. R., Kent, S. B. H., Chu, I. C., and Merrifield, R. B. (1978a). *Anal. Chem.* **50**, 637–640.
Mitchell, A. R., Kent, S. B. H., Engelhard, M., and Merrifield, R. B. (1978b). *J. Org. Chem.* **43**, 2845–2852.
Mizoguchi, T., and Woolley, D. W. (1967). *J. Med. Chem.* **10**, 251.
Mizoguchi, T., Levin, G., Woolley, D. W., and Stewart, J. M. (1968). *J. Org. Chem.* **33**, 903–904.
Mizoguchi, T., Shigezane, K., and Takamura, N. (1969). *Chem. Pharm. Bull.* **17**, 411–412.
Mizoguchi, T., Shigezane, K., and Takamura, N. (1970). *Chem. Pharm. Bull.* **18**, 1465–1474.
Mojsov, S., Mitchell, A. R., and Merrifield, R. B. (1979). *J. Org. Chem.*, in press.
Monahan, M. W., and Gilon, C. (1973). *Biopolymers* **12**, 2513–2519.
Moore, G., and McMaster, D. (1978). *Int. J. Pept. Protein Res.* **11**, 140–148.
Moore, S. (1963). *J. Biol. Chem.* **238**, 235–237.
Moore, S. (1972). *In* "Chemistry and Biology of Peptides" (J. Meienhofer, ed.), pp. 629–653. Ann Arbor Sci. Publ. Ann Arbor, Michigan.
Moore, S., and Stein, W. H. (1963). *In* "Methods in Enzymology" (S. P. Colowick and N. O. Kaplan, eds.), Vol. 6, pp. 819–831. Academic Press, New York.
Moore, S., Spackman, D. H., and Stein, W. H. (1958). *Anal. Chem.* **30**, 1185–1190.
Moore, S., Felix, A. M., Meienhofer, J., Smith, C. W., and Walter, R. (1977). *J. Med. Chem.* **20**, 495–499.
Morley, J. S. (1967). *J. Chem. Soc. C* pp. 2410–2421.
Moroder, L., Wackerle, L., and Wünsch, E. (1976). *Hoppe-Seyler's Z. Physiol. Chem.* **357**, 1647–1650.
Mosbach, K., ed. (1976). "Methods in Enzymology," Vol. 44, Academic Press, New York.
Mukaiyama, T., Ueki, M., Maruyama, H., and Matsueda, R. (1968). *J. Am. Chem. Soc.* **90**, 4490–4491.
Mukaiyama, T., Matsueda, R., and Suzuki, M. (1970). *Tetrahedron Lett.* pp. 1901–1904.
Muramatsu, I., Hirabayashi, T., and Hagitani, A. (1963). *Nippon Kagaku Zasshi* **84**, 855–860; *Chem. Abstr.* **60**, 12100c (1964).
Naider, F., and Bohak, Z. (1972). *Biochemistry* **11**, 3208–3211.
Najjar, V. A., and Merrifield, R. B. (1966). *Biochemistry* **5**, 3765–3770.
Nakaie, C. R., Oshiro, M. E., Goissis, G., and Paiva, A. C. (1977). *Biochim. Biophys. Acta* **495**, 151–158.
Needleman, S. B., ed. (1970). "Protein Sequence Determination." Springer-Verlag, Berlin and New York.
Needleman, S. B., ed. (1975). "Protein Sequence Determination." 2nd revised and enlarged edition. Springer-Verlag, Berlin and New York.
Nefkens, G. H., and Tesser, G. I. (1961). *J. Am. Chem. Soc.* **83**, 1263.
Neubert, K., and Jakubke, H. D. (1975). *J. Prakt. Chem.* **317**, 448–454.
Neumann, N. P., Moore, S., and Stein, W. H. (1962). *Biochemistry* **1**, 68–75.
Niall, H. D., Tregear, G. W., and Jacobs, J. (1972). Ann Arbor Sci. Publ., Ann Arbor, Michigan.

Niederwisser, A., and Pataki, G., eds. (1971). "New Techniques in Amino Acids, Peptides, and Proteins," Ann Arbor Sci. Publ., Ann Arbor, Michigan.
Nishimura, O., and Fujino, M. (1976). *Chem. Pharm. Bull.* **24**, 1568–1575.
Nishimura, O., Kitada, C., and Fujino, M. (1978). *Chem. Pharm. Bull.* **26**, 1576–1585.
Nitecki, D. E., Senyk, G., Williams, E. B., and Goodman, J. W. (1971). *Intra-Sci. Chem. Rep.* **5**, 295–303.
Noble, R. L., Yamashiro, D., and Li, C. H. (1976). *J. Am. Chem. Soc.* **98**, 2324–2328.
Noble, R. L., Yamashiro, D., and Li, C. H. (1977). *Int. J. Pept. Protein Res.* **10**, 385–393.
Noda, K., Terada, S., and Izumiya, N. (1970). *Bull. Chem. Soc. Jpn.* **43**, 1883–1885.
Norris, K., Halstrøm. J., and Brunfeldt, K. (1971). *Acta Chem. Scand.* **25**, 945–954.
Oae, S. (1977). *In* "Organic Chemistry of Sulfur" (S. Oae, ed.), pp. 383–471. Plenum, New York.
Oae, S., Kitao, T., and Kawamura, S. (1963). *Tetrahedron* **19**, 1783–1788.
Offord, R. E., Storey, H. T., Rees, A. R., Hayward, C. F., Johnson, W. H., Pheasey, M. H., and Wightman, D. A. (1976). *Biochem. J.* **159**, 480–486.
Ogawa, H., Sasaki, T., Irie, H., and Yajima, H. (1978). *Chem. Pharm. Bull.* **26**, 3144–3149.
Ohnishi, T., Sugano, H., and Miyoshi, M. (1972). *Bull. Chem. Soc. Jpn.* **45**, 2603–2607.
Ohno, M., Kuromizu, K., Ogawa, H., and Izumiya, N. (1971). *J. Am. Chem. Soc.* **93**, 5251–5254.
Ohno, M., Tsukamoto, S., Makisumi, S., and Izumiya, N. (1972). *Bull. Chem. Soc. Jpn.* **45**, 2852–2855.
Ohno, M., Tsukamoto, S., Sato, S., and Izumiya, N. (1973). *Bull. Chem. Soc. Jpn.* **46**, 3280–3285.
Okuda, T., and Zahn, H. (1969). *Makromol. Chem.* **121**, 87–101.
Okawa, K. (1956). *Bull. Chem. Soc. Jpn.* **29**, 486–489.
Olah, G., Brydon, D. L., and Porter, R. D. (1970). *J. Org. Chem.* **35**, 317–327.
Omenn, G. S., and Anfinsen, C. B. (1968). *J. Am. Chem. Soc.* **90**, 6571–6572.
Omori, Y., Matsuda, Y., Aimoto, S., Shimonishi, Y., and Yamamoto, M. (1976). *Chem. Lett.* pp. 805–808.
Ondetti, M. A., Deér, A., Sheehan, J. T., Pluščec, J., and Kocy, O. (1968). *Biochemistry* **7**, 4069–4075.
Ondetti, M. A., Williams, N. J., Sabo, E. F., Pluščec, J., Weaver, E. R., and Kocy, O. (1971). *Biochemistry* **10**, 4033–4039.
Ontjes, D. A., and Anfinsen, C. B. (1969a). *Proc. Natl. Acad. Sci. U.S.A.* **64**, 428–435.
Ontjes, D. A., and Anfinsen, C. B. (1969b). *J. Biol. Chem.* **244**, 6316–6322.
Ontjes, D. A., and Anfinsen, C. B. (1970). *In* "Peptides: Chemistry and Biochemistry" (B. Weinstein and S. Lande, eds.), pp. 79–98. Dekker, New York.
Ontjes, D. A., Roberts, J. C., Hennessy, J. F., Burford, H. J., and Cooper, C. W. (1973). *Endocrinology* **92**, 1780–1785.
Orlowski, R. C., Walter, R., and Winkler, D. (1976). *J. Org. Chem.* **41**, 3701–3705.
Otsuka, H., Inouye, K., Shinozaki, F., and Kanayama, M. (1966). *Bull. Chem. Soc. Jpn.* **39**, 1171–1178.
Overell, B. G., and Petrow, V. (1955). *J. Chem. Soc.* pp. 232–236.
Park, W. K., and Regoli, D. (1972). *Can. J. Biochem.* **50**, 755–757.
Parr, W., Grohmann, K., and Hägele, K. (1974). *Justus Liebigs Ann. Chem.* pp. 655–666.
Pastuszak, J. J., and Chimiak, A. (1977). *Rocz. Chem.* **51**, 1567–1569.
Patchornik, A., and Sokolovsky, M. (1964). *J. Am. Chem. Soc.* **86**, 1206–1212.
Patchornik, A., Berger, A., and Katchalski, E. (1957). *J. Am. Chem. Soc.* **79**, 6416–6420.
Patchornik, A., Lawson, W. B., Gross, E., and Witkop, B. (1960). *J. Am. Chem. Soc.* **82**, 5923–5927.
Patchornik, A., Fridkin, M., and Katchalski, E. (1967). *In* "Peptides" (H. C. Beyerman, A. van de Linde, and W. Maasen van den Brink, eds.), pp. 91–99. North-Holland Publ., Amsterdam.

Patterson, J. A. (1971). *In* "Biochemical Aspects of Reactions on Solid Supports" (G. R. Stark, ed.), pp. 189–213. Academic Press, New York.
Paul, R. (1963). *J. Org. Chem.* **28**, 236–237.
Paul, R., and Kende, A. S. (1964). *J. Am. Chem. Soc.* **86**, 4162–4166.
Paul, R., Anderson, G. W., and Callahan, F. M. (1961). *J. Am. Chem. Soc.* **26**, 3347–3350.
Peña, C., Stewart, J. M., and Goodfriend, T. C. (1974). *Life Sci.* **14**, 1331–1336.
Penke, B., and Birr, C. (1974). *Justus Liebigs Ann. Chem.* pp. 1999–2002.
Penke, B., Ferenczi, R., and Kovacs, K. (1974). *Anal. Biochem.* **60**, 45–50.
Pepper, K. W., Reichenberg, D., and Hale, D. K. (1952). *J. Chem. Soc.* pp. 3129–3136.
Pereira, W., Close, V., Patton, W., and Halpern, B. (1969). *J. Org. Chem.* **34**, 2032–2034.
Perrin, D. D., Armarego, W. L., and Perrin, D. R. (1966). "Purification of Laboratory Chemicals." Pergamon, Oxford.
Pétra, P. H. (1970). *In* "Methods in Enzymology" (G. E. Perlmann and L. Lorand, eds.), Vol. 19, pp. 460–503. Academic Press, New York.
Pettit, G. R. (1976). "Synthetic Peptides," Vol. 4, pp. 22–32. Elsevier, Amsterdam.
Pfitzner, K. E., and Moffat, J. G. (1965). *J. Am. Chem. Soc.* **87**, 5670–5678.
Photaki, I. (1963). *J. Am. Chem. Soc.* **85**, 1123–1126.
Photaki, I. (1970). *J. Chem. Soc. C* pp. 2687–2689.
Photaki, I. (1976). *In* "Topics in Sulfur Chemistry" (A. Senning, ed.), Vol. 1, pp. 111–183. G. Thieme, Stuttgart.
Photaki, I., and Yiotakis, A. (1976). *J. Chem. Soc., Perkin Trans. 1* pp. 259–264.
Photaki, I., Taylor-Papadimitriou, J., Sakarellos, C., Mazarakis, P., and Zervas, L. (1970). *J. Chem. Soc. C* pp. 2683–2687.
Pietta, P. G., and Brenna, O. (1975). *J. Org. Chem.* **40**, 2995–2996.
Pietta, P. G., and Marshall, G. R. (1970). *J. Chem. Soc. D* pp. 650–651.
Pietta, P. G., Chillemi, F., and Corbellini, A. (1968). *Chem. Ber.* **101**, 3649–3651.
Pietta, P. G., Cavallo, P. F., and Marshall, G. R. (1971). *J. Org. Chem.* **36**, 3966–3970.
Pietta, P. G., Cavallo, P. F., and Marshall, G. R. (1973). *In* "Peptides 1971" (H. Nesvadba, ed.), pp. 172–174. North-Holland Publ., Amsterdam.
Pietta, P. G., Cavallo, P. F., Takahashi, K., and Marshall, G. R. (1974). *J. Org. Chem.* **39**, 44–48.
Pietta, P. G., Biondi, P. A., and Brenna, O. (1976). *J. Org. Chem.* **41**, 703–704.
Piszkiewicz, D., Landon, M., and Smith, E. L. (1970). *Biochem. Biophys. Res. Commun.* **40**, 1173–1178.
Pless, J., and Bauer, W. (1973). *Angew. Chem., Int. Ed. Engl.* **12**, 147–148.
Pless, J., and Boissonnas, R. A. (1963). *Helv. Chim. Acta* **46**, 1609–1625.
Pless, J., and Guttmann, S. (1967). *In* "Peptides" (H. C. Beyerman, A. van de Linde, and W. Maasen van den Brink, eds.), pp. 50–54. North-Holland Publ., Amsterdam.
Poduška, K. (1967). *Collect. Czech. Chem. Commun.* **33**, 3779–3789.
Polzhofer, K. P. (1969). *Tetrahedron* **25**, 4127–4132.
Polzhofer, K. P., and Ney, K. H. (1970a). *Tetrahedron* **26**, 3221–3226.
Polzhofer, K. P., and Ney, K. H. (1970b). *Tetrahedron* **26**, 3787–3790.
Polzhofer, K. P., and Ney, K. H. (1971). *Tetrahedron* **27**, 1997–2001.
Prestidge, R. L., Harding, D. R., and Hancock, W. S. (1976). *J. Org. Chem.* **41**, 2579–2583.
Previero, A., Coletti-Previero, M. A., and Cavadore, J. C. (1967). *Biochim. Biophys. Acta* **147**, 453–461.
Protein Synthesis Group, Shanghai Institute of Biochemistry (1975). *Sci. Sin.* **18**, 745–768.
Ragnarsson, U., Karlsson, S., and Lindeberg, G. (1970a). *Acta Chem. Scand.* **24**, 2821–2825.
Ragnarsson, U., Lindeberg, G., and Karlsson, S. (1970b). *Acta Chem. Scand.* **24**, 3079–3080.
Ragnarsson, U., Karlsson, S., and Sandberg, B. E. (1971). *Acta Chem. Scand.* **25**, 1487–1489.
Ragnarsson, U., Karlsson, S., and Sandberg, B. E. (1974). *J. Org. Chem.* **39**, 3837–3842.

Ragnarsson, U., Karlsson, S., and Hamberg, U. (1975). *Int. J. Pept. Protein Res.* **7**, 307–312.
Ramachandran, J., and Li, C. H. (1962). *J. Org. Chem.* **27**, 4006–4009.
Ramachandran, J., and Li, C. H. (1963). *J. Org. Chem.* **28**, 173–177.
Ramirez, F., Finnan, J. L., and Carlson, M. (1973). *J. Org. Chem.* **38**, 2597–2600.
Raschig, A., and Schneider, F. (1976). *Z. Naturforsch., Teil C* **31**, 418–423.
Ray, W. J., and Koshland, D. E. (1962). *J. Biol. Chem.* **237**, 2493–2505.
Rebek, J., and Feitler, D. (1973). *J. Am. Chem. Soc.* **95**, 4052–4053.
Rebek, J., and Feitler, D. (1974). *J. Am. Chem. Soc.* **96**, 1606–1607.
Rebek, J., and Feitler, D. (1975). *Int. J. Pept. Protein Res.* **7**, 167–169.
Rebek, J., and Gaviña, F. (1975a). *J. Am. Chem. Soc.* **97**, 3221–3222.
Rebek, J., and Gaviña, F. (1975b). *J. Am. Chem. Soc.* **97**, 3453–3456.
Rebek, J., and Trend, J. E. (1979). *J. Am. Chem. Soc.* **101**, 737.
Rebek, J., Brown, D., and Zimmerman, S. (1975). *J. Am. Chem. Soc.* **97**, 4407–4408.
Rees, A. R., and Offord, R. D. (1976). *Biochem. J.* **159**, 467–479.
Reese, C. B. (1973). *In* "Protective Groups in Organic Chemistry" (J. F. W. McOmie, ed.), pp. 95–143. Plenum, New York.
Regen, S. L. (1974). *J. Am. Chem. Soc.* **96**, 5275–5276.
Regen, S. L. (1975a). *J. Am. Chem. Soc.* **97**, 3108–3112.
Regen, S. L. (1975b). *Macromolecules* **8**, 689–690.
Regen, S. L. and Dulak, L. (1977). *J. Am. Chem. Soc.* **99**, 623–624.
Regen, S. L., and Lee, D. P. (1974). *J. Am. Chem. Soc.* **96**, 294–296.
Reid, R. E. (1976). *J. Org. Chem.* **41**, 1027–1031.
Ressler, C. (1956). *J. Am. Chem. Soc.* **78**, 5956–5957.
Ressler, C., and Kashelikar, D. V. (1966). *J. Am. Chem. Soc.* **88**, 2025–2035.
Ressler, C., and Ratzkin, H. (1961). *J. Org. Chem.* **26**, 3356–3360.
Rich, D. H., and Gurwara, S. K. (1973). *J. Chem. Soc., Chem. Commun.* pp. 610–611.
Rich, D. H., and Gurwara, S. K. (1975a). *J. Am. Chem. Soc.* **97**, 1575–1579.
Rich, D. H., and Gurwara, S. K. (1975b). *Tetrahedron Lett.* pp. 301–304.
Rich, D. H., and Tam, J. P. (1975). *Tetrahedron Lett.* pp. 211–212.
Riniker, B., and Schwyzer, R. (1961). *Helv. Chim. Acta* **44**, 674–679.
Riniker, B., Kamber, B., and Sieber, P. (1975). *Helv. Chim. Acta* **58**, 1086–1094.
Rink, H., and Riniker, B. (1974). *Helv. Chim. Acta* **57**, 831–836.
Rivaille, P., and Milhaud, G. (1971). *Helv. Chim. Acta* **54**, 355–357.
Rivaille, P., Raulais, D., Merle, M., Gautron, J. P., Pidoux, E., and Milhaud, G. (1979) "Peptides, 1978" (I. Z. Siemion and G. Kupryszewski, eds.) pp. 519–521. Wroclaw Uni. Press, Wroclaw, Poland.
Rivier, J., Vale, W., Burgus, R., Ling, N., Amoss, M., Blackwell, R., and Guillemin, R. (1973). *J. Med. Chem.* **16**, 545–549.
Rivier, J., Brazeau, P., Vale, W., and Guillemin, R. (1975a). *J. Med. Chem.* **18**, 123–126.
Rivier, J., Brown, M., and Vale, W. (1975b). *Biochem. Biophys. Res. Commun.* **65**, 746–751.
Rivier, J., Brown, M., Rivier, C., Ling, N., and Vale, W. (1976). *In* "Peptides 1976" (A. Loffet, ed.), pp. 427–451. Editions de l'Universite de Bruxelles, Belgium.
Rivier, J., Lazarus, L. H., Derrin, M. H., and Brown, M. R. (1977). *J. Med. Chem.* **20**, 1409–1412.
Robinson, A. B. (1974). *Proc. Natl. Acad. Sci. U.S.A.* **71**, 885–888.
Robinson, A. B., and Tedro, S. (1973). *Int. J. Pept. Protein Res.* **5**, 275–278.
Robinson, P. J., Dunnill, P., and Lilly, M. D. (1971). *Biochim. Biophys. Acta* **242**, 659–661.
Rocchi, R., Benassi, C. A., Tomatis, R., Ferroni, R., and Menegatti, E. (1976). *Int. J. Pept. Protein Res.* **8**, 167–175.
Rocchi, R., Scatturin, A., Moroder, L., Marchiori, F., Tamburro, A. M., and Scoffone, E. (1969). *J. Am. Chem. Soc.* **91**, 492–496.

Roeske, R. (1959). *Chem. Ind. (London)* pp. 1121–1122.
Roeske, R. (1963). *J. Org. Chem.* **28**, 1251–1253.
Roeske, R., and Gesellchen, P. D. (1976). *Tetrahedron Lett.* pp. 3369–3372.
Rosamond, J. D., and Ferger, M. F. (1976). *J. Med. Chem.* **19**, 873–876.
Rosenblatt, M., Goltzman, D., Keutmann, H. T., Treager, G. W., and Potts, J. T. (1976). *J. Biol. Chem.* **251**, 159–164.
Rosenblatt, M., Segre, G. V., and Potts, J. T. (1977a). *Biochemistry* **16**, 2811–2816.
Rosenblatt, M., Keutmann, H. T., Tregear, G. W., and Potts, J. T. (1977b). *J. Med. Chem.* **20**, 1452–1456.
Rothe, M., and Mazánek, J. (1974). *Justus Liebigs Ann. Chem.*, pp. 439–459.
Rothe, M., Schneider, H., and Dunkel, W. (1966). *Makromol. Chem.* **96**, 290–294.
Rothe, M., Rott, H., and Mazánek, J. (1976). *In* "Peptides 1976" (A. Loffet, ed.), pp. 309–318. Editions de l'Université de Bruxelles, Belgium.
Rothe, M., Sander, A., Fischer, W., Mästle, W., and Nelson, B. (1977). *In* "Peptides: Proceedings of the Fifth American Peptide Symposium" (M. Goodman and J. Meienhofer, eds.), pp. 506–509. Wiley, New York.
Rudinger, J. (1963). *Pure Appl. Chem.* **7**, 335–362.
Rudinger, J. (1973). *In* "The Chemistry of Polypeptides" (P. G. Katsoyannis, ed.), pp. 87–123. Plenum, New York.
Rudinger, J., and Buetzer, P. (1975). *In* "Peptides 1974" (Y. Wolman, ed.), pp. 211–219. Wiley, New York.
Rüegg, U. T., and Rudinger, J. (1977). *In* "Methods in Enzymology" (C. H. W. Hirs and S. N. Timasheff, eds.), Vol. 47, pp. 111–116. Academic Press, New York.
Rydon, H. N., and Serrão, F. O. (1964). *J. Chem. Soc.* pp. 3638–3644.
Sakakibara, S. (1971). *In* "Chemistry and Biochemistry of Amino Acids, Peptides, and Proteins" (B. Weinstein, ed.), Vol. 1, pp. 51–85. Dekker, New York.
Sakakibara, S. (1977). *In* "Peptides: Proceedings of the Fifth American Peptide Symposium" (M. Goodman and J. Meienhofer, eds.), pp. 436–447. Wiley, New York.
Sakakibara, S., and Shimonishi, Y. (1965). *Bull. Chem. Soc. Jpn.* **38**, 1412–1413.
Sakakibara, S., Shin, K. H., and Hess, G. P. (1962) *J. Am. Chem. Soc.* **84**, 4921–4928.
Sakakibara, S., Shin, M., Fujino, M., Shimonishi, Y., Inoue, S., and Inukai, N. (1965). *Bull. Chem. Soc. Jpn.* **38**, 1522–1525.
Sakakibara, S., Shimonishi, Y., Kishida, Y., Okada, M., and Sugihara, H. (1967). *Bull. Chem. Soc. Jpn.* **40**, 2164–2167.
Sakakibara, S., Kishida, Y., Nishizawa, R., and Shimonishi, Y. (1968a). *Bull. Chem. Soc. Jpn.* **41**, 438–441.
Sakakibara, S., Kishida, Y., Kikuchi, Y., Sakai, R., and Kakiuchi, K. (1968b). *Bull. Chem. Soc. Jpn.* **41**, 1273.
Sakakibara, S., Fukuda, T., Kishida, Y., and Honda, I. (1970). *Bull. Chem. Soc. Jpn.* **43**, 3322.
Sakami, W., and Toennies, G. (1942). *J. Biol. Chem.* **144**, 203–217.
Sakiyama, F., Okawa, K., Yamakawa, T., and Akabori, S. (1958). *Bull. Chem. Soc. Jpn.* **31**, 926–932.
Salnikow, J., Liao, T. H., Moore, S., and Stein, W. H. (1973). *J. Biol. Chem.* **248**, 1480–1488.
Sandberg, B. E., and Ragnarsson, U. (1974). *Int. J. Pept. Protein Res.* **6**, 111–119.
Sandberg, B. E., and Ragnarsson, U. (1975). *Int. J. Pept. Protein Res.* **7**, 503–504.
Sandberg, B. E., and Ragnarsson, U. (1978). *Int. J. Pept. Protein Res.* **11**, 238–245.
Sanger, F. (1953). *Nature (London)* **171**, 1025–1026.
Sanger, F., and Thompson, E. O. (1963). *Biochim. Biophys. Acta* **71**, 468–471.
Sano, S., and Kawanishi, S. (1975). *J. Am. Chem. Soc.* **97**, 3480–3484.
Sano, S., Tokunaga, R., and Kun, K. A. (1971). *Biochim. Biophys. Acta* **244**, 201–205.

Sarantakis, D., Teichman, J., Lien, E. L., and Fenichel, R. L., (1976). *Biochem. Biophys. Res. Commun.* **73**, 336–342.

Savige, W. E., and Fontana, A. (1977a). *In* "Methods in Enzymology" (C. H. W. Hirs and S. N. Timasheff, eds.), Vol. 47, pp. 442–453. Academic Press, New York.

Savige, W. E., and Fontana, A. (1977b). *In* "Methods in Enzymology" (C. H. W. Hirs and S. N. Timasheff, eds.), Vol. 47, pp. 453–459. Academic Press, New York.

Savige, W. E., and Fontana, A. (1977c). *In* "Methods in Enzymology" (C. H. W. Hirs and S. N. Timasheff, eds.), Vol. 47, pp. 459–469. Academic Press, New York.

Savoie, J. Y., and Barton, M. A. (1974). *Can. J. Chem.* **52**, 2832–2839.

Schafer, D. J., Young, G. T., Elliot, D. F., and Wade, R. (1971). *J. Chem. Soc. C* pp. 46–49.

Schaich, E., Fretzdorff, A. M., and Schneider, F. (1973). *Hoppe-Seyler's Z. Physiol. Chem.* **354**, 897–902.

Schlatter, J. M., Mazur, R. H., and Goodmonson, O. (1977). *Tetrahedron Lett.* pp. 2851–2852.

Schnabel, E., Klostermeyer, H., Dahlmans, J., and Zahn, H. (1967). *Justus Leibigs Ann. Chem.* **707**, 227–241.

Schnabel, E., Stoltefuss, J., Offe, H. A., and Klauke, E. (1971). *Justus Liebigs Ann. Chem.* **743**, 57–68.

Schou, O. (1976). *Acta Chem. Scand., Ser. B* **30**, 991–993.

Schou, O., Brunfeldt, K., Rubin, I., and Hansen, L. (1975). *Hoppe-Seyler's Z. Physiol. Chem.* **356**, 1451–1454.

Schou, O., Bucher, D., and Nebelin, E. (1976). *Hoppe-Seyler's Z. Physiol. Chem.* **357**, 103–106.

Schrauzer, G., and Deutsch, E. (1969). *J. Am. Chem. Soc.* **91**, 3341–3350.

Schreiber, J. (1967). *In* "Peptides" (H. C. Beyerman, A. van de Linde, and W. Maassen van den Brink, eds.), p. 107. North-Holland Publ., Amsterdam.

Schröder, E. (1963). *Justus Liebigs Ann. Chem.* **670**, 127–136.

Schröder, E., and Lübke, K. (1965). "The Peptides" (transl. by E. Gross) 2 Vols. Academic Press, New York.

Schultz, J. (1967). *In* "Methods in Enzymology". (C. H. W. Hirs, ed.), Vol. 11, pp. 255–263. Academic Press, New York.

Schüssler, H., and Zahn, H. (1962). *Chem. Ber.* **95**, 1076–1080.

Schwartz, G. P., and Katsoyannis, P. G. (1973). *J. Chem. Soc., Perkin Trans. 1* pp. 2894–2901.

Schwyzer, R. (1959). *Angew. Chem.* **71**, 742.

Schwyzer, R., and Dietrich, H. (1961). *Helv. Chim. Acta* **44**, 2003–2006.

Schwyzer, R., and Kappeler, H. (1961). *Helv. Chim. Acta* **44**, 1991–2002.

Schwyzer, R., and Sieber, P. (1963). *Nature (London)* **199**, 172–174.

Schwyzer, R., Sieber, P., and Kappeler, H. (1959). *Helv. Chim. Acta* **42**, 2622–2624.

Schwyzer, R., Costopanagiotis, A., and Sieber, P. (1963a). *Helv. Chim. Acta* **46**, 870–889.

Schwyzer, R., Iselin, B., Kappeler, H., Riniker, B., Rittel, W., and Zuber, H. (1963b). *Helv. Chim. Acta* **46**, 1975–1996.

Scoffone, E., Fontana, A., and Rocchi, R. (1968). *Biochemistry* **7**, 971–979.

Scopes, P. M., Walshaw, K. B., Welford, M., and Young, G. T. (1965). *J. Chem. Soc.* pp. 782–786.

Scotchler, J. W., and Robinson, A. B. (1974). *Anal. Biochem.* **59**, 319–322.

Scotchler, J. W., Lozier, R., and Robinson, A. B. (1970). *J. Org. Chem.* **35**, 3151–3152.

Scott, L. T., Rebek, J., Ovsyanko, L., and Sims, C. L. (1977). *J. Am. Chem. Soc.* **99**, 625–626.

Scott, R. P., Chan, K. K., Kucera, P., and Zolty, S. (1971). *J. Chromatogr. Sci.* **9**, 577–591.

Seidel, C., Klauschenz, E., Kaufmann, K. D., and Kunzek, H. (1976). *J. Prakt. Chem.* **318**, 229–247.

Seidl, J., Malinský, J., Dušek, K., and Heitz, W. (1967). *Adv. Polym. Sci.* **5**, 113–213.

Seliger, H., Holupirek, M., and Görtz, H. H. (1978). *Tetrahedron Lett.* pp. 2115–2118.

Seltzman, H. H., and Chapman, T. M. (1974). *Tetrahedron Lett.* pp. 2637–2640.
Semkin, E. P., Smirnova, A. P., and Shchukina, L. A. (1968). *J. Gen. Chem. USSR (Engl. Transl.)* **38**, 2284–2287.
Semmelhack, M. F., and Heinsohn, G. E. (1974). *J. Am. Chem. Soc.* **94**, 5139–5140.
Shaltiel, S. (1967). *Biochem. Biophys. Res. Commun.* **29**, 178–183.
Shaltiel, S., and Fridkin, M. (1970). *Biochemistry* **9**, 5122–5127.
Shapira, R., Chou, F. C., and Kibler, R. F. (1972). *In* "Chemistry and Biology of Peptides" (J. Meienhofer, ed.), pp. 225–226. Ann Arbor Sci. Publ., Ann Arbor, Michigan.
Sharp, J. J., Robinson, A. B., and Kamen, M. D. (1973). *J. Am. Chem. Soc.* **95**, 6097–6108.
Sharts, C. M., and Sheppard, W. A. (1974). *Org. React.* **21**, 125–406.
Sheehan, J. C., and Daves, G. D. (1964). *J. Org. Chem.* **29**, 2006–2008.
Sheehan, J. C., and Frank, V. S. (1949). *J. Am. Chem. Soc.* **71**, 1856–1861.
Sheehan, J. C., and Grenda, V. J. (1962). *J. Am. Chem. Soc.* **84**, 2417–2420.
Sheehan, J. C., and Guziec, F. (1973). *J. Org. Chem.* **38**, 3034–3040.
Sheehan, J. C., and Hess, G. P. (1955). *J. Am. Chem. Soc.* **77**, 1067–1068.
Sheehan, J. C., and Umezawa, K. (1973). *J. Org. Chem.* **38**, 3771–3774.
Sheehan, J. C., Goodman, M., and Hess, G. (1956). *J. Am. Chem. Soc.* **78**, 1367–1369.
Sheehan, J. C., Hasspacher, K., and Yeh, Y. L. (1959). *J. Am. Chem. Soc.* **81**, 6086.
Shemyakin, M. M., Ovchinnikov, Y. A., Kiryushkin, A. A., and Kozhevnikova, I. V. (1965). *Tetrahedron Lett.* pp. 2323–2327.
Sheppard, R. C. (1977). *In* "Molecular Endocrinology" (I. MacIntyre and M. Szelke, eds.), pp. 43–56. Am. Elsevier, New York.
Sheradsky, T., Salemnick, G., and Frankel, M. (1971). *Isr. J. Chem.* **9**, 263–271.
Shimonishi, Y., Sakakibara, S., and Akabori, S. (1962). *Bull. Chem. Soc. Jpn.* **35**, 1966–1970.
Shiori, T., Yokoyama, Y., Kasai, Y., and Yamada, S. (1976). *Tetrahedron* **32**, 2211–2217.
Showell, H. J., Freer, R. J., Zigmond, S. H., Schiffmann, E., Aswanikumar, S., Corcoran, B., and Becker, E. L. (1976). *J. Exp. Med.* **143**, 1154–1169.
Sieber, P. (1977). *Helv. Chim. Acta* **60**, 2711–2716.
Sieber, P., and Iselin, B. (1968a). *Helv. Chim. Acta* **51**, 614–622.
Sieber, P., and Iselin, B. (1968b). *Helv. Chim. Acta* **51**, 622–632.
Sieber, P., Riniker, B., Brugger, M., Kamber, B., and Rittel, W. (1970). *Helv. Chim. Acta* **53**, 2135–2150.
Sieber, P., Kamber, B., Hartmann, A., Jöhl, A., Riniker, B., and Rittel, W. (1974). *Helv. Chim. Acta* **57**, 2617–2621.
Sieber, P., Kamber, B., Eisler, K., Hartmann, A., Riniker, B., and Rittel, W. (1976). *Helv. Chim. Acta* **59**, 1489–1497.
Sieber, P., Kamber, B., Hartmann, A., Jöhl, A., Riniker, B., and Rittel, W. (1977). *Helv. Chim. Acta* **60**, 27–37.
Sieber, P., Eisler, K., Kamber, B., Riniker, B., Rittel, W., Märki, F., and de Gasparo, M. (1978). *Hoppe-Seyler's Z. Physiol. Chem.* **359**, 113–123.
Sievertsson, H., Chang, J. K., von Klaudy, A., Bogentoft, C., Currie, B., Folkers, K., and Bowers, C. Y. (1972). *J. Med. Chem.* **15**, 222–226.
Sifferd, R. H., and du Vigneaud, V. (1935). *J. Biol. Chem.* **108**, 753–761.
Simpson, R. J., Neuberger, M. R., and Liu, T. Y. (1976). *J. Biol. Chem.* **251**, 1936–1940.
Sipos, F., and Gaston, D. W. (1971). *Synthesis* p. 321.
Sklyarov, L. Y., and Shashkova, I. V. (1969). *J. Gen. Chem. USSR (Engl. Transl.)* **39**, 2714–2718.
Skorna, G., Stemmer, R., and Ugi, I. (1978). *Chem. Ber.* **111**, 806–810.
Smith, C. W., Walter, R., Moore, S., Makofske, R. C., and Meienhofer, J. (1978). *J. Med. Chem.* **21**, 117–120.
Smith, C. W., Stahl, G. L., and Walter, R. (1979). *Int. J. Pept. Protein Res.* **13**, 109–112.
Smith, J. A. (1976). *Anal. Biochem.* **74**, 477–483.

Smithers, M. J. (1971). *In* "Peptides 1969" (E. Scoffone, ed.), pp. 44–51. North-Holland Publ., Amsterdam.
Smyth, D. G., Stein, W. H., and Moore, S. (1962). *J. Biol. Chem.* **237**, 1845–1850.
Sondheimer, E., and Holley, R. W. (1954). *J. Am. Chem. Soc.* **76**, 2467–2470.
Southard, G. L. (1971). *In* "Peptides 1969" (E. Scoffone, ed.), pp. 168–170. North-Holland Publ., Amsterdam.
Southard, G. L., Brooke, G. S., and Pettee, J. M. (1969). *Tetrahedron Lett.* pp. 3505–3508.
Southard, G. L., Brooke, G. S., and Pettee, J. M. (1971a). *In* "Peptides 1969" (E. Scoffone, ed.), pp. 95–101. North-Holland Publ., Amsterdam.
Southard, G. L., Brooke, G. S., and Pettee, J. M. (1971b). *Tetrahedron* **27**, 1359–1361.
Southard, G. L., Brooke, G. S., and Pettee, J. M. (1971c). *Tetrahedron* **27**, 2701–2703.
Spackman, D. H., Stein, W. H., and Moore, S. (1958). *Anal. Chem.* **30**, 1190–1206.
Spande, T. F., Witkop, B., Degani, Y., and Patchornik, A. (1970). *Adv. Protein Chem.* **24**, 97–260.
Spangenberg, R., Thamm, P., and Wünsch, E. (1971). *Hoppe-Seyler's Z. Physiol. Chem.* **352**, 655–656.
Sparrow, J. T. (1975). *Tetrahedron Lett.* pp. 4637–4638.
Sparrow, J. T. (1976). *J. Org. Chem.* **41**, 1350–1353.
Spies, J. R., and Chambers, D. C. (1950). *Anal. Chem.* **22**, 1209–1210.
Stammer, C. H. (1961). *J. Org. Chem.* **26**, 2556–2560.
Stark, G. (1971). *In* "Biochemical Aspects of Reactions on Solid Supports" (G. Stark, ed.), pp. 171–188. Academic Press, New York.
Stekol, J. A. (1941). *J. Biol. Chem.* **140**, 827–831.
Stelakatos, G. C., Paganou, A., and Zervas, L. (1966). *J. Chem. Soc C* pp. 1191–1199.
Stevenson, D., and Young, G. T. (1969). *J. Chem. Soc. C* pp. 2389–2398.
Stewart, J. M. (1968). *Fed. Proc., Fed. Am. Soc. Exp. Biol.* **27**, 63–66.
Stewart, J. M. (1976). *J. Macromol. Sci., Chem.* **A10**, 259–288.
Stewart, J. M., and Matsueda, G. R. (1972). *In* "Chemistry and Biology of Peptides" (J. Meienhofer, ed.), pp. 221–224. Ann Arbor Sci. Publ., Ann Arbor, Michigan.
Stewart, J. M., and Woolley, D. W. (1965). *Nature* (*London*) **206**, 619–620.
Stewart, J. M., and Woolley, D. W. (1967). *In* "International Symposium on Vaso-Active Polypeptides: Bradykinin and Related Kinins" (M. Rocha e Silva and H. A. Rothschild, eds.), pp. 7–13. Third International Congress and Sociedade Brasiliera de Farmacologia e de Terapeutica Experimental, São Paulo, Brasil.
Stewart, J. M., and Young, J. D. (1969). "Solid Phase Peptide Synthesis." Freeman, San Francisco, California.
Stewart, J. M., Mizoguchi, T., and Woolley, D. W. (1967). *Abstr., 153rd Natl. Meet., Am. Chem. Soc., 1967* No. O-206.
Stewart, J. M., Knight, M., Paiva, A. C., and Paiva, T. (1972). *In* "Progress in Peptide Research" (S. Lande, ed.), pp. 59–64. Gordon & Breach, New York.
Stewart, J. M., Peña, C., Matsueda, G. R., and Harris, K. (1976). *In* "Peptides 1976" (A. Loffet, ed.), pp. 285–290. Editions de l'Université de Bruxelles, Belgium.
Stirling, C. J. (1977). *In* "Organic Chemistry of Sulfur" (S. Oae, ed.), pp. 473–525. Plenum, New York.
Stoffel, W., and Craig, L. C. (1961). *J. Am. Chem. Soc.* **83**, 145–149.
Storey, H. T., Beacham, J., Cernosek, S. F., Finn, F. M., Yanaihara, C., and Hofmann, K. (1972). *J. Am. Chem. Soc.* **94**, 6170–6178.
Sugano, H., and Miyoshi, M. (1976). *J. Org. Chem.* **41**, 2352–2353.
Sugano, H., Taguchi, Y., Kawai, H., and Miyoshi, M. (1978). *In* "Peptide Chemistry 1977" (T. Shiba, ed.), pp. 23–26. Protein Res. Found., Osaka, Japan.
Suzuki, K., and Endo, N. (1977). *Chem. Pharm. Bull.* **25**, 1143–1146.

Suzuki, K., and Endo, N. (1978). *Chem. Pharm. Bull.* **26**, 2269–2274.
Suzuki, K., Nitta, K., and Endo, N. (1975). *Chem. Pharm. Bull.* **23**, 222–224.
Suzuki, K., Sasaki, Y., and Endo, N. (1976a). *Chem. Pharm. Bull.* **24**, 1–9.
Suzuki, K., Nitta, K., and Sasaki, Y. (1976b). *Chem. Pharm. Bull.* **24**, 3025–3033.
Suzuki, K., Endo, N., and Sasaki, Y. (1977). *Chem. Pharm. Bull.* **25**, 2613–2616.
Suzuki, K., Endo, N., Nitta, K., and Sasaki, Y. (1978). *Chem. Pharm. Bull.* **26**, 2198–2204.
Swan, J. M. (1957). *Nature (London)* **180**, 643–645.
Syrier, J. L., and Beyerman, H. C. (1974). *Recl. Trav. Chim. Pays-Bays* **93**, 117–120.
Syrier, J. L., and Beyerman, H. C. (1975). *In* "Peptides, 1974" (Y. Wolman, ed.), pp. 105–111. Wiley, New York.
Tadema, G., Harryvan, E., Panneman, H. J., and Arens, J. F. (1964). *Recl. Trav. Chim. Pays-Bas* **83**, 345–363.
Takashima, H., du Vigneaud, V., and Merrifield, R. B. (1968). *J. Am. Chem. Soc.* **90**, 1323–1325.
Tam, J. P., Cunningham-Rundles, W. F., Erickson, B. W., and Merrifield, R. B. (1977). *Tetrahedron Lett.* pp. 4001–4004.
Tan, N. H., and Kaiser, E. T. (1976). *J. Org. Chem.* **41**, 2787–2793.
Tanford, C. (1961). "Physical Chemistry of Macromolecules," Wiley, New York.
Tanikaga, R., Nakayama, K., Tanaka, K., and Kaji, A. (1977). *Chem. Lett.* pp. 395–396.
Tarr, G. E. (1977). *In* "Methods in Enzymology" (C. H. W. Hirs and S. N. Timasheff, eds.), Vol. 47, pp. 335–357. Academic Press, New York.
Taylor-Papadimitriou, J., Yovanidis, C., Paganou, A., and Zervas, L. (1967). *J. Chem. Soc. C* pp. 1830–1836.
Tesser, G. I., and Balvert-Geers, I. C. (1975). *Int. J. Pept. Protein Res.* **7**, 295–305.
Tesser, G. I., Buis, J. T., Wolters, E. T., and Bothé-Helmes, E. G. (1976). *Tetrahedron* **32**, 1069–1072.
Teutsch, G. (1976). *Tetrahedron* **32**, 1071–1072.
Theodoropoulos, D. M. and Fruton, J. S. (1962). *Biochemistry* **1**, 933–937.
Tilak, M. A. (1968). *Tetrahedron Lett.*, pp. 6323–6326.
Tilak, M. A., and Hollinden, C. S. (1968a). *In* "Peptides 1968" (E. Bricas, ed.), pp. 173–178. North-Holland Publ., Amsterdam.
Tilak, M. A., and Hollinden, C. S. (1968b). *Tetrahedron Lett.* pp. 1297–1300.
Tjoeng, F. S., Staines, W., St. Pierre, S., and Hodges, R. S. (1977). *Biochim. Biophys. Acta* **490**, 489–496.
Toennies, G., and Kolb, J. J. (1945). *J. Am. Chem. Soc.* **67**, 1141–1144.
Tometsko, A. M. (1973). *Biochem. Biophys. Res. Commun.* **50**, 886–891.
Tometsko, A. M., Garden, J., and Tischio, J. (1971). *Rev. Sci. Instrum.* **42**, 331–336.
Tregear, G. W. (1972). *In* "Chemistry and Biology of Peptides" (J. Meienhofer, ed.), pp. 175–178. Ann Arbor Sci. Publ., Ann Arbor, Michigan.
Tregear, G. W. (1975). *In* "Peptides 1974" (Y. Wolman, ed.), pp. 177–189. Wiley, New York.
Tregear, G. W., Battaerd, H. A., and Lang, G. A. (1966). US Patent #3,700,609, application date 26 April 1966 (Imperial Chemical Industries of Australia and New Zealand, Ltd.); *Chem. Abstr.* **71P**, 508241, (1969).
Tregear, G. W., van Rietschoten, J., Sauer, R., Niall, H. D., Keutmann, H. T., and Potts, J. T. (1977). *Biochemistry* **16**, 2817–2823.
Tritsch, G. L., and Grahl-Nielsen, O. (1969). *Biochemistry* **8**, 1816–1821.
Trudelle, Y., and Birr, C. (1975). *Int. J. Pept. Protein Res.* **7**, 403–410.
Trudelle, Y., and Spach, G. (1972). *Tetrahedron Lett.* pp. 3475–3478.
Tschesche, H., and Jering, H. (1973). *Angew. Chem., Int. Ed. Engl.* **12**, 756.
Tun-Kyi, A. (1978). *Helv. Chim. Acta* **61**, 1086–1090.

Tun-Kyi, A., and Schwyzer, R. (1976). *Helv. Chim. Acta* **59**, 1642–1646.
Turán, A., Patthy, A., and Bajusz, S. (1975). *Acta Chim. Acad. Sci. Hung.* **85**, 327–332.
Tzougraki, C., Makofske, R. C., Gabriel, T. F., Wang, S. S., Kutny, R., Meienhofer, J., and Li, C. H. (1978), *J. Am. Chem. Soc.* **100**, 6248–6249.
Ueki, M., and Ikeda, S. (1974). *Chem. Lett.* pp. 25–28.
Ueki, M., and Ikeda, S. (1976). *Chem. Lett.* pp. 827–830.
Ueki, M., and Shishikura, T. (1973). *Chem. Lett.* pp. 733–736.
Ueki, M., Ikeda, S., and Tonegawa, F. (1977). *In* "Peptides: Proceedings of the Fifth American Peptide Symposium" (M. Goodman and J. Meienhofer, eds.), pp. 546–548. Wiley, New York.
Ugi, I. (1962). *Angew. Chem., Int. Ed. Engl.* **1**, 8–21.
Ugi, I. (1971). *Intra-Sci. Chem. Rep.* **5**, 229–261.
Ugi, I., Aigner, H., Beijer, B., Ben-Efraim, D., Burghard, H., Bukall, P., Eberle, G., Eckert, H., Marquarding, D., Rehn, D., Urban, R., Wackerle, L., and von Zychlinski, H. (1976). *In* "Peptides 1976" (A. Loffet, ed.), pp. 159–181. Editions de l'Université de Bruxelles, Belgium.
Ugi, I., Eberle, G., Eckert, H., Lagerlund, I., Marquarding, D., Skorna, G., Urban, R., Wackerle, L., and Von Zychlinski, H. (1977). *In* "Peptides: Proceedings of the Fifth American Peptide Symposium" (M. Goodman and J. Meienhofer, eds.), pp. 484–487. Wiley, New York.
Upson, D. A., and Hruby, V. J. (1976). *J. Org. Chem.* **41**, 1353–1358.
Uy, R., and Wold, F. (1977). *Science* **198**, 890–896.
van Batenburg, O. D., Raap, J., Kerling, K. E., and Havinga, E. (1976a). *Recl. Trav. Chim. Pays-Bas* **94**, 278–281.
van Batenburg, O. D., Kerling, K. E., and Havinga, E. (1976b). *FEBS Lett.* **68**, 228–230.
van Batenburg, O. D., Voskuyl-Holtkamp, I., Schattenkerk, C., Hoes, K., Kerling, K. E., and Havinga, E. (1977). *Biochem. J.* **163**, 385–387.
van den Akker, M., and Jellinek, F. J. (1967). *Recl. Trav. Chim. Pays-Bas* **86**, 897–906.
van Melik, J. E., and Wolters, E. T. (1972). *Synth. Commun.* **2**, 83–86.
van Rietschoten, J., Tregear, G. W., Leeman, S., Powell, D., Niall, H., and Potts, J. T. (1975a). *In* "Peptides 1974" (Y. Wolman, ed.), pp. 113–115. Wiley, New York.
van Rietschoten, J., Granier, C., Rochat, H., Lissitzky, S., and Miranda, F. (1975b). *Eur. J. Biochem.* **56**, 35–40.
van Rietschoten, J., Pedroso, E. M., and Granier, C. (1977). *In* "Peptides: Proceedings of the Fifth American Peptide Symposium" (M. Goodman and J. Meienhofer, eds.), pp. 522–524. Wiley, New York.
Veber, D. F. (1975). *In* "Peptides: Chemistry, Structure and Biology" (R. Walter and J. Meienhofer, eds.), pp. 307–316. Ann Arbor Sci. Publ., Ann Arbor, Michigan.
Veber, D. F., Varga, S. L., Milkowski, J. D., Joshua, H., Conn, J. B., Hirschmann, R., and Denkewalter, R. G. (1969). *J. Am. Chem. Soc.* **91**, 506–507.
Veber, D. F., Brady, S. F., and Hirschmann. R. (1972a). *In* "Chemistry and Biology of Peptides" (J. Meienhofer, ed.), pp. 315–319. Ann Arbor Sci. Publ. Ann Arbor, Michigan.
Veber, D. F., Milkowski, J. D., Varga, S. L., Denkewalter, R. G., and Hirschmann, R. (1972b). *J. Am. Chem. Soc.* **94**, 5456–5461.
Villemoes, P., Christensen, T., and Brunfeldt, K. (1976). *Hoppe-Seyler's Z. Physiol. Chem.* **357**, 713–719.
Villemoes, P., Christensen, T., and Brunfeldt, K. (1978). *Acta Chem. Scand., Ser. B* **32**, 703–713.
Visser, S., Raap. J., Kerling, K. E., and Havinga, E. (1970a). *Recl. Trav. Chim. Pays-Bas* **89**, 865–875.
Visser, S., Kerling, K. E., and Havinga, E. (1970b). *Recl. Trav. Chim. Pays-Bas* **89**, 880–884.

Vorbrüggen, H., and Krolikiewicz, K. (1975). *Angew. Chem. Int. ed. Engl.* **14**, 818.
Wang, C. T., Kulesha, I. D., Stefko, P. L., and Wang, S. S. (1974). *Int. J. Pept. Protein Res.* **6**, 59–64.
Wang, S. S. (1973). *J. Am. Chem. Soc.* **95**, 1328–1333.
Wang, S. S. (1975). *J. Org. Chem.* **40**, 1235–1239.
Wang, S. S. (1976). *J. Org. Chem.* **41**, 3258–3261.
Wang, S. S., and Kulesha, I. D. (1975). *J. Org. Chem.* **40**, 1227–1234.
Wang, S. S., and Merrifield, R. B. (1969a). *Int. J. Protein Res.* **1**, 235–244.
Wang, S. S., and Merrifield, R. B. (1969b). *J. Am. Chem. Soc.* **91**, 6488–6491.
Wang, S. S., and Merrifield, R. B. (1972). *Int. J. Pept. Protein Res.* **4**, 309–318.
Wang, S. S., Yang, C. C., Kulesha, I. D., Sonenberg, M., and Merrifield, R. B. (1974). *Int. J. Pept. Protein Res.* **6**, 103–109.
Wang, S. S., Gisin, B. F., Winter, D. P., Makofske, R., Kulesha, I. D., Tzougraki, C., and Meienhofer, J. (1977). *J. Org. Chem.* **42**, 1286–1290.
Wang, S. S., Makofske, R., Bach, A. E., and Merrifield, R. B. (1979). *Int. J. Pept. Protein Res.*, unpublished.
Weber, C. J. (1930). *J. Biol. Chem.* **86**, 217–222.
Weber, U., and André, M. (1975). *Hoppe-Seyler's Z. Physiol. Chem.* **356**, 701–714.
Weber, U., and Hartter, P. (1970). *Hoppe-Seyler's Z. Physiol. Chem.* **351**, 1384–1388.
Weber, U., Hartter, P., and Flohé, L. (1970). *Hoppe-Seyler's Z. Physiol. Chem.* **351**, 1389–1394.
Weigele, M., DeBernardo, S. L., Tengi, J. P., and Leimgruber, W. (1972). *J. Am. Chem. Soc.* **94**, 5927–5928.
Weimann, G., and Khorana, H. G. (1962). *J. Am. Chem. Soc.* **84**, 4329–4341.
Weinshenker, N. M., and Shen, C. M. (1972). *Tetrahedron Lett.* pp. 3281–3284.
Weitzel, G., Eisele, K., Schulz, V., and Stock, W. (1973). *Hoppe-Seyler's Z. Physiol. Chem.* **354**, 321–330.
Westall, F. C., Scotchler, J., and Robinson, A. B. (1972). *J. Org. Chem.* **37**, 3363–3364.
Weygand, F. (1968). *In* "Peptides 1968" (E. Bricas, ed.), pp. 183–184. North-Holland Publ., Amsterdam.
Weygand, F., and Csendes, E. (1952). *Angew. Chem.* **64**, 136.
Weygand, F., and Frauendorfer, E. (1970). *Chem. Ber.* **89**, 647–652.
Weygand, F., and Hunger, K. (1962). *Chem. Ber.* **95**, 1–6.
Weygand, F., and Obermeier, R. (1968). *Z. Naturforsch., Teil B* **23**, 1390.
Weygand, F., and Ragnarsson, U. (1966). *Z. Naturforsch., Teil B* **21**, 1141–1144.
Weygand, F., and Reiher, M. (1955). *Chem. Ber.* **88**, 26–34.
Weygand, F., and Steglich, W. (1959). *Z. Naturforsch., Teil B* **14**, 472–473.
Weygand, F., and Zumach, G. (1962). *Z. Naturforsch., Teil B* **17**, 807–810.
Weygand, F., Prox, A., and König, W. (1966a). *Chem. Ber.* **99**, 1446–1450.
Weygand, F., Hoffmann, D., and Wünsch, E. (1966b). *Z. Naturforsch., Teil B* **21**, 426–428.
Weygand, F., Steglich, W., and Pietta, P. (1967). *Chem. Ber.* **100**, 3841–3849.
Weygand, F., Steglich, W., Fraunberger, F., Pietta, P., and Schmid, J. (1968a). *Chem. Ber.* **101**, 923–934.
Weygand, F., Steglich, W., Bjarnason, J., Akhtar, R., and Chytil, N. (1968b). *Chem. Ber.* **101**, 3623–3641.
Weygand, F., Steglich, W., and Bjarnason, J. (1968c). *Chem. Ber.* **101**, 3642–3648.
White, F. H. (1961). *J. Biol. Chem.* **236**, 1353–1360.
White, J. (1934). *J. Biol. Chem.* **106**, 141–144.
Wieland, H. (1912). *Ber. Dtsch. Chem. Ges.* **45**, 484–493.
Wieland, T., and Birr. C. (1967). *In* "Peptides" (H. C. Beyerman, A. van de Linde, and W. Maasen van den Brink, eds.), pp. 103–106. North-Holland Publ., Amsterdam.
Wieland, T., Kern, W., and Sehring, R. (1950). *Justus Liebigs Ann. Chem.* **569**, 117–121.

Wieland, T., Birr, C., and Flor, F. (1969a). *Justus Liebigs Ann Chem.* **727**, 130–137.
Wieland, T., Birr, C., and Wissenbach, H. (1969b). *Angew. Chem., Int. Ed. Engl.* **8**, 764–765.
Wieland, T., Lewalter, J., and Birr, C. (1970). *Justus Liebigs Ann. Chem.* **740**, 31–47.
Wieland, T., Birr, C., and von Dungen, A. (1971a). *Justus Liebigs Ann. Chem.* **747**. 207–212.
Wieland T., von Dungen, A., and Birr, C. (1971b). *Justus Liebigs Ann. Chem.* **752**, 109–114.
Wieland, T., Birr, C., and Flor, F. (1971c). *Angew. Chem. Int. Ed. Engl.* **10**, 336.
Wieland, T., Birr, C., and Fleckenstein, P. (1972a). *Justus Liebigs Ann. Chem.* **756**, 14–19.
Wieland, T., Penke, B., and Birr, C. (1972b). *Justus Liebigs Ann. Chem.* **759**, 71–75.
Wieland, T., Flor, F., and Birr, C. (1973). *Justus Liebigs Ann. Chem.* pp. 1595–1600.
Wilcox, P. E. (1967). *In* "Methods in Enzymology" (C. H. W. Hirs, ed.), Vol. 11, pp. 63–76. Academic Press, New York.
Williams, A. W., and Young, G. T. (1972). *J. Chem. Soc., Perkin Trans. 1* pp. 1194–1200.
Windholz, T. B., and Johnston, D. B. (1967). *Tetrahedron Lett.* pp. 2555–2557.
Windridge, G. C., and Jorgensen, E. C. (1971a). *Intra-Sci. Chem. Rep.* **5**, 375–380.
Windridge, G. C., and Jorgensen, E. C. (1971b). *J. Am. Chem. Soc.* **93**, 6318–6319.
Wissmann, H., and Geiger, R. (1970). *Angew. Chem., Int. Ed. Engl.* **9**, 908–909.
Wissmann, H., König, W., and Geiger, R. (1973). *In* "Peptides 1972" (H. Hanson and H. D. Jakubke, eds.), pp. 158–161. North-Holland Publ., Amsterdam.
Wolfe, S., and Hansan, S. K. (1970). *Can. J. Chem.* **48**, 3572–3579.
Wolters, E. T., Tesser, G. I., and Nivard, R. J. (1974). *J. Org. Chem.* **39**, 3388–3392.
Woodman, D. J. (1974). *In* "Chemistry and Biochemistry of Amino Acids, Peptides, and Proteins" (B. Weinstein, ed.), Vol. 2, pp. 207–296. Dekker, New York.
Woodward, R. B., Olofson, R. A., and Mayer, H. (1961). *J. Am. Chem. Soc.* **83**, 1010–1012.
Woodward, R. B., Heusler, K., Gosteli, J., Naegeli, P., Oppolzer, W., Ramage, R., Raganathan, S., and Vorbrüggen, H. (1966). *J. Am. Chem. Soc.* **88**, 852–853.
Woolley, D. W. (1966). *J. Am. Chem. Soc.* **88**, 2309–2310.
Wünsch, E. (1967). *Z. Natursforsch., Teil B* **22**, 1269–1276.
Wünsch, E., ed. (1974). "Synthese von Peptiden" *In* "Houben-Weyl's Methoden der organischen Chemie" (E. Müller, ed.), 4th ed., Vol. 15, Parts 1 and 2, Thieme, Stuttgart.
Wünsch, E., and Spangenberg, R. (1971). *In* "Peptides 1969" (E. Scoffone, ed.), pp. 30–34. North-Holland Publ., Amsterdam.
Wünsch, E., and Spangenberg, R. (1972). *Chem. Ber.* **105**, 740–742.
Wünsch, E., Fries, G., and Zwick, A. (1958). *Chem. Ber.* **91**, 542–547.
Wünsch, E., Fontana, A., and Drees, F. (1967). *Z. Naturforsch., Teil B* **22**, 607–609.
Wünsch, E., Zwick, A., and Jaeger, E. (1968). *Chem. Ber.* **101**, 336–340.
Wünsch, E., Jaeger, E., Deffner, M., and Scharf, R. (1972). *Hoppe-Seyler's Z. Physiol. Chem.* **353**, 1716–1720.
Wünsch, E., Jaeger, E., Kisfaludy, L., and Löw, M. (1977). *Angew. Chem., Int. Ed. Engl.* **89**, 330–331.
Yajima, H., and Kawatani, H. (1971). *Chem. Pharm. Bull.* **19**, 1905–1913.
Yajima, H., and Kiso, Y. (1974). *Chem. Pharm. Bull.* **22**, 1087–1094.
Yajima. H., Kawatani, H., and Watanabe, H. (1970). *Chem. Pharm. Bull.* **18**, 1333–1339.
Yajima, H., Watanabe, H., and Okamoto, M. (1971). *Chem. Pharm. Bull.* **19**, 2185–2189.
Yajima, H., Fujii, N., Ogawa, H., and Kawatani, H. (1974). *J. Chem. Soc., Chem. Commun.* pp. 407–408.
Yajima, H., Kiso, Y., Ogawa, H., Fujii, N., and Irie, H. (1975). *Chem. Pharm. Bull.* **23**, 1164–1166.
Yajima, H., Takeyama, M., Kanaki, J., and Mitani, K. (1978). *J. Chem. Soc. Chem. Commun.* pp. 482–483.
Yamada, S., Kasai, Y., and Shioiri, T. (1973). *Tetrahedron Lett.* pp. 1595–1598.
Yamada, S., Ikota, N., Shiori, T., and Tachibana, S. (1975). *J. Am. Chem. Soc.* **97**, 7174–7175.

Yamamoto, D. M., Upson, D. A., Linn, D. K., and Hruby, V. J. (1977). *J. Am. Chem. Soc.* **99**, 1564–1570.

Yamashiro, D. (1977). *J. Org. Chem.* **42**, 523–525.

Yamashiro, D., and Blake, J. (1978). *In* "Versatility of Proteins" (C. H. Li, ed.), pp. 81–97. Academic Press, New York.

Yamashiro, D., and Li, C. H. (1972). *Int. J. Pept. Protein Res.* **4**, 181–185.

Yamashiro, D., and Li, C. H. (1973a). *J. Am. Chem. Soc.* **95**, 1310–1315.

Yamashiro, D., and Li, C. H. (1973b). *J. Org. Chem.* **38**, 591–592.

Yamashiro, D., and Li, C. H. (1973c). *J. Org. Chem.* **38**, 2594–2597.

Yamashiro, D., and Li, C. H. (1974). *Proc. Natl. Acad. Sci. U.S.A.* **71**, 4945–4949.

Yamashiro, D., and Li, C. H. (1976). *Biochim. Biophys. Acta* **451**, 124–132.

Yamashiro, D., and Li, C. H. (1978). *J. Am. Chem. Soc.* **100**, 5174–5179.

Yamashiro, D., Blake, J., and Li, C. H. (1972). *J. Am. Chem. Soc.* **94**, 2855–2859.

Yamashiro, D., Noble, R. L., and Li, C. H. (1973). *J. Org. Chem.* **38**, 3561–3565.

Yamashiro, D., Blake, J., and Li, C. H. (1976). *Tetrahedron Lett.* pp. 1469–1462.

Yamashiro, D., Li, C. H., Tseng, L. F., and Loh, H. H. (1978). *Int. J. Pept. Protein Res.* **11**, 251–257.

Yang, C. C., and Merrifield, R. B. (1976). *J. Org. Chem.* **41**, 1032–1041.

Yaron, A., and Schlossman, S. F. (1968). *Biochemistry* **7**, 2673–2681.

Yaron, A., Otey, M. C., Sober, H. A., Katchalski, E., Erlich-Rogozinski, S., and Berger, A. (1972). *Biopolymers* **11**, 607–621.

Yonezawa, H., Takahashi, N., Ohno, M., and Izumiya, N. (1978). *Int. J. Pept. Protein Res.* **11**, 19–27.

Young, G. T. (1967). *In* "Peptides" (H. C. Beyerman, A. van de Linde, and W. Maasen van den Brink, eds.), pp. 55–66. North-Holland Publ., Amsterdam.

Young, J. D., Benjamini, E., Stewart, J. M., and Leung, C. Y. (1967). *Biochemistry* **6**, 1455–1460.

Young, R. W., Wood. K. H., Joyce, R. J., and Anderson, G. W. (1956). *J. Am. Chem. Soc.* **78**, 2126–2131.

Zahn, H., and Fölsche, E. T. (1969). *Chem. Ber.* **102**, 2158–2159.

Zahn, H., and Schmidt, G. (1970). *Justus Liebigs Ann. Chem.* **731**, 91–100.

Zahn, H., and Schnabel, E. (1957). *Justus Liebigs Ann. Chem.* **605**, 212–232.

Zahn, H., Okuda, T., and Shimonishi, Y. (1967). *In* "Peptides" (H. C. Beyerman, A. van de Linde, and W. Maassen van den Brink, eds.), pp. 108–112. North-Holland Publ., Amsterdam.

Zahn, H., Danho, W., Schmidt, G., Dahlmans, J., Costopanogiotis, A., and Engels, E. (1968). In "Peptides 1968" (E. Bricas, ed.), pp. 217–221. North-Holland Publ., Amsterdam.

Zervas, L., and Photaki, I. (1962). *J. Am. Chem. Soc.* **84**, 3887–3897.

Zervas, L., and Theodoropoulos, D. M. (1956). *J. Am. Chem. Soc.* **78**, 1359–1369.

Zervas, L., Benoiton, L., Weiss, E., Winitz, M., and Greenstein, J. P. (1959a). *J. Am. Chem. Soc.* **81**, 1729–1734.

Zervas, L., Otani, T. T., Winitz, M., and Greenstein, J. P. (1959b), *J. Am. Chem. Soc.* **81**, 2878–2884.

Zervas, L., Photaki, I., and Ghelis, N. (1963a). *J. Am. Chem. Soc.* **85**, 1337–1341.

Zervas, L., Borovas, D., and Gazis, E. (1963b). *J. Am. Chem. Soc.* **85**, 3660–3666.

Zervas, L., Photaki, I., Cosmatos, A., and Borovas, D. (1965). *J. Am. Chem. Soc.* **87**, 4922–4933.

Zhukova, G. F., Ravdel, G. A., and Shchukina, L. A. (1970). *J. Gen. Chem. USSR* (*Engl. Transl.*) **40**, 2750–2753.

Zumach, G., and Kühle, E. (1970). *Angew. Chem., Int. Ed. Engl.* **9**, 54–63.

Chapter 2

The Liquid-Phase Method for Peptide Synthesis

MANFRED MUTTER and ERNST BAYER

THE PEPTIDES, VOLUME 2

ISBN 0-12-304202-X

I. INTRODUCTION

A great number of methods for peptide synthesis have been developed in the past, based upon the chemical methodology of classical peptide synthesis but differing in the strategic procedure for building up the peptide chain. Despite considerable achievements in this field, a standardized, unique procedure for synthesizing peptides of any given sequence and length is not available. It seems unlikely that this ultimate goal of peptide synthesis can be realized in the near future in view of the complexity of the problems encountered in peptide synthesis. However, the efficient combination of the strategies at hand, for providing the choice of the appropriate method for a specific synthetic problem, seems to be the most promising approach. The realization of this strategy requires a high degree of flexibility and preparative experience together with an awareness of both the potentialities and the limitations of each method.

The enormous increase in the demand for synthetic peptides has been the main motivation for the development of "rapid procedures" for peptide synthesis. The basic idea of these strategies is to omit the tedious and time-consuming intermediate purification steps, typical of conventional procedures, by forcing the coupling reaction to completion. This is accomplished by the addition of a large excess of the carboxylic component to the growing peptide chain. Thus, the peptide can be built up in a stepwise manner by applying a repetitive, standardized synthesis cycle. The use of an insoluble support as COOH-terminal protecting group, as realized in the solid-phase synthesis by Merrifield (1963), enables easy separation of excess reagents from the polymer-bound peptide but suffers from inherent deficiencies induced by the polymeric, heterogeneous matrix. The use of soluble polymeric supports was devised to eliminate some of the shortcomings of solid-phase synthesis while retaining the enormous advantages of a polymer-mediated synthesis. From the great number of functionalized linear polymers available, polyethylene glycol (PEG) proved to be most compatible with respect to the physical and chemical properties necessary for "liquid-phase" strategy (Mutter *et al.*, 1972; Mutter and Bayer, 1974). The liquid-phase method, as developed by the authors, is therefore primarily based upon the use of PEG, which can be regarded as solubilizing polymeric protecting group, as the COOH-terminal "liquid support." Viewed from the chemical methodology, liquid-phase synthesis is identical to classical solution methods; however, the presence of a solubilizing macromolecular protecting group affords (i) higher solubility of peptides during stepwise synthesis and (ii) facilitated separation of low-molecular-weight compounds from the polymeric peptide ester.

The aim of this chapter is to delineate the potentialities and limitations of liquid-phase synthesis. The specific advantages of this procedure can only be evaluated by comparison with other competing strategies. Therefore, a critical review of the major strategies is given briefly. Then the fundamental principles and experimental procedures of liquid-phase strategy are exemplified for the case of peptide PEG esters. Finally, examples of peptide synthesis according to this strategy are summarized, and some recent applications for the use of soluble polymers in organic chemistry are reviewed.

II. STRATEGIES OF PEPTIDE SYNTHESIS

The strategies of peptide synthesis can be subdivided according to the nature of the COOH-terminal protecting group as classical methods (low-molecular-weight ester group), solid-phase methods (heterogeneous, polymeric ester group), and liquid-phase methods (soluble, polymeric ester group). The major features of these strategies and a critical evalution of their potential are summarized.

A. Classical Methods

1. Segment Condensation

This method of synthesis has been applied to the preparation of many biologically active peptides and can be considered the most promising route to homogeneous sequential peptides containing up to 100 or more amino acid residues (Bodanszky *et al.*, 1976; Finn and Hofmann, 1976). Positive aspects of the method include the ease of purification of the final product and the reliable synthesis of the segments up to about 10 amino acid residues. On the other hand, the method suffers from two limitations, namely, (i) low yield in the segment-coupling step and (ii) low solubility of many sequences.

Furthermore, the procedure is rather time consuming and requires a substantial degree of preparative skill. Since little is known about the relationship between sequence, conformation, and physicochemical properties of sequential polypeptides, it is impossible to predict the efficiency and limits of this strategy. In favorable cases, the solubility of two segments is enhanced after the condensation step. Acyl peptides with COOH-terminal amino acids other than glycine and proline have been coupled by the azide procedure, without racemization in many instances. At present, condensation of medium-sized peptide segments seems to be the most reliable strategy for the synthesis of longer peptides or even of proteins. However, the synthesis of the starting peptides for segment condensation can be accomplished by a number of strategies with much more efficiency than the tedious classical procedure. The use of solubilizing protecting groups offers a possibility for the elimination of solubility problems.

2. *Stepwise Synthesis*

This highly efficient way of building up peptide sequences was successfully applied to the synthesis of oxytocin (Bodanszky and du Vigneaud, 1959). Here, N^{α}-benzyloxycarbonylamino acid *p*-nitrophenyl esters were added in a stepwise manner to the COOH-terminal methyl ester. Protected intermediates were purified by precipitation and washing. Since this promising synthesis, many efforts have undertaken to build up even larger peptides by this procedure. The synthesis of the intestinal hormone secretin, with 27 amino acid residues, has been the most ambitious effort (Bodanszky and Williams, 1967; Bodanszky *et al.*, 1967). When the intermediates in this synthesis are not extensively purified, the procedure is subject to the same requirements as the stepwise synthesis with macromolecular protecting groups, i.e., quantitative reaction in every stage of synthesis. Additionally, solubility problems may prevent a further elongation of the peptide chain. However, the method seems to be very efficient for the synthesis of small- to medium-sized peptides. The synthesis of the protected ACTH(25–39)-peptide (Schwyzer and Sieber, 1966), with an overall yield of 6.85%, and of a large portion of human gastrin (Morley, 1967) are good examples. As protecting group, the N^{α}-benzyloxycarbonyl group in combination with side-chain *tert*-butyl or tosyl groups proved to be the most effective in this strategy as it can be selectively removed by hydrogenation. For activation of the acyl amino acids, trichlorophenyl esters are usually employed. The efficiency of this procedure is demonstrated by the fact that, generally, one new amino acid residue can be introduced into the growing peptide chain every two days. Nevertheless, low solubility of even short peptide sequences can result in insurmountable difficulties. Again, the insertion of solubilizing protecting groups into the peptide chain may provide a solution to this problem inherent in this strategy.

Very useful tools for use in combination with stepwise synthesis of peptides are the polymeric active esters, first introduced by Fridkin *et al.* (1966; see also this volume, Chapter 3). Here, the activated N^{α}-protected amino acid is fixed to an insoluble support and reacted with the amino component in large excess, assuring high yields of reaction. Excess reagent can be readily removed by filtration or centrifugation. The use of these polymeric reagents in peptide synthesis requires good solubility of the reacting peptide ester, thus being subject to the same limitations as classical synthesis. Furthermore, the coupling rate is, due to steric hindrance, expected to decrease with growing chain length of the peptide. Nevertheless, smaller peptides have been built up in excellent yields by this procedure (Fridkin *et al.*, 1968, 1972; Laufer *et al.*, 1968). In a variance of this strategy, polymer-supported triphenylphosphine/carbontetrachloride has been used as condensing agent for peptide synthesis (Appel *et al.*, 1976; Appel and Willms, 1977).

The repetitive excess mixed anhydride procedure involves the stepwise synthesis of peptides using mixed anhydrides of N^{α}-alkyloxycarbonylamino acids and alkyl carbonates as the acylating components (Tilak, 1970; Beyerman *et al.*, 1973; van Zon and Beyerman, 1973; see this volume, Chapter 6). After coupling, the unreacted mixed anhydride is destroyed and the peptide isolated by precipitation or extration. By omitting intermediate purification steps, one coupling cycle can be performed in 6–8 hr. The use of excess acylating reagent guarantees high coupling yields. Thus, the protected 16-peptide(12–27) of secretin was synthesized by this procedure in an overall yield of 31%, corresponding to average coupling yields of 93%. Complications may arise from side reactions of the mixed anhydrides. Whereas the method appears to be very effective for quick synthesis of smaller peptides, the decrease in solubility during chain elongation may prevent its application to the synthesis of larger peptides.

The use of *N*-carboxyamino acid anhydrides (NCAs) for the stepwise synthesis in solution has not found wide application, mainly due to low selectivity of peptide bond formation. Kinetic studies have shown that two competing reactions decrease the yield of the desired coupling product (Bartlett and Jones, 1957; Bartlett and Dittmer, 1957). However, high yields in the preparation of dipeptides could be obtained by using crystalline NCAs (Hirschmann *et al.*, 1971). The stepwise NCA method was also successfully used for the synthesis of a model tripeptide in aqueous solution with polyethyleneimine as the macromolecular protecting group (Blecher and Pfaender, 1973). At present, the stepwise synthesis of longer sequential peptides seems to face size limitations. More promising is the preparation of small segments and subsequent segment condensation, as demonstrated in the synthesis of ribonuclease A *S*-protein (Denkewalter *et al.*, 1969).

B. Solid-Phase Synthesis

The ingenious concept of binding the growing peptide chain to an insoluble support during all steps of synthesis (Merrifield, 1963), thereby allowing a quick and simple strategy for peptide synthesis, has found broad application in peptide chemistry (Erickson and Merrifield, 1976). Despite the enormous accomplishments of solid-phase synthesis, however, the original idea of building up peptides of any desired length and sequence by this efficient strategy has not been realized. The main reason originates in the basic concept itself. As intermediate purification steps are not possible in this strategy, quantitative yields in each step are an absolute prerequisite for the successful application of a multistep solid-phase synthesis. Incomplete reactions result in inhomogeneous products (e.g., failure sequences and truncated peptides) which must be separated from the main product at the end of synthesis. In this case, the undoubted advantages of this strategy cannot be realized. In principle, quantitative reactions in the coupling step should be possible: According to a postulate of Flory (1939), which has been convincingly confirmed, the reactivity of a functional group should not be affected by its attachment to a polymer of high molecular weight. Therefore, differences in the kinetic behavior of low-molecular-weight reagents and their polymeric analogs must arise from inherent effects, e.g., steric hindrance, interaction between polymer and low-molecular-weight reagent, and local solvation effects (Morawetz, 1975), induced by the heterogeneous matrix, most notably by cross-linked polystyrenes.

In most cases, it is impossible to single out specific effects of the polymeric matrix, because little is known about the physical nature of the immediate surroundings of the reaction sites. For example, the observed drop in the coupling yields with increasing chain length in building up the alternating sequence $(\text{Leu-Ala})_n$ (Hagenmaier, 1970) could result from (i) steric effects of the matrix, (ii) insufficient solvation of the growing peptide chain, (iii) a conformational transition of the peptide.

Moreover, reliable comparisons between heterogeneous and homogeneous reactions are impossible, due to the nonreproducibility of specific steric effects in the heterogeneous phase. Thus, systematic studies for optimizing the polymeric support revealed a strong correlation between the physical nature of the solid matrix and the coupling yields (Losse, 1971, 1973; Frank and Hagenmaier, 1974a). The chemical and physical nonequivalence of functional groups within the cross-linked matrix is a source of deviation from linearity in the kinetics of the reactions (Rudinger and Gut, 1967; Rudinger and Buetzer, 1975; Sheppard, 1973; Andreatta and Rink, 1973; Erickson and Merrifield, 1976). In the light of these facts, it is not surprising that comparative investigations of kinetics in heterogeneous and homogeneous systems resulted in rather inconsistant data (Rudinger and Gut, 1967; Gut and Rudinger, 1968a; 1968b; Losse, 1971; Maher *et al.*, 1971). For example, Gut and Rudinger (1968b) found drastic deviations from linearity of the second-order rate constant for yields higher than 50%, whereas Erickson and Merrifield (1976) observed a close fit to second-order kinetics up to 95% completion of reaction. From these results, a drop in the reaction rates in the order of two- to fivefold is indicated for the heterogeneous system dicyclohexylcarbodiimide(DCC)-mediated coupling reactions, the difference in the reaction rates seems to be even smaller. The fact that the reaction rates in solid-phase synthesis are not drastically reduced, compared to homogeneous reactions, indicates that the diffusion of reagent into the polymeric matrix is not a limiting factor of this strategy. This has been confirmed by Andreatta and Rink (1973) in kinetic studies on both functionalized cross-linked and noncross-linked polystyrenes. Consequently, the intrinsic problems of solid-phase synthesis arise from deviations in the linear kinetic course in the final stages of reaction due to nonequivalence of functional groups rather than from reduced reaction rates. Unfortunately, the kinetic studies carried out so far give no information about the course of the final stage of reaction, e.g., for yields >95%. Thus, although such kinetic investigations are extremely important for the optimization of coupling conditions, they give no indication as to whether a quantitative coupling yield can be achieved. The finding that at least 95% of all reactive groups attached to the polymeric matrix follow second-order rate constants, close to those of low-molecular-weight analogs (Frank and Hagenmaier, 1974b; Erickson and Merrifield, 1976), is a strong indication that the above-mentioned postulate of Flory about the chemical equivalence of low- and high-molecular-weight functional groups holds up for high yields of reaction, even in the heterogeneous system; but, to restate this point, the problems of solid-phase synthesis seem to be manifest only in these final stages of reaction.

Numerous analytical investigations have delineated intrinsic effects of the solid support on the coupling reaction (Bayer *et al.*, 1970; Losse and Ulbrich, 1972; Hancock, *et al.*, 1973; Sheppard, 1973; Frank and Hagenmaier, 1974b; Losse and Meisegeier, 1974; Morawetz, 1975; Erickson and Merrifield, 1976). Much effort has been invested in eliminating steric effects of the solid matrix. A great variety of new macromolecular supports were developed (Erickson and Merrifield, 1976). Despite some promising results using the gelatinous beads of copoly(styrene–0.5% divinylbenzene) (Birr, 1973) or the hydrophilic resin poly(*N*,*N*-dimethylacrylamide) (Sheppard, 1973; Atherton *et al.*, 1977) as insoluble supports, the originally introduced cross-linked polystyrene has not yet been replaced on a general basis, although it is not particularly suitable for use in all solid-phase synthesis (Morawetz, 1975). An ideal polymeric support should reduce the problems of peptide synthesis to the purely chemical, as in classical synthesis. The realization of this ultimate goal of support-mediated strategies appears to be incompatible with the use of solid supports. Apart from effects induced by the physical nature of the solid matrix, the chemistry in the heterogeneous system is substantially different from the homogeneous system, as exemplified by the limitations in the chemical methodology of peptide synthesis.

C. Soluble Polymer Techniques

The most rational way to avoid the problems inherent in solid-phase synthesis, while maintaining its major advantages, is by the use of soluble polymeric groups instead of the insoluble matrix. In liquid-phase synthesis, the polymeric support has three functions: (1) It serves as COOH-terminal protecting group of the peptide, (2) it allows a simple and rapid procedure for the removal of excess reagents, and (3) it exerts a solubilizing effect on the growing peptide chain. In order to be compatible with all requirements of a stepwise strategy that omits intermediate purification steps, the polymer chain must be stable during all stages of synthesis, and it should possess chemically and kinetically equivalent functional groups to which the growing peptide chain can be covalently attached. Only a small number of functionalized linear chain molecules exhibit all the requirements of a soluble support for peptide synthesis.

1. Polyfunctional Supports

a. Linear Polystyrene. Shemyakin *et al.* (1965) tried to eliminate the steric complications encountered in solid-phase synthesis by using linear, noncross-linked polystyrene of molecular weight 200,000 as soluble support for the peptide chain. All coupling and deprotection reactions are carried out in solution; for the removal of excess reagent, the polymer-bound peptide is precipitated in water and washed (Scheme 1).

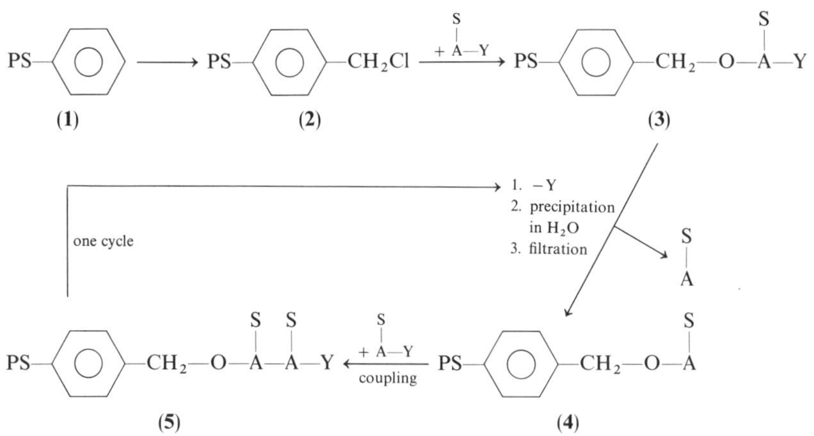

Scheme 1

This polymer-mediated stepwise synthesis in solution was claimed to be superior to solid-phase synthesis because of the greater chance of achieving quantitative coupling reactions. However, investigations of Green and Garson (1969) revealed that the use of soluble polystyrene created new sources of preparative difficulties of which the danger of cross-linking during the various steps of each cycle of synthesis turned out to be most critical. The cross-linking reaction was mainly due to residual chloromethyl groups after the esterification step. Thus, a sample of L-alanyl-polystyrene containing only 0.4% benzyl chloride per gm polymer was sufficient to cause cross-linking, which resulted in phase separation, after neutralization in dimethylformamide(DMF) solution at room temperature. In another example, treatment of Boc-Ser-Ala-polystyrene with a solution of HCl in tetrahydrofuran (THF) at 0°C caused precipitation after a few minutes, due to benzyl ether formation between the free OH-group of L-serine and excess of benzyl chloride residues at the polymer support. Although the authors were successful in reducing the content of benzyl chloride groups to 0.11%, the danger of subsequent cross-linking reactions, e.g., during the deprotection step, limits the applicability of this method. Furthermore, the growth of peptide chains on the polystyrene support influences its physical properties in an undesirable way, owing to the poor solubilizing power of the polymer chain on the peptide. However, when special care was taken to avoid cross-linking reactions, distinct advantages of the soluble polymer technique over heterogeneous synthesis, e.g., greater flexibility in respect to coupling techniques or the elimination of steric effects of the support, could be demonstrated by the synthesis of two model peptides (Green and Garson, 1969).

Andreatta and Rink (1973) used linear polystyrene of molecular weight 20,400 for the synthesis of the pentapeptide Val-Tyr-Val-His-Phe. For separating excess reagent from polymer-bound peptide, gel filtration was used instead of precipitation procedures. The maximal capacity for maintaining satisfactory solubility properties of the peptide polystyrene ester was ca. 0.5 mmol/gm polymer. The addition a Boc-Arg(No_2) residue to the pentapeptide resulted in a drastic decrease in the solubility of the polymer–peptide. Kinetic investigations were carried out in order to delineate differences in the reaction rates of low- and high-molecular-weight peptide esters (Andreatta and Rink, 1973). A result of major impact for the use of polyfunctional supports in general was the finding that the functional groups attached to the polystyrene showed considerable differences in chemical reactivity and kinetic behavior. For example, some anchoring groups proved to be totally resistant to hydrogenation for cleaving the benzyl ester groups between peptide and polystyrene support. Furthermore the kinetic course of the coupling reaction showed significant deviations from linearity due to nonequivalence of functional groups. The enhanced reaction rates of certain groups appear to produce favorable polymer effects, e.g., an increase in the effective concentration or a change in the effective dielectric constant at the reaction site (Morawetz, 1975).

A comparison of the second-order rate constants of aminolysis of N-protected active esters with low- and high-molecular-weight amino components (Table I) shows that the reaction rates of the soluble polymer ester are lower by a factor of about two compared to low-molecular-weight esters, but considerably higher than those of the heterogeneous system.

Although the formation of truncated sequences due to sterically inaccessible reaction sites is unlikely in the homogeneous system, the observed difference in the chemical reactivity of functional groups may severely impede quantitative coupling within reasonable time limits. The random distribution

Table 1. Second Order Rate Constants for the Aminolysis of Z-Ala-ONp with H-Pro-OCH_2-R

R	K_2 (liter·mol^{-1}·sec^{-1})
—C_6H_5	0.081
Polystyrene (noncross-linked)	0.047
Copoly(styrene–*m*-DVP)	0.03

of functional sites along the linear chain molecule must be considered as a possible origin of the heterogeneous kinetic behavior of polyfunctional supports. Thus, functional groups located at the chain ends might be less subject to local effects of the polymer chain than groups near the center of the chain, because of differences in the segment flexibility and in the density of the coiled chain molecule. Also, neighboring effects cannot be ruled out, even when the degree of functionalization is relatively low.

From these results, it must be concluded that the intrinsic problems of solid-phase synthesis, e.g., steric effects of the polymer matrix, are not resolved satisfactorily by using noncross-linked polystyrenes as support. The equivalence of all functional groups attached to a linear chain molecule appears to be an absolute prerequisite for the realization of homogeneous conditions analogous to low-molecular–weight systems.

b. Other Polyfunctional Supports. Polyethyleneimine (PEI) of molecular weight 30,000 was used as water-soluble support for the synthesis of a model tetrapeptide (Blecher and Pfaender, 1973). This polymer is designed for use in combination with the *N*-carboxy anhydride coupling method. For removing excess reagent, the ultrafiltration procedure proved very effective. An enzymatic method was used for removing the peptide from the polymeric support. To this end, arginine was coupled to PEI as the anchoring group and the tetrapeptide was built up in stepwise manner. After cleaving the Arg-PEI bond by digestion with trypsin, the COOH-terminal arginine was removed from the peptide with carboxypeptidase B. In principle, this cleaving procedure could be applied to any water-soluble polymer-bound peptide. Bayer and Geckeler (1975) used copolymers of vinyl pyrrolidinone and vinyl acetate as soluble support for the synthesis of the tetrapeptide Ile-Ala-Val-Gly. With symmetrical anhydrides as coupling reagent, the model peptide was obtained in overall yield of 60%. Owing to the presence of the pyrrolidinone ring, these copolymers exert a very strong solubilizing effect upon the peptide chain in a variety of solvents, including water. The synthesis of the same peptide, in the authors' laboratory, using polyacrylic acid and polyvinyl alcohol gave less satisfactory results, mainly due to poor solubility of the polymers.

None of these polymers has found broad application in liquid-phase synthesis; from a kinetic point of view, the same objections with respect to the equivalence of functional groups may be raised as in the case of polystyrene. However, these deficiencies may not be of major import for the synthesis of small- to medium-sized peptides.

2. Polyethers

Polyethers are linear chain molecules with two hydroxy groups at the chain ends (**6**). The presence of both a hydrophilic and a hydrophobic moiety per monomer unit lends this class of polymers its favourable physical and chemical properties for use in liquid-phase synthesis (Mutter *et al.*, 1972; Mutter and Bayer, 1972, 1974) (see Section III,A and B). So far, mainly

$$HO[(-CH_2)_m-O-]H$$

(**6**)

polyethylene glycols with molecular weights in the range of 2,000 to 20,000 were used as COOH-terminal protecting groups (Mutter *et al.*, 1975). Recently, monofunctional polyethers, e.g., polyethylene glycol monomethyl ether, were used instead of the bifunctional polyethers. In order to obtain higher polyether capacities, block copolymers were synthesized using PEG blocks and diisocyanate derivatives (Scheme 2) (Bayer *et al.*, 1978). When PEG 1000 was used as starting material for the polyaddition reaction, polymers of molecular weight $>20{,}000$ were obtained. As the functional groups are located in equidistant order, neighbouring effects can be excluded in this copolymer.

From theory (Flory, 1939), the reactivity of the two functional groups in PEG should be identical and of the same order as that of the corresponding low-molecular-weight analogs. Whereas the former property is self-evident,

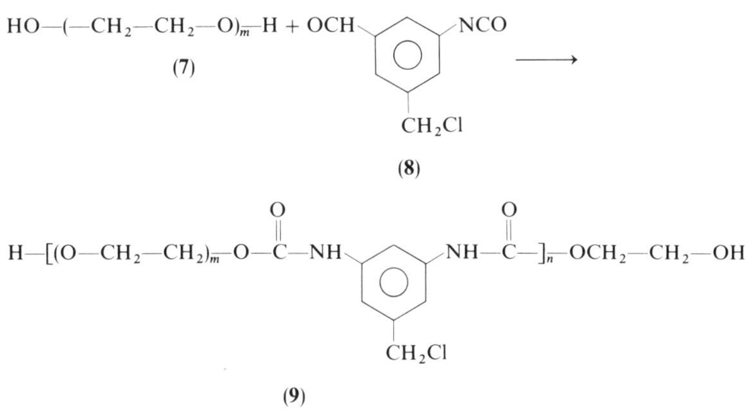

Scheme 2

because of the chemical equivalence of the two OH-groups, a possible influence of the molecular weight of the chain upon the reactivity of the terminal groups had to be delineated by kinetic investigations. To this end, Bayer *et al.* (1974) studied the rate constants of aminolysis of N-protected amino acid esters both with amino acid PEG esters of different molecular weight and with their low-molecular-weight ester analogs. The results are summarized in Table II.

A comparison of the rate constants shows that no significant difference in the reactivity of low- and high-molecular-weight amino acid esters could be detected. Most notably, the low-molecular-weight ester analog of PEG, i.e., 2-methoxyethyl ester shows even lower rate constants than the high-molecular-weight esters. This polymer effect can only be explained by small changes in the effective dielectric constant or differences in solvation in the immediate neighbourhood of the reaction site induced by the polymer chain. The significant increase in the rate constant for H-$(Gly)_3$-OPEG compared to the low-molecular-weight ester analog H-$(Gly)_3$-OEt, which was observed also for other peptide PEG esters, reflects an increase in the nucleophilicity of the terminal amino group due to hydrogen bonding with an oxygen atom of the PEG chain.

On the other hand, no significant influence of the molecular weight of PEG could be detected in the range of 2,000 to 20,000 daltons. Within the investigated range, the reaction between the active amino acid esters and the amino acid PEG esters strictly followed second-order kinetics, from the beginning of the coupling until about 80% completion. Thus, PEG esters showed fully identical behavior to the low-molecular-weight esters, and no

Table II. Rate Constants for the Aminolysis of Boc-Gly-ONp with H-$(Gly)_n$-OR[a]

R	K_2 (liter·mol^{-1}·sec^{-1})
tert-Butyl	0.024
Ethyl	0.013
—$CH_2CH_2OCH_3$	0.0079
PEG 20,000	0.014
PEG 10,000	0.014
PEG 6,000	0.016
PEG 4,000	0.019
PEG 2,000	0.018
Ethyl[b]	0.019
PEG 20,000[b]	0.032

[a] If not otherwise indicated $n = 1$.
[b] $n = 3$.

unfavorable influence of the polymer chain upon the reaction rates could be detected (Bayer *et al.*, 1975b). However, the fact that the reaction rates in classical and liquid-phase system are identical is not a proof for the linearity of the kinetics up to 100% completion of reaction. Although the final stage of reaction is difficult to assess in kinetic investigations, potential deviations from linearity due to sterically heterogeneous functional groups, as observed with solid-phase supports, can be excluded here for several reasons: The functional groups in PEG are equivalent and are consequently in identical "physical surrounding;" in view of the conformational properties of PEG (see Sections III,A and B) local steric effects are unlikely, mainly because of the high flexibility of the chain ends; and diffusion effects are absent in homogeneous solution even when highly viscous (Flory, 1939; Morawetz, 1975).

The results obtained in liquid-phase synthesis using PEG supports confirm these expectations. Most notably in hydrogenation reactions, which revealed considerable heterogeneity of functional groups in noncross-linked polystyrene supports (Andreatta and Rink, 1973), no influence of the PEG group on the reaction rate could be detected and the hydrogenation proceeded to 100% completion (Mutter *et al.*, 1975).

Thus, whereas PEG determines, to a large extent, the physicochemical properties of the peptide attached to its chain end (see Section III,B), it has no significant influence on the reactivity of the peptide. The absence of intrinsic polymer effects is a major prerequisite for a successful application of stepwise strategy, e.g., for obtaining quantitative yields in all reaction steps, and reduces the problems of peptide synthesis to the chemistry of classical procedure. The strategy of liquid-phase synthesis, which takes advantage of these favorable properties of PEG, is outlined in detail in Section III.

D. Liquid–Solid-Phase Techniques

The alternating liquid–solid-phase technique introduced by Frank *et al.* (1975) combines strategic features of both liquid- and solid-phase synthesis and of the polymer reagent technique. The unique feature of this strategy is the possibility of removing unreacted components from the main product after coupling reaction by a simple filtration procedure. To this end, the activated carboxyl component is attached to an insoluble polymeric matrix by an acid-labile urethane bond (Scheme 3).

The temporary immobilization of the growing peptide chain during the coupling reaction eliminates the absolute need for quantitative reactions without losing the practical advantages of solid-phase synthesis. High yields are guaranteed by using an excess of polymeric active ester. Deviations from linear kinetics due to heterogeneous reaction sites are of no concern

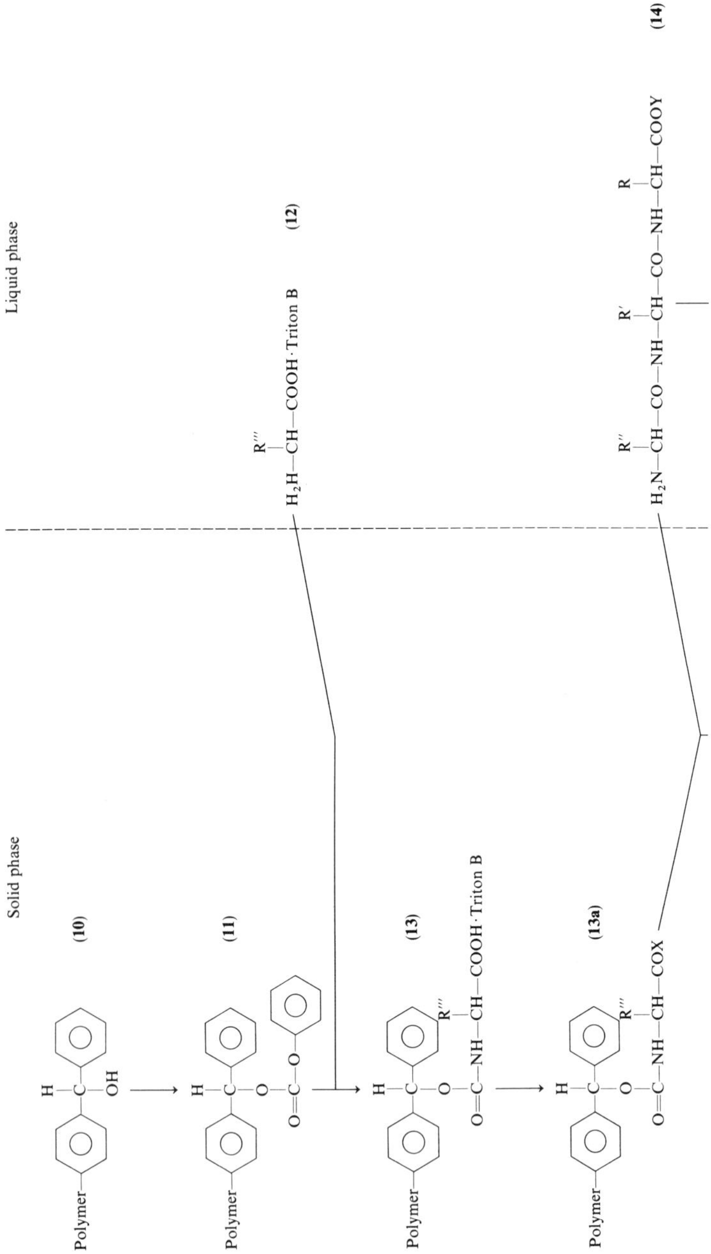
Solid phase
Liquid phase
Polymer
(10)
(11)
(12)
(13)
(13a)
(14)
R'''
H₂H—CH—COOH·Triton B
NH—CH—COOH·Triton B
NH—CH—COX
R''
R'
R
H₂N—CH—CO—NH—CH—CO—NH—CH—COOY

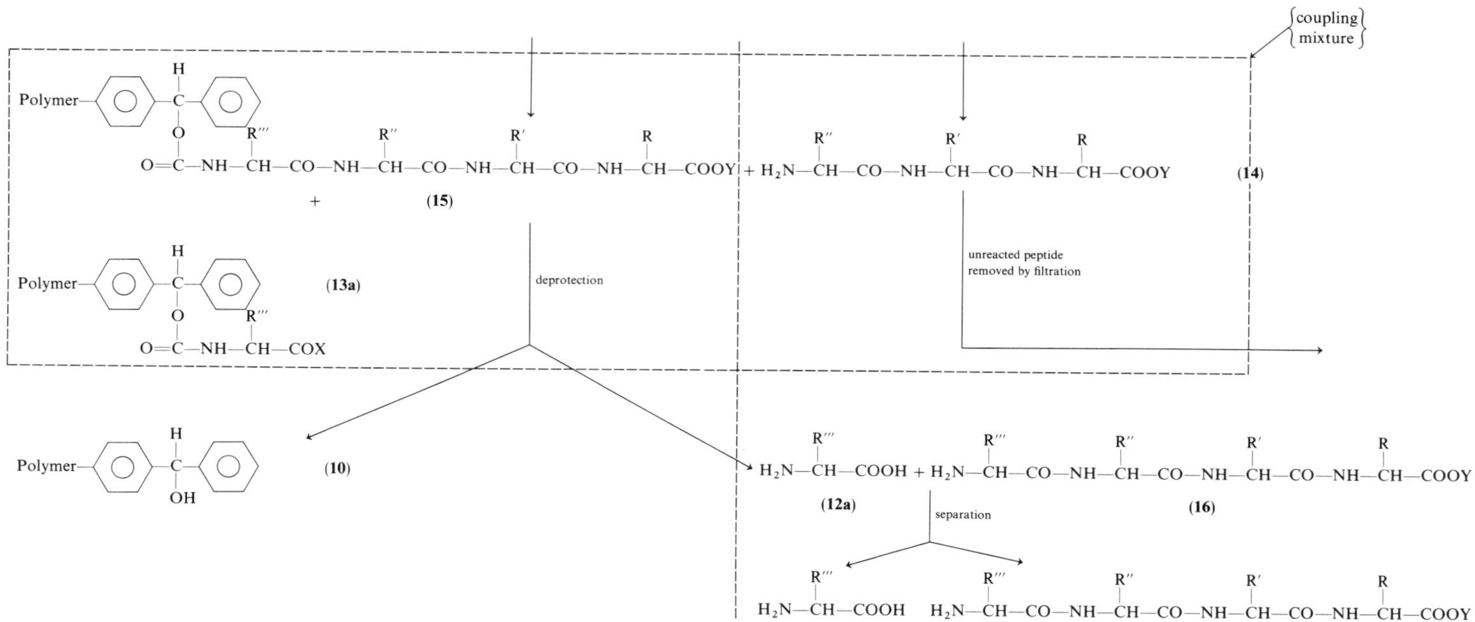

X = activating group
Y = carboxyl protecting group, e.g., monofunctional PEG

Scheme 3

for this strategy. The separation of excess amino acid and peptide, after their cleavage from the heterogeneous matrix by treatment with acid, is simplified by using a macromolecular, solubilizing COOH-terminal protecting group as in liquid-phase synthesis (Frank and Hagenmaier, 1975). Since the main problem of solid-phase synthesis, the formation of failure and truncated sequences due to incomplete coupling reactions, is eliminated in this strategy, products of high purity are obtained as exemplified in the synthesis of LH-RH (Frank *et al.*, 1977).

While this elegant method seems to be very suitable for the synthesis of small- to medium-sized peptides, effects such as a decrease in coupling yields due to steric complications between the macromolecular coupling components or a decrease in solubility of the peptide may prevent the exclusive application of this strategy for the synthesis of longer peptides. However, a major advantage of the strategy is its ease of combination with other stepwise solution methods, most notably with liquid-phase synthesis. For example, by using polyethylene glycol monomethyl ether as COOH-terminal solubilizing protecting group, individual coupling steps can be performed either by the alternating strategy or by the standard liquid-phase procedure (see Section III,D,2). Whenever a particular coupling step is expected to proceed to less than 100% completion within a reasonable time, e.g., in the case of sterically hindered coupling components, the alternating procedure is indicated. Since the diffusion of PEG–peptides into the cross-linked polystyrene matrix proceeds without difficulties up to molecular weights of ca. 6,000 (Frank *et al.*, 1975), the same monofunctional PEG supports can be used as in liquid-phase synthesis.

A combination of liquid-phase strategy and polymer reagent technique for stepwise synthesis of peptides has been described (Jung *et al.*, 1975; Heusel *et al.*, 1977). Polyethylene glycol esters of the COOH-terminal amino acid are reacted with excess active esters covalently bound to cross-linked polystyrene beads. After completion of the coupling step, the excess reagent is removed by a simple filtration step instead of precipitating the PEG–peptide as in liquid-phase synthesis. By the presence of a solubilizing polymeric group, limitations encountered in the classical application of this strategy (Fridkin *et al.*, 1966), such as low solubility of the peptide can be overcome. On the other hand, steric effects between the macromolecular coupling components may have a severe impact on the coupling rates of the growing peptide chain.

III. THE LIQUID-PHASE METHOD

The liquid-phase method for peptide synthesis takes advantage of the efficient concept of stepwise synthesis without intermediate purification procedures. To this end, the COOH-terminal amino acid is covalently attached to a soluble macromolecular group that determines the physical and chemical properties of the growing peptide chain during all stages of synthesis. As a consequence, the polymer-bound peptide can be readily separated from low-molecular-weight reagents so that the synthesis cycle, i.e., the addition of one residue to the peptide chain, is as efficient as in solid-phase synthesis. However, all reactions are carried out in homogeneous solutions, as in classical procedures, eliminating inherent problems of heterogeneous reactions. Compared to classical stepwise strategies (Section II,A), the use of a macromolecular protecting group offers two principal advantages: (1) The solubility of the peptide is strongly enhanced by the polymeric ester group, so that peptides with poor solubility become accessible to the stepwise strategy, and (2) the synthesis cycle is simplified and can be performed according to a standard procedure independent of the physicochemical properties of the peptide.

As intermediate purification steps are difficult in this strategy, the advantages of a polymer-mediated synthesis hold only when quantitative yields of reaction can be achieved in every step of synthesis. Therefore, analytical controls during synthesis are of utmost importance in liquid-phase strategy. So far, polyethylene glycol (also polyoxyethylene for chains with more than about 100 monomer units) has proved to be the most suitable polymer for liquid-phase synthesis, because of favorable physical and chemical properties (Bailey and Koleske, 1967, 1976; Mutter, 1977). Homologous polyethers, such as polymethylene oxide or polypropylene oxide, possess less advantageous solubility and crystallization properties. Moreover, PEG is available in a variety of molecular weights, thus providing a wide spectrum of capacities for the attachment of peptide chains. Most commonly used in liquid-phase synthesis is PEG of molecular weight 2,000 to 20,000. Recently, monofunctional polyglycols were introduced in peptide synthesis.

In addition to its use for peptide synthesis, the liquid-phase technique has found many other promising applications (Section IV). In this section, the fundamentals and experimental procedures of liquid-phase peptide synthesis are presented. Chemical and physical properties of soluble supports, which are the basis of the strategy of liquid-phase synthesis, are exemplified below for the case of PEG.

A. Physicochemical Properties of Polyethylene Glycol (PEG)

The conformation of a polymer both in the solid state and in solution determines its physical properties to a large extent (Flory, 1953). Such aspects as physical handling of solid precipitates, inclusion of low-molecular-weight components, or reaction kinetics of functional groups attached to the polymer are intimately related to the conformational characteristic of the chain molecule. According to X-ray and ir studies (Tadokoro *et al.*, 1964), PEG of molecular weight >2,000 exhibits high crystallinity in the solid state.

The succession of trans, trans, gauche conformations for the bonds C—O—C—C results in a loosely turned 7_2-helix with an identity period of 19.3 Å. The linear chain molecule crystallizes in monoclinic form with 28 monomer units per elementary cell. The degree of crystallization for PEG of molecular weight 3,000 was found to be 98% under optimal crystallization conditions (Neidlinger, 1975; Mutter, 1976); even when precipitated from concentrated solutions by diethyl ether, the degree of crystallization of PEG 20,000 was about 80%, as determined by wide angle X-ray scattering (Mutter, 1976). When PEG is dissolved, the ordered meander-like structure is destroyed, mainly due to rotation around the C—O bonds (Flory, 1969). Using the rotational isomeric state model, which takes into account neighbor-dependent interactions along the chain molecule, conformation-dependent properties of PEG can be adequately reproduced by assuming a random-coil conformation (Abe and Mark, 1976). The conformational parameters deduced from this model indicate that the PEG molecule is very flexible in solution, resulting in high coiling of the chain in theta-like solvents. However, in strongly polar solvents such as water or dimethylformamide, a significant retention of ordered structure is observed (Liu and Parsons, 1969), because of increased steric hindrance of rotation in the polymer chain. Consequently, the average chain dimensions of PEG, with respect to the

Table III. Solubility of PEG 4,000

Solvent	Solubility in % (w/v)[a]
Benzene	10
CCl_4	10
Dioxan	10
Methanol	20
Pyridine	40
$CHCl_3$	47
CH_2Cl_2	53
H_2O	55
EtOH[d]	20
EtOH[c]	1
EtOH[b]	0.1
Diethyl ether	0.01

[a] If not otherwise indicated at 25°C.
[b] 20°C.
[c] 32°C.
[d] 34°C.

average space occupied by the polymer molecule, are considerably increased in polar solvents, so that the chain can be considered as a freely permeable coil under these conditions.

The ethylene units and ether oxygens that alternate along the polymer chain give rise to the hydrophobic and hydrophilic character of PEG. Consequently, PEG is soluble in water and in a large number of organic solvents (Table III).

Whereas PEG is stable toward nucleophilic attack, it is not fully resistant to oxidative degradation. Nevertheless, under the conditions used for liquid-phase synthesis of peptides, no significant degradation was observed. Special precautions against oxidative attack, as e.g., the addition of stabilizers or exclusion of oxygen, proved to be unnecessary.

The molecular weight distribution of PEG with polymerization degree 100 is relatively narrow and corresponds closely to a Poisson distribution (Flory, 1940).

B. Properties of Peptide PEG Esters

The strategy of liquid-phase synthesis is based on the principle that the physical and chemical properties of the peptide are dominated by the COOH-terminal, macromolecular protecting group. Specifically, the crystallization tendency of PEG and its solubility in various organic solvents must be retained after attaching a peptide chain to its chain ends. As expected, the influence of the peptide on these properties of the polymeric ester group depends on the primary sequence, side-chain protection, chain length, and conformation of the growing peptide chain. For example, the degree of crystallization of PEG, as determined by X-ray diffraction, decreased from 80% to about 60% when the homooligopeptide Ala_{10} was bound to the chain termini of the polymer. The high retention of crystalline phases in PEG after the attachment of amorphous peptide blocks is readily explained by a two-phase model (Fig. 1).

According to this model (Mutter, 1976), crystalline PEG layers alternate with amorphous peptide phases; because of the regular folding of the PEG chains (Neidlinger, 1975), the bulky peptide coils are arranged between the PEG blocks without disturbing the crystalline structure of PEG. The high degree of ordered structure in PEG–peptides is of practical relevance for liquid-phase strategy—the precipitated peptide ester is easy to handle owing to its partially crystalline state; moreover, the danger of inclusions of low-molecular-weight components, as observed in amorphous precipitates (Shemyakin *et al.*, 1965), is considerably reduced. Consequently, a single precipitation step of PEG–peptide is often sufficient to remove excess reagent after the coupling reaction. Even when the ratio of the molecular

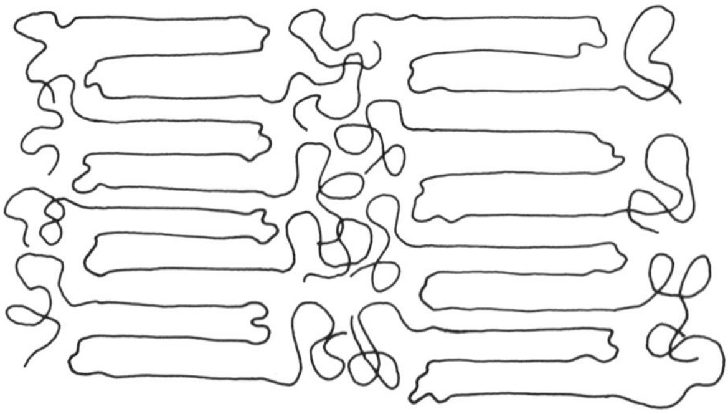

Figure 1. Two-phase model for PEG-bound peptides in the solid state. (a) Crystalline PEG phase. (b) Amorphous peptide phase.

weights of PEG and peptide was close to unity, the PEG–peptide could be precipitated quantitatively from organic solvents and showed partially crystalline structure.

The solubility of PEG–peptides is generally similar to that of the free polymer. However, changes in solubility and viscosity are difficult to predict. For example, the solubility of the PEG esters of the hydrophobic homo-oligomers Val_n and Ile_n decreased considerably for chain lengths $n > 6$. Nevertheless, the solubilizing power of PEG was sufficient to enable the synthesis of oligomers with chain lengths up to 12 residues by the stepwise liquid-phase method. The conformation of oligomers whose low-molecular-weight esters are insoluble could be investigated in the form of PEG esters in such solvents as trifluoroethanol (TFE), methylene chloride (CH_2Cl_2), and water (Section IV,A). A major reason for the exceptionally low solubility of these homooligopeptides is their tendency to form aggregated β-structures with intermolecular hydrogen bonds (Toniolo *et al.*, 1976). The intimate relationship between conformation and physical properties was also clearly manifested in the case of the PEG esters of oligo[LGlu (OBzl)] (Mutter, 1977). The lower oligomers, which form intermolecular β-structures in organic solvents such as methylene chloride (Imae and Ikeda, 1972), show decreasing solubility with increasing chain length. By way of contrast, a continuous increase in solubility and decrease in viscosity was found when the PEG–peptide was further elongated. This unusual behavior is the result of a conformational transition from aggregated β-structures to a helical conformation. Similar observations were made during stepwise synthesis of sequential polypeptides (Bayer and Mutter, 1974; Jung *et al.*, 1977). Variation of the molecular weight of PEG in range of 2,000 to 20,000 showed no dramatic effect on the solubility of the attached peptides. Molecular weights higher than 20,000 have very low capacity, whereas the lower polyethers are no longer amenable to crystallization. Low-molecular-weight PEG may, however, be used as solubilizing group for side-chain protection of peptides (Anzinger *et al.*, 1979).

From the present data, the following conclusions can be drawn. The liquid-phase strategy of peptides has no absolute limitations with respect to the size and sequence of the peptide to be synthesized. The physicochemical properties of PEG-bound peptides are governed principally by the polymeric ester group up to chain lengths of 10–15 amino acid residues, independent of the nature of the attached peptide. The compatibility of the liquid-phase synthesis for the preparation of longer peptides (more than 20 residues) strongly depends upon primary sequence, side-chain protection, and conformation of the peptide. When conformational transitions from aggregated structures to helical conformations occur with growing peptide chain, the solubility is generally significantly enhanced. The limitations of the stepwise strategy in the homogeneous phase produced by the insolubility of the peptide are far beyond those of classical methods.

C. Strategy of the Liquid-Phase Method

The standard procedure of liquid-phase synthesis combines strategic elements of classical and solid-phase technique. The factors operative in the selection of protecting groups and coupling methods are the same as in classical synthesis. Thus, the "chemistry" of peptide synthesis in liquid-phase and classical methods is identical. On the other hand, through the presence of a macromolecular protecting group at the COOH-terminus of the peptide chain, the advantage of a polymer-mediated synthesis—the efficient stepwise synthesis without intermediate purification procedures—are preserved. In liquid-phase synthesis, the COOH-terminal N-protected amino acid is esterified using PEG as the alcoholic component (Scheme 4). By attaching an anchoring group to the chain ends of PEG prior to esterification, the chemical stability of the peptide–polymer bond can be modified. After removing the protecting group at the NH_2-terminus, the next N-protected amino acid is coupled in excess, analogous to the classical stepwise synthesis. The yield of the coupling reaction is monitored analytically until complete reaction is assured. Subsequently, excess reagents are removed by quantitative precipitation of the PEG–peptide from organic solution. Components not covalently bound to the polymer remain in solution and are removed by filtration. For elongation of the peptide chain, the steps—deprotection of the NH_2-terminus, coupling reaction, analytical control for completeness of reaction, and removal of excess reagent—are repeated until the entire sequence is built up. Finally, the polymeric ester group is cleaved from the

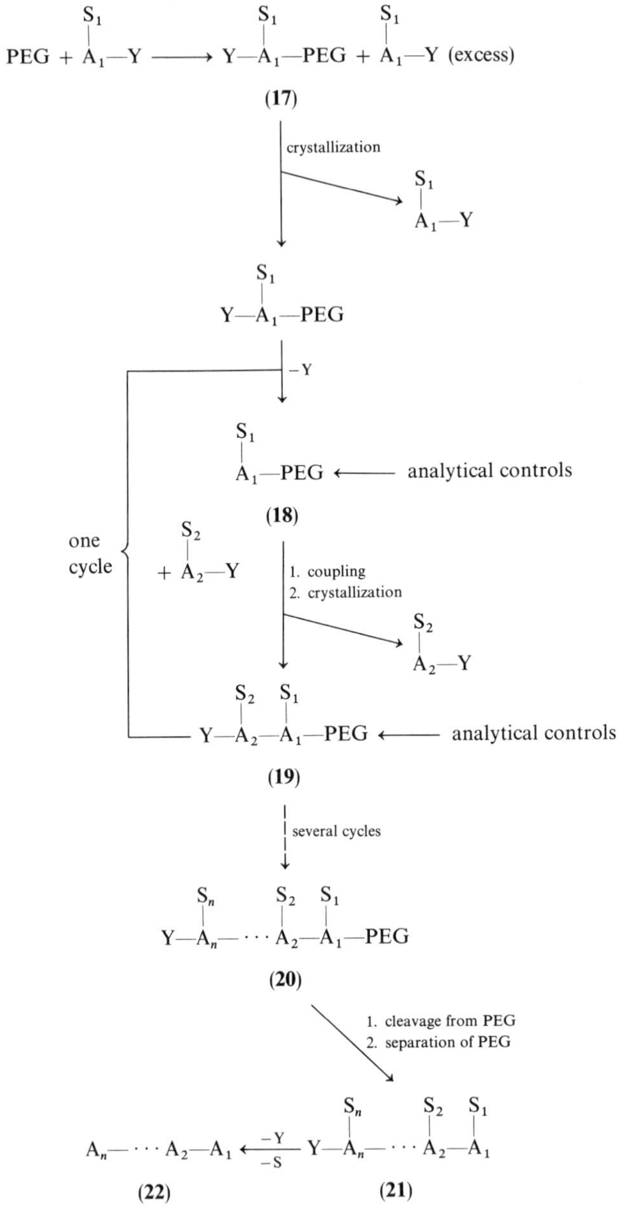
PEG + A1—Y → Y—A1—PEG + A1—Y (excess)
(17)
crystallization
Y—A1—PEG
−Y
A1—PEG ← analytical controls
(18)
one cycle
+ A2—Y
1. coupling
2. crystallization
Y—A2—A1—PEG ← analytical controls
(19)
several cycles
Y—An— · · · A2—A1—PEG
(20)
1. cleavage from PEG
2. separation of PEG
An— · · · A2—A1 ← Y—An— · · · A2—A1
(22)
(21)

Scheme 4

peptide and removed by selective precipitation or chromatographic procedures.

Alternative strategies are indicated in cases where (i) the coupling or deprotection step does not proceed to completion, (ii) analytical tests indicate deletion in the peptide chain, and (iii) the solubility of the polymeric peptide ester is too low to allow further prolongation of the peptide chain. Principally, the stepwise, polymer-mediated strategy does not permit intermediate purification of the peptide, e.g., by crystallization of the peptide. A number of alternative strategies have been proposed to overcome this problem. The most rational method would be the cleavage of the side-chain protected peptide from the polymeric ester group and purification of the desired peptide. Subsequently, the synthesis could be continued by the stepwise strategy after reattaching the peptide to the polymer. So far, no satisfactory procedure is available that would allow the reattachment of a peptide to a polymeric group, not even in homogeneous solution. Therefore, the continuation of the stepwise strategy according to the liquid-phase procedure, after intermediate purification of the peptide, is difficult to achieve. However, as long as the peptide is soluble in organic solvents (which is a prerequisite for an efficient purification procedure) the stepwise synthesis can be continued using a low-molecular-weight COOH-terminal protecting group as in classical synthesis. In this case, the cleavage from the PEG group is accomplished by transesterification. By using low-molecular-weight polyglycols as solvent, e.g., diethylene glycol monomethyl ether, the solubilizing effect of the COOH-terminal ester group can be retained to a certain degree (Anzinger *et al.*, 1979).

Alternatively, the synthesis is continued by segment condensation. The stepwise synthesis of small- to medium-sized peptides, with subsequent segment condensation, appears to be the most promising strategy for building up longer peptide chains. The liquid-phase method offers specific advantages in respect to this strategy: (1) The synthesis of the segments is very efficient and high purity of the products is guaranteed by the analytically, fully controlled synthesis; (2) intermediate solubility problems are eliminated by the solubilizing polymeric ester group; (3) the same tactics can be used for side-chain protection as in classical synthesis, and the cleavage of the PEG ester can be performed under mild conditions; and (4) PEG-bound peptides can be used directly for segment condensation, thus allowing easy separation of the reaction products.

Several methods have been developed to avoid the formation of failure sequences resulting from incomplete coupling reactions. In cases where the coupling reaction proceeds very slowly due to steric hindrance of the coupling components, quantitative yields can be achieved only by applying large excesses of reagent and long reactions times. This may result either in complications from side reactions or in racemization. The alternating liquid–solid-phase procedure developed by Frank and Hagenmaier (1975) was devised to overcome this problem (see Section II,D). The combination of this strategy with the standard procedure of liquid-phase synthesis is readily accomplished when monofunctional polyethers are used as COOH-terminal protecting groups.

Since the influence of the NH_2-terminus upon the physical properties of PEG–peptides turned out to be substantial, the direct removal of unreacted (N-unprotected) product from the desired peptide (N-protected) is possible in the case of the monomethoxy PEG esters.

A less-satisfying procedure for preventing the formation of failure sequences involves the irreversible blocking of residual amino groups. Blocking groups such as 3-nitrophthalic anhydride (Wieland *et al.*, 1969), 2-sulfobenzoic acid anhydride (Birr *et al.*, 1973), and 3-sulfopropionic anhydride (Wissmann and Geiger, 1970)—all first introduced in solid-phase synthesis—result in truncated peptides which can be separated more easily from the desired product in the final purification of the peptide. By using highly reactive blocking groups that are bound to an insoluble matrix, unreacted peptide ester can be immobilized and removed by filtration. Cross-linked polystyrene and glass beads carrying isocyanate groups proved to be most effective.

D. Experimental Procedures

1. *Attachment of the α-Carboxyl Group to PEG*

At present, a number of COOH-terminal protecting groups, possessing widely differing chemical stability, are available. In principle, any protecting group used in classical procedures can be attached to the chain ends of PEG. To this end, the hydroxyl groups of PEG are modified in appropriate manner. Thus far, aliphatic and benzyl ester bonds have been used predominantly for liquid-phase synthesis. However the accessibility of a larger number of anchoring groups is highly desirable in view of the complexity of peptide synthesis.

a. Alkyl Ester Links. The direct esterification of PEG with an N-protected amino acid results in an ester bond with chemical stability similar that of low-molecular-weight aliphatic esters. By applying a large excess of activated COOH-terminal amino acid, high yields are obtained. For example, Boc-Gly-OH and Boc-Ala-OH can be quantitatively attached to the hydroxyl end groups of PEG using fourfold excess of symmetrical amino acid anhydride: the esterification of sterically hindered amino acid derivatives most notably trifunctional residues, proceeds with somewhat lower yields.

The use of dicyclohexylcarbodiimide in the presence of 1-hydroxybenzotriazole (DCC–HOBt) for activation of the carboxylic component resulted in formation of considerable amounts of PEG–HOBt, thus reducing the esterification yield (Hemmasi and Bayer, 1977). In order to avoid esterification of unreacted hydroxylic groups in later stages of synthesis, residual groups are acetylated before removal of the N-protective group of the amino acid PEG ester. The total capacity of hydroxylic groups per weight of PEG can be determined spectroscopically by reacting the polymer with phenyl isocyanate (Bayer *et al.*, 1978); from these data, the degree of esterification and the average number molecular weight of PEG can be calculated.

During the stepwise synthesis, a slight decrease in the capacity of PEG has been observed, mainly in the first stages of synthesis. With growing peptide chain, the nucleophilic attack of the COOH-terminal ester bond is apparently rendered more difficult by steric hindrance; this is exemplified by the fact that the COOH-terminal alanine analogue of the insulin A-chain could be cleaved from PEG only in poor yields, even when drastic alkaline reaction conditions were applied (Woiwode, 1977). The aliphatic ester could, however, be hydrolyzed under mild alkaline reaction conditions in most cases. Because of the danger of side reactions during alkaline hydrolysis, most notably racemization, the use of an aliphatic ester bond cannot be recommended unconditionally. Attempts to cleave the peptide from PEG under acidic reaction conditions were not successful. On the other hand, high yields of reaction are obtained when the peptide was cleaved from PEG by transesterification, hydrazinolysis, or aminolysis.

The experimental conditions for the esterification of polyethylene glycol with *tert*-butyloxycarbonylglycine are described as follows. Twelve grams of PEG 6,000 are dried over P_4O_{10} at 1 Torr and 80°C for two hours and dissolved in 80 ml of CH_2Cl_2. In a separate vessel, 32 mmol Boc-Gly-OH and 16 mmol DCC are reacted in 20 ml of CH_2Cl_2 for 30 min at 25°C. The resulting precipitate of *N,N′*-dicylohexylurea (DCU) is filtered off and the clear solution of Boc-Gly annhydride is combined with the PEG solution. After adding 5 ml of dry pyridine, the solution is concentrated in *vacuo* to about 40 ml and kept at 25°C for 4–6 hr. The PEG ester is precipitated by slowly adding diethyl ether (about 200 ml) to the vigorously stirred solution. The mixture is stirred at 0°C for about 15 min, filtered through a glass funnel, and washed several times with ether. In order to remove small traces of excess Boc-Gly-OH, the precipitate is redissolved in a little CH_2Cl_2, DMF, or EtOH (by warming to 40°C) and precipitated as before. For the esterification of trifunctional amino acids with PEG, the reaction time and excess of reagent has to be increased. The individual yields of esterification must be determined before each synthesis, by amino acid analysis.

b. Benzyl Esters. The benzyl ester link, which is predominantly used in solid-phase synthesis, can be cleaved under acidic reaction conditions and by catalytic hydrogenation. The benzyl group is attached to the chain ends of PEG by reacting excess *p*-chloromethyl benzoic acid and dicyclohexylcarbodiimide (Künzi, 1976) or *p*-bromomethylbenzoic acid chloride (Albrecht, 1978) with PEG. Instead of PEG, its amino derivative NH_2—$(CH_2CH_2O)_n$—CH_2CH_2—NH_2 can be used, resulting in a chemically more stable amide bond between PEG chain and anchoring group. The attachment of the COOH-terminal amino acid to the polymer is achieved by reacting its cesium salt with the chain-terminal bromomethyl groups (Scheme 5).

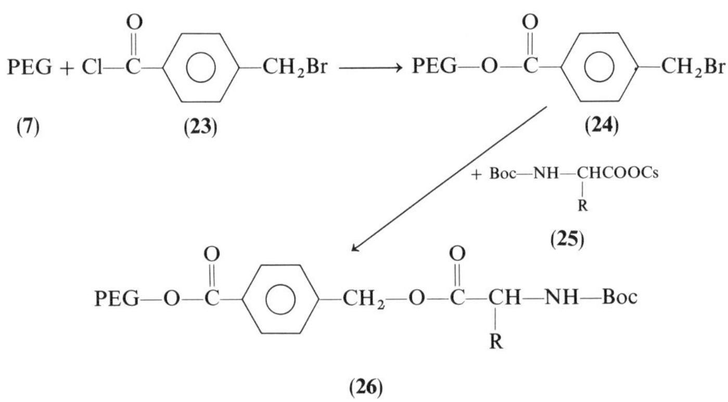

Scheme 5

According to this procedure, esterification yields of 80–100% could be obtained. No significant loss of peptide was observed during liquid-phase syntheses of a model 12-peptide (Gatfield, 1975), bradykinin, and LH-RH (Künzi, 1976). Obviously, the presence of an ester bond in the *p*-position to the benzyl ester link exerts a stabilizing effect on the hydrolytic cleavage of this linkage. Similar observations were made in solid-phase synthesis when alkyl groups were inserted in a position *para* to the benzyl group (Mitchell *et al.*, 1976). The cleavage of the benzyl ester bond is accomplished by treatment of the PEG–peptide with liquid HF, HBr in TFA, or by catalytic hydrogenation.

c. Other Anchoring Groups. A photosensitive anchoring group was attached to PEG by Tjoeng *et al.* (1977). Polyethylene glycol of molecular weight 6000 was esterified with 3-nitro-4-bromomethylbenzoic acid (**27**) using dicyclohexylcarbodiimide for activation, and the COOH-terminal N^{α}-*tert*-butyloxycarbonyl protected amino acid was subsequently linked to

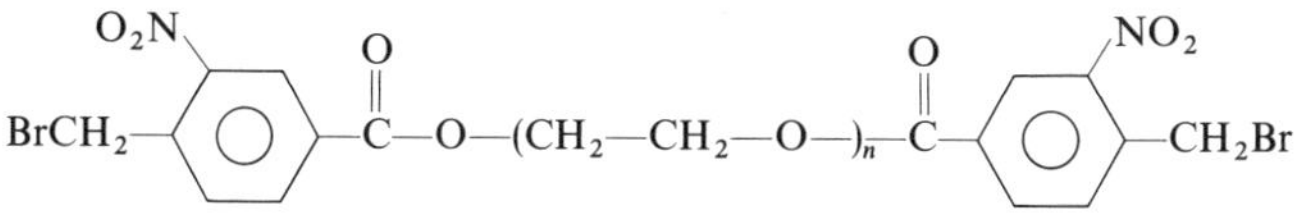

(27)

the bromomethyl group by reaction with its DEA salt. A tetrapeptide was built up stepwise and released from the PEG by photolytic cleavage. The cleavage reaction proceeded to 98% within 24 hr in the homogeneous system of liquid-phase synthesis and was considerably faster than the analogous heterogeneous reaction. This elegant and mild cleavage method, devised originally for solid-phase synthesis by Rich and Gurwara (1975), leaves aromatic residues and both acid- and base-labile side-chain protecting groups intact; it deserves further attention. The yields of attaching the anchoring group to PEG can be considerably increased by using PEG with terminal amino groups (Pillai *et al.*, 1979).

The most commonly used COOH-terminal protecting group in classical synthesis is the *tert*-butyl ester, because of its ease of cleavage by acidic reagents and its resistance toward nucleophilic attack. The attachment of a tertiary alcohol group to the chain end of monofunctional PEG was achieved according to Scheme 6 (Anzinger *et al.*, 1979). However, the sub-

$$CH_3{-}(O{-}CH_2{-}CH_2)_n{-}OH$$

(28)

1. $CH_2{=}CH{-}C(=O){-}CH_3$
2. $CH_3{-}MgJ$

$$CH_3{-}(O{-}CH_2{-}CH_2)_n{-}OCH_2{-}CH_2{-}C(CH_3)_2{-}OH$$

(29)

Scheme 6

sequent esterification of the COOH-terminal amino acid with the polymeric tertiary alcohol proceeded in poor yields. Recently, the replacement of the terminal OH-groups of the PEG chain by various functional groups (**30**) has been successfully realized. As the stability of the bond between PEG

$$X({-}CH_2{-}CH_2{-}O)_n{-}CH_2{-}CH_2{-}X$$

$$X = Cl, Br, I, NH_2$$

(30)

and COOH-terminal amino acid of the peptide sets limits upon the choice of side-chain protecting groups, tailoring of the anchoring group should allow greater flexibility in the tactical strategy of peptide synthesis.

2. *Synthesis Cycle*

a. The Coupling Step. The COOH-terminal PEG chain exerts a strong solubilizing effect on the attached peptide chain in all solvents commonly used for peptide synthesis. Usually the coupling reaction is carried out in methylene chloride, but such solvents as dimethylformamide or dimethylsulfoxide (DMSO) can also be used in liquid-phase synthesis. In the coupling steps, the neutralization of the protonated terminal amino groups of the peptide PEG ester by a tertiary base is performed subsequent to the addition of the activated carboxyl component in order to avoid the presence of large numbers of free amino groups at any time. This procedure is of utmost importance, at the stage of dipeptide synthesis, for suppressing the formation of diketopiperazines. In cases where long reaction times are necessary to achieve quantitative reaction, the presence of salts during the coupling reaction may give rise to some degree of racemization. Neutralization and removal of salts by precipitating the N-deprotonated PEG–peptide prior to the addition of the activated carboxyl component might be advantageous in these instances. For the formation of the peptide bond, all commonly used methods of classical synthesis can be applied. The use of symmetrical anhydrides, prepared separately by reacting two moles of N-protected amino acid with one mole of dicyclohexylcarbodiimide (Hagenmaier and Frank, 1972) proved to be convenient, since insoluble *N*,*N*′-dicyclohexylurea can be removed prior to coupling. Usually, two- or threefold excess of carboxylic components is applied. Depending upon the molecular weight of PEG, the coupling reaction is carried out in 10^{-1} to 10^{-2} molar solution, which is about 10 times higher than in solid-phase synthesis. The increase in viscosity due to the polymeric ester group has no influence upon the reaction rate of the coupling step (Flory, 1939; Morawetz, 1975). A gradual increase in concentration during the coupling reaction (e.g., by continuous evaporation of the solvent) is advantageous for maintaining reasonable reaction rates in the final stages of the coupling reaction. In this way, the use of very large excesses of reagent, which may result in side reactions, can be avoided. By carrying out the coupling reaction in concentrated solutions, the chances of obtaining quantitative reactions in each step of synthesis are considerably enhanced.

After completion of the coupling reaction, the peptide PEG ester is precipitated by adding diethyl ether to the reaction mixture. Excess reagent remains in solution and is removed by filtration. Before splitting off the NH_2-terminal protecting group, the coupling yield is determined quantitatively (Section III,B). When the analytical tests indicate incomplete reaction, the coupling step is repeated by the same or by another coupling method. The reaction time for obtaining quantitative coupling yields depends mainly on steric factors of the coupling components; however, effects such as solvation and conformation of the peptide chain may also influence the reaction rates. Therefore, optimal reaction conditions in terms of activation method, excess reagent, solvent, and reaction time strongly depend on the primary sequence of the peptide and must be determined for each specific sequence. Valuable advanced information can be deduced from competition experiments (Section IV,B). Reaction times for quantitative coupling in the range of a few minutes to several hours were found. Consequently, the stipulation of a time interval for the addition of one residue to the peptide chain is not possible.

b. Removal of Excess Reagents. The simplification of the procedure for the removal of excess reagents after the coupling reaction is the main motivation for using polymeric supports in peptide synthesis. In liquid-phase synthesis, the presence of a COOH-terminal macromolecular protecting group allows the selective and quantitative precipitation of the peptide ester; thus, low-molecular-weight reagents can be removed by filtration. Owing to the highly crystalline state (Section III,B), solid peptide PEG esters are obtained as granular powder and can be easily filtered and dried. Depending upon the solubility of the N-protected carboxyl component, the precipitation of peptide PEG esters is accomplished either by (i) adding diethyl ether to the coupling solution or by (ii) recrystallization from ethanolic or methanolic solution. Alternatively, excess reagents in homogeneous solution can be removed by ultrafiltration or chromatographic procedures, taking advantage of the difference in the molecular weights (Mutter, 1972). In exceptional cases, the N-protective group is removed after completion of the coupling step and excess amino acid is extracted from organic solution (e.g., CH_2Cl_2, $CHCl_3$, benzene) with water.

c. Deprotection Step. The *N-tert*-butyloxycarbonyl group is cleaved by dissolving the PEG–peptide in TFA/CH_2Cl_2 (1:1). Peptide PEG esters synthesized to date were readily soluble in this mixture. When HCl in organic solvent is used for deprotection, the concentration of HCl must be increased by about 10% in order to compensate for the formation of oxonium salts of PEG. The peptide PEG ester is isolated by adding diethyl ether to the reaction mixture. The precipitation step is repeated from ethanolic solution in order to remove the last traces of acid. N^{α}-Benzyloxycarbonyl peptides are subject to catalytic hydrogenation analogous to the classical procedure. So far, no kinetic investigations of the hydrogenation reaction have been carried out. The reaction rates seem to be of similar order to low-molecular-weight peptide ester. The completeness of the deprotection step can be determined by titration and by ir and nmr spectroscopy of the resulting peptide PEG ester (Section III,D,3). The experimental procedure for adding one amino acid residue to the growing peptide chain according to the liquid-phase method is described in the following paragraph.

For cleavage of the Boc group, 12 gm N^{α}-Boc protected peptide PEG ester (PEG 6,000) are dissolved in 80 ml of a mixture of CH_2Cl_2/TFA (1:1) and kept under exclusion of moisture at 25°C for 30 min. The clear solution is concentrated *in vacuo* to ca. 30 ml (bath temperature ca. 30°C); from this solution, the peptide PEG ester is precipitated by adding diethyl ether (ca. 200 ml) under vigorous stirring; the mixture is stirred for ca. 15 min in an ice bath and the precipitate filtered and washed with ether. After reprecipitation from ethanolic solution, the N-deprotected PEG peptide is dried *in vacuo* over P_4O_{10}. For coupling, the NH_2-terminal protonated peptide PEG ester (12 gm = 4 mmol peptide/PEG) is dissolved in 100 ml of CH_2Cl_2 (or in 50 ml of DMF). The symmetrical anhydride of the N^{α}-Boc amino acid is prepared in a separate reaction vessel by adding 10 mmol DCC to 20 mmol Boc amino acid in CH_2Cl_2 (Hagenmaier and Frank, 1972). The NH_2-terminal amino group of the peptide is deprotonized by adding *N*-methylmorpholine (1 *M* solution in CH_2Cl_2) to the solution of the peptide PEG ester, shortly before or after addition of the symmetrical anhydride, until moist pH-paper shows pH 7 (the precipitated DCU is removed by filtration of the anhydride solution). The coupling solution is concentrated to ca. 50% (w/v; after quantitative reaction (the coupling time may very between 15 min and several hr), the PEG-bound peptide is precipitated by adding diethyl ether to the coupling solution, and the product is worked up as described before. When traces of excess reagents are indicated by a thin-layer chromatogram, the PEG peptide is reprecipitated as before or by recrystallization from ethanolic or methanolic solution (5% w/v).

d. Isolation of the Peptide. After the peptide sequence has been built up, the COOH-terminal polymeric protecting group is cleaved from the peptide chain either by hydrolysis, aminolysis, hydrazinolysis, transesterification, photolysis, or hydrogenolysis. Since the polymeric support is readily soluble in liquid-phase synthesis, the separation of the two components must be achieved by taking advantage of the differences either in their chemical and physical properties or in their molecular weights.

In cases where the peptide is insoluble after splitting off the solubilizing COOH-terminal protecting group, it can be separated simply by filtration. For example, side-chain protected substance P precipitated immediately after aminolysis of its PEG ester (Bayer and Mutter, 1974). The most general method for the separation of peptide and PEG is the selective precipitation of the polymer from organic solution. This procedure can be applied in cases where the peptide is sufficiently soluble in solvents such as methanol, ethanol, or in mixtures of polar organic solvents and diethyl ether. In these solvents, PEG is quantitatively precipitated at low temperatures and can be removed by filtration. Similarly, PEG can be precipitated selectively by formation of insoluble complexation compounds with metal salts (Springer and Weigand, 1959; Shaffer and Critchfield, 1947) or with soluble polymers such as polyacrylic acid (Bailey and Koleske, 1967). These methods, which are also used for the analytical identification of PEG, are especially useful in removing traces of PEG from the peptide. The separation of peptide and PEG by gel chromatography or ultrafiltration is recommended in cases where both components show similar chemical properties but differ considerably in molecular weight. Chromatography on silica gel also proved efficient for this purpose.

3. *Analytical Methods*

For the direct application of analytical control steps to the polymer-bound peptide, the following aspects are of fundamental importance.

1. The COOH-terminal polymeric ester group (i.e., PEG) exerts no in fluence upon reactivity of the attached peptide; thus, the reaction kinetics in the liquid-phase system corresponds to that of homogeneous, low-molecular-weight systems (compare Section II,2). Since no inaccessible reaction sites are extant in homogeneous solution, the analytical controls are not subject to the limitations of solid-phase synthesis. For example, the tests for free amino groups after coupling reaction give information as to the absolute (and not the accessible) amount of unreacted amino groups. Consequently, the objection brought up in solid-phase synthesis—that the analytical controls are subject to the same limitations as the reaction to be tested—does not hold for the liquid-phase system.
2. PEG does not interfere with the analytical methods applied for the control of synthesis. For example, PEG is optically transparent to 200 nm, shows no ir absorption in the range of the amide band and its ^{1}H- and ^{13}C-nmr spectra are extremely simple (Mutter *et al.*, 1977; Leibfritz *et al.*, 1978). Thus, chemical and physical methods, e.g., nmr, ir, or CD spectroscopy, can be applied directly to peptide PEG esters.

The completeness of reactions in stepwise liquid-phase synthesis can thus be monitored with high reliability by several independent chemical and physical analytical methods.

So far, the fluorescamine and the ninhydrin methods have been used predominantly for control of the coupling reaction (Hagenmaier and Mutter,

1974). The dansyl method was introduced for control in liquid-phase synthesis by Stein (1975). For a rapid qualitative test, a sample of the reaction mixture (e.g., 20 μmol peptide PEG ester) is precipitated with ether, filtered, and heated in ninhydrin reagent (Moore *et al*., 1959). By comparing the color in the test tube with that of a sample of the uncoupled product developed analogously (e.g., 0.5 μmol peptide PEG ester), the yield can be estimated roughly (e.g., $\approx 98\%$ when the same intensity of the color reaction appears). Less than 0.1% of unreacted amino groups can be detected when a scanner or an amino acid analyzer are used for quantitative evaluation. The fluorescamine method and ir spectroscopy, in conjunction with stop-flow technique, was used for monitoring the kinetic course of the coupling reaction (Hagenmaier and Mutter, 1974; Mutter and Hagenmaier, 1974). Since PEG has a strong solubilizing effect in water, titration of residual amino groups offers a useful analytical tool (Schattenberg, 1975).

The quantitative removal of the NH_2-terminal protecting group can be tested by titration, ninhydrin reaction, or spectroscopy (Mutter, 1972; Hagenmaier and Mutter, 1974). Recently ^{13}C-nmr spectroscopy was proposed as a tool for control of liquid-phase synthesis (Albert *et al*., 1978; Leibfritz *et al*., 1978). ^{19}F nmr was applied for the detection of failure peptides and truncated sequences (Bayer *et al*., 1972). For amino acid analysis, the peptide PEG ester is subject to hydrolysis and analyzed directly; the soluble polymer is eluted at the front. Similarly, the total number of amino groups can be detected quantitatively without cleaving the PEG support from the peptide by subjecting the NH_2-terminal unprotected peptide PEG ester to amino acid analysis.

4. *Automation*

Automation is the ultimate goal of any polymer-mediated stepwise strategy for peptide synthesis. The automation of the main operations in the synthesis cycle of liquid-phase strategy (precipitation, evaporation, and filtration) has been successfully realized (Bayer *et al*., 1975a; Heitmann, 1978). Rapid synthesis, elimination of operative errors, and increase in total yields are major advantages of automatized procedures. Owing to the equivalence of the functional sites in PEG, high reproducibility is guaranteed in the standardized procedure of liquid-phase synthesis. However, the change in the physicochemical properties of the peptide PEG ester with growing peptide chain represents a potential limitation for the synthesis of larger peptides. Automation of synthesis is advantageous when the optimal reaction conditions for the synthesis of a specific peptide have been elucidated in advance. In this case, the analytical controls of synthesis can even be omitted. The synthesis of small- to medium-sized peptides appears presently to be within the scope of automated liquid-phase synthesis.

E. Results and Discussion

A number of model peptides have been synthesized in order to demonstrate the feasibility of the liquid-phase synthesis. The tetrapeptides Leu-Ala-Gly-Val and Ile-Ala-Val-Gly were prepared with PEG of molecular weights 4,000 to 20,000 as COOH-terminal protecting group by applying various coupling methods and N-protecting groups (Mutter, 1972; Mutter *et al.*, 1972); overall yields of ca. 70% were obtained when mixed anhydrides, symmetrical anhydrides, or the dicyclohexylcarbodiimide method were used for coupling. Quantitative coupling reactions were achieved within less than one hour by applying a two- to threefold excess of carboxyl component. Reaction times of several hours were necessary for completion of the coupling step when *p*-nitrophenyl esters were used. The removal of excess reagent was accomplished by either precipitation, recrystallization, ultrafiltration, or extraction with water. Taking advantage of the strong solubilizing power of PEG in water, a model tetrapeptide was synthesized in aqueous solution using *N*-2-(cyclohexyliminomethyleneamino)ethyl-*N*-methylmorpholino-*p*-toluene sulfonate (**31**) as water soluble coupling reagent (Mutter, 1972). The coupling rates in aqueous solution were considerably lower compared to analogous reactions in organic solution. On the other hand, manipulation

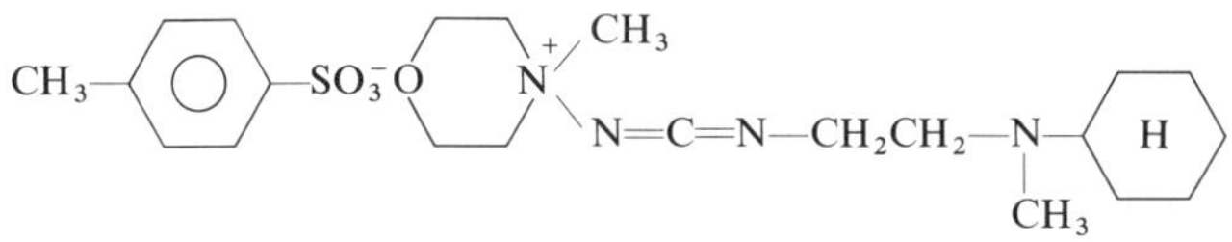

(**31**)

of the synthesis cycle is greatly facilitated when the coupling reaction is carried out in an ultrafiltration cell and when membrane filtration is used for removing excess reagent. This elegant strategy appears to be practicable for the synthesis of smaller peptides that are readily soluble in water, in the form of their PEG esters.

The 12-peptide (Leu-Ala)$_6$ was synthesized using PEG 6,000 as soluble support (Gatfield, 1975). The absence of the dipeptides Leu-Leu or Ala-Ala in the partial hydrolysate proved that no deletion peptides were formed during stepwise liquid-phase synthesis. The COOH-terminal pentapeptide of calcitonin with the sequence Ala-Gly-Leu-Gly-Pro was built up within 2 days using PEG 10,000 and symmetrical anhydrides for coupling. Less than 1% of deleted peptide could be detected in the raw material, applying an optimized chromatographic procedure for analytical investigation (Frank and Hagenmaier, 1974a). Hagenmaier (1975) carried out systematic investigations for the optimization of polymer-supported peptide synthesis. To th end, deaminocalcitonin M(19–32)-peptide was synthesized using cross-linked polystyrene supports and soluble macromolecular protecting groups, i.e., PEG. In liquid-phase synthesis, the peptide was built up with prolyl PEG ester of molecular weight 6,000 as starting material using N^{α}-*tert*-butyloxycarbonyl protected derivatives and symmetrical anhydrides for coupling. The fully protected 14-peptide was readily soluble in dimethylformamide and precipitated after alkaline hydrolysis in H_2O. Preparative ion-exchange-chromatography resulted in a fraction A comprising 78% and a fraction B comprising 15% of the raw material. Further investigation proved that this inhomogeneity was due to partial cleavage of the benzyl ether group of tyrosine in fraction B. Thus, the correct sequence of the 14-peptide was established in yields of 92%. When the average loss of 0.5% per precipitation step and the low yield of the cleavage reaction from PEG (64.6%) were taken into account, the overall yield of analytically pure 14-peptide was 32.7%, with respect to the COOH-terminal prolyl PEG ester.

Compared to an optimized synthesis of the same peptide using a solid support (Frank and Hagenmaier, 1974b), the purity of the liquid-phase product was considerably higher. The heterogeneity of the solid-phase peptide was explained by inhomogeneous reaction sites in the cross-linked polystyrene matrix that resulted in incomplete coupling reactions with growing peptide chain. The high purity of the liquid-phase product was evaluated as proof of the absence of such polymer effects in PEG-supported synthesis. The strong solubilizing strength of PEG was used for the synthesis of a number of hydrophobic peptides that could be subjected to conformational studies in the form of their PEG esters (Section IV,A). For example, the homooligopeptides Ala_{10}, Val_8, and Ile_8 (Bonora and Toniolo, 1978; Mutter *et al.*, 1977; Mutter, 1977, Mutter, 1979b) and the alamethicin model peptide $(Ala\text{-}Aib)_7$ (Jung *et al.*, 1977; Mayr *et al.*, 1979) were obtained in good yield and high purity according to the standard procedure of liquid-phase strategy.

In all series, a gradual decrease in solubility with growing peptide chain was observed; polar solvents such as DMF and DMSO had to be used to obtain homogeneous solutions of the peptide PEG esters. The corresponding peptide analogs with low-molecular-weight ester groups proved to be insoluble in these solvents. The crystallization tendency of the peptide PEG esters however did not change significantly. Solutions of NH_2-terminal unprotected peptide esters were considerably less viscous in CH_2Cl_2 than the corresponding N-protected analogs. A pronounced influence of the conformation of the growing peptide chain on the physicochemical properties was observed during the stepwise synthesis of the homooligopeptide $[Glu(OBzl)]_{20}$–PEG (Mutter, 1977).

Several peptides with biological activity have been synthesized by the liquid-phase strategy. The 11-peptide, substance P, was obtained in overall yield of 30% with full biological activity (Bayer and Mutter, 1974). Symmetrical anhydrides were used throughout for coupling. PEG 15,000 solubilized the fully protected peptide, which is sparingly soluble as low-molecular-weight analogue in methylene chloride, in dimethylformamide, and in trifluoroethanol. The unexpected increase in solubility with growing peptide chain was attributed to the onset of a helical structure (Section IV,A).

Oxytoxin was synthesized using PEG 10,000 as polymeric support (Schattenberg, 1975). Somatostatin was built up by segment condensation of the COOH-terminal heptapeptide PEG ester with a fully protected heptapeptide and subsequent oxydative cyclization (Schattenberg, 1975). Künzi (1976) used a benzyl anchoring group attached to PEG 4,000 for liquid-phase synthesis of bradykinin. The fully protected peptide PEG ester, which was built up using symmetrical anhydrides for coupling, was readily soluble in mixtures of methylene chloride and dimethylformamide. The overall yield of analytically pure bradykinin was 20%. By a similar strategy, the peptide hormone LHRH was synthesized using symmetrical anhydrides of N^{α}-*tert*-butyloxycarbonylamino acids for coupling (Künzi, 1976). After aminolysis of the side-chain protected nonapeptide PEG ester, the peptide was separated from the polymer by chromatographic procedures, with a yield of 16.5% with respect to the starting glycyl PEG ester, and a peptide was obtained which behaved identically to a product synthesized by classical methods. The use of PEG 20,000 with an aliphatic COOH-terminal ester bond resulted in less satisfying yields of LHRH (Stein, 1975).

Fully side-chain protected segments of secretin suitable for usage in classical segment condensation have been prepared by Goehring and Jung (1975a,b,c). Comparative investigations showed that the coupling yields in liquid-phase synthesis with PEG 10,000 as polymeric support were considerably higher than in solid-phase synthesis.

ACTH(1–10)-peptide and ACTH(11–23)-peptide have been built up using PEG 15,000 as solubilizing protecting group (Mutter, 1978). During the stepwise synthesis, the conformation of the peptide PEG esters was investigated by CD measurements in a variety of solvents, including water (Section IV,A). Conformational studies were also carried out during liquid-phase synthesis of myoglobin(66–73)-peptide PEG ester (Uhmann, 1971; Mutter *et al.*, 1976). The stepwise synthesis and conformational studies of the NH_2-terminal α-helical partial sequence of the membrane-modifying peptide antibiotic alamethicin were described by Mayr *et al.* (1979). The peptide was prepared using N^{α}-*tert*-butyloxycarbonylamino acid 1-hydroxybenzotriazole esters for coupling and PEG 10,000 as solubilizing ester group.

Coupling reactions in which the sterically hindered residue α-aminoisobutyric acid is involved required large excess of acylating component for completion of reaction.

Recently, insulin A-chain was synthesized using N^{α}-*tert*-butyloxycarbonylamino acid 1-hydroxybenzotriazole esters for coupling and PEG 20,000 as support (Woiwode, 1977). The analysis of the fully side-chain protected peptide PEG ester indicated that the coupling yields were very high. However, the cleavage of the peptide from PEG turned out to be extremely difficult. Drastic alkaline reaction conditions resulted in considerable damage to the peptide. Preliminary results indicate that this problem can be eliminated by using an acid-labile or photosensitive anchoring group.

Weber and Hartter (1979) investigated the segment condensation of insulin A- and insulin B-chain on solid and liquid supports. The stepwise condensation of five segments that were synthesized by classical procedures (Weber, 1975) resulted in overall yields of 15–20% of biologically active insulin A-chain when PEG 10,000 was used as COOH-terminal solubilizing group. Unreacted product could be removed by recrystallization of the PEG components, which were soluble in mixtures of methylene chloride and dimethylformamide. The fully protected PEG ester of insulin A-chain was readily hydrolized under mild alkaline conditions. The yields for the analogous synthesis using 2% cross-linked polystyrene as insoluble COOH-terminal protecting group were significantly lower, mainly because of lower coupling rates in the segment condensation step and incomplete cleavage of the final product from the resin. For similar reasons, the overall yield in the synthesis of insulin B-chain by segment condensation of six partial sequences was below 2% in the solid-phase system. In this case, the use of a solubilizing polymeric protecting group, e.g., PEG 10,000 or PEG 20,000, did not improve the results of the heterogeneous system. Here, the poor solubility of the segments caused difficulties in the separation of the condensation components. However, the low biological activity of the liquid-phase product was mainly attributed to damage of the peptide during alkaline hydrolysis of its PEG ester. Again, the use of an acid-labile anchoring group should provide a solution to this problem.

From the present data, the following conclusions can be drawn.

1. Liquid-phase synthesis has proved very efficient for the rapid synthesis of medium-sized peptides. Due to the fully analytically controlled, stepwise synthesis, the formation of deleted sequences due to incomplete reactions is avoided. By the use of PEG as soluble support, the problems associated with heterogeneous polymer-mediated synthesis can be eliminated.

2. Owing to the solubilizing power of PEG, peptides with low solubility become accessible for synthesis and conformational investigation. The solubilizing effect on the peptide does not differ significantly for PEG with

molecular weights of 2,000 to 20,000. However, the crystallization of peptide PEG esters is more readily achieved for the higher molecular weights. The use of PEG 6,000 with a capacity of 0.33 mmol/gm is recommended in most cases. From the practical side, starting amounts up to 50 gm of PEG can be easily handled in the laboratory.

3. For stepwise elongation of the peptide chain the use of N^{α}-*tert*-butyloxycarbonylamino acids and symmetrical anhydrides for coupling has proved most effective. Two- to threefold excess of acylating agent and reaction times of less than two hours are generally sufficient for quantitative coupling.

4. For the synthesis of biologically active peptides, anchoring groups other than the aliphatic esters are highly preferable. The availability of a variety of anchoring groups with different chemical stability at the chain ends of PEG for the attachment of COOH-terminal amino acids is under extensive investigation.

IV. APPLICATIONS OF THE LIQUID-PHASE METHOD

A. Conformational Studies of PEG-Bound Peptides

Conformational studies have become a major motivation for peptide synthesis. However, the investigation of the conformational behavior of peptides in solution is limited by the low solubility of even relatively short peptides. This limitation can often be overcome by the use of liquid-phase synthesis, since the solubility of the peptide is greatly enhanced by the presence of a macromolecular COOH-terminal protecting group. For the direct investigation of polymer-bound peptides, it is essential that the polymeric group have no influence upon the conformation of the peptide and that it does not disturb the application of the spectroscopic methods used in conformational studies. PEG, which has been used exclusively in liquid-phase synthesis, fulfils these requirements to a high degree. For example, commonly used investigational methods such as nmr, CD, ir, and uv spectroscopy can be applied to peptide PEG esters (Mutter *et al.*, 1977; Bonora *et al.*, 1979; Toniolo *et al.*, 1979a,b,c). Systematic studies have been carried out in order to determine the influence, if any, of the COOH-terminal protecting group on the conformation of the attached peptide (Mutter, 1977, Bonora *et al.*, 1978). To this end, several homooligopeptides were synthesized by the liquid-phase procedure using PEG of different molecular weights as soluble support. The CD was measured for various chain lengths of the peptide and compared to that of the corresponding low-molecular-weight ester analogs. In these studies, no significant influence of the PEG group on the conformation of the homooligomers could be detected. Most notably, the presence of the polymeric group did not interfere with the formation of intermolecularly ordered β-structures. This finding is readily explained by the conformational properties of the PEG chain in solution (Section III,A).

The possibility of investigating the conformation of solubilized peptide esters during liquid-phase synthesis provides a deeper insight into the relationship between conformation and chain length, primary sequence, protecting groups, and solvent. This has been demonstrated during liquid-phase synthesis of a number of peptides possessing interesting conformational features. For example, a close relationship between conformation and physicochemical property of peptides could be established in the case of oligo[LGlu(OBzl)] (Mutter, 1977) and substance P (Bayer and Mutter, 1974). Circular Dichroism investigations of PEG-bound ACTH sequences revealed a strong tendency for ordered structures at the NH_2-terminus of the peptide (Mutter, 1978). A significant increase in helical structure was observed when the helix-promoting sequence Pro-Ala-Ala was inserted at the NH_2-terminus. Transitions from random-coil conformation to partially helical structures, at chain lengths of about eight to ten residues, were observed during the stepwise liquid-phase synthesis of myoglobin(66–73), substance P (Mutter *et al.*, 1975) homo-L-oligomethionine (Toniolo *et al.*, 1978b,c) and model peptides for alamethicin (Mayr *et al.*, 1979). The strong helicogenic character of the rarely occuring α-aminoisobutyric acid became apparent in the study of the alamethicin models; thus, despite the lower intrinsic chirality, the model peptide $(\text{Ala-Aib})_5$ exhibited ellipticities in the CD spectra similar to those of Ala_{10}. In the case of the PEG-bound hydrophobic homo-oligopeptide Ala_n, a transition from unordered structures to β-conformation occured at a chain length of seven to eight residues in trifluoroethanol and in water (Fig. 2) (Mutter *et al.*, 1977; Toniolo *et al.*, 1979a; Bonora *et al.*, 1979). A transition from random-coil to β-structure was also observed for PEG-bound oligo[LVal] and oligo[LIle] (H. Mutter, 1978; Bonora and Toniolo, 1979). For oligo[LAla], a pronounced influence of the NH_2-terminus upon the conformation of the peptide was observed. Thus, the N^α-*tert*-Butyloxycarbonyl protected 10-peptide PEG ester showed CD spectra typical for partially helical structures in trifluoroethanol, whereas its N-unprotected, protonated form exhibited a pure β-structure in the same solvent.

Preliminary results indicate that the reactivity of the NH_2-terminal amino group in coupling reactions is strongly influenced by the conformation of the peptide (Mutter, 1979a). For example, during stepwise synthesis of oligo [LVal] PEG ester, a significant decrease in the reaction rates paralleled by a change from random-coil conformation to β-like structures was registered. The systematic elucidation of these effects represents an interesting

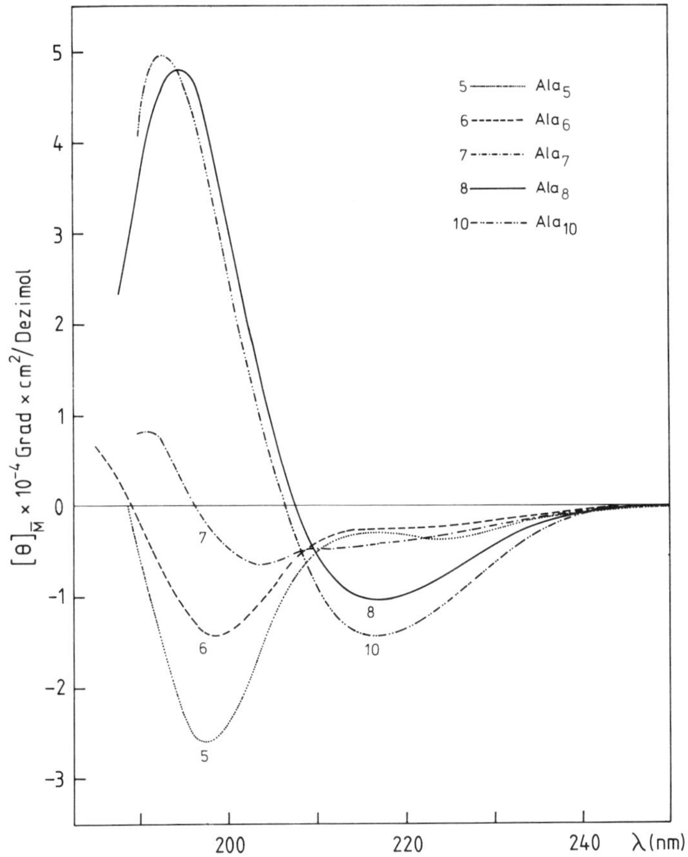

Figure 2. Circular dichronium spectra of H-(Ala)$_n$-OPEG (Mol wt 6×10^3) in trifluoroethanol (0.5 mg peptide/ml) for various chain lengths *n*.

challenge for liquid-phase synthesis and is at present under investigation. ^{13}C and ^{1}H nmr studies (Leibfritz *et al.*, 1978; Goodman *et al.*, 1979) of PEG-bound peptides have been performed. In ^{13}C nmr spectra, PEG exhibits one signal at 71.5 ppm for the inner chain carbons. In conformational studies of model compounds of alamethicin, this signal has been eliminated from the spectrum in a double resonance experiment by selective saturation of the methylene carbons. High resolution revealed distinct differences in the chemical shifts of the peptide carbons between random-coil and α-helical conformations. Owing to the strong solubilizing effect of PEG in solvents such as $CDCl_3$ and D_2O, the high concentrations necessary for effective application of nmr spectroscopy to peptide PEG esters can be achieved readily (Goodman *et al.*, 1979) These examples give only a limited indication of the potentialities of the new method. Since liquid-phase synthesis allows efficient synthesis of peptides in very short periods of time, much more information about the relationship between conformational and physicochemical properties of peptides should be available in the near future.

B. Competition Experiments

For the optimization of stepwise synthesis, a knowledge of the reaction rates for the condensation of individual amino acids is of exceptional value. At present, only a limited amount of such data is available (Ragnarsson *et al.*, 1971, 1974; Bayer *et al.*, 1975a). Consequently, standardized reaction conditions are normally chosen for each coupling step, without taking into account the specific requirements of each individual step. Competitive coupling reactions afford a simple and effective method for the evaluation of relative reaction rates for individual amino acids. Furthermore, effects on the coupling rate of solvent, activation method, and protecting groups can be studied. Liquid-phase synthesis offers a convenient and efficient method for the realization of such experiments. In a typical model system, a PEG-bound NH_2-terminal amino acid is reacted with equimolar amounts of several N-protected activated amino acids (Scheme 7).

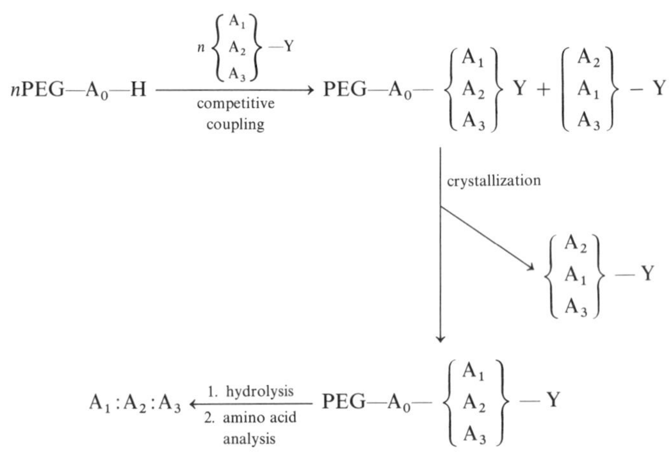

Scheme 7

After competitive coupling, excess reagents are removed by precipitating the PEG–dipeptides. Direct hydrolysis of the PEG–peptides and amino acid analysis give the amount of incorporation, i.e., the relative reaction rates for the individual amino acids. As the dipeptide PEG esters are subjected to analysis without cleavage from the polymer, this procedure is extremely rapid and guarantees reproducible results. Once the absolute reaction rate is known for one competing amino acid, the time for quantitative reaction of any other component can be predicted from the experiment, as linear kinetics are assured in liquid-phase synthesis (Bayer *et al.*, 1974). Preliminary results from such experiments show that the relative reaction rates for individual amino acids differ by factors of up to 50 (Mutter, 1979a). A significant dependence of the relative reaction rates upon the nature of the amino component, the solvent, and the protecting group of the carboxyl components is indicated (Bayer *et al.*, 1975a; Mutter, 1976, 1979a). Absolute and relative reaction rates in liquid-phase synthesis have been determined also by ir and uv spectroscopic investigations of the aminolysis of amino acid anhydrides and active esters, respectively (Bayer *et al.*, 1974; Mutter and Hagenmaier, 1974).

C. Polymer Reagents

The main motivation for using polymer-bound reagents in organic synthesis is the ease with which such reagents can be manipulated and removed from the desired reaction product. The use of soluble polymers offers two distinct advantages over solid supports: (1) all reactions are carried out in homogeneous solution and (2) insoluble reactants are solubilized in a variety of solvents. For the removal of the polymer-bound reagent from the reaction mixture, several procedures that take advantage of the difference in the physical and chemical properties between polymeric and low-molecular-weight component are available.

By analogy to solid polymer reagents devised for peptide synthesis (Fridkin *et al.*, 1966; Heusel *et al.*, 1977), PEG-bound active amino acid esters have been synthesized (**32** and **33**). The application of these polymer reagents,

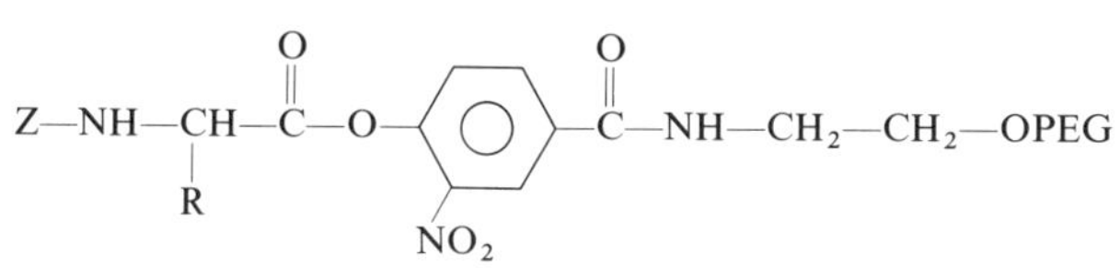

(32)

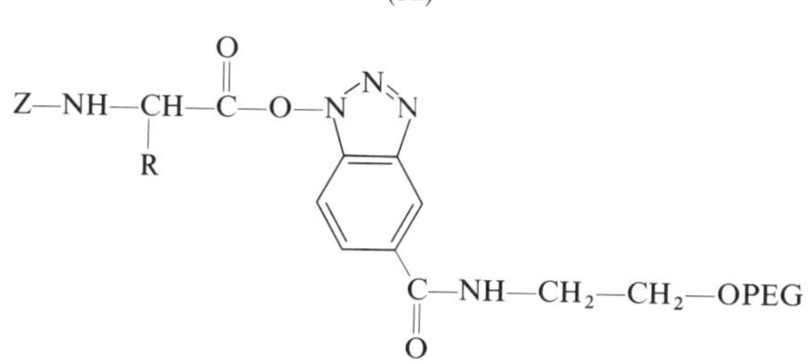

(33)

which are readily soluble in all solvents commonly used in peptide synthesis, is advantageous for the stepwise, classical synthesis (Mutter, 1978) (Section II,A,2). The polymer-bound reagent can be used in large excess and is readily removed by crystallization from MeOH and EtOH or by direct precipitation from the reaction solution with diethyl ether. As long as the peptide is adequately soluble in these solvents, this method represents one of the most efficient and rapid procedures for building up peptides. Thus, several model peptides could be obtained in high yield and purity in a time comparable to polymer-supported synthesis. The ease of accessibility of the PEG-bound amino acid derivatives, the flexibility in the choice of solvents, and the higher reaction rates compared to the analogous solid-phase reagents provide a wide field of application for this strategy.

D. PEG–Peptides as Models for Biologically Active Compounds

One of the most ambitious applications of synthetic polymers involves their use as a replacement for the protein component in biological active molecules, e.g., in metalloproteins. First results on model systems for the reversible binding of oxygen in synthetic hemopolymers indicate that this concept is not unrealistic (Bayer and Holzbach, 1977). For example, heme-containing functionalized polymers, e.g., derivatives of polyethylene glycol or polyvinyl pyrrolidinone, for the reversible transport of oxygen have been prepared with properties similar to those of the natural oxygen carriers hemoglobin and myoglobin. To this end, H-His-Gly-PEG ester was synthesized according to the procedure of liquid-phase synthesis and heme covalently attached by formation of an amide bond between the amino group of histidine and one of the carboxylic groups of heme. The role of the peptide PEG ester is threefold in this case. It must afford good solubility of the heme part in water, hindrance of irreversible oxidation of the oxygen complex, and imitation of the distal imidazole.

Reversible oxygenation of the Fe(III)-complex of this model compound was possible several times in aqueous solution at room temperature under reduced pressure. The oxygen partial pressure at half saturation of these compounds equalled the values of myoglobin.

Recently PEG–peptides have been used to synthesize models for the visual pigment, rhodopsin. It is known that in rhodopsin an ε-amino group of a lysyl residue of the protein opsin is bound via Schiff's base link to vitamin A aldehyde. Attempts to imitate the protein part with low-molecular-weight Schiff's bases of retinal failed to reproduce the spectroscopic properties of the natural compound. With PEG–peptide retinal Schiff's bases like PEG-Ala-Lys-Tyr(Boc)-retinalimine, the spectral properties of rhodopsin could be closely approached (Hannak, 1977; Das *et al.*, 1979).

E. Oligonucleotide Synthesis

The strategy of liquid-phase synthesis has also been applied to the synthesis of oligonucleotides. The efficient stepwise synthesis requires the formation of phosphate ester bonds in high yields (Agarwal *et al.*, 1972; Schott *et al.*, 1973; Brandstetter *et al.*, 1973; Koester, 1972). The originally applied linear polystyrenes as soluble support (Hayatsu and Khorana, 1966; Cramer *et al.*, 1966) proved not to be ideal for this purpose, mainly because of a considerable decrease in solubility during synthesis. Subsequently, polymers with more favorable physicochemical properties, such as polyethylene glycol, and with high capacity, such as polyvinyl alcohol and polyethyleneimine, were used (Schott *et al.*, 1973; Schott, 1973). In the case of the more efficient strategy of segment condensation of smaller, fully protected segments to larger oligonucleotide chains, polymers with low capacity, e.g., polyethylene glycol, proved most effective. The first nucleotide was attached to the polymer by phosphoramidate and thiophosphoramidate bonds in yields of 60–90% (Brandstetter, 1972; Ohtsuka *et al.*, 1970; Cook, 1970). After the condensation step, excess reagents were removed by ultrafiltration or chromatography on DEAE-cellulose, when PEG was used as soluble support. The yields in subsequent steps decreased considerably with chain length, resulting in overall yields of 14% and 8% for the trinucleotide d(pA-A-A) and for the pentanucleotide $d(T)_5$, respectively.

F. Other Applications

In contrast to heterogeneous catalysis, homogeneous catalysts often exhibit higher substrate selectivity, better reproducibility, and better reactivity under milder conditions. The frequently encountered problem of separating low-molecular-weight catalysts from reaction products can be circumvented by binding the catalyst to soluble polymers. To this end, noncross-linked linear polystyrene was chloromethylated and converted into the soluble polydiphenyl(styrylmethyl)phosphane by treatment with potassium diphenylphosphide (Bayer and Schrig, 1975). The soluble macromolecular metal complex, which catalyses the homogeneous hydrogenation or hydroformylation of alkenes under mild conditions, was prepared by addition or by ligand exchange reaction of transition metal derivatives.

Continuous homogeneous catalysis (or cyclic catalyst reused) is achieved by membrane filtration, which separates the polymeric catalyst from low-molecular-weight products. Hydrogenation of 1-pentene with soluble polymer-attached Wilkinson catalyst resulted in quantitative yield of *n*-pentane. A variety of other catalysts have been attached to functionalized polystyrenes (Bayer and Schurig, 1975, 1976). In addition to linear polystyrenes, polyethylene glycols, polyvinyl pyrrolidinones, and polyvinyl chlorides were used for liquid-phase catalysis. A broad field of application is opened by PEG functionalized with terminal metal centers (Bayer and Kutubuddin, 1979).

Macromolecular soluble complexing agents have been prepared for selective concentration of metals from diluted solutions (Lange, 1975; Jaeger, 1975). Highly specific low-molecular-weight metal-complexing agents were covalently bound to soluble polymers (such as polyacrylic acid, polyvinyl pyrrolidinone, polyethyleneimine, or polyethylene glycols) that form very stable metal complexes.

Such polymeric complexing agents afford distinct advantages over their low-molecular-weight analogs. By using efficient ultrafiltration systems, the very large quantities neccessary for industrial use can be handled. Recently, the selective concentration of plutonium was successfully accomplished by this method. Also, the use of pharmacologically active polymers has recently attracted attention (Orth and Bruemmer, 1972; Batz *et al.*, 1973; Abuchowski *et al.*, 1977).

V. CONCLUSIONS

Macromolecular supports suitable for peptide synthesis must dominate the physicochemical properties of the growing peptide chain during all stages of synthesis, but at the same time, the supports should not influence the chemical reactivity of the peptide. Obviously, the fulfillment of both properties is difficult to achieve. However, the use of polyethylene glycols as "liquid support" offers an acceptable compromise: Peptide PEG esters possess physicochemical properties that allow easy separation from low-molecular-weight reagents, are readily soluble in a large variety of solvents, and do not differ in the chemical reactivity from their low-molecular-weight ester analogs. At present, the rapid synthesis of medium-sized peptides, with subsequent segment condensation to larger units, appears to be the most straightforward approach to the synthesis of long peptide chains.

The stepwise synthesis is limited in many cases by a slow decrease in the solubility with growing peptide chain and by formation of deletion peptides as a consequence of side reactions. The use of additional solubilizing groups for both side-chain protection and further improvements in the chemical methodology of peptide synthesis may render the stepwise liquid-phase synthesis feasible for the synthesis of longer peptide chains.

Whereas a number of competing strategies are available for the rapid synthesis of medium-sized peptides, the use of liquid-phase synthesis offers specific features which other methods lack. Most important is its use for the synthesis of peptides with low solubility. The direct accessibility of solubilized peptide PEG esters to conformational investigations should provide more information about the relationship between conformation and physico-chemical properties of the peptide in the near future. Soluble polymers can be used to advantage as both reagent and support in many organic reactions. Finally, with peptide PEG esters as model compounds, it is possible to simulate properties of biologically active proteins.

REFERENCES

Abe, A., and Mark, J. E. (1976). *J. Am. Chem. Soc.* **98**, 6468–6476.

Abuchowski, A., van Es, T., Palczuk, N. C., and Davis, F. (1977). *J. Biol. Chem.* **252**, 3578–3581.

Agarwal, K. L., Yamazaki, A., Cashion, P. J., and Khorana, H. G. (1972). *Angew. Chem.* **84**, 489–498.

Albrecht, H. (1978). Ph.D. Thesis, University of Tuebingen, Germany.

Andreatta, R. H., and Rink, H. (1973). *Helv. Chim. Acta* **56**, 1205–1218.

Anzinger, H., Mutter, M., and Bayer, E. (1979). *Angew. Chem.*, **91**, 747–748.

Appel, R., and Willms, L. (1977). *J. Chem. Res.* (*S*) p. 84.

Appel, R., Struever, W., and Willms, L. (1976). *Tetrahedron Lett.* pp. 905–906.

Atherton, E., Caviezel, M., Over, H., and Sheppard, R. C. (1977). *J. Chem. Soc., Chem. Commun.* **22**, 819–821.

Bailey, F. E., and Koleske, J. V. (1967). *In* "Nonionic Surfactants" (M. J., Schick, ed.), pp. 794–821. Dekker, New York.

Bailey, F. E., Jr., and Koleske, J. V. (1976). "Poly(Ethylene Oxide)." Academic Press, New York.

Bartlett, P. D., and Dittmer, D. C. (1957). *J. Am. Chem. Soc.* **79**, 2159–2160.

Bartlett, P. D., and Jones, R. H. (1957). *J. Am. Chem. Soc.* **79**, 2153–2159.

Batz, H. G., Franzmann, G., and Ringsdorf, H. (1973). *Makromol. Chem.* **172**, 27–47.

Bayer, E., and Geckeler, K. (1975). *Justus Liebigs Ann. Chem.* **1975**, 1671–1674.

Bayer, E., and Holzbach, G. (1977). *Angew. Chem., Int. Ed. Engl.* **16**, 117–118.

Bayer, E., and Kutubuddin, M. (1979). In preparation.

Bayer, E., and Mutter, M. (1974). *Chem. Ber.* **107**, 1344–1352.

Bayer, E., and Schurig, V. (1975). *Angew. Chem., Int. Ed. Engl.* **14**, 493–494.

Bayer, E., and Schurig, V. (1976). *CHEMTECH* **6**, 212–214.

Bayer, E., Eckstein, H., Haegele, K., Koenig, W. A., Bruening, W., Hagenmaier, H., and Parr, W. (1970). *J. Am. Chem. Soc.* **92**, 1735–1738.

Bayer, E., Hunziker, P., Mutter, M., Sievers, R. E., and Uhmann, R. (1972). *J. Am. Chem. Soc.* **94**, 265–268.

Bayer, E., Mutter, M., Uhmann, R., Polster, J., and Mauser, H. (1974). *J. Am. Chem. Soc.* **96**, 7333–7336.

Bayer, E., Mutter, M., and Holzer, G. (1975a). *In* "Peptides: Chemistry, Structure and Biology" (R. Walter, and J., Meienhofer, eds.), pp. 425–432. Ann Arbor Sci. Publ., Ann Arbor, Michigan.

Bayer, E., Mutter, M., Polster, J., and Uhmann, R. (1975b). *In* "Peptides 1974" (Wolman, Y., ed.), pp. 129–136. Wiley, New York.

Bayer, E., Gatfield, I., Mutter, H., and Mutter, M. (1978). *Tetrahedron* **34**, 1829–1831.

Beyerman, H. C., de Leer, E. W. B., and Floor, J. (1973). *Recl. Trav. Chim. Pays-Bas* **92**, 481–492.

Birr, C. (1973). *In* "Peptides 1972" (Hanson, H. and Jakubke, H. D., eds.), pp. 72–77. North-Holland Publ., Amsterdam.

Birr, C., Flor, F., Fleckenstein, P., and Wieland, T. (1973). *In* "Peptides 1971" (Nesvadba, H., ed.), pp. 175–184. North-Holland Publ., Amsterdam.

Blecher, H., and Pfaender, P. (1973). *Justus Liebigs Ann. Chem.* **1973**, 1263–1268.

Bodanszky, M., and du Vigneaud, V. (1959). *J. Am. Chem. Soc.* **81**, 5688–5691.

Bodanszky, M., and Williams, N. J. (1967). *J. Am. Chem. Soc.* **89**, 685–689.

Bodanszky, M., Ondetti, M. A., Levine, S. D., and Williams, N. J. (1967). *J. Am. Chem. Soc.* **89**, 6753–6757.

Bodanszky, M., Klausner, Y. S., and Ondetti, M. A. (1976). "Peptide Synthesis," 2nd ed. Wiley, New York.

Bonora, G. M., and Toniolo, C. (1979). *Makromol. Chem.*, **180**, 2095–2099.

Bonora, G. M., Toniolo, C., and Mutter, M. (1978). *Polymer* **19**, 1382–1386.

Bonora, G. M., Palumbo, M., Tonido, C., and Mutter, M. (1979). *Makromol. Chem.*, in press.

Brandstetter, F. (1972). Ph.D. Thesis, University of Tuebingen, Germany.

Brandstetter, F., Schott, H., and Bayer, E. (1973). *Tetrahedron Lett.* pp. 2997–3000.

Cook, A. F. (1970). *J. Am. Chem. Soc.* **92**, 190–195.

Cramer, F., Hettler, H., Seliger, H., Helbig, R., and Scheit, K. H. (1966). *Angew. Chem.* **78**, 640–641.

Das, P. K., Becker, R. S., Hannak, D., and Bayer, E. (1979). *J. Am. Chem. Soc.* **101**, 239–240.

Denkewalter, R. G., Veber, D. F., Holly, F. W., and Hirschmann, R. (1969). *J. Am. Chem. Soc.* **91**, 502–503.

Erickson, B. W., and Merrifield, R. B. (1976). *In* "The Proteins" 3rd ed. (H. Neurath, and R. L., Hill, eds.) Vol. 2, pp. 255–527. Academic Press, New York.

Finn, F. M. and Hofmann K. (1976). *In* "The Proteins" 3rd ed. (H. Neurath and R. L., Hill, eds.), Vol. 2, pp. 105–253. Academic Press, New York.

Flory, P. J. (1939). *J. Am. Chem. Soc.* **61**, 3334–3340.

Flory, P. J. (1940). *J. Am. Chem. Soc.* **62**, 1561–1565.

Flory, P. J. (1953). "Principles of Polymer Chemistry." Cornell Univ. Press, Ithaca, New York.

Flory, P. J. (1969). "Statistical Mechanics of Chain Molecules." Wiley (Interscience), New York.

Frank, H., and Hagenmaier, H. (1974a). *Beckman Rep.* pp. 12–17.

Frank, H., and Hagenmaier, H. (1974b). *Tetrahedron* **30**, 2523–2531.

Frank, H., and Hagenmaier, H. (1975). *Experientia* **31**, 131–132.

Frank, H., Meyer, H., and Hagenmaier, H. (1975). *In* "Peptides: Chemistry, Structure and Biology" (R. Walter, and J., Meienhofer, eds.), pp. 439–445. Ann Arbor Sci. Publ., Ann Arbor, Michigan.

Frank, H., Hagenmaier, H., Bayer, E., and Desiderio, D. M. (1977). *In* "Peptides, Proceedings of the Fifth American Peptide Symposium" (M. Goodman and J., Meienhofer, eds.), pp. 514–517. Wiley, New York.

Fridkin, M., Patchornik, A., and Katchalski, E. (1966). *J. Am. Chem. Soc.* **88**, 3164–3165.

Fridkin, M., Patchornik, A., and Katchalski, E. (1968). *J. Am. Chem. Soc.* **90**, 2953–2957.

Fridkin, M., Patchornik, A., and Katchalski, E. (1972). *Biochemistry* **11**, 466–471.

Gatfield, I. (1975). Ph.D. Thesis, University of Tuebingen, Germany.

Goehring, W., and Jung, G. (1975a). *Justus Liebigs Ann. Chem.*, **1975**, 1776–1780.

Goehring, W., and Jung, G. (1975b). *Justus Liebigs Ann. Chem.*, **1975**, 1781–1789.

Goehring, W., and Jung, G. (1975c). *Justus Liebigs Ann. Chem.*, **1975**, 1965–1975.

Goodman, M., Ribeiro, A., and Mutter, M. (1979). in preparation.

Green, B., and Garson, L. R. (1969). *J. Chem. Soc. C* pp. 401–405.

Gut, V., and Rudinger, J. (1968a). *In* "Peptides 1968" (Bricas, E., ed.), pp. 185–186. North-Holland Publ., Amsterdam.

Gut, V., and Rudinger, J. (1968b). *Colloq. Int. C. N. R. S.* **175**, 185–188.

Hagenmaier, H. (1970). *Tetrahedron Lett.* pp. 283–286.

Hagenmaier, H. (1975). *Hoppe-Seyler's Z. Physiol. Chem.* **356**, 777–785.

Hagenmaier, H., and Frank, H. (1972). *Hoppe-Seyler's Z. Physiol. Chem.* **353**, 1973–1976.

Hagenmaier, H., and Mutter, M. (1974). *Tetrahedron Lett.* pp. 767–770.

Hancock, W. S., Prescott, D. J., Vagelos, P. R., and Marshall, G. R. (1973). *J. Org. Chem.* **38**, 774–781.

Hannak, D. (1977). Ph.D. Thesis, University of Tuebingen, Germany.

Hayatsu, H., and Khorana, H. G. (1966). *J. Am. Chem. Soc.* **88**, 3182–3183.

Heitmann, U. (1978). Ph.D. Thesis, University of Tuebingen, Germany.

Hemmasi, B., and Bayer, E. (1977). *Tetrahedron Lett.* pp. 1599–1602.

Heusel, G., Bovermann, G., Goehring, W., and Jung, G. (1977). *Angew. Chem., Int. Ed. Engl.* **16**, 642–643.

Hirschmann, R., Schwam, H., Strachan, R. G., Schoenewaldt, E. F., Barkemeyer, H., Miller, S. M., Conn, J. B., Garsky, V., Veber, D. F., and Denkewalter, R. G. (1971). *J. Am. Chem. Soc.* **93**, 2746–2754.

Imae, T., and Ikeda, S. (1972). *Biopolymers* **11**, 509–517.

Jaeger, W. (1975). Ph.D. Thesis, University of Tuebingen, Germany.

Jung, G., Bovermann, G., Goehring, W., and Heusel, G. (1975). *In* "Peptides: Chemistry, Structure and Biology" (R. Walter, and J., Meienhofer, eds.), pp. 433–437. Ann Arbor Sci. Publ., Ann Arbor, Michigan.

Jung, G., Dubischar, N., Irmscher, G., Mayr, W., and Oekonomopulos, R. (1977). *Chem.-Ztg.* **101**, 196–201.

Koester, H. (1972). *Tetrahedron Lett.* pp. 1535–1538.

Künzi, H. (1976). German Off. 2,435,642; Patent Appl. 2,435,642.1.

Lange, G. (1975). Ph.D. Thesis, University of Tuebingen, Germany.

Laufer, D. A., Chapman, T. M., Malborough, D. I., Vaidya, V. M., and Blout, E. R. (1968). *J. Am. Chem. Soc.* **90**, 2696–2698.

Leibfritz, D., Mayr, W., Oekonomopulos, R., and Jung, G. (1978). *Tetrahedron* **34**, 2045–2050.

Liu, K., and Parsons, J. L. (1969). *Macromolecules* **2**, 529–533.

Losse, A. (1971). *Tetrahedron Lett.* pp. 4989–4992.

Losse, A. (1973). *Tetrahedron* **29**, 1203–1208.

Losse, G., and Meisegeier, B. (1974). *J. Prakt. Chem.* **316**, 629–637.

Losse, G., and Ulbrich, R. (1972). *Tetrahedron* **28**, 5823–5829.

Maher, J. J., Furey, M. E., and Greenberg, L. J. (1971). *Tetrahedron Lett.* pp. 27–28.

Mayr, W., Oekonomopulos, R., and Jung, G. (1979). *Biopolymers* **18**, 425–450.

Merrifield, R. B. (1963). *J. Am. Chem. Soc.* **85**, 2149–2154.

Mitchell, A. R., Erickson, B. W., Ryabtsev, M. N., Hodges, R. S., and Merrifield, R. B. (1976). *J. Am. Chem. Soc.* **98**, 7357–7362.

Moore, S., Spackman, D. H., and Stein, W. H. (1959). *Anal. Chem.* **30**, 1185–1190.

Morawetz, H. (1975). *In* "Peptides: Chemistry, Structure and Biology" (R. Walter, and J., Meienhofer, eds.), pp. 385–394. Ann Arbor Sci. Publ., Ann Arbor, Michigan.

Morley, J. S. (1967). *J. Chem. Soc. C* pp. 2410–2421.

Mutter, H. (1978). Ph.D. Thesis, University of Tuebingen, Germany.

Mutter, M. (1972). Ph.D. Thesis, University of Tuebingen, Germany.

Mutter, M. (1976). Habilitationsschrift, University of Tuebingen, Germany.

Mutter, M. (1977). *Macromolecules* **10**, 1413–1414.

Mutter, M. (1978). *Tetrahedron Lett.*, pp. 2843–2846.

Mutter, M. (1979a). *Int. J. Peptide Protein Res.* **13**, 274–277.

Mutter, M. (1979b). *Biopolymers*, submitted.

Mutter, M., and Bayer, E. (1972). *Nature (London)* **237**, 512–513.
Mutter, M., and Bayer, E. (1974). *Angew. Chem., Int. Ed. Engl.* **13**, 88–89.
Mutter, M., and Hagenmaier, H. (1974). *Angew. Chem., Int. Ed. Engl.* **13**, 149.
Mutter, M., Hagenmaier, H., and Bayer, E. (1972). *Angew. Chem., Int. Ed. Engl.* **10**, 811–812.
Mutter, M., Uhmann, R., and Bayer, E. (1975). *Justus Liebigs Ann. Chem.* **1975**, 901–915.
Mutter, M., Mutter, H., Uhmann, R., and Bayer, E. (1976). *Biopolymers* **15**, 917–927.
Mutter, M., Mutter, H., and Bayer, E. (1977). *In* "Peptides: Chemistry, Structure and Biology" (M. Goodman and J. Meienhofer, eds.) pp. 403–406. Wiley, New York.
Neidlinger, H. (1975). Ph.D. Thesis, University of Mainz, Germany.
Ohtsuka, E., Murao, K., Ubasawa, M., and Ikehara, M. (1970). *J. Am. Chem. Soc.* **92**, 3441–3445.
Orth, H. D., and Bruemmer, W. (1972). *Angew. Chem.* **84**, 319–329.
Pillai, V. N. R., Mutter, M., and Bayer, E. (1979). *Tetrahedron Lett.* pp. 3409–3412.
Ragnarsson, U., Karlsson, S., and Sandberg, B. (1971). *Acta Chem. Scand.* **25**, 1487–1489.
Ragnarsson, U., Karlsson, S. M., and Sandberg, B. (E. B.) (1974). *J. Org. Chem.* **39**, 3837–3841.
Rich, D. H., and Gurwara, S. K. (1975). *J. Am. Chem. Soc.* **97**, 1575–1578.
Rudinger, J., and Buetzer, P. (1975). *In* "Peptides 1974". (Y., Wolman, ed.), pp. 211–219. Wiley, New York.
Rudinger, J., and Gut, V. (1967). *In* "Peptides 1967" (H. C., Beyerman, van de Linde and Maassen, W. van den Brink, eds.), pp. 89–90. North-Holland Publ., Amsterdam.
Schattenberg, M. (1975). Ph.D. Thesis, University of Tuebingen, Germany.
Schott, H. (1973). *Angew. Chem., Int. Ed. Engl.* **12**, 246.
Schott, H., Brandstetter, F., and Bayer, E. (1973). *Makromol. Chem.* **173**, 247–251.
Schwyzer, R., and Sieber, P. (1966). *Helv. Chim. Acta* **49**, 134–158.
Shaffer, C. B., and Critchfield, F. H. (1947). *Anal. Chem.* **19**, 32–34.
Shemyakin, M. M., Ovchinnikow, Y. A., Kiryushkin, A. A., and Kozhevnikova, I. V. (1965). *Tetrahedron Lett.* pp. 2323–2327.
Sheppard, R. C. (1973). *In* "Peptides 1971" (H., Nesvadba, ed.), pp. 111–123. North-Holland Publ., Amsterdam.
Springer, R., and Weigand, H. (1959). *Pharm. Ztg.* **104**, 1018–1028.
Stein, W. (1975). Ph.D. Thesis, University of Tuebingen, Germany.
Tadokoro, H., Chatani, Y., Yoshihara, T., Tahara, S., and Murahashi, S. (1964). *Makromol. Chem.* **73**, 109–126.
Tilak, M. A. (1970). *Tetrahedron Lett.* pp. 849–854.
Tjoeng, F.-S., Staines, W., St-Pierre, S., and Hodges, R. S. (1977). *Biochim. Biophys. Acta* **490**, 489–496.
Toniolo, G., Bonora, G. M., Palumbo, M., and Pysh, E. S. (1976). *In* "Peptides 1976" (Loffet, A., ed.), pp. 597–600. Editions de l'Université de Bruxelle, Belgium.
Toniolo, C., Bonora, G. M., and Mutter, M. (1979a). *J. Am. Chem. Soc.*, in press.
Toniolo, C., Bonora, G. M., Salardi, S., and Mutter, M. (1979b). *Polymer*, submitted.
Toniolo, C., Bonora, G. M., and Mutter, M. (1979c). *Macromolecules*, in press.
Uhmann, R. (1971). Ph.D. Thesis, University of Tuebingen, Germany.
van Zon, A., and Beyerman, H. C. (1973). *Helv. Chim. Acta* **56**, 1729–1740.
Weber, U. (1975). *Hoppe-Seyler's Z. Physiol. Chem.* **356**, 701–714.
Weber, U., and Hartter, P. (1979). in preparation.
Wieland, T., Birr, C., and Wissenbach, H. (1969). *Angew. Chem., Int. Ed. Engl.* **8**, 764–765.
Wissmann, H., and Geiger, R. (1970). *Angew. Chem., Int. Ed. Engl.* **9**, 908–909.
Woiwode, W. (1977). Ph.D. Thesis, University of Tuebingen, Germany.

Chapter 3

Polymeric Reagents in Peptide Synthesis

MATI FRIDKIN

I. INTRODUCTION

Ever since the pioneering work of Emil Fischer (1906) on the controlled stepwise synthesis of polypeptides with predetermined sequences, the chemical preparation of naturally occurring and model peptides and polypeptides

THE PEPTIDES, VOLUME 2

ISBN 0-12-304202-X

has been a prime challenge for chemists and a matter of major experimental importance. Scientists are motivated by a deep desire to elucidate the detailed architecture of these remarkable molecules, to explore their various physiological tasks and modes of action in living cells, to evaluate their therapeutical importance, and to no lesser extent, to examine as well as to satisfy the intellectual and experimental skill of the chemists themselves. With the ongoing discovery of many new naturally occurring peptides and polypeptides that exhibit a wide spectrum of biological activities (e.g., enzymatic, hormonal, and antibiotic), stimuli for improving peptide synthesis have gradually and constantly increased. Thus, in order to cope with new challenges, the methodology of peptide synthesis has been markedly upgraded, with a concomitant increase in sophistication. The efficient, unambiguous synthesis of complicated proteins that not too long ago seemed impossible is now much less chimerical and far closer to implementation.

A major disadvantage of all multistep syntheses is the sometimes dramatic diminishing of yields and often of reaction rates as the synthesized molecule increases in size and complexity. In the preparation of polypeptides, a classic multistep synthesis, this problem may be attributed to (1) decreasing solubilities, namely of effective molar concentrations of reactants (i.e., the amine and carboxyl components), with growth of the synthesized peptide chains; (2) increasing steric hindrance and its negative effects upon the efficiency of couplings; and (3) the enhancement of side reactions because of the slow progress of the main reaction, the formation of the desired peptide bond. In order to achieve optimal results in the synthesis of a long and complicated polypeptide, it is therefore essential to overcome the above-mentioned difficulties and several solutions have been proposed. One of them is the step-by-step assembly of the polypeptide through its NH_2-terminal using a large excess of acylating agents, activated N-blocked amino acids. The excess of reagent, which markedly enhances the course and efficiency of synthesis, may however, complicate the purification of desired products. The polymeric reagent approach (Fridkin *et al.*, 1965a, 1966; Wieland and Birr, 1966, 1967a,b; Sklyarov *et al.*, 1966) as well as several other techniques (e.g., Merrifield, 1963; Bodanszky and Bodanszky, 1967; Bodanszky, 1971; Tilak *et al.*, 1972) have been designed and developed to overcome this difficulty.

This chapter reviews peptide synthesis using the polymeric reagent approach and discusses its scope and limitations. The subject has been reviewed previously by Patchornik *et al.* (1973a) and briefly covered in several articles and reviews (e.g., Wünsch, 1971; Fridkin and Patchornik, 1974; Jones and Ridge, 1974; Patchornik and Kraus, 1976).

II. POLYMERIC REAGENTS IN ORGANIC CHEMISTRY: GENERAL ASPECTS

The application of polymers, primarily insoluble ones, as tools for facilitating organic synthesis has recently attracted much attention and is still gaining appreciable momentum. This interest stems both from significant technological advantages as well as from unique effects such as the high selectivity of certain chemical reactions when using polymers (Patchornik *et al.*, 1973a; Leznoff, 1974; Patchornik and Kraus, 1976). In classical organic chemistry, polymers have been mainly used as (1) supports on which molecules such as polypeptides (Barany and Merrifield, see this volume Chapter 1) or oligonucleotides (Hayatsu and Khorana, 1967; Cramer and Köster, 1968) were assembled in a repetitive, sequential manner; (2) supports for selective modification of various reactive molecules (Patchornik and Kraus, 1976) or for selective single reactions on symmetrical bifunctional compounds (Leznoff, 1974); (3) as supports for other miscellaneous reactions (Leznoff, 1974; Overberger and Sannes, 1974; Patchornik and Kraus, 1976); and (4) as catalysts (Helfferich, 1962; Grubbs, 1977) and reagents (Patchornik *et al.*, 1973a; Mathur and Williams, 1976; Patchornik and Kraus, 1976).

In our terminology, "polymeric reagent" refers to polymers used to modify other molecules; in turn, the polymeric reagents themselves are chemically altered as a result of the reaction. It should be emphasized, however, that this definition of polymeric reagent is rather arbitrary and restrictive. Thus, for example, polymeric catalysts which are recovered unchanged from reaction mixtures are *not* regarded as polymeric reagents according to this definition.

A general reaction of a polymeric reagent applied in synthesis is shown in Scheme 1. In this type of reaction, a low-molecular-weight reagent R, which is attached to a polymeric carrier Ⓟ to form the polymeric reagent Ⓟ—R, reacts with a substrate S, yielding a product S′ and a spent, modified polymer Ⓟ—R′.

$$\text{Ⓟ}\text{—R} + \text{S} \longrightarrow \text{Ⓟ}\text{—R}' + \text{S}'$$

Scheme 1. General scheme for synthesis using polymeric reagents. Ⓟ—, polymeric support.

The action of a family of polymeric reagents that is of a particular relevance to peptide synthesis and to our present chapter, i.e., the polymeric transfer agents, is outlined generally in Scheme 2. According to this scheme compound A, which is attached in large numbers (*n*) at many points to the polymer Ⓟ

$$\text{(P)}\text{—}A_n + B \longrightarrow \text{(P)}\text{—}A_{n-1} + A\text{—}B$$

Scheme 2. General scheme for synthesis using polymeric transfer agents. (P)—, polymeric support; A, activated reagent; B, acceptor molecule.

through chemically active bonds, is transferred during chemical reaction to an acceptor molecule B, to form A—B.

Since the A molecules are always low-molecular-weight compounds, one would expect that the physicomechanical properties of the high-molecular-weight carrier (P) (e.g., complete insolubility in all common solvents and hence simple separation techniques if the polymer is cross-linked) would determine the properties of the polymeric reagent (P)—A_n.

The potential advantages of using such polymeric reagents are of four principal types:

1. Synthetic procedures are facilitated and yields are increased through the use of a large excess of either insoluble or soluble reagents, which may be conveniently removed at the end of synthesis by filtration or differential solubility characteristics of reagents and products.
2. High purity of products may be achieved as a result of the technical advantages intrinsically associated with the simplified synthetic procedures, Moreover, by-products of the reagent, if formed, remain attached to the polymeric support.
3. Continuous columnwise or batchwise synthesis and automation of synthesis are possible when insoluble polymeric reagents are employed. Simplified procedures for the regeneration of reagents are feasible.
4. The polymeric matrix constitutes a special microenvironment that may introduce selectivity into a chemical reaction due to steric effects, controlled polymeric pore sizes, dilution on the polymer chain, or specific affinity of reactant B for groups on polymer (P).

A considerable number of organic reactions such as halogenation, oxidation–reduction, nucleotide synthesis, or Wittig-reaction (for review, see Leznoff, 1974; Patchornik and Kraus, 1976; Mathur and Williams, 1976) have been carried out according to Schemes 1 and 2. In these reactions, some or even most of the above-mentioned charactertistics were taken advantage of. In the following sections we describe in some detail the application of polymeric reagents, namely polymeric active esters, to peptide synthesis.

III. POLYMERIC REAGENTS IN PEPTIDE SYNTHESIS

The optimal synthesis of relatively high-molecular-weight polypeptides with well-defined amino acid sequence requires that: (1) the yield of every step of the synthesis be nearly quantitative; (2) the main reaction products be easily separable from reagent excesses, starting materials, and side products; (3) product purification be carried out conveniently; and (4) preferably, that automation of synthesis be practicable. Utilization of insoluble polymeric reagents as synthetic tools has been introduced in response to some of the above considerations.

A. Polymeric Active Esters

1. General Considerations

Polymeric active esters may be prepared by attachment of N-protected amino acids or peptides to a suitable insoluble polymer (Scheme 3). Potentially useful polymeric supports, Ⓟ—OH, should carry free hydroxyl groups to which the carboxyl group of amino acid derivatives may be coupled,

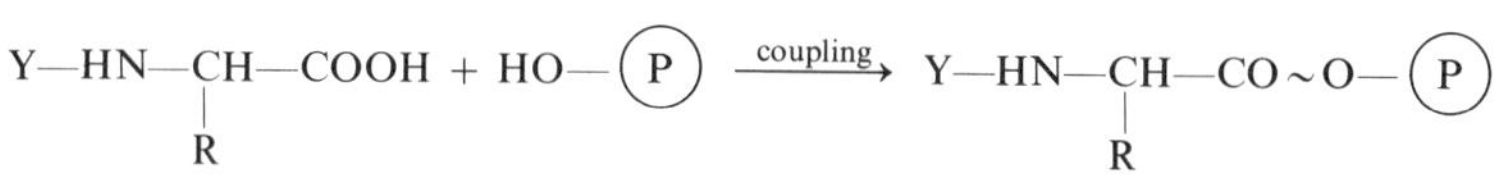

Scheme 3. Preparation of polymeric active esters. Y, α-amine protecting group; Ⓟ—, polymeric carrier; wavy line, ~, denotes an active ester bond.

yielding active ester bonds. Most of the polymer groups which have been prepared and investigated so far are illustrated in Scheme 4. (**1** to **9**).

Various conventional synthetic procedures have been used to couple the polymer Ⓟ—OH and the carboxylic derivatives. The reagent most commonly used has been *N,N'*-dicyclohexylcarbodiimide (DCC) (Fridkin *et al.*, 1966, 1968, 1972; Wieland and Birr, 1966, 1967a,b; Sklyarov *et al.*, 1966; Laufer *et al.*, 1968; Narita *et al.*, 1972; Rogozhin *et al.*, 1973b; Kalir *et al.*, 1974, 1975; Teramoto *et al.*, 1977). But mixed anhydrides (Wieland and Birr, 1966; Sklyarov *et al.*, 1966; Patchornik *et al.*, 1967; Rogozhin *et al.*, 1975), thionyl chloride (Wieland and Birr, 1966; Fridkin, 1967), trifluoroacetic anhydride (Rogozhin *et al.*, 1973b, 1975), *N*-ethyloxycarbonyl-2-ethyloxy-1,2-dihydroquinoline (EEDQ) (Rogozhin *et al.*, 1975), and phosgene (Wieland and Birr, 1966; Fridkin, 1967) were also utilized. The couplings were carried out in appropriate organic solvents such as *N,N*-dimethylformamide (DMF), dichloromethane, chloroform, or tetrahydrofuran. At the completion of reaction, the polymer was washed free of excess reagents and was ready for peptide synthesis. Several analytical procedures were used to evaluate the content of N-blocked amino acid derivatives attached to the polymer (Patchornik *et al.*, 1973a): (1) Determination of the weight increase of the

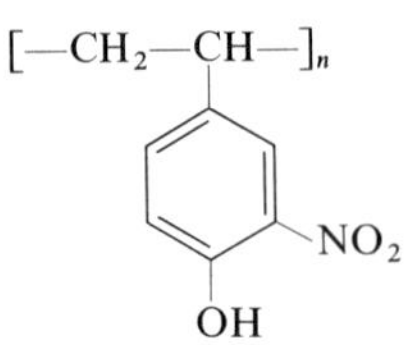

(1)

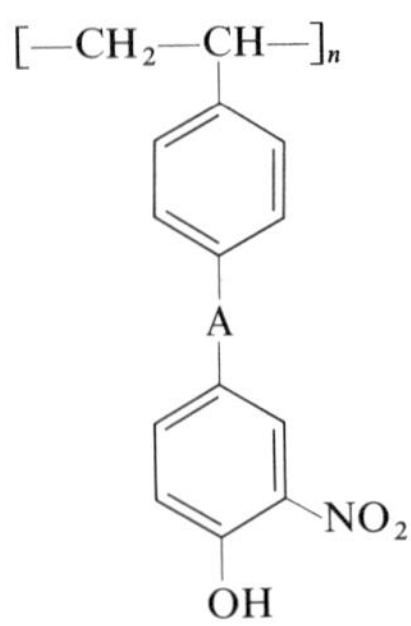

(2)

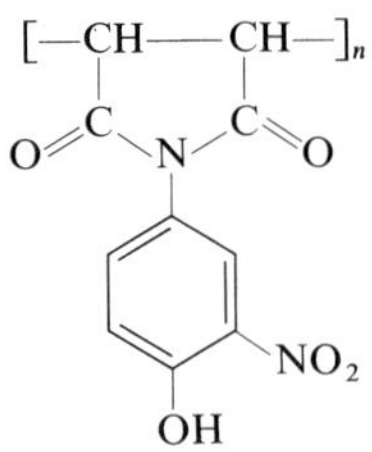

(3)

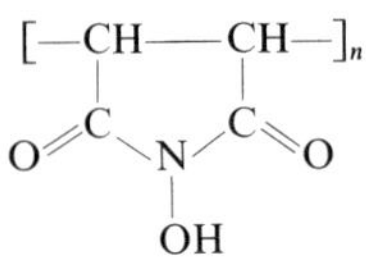

(4)

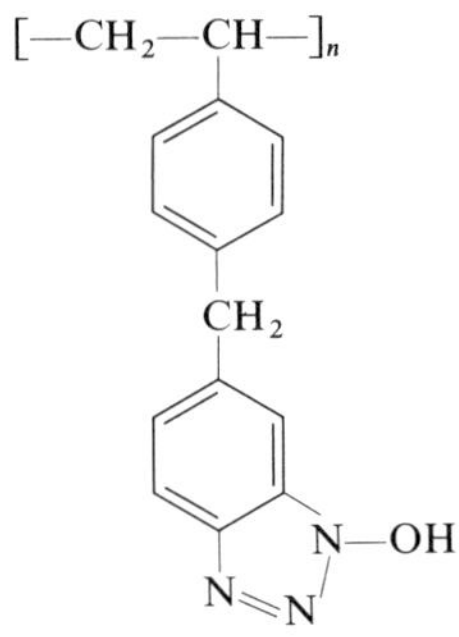

(5)

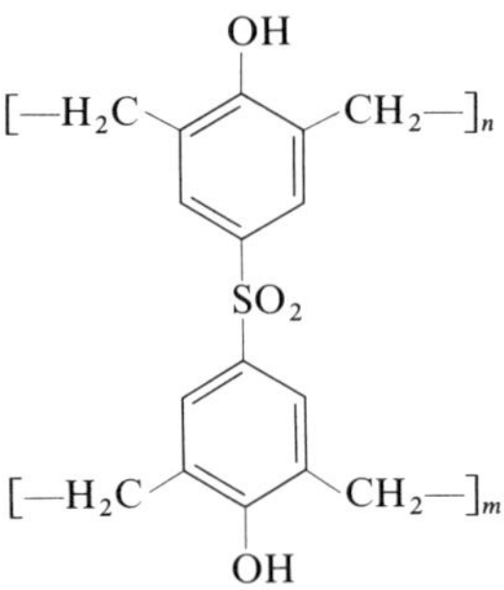

(6)

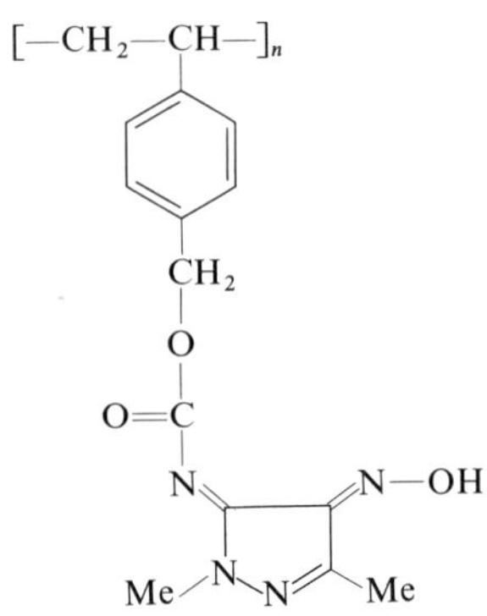

(7)

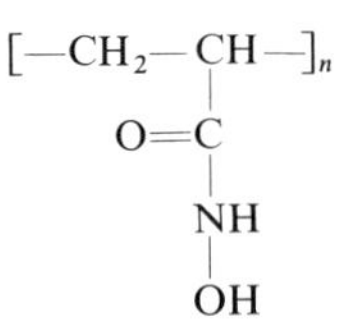

(8)

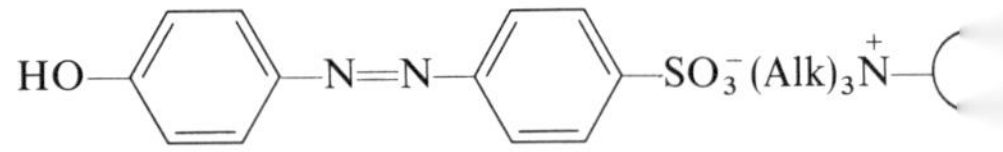

(9)

Scheme 4. Main polymers used for preparation of polymeric active esters. Alk, alkyl; Ⓟ—, polymeric carrier; A, extension arm.

parent polymer due to the addition of the carboxylic component; (2) determination of amino acid liberated upon exhaustive acid hydrolysis; (3) spectrophotometric determination of the N-protected amino acid liberated on treatment with NaOH; (4) elemental analysis (sulfur analysis was of particular value for *o*-nitrophenylsulfenyl, cysteine, and methionine derivatives); (5) determination of yields of cyclohexylamides, produced by treating the the activated polymers with cyclohexylamine (Teramoto *et al.*, 1977); and (6) reaction of the polymeric active ester with an excess of benzylamine and back titration of the latter with perchloric acid in dioxane using thymol blue as an indicator. This analytical method (Kalir *et al.*, 1974) is highly accurate and reproducible and is routinely used in our laboratory (see Scheme 5).

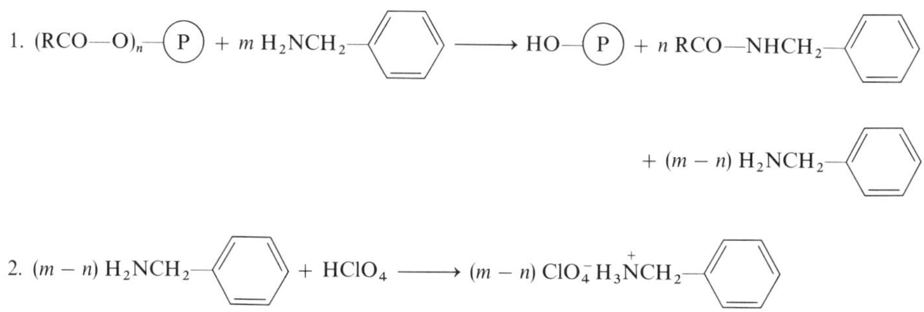

Scheme 5. Nonaqueous titrimetric analysis of polymeric active esters (Kalir *et al.*, 1974). Ⓟ—, Polymeric carrier.

Analyses of various polymeric esters prepared reveal that the extent of loading of the parent polymer Ⓟ—OH largely depends on the type of polymer, the coupling procedure used, the molar ratio between available polymeric hydroxyl groups and soluble reactants, the nature of the attached N-blocked carboxylic derivative, and the solvent used. Sometimes nearly quantitative loading may be attained by selecting appropriate reaction conditions. The polymeric active esters used for peptide synthesis usually contained 1–2 mmol of N-blocked amino acid per 1 gm of reagent.

Utilization of polymeric active esters in a step-by-step peptide synthesis is illustrated in Scheme 6. The polymeric reagent (**10**) (containing 1.25–3 equiv of the desired activated amino acid) suspended in an organic solvent such as dimethylformamide, dichloromethane, or chloroform is coupled with one equivalent of the appropriate soluble amino acid ester containing a free α-amino group (**11**), yielding the amine- and carboxyl-blocked soluble dipeptide (**12**). The product is then isolated by filtering spent polymeric reagent and evaporating the reaction solvent. Synthesis may be carried further

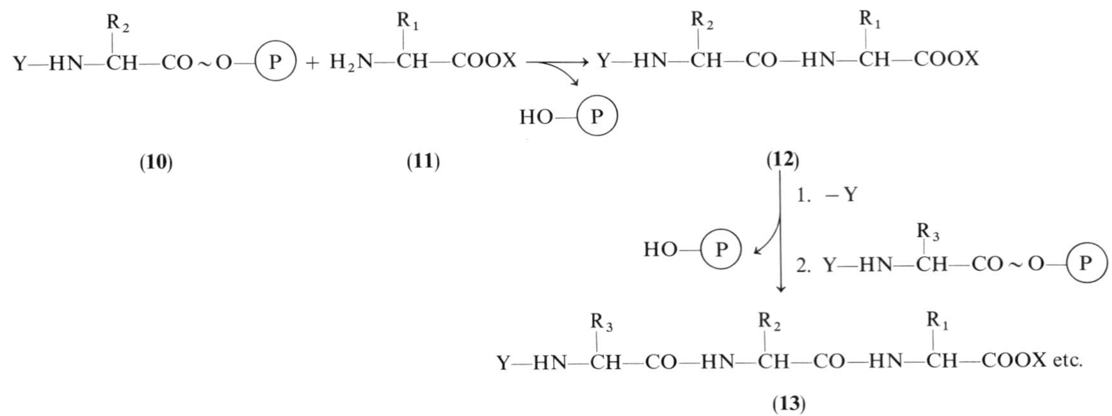

Scheme 6. General scheme for a stepwise peptide synthesis utilizing polymeric active esters. Wavy line, ~, denotes an active ester bond; Ⓟ —, polymeric carrier; X, α-carboxyl protecting group; Y, N^{α}-protecting group.

through selective removal of the NH_2-terminal protecting group of **12** and acylation of the newly exposed free α-amino group by the suitable polymer-attached activated N-blocked amino acid. Further repetition of this reaction scheme leads to the desired peptide having the required chain length and amino acid sequence.

Several features are prerequisites for ideal polymeric active esters: mechanical stability, i.e., integrity of polymeric particles over many repetitive synthesis and reloading cycles; good swelling in the organic solvents commonly used in peptide synthesis; accessibility of polymer-attached activated amino acid moieties to incoming amine components to achieve quantitative acylation; minimal dependence of acylations on steric hindrance; high reactivity, i.e., fast formation of new peptide bonds; possibility of loading relatively large amounts of activated N-blocked amino acids; and stability during storage. These were the guidelines for preparation of the polymers (**1** to **9**) illustrated in Scheme 4 as well as the corresponding polymeric active esters.

2. *Esters Based on Nitrophenol*

Two methods were used to prepare polymers bearing nitrophenol residues (**1** to **3**); the first involved polymerization of the appropriate monomers while the second involved covalent attachment of nitrophenol-containing moieties to polystyrene. Polymers of the former class include:

1. Insoluble poly(4-hydroxy-3-nitrostyrene) (**1**) was obtained either by copolymerization of 4-acetoxystyrene with divinylbenzene (DVB) to yield cross-linked poly-4-acetoxystyrene followed by nitration and deacetylation (Fridkin *et al.*, 1968), or by copolymerization of 4-methoxystyrene with DVB and subsequent demethylation and nitration of the copolymer thus obtained (Williams, 1972). Insoluble active esters derived from this polymer were used in the synthesis of a variety of peptides among them oxidized glutathione in 90% overall yield (Fridkin *et al.*, 1966; Patchornik *et al.*, 1967) and the nonapeptide bradykinin, which possessed full biological activity, in 39% overall yield (Fridkin *et al.*, 1968).
2. Poly(2-hydroxy-5-nitro-*m*-phenylenedimethylene) was obtained by copolymerization of *p*-nitrophenol, phenol, and bischloromethyl ether and utilized for the synthesis of several di- and tripeptides (Sklyarov *et al.*, 1966).
3. Polymers containing *N*-(4-hydroxy-3-nitrophenyl)succinimide residues (**3**) were obtained by copolymerization of *N*-(4-acetoxyphenyl)maleimide with styrene and DVB followed by deacetylation and nitration, or by reaction of a copolymer of maleic anhydride and styrene with *p*-aminophenol, followed by nitration—a method which is in fact a polymer-modification procedure (Teramoto *et al.*, 1977). Polymeric active esters derived from these

polymers proved to be most efficient acylating agents and were used to make several dipeptides containing glycine and leucine in nearly quantitative yields.

Preparation of polymers by the above-described methods is sometimes rather laborious and time consuming. Although they do yield useful active esters, the polymeric products are often powders with poor physicomechanical properties that cause major difficulties in filtration and washings. In addition, these polymers will often swell in a limited number of highly polar solvents only, whereas a broader choice of solvents is desirable for versatile peptide synthesis. Moreover, reactions utilizing these polymers were occasionally found to be slow and inefficient (Fridkin *et al.*, 1971; Kalir *et al.*, 1974), perhaps due to the close proximity of reacting ester groups to the polymer backbone. Improved polymers may be prepared by using preformed bead polymers as starting material. Cross-linked polystyrene is particularly suitable for this purpose since it may be easily and efficiently modified by a variety of chemical reactions and since its products have the excellent mechanical properties of the parent polymer. Thus, reaction of chloromethylated polystyrene with 4-hydroxy-3-nitrobenzoic acid (Panse and Laufer, 1970) or with 4-nitroresorcinol (Fridkin *et al.*, 1971) yielded polymers **2a** and **2b**, respectively (derived from **2**, Scheme 4). Both **2a** and **2b** provided moderately

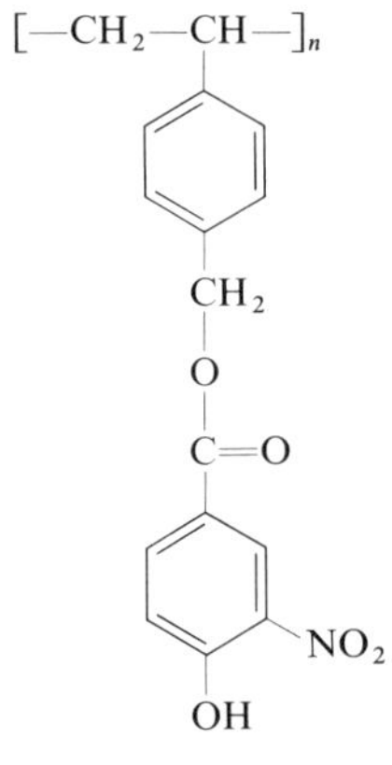

(2a)

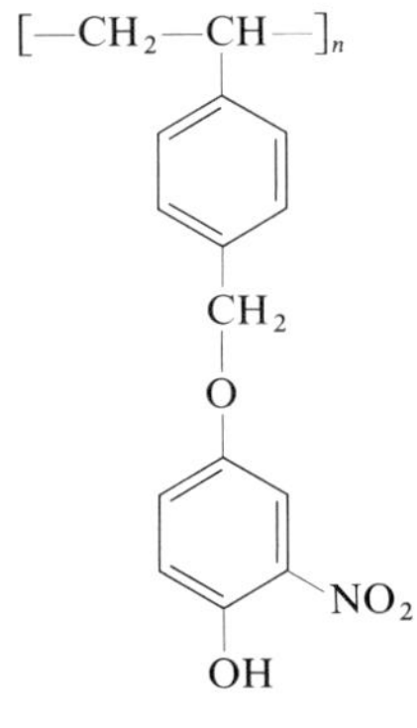

(2b)

efficient acylating agents, clearly demonstrating the feasibility of the general approach. The most successful polystyrene-derived carrier is (4-hydroxy-3-nitro)benzylated polystyrene (structure **2c**, Scheme 7), the product of a Friedel–Crafts alkylation of beads of copoly(styrene–2%-divinylbenzene) (**14**) by 4-hydroxy-3-nitrobenzyl chloride, (**15**) (Kalir *et al.*, 1974). This carrier retains the good mechanical properties of the starting polymer, swells in a

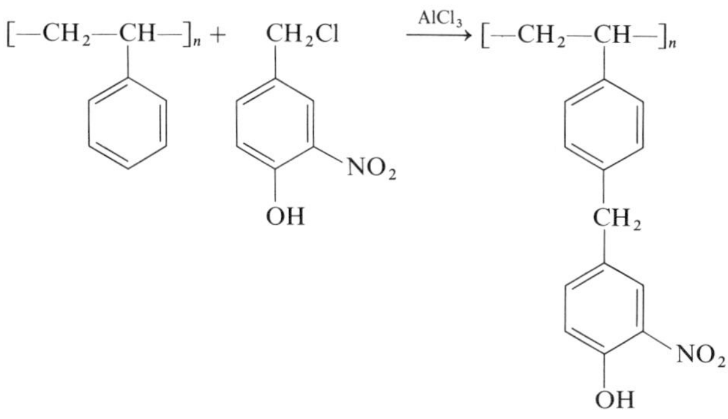

(**14**) (**15**) (**2c**)

Scheme 7. Preparation of (4-hydroxy-3-nitro)benzylated polystyrene by a Friedel–Crafts alkylation of polystyrene (Kalir *et al.*, 1974).

large number of solvents (including dimethylformamide, methylene chloride, chloroform and benzene), is easy to filter and wash, is efficiently loaded with N-blocked amino acid, and has acylation rates with the derived active esters similar to those of analogous reactions in homogeneous phase. Using this reagent, many peptides were prepared in high coupling yields and of high purity, including substrates for porcine elastase (Atlas *et al.*, 1975) and the

Thr-Lys-Pro-Arg

(**16**)

naturally occurring peptides tuftsin (**16**) in 87% overall yield (Fridkin *et al.*, 1977b) and luliberin (LHRH) in 40% overall yield (Fridkin *et al.*, 1977a), both of which possessed high biological activity. The synthesis of LHRH (**17**) is illustrated in Scheme 8.

3. Esters Based on N-Hydroxysuccinimide

Polymers containing *N*-hydroxysuccinimide residues (**4**) may be prepared in a manner similar to that of the previous group of polymers, either by polymerization of appropriate monomers or by modification of preformed polymers. Thus, *N*-benzyloxymaleimide (**18**) (Narita *et al.*, 1972), *N*-acetoxymaleimide (Narita *et al.*, 1972), and *N*-benzyloxymaleamic acid (**19**) (Akiyama *et al.*, 1969) were copolymerized with styrene and DVB, and the polymeric products (**20**) converted by hydrolysis into the corresponding *N*-hydroxysuccinimide polymer (**21**). *N*-Benzyloxyisomaleimide was similarly copolymerized and converted into the *N*-hydroxysuccinimide polymer (Akiyama *et al.*, 1969). The insoluble active esters derived from these polymers served

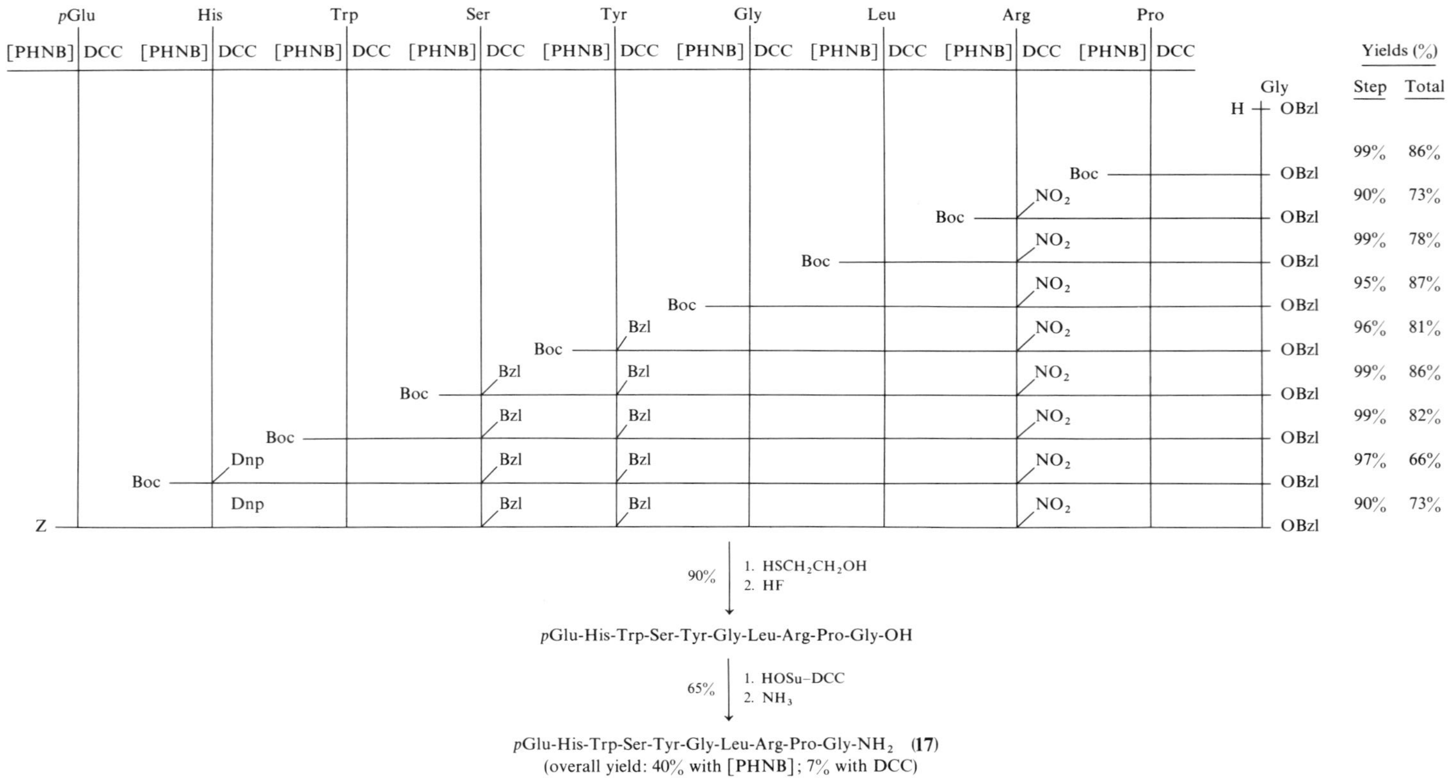

Scheme 8. Comparative stepwise synthesis of luliberin using insoluble active esters derived from (4-hydroxy-3-nitro)benzylated polystyrene, [PHNB], and *N*,*N*-dicyclohexylcarbodiimide (DCC) as reagents (Fridkin *et al.*, 1977a). Dnp, 2,4-dinitrophenyl; HOSu, *N*-hydroxysuccinimide; pGlu, pyroglutamic acid.

for a high yield synthesis of dipeptides containing glycine and leucine (Narita *et al.*, 1972) and for acetylation of cyclohexylamine (Akiyama *et al.*, 1969).

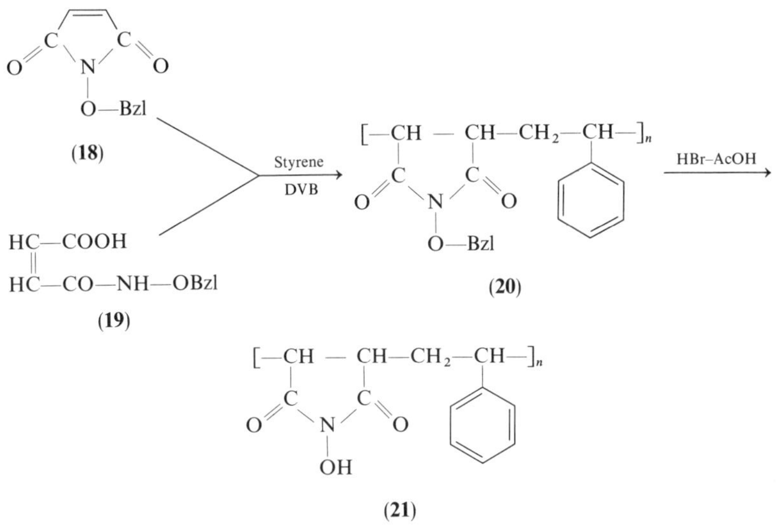

Scheme 9. Preparation of polymers containing *N*-hydroxysuccinimide residues through polymerization of monomers. Compound **18**: Narita *et al.*, 1972, compound **19**: Akiyama *et al.*, 1969. DVB, Divinylbenzene.

Copoly(ethylene-*N*-hydroxymaleimide), both soluble (Wildi and Johnson, 1968; Laufer *et al.*, 1968) and insoluble, cross-linked either by high-energy electron irradiation (Laufer *et al.*, 1968) or by various diamines (Patchornik *et al.*, 1967; Fridkin *et al.*, 1972) was prepared by reacting hydroxylamine with copoly(ethylene–maleic anhydride) as a starting polymer. Similarly, copoly(styrene-*N*-hydroxymaleimide) cross-linked with benzidine was prepared from copoly(styrene–maleic anhydride) (Rogozhin *et al.*, 1973a). Many soluble and insoluble polymeric active esters, carrying a wide variety of N-protected amino acid derivatives, have been prepared from this group of polymers. These reagents were found to be highly reactive and most efficient in peptide synthesis (Laufer *et al.*, 1968; Fridkin *et al.*, 1972; Rogozhin *et al.*, 1973b; Davidovich and Gut, 1976). Their application to the α-amino group acylation of free amino acids and peptides is noteworthy (Andreev *et al.*, 1977). A step-by-step synthesis of bovine carboxypeptidase A-(159–165)-peptide (**22**) using polymeric *N*-hydroxysuccinimide esters is illustrated in Scheme 10.

Cross-linked polystyrene in bead form was also used as a starting material for the preparation of a polymer with *N*-hydroxysuccinimide residues (**24**). The product possessed excellent mechanical and chemical properties but has

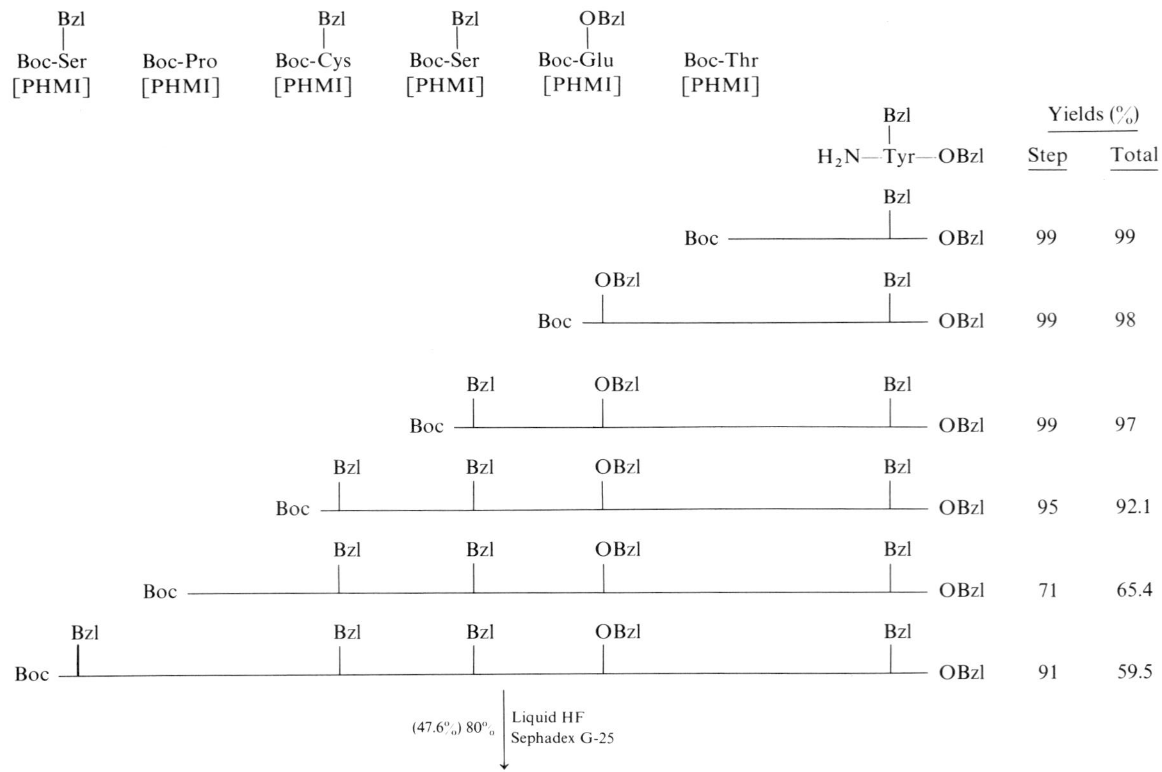

H-Ser-Pro-Cys-Ser-Glu-Thr-Tyr-OH **(22)**

Scheme 10. Synthesis of bovine carboxypeptidase A-(159–165)-peptide using polymeric active esters derived from cross-linked copoly(ethylene-*N*-hydroxymaleimide) [PHMI]. Reprinted with permission from Fridkin *et al.*, 1972. Copyright by the American Chemical Society.

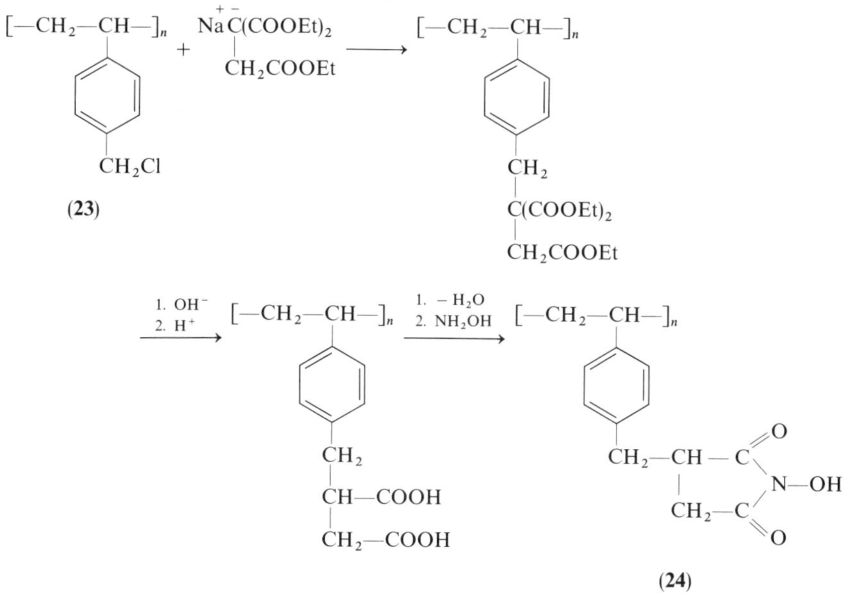

Scheme 11. Preparation of *N*-hydroxysuccinimide derivative of polystyrene (Kalir, 1975).

been used so far only for preparation of simple di- and tripeptides (Kalir, 1975).

4. Esters Based on 1-Hydroxybenzotriazole

At the end of Section III,A,1, we outlined several features which, characterize the ideal polymeric reagents. Our laboratory experience indicates that most of the above-mentioned criteria are successfully met by polymeric active esters derived from cross-linked (4-hydroxy-3-nitro)benzylated polystyrene (Kalir *et al.*, 1974) (see Section III,A,2). However, rates of peptide bond formation with these reagents are not always fast enough for experimental purposes and may even be very slow indeed. This is especially evident when either the insolubilized carboxyl component or the terminal residue of the soluble amine component is a derivative of a bulky amino acid such as leucine or valine (Kalir *et al.*, 1974). It is thus obvious that these polymeric reagents are not ideally suited for the synthesis of long and complicated peptides.

The well-known high reactivity of 1-hydroxybenzotriazole esters of N-blocked amino acids and peptides, and the consequent remarkable facilitation of peptide synthesis when these derivatives are used as coupling agents (König and Geiger, 1970a,b), have motivated preparation of the corresponding polymeric analogues (Kalir *et al.*, 1975). The synthesis of insoluble

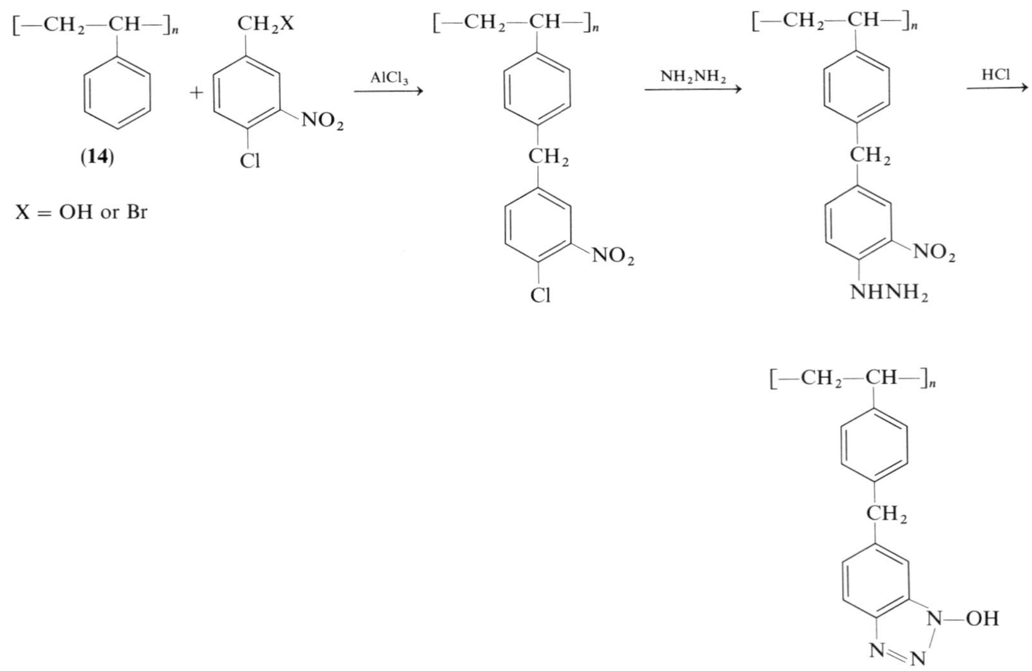

Scheme 12. Preparation of polystyrene-bound 1-hydroxybenzotriazole (Kalir *et al.*, 1975).

polystyrene-bound 1-hydroxybenzotriazole (**5**) is illustrated in Scheme 12. The beads of polystyrene used as the starting polymer (**14**) were either macroporous-type or co(polystyrene–2%-divinylbenzene). The polymeric 1-hydroxybenzotriazole esters contained a relatively large amount (0.6–1.3 mmol/gm) of activated N-blocked amino acid derivatives, and were stable for at least 2 months when stored in a dry state in the desiccator. They proved to be highly active acylating agents. Peptide bond formation with these esters was generally very rapid, often a matter of minutes (Kalir *et al.*, 1975), and they were used for high-yield syntheses of many peptides, including various analogs of tuftsin (Fridkin *et al.*, 1977b), thyrotropin-releasing hormone (in 80% overall yield) and Boc-LLeu-LLeu-LVal-LTyr(Bzl)-OBzl. It is pertinent to note that the three coupling steps involved in the synthesis of this latter bulky tetrapeptide took as little as 60 min (i.e., 20 min each) and gave an overall yield of 91%. A comparative synthesis of this peptide with (4-hydroxy-3-nitro)benzylated polystyrene took 112 hr with an overall yield of 82% (Fridkin *et al.*, 1975; Kalir *et al*, 1975). Our experience so far with this polymer strongly indicates a potential for its wider application in peptide synthesis. Moreover, the racemization-suppressed coupling of peptide segments using a combination of DCC and 1-hydroxybenzotriazole (König and Geiger, 1970a,b) suggests that an insoluble 1-hydroxybenzotriazole polymer may be equally useful in segment condensations.

5. Esters Based on Miscellaneous Hydroxyl Groups

In the following section we briefly describe several hydroxyl-containing polymers which do not belong to the above families of polymers but have also been used to prepare insoluble active esters. Although these polymers have found a rather limited use in peptide synthesis, we decided to mention them in order to demonstrate the wide variety of potentially useful reagents.

Insoluble *p,p′*-dihydroxydiphenyl sulfone (polymer **6**) was prepared by Wieland and Birr (1966) by condensing *p,p′*-dihydroxydiphenyl sulfone with formaldehyde. The insoluble active esters derived from this polymer were utilized in the synthesis of several di and tripeptides in moderate yields.

In a most attractive approach suggested by Wieland and Birr (1967a) (Scheme 13), an N-protected amino acid active ester bearing a free sulfonic functional group—the active ester of *N*-3,5-dimethyloxybenzyloxycarbonyl-L-phenylalanine and 4-hydroxyazo-4′-sulfonic acid—was attached through ionic bonds to the ion exchanger Dowex 1 (**9**). The ionically bound active ester was then reacted with L-alanylglycine methyl ester, to produce the corresponding blocked tripeptide in 69% yield.

Peptide synthesis by means of insoluble active esters derived from *N*-5(polystyryl-4-methoxycarbonyl)-imino-4-oximino-1,3-dimethyl 2-pyrazoline (**7**) and N-blocked amino acids was described by Guarneri *et al.* (1972).

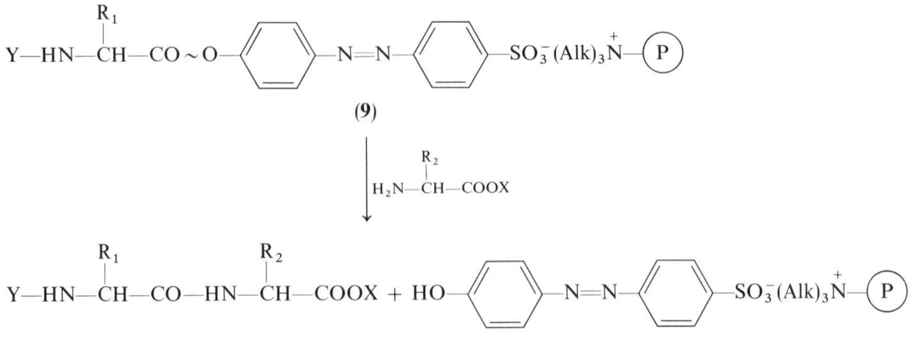

Scheme 13. Peptide synthesis using an insoluble, ionically bound active ester (Wieland and Birr, 1967a). Alk, Alkyl; Ⓟ— polymeric carrier; X, α-carbonyl protecting group; Y, N^{α}-protecting group.

The reagents were used to prepare several dipeptides, as well as the NH_2-terminal hen egg-white lysozyme tetrapeptide, Z-Lys(Z)-Val-Phe-Gly-OEt in 80% overall yield.

A very interesting variation of the polymeric reagent approach is one in which a stable anchoring bond linking the N-blocked amino acid or peptide derivative to a polymeric carrier is activated at will; thus the polymer becomes a transfer agent. A 4-methylthiophenol derivative of polystyrene (**25**) was used by Flanigan and Marshall (1970) and by Marshall and Liener (1970) as support for a stepwise synthesis of peptides following the Merrifield method. Oxidation of the thioether to a sulfone (**26**), however, activates the COOH-terminus of the peptide and allows easy coupling to free α-amino groups (see Scheme 14, Pathway A). Only a limited number of peptides, in moderate yields, have been synthesized so far by this method.

Two additional polymers utilizing the principle of subsequent activation were described. In the first, the α-carboxyl group of an *N*-blocked amino acid or a peptide derivative is attached to a polystyrene-derived carrier through a phenylhydrazide bond which is activated by oxidation with *N*-bromosuccinimide to the corresponding azo bond (Wieland *et al.*, 1970). In the second, the carboxyl is attached to the polymer by an ester bond through 2,2-diphenylethylene glycol residues. Activation is accomplished by conversion of the glycol–ester bond to an enol–ester bond effected by elimination of a water molecule with trifluoroacetic acid (Wieland *et al.*, 1970, 1972). These two potentially useful polymers have found only limited use in actual peptide synthesis so far.

The preparation and use in peptide synthesis of poly(5-vinyl-8-hydroxyquinaldine) (Manecke and Haake, 1968), of poly(5-vinyl-8-hydroxyquinoline) (Kalir, 1975), of poly(4-thiostyrene) (Sklyarov *et al.*, 1966; Patchornik *et al.*, 1967), of a polymer containing *N*-(4-hydroxy-3,5-dinitrophenyl)succinimide

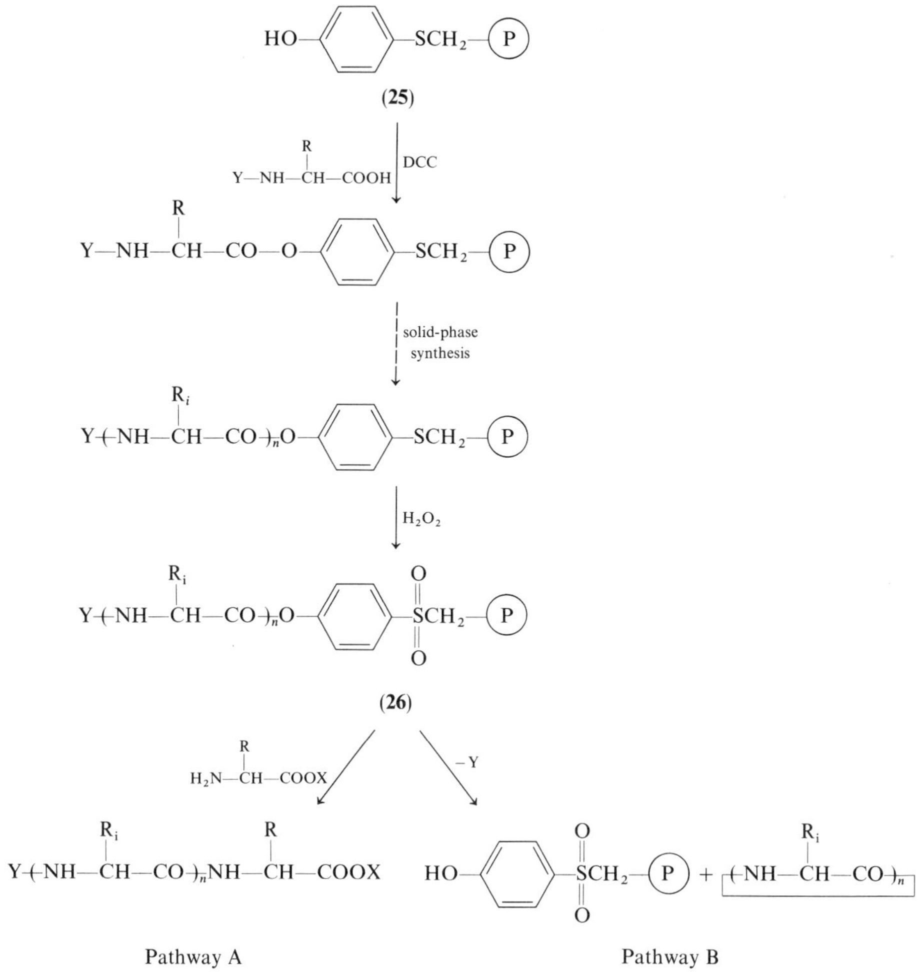

Scheme 14. Synthesis of linear (A) and cyclic (B) peptides using the principle of "subsequent-activation" of polymeric esters (Marshall and Liener, 1970; Flanigan and Marshall, 1970). DCC, dicyclohexylcarbodiimide; Ⓟ—, polymeric carrier; OX, ester protective group; R_i, individual side-chain residues; Y, N^{α}-protecting group.

residues (Teramoto *et al.*, 1977), and of a polymer containing hydroxamic acid residues (polymer **8**) (Narita *et al.*, 1972) were also reported. However, these reports include very little information.

6. Racemization during Use of Polymeric Active Esters

As might have been expected from reactions with the corresponding soluble monomers, racemization is insignificant in the stepwise synthesis of peptides in which a single N-protected amino acid is transferred from a polymeric donor to a free α-amino group acceptor. Thus, for example, it was found that charactertistic properties such as optical rotations, melting points, and chromatographic mobilities of various peptides prepared by different polymeric reagents were identical with the values of authentic markers prepared by conventional and "racemization free" procedures (Fridkin *et al.*, 1972; Kalir *et al.*, 1974; Teramoto *et al.*, 1977). Using polymeric *N*-hydroxysuccinimide esters, Andreev *et al.* (1977) have synthesized a large number of diastereomeric dipeptides containing L- and D-isomers of alanine, aspartic and glutamic acid, proline, leucine, and valine sequences in various combinations. Careful analysis of these peptides by gas liquid chromatography and ion exchange chromatography on the amino acid analyzer revealed high optical purity. Moreover, the synthesis of the dipeptide Z-LVal-LVal-OBzl by reaction of a polymeric *N*-hydroxysuccinimide Z-LVal ester with H-LVal-OBzl at 70°C was found to be free of racemization (Fridkin *et al.*, 1972). Finally, the high biological potencies of synthetic bradykinin (Fridkin *et al.*, 1968), thyroliberin (Kalir *et al.*, 1975), and luliberin (Fridkin *et al.*, 1977a), as well as their physicochemical identity with the authentic peptides, strongly support the above conclusions.

On the other hand, as may have been anticipated, the transfer of a peptide from a polymeric reagent to an α-amino group will probably lead to racemization, the extent of which will be dependent upon the particular reagent used. Lauren and Williams (1972) have studied the racemization accompanying the synthesis of Z-Gly-L(D)Ala-LLeu-OBzl—the Izumiya racemization model—using *N*-hydroxysuccinimide and *o*-nitrophenyl polymeric esters of Z-Gly-LAla (Izumiya *et al.*, 1971). Coupling of the former reagent with L-leucine benzyl ester led to racemization of 6.4% of the alanine residue whereas the latter reagent led to a racemization of as high as 37.4%. A polymeric *N*-ethyloxycarbonyl-2-ethyloxy-1,2-dihydroquinoline derivative (poly-EEDQ; see Section III,B) exhibited a racemization value of 6%. However, coupling of the insoluble active ester formed between polystyrene-bound 1-hydroxybenzotriazole and Z-LAla-LAla-OH with H-DAla-ONb was found in our laboratory (M. Fridkin and R. Kalir, unpublished results) to preserve optical purity very well (>99%), as revealed by digestion with leucine aminopeptidase followed by amino acid analysis (Bosshard *et al.*, 1973) or by high-voltage paper electrophoresis at pH 3.5 (Burstein *et al.*, 1970).

B. Polymeric Coupling Agents

Insoluble coupling or condensing agents belong to a family of polymeric reagents which are not transfer agents. These reagents join the free α-amino group of an amino acid or peptide ester with the free α-carboxyl group of an N-blocked amino acid with the elimination of water and the formation of an elongated amine- and carboxyl-blocked peptide. In spite of their potentiality, they have so far found limited use in peptide synthesis.

Insoluble polyhexamethylene carbodiimide

$$-(CH_2)_6-N{=}C{=}N-[(CH_2)_6-N{=}C{=}N]_n-(CH_2)_6-$$

was prepared by Wolman *et al.* (1967) and used for several condensations, including that of N^α,*S*-dibenzyloxycarbonyl-L-cysteine and glycine benzyl ester to form Z-Cys(Z)-Gly-OBzl in 93% yield. A series of insoluble polymeric carbodiimides (**27**), derived either from polystyrene or from polyvinylamine, was also prepared (Fridkin *et al.*, 1971) and applied to the successful synthesis of several di- and tripeptides. It is also interesting to note that polymeric diimides have been used successfully in the preparation of ketones and aldehydes by the Moffatt oxidation procedure (Weinshenker and Shen, 1972a,b).

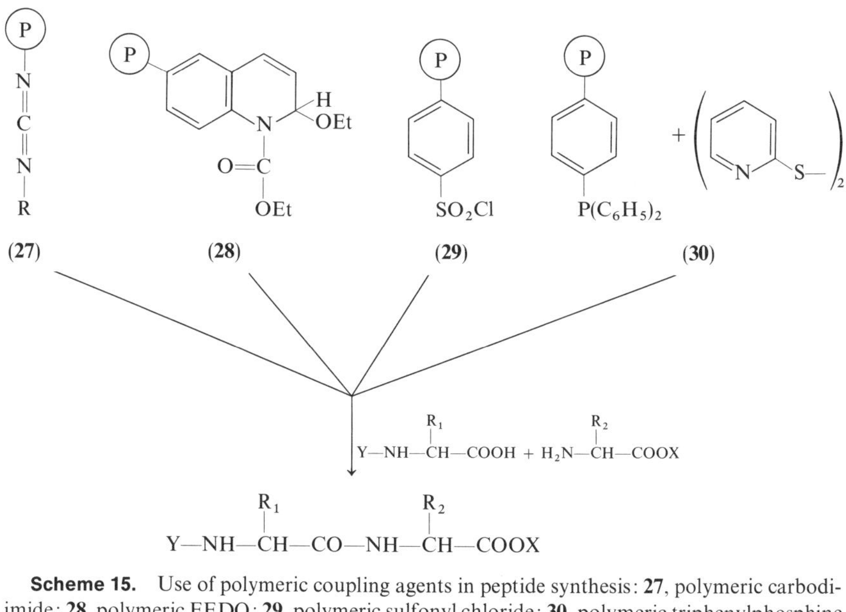

Scheme 15. Use of polymeric coupling agents in peptide synthesis: **27**, polymeric carbodiimide; **28**, polymeric EEDQ; **29**, polymeric sulfonyl chloride; **30**, polymeric triphenylphosphine and 2,2′-dipyridyl disulfide; Ⓟ —, polymeric carrier; X, α-carboxyl protecting group; Y, N^α-protecting group.

Insoluble chlorosulfonated polystyrene (**29**) was prepared and used as a condensing agent for the synthesis of several di- and tripeptides, such as Z-Cys(Z)-Gly-OBzl (99% yield) and Z-Gly$_3$-OBzl (30% yield) (Fridkin *et al.*,

1971). A related polymer, poly(3,5-diethylstyrene)sulfonyl chloride, was used in oligonucleotide synthesis (Rubinstein and Patchornik, 1972).

An insoluble regenerable polymeric derivative of *N*-ethyloxycarbonyl-2-ethyloxy-1,2-dihydroquinoline (EEDQ) (**28**) was synthesized and used in a rather efficient synthesis (83.5% yield) of Z-Phe-Gly-OEt. Its potential application in an automated peptide synthesis has also been discussed (Brown and Williams, 1971; Williams *et al.*, 1972).

Peptide synthesis by means of a combination of insoluble, polystyrene-derived triphenylphosphine and soluble 2,2′-dipyridyl disulfide (**30**) was recently described. Z-Val-Gly-OEt and Z-Phe-Gly-OEt were prepared in reasonable yields using this method (Horiki, 1976).

C. Toward Mechanized Peptide Synthesis

Although many peptides have been successfully synthesized by the polymeric reagent technology, one of the method's major potentials, its being applied in mechanized and even automated peptide synthesis, has not yet been realized.

The most obvious way of using polymeric reagents in mechanized peptide synthesis would involve the passage of the amine component to be acylated, dissolved in an appropriate solvent, through a column packed with a desired insoluble, activated carboxyl component and subsequent collection of product in the eluent. The NH_2-terminal protecting group would then be selectively removed from the newly formed peptide, preferably with the formation of an unprotonated amine function, and the product passed through yet another column containing a suitable polymeric reagent, yielding further elongation. The ideal process would be carried out continuously, using the same solvent throughout all operations, including removal of N-blocking groups. A general process using columns is outlined in Scheme 16. Columns packed with insoluble polymeric active esters have indeed been used in peptide synthesis, although only for single acylation steps rather than for stepwise operation. Thus, for example, Wieland and Birr (1967b) have used columns packed with insoluble active esters of *N*-blocked amino acids and crosslinked *p,p*′-dihydroxydiphenylsulfone (**6**) to synthesize several peptides in nearly quantitative yields. These investigators have also proposed the use of photolabile protective groups for the α-amino groups in such a stepwise procedure. Photocleavable protecting groups are in principle ideal for continuous applications, since they may generate the free nonprotonated amino functions by simple irradiation. Further progress using this interesting approach has not been reported.

Other investigators have also used polymeric columns for single acylation steps. For example, Patchornik *et al.* (1973b) have synthesized several

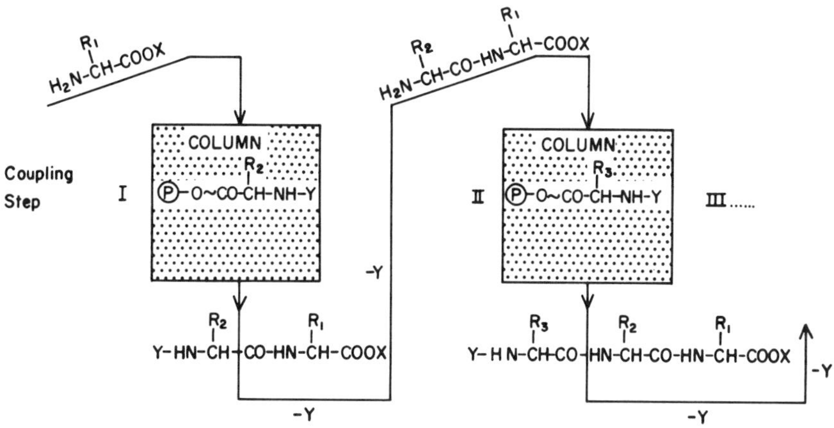

Scheme 16. A general scheme for a continuous columnwise synthesis of peptides utilizing polymeric active esters. Based on scheme of Wieland and Birr (1967b). Ⓟ—, polymeric carrier; X, α-carboxyl protecting group; Y, N^{α}-protecting group.

penicillins in high yields by passage of 6-aminopenicillanic acid through columns of insoluble active esters derived from various carboxylic acids and cross-linked nitrophenol resin. Kalir (1975) has used polymeric nitrophenyl, *N*-hydroxysuccinimide, and 1-hydroxybenzotriazole esters for the efficient synthesis of a few dipeptides. Rogozhin *et al.* (1973b) employed a column of polymeric *N*-hydroxysuccinimide ester of Z-L-leucine to synthesize Z-Leu-Gly-OMe in nearly quantitative yields.

A batch-processing approach to mechanized peptide synthesis using polymeric reagents was recently investigated in our laboratory (Stern *et al.*, 1977). It is comprised of three operative steps, performed consecutively without isolation of intermediate compounds: (1) coupling of the free α-amino group of an amino acid ester with a polymeric active ester to yield the amine- and carboxyl-blocked dipeptide; (2) selective removal by acidolysis of the N-protecting group to yield the corresponding ammonium salt of the dipeptide ester; and (3) neutralization of the ammonium salt with a weakly basic polymer concomitant to the coupling of the newly exposed α-amino function with a suitable polymeric reagent to afford the amine- and carboxyl-blocked tripeptide.

The main features of this method are illustrated in Scheme 17. It was used to prepare the peptides Boc-Pro-Val-Lys(NO_2z)-Val-Tyr(Dnp)-Pro-OBzl [corresponding to human ACTH(19–24)-peptide] and Boc-Lys(NO_2z)-Lys(NO_2z)-Arg(NO_2)-Arg(NO_2)-OBzl [corresponding to human ACTH-(15–18)-peptide] in 63% and 70% overall yields, respectively. As shown in Scheme 17, the *tert*-butyloxycarbonyl group was used for α-amine protection

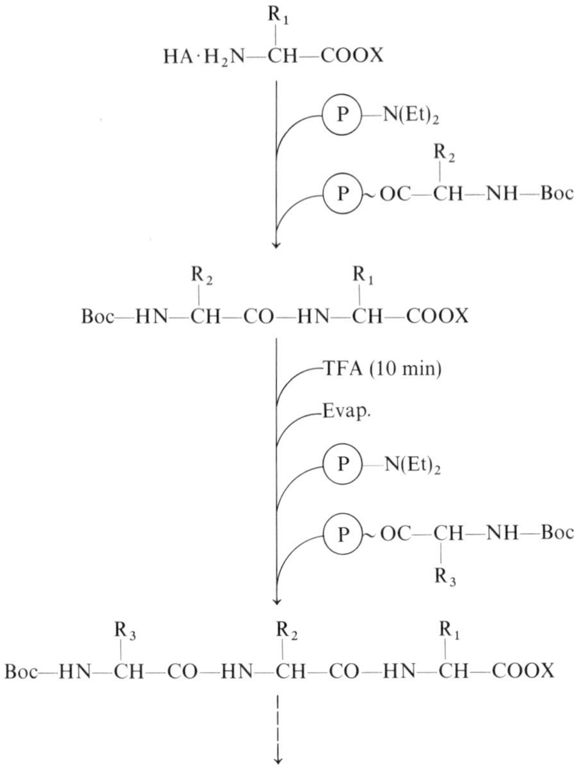

Scheme 17. A general scheme for a batchwise synthesis of peptides utilizing polymeric active esters (Stern *et al.*, 1977). HA, CF_3COOH, HCl; TFA, trifluoroacetic acid; X, α-carboxyl protecting group; Ⓟ—, polymeric carrier.

throughout synthesis. It was cleaved by anhydrous trifluoroacetic acid, the excess of which was removed from the ammonium salt product at the end of the reaction by repeated evaporation (3–4 times) with dichloromethane. Final neutralization was carried out using commercial bead-form macroporous polyvinyl-*N*,*N*-diethylbenzylamine, or a bead-form gel-type polymer prepared by reaction of diethylamine with chloromethylated copolystyrene-2%-divinylbenzene. The entire procedure was conducted very simply and smoothly in two regular round-bottom flasks in alternate succession of couplings and deprotections (Stern *et al.*, 1977).

The experiments using columnwise or batchwise procedures described above are undoubtedly exploratory in nature but, nevertheless, encouraging. It is clear that optimal polymeric reagents are needed to meet the many demands of a longer and more complicated mechanized process. Polymeric active esters derived from insoluble 1-hydroxybenzotriazole (Kalir *et al.*,

1975) seem particularly useful for this purpose, especially in columnwise processes where high reactivity of acylating agents is crucial.

D. Cyclic Peptides

Cyclization of linear peptides, as with many other molecules, usually takes place in high dilution in order to diminish intermolecular and enhance intramolecular condensation (for a review, see Kopple, 1972). Often, however, yields of cyclic peptides are low and they are accompanied by oligomerization products, i.e., oligo- and polypeptides. This undesirable side reaction might, in principle, be prevented by using polymers, preferably rigid ones, as carriers for the peptides to be cyclized. If the polymer-attached molecules are sufficiently distant from one another, only intramolecular cyclization should result as a consequence of their activation.

As no real progress has been made since the cyclization of peptides on polymeric supports was last reviewed (Patchornik *et al.*, 1973a), only a very brief account of this topic is given.

Cyclization on polymeric carriers may be divided into two categories, as shown in Scheme 18. In reactions of type I, cleavage of product from the polymer occurs simultaneously with cyclization. In reactions of type II, the linear peptide attached to the polymer is first cyclized and then cleaved.

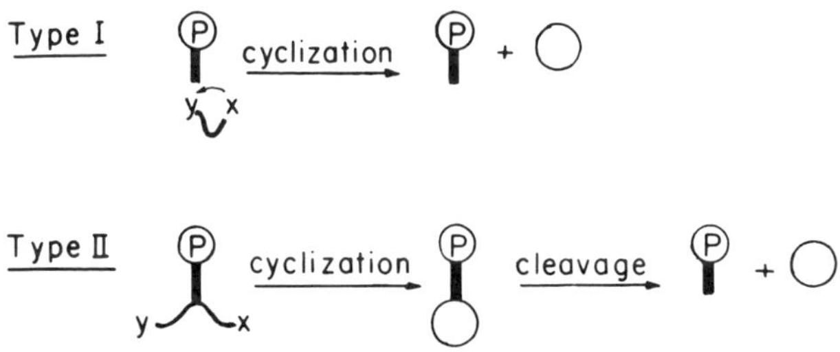

Scheme 18. General scheme for cyclization on polymeric carriers. Ⓟ—, polymer; X and Y, NH_2- and COOH-termini of activated polymer-attached linear molecules.

Type I cyclization was first achieved (Fridkin *et al.*, 1965b; Patchornik *et al.*, 1967) with polymeric esters derived from poly(4-hydroxy-3-nitrostyrene), as illustrated in Scheme 19. According to this procedure several

$$\text{Y—NH—Pep—COOH} + \text{H—[PNP]} \longrightarrow \text{Y—NH—Pep—CO—[PNP]} \xrightarrow{-\text{Y}}$$

$$\text{HA}\cdot\text{H}_2\text{N—Pep—CO—[PNP]} \xrightarrow{\text{Et}_3\text{N}} \underbrace{\text{NH—Pep—CO}} + \text{H—[PNP]}$$

Scheme 19. Synthesis of cyclic peptides utilizing insoluble poly(4-hydroxy-3-nitrostyrene) as a carrier. HA, HBr, HCl; [PNP], 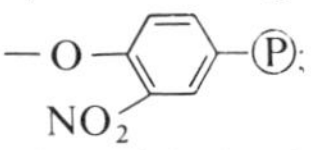; Y, N^{α}-protecting group. Reprinted with permission from Fridkin *et al.*, 1965b. Copyright by the American Chemical Society.

diketopiperazines and cyclic lactams as well as cyclo(tetra-Ala) (65% yield) and cyclo(Ala-Gly-Ala-Ala) (56% yield) were synthesized. Similarly, Flanigan and Marshall (1970) have utilized their polystyrene-derived 4-methylthiophenol support for the build-up and subsequent cyclization of attached linear peptides. Prior to cyclization, the polymeric ester bond is activated by oxidation, as shown in Scheme 14, Pathway B. Cyclic di-, tetra-, and hexapeptides were prepared by this method in reasonable yields. Preliminary reports on the attempted, although not yet successful, cyclization utilizing the attractive polystyrene-derived phenylhydrazine and 2,2-diphenyl glycol supports were also recorded (Wieland *et al.*, 1970).

Cyclization of type II is illustrated in Scheme 20 for the synthesis of a cystine-containing tripeptide (Bondi *et al.*, 1968). Following a similar procedure, Inukai *et al.* (1968) were able to prepare biologically active oxytocin.

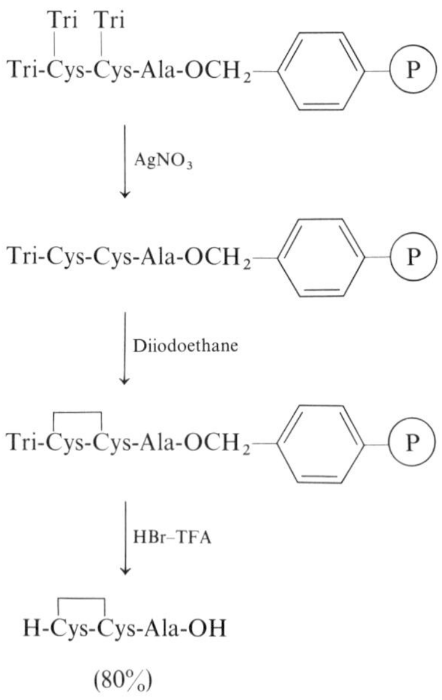

Scheme 20. Synthesis of a tripeptide cyclic disulfide on a polystyrene carrier (Bondi *et al.*, 1968). Ⓟ—, polymeric carrier; TFA, trifluoroacetic acid; Tri, trityl.

The synthesis of a glycine analog of gramicidin S (Sklyarov and Shashkova, 1969) is another example for cyclization of type II. The corresponding linear peptide was assembled on a benzyloxycarbonyl derivative of polystyrene to which a δ-amino group of an ornithine residue of the peptide was attached. Cyclization of the bound linear peptide followed by treatment with HBr in acetic acid afforded the desired cyclic product in solution, in 20% overall yield.

We strongly feel that the advantages associated with the use of polymers as supports for cyclization of peptides may lead to greatly improved synthetic procedures. Thus, there is a wide variety of recently synthesized polymers, some of which possess excellent physicomechanical and chemical properties, that are waiting to be used for this purpose.

IV. CONCLUDING NOTES

This chapter illustrated the successful utilization in peptide synthesis of insoluble polymeric reagents, namely, high-molecular-weight active esters derived from N-blocked amino acids and various cross-linked hydroxyl-containing resins. Most of the advantages anticipated *a priori* from such reagents were indeed experienced in practical work. The principal advantages were the convenient, facilitated operative procedures. When excess reagent was employed, it provided accelerated reaction rates, high, often nearly quantitative coupling yields, and high product purity. These advantageous features were common to polymeric reagents derived from various kinds of "parent" carriers including soluble polymers and insoluble-amorphous polymers, but mainly from cross-linked bead-form polymers of either gel or rigid-macroporous type. The superiority of these latter two types of supports was due to their excellent physicomechanical properties.

The polymeric reagent technique has been used so far for the synthesis of a large number and variety of peptides, though of the relatively short chain length of not more than 10 amino acid residues. The practicability of the method for longer peptides (~15–40 amino acid residues) has not yet been evaluated. Successful preparation of such compounds will undoubtedly depend on the answers to two key questions: first, will increasing steric hindrance, an obvious result of chain lengthening, diminish coupling rates to impractically low values; and second, how does one overcome similar kinetic difficulties which are expected from the decreased solubilities, i.e., concentrations, of the longer peptides?

The answer to the first question is perhaps supplied by known experimental data. It has been found that as long as the amine to be acylated is soluble in an appropriate organic solvent, its coupling with polymer-activated amino acids is still efficient. When dissolved in *N*,*N*-dimethylformamide, both insulin and poly-ε-benzyloxycarbonyllysine (of an average molecular weight ~10,000) for example, were quantitatively alanylated by polymeric *N*-hydroxysuccinimide ester of *tert*-butyloxycarbonylalanine at rates comparable to those of low-molecular-weight amines (Fridkin *et al.*,

1972). Moreover, Jung *et al.* (1975) and Heusel *et al.* (1977) have shown that it is possible to carry out a stepwise synthesis of peptides on soluble polymeric supports (polyethylene glycol) using polymeric active esters as reagents. This original and most attractive approach, in which the advantageous features of the liquid-phase peptide synthesis (Mutter and Bayer, 1974; this volume, Chapter 2) and those of the polymeric reagent method are combined, may lead to significant achievements.

The answer to the second question may be either: (1) development of new protecting groups, primarily of a hydrophobic nature, which will enhance the solubility of large oligopeptides in organic solvents; or (2) peptide synthesis in aqueous media, with minimum protection of side-chain functional groups using polymeric acylating agents, which has never been described but is worth exploring; or (3) limiting polymeric reagent synthesis to medium-sized fragments ($\sim$10–20 amino acid residues), followed by their subsequent assembly into the desired final peptide. This segment condensation may be carried out in homogeneous phase according to commonly used technology, or by using polymers, such as insoluble polystyrene-bound 1-hydroxybenzotriazole, as additives or even as carriers for activated peptide fragments.

Comparison of the polymeric reagent method with Merrifield's solid phase method, which uses the polymer as a support for the growing peptide, is almost irresistible, particularly in connection with the possibility of sequential-automatic processing. As assembly of the peptide using the solid-phase technology involves a single polymeric carrier, it might as well be performed in a single reactor and easily automated. Indeed, there are several useful variations of automatic peptide synthesizers, based on the principles of the Merrifield synthesis (Merrifield, 1968; Stewart and Young, 1969; Barany and Merrifield, this volume, Chapter 1); their applications have led to some impressive results. Polymeric reagent synthesis, on the other hand, requires several different insoluble polymers, creating considerable technological difficulties when automation is attempted. Since, however, both intermediate and final peptides formed by this method are always soluble in the reaction medium, the following principal advantages are obvious: the course of coupling reactions may be easily and continuously monitored by nondegradative analyses, e.g., by simple spectroscopic techniques; if necessary, synthesis may be accelerated in midreaction by addition of further excess of reagents; synthesis may be stopped at will and the desired product readily separated from accompanying contaminations; and efficient and mild final deblocking may be accomplished with properly selected protecting groups.

We have previously described (Section III,C) some preliminary experiments aimed at mechanized synthesis by polymeric reagents. Attempts in this direction will undoubtedly continue to encourage the development of new polymeric supports, perhaps more suitable for peptide synthesis. In view of the ease of regenerating these polymers after use and of reloading them with new N-blocked amino or other carboxylic acids (e.g., Kalir, 1975; Mathur and Williams, 1976), a fact which has significant industrial implications, the chemical and mechanical stability of these polymers will be of particular importance. No less important will be the need for appreciable solvation of polymers in as wide a variety of solvents as possible, including aqueous ones. Availability of such polymers will increase the versatility of peptide synthesis and allow the use of various approaches to side-chain protection.

In this chapter we have attempted to evaluate briefly a method which is not yet "conventional" for peptide synthesis. It is our strong belief and hope that its application, alone or combined with other existing technologies, will help in the near future to improve further the capabilities of synthetic peptide chemistry.

ACKNOWLEDGMENT

Financial support from the Camille and Henry Dreyfus Foundation which helped in the preparation of this chapter is gratefully acknowledged.

REFERENCES

Akiyama, N., Yanagisawa, Y., and Okawara, M. (1969). *J. Polym. Sci., Part A-1* **7**, 1905–1912.
Andreev, S. M., Tsiryapkin, V. M., Samoilova, N. A., Mironova, N. V., Davidovich, Yu. A., and Rogozhin, S. V. (1977). *Synthesis* pp. 303–304.
Atlas, D., Kalir, R., and Patchornik, A. (1975). *Tetrahedron Lett.* pp. 179–180.
Bodanszky, M. (1971). *In* "Prebiotic and Biochemical Evolution" (A. P. Kimball and J. Oro, eds.), pp. 217–222. North-Holland Publ., Amsterdam.
Bodanszky, M., and Bodanszky, A. (1967). *Am. Sci.* **55**, 185–196.
Bondi, E., Fridkin, M., and Patchornik, A. (1968). *Isr. J. Chem.* **6**, 22p.
Bosshard, H. R., Schechter, I., and Berger, A. (1973). *Helv. Chim. Acta* **56**, 717–723.
Brown, J., and Williams, R. E. (1971). *Can. J. Chem.* **49**, 3765–3766.
Burstein, Y., Saltoun, G., and Patchornik, A. (1970). *Isr. J. Chem.* **8**, 175p.
Cramer, F., and Köster, H. (1968). *Angew. Chem., Int. Ed. Engl.* **7**, 473–474.
Davidovich, Yu. A., and Gut, V. (1976). *Collect. Czech. Chem. Commun.* **41**, 1805–1810.
Fischer, E. (1906). *Ber. Dtsch. Chem. Ges.* **39**, 530–610.
Flanigan, E., and Marshall, G. R. (1970). *Tetrahedron Lett.* pp. 2403–2406.
Fridkin, M. (1967). Ph.D. Thesis, Weizmann Institute of Science, Rehovot, Israel.
Fridkin, M., and Patchornik, A. (1974). *Annu. Rev. Biochem.* **43**, 419–443.
Fridkin, M., Patchornik, A., and Katchalski, E. (1965a). *Isr. J. Chem.* **3**, 69–70p.
Fridkin, M., Patchornik, A., and Katchalski, E. (1965b). *J. Am. Chem. Soc.* **87**, 4646–4648.
Fridkin, M., Patchornik, A., and Katchalski, E. (1966). *J. Am. Chem. Soc.* **88**, 3164–3165.

Fridkin, M., Patchornik, A., and Katchalski, E. (1968). *J. Am. Chem. Soc.* **90**, 2953–2957.
Fridkin, M., Patchornik, A., and Katchalski, E. (1971). *In* "Peptides 1969" (E. Scoffone, ed.), pp. 164–167. North-Holland Publ., Amsterdam.
Fridkin, M., Patchornik, A., and Katchalski, E. (1972). *Biochemistry* **11**, 466–471.
Fridkin, M., Kalir, R., Warshawsky, A., and Patchornik, A. (1975). *In* "Peptides: Chemistry, Structure and Biology" (R. Walter and J. Meienhofer, eds.), pp. 395–401. Ann Arbor Sci. Publ., Ann Arbor, Michigan.
Fridkin, M., Hazum, E., Kalir, R., Rotman, M., and Koch, Y. (1977a). *J. Solid-Phase Biochem.* **2**, 175–182.
Fridkin, M., Stabinsky, Y., Zakuth, V., and Spirer, Z. (1977b). *Biochim. Biophys. Acta* **496**, 203–211.
Grubbs, R. H. (1977). *Chemtech.* **7**, 512–518.
Guarneri, M., Ferroni, R., Giori, P., and Benassi, C. A. (1972). *In* "Chemistry and Biology of Peptides" (J. Meienhofer, ed.), pp. 213–216. Ann Arbor Sci. Publ., Ann Arbor, Michigan.
Hayatsu, H., and Khorana, H. G. (1967). *J. Am. Chem. Soc.* **89**, 3880–3887.
Helfferich, F. (1962). "Ion Exchange." McGraw-Hill, New York.
Heusel, G., Bovermann, G., Gohring, W., and Jung, G. (1977). *Angew Chem. Int. Ed. Engl.* **16**, 642–643.
Horiki, K. (1976). *Tetrahedron Lett.* pp. 4103–4106.
Inukai, N., Nakano, K., and Murakami, M. (1968). *Bull. Chem. Soc. Jpn.* **41**, 182–186.
Izumiya, M., Muraoko, M., and Aoyagi, H. (1971). *Bull. Chem. Soc. Jpn.* **44**, 3391–3395.
Jones, J. H., and Ridge, B. (1974). *In* "Amino Acids, Peptides and Proteins" (R. C. Sheppard, ed.), Vol. 5, pp. 258–350. The Chemical Society, London.
Jung, G., Bovermann, G., Gohring, W., and Heusel, G. (1975). *In* "Peptides: Chemistry, Structure and Biology" (R. Walter and J. Meienhofer, eds.), pp. 433–437. Ann Arbor Sci. Publ., Ann Arbor, Michigan.
Kalir, R. (1975). Ph.D. Thesis, Weizmann Institute of Science, Rehovot, Israel.
Kalir, R., Fridkin, M., and Patchornik, A. (1974). *Eur. J. Biochem.* **42**, 151–156.
Kalir, R., Warshawsky, A., Fridkin, M., and Patchornik, A. (1975). *Eur. J. Biochem.* **59**, 55–61.
König, W., and Geiger, R. (1970a). *Chem. Ber.* **103**, 788–798.
König, W., and Geiger, R. (1970b). *Chem. Ber.* **103**, 2034–2040.
Kopple, K. D. (1972). *J. Pharm. Sci.* **61**, 1345–1356.
Laufer, D. A., Chapman, T. M., Marlborough, D. I., Vaidya, V. M., and Blout, E. R. (1968). *J. Am. Chem. Soc.* **90**, 2696–2698.
Lauren, D. R., and Williams, R. E. (1972). *Tetrahedron Lett.* pp. 2665–2668.
Leznoff, C. C. (1974). *Chem. Soc. Rev.* **3**, 65–85.
Manecke, G., and Haake, E. (1968). *Naturwissenschaften* **55**, 343.
Marshall, D. L., and Liener, I. E. (1970). *J. Org. Chem.* **35**, 867–868.
Mathur, N. K., and Williams, R. E. (1976). *J. Macromol. Sci., Rev. Macromol. Chem.* **15**, 117–142.
Merrifield, R. B. (1963). *J. Am. Chem. Soc.* **85**, 2149–2154.
Merrifield, R. B. (1968). *Sci. Am.* **218**, 56–74.
Mutter, M., and Bayer, E. (1974). *Angew Chem.* **86**, 97.
Narita, M., Teramoto, T., and Okawara, M. (1972). *Bull. Chem. Soc. Jpn.* **45**, 3149–3155.
Overberger, C. G., and Sannes, K. N. (1974). *Angew. Chem., Int. Ed. Engl.* **13**, 99–104.
Panse, G. T., and Laufer, D. A. (1970). *Tetrahedron Lett.* pp. 4181–4184.
Patchornik, A., and Kraus, M. A. (1976). *Encycl. Polym. Sci. Technol.* **1**, Suppl., 468–492.
Patchornik, A., Fridkin, M., and Katchalski, E. (1967). *In* "Peptides" (H. C. Beyerman, A. Van de Linde, and W. Maassen van den Brink, eds.), pp. 91–99. North-Holland Publ., Amsterdam.

Patchornik, A., Fridkin, M., and Katchalski, E. (1973a). *In* "The Chemistry of Polypeptides" (P. G. Katsoyannis, ed.), pp. 315–333. Plenum, New York.

Patchornik, A., Fridkin, M., and Katchalski, E. (1973b). British Patent 1,268,631.

Rogozhin, S. V., Davidovich, Yu. A., Andreev, S. M., and Yurtanov, A. I. (1973a). *Dokl. Akad. Nauk SSSR* **211**, 1356–1358.

Rogozhin, S. V., Davidovich, Yu. A., Andreev, S. M., and Yurtanov, A. I. (1973b). *Dokl. Akad. SSSR* **212**, 108–111.

Rogozhin, S. V., Davidovich, Yu. A., Samoilova, N. A., Mironova, N. V., Yurtanov, A. I., and Andreev, S. M. (1975). *Izv. Akad. Nauk SSSR, Ser. Khim.* pp. 428–433.

Rubinstein, M., and Patchornik, A. (1972). *Tetrahedron Lett.* pp. 2881–2884.

Sklyarov, L. Yu., and Shashkova, I. V. (1969). *Zh. Obshch. Khim.* **39**, 2778–2783.

Sklyarov, L. Yu., Gurbunov, V. I., and Shchukina, L. A. (1966). *Zh. Obshch. Khim.* **36**, 2220–2221.

Stern, M., Kalir, R., Patchornik, A., Warshawsky, A., and Fridkin, M. (1977). *J. Solid-Phase Biochem.* **2**, 131–139.

Stewart, J. M., and Young, J. D. (1969). "Solid Phase Peptide Synthesis." Freeman, San Francisco, California.

Teramoto, T., Narita, M., and Okawara, M. (1977). *J. Polym. Sci., Polym. Chem. Ed.* **15**, 1369–1379.

Tilak, M. A., Hendricks, M. L., and Wedel, D. S. (1972). *In* "Progress in Peptide Research" (S. Lande, ed.), Vol. II, pp. 351–359. Gordon and Breach, New York.

Weinshenker, N. M., and Shen, C. M. (1972a). *Tetrahedron Lett.* pp. 3281–3284.

Weinshenker, N. M., and Shen, C. M. (1972b). *Tetrahedron Lett.* pp. 3285–3288.

Wieland, T., and Birr, C. (1966). *Angew. Chem., Int. Ed. Engl.* **5**, 310.

Wieland, T., and Birr, C. (1967a). *Chimia* **21**, 581–582.

Wieland, T., and Birr, C. (1967b). *In* "Peptides 1966" (H. C. Beyerman, A. Van de Linde and W. Maassen van den Brink, eds.), pp. 103–106. North-Holland Publ., Amsterdam.

Wieland, T., Lewalter, J., and Birr, C. (1970). *Justus Liebigs Ann. Chem.* **740**, 31–47.

Wieland, T., Birr, C., and Fleckenstein, P. (1972) *Justus Liebigs Ann. Chem.* **756**, 14–19.

Wildi, B. S., and Johnson, J. H. (1968). *Abstr. Pap., 155th Natl. Meet. Am. Chem. Soc.*, Abstract A-8.

Williams, R. E. (1972). *J. Polym. Sci., Part A-1* **10**, 2123–2127.

Williams, R. E., Brown, J., and Lauren, D. R. (1972). *Polym. Prepr.* **13**, 823–827.

Wolman, Y., Kivity, S., and Frankel, M. (1967). *Chem. Commun.* pp. 629–630.

Wünsch, E. (1971). *Angew. Chem., Int. Ed. Engl.* **10**, 786–795.

Chapter 4

The Four Component Synthesis

IVAR UGI

I. THE FOUR COMPONENT CONDENSATION AS A BASIS OF PEPTIDE SYNTHESIS

The four component condensation (4CC) (Ugi, 1962, 1969, 1971; Gokel *et al.*, 1971a) of suitably chosen acid components (**1**) with amines (**2**), carbonyl compounds (**3**), and isonitriles (**4**) leads via generally labile α-adducts (**5**) to stable α-amino acid derivatives (**6**) (Scheme 1). In one case an α-adduct could be isolated (Marquarding, 1973).

THE PEPTIDES, VOLUME 2

ISBN 0-12-304202-X

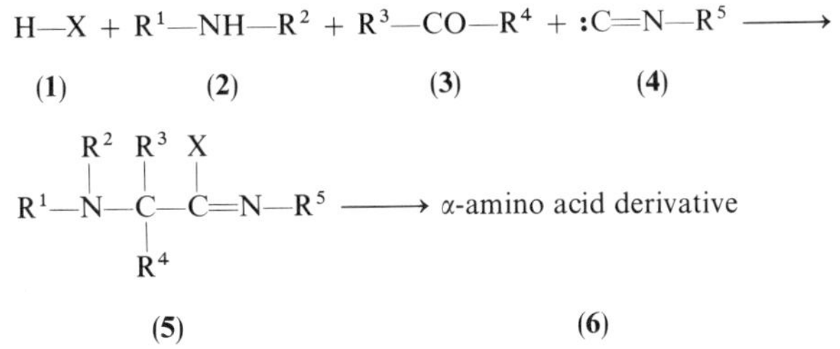

Scheme 1

The term "four component condensation"* is also used for related reactions which involve instead some condensation products (imines, enamines, etc. Schutz and Ugi, 1979a,b) of amines (**2**) and carbonyl compounds (**3**). The chemical nature of the spontaneous rearrangement of the α-adduct **5** and the general constitutional features of the resulting α-amino acid derivatives **6** are primarily determined by the acid components. Their incorporation in the four component condensation products is illustrated by structures **7** to **14**.

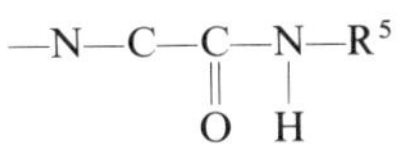

(**7**) (1 = H_2O)

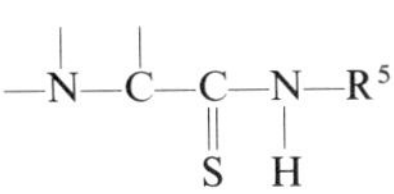

(**8**) (1 = $H_2S_2O_3$)

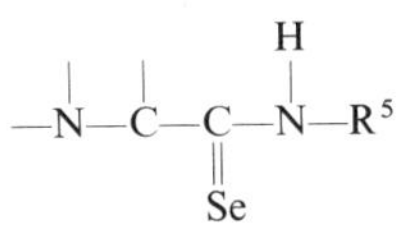

(**9**) (1 = H_2Se)

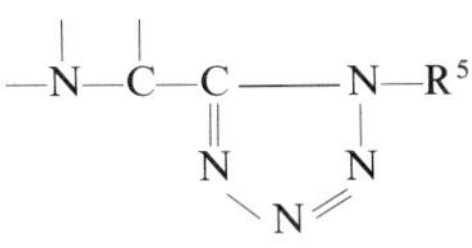

(**10**) (1 = HN_3)

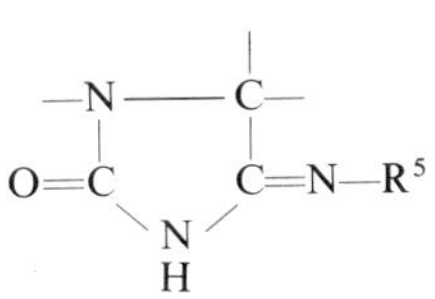

(**11**) (1 = HNCO)

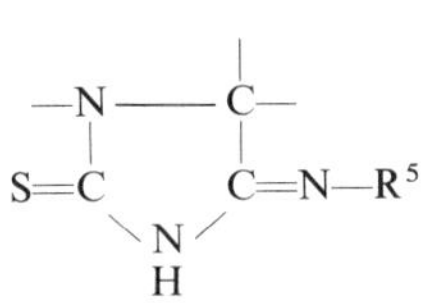

(**12**) (1 = HNCS)

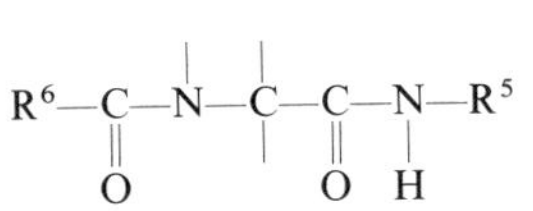

(**13**) (1 = R^6CO_2H; 2 = primary amine)

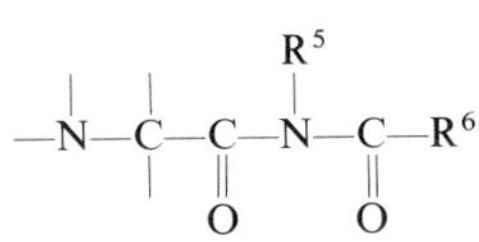

(**14**) (1 = R^6CO_2H; 2 = secondary amine)

* Some other names have also been used for this reaction. (see Kabbe 1971, and Gokel *et al.*, 1971a)

Whenever a target compound of a synthesis can be prepared by a four component condensation, its synthesis by this procedure is likely to be among the most effective of suitable synthetic approaches. Generally, the result of a four component condensation can only be achieved by a multistep sequence of other synthetic reactions. As a rule, there exist some conditions under which a considered four component condensation affords the desired product in satisfactory, often excellent, yield. Due to the special nature of the four component condensation, the impurities of the product (unreacted starting materials, co-products, by-products from side reactions) are, as a rule, chemically quite different from the product itself and easy to remove.

The four component condensation with carboxylic acids (as **1**) and ammonia or primary amines yields α-acylamino carbonamides (**13**). Since these also comprise the peptide derivatives, this type of four component condensation (Ugi and Steinbrückner, 1961; Steinbrückner, 1961) is of special interest to peptide synthesis. By virtue of the constitutional features of the starting materials and mechanistic peculiarities, such four component condensations are closely related to other multicomponent reactions such as the Mannich condensation, the Strecker synthesis, or the Bucherer–Bergs reaction (Hellmann and Opitz, 1960). Four component condensations appear to be closely similar to the Passerini reaction (Passerini *et al.*, 1931); however, they differ from it by the absence of an amine component in those reactions. In contrast to the Passerini reaction, a great variety of acid components (**1**) other than carboxylic acids can participate in the four component condensation. In many cases the four component condensation and the Passerini reaction do not have much in common with respect to either the reaction mechanism or the chemical nature of the products. With certain reactants and under certain (avoidable) conditions (Skorna, 1978; Ugi *et al.*, 1977), the Passerini reaction and the four component condensation with a carboxylic acid as **1** compete as parallel reactions and lead to a mixture of the four component condensation and Passerini reaction products.

There are three different types of peptide syntheses that involve four component condensation with N^{α}-protected α-amino acids or peptides (**15**) as the acid component (Schemes 2–4).

Scheme 2 represents asymmetric syntheses of peptide segments (Ugi, 1969, 1971; Ugi *et al.*, 1975, 1977a,b; Gokel *et al.*, 1971b; Urban *et al.*, 1978) (**20**) by stereoselective four component condensation with a chiral amine (**16**), an aldehyde (**17**), and an isocyanide (**18**) whose isocyano group replaces the α-amino group of a COOH-terminally protected amino acid or peptide (Urban *et al.*, 1977; Skorna and Ugi, 1977), followed by a cleavage of the R*—N bond (see **20** → **21** + **22**) (Gokel *et al.*, 1971b, 1972; Ugi and Offermann, 1964; Marquarding *et al.*, 1970).

Scheme 3 describes the coupling of suitably protected peptides, i.e., a carboxyl component (**15**) and an amine component (**23**), by four component

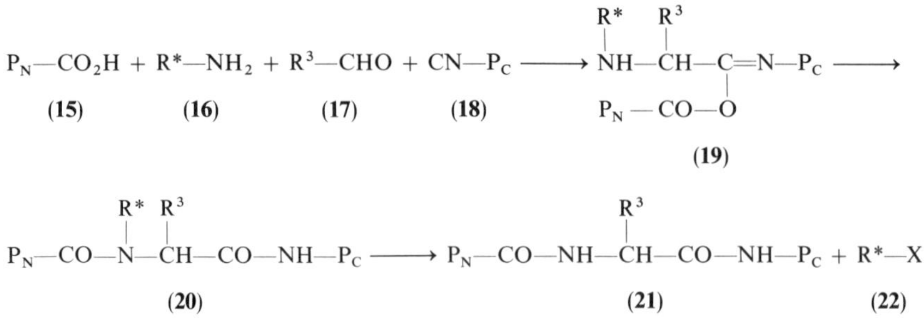

Scheme 2

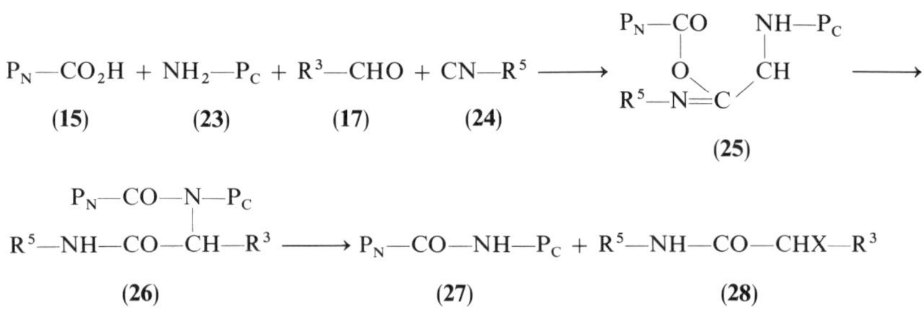

Scheme 3

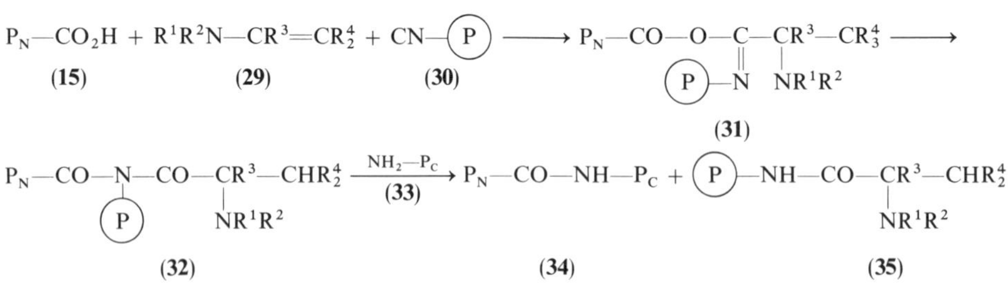

Scheme 4

condensation (Gokel *et al.*, 1971b; von Zychlinski *et al.*, 1974; Wackerle and Ugi, 1975 Ugi *et al.*, 1975, 1977a,b; Zahn and Ugi, 1979) and subsequent conversion of the product (**26**) into a peptide derivative (**27**).

In Scheme 4 (Ugi and Steinbrückner, 1961; Steinbrückner, 1961; Skorna and Ugi, 1977; Opitz and Merz, 1962) an N^{α}-protected amino acid or peptide derivative (**15**) is reacted with an enamine (**29**) and a functionalized polymer (**30**) with isocyano groups (Skorna *et al.*, 1978; Skorna and Ugi, 1978, 1979). The resulting four component condensation product (**32**), again a functionalized polymer, corresponds to a COOH-terminally activated derivative of **15**, and may be used for the acylating transfer of the P_N-CO group in peptide (**34**) synthesis (Skorna and Ugi, 1978, 1979; Skorna, 1978).

II. ADVANTAGES OF THE FOUR COMPONENT SYNTHESIS

The synthesis of peptide segments by stereoselective four component condensation (Scheme 2) proceeds with the formation of two bonds belonging to the peptide chain and a new amino acid unit from an aldehyde (**17**) as a connecting moiety within the peptide chain of **21**. In the four component condensation product, **20**, the newly formed amino acid carries an N-substituent (Gokel *et al.*, 1971b; Failli *et al.*, 1973, 1978; Immer *et al.*, 1973; Zinner and Kliegel, 1966; Urban *et al.*, 1978; Urban, 1979).

Such peptide segment syntheses have several advantages over comparable syntheses by conventional methods. In general, the number of steps required is substantially lower. The procedure permits the direct synthesis—as part of a peptide—of certain amino acids, which are otherwise not readily available or difficult to manipulate, in desired configuration. Such amino acids include those of D-configuration, others not derived from protein (Immer *et al.*, 1973; Failli *et al.*, 1973, 1978), and isotope-labeled ones; and those difficult to incorporate into peptides by conventional methods include N-substituted (e.g., *N*-alkyl) (Zinner and Kliegel, 1966; Immer *et al.*, 1973; Failli *et al.*, 1973, 1978) and α-substituted amino acids (Maia *et al.*, 1973; Urban *et al.*, 1979). Probably the most important advantage of four component peptide segment syntheses is the generally observed enhancement of the solubility of peptide derivatives by the N-substituents (R* in structure **20** and R^5-NH-CO-CHR^3 in **16**) which are typical of four component peptide segment syntheses (Wackerle and Ugi, 1975, also unpublished; Urban and Ugi, 1975 Urban *et al.*, 1978). Preliminary results suggest that the derivatives of larger peptides with two or more such N-substituents are also much more soluble in most organic solvents than the corresponding non-N-substituted compounds (L. Wackerle and I. Ugi, unpublished). Predictably, the frequently encountered failures in conventional synthesis of large peptides due to the sparing solubility of essential intermediates will be avoidable through the use of four component condensation products, as segments, or intermediates.

The synthesis of large peptides through four component segment coupling (Gokel *et al.*, 1971b; von Zychlinski *et al.*, 1974; Wackerle and Ugi, 1975 Ugi *et al.*, 1975, 1977a,b; Eckert *et al.*, 1979) (see Scheme 3) also has some attractive features. In contrast to segment coupling by conventional methods, coupling of peptide segments by four component condensation does not involve α-carboxyl activation of peptide COOH-termini and acylation of the amine component by a second order reaction in competition with racemization by pseudo first-order processes (Ugi, 1971). The α-adduct **25** is the only activated, racemization-prone derivative of the COOH-terminal amino acid unit of **15**. The formation of the peptide bond occurs by a fast, spontaneous intramolecular rearrangement. Racemization of the COOH-terminal amino acid unit of **25** via a 5(4*H*)-oxazolone would not lead to an isomer of **26** because the formation of the 5(4*H*)-oxazolone (Goodman and Glaser, 1969) of the carboxyl component **15** from the intermediate **25** would lead to nonisomeric fragments of the latter (Nah, 1973).

Similar to the four component condensation intermediates (**20**) in the synthesis of peptide segments (**21**), the intermediates (**26**) of four component peptide segment coupling are generally more soluble in organic solvents than are the corresponding final products (**27**). This feature of four component segment coupling is potentially useful for the synthesis of large peptides that would be difficult to prepare by other methods involving sparingly soluble intermediates.

Four component segment coupling does not lead to the hard-to-separate, contaminating coproducts and side products that are occasionally formed during some of the conventional peptide segment couplings (e.g., by the carbodiimide or azide methods).

Since four component peptide syntheses according to Schemes 2 and 3 do not involve the customary activated derivatives of amino acids that are reactive toward nucleophilic solvents, they can also be carried out in highly polar and nucleophilic solvents, e.g., methanol.

The resin-bound diacylimine derivatives of amino acids and peptides (**32**) are formed by a practically quantitative four component condensation from an enamine (**29**), an isocyano resin (**30**) (Skorna *et al.*, 1978), and the acid component (**15**) (Skorna, 1978). The synthetic usefulness of compound **32** is similar in scope and limitations to some other resin-bound, activated amino acid derivatives (Fridkin *et al.*, 1965; Kalir *et al.*, 1974) (see Chapter 3, this volume). If **32** is derived from a peptide (**30**) with a suitable NH_2-terminal protective group, deblocking may lead to the exclusive formation of the corresponding cyclic peptide through "intramolecular" cyclization.

III. PROBLEMS

A. Stereoselective Synthesis of Peptide Segments

The four component condensation according to Scheme 2 yields a mixture of the stereoisomeric peptide derivatives **20** (p) and **20** (n) [p = (R),(R) or (S),(S) and n = (R),(S) or (S),(R); Ugi, 1965a,b] with a newly formed L- or D-amino acid unit, respectively. In order to obtain the desired stereoisomer of the product, **21** (L) or **21** (D), respectively, in high yield and purity, a high degree of stereoselectivity is needed for the essential steps of the synthesis. A chiral amine component (**16**) with strong asymmetric inducing power (Kopp *et al.*, 1980; Dugundji *et al.*, 1978; Urban *et al.*, 1978) is needed as a chiral template whose participation in the four component condensation under suitable conditions leads to the preferred formation of the desired stereoisomer of compound **20**. A replacement of the R* group of the four component condensation product **20** by hydrogen is needed in order to obtain a peptide derivative **21**. Thus, the chiral residue R* of the amine component (**16**) must have constitutional features which make it possible to cleave the R*-N bond in compound **20** under conditions that do not destroy or racemize peptides (Gokel *et al.*, 1971b; Marquarding *et al.*, 1970; Urban *et al.*, 1978). The chiral templates of asymmetric syntheses are generally relatively difficult to obtain. If the synthesis of peptide segments by four component condensation proceeded with a total loss of the chiral amine component (Eberle and Ugi, 1976; Urban *et al.*, 1976, 1978) it would be of only limited usefulness. Therefore, the amine component (**16**) must not only contain a "cleavable" R*-N bond, but, moreover, the cleavage reaction (**20** → **21** + **22**) ought to produce a coproduct (**22**) from which the chiral amine (**16**) can be resynthesized with full retention (Gokel *et al.*, 1972; Urban *et al.*, 1976, 1978) of its configuration and in good yield.

The ratio of the relative amounts of any two stereoisomeric products of a kinetically controlled productive stereoselective reaction (Dugundji *et al.*, 1978), as **15** + **16** + **17** + **18** → **20**, has a positive lower bound and a finite upper bound (Gokel *et al.*, 1971a; Kopp *et al.*, 1980; Dugundji *et al.*, 1978; Kaufhold and Ugi, 1967). Both of these bounds are the ratios of two rate constants. Since such ratios are determined by a free enthalpy difference of stereoisomeric transition states and are due to differences in nonbonded interactions Ruch and Ugi, 1969), the stereoselectivity of productive reactions is generally relatively low (mostly 50–90% relative yield of the preferred product). Products with an isomer purity required for many synthetic peptides (99.99% and better) can thus only be obtained from productive stereoselective reactions after cumbersome purification of the product, as a rule by chromatographic methods (Jaeger, 1974). This disadvantage of productive asymmetric syntheses can however be overcome if the latter are combined with a subsequent destructive stereoselective reaction (Dugundji *et al.*, 1978; Urban *et al.*, 1976; 1978; Brandt *et al.*, 1977). In a peptide synthesis by stereoselective four component condensation according to Scheme 5, the limited stereoselectivity of the productive pair of mutually dependent parallel reactions (Eq. 1) is multiplied by the unlimited nonlinear stereoselectivity

$$\mathbf{15} + \mathbf{16} + \mathbf{17} + \mathbf{18} \longrightarrow \mathbf{20}(\text{L}) + \mathbf{20}(\text{D}) \qquad (1)$$

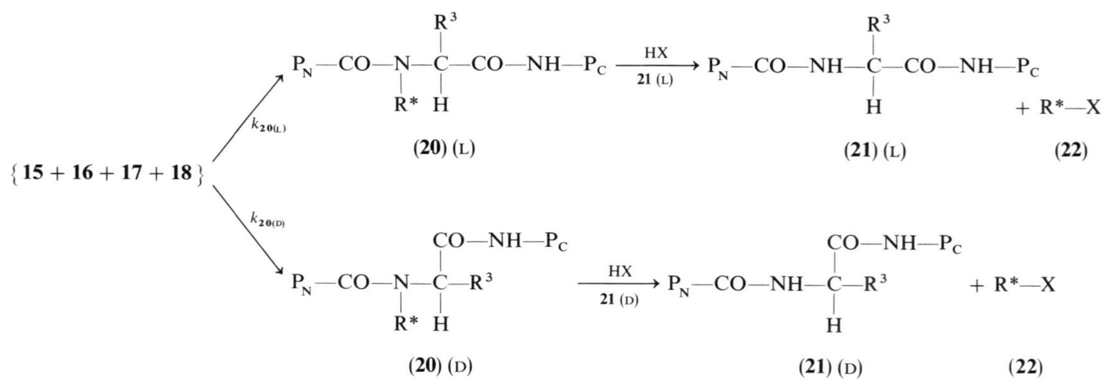

Scheme 5

of a destructive pair of independent parallel reactions, **20** (L) → 21 (L) + **22** together with **22** (D) → **21** (D) + **22**. Let **21** (L) be the desired product. Its synthesis by Scheme 5 is only effective if a chiral amine component can be found which obeys the correlations $k_{\mathbf{20}\,(\mathrm{L})} \gg k_{\mathbf{20}\,(\mathrm{D})}$ and $k_{\mathbf{21}\,(\mathrm{L})} \ll k_{\mathbf{21}\,(\mathrm{D})}$ under suitable conditions. In such a synthesis one produces, first, a mixture of stereoisomers in which **20** (L) prevails over **20** (D) and then subjects this mixture in a second step to partial cleavage **20** (L) → **21** (L) + **22** and **20** (D) → **21** (D) + **22** by which only a minor fraction of **20** (L) is affected, while **20** (D) is almost totally destroyed. The subsequent separation of the nonisomeric compounds **20** and **21** is generally easy to accomplish and also yields the semifinal product **20** (L) in a high degree of purity. Cleavage of the R*—N bond of **20** (L) yields practically pure peptide **21** (L) (Urban *et al.*, 1978).

Further problems associated with peptide segment syntheses by stereoselective four component condensation are the avoidance of racemization at any one of the amino acid units (Charles *et al.*, 1975) during any of the required operations, the synthesis of the aldehyde and isocyanide components needed, and the necessity to find substitutes (P. Seidel, and I. Ugi, unpublished) for some aldehydes that do not undergo four component condensations sufficiently well, due to their proneness to react in some undesirable manner (e.g., the extreme tendency of phenyl acetaldehyde to undergo aldol-type condensations).

B. Segment Coupling

The effectiveness of peptide segment coupling by four component condensation (Scheme 3) depends primarily on the properties of the aldehyde component (**17**). It is critically important that **17** is capable of forming a four component condensation product (**26**) in good yield from suitably protected peptide segments (**15** and **23**) without racemization of the COOH-terminal amino acid of the carboxyl component (**15**) (L. Wackerle, unpublished). The conversion of the four component condensation product **26** into a peptide derivative **27** is a necessary condition for peptide segment coupling by four component condensation. The aldehyde **17** must contain a group R^3 which facilitates the necessary C—N cleavage **26** → **27** + **28** (von Zychlinski *et al.*, 1974; Eckert *et al.*, 1979) so that this can be achieved under conditions which do not damage a peptide. Further, the coproduct **28** of this cleavage should be easy to remove from the target peptide (**27**).

C. Polymer-Bound Activated Amino Acid Derivatives

After it had been demonstrated that the activation of amino acid derivatives according to Scheme 4 with aliphatic low molecular-weight isocyanides (Ugi and Steinbrückner, 1961; Steinbrückner, 1961) does not succeed (Skorna, 1978), polymeric compounds with isocyano groups [(**30**), e.g., isocyanomethyl polystyrene (Skorna *et al.*, 1978)] were used, leading to activated amino acid derivatives (**32**) that are attached to the polymer **32** (Skorna, 1978), which is easy to purify. Here, the overall yield and isomer purity of the peptide derivatives that are obtained from **32** by treatment with amine components, i.e., protected amino acid and peptide derivatives with blocked COOH-terminus, is satisfactory.

Since the diacylimides (**32**) contain two acylating electrophilic carbonyl groups, it is important to find enamines (**29**) leading to four component condensation product (**32**) in which the undesirable acyl transfer is suppressed. This is achieved through selective steric hindrance.

IV. RECENT ADVANCES

A. Synthesis of Optically Active Isocyanides

A comparison (Hoffmann, *et al.*, 1971 Urban *et al.*, 1977; Skorna and Ugi, 1977) of the various methods to prepare isonitriles **37** showed that, in general, the dehydration of N-monosubstituted formylamines (**36**) by phosgene in the presence of tertiary amines is particularly well suited for the preparation of isonitriles which are needed in four component peptide syntheses.

However, it has been found (Skorna and Ugi, 1977) that it is advantageous in many cases to prepare isonitriles (**38**) from N-monosubstituted *N*-formylamines (**36**) with "diphosgene" (**37**) as the dehydrating agent (Eq. 2). In the

$$\underset{(\mathbf{36})}{2\,R^5\text{—NH—CHO}} \xrightarrow[\text{tertiary amine}]{Cl_3C\text{—OCOCl}\ (\mathbf{37})} \underset{(\mathbf{38})}{R^5\text{—N}{=}C\colon} \tag{2}$$

preparation of isocyanomethyl polystyrene **30**, a two-phase procedure gives the best conversion of *N*-formylamino groups to isocyano groups (Skorna *et al.*, 1978).

When chiral isocyano derivatives of amino acids and peptides with blocked COOH-termini (**18**) are needed for the synthesis of peptide segments (**21**) by stereoselective four component condensation, the isocyano derivatives of dipeptide esters are prefered to the corresponding chiral α-isocyano esters, because, with few exceptions, the latter have a pronounced tendency to racemize (Hoffmann *et al.*, 1971; Urban *et al.*, 1977). This is due to the combination of C—H acidifying ester and isocyano groups.

In a systematic investigation of the relative reactivities of isocyanides in either four component condensation or the Passerini reaction it was found that cyclohexyl isocyanide is particularly reactive in the above context (H.

Aigner, and I. Ugi, unpublished). Cyclohexyl isocyanide reacts in the four component condensation about 10 times faster than *tert*-butyl isocyanide, due to a particularly advantageous combination of inductive effects and steric factors.

B. Stereoselective Four Component Condensation

1. Amine Components

The synthesis of peptide segments according to Scheme 2 requires chiral primary amine components (**16**) that have a strong asymmetric inducing power (Kopp *et al.*, 1980) in stereoselective four component condensation and whose products, **20**, contain a C—N bond which is cleavable under mild conditions (Gokel *et al.*, 1971b). Recently we found that α-aminoalkylferrocenes of the type **39** (Marquarding *et al.*, 1977a) fulfill the requirements.

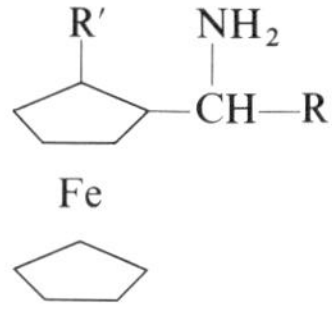

R = *sec*-alkyl

(**39**) R′ = H or small alkyl (methyl, ethyl)

In particular, the chiral amine (**39**) with R = isopropyl and R′ = H has been used successfully in this context (Urban and Ugi, 1975; Eberle *et al.*, 1978; Marquarding *et al.*, 1977a,b; Urban *et al*; 1978). Not only does it participate in stereoselective four component peptide syntheses under suitable conditions with satisfactory stereoselectivity, but it can also be resynthesized from **22** after the cleavage step **20** → **21** + **22**) if thioglycolic acid is used as a scavenger of chiral α-ferrocenylalkylcarbenium ions (Eberle and Ugi, 1976; Ratajczak and Misterkiewicz, 1975; Urban *et al.*, 1978). Other promising chiral amine components are **39**, with R = c-Hex, R = H (Ettengruber and Ugi, unpubl.), and **40** (Stüber and Ugi, unpubl.). The α-ferrocenyl alkylamines (**39**, R′ = H) are now readily available from ferrocene, aldehydes RCHO and ammonia (Hermann and Ugi, 1978, 1979).

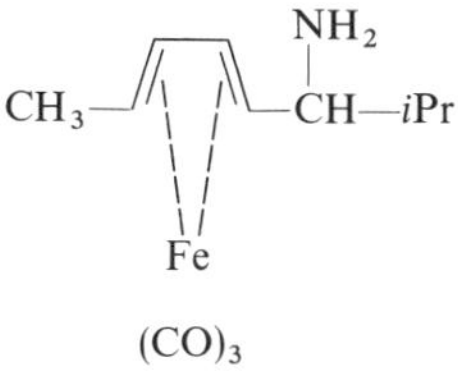

(**40**)

2. *Reaction Conditions*

With the knowledge of the reaction mechanism (Kaufhold and Ugi, 1967 Gokel *et al.*, 1971), which represents a kinetic scheme of four competing pairs of corresponding reactions (Ruch and Ugi, 1969; Ugi, 1965a,b), conditions were found under which the asymmetrically induced synthesis of peptide segments by four component condensation takes place with a high degree of stereoselectivity (Urban *et al.*, 1978; Urban and Ugi, 1975; Rehn, 1976), with a stereoisomer ratio of up to 1:200 for **20** (p): **21** (n).

Since peptides are needed in an isomer purity which is much higher than the purity that can be attained through the direct formation of **20** by a "productively" stereoselectivity, this step is followed by a "destructively" stereoselective step **20** → **21** + **22** (see Scheme 5) that is endowed with unlimited stereoselectivity (Kopp *et al.*, 1980; Dugundji *et al.*, 1978; Brandt *et al.*, 1977). Thus, the combination of a productively stereoselective four component condensation with a subsequent destructively stereoselective partial acidolysis leads to peptide segment derivatives **20** of high isomer purity (>99.99%) with a relatively good overall yield (Urban *et al.*, 1978, 1979; Urban, 1979).

3. *Solubility Effects and New Protective Groups*

In order to be able to make use in subsequent synthetic steps of the fact that N^{α}-ferrocenylalkyl peptide (Urban and Ugi, 1975) derivatives are more soluble in a variety of required solvents than the corresponding N-unsubstituted peptide derivatives, a new proptective group technique has been developed (Eckert and Ugi, 1975; 1976a,b; 1979; Eckert *et al.*, 1975, 1977, 1978, 1979; Eckert, 1977) which involves β-halogenated alkyl groups that are stable towards the usual acidolytic deprotection conditions, against which the N^{α}-ferrocenylalkyl groups are rather sensitive.

C. Segment Coupling

The development of coupling methods for peptide segments by four component condensation according to Scheme 3 was at first very much determined by the search for aldehyde components (**17**) whose four component condensation products **26** were acidolytically cleavable (von Zychlinski *et al.*, 1974; Ugi *et al.*, 1975, 1977a,b). This effort led to the use of 1-*tert*-butyloxycarbonyl-3-formyl indole (**41**) (Wackerle and Ugi, 1975) as an aldehyde

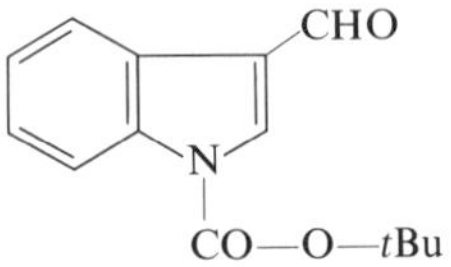

(**41**)

component (**17**) that fulfilled the expectations quite well, in particular, if the cleavage (**26** → **27** + **28**) was carried out in one step. This is possible through the use of trifluoroacetic acid in combination with trifluoroethanol. In the latter acidic medium the acidolytic removal of the *tert*-butyl group, the N-decarboxylation, and the cleavage of the resulting intermediate (**43**) all proceed in a satisfactory manner.

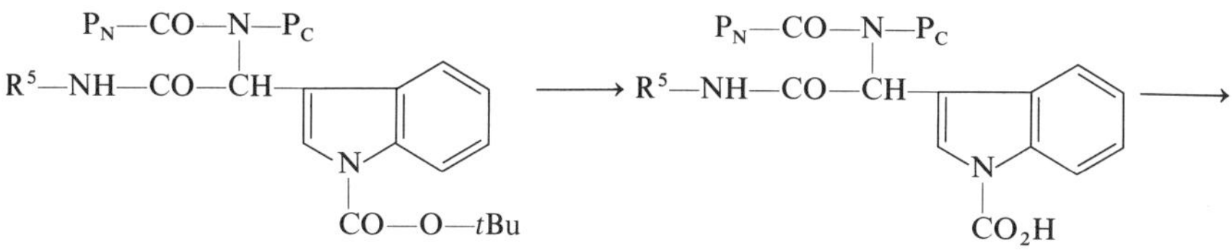

(**42**) (**43**)

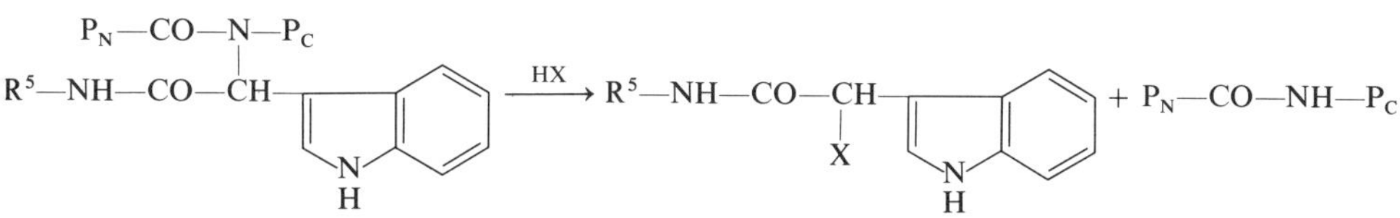

(**44**)

If the NH_2-terminal protective group of the carboxyl component **15** is chosen to be also acid labile, the final product **27** is protonated at the NH_2-terminus and thus easy to separate from the by-product **45**, e.g., by anhydrous ether extraction (Wackerle, L., unpubl.).

The potential of such fragment coupling was recently studied by Waki and Meienhofer (1977), who also found that 4-pyridinecarboxaldehyde (**46**) is recommendable as an aldehyde component (**17**), since it undergoes the four component condensation in satisfactory yield and its condensation product (**26**) are readily cleaved by electrolysis. Recently, it was observed that the latter compounds are also cleavable by photolysis (Bukall and Ugi, 1978), and by treatment with acetic anhydride in the presence of 4-dimethylamino-pyridine (Bukall and Ugi, unpubl.).

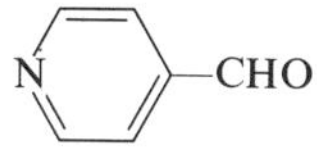

(**46**)

In four component condensations with (**41**), according to Scheme 3, there is not only some degree of racemization of the COOH-terminal amino acid unit of the acid component (**15**), but experimental evidence (Charles *et al.*, 1975; Nah, 1973) indicates that the amine component (**23**) is also somewhat racemized if a Schiff base (**47**) is preformed before the four component condensation is carried out. This is particularly prevalent in apolar solvents and can, thus, be avoided if the four component condensation is carried out

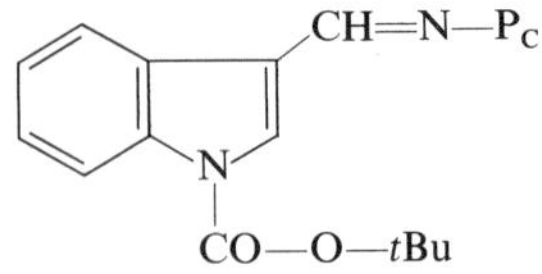

(47)

in strongly polar solvents such as methanol without preformation of the respective Schiff base (L. Wackerle, I. Ugi and E. Gil-Av, unpubl.). The results of some model experiments (Zahr, 1978) indicate that **48** (Zahr and Ugi, 1979; Seidel and Ugi, 1979) is particularly well-suited as the aldehyde component (**17**) of peptide segment coupling by four component condensations according to Scheme 3.

$$\mathrm{CCl_3{-}CH{=}CH{-}CHO}$$

$$(\mathbf{48} = \mathbf{17},\ \mathrm{R}^3 = \mathrm{CCl_3{-}CH{=}CH})$$

The aldehyde **48** reacts generally fast and in good yield with **15**, **23**, and **24**. The products (**26**, $\mathrm{R}^3 = \mathrm{CCl_3{-}CH{=}CH} = \mathbf{49}$) are readily cleaved by the supernucleophilic Co(I)-phthalocyanine anion $\mathrm{PcCO(I)^-}$ to form **27** and **50**.

$$\underset{(\mathbf{49})}{\mathrm{R^5{-}NH{-}CO{-}\underset{\displaystyle|}{\overset{\mathrm{P_N{-}CO{-}N{-}P_C}}{CH}}{-}CH{=}CH{-}CCl_3}} \xrightarrow{\mathrm{2PcCO(I)^-}} \mathbf{27} + \underset{(\mathbf{50})}{\mathrm{CCl_2{=}CH{-}CH{=}CH{-}CO{-}NH{-}R^5}}$$

This method of segment coupling is most promising with respect to racemization (Eckert *et al.*, 1979).

It is interesting to note that peptide segments (**15** and **23**) can also be coupled by the isonitrile (**24**) in the absence of an aldehyde component (**17**) (Wackerle 1979, Aigner and Marquarding, 1978) to yield directly the final produce (**27**). This succeeds particularly well with *N*-(β-isocyanoethyl) morpholine (Aigner and Marquarding, 1978), and in the presence of *N*-hydroxysuccinimide (Wackerle, 1979).

V. PERSPECTIVES

Those problems whose solution was necessary for the use of four component condensations in the synthesis of peptides have been attacked in the recent past with rather encouraging results. The syntheses of some natural products, e.g., glutathione (Urban, 1979) and alamethicine (Reusser, 1967, Urban, 1979; Urban *et al.*, 1979), and the results of some preliminary studies on the synthesis of the enkephalins (P. Seidel and I. Ugi, unpublished) and an N-terminal part of the B-chain of insulin (Paech, 1978) indicate that the development has already reached a stage at which one can see that the four component condensation and related methods will soon be a useful supplement to the present routine methods of peptide chemistry. In some cases, our methods will have definite advantages over the conventional ones for the achievement of certain goals in peptide chemistry.

The rather complex, detailed information that must be gathered before any one of the considered (then experimentally rather easy to perform) four component condensations can be carried out well enough to make full use of its versatility and inherent advantages is only justified if a given compound is synthesized repeatedly: Still, the four component condensation methods are rather worthwhile to apply. In addition to the progress made with respect to peptide synthesis, the development of the methods presented here has led to many new approaches to the chemistry of the isonitriles, the ferrocene derivatives, the supernucleophiles, the β-lactam antibiotics, and protecting group techniques that are also useful outside peptide synthesis, and above all, to new concepts and techniques in stereochemistry.

ACKNOWLEDGMENT

Collaborators in these studies include H. Aigner, W. Breuer, P. Bukall, G. Eberle, H. Eckert, R. Hermann, I. Lagerlund, D. Marquarding, G. Skorna, P. Seidel, S. Stüber, R. Urban, L. Wackerle, S. Zahr, and H. von Zychlinski.

REFERENCES

Aigner, H. and Marquarding, D. (1978). *Tetrahedron Lett.* pp. 3325–3326.

Brandt, J., Jochum, C., Ugi, I., and Jochum, (1977). *Tetrahedron* **33**, 1353–1363.

Bukall, P. and Ugi, I. (1978). *Heterocycles (Sendai)* **11**, 467–477.

Charles, R., Feibush, B., and Gil-Av, E. (1975). *In* "Peptides 1974" (Y. Wolman, ed.), pp. 93–96. Wiley, New York. Dugundji, J., Kopp, R., Marquarding, D., and Ugi, I. (1978) *Top. Curr. Chem.* **75**, 165–180.

Eberle, G., and Ugi, I. (1976). *Angew. Chem.* **88**, 509–510; *Angew. Chem. Int. Ed. Engl.* **15**, 492–493 (1976).

Eberle, G., Lagerlund, I., Ugi, I., and Urban, R. (1978). *Tetrahedron* **34**, 977–980.

Eckert, H. (1977). *Synthesis* p. 332.

Eckert, H., and Ugi, I. (1975). *Angew. Chem.* **87**, 847; *Angew. Chem. Int. Ed. Engl.* **14**, 825.

Eckert, H., Schrauzer, G. N., and Ugi, I. (1975). *Tetrahedron* **31**, 1399.

Eckert, H., and Ugi, I. (1976a). *J. Organomet. Chem.* **118**, C55 and C59.

Eckert, H. and Ugi, I. (1976). *Angew. Chem.* **88**, 717, *Angew Chem. Int. Ed. Engl.* **15**, 681.

Eckert, H., Lagerlund, I. and Ugi, I. (1977). *Tetrahedron* **33,** 2243.

Eckert, H., Lenoir, D., and Ugi, I. (1977), *J. Organomet. Chem.* **141,** C23-C27.

Eckert, H., Listl, M. and Ugi, I. (1978). *Angew. Chem.* **90**, 388–389, *Angew. Chem. Int. Ed. Engl.* **17**, 361–362.

Eckert, H., and Ugi, I. (1979). *Justus Liebigs Ann. Chem.* 278–295.

Eckert, H., Breuer, W., Geller, J., Listl, M., Marquarding, D., Stüber, S., Ugi, I., Zahr, S. and von Zychlinski, H. (1979). *Pure and Appl. Chem.*, **51**, 1219–1233.

Failli, A., Nelson. V., Immer, H., and Götz, M. (1973). *Can. J. Chem.* **51,** 2769–2775.

Failli, A., Sestanj, K., Immer, H., U., and Götz, M. (1978). *ArzneimForsch.* submitted for publication.

Fridkin, M., Patchornik, A., and Katchalski, E. (1965). *J. Am. Chem. Soc.* **87**, 4646–4648.

Gokel, G., Lüdke, G., and Ugi, I. (1971a). *In* "Isonitrile Chemistry" (I. Ugi, ed.), Chapter 8. Academic Press, New York.

Gokel, G., Hoffmann, P., Kleimann, H., Klusacek, H., Lüdke, G., Marquarding, D., and Ugi, I. (1971b). *In* "Isonitrile Chemistry" (I. Ugi, ed.), Chapter 9. Academic Press, New York.

Gokel, G., Marquarding, D., and Ugi, I. (1972). *J. Org. Chem.* **37**, 3052–3058.

Goodman, M., and Glaser, C. B. (1969). *Tetrahedron Lett.* pp. 3473–3475.

Hellmann, H., and Opitz, G. (1960). *In* "α-Aminoalkylierung," Verlag Chemie, Weinheim.

Herrmann, R., and Ugi, I. (1978). *Angew. Chem.* **90**, 134–135, *Angew. Chem. Int. Ed.* **17**, 689–699.

Herrmann, R. and Ugi, I. (1979). *Angew. Chem.*, in press.

Hoffmann, P., Gokel, G., Marquarding, D., and Ugi, I. (1971). *In* "Isonitrile Chemistry" (I. Ugi, ed.), Chapter 2. Academic Press, New York.

Immer, H., Nelson, V., Robinson, W., and Götz, M. (1973). *Justus Liebigs Ann. Chem.* **1973**, 1789–1796.

Jaeger, E. (1974). *In* "Houben-Weyl's Methoden der organischen Chemie" (E. Wünsch, ed.), Vol. 15, Part 2, p. 462. Stuttgart.

Kabbe, H. J. (1971). *In* "Isonitrile Chemistry" (I. Ugi, ed.), Chapter 5, footnote on p. 95, Academic Press, New York.

Kalir, R., Fridkin, M., and Patchornik, A. (1974). *Eur. J. Biochem.* **42**, 151–156.

Kaufhold, G., and Ugi, I. (1967). *Justus Liebigs Ann. Chem.* **709**, 11.

Kopp, R., Lemmen, P., Marquarding, D., and Ugi, I. In: "Chiralität und Stereoselektivität." G. Thieme Verlag, Stuttgart, in preparation.

Lagerlund, I., Eckert, H., and Ugi, I. (1979). *Tetrahedron* (in press).

Maia, H. L., Ridge, B., and Rydon, H. N. (1973). *J. Chem. Soc., Perkin Trans. 1* pp. 98–105.

Marquarding, D., Hoffmann, P., Heitzer., and Ugi, I. (1970). *J. Am. Chem. Soc.* **92**, 1969–1971

Marquarding, D. (1973). *Angew. Chem.* **85**, 92–93.

Marquarding, D., Burghard, H., Ugi, I., Urban, R., and Klusacek, H. (1977a). *J. Chem. Res.* (M) p. 0915. *J. Chem. Res.* (*S*) pp. 82–83.

Nah, H. S., (1973). Ph.D. Thesis, Univ. So. Cal.

Opitz, G., and Merz, W. (1962). *Justus Liebigs Ann. Chem.* **652**, 163–175.

Passerini, M., Ragni, G., and Simone, L. (1931). *Gazz. Chim. Ital.* **61**, 964–969, and preceding articles.

Paech, C. (1978). Diplomarbeit, Technische Univers. München.

Ratajczak, A., and Misterkiewicz, B. (1975). *J. Organomet. Chem.* **91**, 73–79.

Rehn, D. (1976). Doctoral Thesis, Technische Universität München.

Reusser, F. (1967), *J. Biol. Chem.* **242**, 243–247.

Ruch, E., and Ugi, I. (1969). *Top. Stereochem.* **4**, 99.

Schutz, A., and Ugi, I. (1979). *J. Chem. Res.*, (*s*) **157** (M), 2064.

Schutz, A., and Ugi, I. (1979). *Z. Naturforsch.*, in press.

Seidel, P., and Ugi, I. (1979). *Z. Naturforsch.*, in preparation.

Skorna, G., and Ugi, I. (1977). *Angew. Chem.* **89**, 267–268; *Angew. Chem., Int. Ed. Engl.* **16**, 259–260 (1977).

Skorna, G., Stemmer, R. and Ugi, I. (1978). *Chem. Ber.* **111**, 806.

Skorna, G., and Ugi, I. (1978). *Chem. Ber.* **111**, 3965–3968.

Skorna, G., and Ugi, I. (1979). *Chem. Ber.* **112**, 776–777

Skorna, G. (1978). Doctoral Thesis, Technische Universität München.

Steinbrückner, C. (1961). Doctoral Thesis, Ludwig-Maximilian-Universität München.

Ugi, I. (1962). *Angew. Chem.* **74**, 9–22; *Angew. Chem., Int. Ed. Engl.* **1**, 8–21 (1962).

Ugi, I. (1965a). *In* "Jahrbuch der Akademie der Wissenschaften in Göttingen für das Jahr," p. 21. Vandenhoeck & Ruprecht, Göttingen.

Ugi, I. (1965b). *Z. Naturforsch.* **20B**, 405.

Ugi, I. (1969). *Recent Chem. Prog*, **30**, 289–311

Ugi, I. (1971). *Intra-Sci. Chem. Rep.* **5**, 229–261.

Ugi, I., and Offermann, K. (1964). *Chem. Ber.* **97**, 2996–3007.

Ugi, I., and Steinbrückner, C. (1961). *Chem. Ber.* **94**, 2802–2814.

Ugi, I., Arora, A., Burghard, H., Eberle, G., Eckert, H., George, G., Gokel, G., Herlinger, H., V. Hinrichs, E., Hoffmann, P., Kleimann, H., Klusacek, H., Lam, H. L., Marquarding, D., Nah, H. S., Offermann, K., Rehn, D., Stüber, S., Tamasi, M., Urban, R., Wackerle, L., Zahr, S., and von Zychlinski, H. (1975). *In* "Peptides 1974" (Y. Wolman, ed.), pp. 71–91. Wiley, New York.

Ugi, I., Aigner, H., Beijer, B., Ben-Efraim, D., Burghard, H., Bukall, P., Eberle, G., Eckert, H., Marquarding, D., Rehn, D., Urban, R., Wackerle, L., and von Zychlinski, H. (1977a). *In* "Peptides 1976" (A. Loffet, ed.), pp. 159–180. Editions de l'Universite de Bruxelles, Belgium.

Ugi, I., Eberle G., Eckert, H., Lagerlund, I., Marquarding, D., Skorna, G., Urban, R., Wackerle, L., and von Zychlinski, H. (1977b). *In* "Peptides, Proceedings of the Fifth American Peptide Symposium (M. Goodman and J. Meienhofer, eds.), pp. 484–487. Wiley, New York.

Urban, R., and Ugi, I. (1975). *Angew. Chem.* **87**, 67–69; *Angew. Chem., Int. Ed. Engl.* **14**, 61–62.

Urban, R., Eberle, G., Marquarding, D., Rehn, D., Rehn, H., and Ugi, I. (1976). *Angew. Chem.* **88**, 644–646; *Angew. Chem., Int. Ed. Engl.* **15**, 627–628.

Urban, R., Marquarding, D., Seidel, P., Ugi, I., and Weinelt, A. (1977). *Chem. Ber.* **110**, 2012–2015.

Urban, R., Marquarding, D. and Ugi, I. (1978). *Hoppe Seyler's Z. Physiol. Chem.* **359**, 1541–1552.

Urban, R. (1979). *Tetrahedron* **35**, 1841–1843.

Urban, R., Marquarding, D. and Ugi, I. (1979). *In* "Peptides, Structure and Biological Function" (E. Gross and J. Meienhofer, eds.). Pierce Chemical Co., Rockport, IL.

Wackerle, L. (1979). *Synthesis* pp. 197–198.

Wackerle, L., and Ugi, I. (1975). *Synthesis* pp. 598–599.

Waki, M., and Meienhofer, J. (1977). *J. Am. Chem. Soc.* **99**, 6075–6082. See also Waki, M., Minematsa, Y., Meienhofer, J., and Izumiga, N. (1979). *Chem. Lett.* 823–824.

Waki, M., and Meienhofer, J. (1977). *J. Am. Chem. Soc.* **99**, 6075–6082.

Zahr, S. (1978). Doctoral Thesis, Technische Univ. München.

Zahr, S., and Ugi, I. (1979). *Synthesis* pp. 266–267.

Zinner, G., and Kliegel, W. (1966). *Arch. Pharm.* (*Weinheim, Ger.*) **299**, 746–756.

von Zychlinski, H., Ugi, I., and Marquarding, D. (1974). *Angew. Chem.* **86**, 517–518; *Angew. Chem., Int. Ed. Engl.* **13**, 473–474. (1974).

Chapter **5**

The Oxidation–Reduction Condensation

TERUAKI MUKAIYAMA, REI MATSUEDA, and MASAAKI UEKI

THE PEPTIDES, VOLUME 2

ISBN 0-12-304202-X

I. INTRODUCTION

Peptide synthesis is a typical condensation reaction. Although in some cases activated carboxyl derivatives are prepared prior to coupling, a new peptide bond is formed by abstracting an OH group from a carboxyl component and an H from an amine component. The most characteristic feature of the oxidation–reduction condensation in accomplishing dehydration is the combined use of the reductive potential of the phosphine to phosphine oxide conversion and the oxidative potential of the conversion of sulfenium compounds to thiols ($RS^+ \rightarrow RS^-$) [see Mukaiyama (1976) for a historical review]. Sulfenyl halides, sulfenate esters, disulfides, and sulfenamides have all found practical application in this capacity.

The success of the oxidation–reduction condensation can be attributed to the specific reaction patterns of organophosphorus and organosulfur compounds outlined in Scheme 1. Trivalent phosphorus compounds (**1**) react with many oxidizing reagents (**2**) to give, initially, phosphonium salts or pentacovalent phosphoranes (structures **3**). In these states ligand exchange reactions occur by which carboxyl compounds (**4**) are incorporated (**5**). Reaction of **5** with an amine (**7**) results in peptide bond (**8**) formation.

$$\underset{(\mathbf{1})}{R_3P:} + \underset{(\mathbf{2})}{X^+{-}Y^-} \rightleftharpoons \underset{(\mathbf{3})}{\left[R_3P\begin{smallmatrix} X \\ Y \end{smallmatrix} \rightleftharpoons (R_3\overset{+}{P}{-}X)Y^- \right]}$$

$$\mathbf{3} + \underset{(\mathbf{4})}{HO{-}COR^1} \rightleftharpoons \underset{(\mathbf{5})}{\left[R_3P\begin{smallmatrix} O{-}COR^1 \\ Y \end{smallmatrix} \rightleftharpoons (R_3\overset{+}{P}{-}O{-}COR^1)Y^- \right]} + \underset{(\mathbf{6})}{HX}$$

$$\mathbf{5} + \underset{(\mathbf{7})}{H_2NR^2} \longrightarrow \underset{(\mathbf{8})}{R^1CO{-}NHR^2} + \underset{(\mathbf{9})}{R_3P{=}O} + HY$$

Scheme 1

Disulfides are the preferred oxidizing reagents because they can convert trivalent phosphorus compounds to pentacovalent states without affecting amino acids sensitive to oxidation. On the contrary, disulfides are usually not sufficient for the completion of the reaction. This defect is overcome by using thiol scavengers or by the thiol–thione tautomerism. Thus, the oxidation–reduction condensation shown in Scheme 2 is accomplished with the use of triphenylphosphine (**10**) and 2,2′-dipyridyl disulfide (**11**), i.e., by

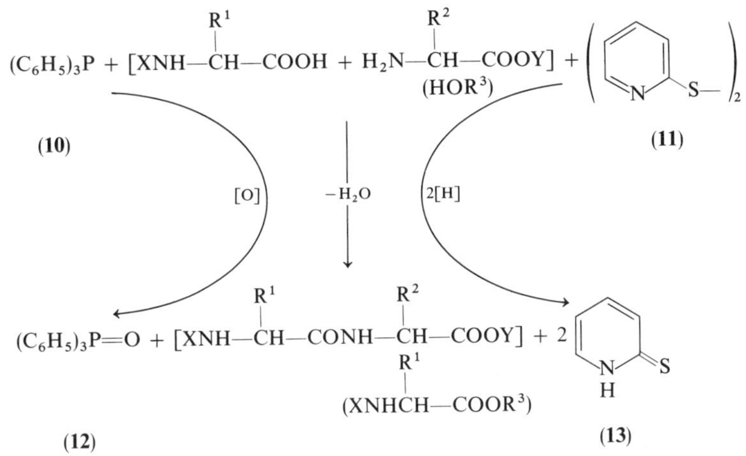

Scheme 2

a combination of the strong deoxygenating power of trivalent phosphorus compounds, the mild oxidizing power of disulfides, and the effective conversion of thiol to inactive thione.

II. PROCEDURES OF PEPTIDE BOND FORMATION

A. Coupling of Amino Acid or Peptide Carboxyl and Amine Components

The most common procedure of peptide bond formation is characterized by the condensation of an N^{α}-protected carboxyl component and an α-carboxyl protected amine component. The oxidation–reduction condensation with the use of triphenylphosphine and 2,2′-dipyridyl disulfide (**11**) shown in Scheme 2 (Mukaiyama *et al.*, 1970b) proceeds under mild conditions without the need for basic or acidic catalysts. This method has the advantages of minimizing racemization of carboxyl components and side reactions of certain amino acids. The active dehydrating agent in the reaction, phosphorane, is generated simply by mixing the carboxyl component with triphenylphosphine (mp 80°C) and 2,2′-dipyridyl disulfide (mp 58°C).* It has been proposed that peptide bond formation may proceed through a cyclic transition state (**14**) as depicted in Scheme 3.

* Contrary to dicyclohexylcarbodiimide, the reagents used in oxidation–reduction condensation do not cause any contact dermatitis.

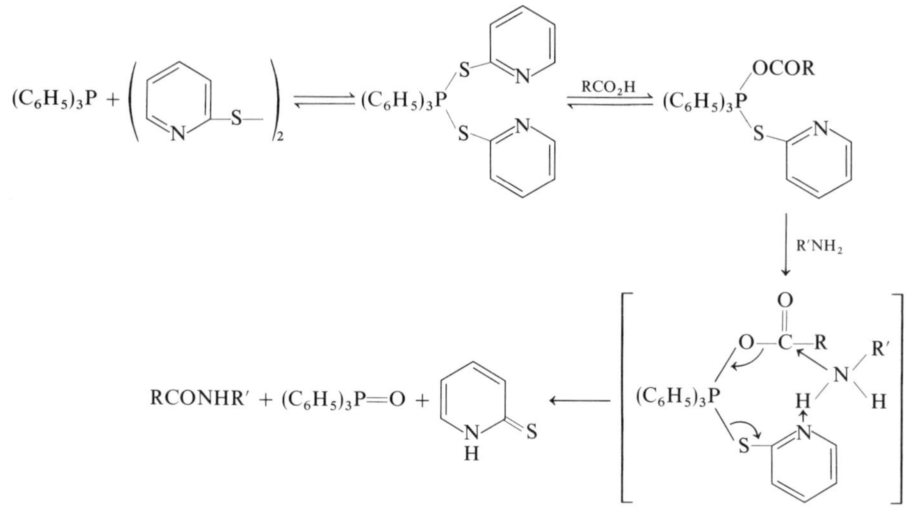

(14)

Scheme 3

In this reaction, the peptide is produced along with triphenylphosphine oxide (**12**, an oxidized product) and 2-pyridinethiol (a hydrogenated product), which isomerizes to the stable thione form (**13**). According to the Young test (Williams and Young, 1963), racemization is effectively suppressed in this reaction system and a dipeptide with high optical purity is obtained. For example, synthetic Bz-Leu-Gly-OEt was shown to contain 96% of the L isomer by this test. Other disulfides, such as bis(5-nitro-2-pyridyl), di(2-benzothiazolyl), and di(2-benzimidazolyl) disulfides, that take up hydrogen and similarly isomerize to thiones are suitable for peptide synthesis, but 4,4′-dipyridyl disulfide, which cannot participate in the cyclic transition state [(**14**) shown in Scheme 3], is not. The reaction can be carried out in a wide variety of solvent systems, but some modifications are necessary when acetonitrile or dimethylformamide is used.

Racemization is largely suppressed when 1 equiv of 2-pyridinethiol is added at room temperature in acetonitrile or when 2 equiv of 2-pyridinethiol are added at $-30°C$ in dimethylformamide at the beginning of the reaction. The presence of water has little effect on the reaction if 1 equiv of 2-pyridinethiol is added intially. Furthermore, the reaction can be carried out over a wide range of temperatures (from $-70°C$ to the boiling point of methylene chloride). No difficulties are encountered when N^{α}-protected methionine, cysteine, tryptophan, and asparagine carboxyl groups are coupled with amine components. A special advantage of this method lies in its compatibility with several side-chain functionalities; i.e., N^{α}-protected serine, threonine, and tyrosine can be coupled without hydroxyl group protection. Lactam formation by nitroarginine was not observed. In a synthesis of Z-Asn-Cys(Bzl)-OMe, dehydration of the asparagine β-carboxamide to form the nitrile amounted to less than 1% (see Table I), whereas Gish *et al.* (1956) obtained the undesired nitrile in 28% yield by the dicyclohexylcarbodiimide method. This advantage of the oxidation–reduction condensation has been confirmed by Wang (1975) who did not detect any nitrile formation during *tert*-butyloxycarbonylglutamine incorporation by oxidation–reduction condensation in a solid phase synthesis of pGlu-Gln-Ala-NH_2.

Table I. Comparison of Dicyclohexylcarbodiimide Coupling and Oxidation–Reduction Condensation in Syntheses of Z-Asn-Cys(Bzl)-OMe[a]

	Yield (%)	
Coupling method	Z-Asn-Cys(Bzl)-OMe	Nitrile formation (side product)
Oxidation–reduction condensation	73	trace < 1
Dicyclohexylcarbodiimide	35	28
Literature values[b]	39	26

[a] Abbrev: Bzl, benzyl; OMe, methyl ester; Z, benzyloxycarbonyl.
[b] Gish *et al.* (1956).

Experimentally, several precautions should be taken in the use of the oxidation–reduction condensation: Triphenylphosphine and 2,2′-dipyridyl disulfide must be added separately to the dry solution of the mixture of the carboxyl and the free (unprotonated) amine component. If the intermediate phosphorane generated from triphenylphosphine and 2,2′-dipyridyl disulfide were to suffer partial hydrolysis, its dehydrating activity would be lost. When the amine component is used as a hydrochloric acid salt, a more reactive intermediate, acyloxyphosphonium chloride, may be formed in part and cause partial racemization during segment condensations (see Section II,E).

B. Coupling with the Use of N^{α}-*o*-Nitrophenylsulfenyl Peptides as Amine Components

Many side reactions in peptide synthesis occur during acidolytic deprotection because highly reactive cationic species are produced by exposure to strong acid. Special care should be taken in syntheses of peptides which contain tryptophan, tyrosine, cysteine, or methionine residues that are susceptible to alkylation. In such cases, peptide synthesis with the use of N^{α}-*o*-nitrophenylsulfenyl (Nps) derivatives as amine components offers the advantage that cleavage of the N^{α}-protecting groups is not required. Instead, an N^{α}-*o*-nitrophenylsulfenyl amino acid ester (**16**), a kind of sulfenamide, can be used directly for peptide bond formation by treatment with the copper(II) salt of an N^{α}-protected amino acid (**15**) and triphenylphosphine [see Eq. (1)] or with an N^{α}-protected amino acid (**4**) and triphenylphosphine in the presence of a thiol scavenger [see Eq. (2)] (Mukaiyama *et al.*, 1968a).

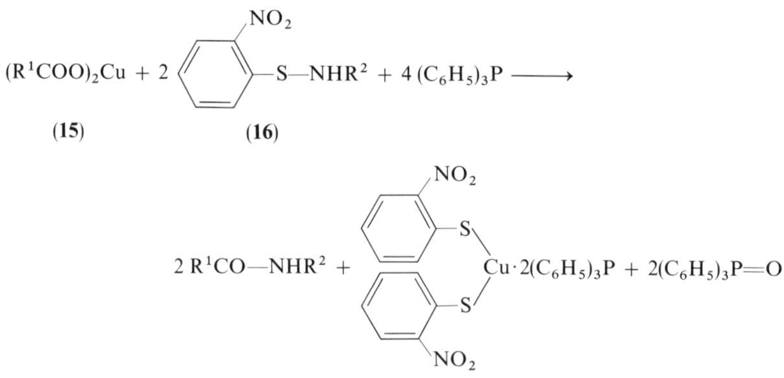

(1)

(**17**)

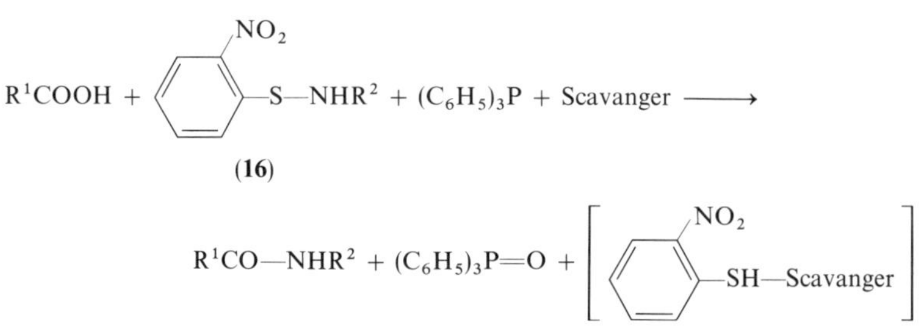

(2)

The reaction symbolized by Eq. (1) has been shown by the Young test (Williams and Young, 1963) to be accompanied by partial racemization, probably due to the generation of strongly basic amide anion. This can be considerably suppressed by the addition of acid. Accordingly, racemization is minimized in the latter reaction (Ueki *et al.*, 1971).

The synthetic utility of this type of reaction was shown in the synthesis of a glutamine-containing peptide. Amino-terminal glutamine tends to cyclize to pyroglutamic acid during deprotection depending on the reagents used. This problem can be avoided by direct coupling of *o*-nitrophenylsulfenylglutamine derivatives, as shown in the synthesis of Z-Ile-Gln-Asn-O*t*Bu [oxytocin(3–5), **18**], in 72% yield.

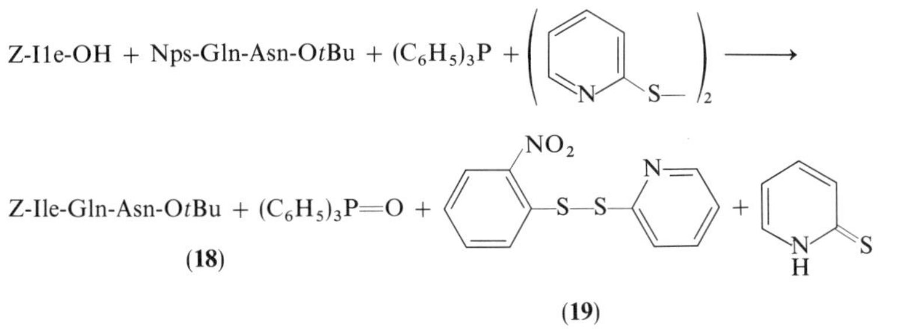

(3)

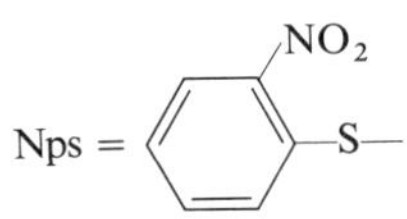

When dipyridyl disulfide is used as a thiol scavenger, two peptide bonds can be formed successively without isolation of the first coupling product (e.g., **20**). For example, Z-Pro-Leu-Gly-OEt (**21**) can be obtained in 72% yield by triphenylphosphine–2,2′-dipyridyl disulfide mediated coupling of Nps-Leu-OH and H-Gly-OEt [Eq. (4)] followed by coupling of Z-Pro-OH [Eq. (5)] after removal of unreacted amino acid derivatives by acid and base washing (Mukaiyama *et al.*, 1970d).

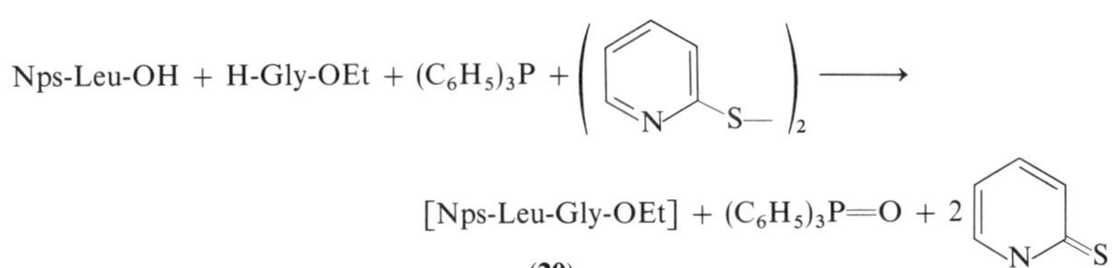

(4)

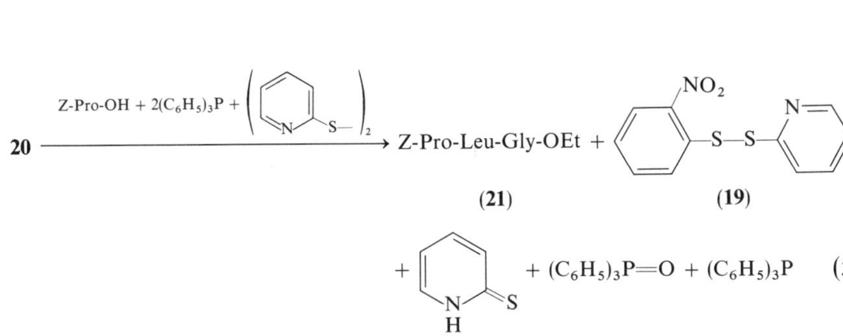

(5)

C. Active Ester Synthesis

Active esters of alkyloxycarbonylamino acids have also been prepared by the oxidation–reduction condensation (Mukaiyama *et al.*, 1970c,d) according to Eq. (6). The results are shown in Table II. The pentachlorophenyl

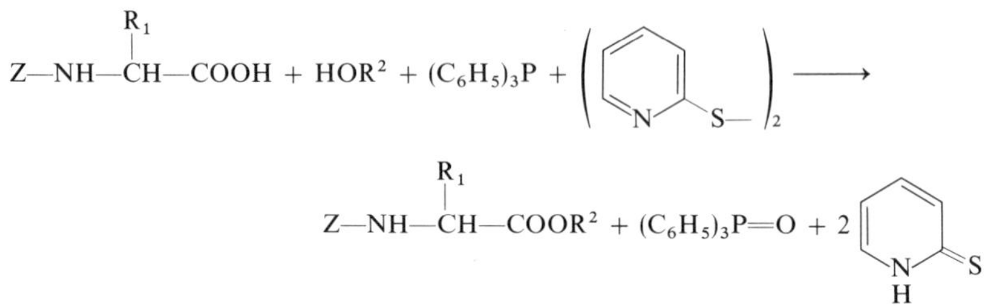

(6)

Table II. Active Esters Synthesized by Oxidation–Reduction Condensation[a]

Active ester	Reagent[b]	Yield (%)	Mp (°C) (lit.)	$[\alpha]_D$C (lit.)		Concentration
Nps-Gln-OPcp	A	74	146–147 (142–146)	$[\alpha]_D^{23}$	−59.8° (61.3°)	(*c* 2, DMF)
Z-Gln-OSu	A	72	182–184.5 (182–187)	$[\alpha]_D^{20}$	−17.3° (11.0°)	(*c* 1, DMF)
Z-Asn-OPcp	A	74	167–168.5 (—)	$[\alpha]_D^{20}$	−26.7° (—)	(*c* 1, DMF)
Z-Arg(NO_2)-OPcp	A	71	109–111 (109–111.5)	$[\alpha]_D^{20}$	−14.0° (−14.2°)	(*c* 1, DMF)
Z-Phe-OSu	B	75	136–137.5 (134–136)	$[\alpha]_D^{24}$	−54.2° (−53.8°)	(*c* 1, DMF)
Z-Val-OSu	B	61	116–117.5 (116–117)	$[\alpha]_D^{25}$	−25.8° (−25.1 ± 2.4°)	(*c* 2, dioxane)
Z-Phe-OPcp	B	86	157.5–158 (157)	$[\alpha]_D^{24}$	−51.4° (−49.4°)	(*c* 1, DMF)
Z-Val-OPcp	B	82	140 (140–141)	$[\alpha]_D^{24}$	−22.6° (−22.8°)	(*c* 1, DMF)
Z-Phe-OQ	B	70	138.5–139.5 (139.5–140)	$[\alpha]_D^{25}$	−70.6° (−70.0°)	(*c* 2, DMF)
Z-Ala-OQ	B	72	101–102.5 (102–103)	$[\alpha]_D^{26}$	−67.7° (−67.9°)	(*c* 1, DMF)
Bz-Leu-OSu	B	59	168–169 (167–169)		— (—)	

[a] Mukaiyama *et al.* (1970c). Abbrev: Bz, benzoyl; Nps, *o*-nitrophenylsulfenyl; OPcp, pentachlorophenyl ester; OQ, 8-quinolyl ester; OSu, *N*-hydroxysuccinimide ester; Z, benzyloxycarbonyl.

[b] A: Triphenylphosphine + 2,2′-dipyridyl disulfide; B: $[(C_6H_5)_2PC_6H_4SO_3]_2Hg$ + 2,2′-dipyridyl disulfide.

esters of N^{α}-benzyloxycarbonylglutamine, -asparagine, and -nitroarginine were precipitated from the reaction mixture, and analytically pure, crystalline products were obtained after simple filtration and washing. More soluble active esters were prepared using the modified phosphine (**22**), mercuric bis(diphenylphosphinobenzene sulfonate). The phosphine oxide that is formed from compound **22** was removed by washing with water, and pure active esters were obtained after one recrystallization.

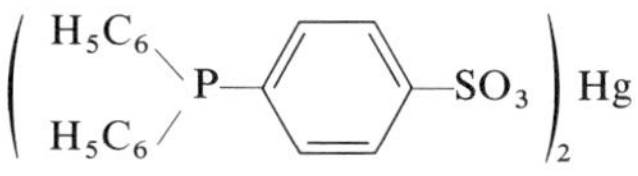

(**22**)

D. 2-Pyridinethiol Ester Synthesis

2-Pyridinethiol esters of N^{α}-alkyloxycarbonylamino acids, introduced by Lloyd and Young (1971), are effective aminolysis agents through the assistance of intramolecular catalysis. This type of active ester is easily prepared by the oxidation–reduction condensation using triphenylphosphine and 2,2′-dipyridyl disulfide (Mukaiyama *et al.*, 1972, 1973a) according to Eq. (7), provided that the reaction is carried out in the absence of nucleophiles such as amine or alcohol components, or in the presence of an alcohol with weak nucleophilicity. This thiolester synthesis has been successfully applied to the syntheses of macrolides by Corey *et al.* (1975a,b,c); it will be described later (Section V,B). Thiolesters such as **23** seem not to be the major intermediates of the usual oxidation–reduction condensation (described in

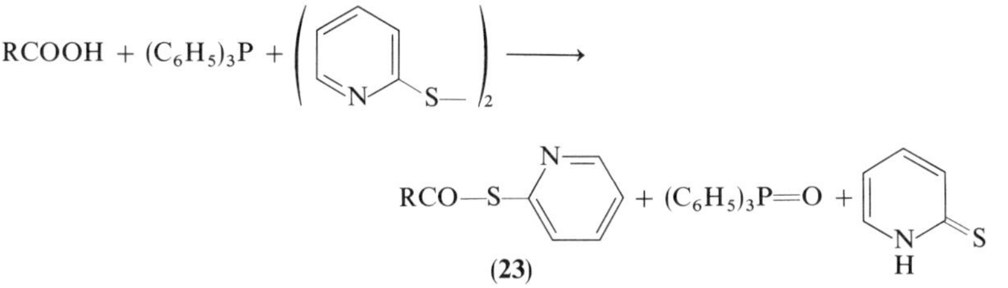

(**23**) (7)

Section II,A and B) since the rate of alcoholysis of Z-glycine 2-pyridine thiolester with benzyl alcohol was slower (yield of Z-Gly-OBzl was 9% after 5 hr) than the esterification reaction by oxidation–reduction condensation: The reaction of Z-Gly-OH, benzyl alcohol, triphenylphosphine, and 2,2′-dipyridyl disulfide produce Z-Gly-OBzl in 79% yield after 3 hr (Mukaiyama *et al.*, 1969b).

E. Selection of Oxidant and Reductant in Oxidation–Reduction Condensation

The acylation reaction for peptide bond or ester bond formation is widely variable due to the nature of carboxyl, amine, and alcohol components and to reaction conditions such as temperature, solvent, concentration. This is especially so in peptide segment condensation since each segment differs from others with respect to amino acid residue, solubility, reactivity, steric hindrance, etc. Therefore, the availability of a wide range of reagents of different reactivity (or degree of activation) would seem to be necessary to obtain efficient coupling in each case. The oxidation–reduction condensation has been developed to the point where it is possible to select a large variety of reactivities by the choice of suitable oxidizing and reducing agents. Representative oxidants and reductants are listed in Table III, in the order of their reactivity.

In the most reactive combination, a halogen would be used as an oxidant and a phosphine with electron-donating substituents as a reductant, e.g., tri-*n*-butylphosphine [24, Eq. (8)] (Ueki and Mukaiyama, 1968; Mukaiyama *et al.*, 1971a).

$$R^1COOH + H_2HR^2 + \underset{\mathbf{(24)}}{(nC_4H_9)_3P} + \underset{\mathbf{(25)}}{I_2} + 2\,Alk_3N \longrightarrow R^1CONHR^2 + \underset{\mathbf{(26)}}{(nC_4H_9)_3P{=}O} + 2\,Alk_3\overset{+}{N}H\,I^- \quad (8)$$

This is a useful acylation reaction for the preparation of penicillin derivatives such as dichloxacillin (**27**), which is difficult to prepare by commonly

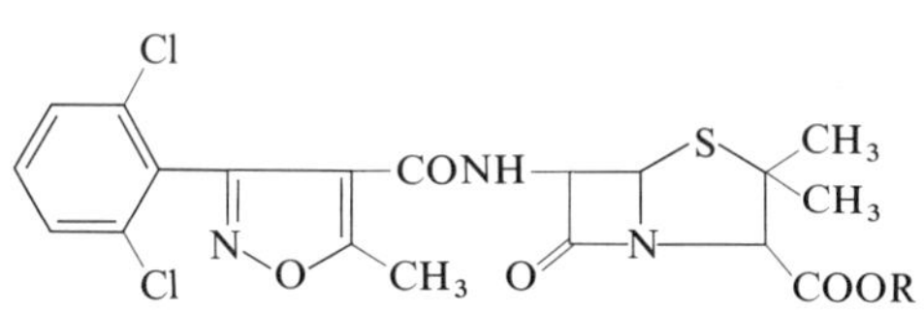

(**27**)

used acylation procedures such as the dicyclohexylcarbodiimide, mixed anhydride, or active ester methods (Mukaiyama and Maruyama, 1972). However, this reaction is accompanied by racemization, as shown by the Young test (Williams and Young, 1963), due probably to the formation of an unstable intermediate onium structure (i.e., an acyloxyphosphonium halogenate) rather than a pentacovalent structure. When the usual sulfenamides

Table III. The Order of Representative Oxidants and Reductants in Terms of Reactivity

	Order of reactivity
Oxidant	X_2[a], RSX[b], RS—NR'_2[b], RS—OR'[c] (halogen) 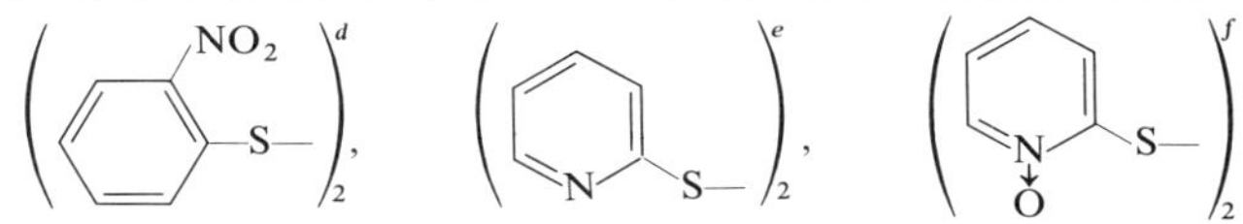
Reductant	$(n—C_4H_9)_3P$[g], 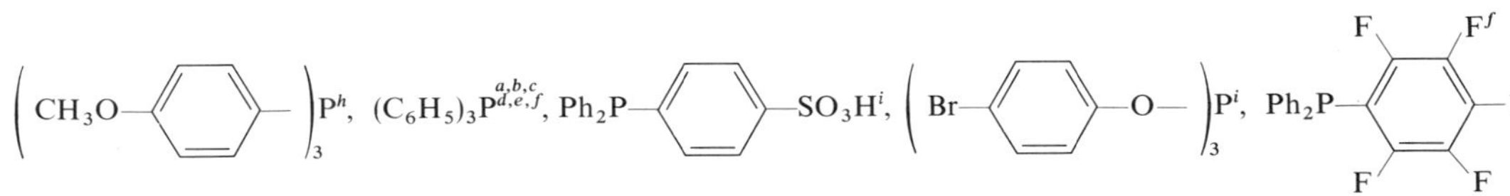
	large ← Reactivity → small
	small ← Selectivity → large

[a] Ueki and Mukaiyama (1968). Mukaiyama *et al.* (1971a). Similar reactions using CCl_4 were subsequently developed (Wieland and Seeliger. 1971; Yamada and Takeuchi, 1971).

[b] Mukaiyama *et al.* (1971b, 1974).

[c] Ueki *et al.* (1971), Matsueda *et al.* (1971).

[d] Mukaiyama *et al.* (1968b, 1969a).

[e] Mukaiyama *et al.* (1970a,b).

[f] Matsueda *et al.* (1977).

[g] Ueki and Ikeda (1974).

[h] Birr *et al.* (1975).

[i] Mukaiyama *et al.* (1969b, 1970c).

[j] Ueki *et al.* (1973).

(**28**), sulfenates, (**29**) or disulfides (**31**) are used as oxidants, the resultant thiols, which have no tendency to isomerize to thione form, have to be scavenged for the completion of peptide bond formation. Sulfenate (**29**), dihydropyran (**32**), soft metal salts (Pearson, 1962, 1963) such as the mercuric salts of urea and succinimide, and bis(*p*-anisyl)mercury (**34**) are useful as thiol scavengers (Mukaiyama *et al.*, 1969a, 1979a) (see Scheme 4 with several examples, e.g., compounds **30**, **33**, and **35**).

Soft oxidants such as 2,2′-dipyridyl disulfide and 2,2′-dithiodipyridine 1,1′-dioxide and reductants with electron-withdrawing substituents such as triphenylphosphine, tris(*p*-bromophenyl)phosphite and pentafluorodiphenylphosphine (listed in Table III) increase the stability of the intermediate pentacovalent structure in the reaction and result in decreasing racemization.

Very recently, the coupling of Boc-Leu-Ile-OH and H-Asn-Leu-O*t*Bu in dimethylformamide has been utilized as a sensitive test to detect racemization in azide, dicyclohexylcarbodiimide–hydroxybenzotriazole (DCC–HOBt; König and Geiger, 1970) and other coupling methods (Fujino, 1976) which seem to be more stringent than the Young test. The results obtained with this racemization test by oxidation–reduction condensations in solution and solid phase synthesis are shown in Tables IV and V, respectively Matsueda *et al.*, 1977).

The levels of reactivity obtained by the use of triphenlphosphine and 2,2′-dithiodipyridine 1,1′-dioxide at room temperature or triphenylphosphine and 2,2′-dipyridyl disulfide in the presence of *N*-hydroxy-5-norbornene-endo-2,3-dicarboximide (HONb)* proved to be more effective in the suppression of racemization in this test than that of triphenylphosphine and 2,2′-dipyridyl disulfide at $-30°C$.

Recently, a Chinese group (Protein Synthesis Group, Shanghai, 1975) has succeeded in a synthesis of crystalline glucagon by segment condensation on a solid support with use of DCC–HOBt as a coupling reagent which, according to Fujino's test (Fujino, 1976), can cause racemization to the extent of $7.5 \pm 3\%$. If the reactivity of the (intermediate) 1-hydroxybenzotriazole ester was considered to be universally sufficient for each segment condensation on a solid support in the Chinese glucagon synthesis, the combinations of triphenylphosphine, 2,2′ dipyridyl disulfidc and 1-hydroxybenzotriazole as well as triphenylphosphine, 2,2′-dipyridyl disulfide and *N*-hydroxy-5-norbornene-endo-2,3-dicarboxamide (see Table IV) should be

* *N*-Hydroxy-5-norbornene-endo-2,3-dicarboximide (HONb) has been the most effective racemization-suppressing additive to dicyclohexylcarbodiimide coupling of Boc-Leu-Ile-OH and H-Asn-Leu-O*t*Bu (Fujino test). However, the question of which racemization test resembles most closely practical peptide synthesis in these examples or other oxidation–reduction condensations remains open.

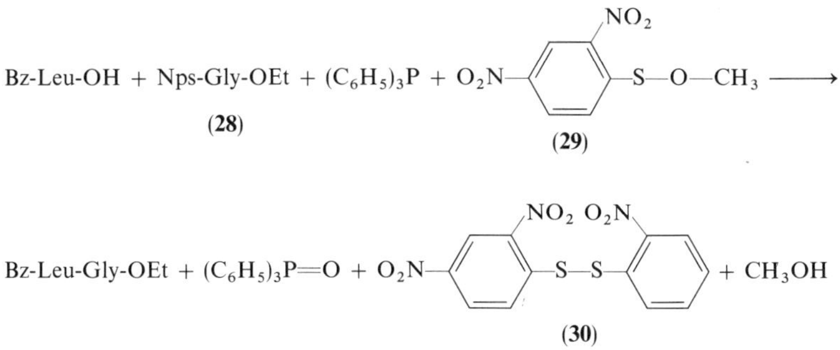

[Mukaiyama *et al.*, 1969b, 1970a]

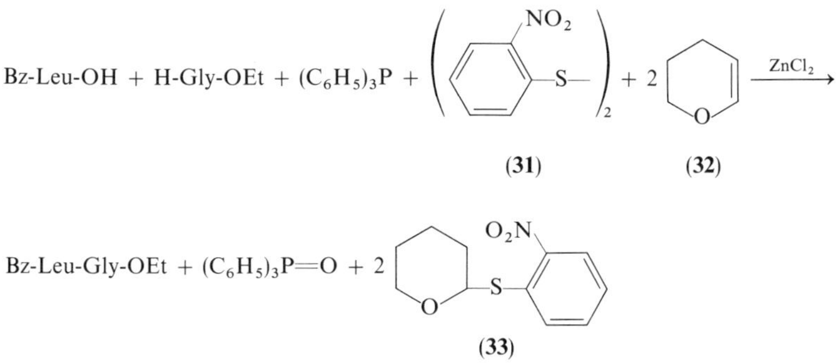

[Mukaiyama *et al.*, 1969b, 1970a]

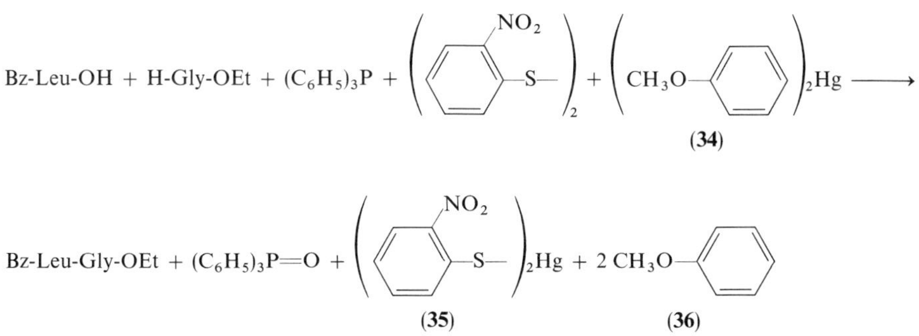

Mukaiyama *et al.*, 1968b, 1969a

Scheme 4

Table IV. Racemization Test Using the Coupling of Boc-Lec-Ile-OH and H-Asn-Leu-O*t*Bu in Dimethylformamide[a]

Reaction conditions	Yield (%)	Extent of racemization[b] (%)
1. $(C_6H_5)_3P$ + 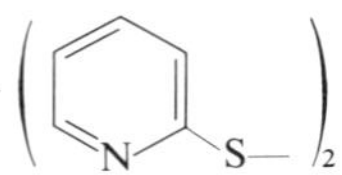 + 2 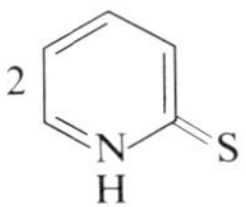 (10 hr at −30°C)	43.3	10.2[c] (9.27)
2. $(C_6H_5)_3P$ + 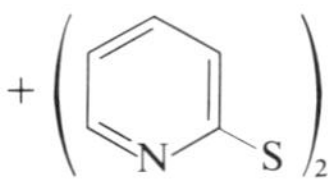 + 2 HONb[d] (2 hr at −15°C, 5 hr at room temp.)	70.2	3.59 (2.66)
3. $(C_6H_5)_3P$ + 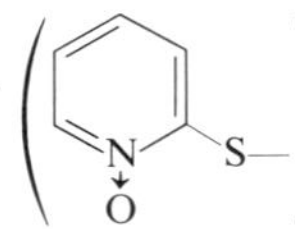 + 2 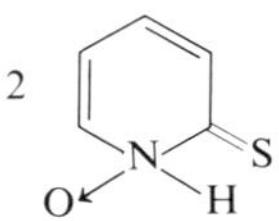 (0.5 hr at 0°C, 5 hr at room temp.)	78.2	1.82 (0.89)
4. Stepwise synthesis[e]	—	0.93
5. Azide method		
Rudinger	47.8[f]	1.4 ± 0.4[f]
Curtius	59.3[f]	11.2 ± 0.8[f]
Dicyclohexylcarbodiimide–hydroxybenzotriazole	45.6[f]	7.5 ± 3[f]

[a] Matsueda *et al.* (1977); Abbrev: Boc, *tert*-butyloxycarbonyl; O*t*Bu, *tert*-butyl ester.

[b] Hydrolyses of samples were performed simultaneously in 12 *N* HCl at 110°C for 24 hr and extent of racemization is expressed by (*a*Ile/Ile + *a*Ile) × 100%. Values in parentheses are calculated by deduction of 0.93 of racemization occurring during acid hydrolysis; these values correspond to the data marked *f*.

[c] The difference between this value and the reported value of $24.7 \pm 2.1\%$ (Fujino, 1976) may be due to lack of decomposition of the remaining phosphorane intermediate in the latter case since an experiment without the decomposition procedure gave 20.6%. Moreover, it was learned from personal communication that Fujino's experiment (at −5°C) did not satisfy our reported conditions of −30°C in dimethylformamide (Mukaiyama *et al.*, 1970b).

[d] *N*-hydroxy-5-norbornene-endo-2,3-dicarboxiimide (HONb) was found to be the best additive to dicyclohexylcarbodiimide coupling according to this racemization test.

[e] Starting with H-Asn-Leu-O*t*Bu, the tetrapeptide was prepared by coupling successively with Z-Ile-OH and Boc-Leu-OH by triphenylphosphine and 2,2′-dipyridyl disulfide.

[f] Literature data (Fujino, 1976).

Table V. Racemization Test in Dimethylformamide by Two Types of Chain Elongation[a] on Solid Support[b]

Coupling conditions[c]	Coupling yield (%)	Extent of racemization (%)
Type A chain elongation		
1. I	37	4.97
2. I repeated 3 times	Quantitative	5.72
3. II	Quantitative	1.89
4. Coupling of Boc-Leu-Ile-OH with H-Val-resin by II	Quantitative	2.02
Blank		
5. Stepwise synthesis[d]	Quantitative	1.42
Type B chain elongation		
6. I repeated twice	~Quantitative	7.65
7. II	Quantitative	2.02

[a] Type A chain elongation: coupling reaction of Boc-Leu-Ile-OH and H-Asn-Leu-Ⓟ. Type B chain elongation: coupling reaction of Ⓟ-Leu-Ile-OH and H-Asn-Leu-O*t*Bu.

[b] Matsueda *et al.* (1977). Abbrev: Boc, *tert*-butyloxycarbonyl; O*t*Bu, *tert*-butyl, ester.

[c] I: 3 equiv of carboxyl or amine component, triphenylphosphine, 2,2′-dipyridyl disulfide and 6 equiv of 2-pyridinethiol, 8 hr at −30°C.

II: 3 Equiv of carboxyl or amine component triphenylphosphine, 2,2′-dithiodipyridine 1,1-dioxide and 6 equiv of pyridinethiol 1-oxide, 2 hr at 0°C and 22 hr at room temperature.

[d] Starting with H-Asp(NH_2)-Leu-resin, Boc-Ile-OH and Boc-Leu-OH were successively coupled in CH_2Cl_2 by triphenylphosphine and 2,2′-dipyridyl disulfide.

more advantageous than the use of DCC–HOBt for the following reasons: (a) they do not cause side reactions such as the formation of *N*-acylurea derivatives and (b) they are not accompanied by any insoluble material such as *N*,*N*′-dicyclohexylurea. From the standpoint of general utility and (commercial) availability, the combination of triphenylphosphine and 2,2′-dipyridyl disulfide is the most highly recommended one for practical peptide synthesis both in solution and on a solid support and by either the stepwise chain elongation strategy or segment condensation. The use of 2,2′-dithiodipyridine 1,1′-dioxide as an oxidant in place of 2,2′-dipyridyl disulfide is effective in segment condensation in dimethylformamide; and the use of tri-*p*-anisylphosphine as a reductant in place of triphenylphosphine promotes the coupling reaction in stepwise synthesis on solid supports (Birr *et al.*, 1975; see also Section III,A and B).

III. SOLID-PHASE SYNTHESIS BY OXIDATION–REDUCTION CONDENSATION

A. Attachment of N^{α}-*tert*-Butyloxycarbonyl (Boc) Amino Acid or Peptide to the Hydroxymethyl Resin

While peptide bond formation in solid-phase synthesis is carried out under mild conditions, rather drastic conditions are often used to form the covalent benzyl ester anchorage of the first N^{α}-protected amino acid to the resin support, i.e., reaction with chloromethylated resin in the presence of triethylamine for 24 ~ 50 hr in refluxing ethanol or other solvents (Stewart and Young, 1969). This procedure causes problems in the attachment of Boc-Met-OH, Boc-His(Bzl)-OH, Boc-Asn-OH, Boc-Gln-OH, and Boc-peptide segments. The esterification reaction by the oxidation–reduction condensation shown in Eq. (9) can solve these problems (Matsueda *et al.*, 1972; Kitazawa *et al.*, 1971). Boc-Met-OH, Boc-His(Bzl)-OH, Boc-Asn-OH, Boc-Gln-OH, and Boc-peptide segments are easily attached to the resin (**38**) in desirable amounts (0.1 ~ 0.5 mmol/gm) without side reactions by treating the hydroxymethyl resin (**37**) with 3 equiv of reagents for 24 hr at room temperature (see Table VI for a compilation). The use of tri-*n*-butylphosphine

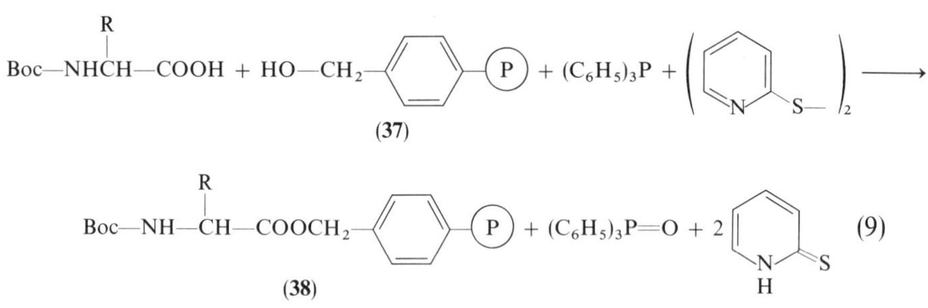

(9)

or tri-*p*-anisylphosphine in place of triphenylphosphine can speed up the reaction such that a reaction time of 1 hr is sufficient for the incorporation of N^{α}-*tert*-butyloxycarbonyl amino acid into the hydroxymethyl resin, with the use of 2 equiv of reagents. However, this procedure is not recommended for the incorporation of *tert*-butyloxycarbonyl peptides since it is accompanied by partial racemization, as shown by the Young test (Ueki and Ikeda, 1974; see also Section II,E).

B. Stepwise Chain Elongation

The oxidation–reduction condensation is applicable to the standard stepwise solid phase synthesis simply by replacing the addition of dicyclohexyl-

Table VI. Attachment of *tert*-Butyloxycarbonylamino Acid or Peptide to the Hydroxymethyl Resin by Oxidation–Reduction Condensation[a]

Amino acid	Solvent	Content (mmol/gm)
Boc-Asn-OH	DMF	0.19 (Asn)
Boc-Gln-OH	DMF	0.21 (Gln)
Boc-Arg(NO_2)-OH	DMF	0.12
Boc-Trp-OH	DMF	0.12
Boc-Cys(Bzl)-OH	CH_2Cl_2	0.18
Boc-Met-OH	CH_2Cl_2	0.15
Boc-His(Bzl)-OH	DMF	0.44 [His(Bzl)]
Boc-Gly-OH	CH_2Cl_2	0.44
Boc-Ala-OH·1/2 H_2O	CH_2Cl_2	0.25
Boc-Phe-OH	CH_2Cl_2	0.38
Boc-Pro-OH	CH_2Cl_2	0.16
Boc-Val-OH	CH_2Cl_2	0.12
Boc-Ile-OH·1/2 H_2O	CH_2Cl_2	0.11
Boc-Ala-Ala-OH·H_2O	CH_2Cl_2	0.11 (Ala-Ala)
Boc-Lys(Z)-Phe-OH	CH_2Cl_2	0.26 (Lys, Phe)

[a] Matsueda *et al.* (1972). Boc, *tert*-butyloxycarbonyl; Bzl, benzyl; DMF, dimethylformamide; Z, benzyloxycarbonyl.

carbodiimide by two separate additions of excess triphenylphosphine and 2,2′-dipyridyl disulfide in methylene chloride. Washing procedures after the coupling can be shortened because both reagents (i.e., the remaining excess) and the respective by-products (triphenylphosphine oxide and 2-pyridinethiol) have good solubility in methylene chloride. Storage of triphenylphosphine and 2,2′-dipyridyl disulfide as a mixed solution must be avoided. While ^{31}p nmr spectra do not indicate the existence of any active species such as a phosphorane or a phosphonium salt, the mixture is liable to slow hydrolytic decomposition.

Wang (1975) synthesized Boc-Gly-Phe-Phe-Tyr(Bzl)-Thr(Bzl)-$NHNH_2$ in a stepwise manner on the usual benzyl ester resin in order to evaluate the effectiveness of the oxidation–reduction condensation. By using a 2.6-fold excess of *tert*-butyloxycarbonyl amino acids and coupling reagents, triphenylphosphine and 2,2′-dipyridyl disulfide and a reaction time of 3 hr, the protected peptide hydrazide was obtained in 57.7% yield. This result was quite similar to those produced by other methods including the dicyclohexylcarbodiimide (2 hr reaction time), the DCC–HOBt (22 hr), and the *N*-ethyloxycarbonyl-2-ethyloxy-1,2-dihydroquinoline (EEDQ, 22 hr) procedures. The absence of nitrile formation during Boc-Gln-OH coupling was confirmed in this study by a preparation of pGlu-Gln-Ala-NH_2.

Racemization via 5(4*H*)-oxazalone* formation does not present a problem in stepwise synthesis as long as the α-amino groups are protected by urethane or other racemization-suppressing groups. Accordingly, more reactive phosphines (see Table III) that have a stronger nucleophilicity than tri-phenylphosphine may be used in stepwise synthesis. For example, tri-*p*-anisylphosphine was used in the synthesis of the secretin(18–27)-peptide by the stepwise incorporation of α,α-dimethyl-3,5-dimethyloxybenzyloxycarbonyl (Ddz) amino acids. From quantitative monitoring of couplings by uv and ir absorption measurements, reactions seemed to be almost complete within the first 15 minutes (Birr *et al.*, 1975).

The oxidation–reduction condensation itself has no limitation with regard to the selection of suitable protecting groups for α-amino and side-chain functions, and it causes no side reactions. But, much care should be taken in the use of N^{α}-*tert*-butyloxycarbonyl tryptophan. Recently the diphenyl-phosphinothioyl group (Ppt, see **40**; Ueki and Ikeda, 1975) was shown to be advantageous in the synthesis of tryptophan-containing peptides. The N^{α}-diphenylphosphinothioyl group could be removed by treatment with

(40)

triphenylphosphine dihydrochloride in methylene chloride. Since the cleavage by-product, diphenylphosphinothioyl chloride, did not affect the tryptophan indole moiety, pure Ppt-Trp_3-OMe was obtained in good yield (Ueki and Ikeda, 1977; Ueki *et al.*, 1977). Still milder cleavage conditions [0.5 *N* HCl–$(C_6H_5)_3P$–CH_2Cl_2] were applied to N^{α}-dimethylphosphinothioyl

* Nomenclature according to *Chem. Abstr.* 1977 *Index Guide*, pp. 836G, 837G. Example: Structure (**39**) is designated as 2-methyl-4-(phenylmethyl)-5(4*H*)-oxazolone in *Chem. Abstr.*

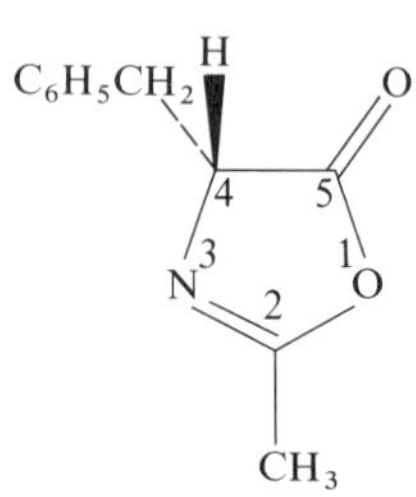

(derived from N^{α}-acetyl-L-phenylalanine)

(39)

Chemical Substance Index (**83**, 3633CS (1975), as a reference to Siemion and Baran (1975): "Conformational investigation on azlactones of aromatic *N*-acylamino acids." The term 5(*H*)-oxazolone replaces all earlier terms such as azlactone, 4,5-dihydro-1,3-oxazol-5-one, oxazolinone, 1,3-oxazolin-5-one, and 2-oxazolin-5-one, whose use should be discontinued.

(Mpt) protected amino acids and peptides, such that loss of growing peptide chain from the resin could be completely prevented. Couplings of Ppt-Trp-OH and Mpt-Trp-OH were mediated effectively with tri-*p*-anisylphosphine and 2,2′-dipyridyl disulfide (Ueki *et al.*, 1977).

C. Segment Condensation Method

The coupling of protected segments to the growing resin-bound peptide chain (segment condensation) appears to be an ideal strategy for syntheses of large polypeptides, since potential microheterogeneous contamination—consisting of countless, closely related failure sequences as in stepwise solid phase synthesis—is avoided (Meienhofer, 1973). In this strategy, two types of chain elongation are considered: (1) chain elongation from the COOH-terminal toward the NH^2-terminal amino acid (type A chain elongation) and (2) that from the NH^2-terminal to the COOH-terminal amino acid (type B chain elongation).

1. *Chain Elongation Starting from the COOH-Terminal Amino Acid (Type A Chain Elongation)*

Many different procedures (Weygand and Ragnarsson, 1966; Omenn and Anfinsen, 1968; Yajima *et al.*, 1972; Matsueda *et al.*, 1972) have been explored in segment condensations starting from the COOH-terminal amino acid and elongating the peptide chain in the direction of the NH_2-terminal amino acid (type A chain elongation). Coupling methods other than the oxidation–reduction condensation include the dicyclohexylcarbodiimide–*N*-hydroxysuccinimide (DCC–HOSu; Weygand *et al.*, 1966), the *N*-ethyl-5-phenylisoxazolium 3′-sulfonate, and the EEDQ procedures. The coupling of segments with resin-bound peptides is usually less complete than that of *tert*-butyloxycarbonylamino acids, but this is counterbalanced by simplification of purification problems. Omenn and Anfinsen (1968) reported that a COOH-terminal valine residue and an NH_2-terminal aspartic acid residue gave poor coupling yields (20 ~ 30% and 40%, respectively) by the DCC–HOSu method in syntheses of ribonuclease T segments. The most important requirements in these strategies are efficiency of the coupling reaction and suppression of racemization of the carboxyl component. It would seem ideal to have a wide variety of reactivities, so that one suitable for obtaining a high yield might be selected, rather than one uniform level of reactivity as with the DCC–HOSu method which is not adequate for all cases. The oxidation–reduction condensation can offer a wide variety of reactivities by the choice of oxidant and reductant (as discussed in Section II,E).

As an example, a synthesis of luliberin(2–10) by the use of the most common combination, triphenylphosphine and 2,2′-dipyridyl disulfide (Matsueda *et al.*, 1973a), is shown in Fig. 1A. No problems were observed in this case.

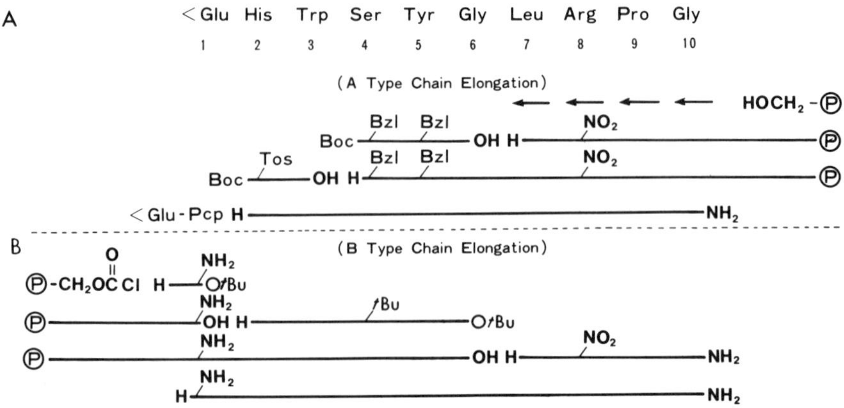

Figure 1. Solid-phase syntheses of luliberin by oxidation–reduction condensation procedures (Matsueda *et al.*, 1973a). A, Type A chain elongation starting from the COOH-terminus. B, Type B chain elongation starting from the NH_2-terminus.

However, for difficult syntheses suitable choices of oxidants and reductants (listed in Table III) may be required to solve the problems (see also Section II,E). Suitable reaction conditions may be selected as follows: (1) In the first experiment, the most commonly used reaction conditions, e.g., use of 3 equiv each of peptide segment, triphenylphosphine, 2,2′-dipyridyl disulfide, and 2-pyridinethiol should be tried, initially at $-15°C$, followed by room temperature in methylene chloride or $-30°C$ in dimethylformamide; (2) if the coupling reaction is inefficient in methylene chloride it should be repeated by adding an excess ($10 \sim 30$ equiv) of triphenylphosphine, 2,2′-dipyridyl disulfide, and 2-pyridinethiol or by using tri-*p*-anisylphosphine in place of triphenylphosphine initially at $-30°C$; (3) when the coupling reaction is inefficient or racemization is observed in dimethylformamide, the coupling reaction should be tried using 2,2′-dithiodipyridine 1,1′-dioxide and 2-pyridinethiol 1-oxide in place of 2,2′-dipyridyl disulfide and 2-pyridinethiol, initially at $-15° \sim 0°C$ and then at room temperature, and followed by repeated coupling; and (4) if the coupling reaction is inefficient at COOH-terminal glycine or proline residues that are free from racemization, more reactive combinations such as triphenylphosphine and halogen (see Table III) are also available for the reaction.

2. *Chain Elongation Starting from the NH_2-Terminal Amino Acid (Type B Chain Elongation)*

Letsinger and Kornet (1963) investigated the chain elongation starting from the NH_2-terminal amino acid and proceeding to the COOH-terminal amino acid (type B chain elongation). This strategy has not been followed up because of the danger of racemization, both in the deprotection and in the

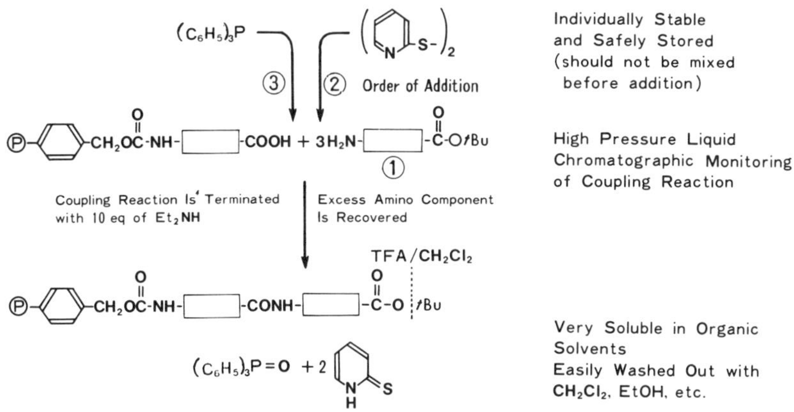

Figure 2. General procedures for type B chain elongation by oxidation–reduction condensation (Matsueda *et al.*, 1975b).

coupling steps. Recently, a modification of this method was studied by the use of azide coupling (Felix and Merrifield, 1970), but it was found not to be a generally useful procedure because of low yields and restrictions in the availability of suitable solvents. The most severe problem in this strategy is the high degree of termination because of side reactions of the activated carboxyl component.

An effective approach to chain elongation starting from the NH_2-terminus has been developed (Matsueda *et al.*, 1973a,b; 1975a,b) by combining the use of the oxidation–reduction condensation, which couples with a minimum incidence of side reactions and racemization, with *tert*-butyl ester protection of the α-carboxyl groups. The general reaction scheme of this method is shown in Fig. 2 and a synthesis of luliberin (LHRH) is shown in Fig. 1. The wide variety of reactivities available in the oxidation–reduction condensation by the choice of suitable oxidants and reductants (as discussed in Sections III,C,1 and II,E) can serve to increase further the efficiency of coupling reactions. The excess segment (amine component) is recoverable in this method since it is not changed during the coupling reaction. Some of the particular merits of this method of peptide synthesis are the following: (1) recovered amine component segment can be recycled in the chain elongation*; (2) the number of side-chain protecting groups can be reduced; (3) segments can be synthesized by solution methods from readily available N^{α}-benzyloxycarbonylamino acids; (4) diacylations are avoided since excess amine component is used; (5) the end product is not contaminated by closely related failure sequences; (6) the frequency of acidolysis is reduced—which

* This is in contrast to type A chain elongation where excess carboxyl component cannot be recovered in unaltered form.

is of importance for the acid-sensitive tryptophan residue; and (7) the progress and completeness of each coupling reaction is easily monitored by high-pressure liquid chromatography (HPLC), which will be described below.

D. Related Techniques

1. *High Pressure Liquid Chromatographic (HPLC) Techniques*

The purification of peptide segments and completed peptides is laborious, especially with the products of solid phase synthesis. Gel filtration, countercurrent distribution, ion-exchange chromatography, and silica gel chromatography are usually used for this purpose; however, such methods are time consuming. High pressure liquid chromatographic (HPLC) techniques (Takahagi *et al.*, 1974; Matsueda and Takahagi, 1976; Matsueda *et al.*, 1975a,b; Takahagi and Matsueda, 1975) are effective for purification of free and protected peptides within short periods of time. Porous polymer Hitachi gels and TSK gels of polymerized styrene–divinylbenzene have been shown

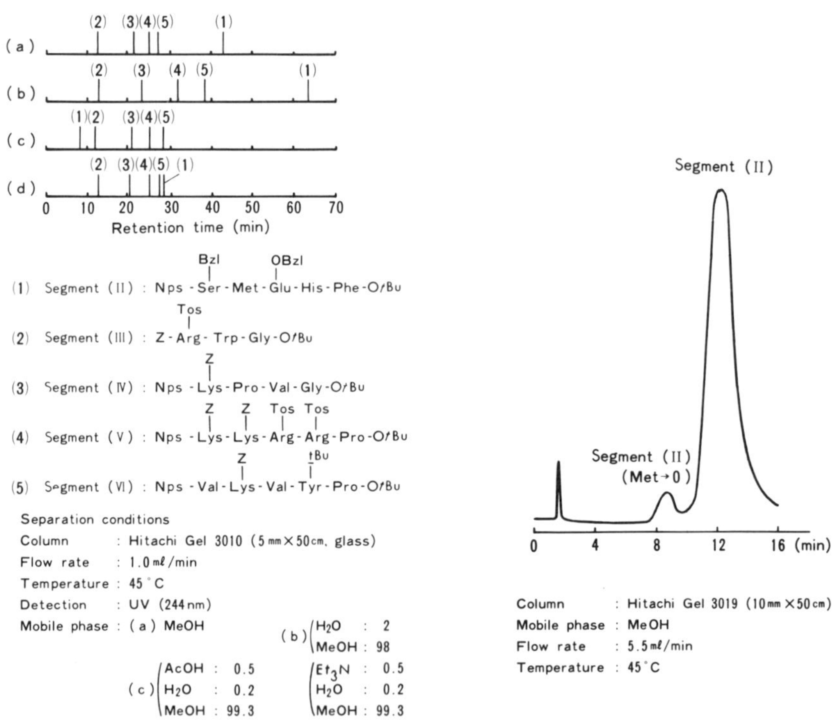

Figure 3. High-pressure liquid chromatographic separation of protected segments and an example of the purity test of fragment II prepared by ordinary methods.

to be useful packing materials for HPLC (Takahagi and Seno, 1974, 1975). The individual bead of these polymers has a lipophilic surface which retains polar compounds less well; thus, the separation properties are somewhat similar to those of reversed-phase partition chromatography. The advantages of the porous polymer packing over other packings for reversed-phase chromatography are 1) the packing has excellent chemical and physical durability and thus permits long-term operation in the separation; 2) with acidic or basic mobile phase, effective separation, especially for polar compounds, can be obtained; and 3) the packing material has a surface area of 400 ~ 500 m^2/gm and is suitable for preparative use. This technique has been successfully applied to purification of protected peptide segments as well as final free peptides, and analytically, to testing of the purity of protected peptide segments and the final product in the synthesis of ACTH(1–24) (Takahagi and Matsueda, 1975). Examples of high pressure liquid chromatographic separations and purity tests are shown in Figs. 3 and 4.

In common solid-phase synthesis the investigator is usually not able to monitor the progress of the coupling reaction until the very end when the residual amine may be determined by titrimetric (Dorman, 1969) or colorimetric methods (Kaiser *et al.*, 1970), which consume part of the peptide resin. High-pressure liquid chromatography techniques are applicable to the monitoring of solid-phase coupling reactions (Takahagi *et al.*, 1974; Matsueda

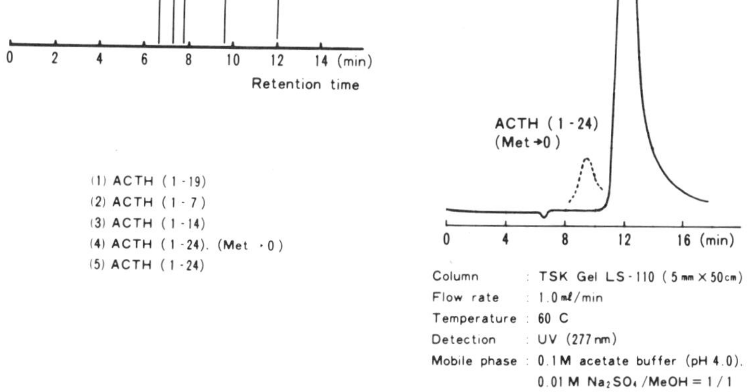

Figure 4. Separation of ACTH(1–24)-peptide from its related compounds. Left: retention time; right: chromatogram of the purified ACTH(1–24)-peptide.

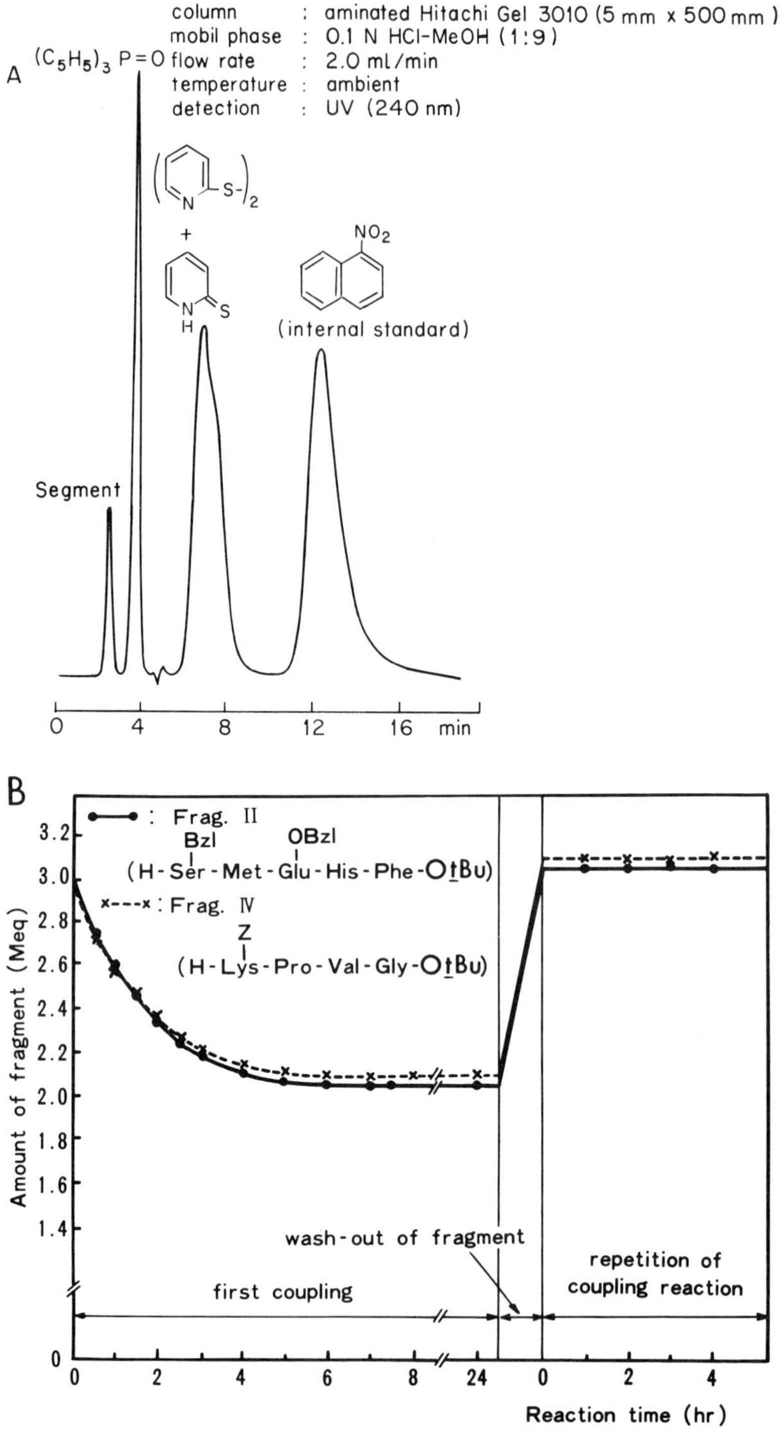

Figure 5. Monitoring of coupling reaction in type B chain elongation by HPLC (Matsueda *et al.*, 1975b). A, Elution profile of HPLC column. B, Monitoring of solid-phase coupling reaction.

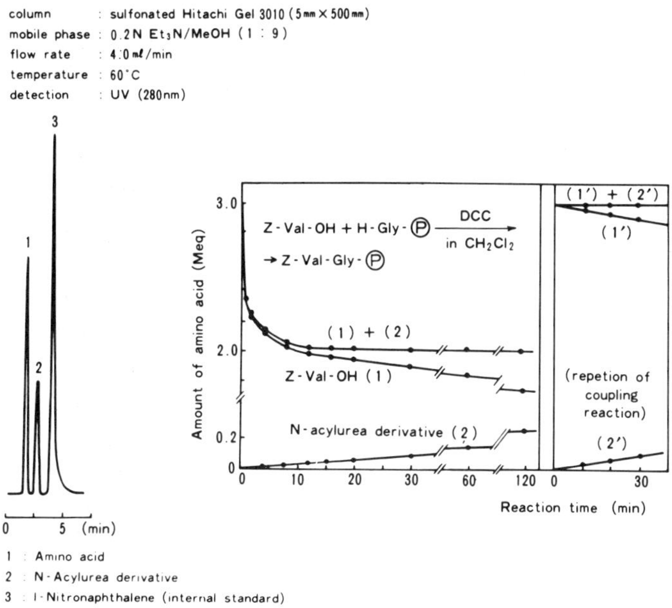

Figure 6. Monitoring of a dicyclohexylcarbodiimide-mediated coupling reaction using type B chain elongation (Merrifield synthesis).

et al., 1975a,b, 1976) (see Figs. 5 and 6 for examples with a segment condensation via type B chain elongation) and to the usual stepwise Merrifield synthesis, respectively. Coupling rates are monitored by measuring the amount of peptide segment or amino acid in solution by HPLC with Hitachi gels with, for example, a uv or other applicable detector. The completion of a reaction is indicated by the plateauing of the kinetic curve that occurs when no more peptide segment or amino acid is consumed. In addition, couplings are repeated until no more uptake is detectable. Further confirmation may be obtained by previously known methods, such as amino acid analysis or titrimetric methods. The HPLC technique offers the following advantages: 1) It does not consume the resulting peptide resin; 2) the analytical procedure is fast (about 5–15 minutes) and, 3) the entire progress of a coupling reaction is monitored by the kinetic curve. The procedure may also be utilized for process control in automated synthesis.

2. *Cleavage of the N-o-Nitrophenylsulfenyl Protecting Group under Neutral Conditions*

The *o*-nitrophenylsulfenyl protecting group (Zervas *et al.*, 1963) is well known as an acid-labile protecting group of amine functions. It is usually difficult to remove selectively the *o*-nitrophenylsulfenyl group from peptides that contain other acid-labile protecting groups such as the *tert*-butyl ester or ether or the *tert*-butyloxycarbonyl group, but this can be achieved readily via phosphonium salts (**41**) obtained with triphenylphosphine and active hydrogen compounds such as phenol, alcohol, or water in neutral solution at room temperature (see Scheme 5; Matsueda *et al.*, 19751). This procedure

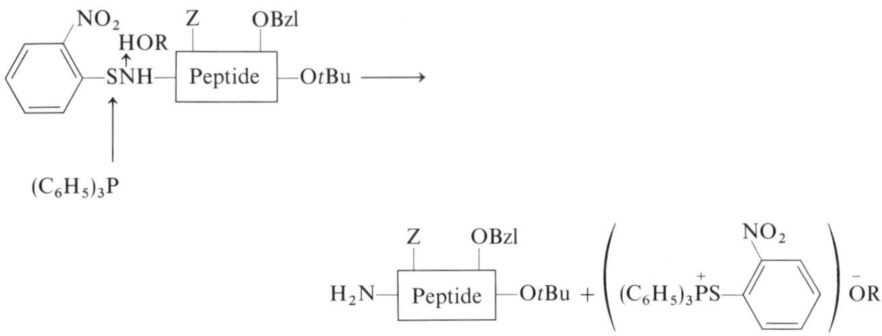

(**41**)

Scheme 5

expands the utility of the *o*-nitrophenylsulfenyl group to a protecting group, cleavable not only under acidic but also under neutral conditions, in addition to its activation for coupling (described in Section II,B).

IV. APPLICATIONS TO THE SYNTHESIS OF BIOLOGICALLY ACTIVE PEPTIDES

Peptide synthesis by the oxidation–reduction condensation with triphenylphosphine and 2,2′-dipyridyl disulfide can be applied in three types of reactions [Eqs. (10) to (12)]

Type A (see Section II, A)

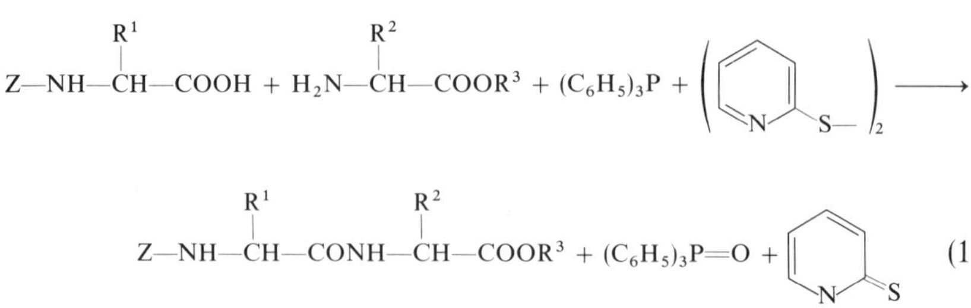

(10)

Type B (see Section II, B)

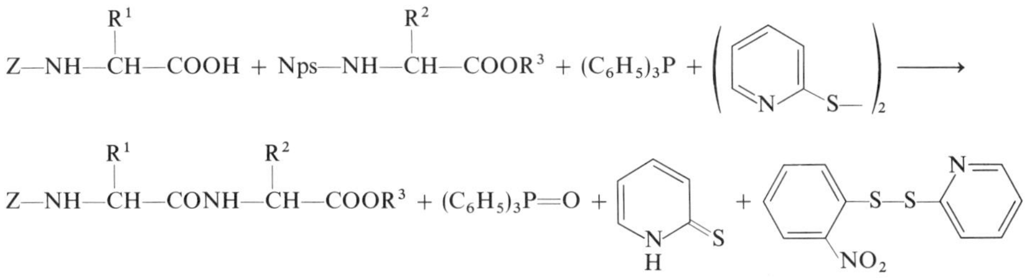

(11)

Type C (see Section II,C)

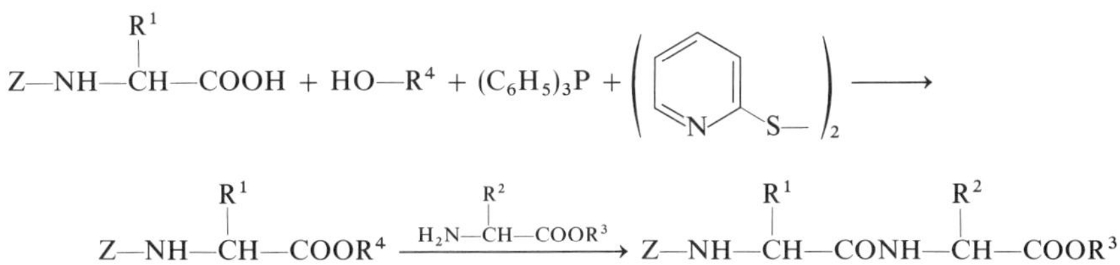

(12)

These reactions have been successfully applied to the synthesis of biologically active peptides in solution and on solid supports. Protected oxytocin (Mukaiyama *et al.*, 1970d), fully bioactive [Phe^2]-lysine-vasopressin [(2 + 3) + 4 condensation, R. Matsueda, H. Maruyama, and E. Kitazawa, unpublished], and luliberin (LHRH; 6 + 4 condensation; Mukaiyama *et al.*, 1973b) were obtained by solution synthesis via segment condensation. An example of a synthesis of protected oxytocin, involving three different types of reactions, is shown in Fig. 7. Pyroglutamic acid formation at the glutamine residue (Takashima *et al.*, 1968) and nitrile formation at the asparagine residue do not occur in the oxidation–reduction process. [Phe^2]-Lysine-vasopressin was also prepared by solid-phase synthesis both by the stepwise method and by segment condensation [2 + (3 + 4)] with type A chain elongation (Matsueda and Mukaiyama, 1970).

Luliberin (LHRH) was synthesized on solid support by two types of stepwise chain elongation (Figs. 1A and 1B). In the COOH to NH_2 strategy (type A), the product was obtained in 18% yield after the final coupling with pyroglutamic acid pentachlorophenyl ester; the NH_2 to COOH approach (type B) yield was 51%, after cyclization of the NH_2-terminal glutamine residue. Both yields were based on the amount of the first amino acid attached to the resin (Matsueda *et al.*, 1973a; Maruyama *et al.*, 1972).

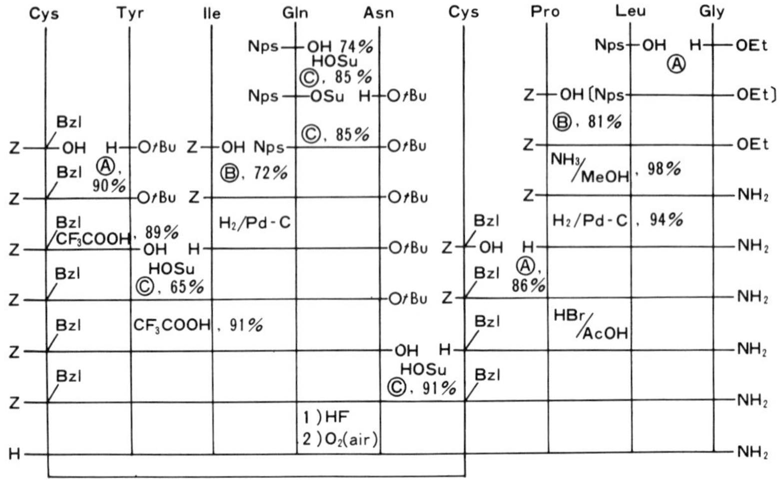

Figure 7. An example of a synthesis of protected oxytocin using 3 types of oxidation–reduction condensation reactions.

An example of a synthesis of ACTH(1–24) by a type B segment chain elongation (Matsueda *et al.*, 1975a; Maruyama *et al.*, 1976) is shown in Fig. 8. The following precautions were taken in planning this synthesis: 1) the number of amino acid residues per segment was held below 6 to facilitate the preparation of these intermediates; 2) each segment contained a unique, diagnostic amino acid residue which made it possible to check the success of

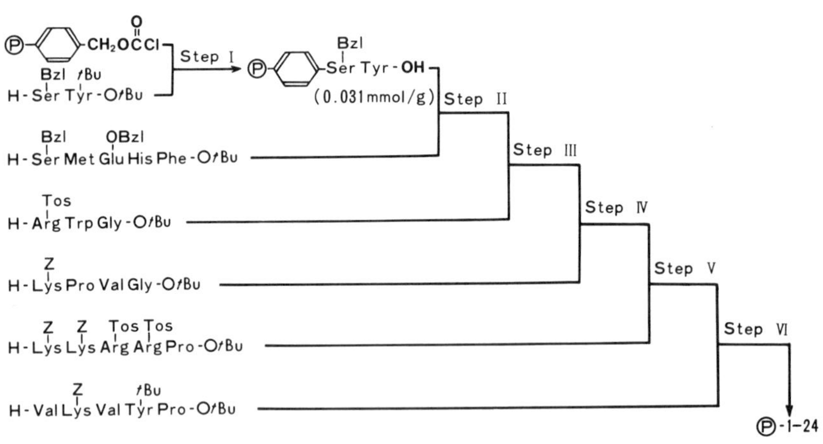

Figure 8. Synthesis of ACTH(1–24)-peptide by type B chain elongation on solid support (Matsueda *et al.*, 1975a).

segment condensations by amino acid analysis; 3) each segment had a uv-absorbing moiety to facilitate the monitoring of the coupling reaction by HPLC (see Section III,D,1); and 4) a binding site was included in one segment (V) so that its incorporation could be shown from the bioactivity. Each protected segment was synthesized in a stepwise manner by the oxidation–reduction condensation in solution using readily available N^{α}-benzyloxycarbonylamino acids. However, N^{α}-*o*-nitrophenylsulfenylamino acid residues were coupled by either the dicyclohexylcarbodiimide or the mixed anhydride methods. The technique of selective removal of the *o*-nitrophenylsulfenyl protecting group (see Section III,D,2) was successfully applied in the syntheses of segments. An example of the synthesis of segment IV is shown in Fig. 9.

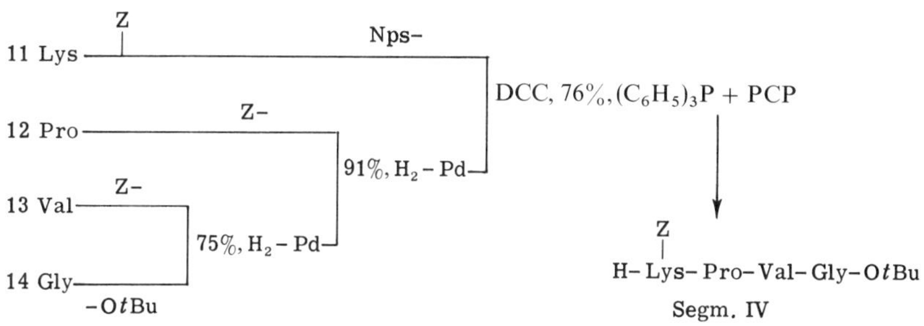

Figure 9. An example of a synthesis of segment IV in ACTH(1–24)-peptide (Maruyama *et al.*, 1976). DCC, dicyclohexylcarbodiimide; PCP, pentachlorophenol.

Following the completion of the type B segment condensation (Fig. 8), the peptide was cleaved from the resin by liquid hydrogen fluoride treatment. The crude product was purified by Sephadex G-25 gel filtration and carboxymethyl cellulose ion-exchange chromatography. High-pressure liquid chromatography (Section III,D,1) were also effective in rapid purification of the product (Takahagi and Matsueda, 1975). Pure corticotropin(1–24)·7 AcOH·9 H_2O was obtained in 25% yield, based on the starting resin segment (**42**).

(**42**)

Amino acid ratios determined after complete enzymatic digestion and acid hydrolysis were in good agreement with expected values. The ACTH activity of the purified peptide was assayed by the ascorbic acid depletion test according to the U.S. Pharmacopoeia (XVIII). Based on the 2nd international standard, it exhibited an activity of 96 (83 ~ 128) IU/mg. This value agrees with that given in the literature.

The versatility of the approach of segment condensation on solid support, combined with solution synthesis, opens up new possibilities for facile preparation of pure long-chain peptides and significantly expands the scope of standard solid phase synthesis.

V. RELATED APPLICATIONS

A. Nucleotide Synthesis

Phosphorylation reactions are typical condensation reactions similar to peptide synthesis. The oxidation–reduction condensation method is also useful for the synthesis of all types of phosphoric acid derivatives.

While in peptide synthesis no active acylating intermediate could be identified, in phosphorylation, active intermediates are stable in solution and storable for a few days without significant decomposition. Although isolation was not successful, structures of these intermediates could be elucidated as *S,S*-bis(2-pyridyl) phosphorodithiolates by ^{31}P nmr studies of model compounds (Hashimoto *et al*., 1976). This difference in mechanism of acylation and phosphorylation probably depends on the relative stability orders of —C(O)—S— < —C(O)—O— < —C(O)—N in the acyl series, and of =P(O)—N= < =P(O)—O— < =P(O)—S— in the phosphoryl series. Formation of symmetrical anhydrides (pyrophosphates), the main problem in nucleotide synthesis, was not encountered in the oxidation–reduction condensation because the (above) triester-type intermediates are produced preferentially to pyrophosphates.

This method has been successfully applied to the synthesis of coenzyme A (Hashimoto and Mukaiyama, 1972), oligothymidilates and nucleotide cyclic phosphates (Mukaiyama and Hashimoto, 1972), and nucleotides from O^2,2′-cyclouridine (Ogilvie and Iwacha, 1974).

B. Macrolide Synthesis

2-Pyridinethiol ester formation by the oxidation–reduction condensation (see Section II,D) has been successfully applied by Corey *et al.* (1975a,b,c; Corey and Nicolaou, 1974) to the internal esterification of hydroxy acids to form medium- and large-ring lactones. The properties of the 2-pyridinethiol ester, having simultaneously activated carboxyl and hydroxyl groups by proton transfer from the hydroxyl to the carboxyl as shown in Scheme 6, are reported to be favorable for cyclization.

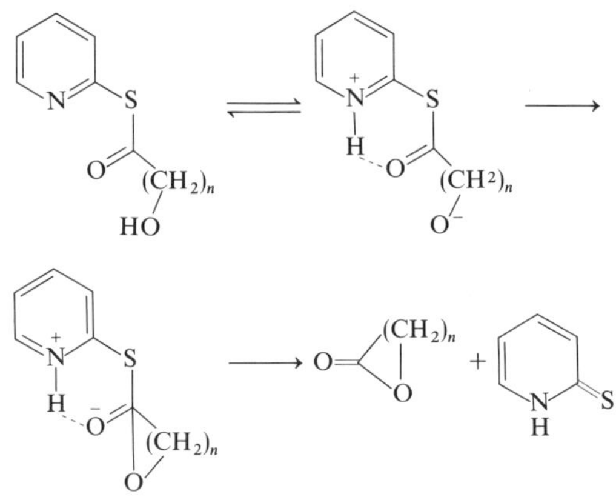

Scheme 6

This method proved to be highly effective and yet mild enough to be used with complex and polyfunctional substrates. A series of novel macrocyclic lactones such as (±)-zearalene, prostaglandin $F_{2\alpha}$1,9-lactone, prostaglandin $F_{2\alpha}$1,15-lactone, the lactone of the polyether antibiotic monensin, brefeldin A, Carpaine, vertaline, erythronolide B, and (±) Vermiculine were prepared from the hydroxy acid derivatives by Corey *et al.* (1975a,b,c; Corey and Nicolaou, 1974). With some modification, Gerlach *et al.* (1975) also succeeded in the synthesis of nonactins by this method, using $AgClO_4$ as thiol scavenger (see Mukaiyama *et al.*, 1969a, 1971b).

C. Polymeric Reagent

Horiki (1976) has prepared a polymeric triphenylphosphine-type reagent and synthesized Z-Val-Gly-OEt and Z-Phe-Gly-OEt by oxidation–reduction condensation with the use of this polymeric reagent as the triphenylphosphine component (see Scheme 7). The polymeric triphenylphosphine

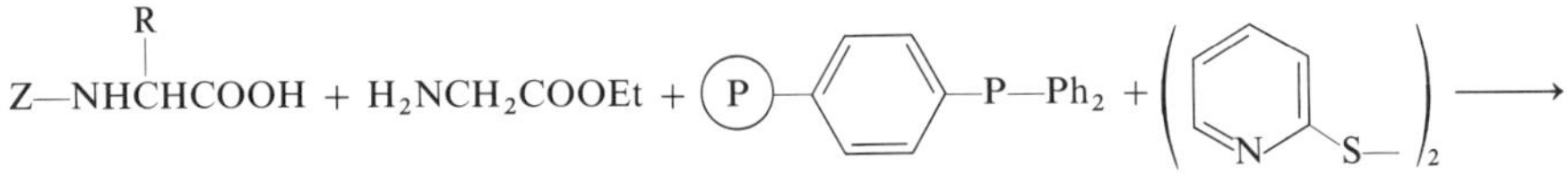

(43)

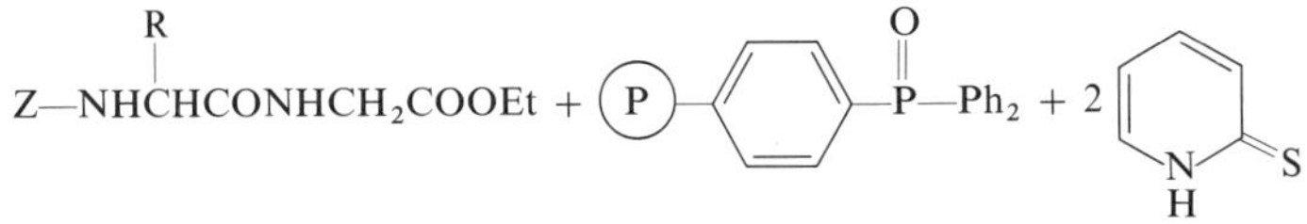

(44)

Scheme 7

oxide produced along with the peptide is easily removed by filtration and can be reconverted—by reductive treatment with trichlorosilane in benzene, followed by aqueous NaOH—for use in peptide synthesis.

VI. CONCLUDING REMARKS

The oxidation–reduction condensation is remarkably distinguished from other coupling methods in providing a wide range of reactivity levels by judicious selection of suitable oxidizing and/or reducing reagents. Indeed, preparations of peptides, nucleotides, and macrolides proceed in high yields with choice of optimum oxidant and reductant. Consequently, it would appear that the method can be resorted to in numerous synthetic schemes.

REFERENCES

Birr, C., Ueki, M., and Frank, R. (1975). *In* "Peptides, Chemistry, Structure and Biology" (R. Walter and J. Meienhofer eds.), pp. 409–418. Ann Arbor Sci. Publ., Ann Arbor, Michigan.

Corey, E. J., and Nicolaou, K. C. (1974). *J. Am. Chem. Soc.* **96**, 5614–5616.

Corey, E. J., Nicolaou, K. C., and Melvin, L. S., Jr. (1975a). *J. Am. Chem. Soc.* **97**, 653–654.

Corey, E. J., Nicolaou, K. C., and Melvin, L. S., Jr. (1975b). *J. Am. Chem. Soc.* **97**, 654–655.

Corey, E. J., Nicolaou, K. C., and Toru, T. (1975c). *J. Am. Chem. Soc.* **97**, 2287–2288.

Dorman, L. C. (1969). *Tetrahedron Lett.* pp. 2319–2321.

Felix, A. M., and Merrifield, R. B. (1970). *J. Am. Chem. Soc.* **92**, 1385–1391.

Fujino, M. (1976). *In* "PRF," 2, p. 21. Protein Res. Found., Minoh City, Osaka.

Gerlach, H., Oertle, K., Thalmann, A., and Servi, S. (1975). *Helv. Chim. Acta* **58**, 2036–2043.

Gish, D. T., Katsoyannis, P. G., Hess, G. P., and Stedman, R. J. (1956). *J. Am. Chem. Soc.* **78**, 5954.

Hashimoto, M., and Mukaiyama, T. (1972). *Chem. Lett.* pp. 595–598.

Hashimoto, M., Ueki, M., and Mukaiyama, T. (1976). *Chem. Lett.* pp. 157–160.

Horiki, K. (1976). *Tetrahedron Lett.* pp. 4103–4106.

Kaiser, E., Colescott, R. L., Bossinger, C. D., and Cook, P. I. (1970). *Anal. Biochem.* **34**, 595–598.

Kitazawa, E., Matsueda, R., Maruyama, H., Takahagi, T., and Mukaiyama, T. (1971). *In* "The 9th Symposium on Peptide Chemistry 1971" (N. Yanaihara, ed.), pp. 23–27. Protein Res. Found., Osaka.

König, W., and Geiger, R. (1970). *Chem. Ber.* **103**, 788–798.

Letsinger, R. L., and Kornet, M. J. (1963). *J. Am. Chem. Soc.* **85**, 3045–3046.

Lloyd, K., and Young, G. T. (1971). *J. Chem. Soc. C* pp. 2890–2896.

Maruyama, H., Matsueda, R., Kitazawa, E., Takahagi, H., and Mukaiyama, T. (1972). *In* "The 10th Symposium on Peptide Chemistry 1972" (J. Noguchi, ed.), pp. 60–65. Protein Res. Found., Osaka.

Maruyama, H., Matsueda, R., Kitazawa E., Takahagi, H., and Mukaiyama, T. (1976). *Bull. Chem. Soc. Jpn.* **49**, 2259–2267.

Matsueda, R., and Mukaiyama, T. (1970). *In* "The 8th Symposium on Peptide Chemistry 1970" (T. Kaneko, ed.), pp. 115–120. Protein Res. Found., Osaka.

Matsueda, R., and Takahagi, H. (1976). Japanese Patent Publication 76-46194.

Matsueda, R., Maruyama, H., Ueki, M., and Mukaiyama, T. (1971). *Bull. Chem. Soc. Jpn.* **44**, 1373–1378.

Matsueda, R., Kitazawa, E., Maruyama, H., Takahagi, H., and Mukaiyama, T. (1972). *Chem. Lett.* pp. 379–384.

Matsueda, R., Kitazawa, E., Maruyama, H., Takahagi, H., and Mukaiyama, T. (1973a). *Bull. Chem. Soc. Jpn.* **46**, 3240–3247.

Matsueda, R., Maruyama, H., Kitazawa, E., Takahagi, H., and Mukaiyama, T. (1973b). *In* "The 11th Symposium on Peptide Chemistry 1973" (H. Kotake ed.), pp. 40–46. Protein Res. Found., Osaka.

Matsueda, R., Maruyama, H., Kitazawa, E., Takahagi, H., and Mukaiyama, T. (1975a). *J. Am. Chem. Soc.* **97**, 2573–2575.

Matsueda, R., Maruyama, H., Kitazawa, E., Takahagi, H., and Mukaiyama, T. (1975b). *In* "Peptides, Chemistry, Structure and Biology" (R. Walter and J. Meienhofer, eds.), pp. 403–407. Ann Arbor Sci. Publ., Ann Arbor, Michigan.

Matsueda, R., Takahagi, H., and Mukaiyama, T. (1977). *Chem. Lett.* pp. 719–722.

Meienhofer, J. (1973). *In* "Hormonal Proteins and Peptides" (C. H. Li, ed.), Vol. 2, pp. 45–267. Academic Press, New York.

Mukaiyama, T. (1976). *Angew. Chen., Int. Ed. Engl.* **15**, 94–103.

Mukaiyama, T., and Hashimoto, M. (1972). *J. Am. Chem. Soc.* **94**, 8528–8532.

Mukaiyama, T., and Maruyama, H. (1972). Japanese Patent Kokai 72-20190.

Mukaiyama, T., Ueki, M., Maruyama, H., and Matsueda, R. (1968a). *J. Am. Chem. Soc.* **90**, 4490–4491.

Mukaiyama, T., Ueki, M., Maruyama, H., and Matsueda, R. (1968b). *In* "The 6th Symposium on Peptide Chemistry 1968", pp. 29–33. Protein Res. Found., Osaka.

Mukaiyama, T., Matsueda, R., Ueki, M., and Maruyama, H. (1969a). *J. Am. Chem. Soc.* **91**, 1554–1555.

Mukaiyama, T., Ueki, M., Matsueda, R., Maruyama, H., and Goto, K. (1969b). *In* "The 7th Symposium on Peptide Chemistry 1969" (S. Akabori, ed.), pp. 25–30. Protein Res. Found., Osaka.

Mukaiyama, T., Matsueda, R., and Maruyama, H. (1970a). *Bull. Chem. Soc. Jpn.* **43**, 1271.

Mukaiyama, T., Matsueda, R., and Suzuki, M. (1970b). *Tetrahedron Lett.* pp. 1901–1904.

Mukaiyama, T., Goto, K., Matsueda, R., and Ueki, M. (1970c). Tetrahedron Lett. pp. 5293–5296.

Mukaiyama, T., Goto, K., Matsueda, R., Hayashida, A., and Ueki, M. (1970d). *In* "The 8th Symposium on Peptide Chemistry 1970" (T. Kaneko, ed.), pp. 110–114. Protein Res. Found., Osaka.

Mukaiyama, T., Ueki, M., Maruyama, H., and Matsueda, R. (1971a). Japanese Patent 642,604.

Mukaiyama, T., Ueki, M., Maruyama, H., and Matsueda, R. (1971b). Japanese Patents 622,838 and 622,840.

Mukaiyama. T., Ueki, M., Matsueda, R., and Maruyama, H. (1972). Japanese Patent 687,121.

Mukaiyama, T., Araki, M., and Takei, H. (1973a). *J. Am. Chem. Soc.* **95**, 4763–4765.

Mukaiyama, T., Matsueda, R., Kitazawa, E., and Maruyama, H. (1973b). Japanese Patent Kokai 73-40769.

Mukaiyama, T., Ueki, M., Maruyama, H., and Matsueda, R. (1974). Japanese Patent 720,113.

Ogilvie, K. K., and Iwacha, D. J. (1974). *Can. J. Chem.* **52**, 1787–1797.

Omenn, G. S., and Anfinsen, C. B. (1968). *J. Am. Chem. Soc.* **90**, 6571–6572.

Pearson, P. G. (1962). *J. Am. Chem. Soc.* **84**, 16–24.

Pearson, P. G. (1963). *J. Am. Chem. Soc.* **85**, 3533–3539.

Protein Synthesis Group, Shanghai (1975). *Sci. Sin.* **18**, 745–768.

Siemion, I. Z., and Baran, G. (1975). *Bull. Acad. Prl. Sci., Ser. Sci. Chim.* **23**, 317–322.

Stewart, J. M., and Young, J. D. (1969). "Solid Phase Peptide Synthesis." Freeman, San Francisco, California.

Takahagi, H., and Matsueda, R. (1975). *In* "The 13th Symposium on Peptide Chemistry 1975" (S. Yamada, ed.), pp. 88–93. Protein Res. Found., Osaka.

Takahagi, H., and Seno, S. (1974). *J. Chromatogr. Sci.* **12**, 507–511.

Takahagi, H., and Seno, S. (1975). *J. Chromatogr.* **108**, 354–357.

Takashima, H., du Vigneaud, V., and Merrifield, R. B. (1968). *J. Am. Chem. Soc.* **90**, 1323–1325.

Takahagi, H., Matsueda, R., Maruyama, H., and Mukaiyama, T. (1974). *In* "The 12th Symposium on Peptide Chemistry 1974" (H. Yajima, ed.), pp. 50–56. Protein Res. Found., Osaka.

Ueki, M., and Ikeda, S. (1974). *Chem. Lett.* pp. 25–28.

Ueki, M., and Ikeda, S. (1975). *Chem. Lett.* pp. 827–830.

Ueki, M., and Ikeda, S. (1977). *Chem. Lett.* pp. 869–870.

Ueki, M., and Mukaiyama, T. (1968). *In* "The 6th Symposium on Peptide Chemistry 1968," pp. 34–38. Protein Res. Found., Osaka.

Ueki, M., Maruyama, H., and Mukaiyama, T. (1971). *Bull. Chem. Soc. Jpn.* **44**, 1108–1111.

Ueki, M., Shishikura, T., Hayashida, A., and Mukaiyama, T. (1973). *Chem. Lett.* pp. 733–736.

Ueki, M., Ikeda, S., and Tonegawa, F. (1977). *In* "Peptides, Proceedings of the Fifth American Peptide Symposium" (M. Goodman and J. Meienhofer, eds.) pp. 546–548. Wiley, New York.

Wang, S. S. (1975). *J. Org. Chem.* **40**, 1235–1239.

Weygand, F., and Ragnarsson, U. (1966). *Z. Naturforsch., Teil B* **21**, 1141–1144.

Weygand, F., Hoffmann, D., and Wünsch, E. (1966). *Z. Naturforsch., Teil B* **21**, 426–428.

Wieland, T., and Seeliger, A. (1971). *Chem. Ber.* **104**, 3992–3994.

Williams, M. W., and Young, G. T. (1963). *J. Chem. Soc.* pp. 881–889.

Yajima, H., Kawatani, K., and Watanabe, H. (1972). *Chem. Pharm. Bull.* **18**, 1333–1339.

Yamada, S., and Takeuchi, Y. (1971). *Tetrahedron Lett.* pp. 3595–3598.

Zervas, L., Borovas, D., and Gazis, E. (1963). *J. Am. Chem. Soc.* **85**, 3660–3666.

Chapter 6

Repetitive Methods in Solution

LAJOS KISFALUDY

THE PEPTIDES, VOLUME 2

ISBN 0-12-304202-X

I. INTRODUCTION

It is highly probable that the four methods of peptide synthesis to be discussed in this chapter would not have been developed if the solid-phase method had satisfied most expectations. Within a few years after the introduction of this otherwise ingenious method, it became apparent that some of its inherent procedural defects could not be easily eliminated. Thus, the problem of an effective method for the synthesis of longer peptides and small proteins remains unsolved. Those peptide chemists who continued using solution methods aimed to improve the existing procedures or to find new ones. However, one cannot ignore the fact that the advances and results of the two approaches mutually influenced each other. Bodanszky (1960) supposed that the repetitive nature of the stepwise strategy would probably lead to a considerable simplification of peptide synthesis. The aim of simplification was successfully implemented both in the solid phase procedure and in the methods to be described here. The basic difference is that solid-phase methods are carried out on a polymer support in a heterogeneous phase, while repetitive methods in solution always involve the application of a homogeneous phase. Because one of the repetitive reactions is a nucleophilic substitution with an S_N2 mechanism and is generally sensitive to steric hindrance, the completion of the acylation reaction is far more likely to be accomplished in solution.

A. Common Features

From the foregoing, it will be appreciated that repetitive methods in solution have several features in common.

1. General Aims

In each approach the idea, which did not necessarily mean a decrease in experimental effort, was to facilitate the synthetic work. The main aim was, indeed, to control every reaction step and avoid the problems arising from inhomogeneity of the final products, because complete removal of closely related contaminants by purification at the end of a synthesis is often impracticable. Special emphasis was placed on decreasing the time required, without sacrificing the quality of the final product.

2. Reactions in Solution

The advantages of carrying out acylation reactions in solution have already been mentioned. It is also expedient to use solution methods during the removal of the NH_2-terminal protecting groups, preferably in a solvent different from that used for acylation. This use of different solvents aids in the removal of excess acylating agent, by-products, salts, etc. The increasing chain length and decreasing solubility during the progress of a synthesis result in a continuous decrease of the molar concentration of the reaction partners and, consequently, in a diminished reaction rate. Thus, solubility plays a decisive part in determining the maximum chain length of peptide that can be synthesized and the reaction time required.

3. *Stepwise Strategy*

The application of the stepwise strategy in solution provides for considerable control of every reaction step. In this way the necessity for isolating each intermediate is eliminated, and consequently, considerable time can be saved. On the other hand, when contamination of an intermediate is detected during the monitoring procedure, it is possible to stop at any stage of the synthesis and purify the intermediate, if necessary. This is the main advantage over the solid phase method. The acylating agent is always present in excess during the coupling reactions; the extent of this excess depends on chain length and steric factors. The routine use of the same excess of acylating component and the same reaction time is unfavorable, considering that both the acylating agent and the amine component vary in every step.

The employment of these main principles lends the repetitive methods a very high degree of flexibility and permits optimization of procedures, depending on the character of each individual peptide to be synthesized. Naturally, in addition to these common features, each method has its own special characteristics with respect to coupling methods, combination of the protecting groups, and purification.

II. THE FUNCTIONAL HANDLE METHOD

A. General Considerations

The functional handle principle was introduced to peptide chemistry by the Oxford group (Camble *et al.*, 1968). The essence of this method (Young, 1973) is the incorporation into the peptide of one or several functional groups which are stable under the conditions of the synthesis and which make it possible to separate the main product from the by-products and side products in a simple way. The 4-picolyl group lends itself to this purpose because it shows a high degree of acid resistance, due to its basic nitrogen atom, and at the same time it offers the possibility of separation by salt formation. Thus, in addition to the general principles just described, the main objective of the method is efficient *purification*. The picolyl group is usually incorporated into the COOH-terminus, as an ester, but occasionally into side-chain carboxyl groups as well. The additional "handles" formed in this manner further facilitate the purification. In the same way that the solubilizing character of a hydroxyl function decreases with the increasing carbon chain length in aliphatic alcohols, the solubilizing effect of a single picolyl ester group also diminishes with increasing peptide chain length. Usually, di- to tetrapeptide picolyl esters can be extracted with aqueous citric acid from a water-immiscible solvent, but beyond this stage, it is expedient to carry out the salt formation by means of an ion-exchange resin. Subsequently, all nonbasic contaminants can easily be removed from the main product by a simple washing procedure. The main product is then eluted from the resin with basic solvents and, often, obtained in analytically pure form. The characteristics of this method have been summarized by the authors as follows: "The advantage over the conventional 'solution' method lies in the ease with which pure product can be isolated at each step; in comparison with 'solid phase' synthesis, the time and the labour required are no doubt greater, but the isolation and characterization of each protected intermediate provide in the synthesis the all-important cumulative evidence for the structure of the final product (Young, 1973)." Scheme 1 demonstrates one cycle of the synthesis.

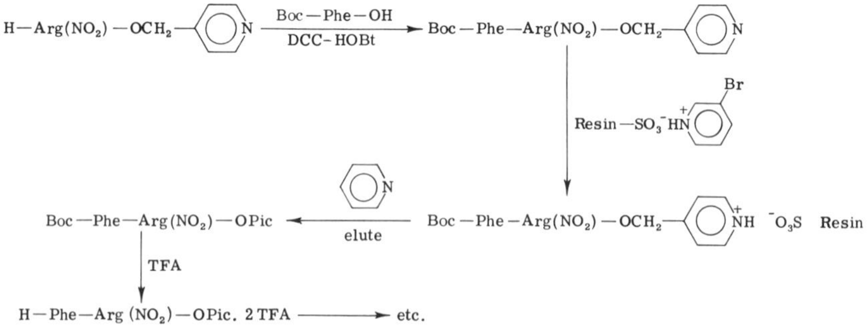

Scheme 1. First stages of the synthesis of a bradykinin analogue by the picolyl ester method.

The starting 4-picolyl ester of N^{α}-*tert*-butyloxycarbonyl- and benzyloxycarbonylamino acids can be prepared conventionally. These, in contrast to benzyl esters, are relatively acid stable; thus either their trifluoroacetic or hydrobromic salts are easily available. On the other hand, they may be cleaved under reductive conditions such as hydrogenation, sodium in liquid ammonia, electrolytic reduction at a mercury cathode, or cold alkali. This feature of the 4-picolyl esters offers the use of protecting-group combinations where all protecting groups can be removed in a single step.

B. Problems with the Coupling Step

In the early days of the method the authors took no special precautions during coupling, the only restriction being the avoidance of acylating agents that form basic by-products, such as quinolyl or piperidyl esters. Later on, Fletcher *et al.* (1973) reported that the use of dicyclohexylcarbodiimide to form peptide bonds in the presence of trifluoroacetate anion can cause substantial trifluoroacetylation (up to 40%) of the amine component. Such undesired acylation was also observed to a lesser extent during peptide bond formation by certain active esters (Schnabel *et al.*, 1971; Fletcher *et al.*, 1973; Bodanszky *et al.*, 1974b). Alternatives to bypass this undesired side reaction include removal of the *tert*-butyloxycarbonyl group by hydrochloric acid in dioxane. However, some protected peptide picolyl esters were found to precipitate as monohydrochlorides along with the dihydrochloride of the N^{α}-deblocked peptide, resulting in an incomplete deprotection. Another alternative would be to replace the trifluoroacetate anion by a less nucleophilic anion, e.g., pivalate. It was subsequently found that dicyclohexylcarbodiimide coupling in the presence of 1-hydroxybenzotriazole (DCC–HOBt, König and Geiger, 1970) very nearly eliminates this side reaction. It is of interest that no trifluoroacetylation was observed using 2,4,5-trichlorophenyl esters. The side reaction presumably takes place through a mixed anhydride intermediate and is influenced mainly by the nucleophilic character of the acylate anion and the degree of activation of the active esters. Thus, in the presence of formate, and to a lesser extent acetate ions, transacylation could be observed even in the case of moderately active esters.

Another problem during the coupling step arose from unacceptably slow coupling rates in some cases where 2,4,5-trichlorophenyl and pentachlorophenyl esters were used. For example, in some steps of a simplified bradykinin synthesis (Schafer *et al.*, 1971), the acylation time was as long as 11–12 days, and in the first synthesis of bradykinin by the picolyl ester handle procedure it even amounted to 42 days! Probable reasons for these slow reaction rates may be either the considerable decrease of molar concentrations or partial protonation of the amine component by the acidic phenol derivatives released during the acylation reaction. Therefore, the DCC–HOBt procedure (König and Geiger, 1970), with reaction times of 15–20 hr, was adopted for general use. In the first stage of the synthesis it is important to avoid the formation of dioxopiperazines because this reaction liberates 4-picolyl alcohol which, due to its basic character, remains with the main product. Consequently, in the next step 4-picolyl ester can be formed that, in turn, will produce peptides truncated at the COOH-terminus.

C. Purification by the "Handle" Principle

As already mentioned, the cardinal point of the method is the ability to purify intermediates by the "handle" principle. Therefore, the inventors placed special emphasis on the development of this procedure, and the optimal conditions were determined with great care. As long as the partition of the protected peptide—provided with a handle—between aqueous citric acid and a water-immiscible solvent is satisfactory, the purification is very simple. Adequate partition of this kind generally extends to tetrapeptides. Beyond this stage the salt formation was initially carried out by means of moist Sulfoethyl-Sephadex C-25 in the H-form. The use of the far more effective macroreticular Amberlyst-15 cation-exchange resin was introduced by Burton *et al.* (1971). This resin has to be presaturated with 3-bromopyridine to avoid inadvertent cleavage of acid-sensitive protecting groups (see Scheme 1). The time required for absorption does not exceed 0.5–1 hr and the resin is then washed with organic solvent and the product eluted with pyridine in dimethylformamide. The 3-bromopyridine can be removed after evaporation by trituration of the residue with anhydrous ether. The entire purification step requires skillful and careful work. To illustrate the efficiency of the method, results in the synthesis of a bradykinin analog are presented in Table I (Pinker *et al.*, 1976).

Table I. Synthesis of Boc-Pro-Pro-Gly-Phe-Ser(Bzl)-Pro-Phe-Orn(Picoc)-OPic

Protected peptide	Amine component (mmol)	Acylating component (mmol)	DCC–HOBt (mmol)	DMF (ml)	Isolation[a]	Yield[b] (%)
Dipeptide(8–9)	17.8	23	23	30	A	91
Tripeptide(7–9)	15.0	20	20	30	A	96
Tetrapeptide(6–9)	6.7	8.7	8.7	15	A	89
Pentapeptide(5–9)	5.75	7.47	7.47	20	B	84
Hexapeptide(4–9)	2.6	3.47	3.47	16	B	91
Heptapeptide(3–9)	1.83	2.4	2.4	10	B	92.5
Octapeptide(2–9)	1.41	2.2	2.2	11	B	96

[a] A, Isolation by extraction into aqueous citric acid; B, isolation by means of Amberlyst-15 resin.

[b] The overall yield of the protected octapeptide was 59.4%.

D. Improvements and Further Development of the Method

a. For the preparation of N^{α}-*tert*-butyloxycarbonylamino acid 4-picolyl esters from the acids and alcohol, a modified procedure was recommended (Pinker *et al.*, 1976) by which considerably improved yields were obtained through the gradual addition of dicyclohexylcarbodiimide to the solution of the acid and alcohol at −3°C.

b. To avoid dioxopiperazine formation by the free amine component, the use of diisopropylethylamine was generally adopted (Pinker *et al.*, 1976).

c. The use of an adequate coupling method is of the utmost importance. The best reagent proved to be DCC–HOBt which is used in a 1:1 molar ratio with the carboxyl component, after "preactivation" for one hour at 0°C and one hour at room temperature.

d. The synthesis of [Val^5]-angiotensin II was the first example to demonstrate the advantage of using more than one handle for efficient purification. Thus, if the β-carboxyl group of the aspartic acid-1 residue is protected by a 4-picolyl ester group, the protected octapeptide Z-Asp(OPic)-Arg(NO_2)-Val-Tyr(Bzl)-Val-His(Bzl)-Pro-Phe-OPic is readily soluble in aqueous citric acid. In the case of longer peptides, not only the acid solubility but the general solubility plays a very important role. For instance, the well-known insolubility of the insulin B-chain was considerably ameliorated when its reduced form was converted to the *S*-picolyl derivative (Seely *et al.*, 1973). In the synthesis of gastric inhibitory polypeptide(1–13) the side-chain functionalities of the glutamic acid and tyrosine side chains were protected by picolyl groups, and although there was no significant improvement in the general solubility, the absorption by resin was facilitated (Macrae and Young, 1975).

e. The method was further extended by the use of 4-picolyloxycarbonylhydrazides (Macrae and Young, 1974). Reaction of the starting material, 4-picolyl *p*-nitrophenyl carbonate (prepared by the reaction of bis-*p*-nitrophenyl carbonate with 4-picolyl alcohol in the presence of *N*-methylmorpholine) with hydrazine hydrate produced 4-picolyloxycarbonyl hydrazide nearly quantitatively. This compound can be used in a manner similar to other protected hydrazides, and the 4-picolyloxycarbonyl group can be as easily cleaved as other picolyl groups. The utility of this "handle" was well demonstrated in the synthesis of gastric inhibitory polypeptide(1–13) hydrazide (Macrae and Young, 1975).

f. The marked differences in properties between 4-picolyl and benzyl esters led to the introduction of other protecting groups having tertiary amine substituents. Such a basic residue greatly increases the stability to acid, presumably by distracting the attacking proton onto a nonproductive site. For instance, whereas *S*-tritylcysteine is rapidly cleaved by acid, *S*-(diphenyl-4-pyridylmethyl)cysteine is stable under acidic conditions (Coyle and Young, 1976). The same group provides improved protection of the imidazole group of histidine. Similar advantages were obtained by using the 4-picolyl group for protection of side-chain functionalities of tyrosine and cysteine (Gosden *et al.*, 1977). The utility of these protecting groups in actual synthesis was ascertained and further improvements in the solubility of longer peptides were gained with the use of the tyrosine picolyl ether residues (Macrae and Young, 1975).

E. Examples

The biologically active peptides or their segments, synthesized so far by the "functional handle" method, are listed in Table II.

Table II. Protected Peptides Synthesized by Functional Handle Method[a]

Protected peptide	Coupling methods	Isolation procedure	Overall yield (%)	Reference
[Val^5]-Angiotensin II	Active ester	Citric acid SE-Sephadex	38	Garner and Young (1969, 1971)
Bradykinin	Active ester	Citric acid SE-Sephadex	42	Schafer *et al.* (1971)
Bradykinin analogs	DCC–HOBt	Citric acid Amberlyst-15	14–42	Fletcher and Young (1972)
GIP(1–13)-peptide	DCC–HOBt	Citric acid Amberlyst-15	16	Macrae and Young (1975)
Bradykinin analogs	DCC–HOBt	Citric acid	52	Pinker *et al.* (1976)

[a] DCC–HOBt, Dicyclohexylcarbodiimide coupling in the presence of 1-hydroxy-benzotriazole; GIP, gastric inhibitory polypeptide.

III. THE REPETITIVE EXCESS MIXED ANHYDRIDE (REMA) METHOD

A. Mixed Anhydrides

The *r*epetitive *e*xcess *m*ixed *a*nhydride procedure is an excellent example of the further development of a long-standing method. The mixed anhydride method proved to be a valuable tool for peptide chemists when (in 1951) three research groups (Boissonnas, 1951; Wieland and Bernhard, 1951; Vaughan, 1951) independently proposed the use of mixed anhydrides prepared from *N*-alkyloxycarbonylamino acids and alkyl carbonates. A barrier to the extensive use of this method was the fact that these compounds are rather "sensitive"—they are stable only in solution at a low temperature—and consequently, liable to undergo various side reactions. These are disproportionation; ester, urethane, and, 5(4*H*)-oxazolone formation; reactions with solvent and tertiary bases; and diacylation (Albertson, 1962). Some of these undesirable side reactions, e.g., the time- and temperature-dependent disproportionation, could be easily suppressed. It is also possible to find the conditions under which racemization could be kept to a minimum (Determann and Wieland, 1963; Anderson *et al.*, 1967). It was quite fortunate that the optimal conditions, including the molar ratios, aided the suppression of nearly every side reaction. However, there is still a great need for a thorough kinetic study of the aminolytic reaction of mixed anhydrides. The high reactivity of these compounds, however, is apt to cause difficulties that may discourage kinetic studies.

At present the formation of urethanes (Scheme 2) seems to be the main unsolved problem in the use of the method (see Vol. 1, Chapter 6). The attack of the amine component on the carbonyl of the alkyloxycarbonyl-amino acid leads to the desired peptide (Scheme 2, path a), whereas the competitive attack on the "wrong site" of the mixed anhydride leads to the permanently urethane-blocked amine component (Scheme 2, path b), resulting in truncated sequences. The reaction is greatly influenced both by the nucleophilicity of the amine component and by steric factors. Thus, the mixed anhydride of Z-Pro-OH and ethyl chloroformate was reported to react with H-Pro-Lys(For)-Asp-OH to produce 61% of EtOCO-Pro-Lys-(For)-Asp-OH and very little of the desired peptide (Yajima *et al.*, 1969b). Similar results, perhaps due to steric hindrance, were observed during the formation of a Leu-Val bond in a synthesis of insulin B-chain (Gillessen *et al.*, 1963). Urethane formation of this kind was also observed by Vaughan (1955; Vaughan and Eichler, 1953), Leister and Tarbell (1958), Ondetti and Thomas (1965), Stewart (1965), and Yajima *et al.* (1968, 1969a). Only recently has urethane side-product formation been systematically studied (Bodanszky and Tolle, 1978):

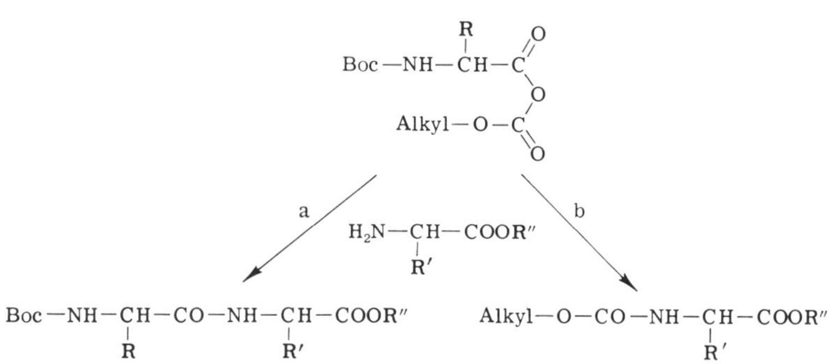

Scheme 2. Competitive aminolytic reaction of mixed anhydrides.

B. The Use of Excess Mixed Anhydrides

In view of the above information, Tilak's communication (1970); Tilak *et al.*, 1972), which disclosed two new findings, aroused considerable interest.

The first was the use of an *excess* of mixed anhydride, previously always used in a 1:1 molar ratio to prevent certain undesired side reactions. With the use of excess, practically quantitative acylation could be effected within short reaction periods. The second significant observation was that the hydrolysis of the excess mixed anhydride with $KHCO_3$–water at 0°C was complete within half an hour. At the same time the desired protected peptide precipitated. When the N^{α}-benzyloxycarbonyl protecting group is used, the coupling is followed by catalytic hydrogenolysis, also a quantitative reaction. Since purification of the intermediates is usually not necessary, one cycle (coupling and deprotection) can be completed in one day. A hexapeptide, Z-Tyr(*t*Bu)-Ser(*t*Bu)-Lys(Boc)-Tyr(*t*Bu)-Leu-Asp(O*t*Bu)-OEt was synthesized in one week with an overall yield of 97% (!). This unusually high yield indicated that greater than 99% yields were achieved at each of the coupling and deblocking steps. From the results obtained by the synthesis of several peptides, including a nonapeptide, Tilak drew the following conclusions. The *advantages* are (1) rapidity; (2) quantitative coupling, the 0.6 equivalent excess of acylating agent being recoverable and reusable; (3) high analytical purity of products without intermediate purification; (4) a capacity for intermediate purification when necessary; (5) mild reaction and work-up conditions; (6) high yields; (7) easy activation and work-up; (8) the ability to use N^{α}-benzyloxycarbonyl derivatives; and (9) the ability to use other protecting groups.

The *limitations* include (1) the necessity for side-chain protection of asparagine to prevent cyclization, (2) side reactions of benzyloxycarbonylglycine (however, glycine may be used advantageously as a COOH-terminus), (3) necessity for methionine and cysteine residues to use N^{α}-*tert*-butyloxycarbonyl protection or to be located at the NH_2-terminus (because of catalyst poisoning), and (4) potential side reactions of the Pro-Pro sequence.

C. Improvements and Further Development

Further improvement was reported by Beyerman *et al.* (1973) and the method named REMA at that time. In the synthesis of the human growth hormone(1–10) segment, the N^{α}-*tert*-butyloxycarbonyl group was used throughout. It was removed either by trifluoroacetic acid in methylene chloride or by 1 *N* hydrogen chloride in acetic acid. In this case, in contrast to Tilak's experiments, acid-stable benzyl-type side-chain protecting groups were used. Unfortunately, even this improved synthesis did not overcome the main problem, the elimination of urethane formation. Thus, when *tert*-butyloxycarbonylisoleucine was coupled to H-Pro-Leu-Ser(Bzl)-Arg(NO_2)-Leu-Phe-OMe, the *only* product isolatable was the isobutyloxycarbonyl derivative of the hexapeptide. In the successful synthesis of the human growth hormone(1–10) segment, this problem was avoided by coupling peptides, namely, Boc-Ile-Pro-OH to the pentapeptide. Thus, an excellent overall yield of 59% was obtained with high purity of the intermediates and final product. Beyerman concluded that "it is possible to extend by the REMA synthesis a peptide sequence by one amino acid every day i.e., in the order of magnitude of the solid-phase method. In contrast, it is possible to check the correct progress of the synthesis by analysis of each intermediate peptide; this is a considerable advantage (Beyerman *et al.*, 1973)." In spite of these results, some concerns about the use of mixed anhydrides in stepwise synthesis of longer chains have been mentioned in the literature (Bodanszky *et al.*, 1973). In recent model experiments, a low degree (<1%) of urethane formation could be detected in all cases, but larger amounts (6–8%) were formed with hindered carboxyl components (Bodanszky and Tolle, 1978).

D. Examples

A large effort was made in Beyerman's laboratory to improve the method and to determine its limitations. As a result of this work, a great achievement was obtained with this method, a total synthesis of secretin (van Zon and Beyerman, 1976), starting from the secretin(1–16) peptide described previously (van Zon and Beyerman, 1973). The detailed procedure and tactics of this synthesis are shown in Fig. 1. In coupling reactions, a 100% excess of mixed anhydride was used, i.e., 2.0 equiv of isobutyl chloroformate, 2.2 equiv of N^{α}-protected amino acid, and 2.2 equiv of *N*-methylmorpholine. The progress of the coupling reactions was monitored using fluorescamine, which permitted the disappearance of the amine component to be determined with an accuracy of 99.8–99.9%. It is of interest that no gradual increase in the reaction time with growing chain length of the amine component was observed (see Fig. 2). The authors concluded that the (unknown) conformation of the amine component is of greater importance than its chain length as a determining factor for coupling time. The cleavage of the N^{α}-protecting group in each step was carried out either by trifluoroacetic acid or by 2 *N* hydrogen bromide in acetic acid, and in the final step in one operation by catalytic hydrogenolysis. The purity of all intermediates was checked by thin-layer chromatography and amino acid analysis. Figure 2 shows the extent of incorporation of each amino acid and a reaction time during the "main" synthesis that did not seem to correlate with each other. For example, the incorporation of the threonine-5 residue remained incomplete even after a relatively long reaction time, while in the first steps of the synthesis complete incorporation took place within a few minutes. Furthermore, due to incomplete incorporation in some steps, the formation of truncated sequences was expected. The removal of these side products was actually

Deprotection	N^{α}-Protection	Sequence	Side-chain protection	Yield (%)
H_2/Pd	Z	His	Z	80
	Boc	Ser	Bzl	94
	Boc	Asp	Bzl	94
	Boc	Gly		91
	Boc	Thr	Bzl	96
	Boc	Phe		92
	Boc	Thr	Bzl	96
	Boc	Ser	Bzl	86
TFA	Boc	Glu	Bzl	94
	Boc	Leu		97
	Boc	Ser	Bzl	93
	Boc	Arg	NO_2	97
	Boc	Leu		93
	Boc	Arg	NO_2	99
	Boc	Asp	Bzl	85
	Boc	Ser	Bzl	89
	Z	Ala		93
	Z	Arg	NO_2	94
2 *N* HBr/AcOH	Z	Leu		87
	Z	Gln		98
	Z	Arg	NO_2	93
	Z	Leu		91
	Z	Leu		89
	Z	Gln		98
TFA	Boc	Gly		97
2 *N* HBr/AcOH	Z	Leu		96
	Z	Val		90

Figure 1. Detailed procedure for synthesis of secretin. Yields were calculated on preceding protected peptide; the 80% value for Z is associated with *o*-nitrophenyl ester coupling.

possible at the final stage by means of ion-exchange chromatography, which showed the presence of about 40% of contaminating truncated sequences. The yield of purified secretin, which was identical in every respect with an authentic sample, amounted to 5%. The main conclusion was that the synthesis did not give rise to any fundamental difficulties at either a 0.2 mmol or a 2 mmol level. At present the size limit of a purifiable peptide that can be synthesized by the REMA method is about 30 amino acid residues. Table III summarizes the protected, biologically active peptides or their fragments prepared so far.

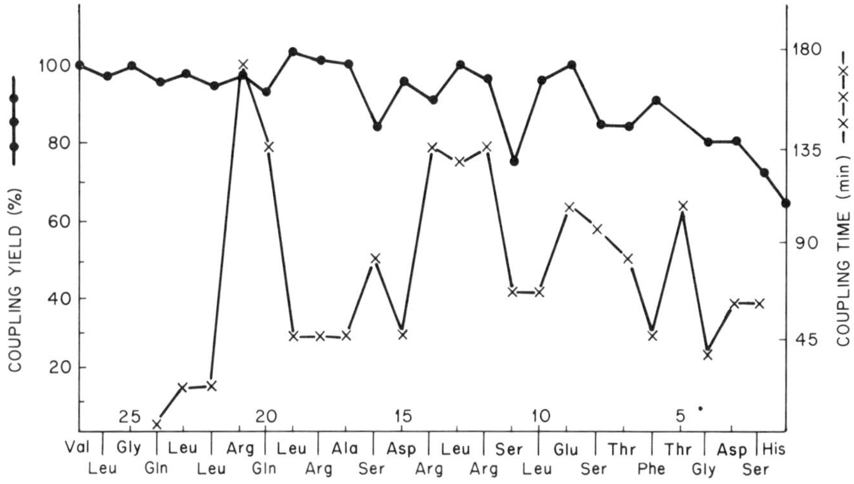

Figure 2. Incorporation of amino acids and reaction time of couplings in the synthesis of secretin by the REMA method.

Table III. Protected Peptides Synthesized by the REMA Method

Protected peptide[a]	Overall yield (%)	Reference
Glucagon(7–15)-peptide	79	Tilak (1970)
HGH(1–10)-peptide	59	Beyerman *et al.* (1973)
Secretin(12–27)-peptide	31	van Zon and Beyerman (1973)
Secretin	13	van Zon and Beyerman (1976)
LHRH	22	Beyerman *et al.* (1976)
LHRH(3–9) analogs	—	Beddell *et al.* (1975)
Substance P	18	Izebond and Beyerman (1978)
Human calcitonin	—	Hirt *et al.* (1979)

[a] HGH, Human growth hormone.
[b] LHRH, luliberin.

IV. PEPTIDE SYNTHESIS *in Situ*

A. General Features

In 1973 Bodanszky published a new technique by which the stepwise synthesis could be considerably facilitated (Bodanszky *et al.*, 1973). A reaction vessel was constructed that permits mixing, evaporation, centrifugation, and drying. When a cycle is started with the removal of the N^{α}-*tert*-butyloxycarbonyl protecting group, the successive steps are (1) trifluoroacetic acid treatment, (2) evaporation, (3) trituration with ether, (4) centrifugation, and

(5) drying. The subsequent coupling step includes (1) dissolving of the trifluoroacetate salt of the amine component in an appropriate solvent and adding the acylating component (completion of the acylation is ascertained by a spot test with ninhydrin); (2) evaporation of the solvent, followed by the addition of a "nonsolvent"; (3) centrifugation; (4) washing with nonsolvent; and (5) drying. An essential feature of this technique is that intermediates of the synthesis remain in the centrifuge tube throughout the operations and throughout the series of chain-lengthening steps. As with other methods described in this chapter, the intermediates can be removed from the apparatus and purified if necessary. The utility of this technique was first demonstrated by the synthesis of a protected nonapeptide, Boc-Leu-Gln-Lys(Dcz)-Leu-Leu-Gln-Gly-Leu-Val-NH_2, corresponding to the COOH-terminal segment of a secretin analog (Bodanszky *et al.*, 1973).

B. Characteristics of *o*-Nitrophenyl Esters

In the above synthesis the moderately active *o*-nitrophenyl esters of amino acids were used as acylating components. Some interesting features of *o*-nitrophenyl esters, such as less solvent-dependent rates of aminolysis and diminished conformational freedom compared with the *p*-nitrophenyl esters, led to an extensive comparative study (Bodanszky *et al.*, 1974a). The ir and nmr spectra of pairs of active esters provided convincing evidence for higher activation in the *ortho*-substituted series. The lower solvent sensitivity of *o*-nitrophenyl esters should be due in part to the proximity of the nitro and ester groups. The higher optical activity, together with the difference in the relative temperature coefficients, pointed to a cyclic conformation of *o*-nitrophenyl esters that could not be established. A possible explanation for the involvement of the *o*-nitro group is an intramolecular dipole–dipole interaction with the amide groupings of the protected amino acids (see Fig. 3).

In this context, kinetic studies (Kömives *et al.*, 1977) suggest that aminolytic reactivities of *o*-nitrophenyl esters are, at least in part, due to intramolecular catalysis by the *o*-nitro group. A frequently discussed question concerns the optimal activation of amino acid active esters. Moderately

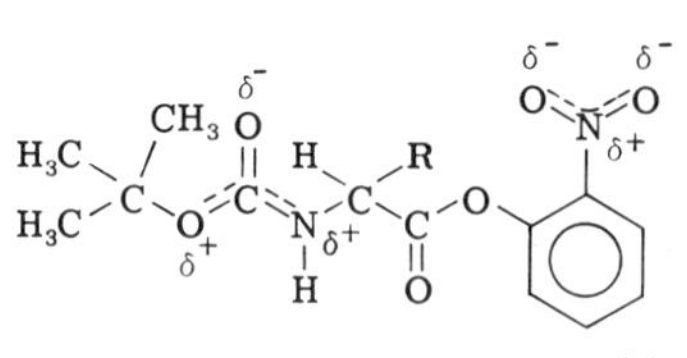

Figure 3. Polar groups in *o*-nitrophenyl esters of protected (*tert*-butyloxycarbonyl) amino acids, as proposed by Bodanszky *et al.* (1974a). Copyright by the American Chemical Society. Reprinted with permission of the copyright owners.

active esters (e.g., *o*- or *p*-nitrophenyl) provide a more selective but slower acylation, while with highly activated esters (e.g., pentafluorophenyl) acylation is much faster, but the possibility of side reactions (e.g., side-chain acylation or racemization) is also enhanced. The above synthesis of a protected nonapeptide exemplifies some of the unexpected difficulties that may be encountered. In the formation of the Leu-Leu bond, using 20–50% excess of acylating agent, the completion of the coupling reaction required one week. The problem was overcome by the application of the active ester in 100% excess and by catalysis with 1-hydroxybenzotriazole. Under these conditions acylations were complete in less than one day (see this volume, Chapter 8).

C. Purification with Nonsolvents

The *in situ* technique requires a so-called "nonsolvent," which means that after completion of the coupling and evaporation, a solvent which precipitates the desired peptide while keeping the by-products in solution, was added to the residue. Ethyl acetate or, in the case of very insoluble peptides, 95% ethanol are particularly useful for this purpose. The considerable solubility of many small peptides in these "nonsolvents" permits the application of this technique only at an advanced stage of the synthesis. This was the situation in the synthesis of [Nle^{17}]vasointestinal peptide(14–28) where the application of the *in situ* technique started at the octapeptide stage. For acylation, *p*-nitrophenyl or *o*-nitrophenyl esters were both used in about 100% excess. The danger of undesirable transacylation during coupling, e.g., trifluoroacetylation, has been mentioned previously (see Section II,B). In the *in situ* technique, the application of such acylating agents as *ortho*- or *para*-nitrophenyl esters virtually eliminates any trifluoroacetylation. If, however, the stronger nucleophilic acetate ions should be present, the transacylation side reaction may occur (see footnote in lit., Bodanszky *et al.*, 1974b).

D. Examples

In the synthesis of *S,S*′-dibenzyloxytocein (Bodanszky *et al.*, 1974c), the starting material was the COOH-terminal tetrapeptide since the shorter intermediates were too soluble for effective precipitation in suitable nonsolvents (see Section V,C) which could keep the by-products in solution. All reactions were carried out in the same centrifuge tube without removal of any of the intermediates. The results of this synthesis, as compared with those obtained by conventional technique, are summarized in Table IV.

It seems that the most successful application of the *in situ* technique so far has been the synthesis of analogs of secretin(5–27) (Fink and Bodanszky,

Table IV. *In situ* Synthesis of Z-Cys(Bzl)-Tyr-Ile-Gln-Asn-Cys(Bzl)-Pro-Leu-Gly-NH_2

	Yield (%)		Melting point (°C)	
Protected peptide	*In situ*	Conventional	*In situ*	Conventional
Oxytocin(5–9)-peptide	82	79	211–213	211–212
Oxytocin(4–9)-peptide	95	80	202–203	210–212
Oxytocin(3–9)-peptide	79	80	229–232	228–230
Oxytocin(2–9)-peptide	96	92	237–239	240–242
Oxytocein	92	92	239–242	239–242

1976). The procedure used in the first synthesis of secretin was followed, except that *o*-nitrophenyl rather than *p*-nitrophenyl esters were employed. In some steps, acylation was catalyzed by 1-hydroxybenzotriazole. Parallel experiments demonstrated the efficacy of this catalyst, since without catalysis several hours were needed for completion of acylation. In the presence of 1-hydroxybenzotriazole the ninhydrin reaction was negative after 0.5 hr. Beginning with the decapeptide stage, the synthesis was facilitated by the application of the *in situ* technique to the extent that it became possible to carry out the preparation of four separate 23-peptides *in parallel*. Biologically active peptides and their fragments prepared so far are compiled in Table V.

Table V. Protected Peptides Synthesized by *in Situ* Technique

Protected peptide[a]	Starting compound	Yield (%)	Reference
Secretin(19–27)-peptide	Boc-Val-NH_2	66	Bodanszky *et al.* (1973)
S,S'-Dibenzyloxytocein	Tetrapeptide	54	Bodanszky *et al.* (1974c)
[Nle^{17}]VIP(14–28)-peptide	Heptapeptide	—	Bodanszky *et al.* (1974b)
Secretin(5–27) analogs	12-Peptide	—	Fink and Bodanszky (1976)
VIP(14–28)-peptides	Tetrapeptide	—	Bodanszky *et al.* (1976)
[Asp^9, Lys^{21}]secretin(5–27)-peptide	Pentapeptide	—	Bodanszky and Fink (1976)

[a] Nle, Norleucine; VIP, vasointestinal peptide.

V. THE PENTAFLUOROPHENYL ESTER METHOD

A. General Considerations

In the development of the pentafluorophenyl (Pfp) ester method, the general considerations are the same as those previously described. Two factors played a decisive role in selecting the pentafluorophenyl esters: (1) their high reactivity in aminolytic reactions and (2) their excellent solubility in common solvents. From the experience gained in earlier studies by Kisfaludy *et al.* (1970, 1973a) it was expected that the high reactivity would assure not only very short reaction times, but also suppression of time-dependent side reactions. This was demonstrated by the study of a competitive intramolecular rearrangement of Z-Asn-OPfp (Kisfaludy *et al.*, 1975) where the half-life of the rearrangement was found to be 117 min, whereas under similar conditions, that of the coupling was 1 min. Thus, the rearrangement cannot compete with the acylation reaction. The excellent solubility of the pentafluorophenyl esters enables the preparation of concentrated solutions—depending on the amine component—and more importantly, the removal of their excess with ether or ether–hexane mixtures (as nonsolvents).

B. Structural Characteristics of Pentafluorophenyl Esters

One cycle of the so-called rapid synthesis (Kisfaludy *et al.*, 1973b, 1974) consists of three main steps: coupling, purification, and deprotection. In the coupling step the acylating component is always a *tert*-butyloxycarbonyl-amino acid pentafluorophenyl ester, which can be prepared conventionally. These esters are quite stable and may be stored in pure form in a desiccator over prolonged periods of time. The thorough kinetic studies of Kovacs *et al.* (1970, 1972, 1973; see also this volume, Chapter 8) have shown that the relative rates of aminolysis of benzyloxycarbonylphenylalanine *p*-nitrophenyl, pentachlorophenyl, and pentafluorophenyl esters were 1:10:400. This outstanding reactivity of pentafluorophenyl esters has not as yet been explained. Therefore, experiments were made to separate the electronic, steric, and possible intramolecular catalytic effects by neighboring halogen substituents in aminolytic reactions (Kisfaludy *et al.*, 1976a). The difference in reactivity between *p*-nitrophenyl and pentahalogenophenyl esters can be explained by electronic effects. However, the latter cannot account for the difference in reactivity of the two pentahalogenophenyl esters. The difference in steric effects is clear from the difference in Van der Waals atomic radii of the two halogen atoms. The question was whether the steric effects or a possible neighboring group participation has greater influence on the aminolytic reaction. The kinetic behavior of different sterically hindered esters and amines indicated that the high reactivity of pentafluorophenyl esters can mainly be attributed to *steric factors* rather than to anchimeric assistance.

The conformation of pentafluorophenyl esters in the solid state was studied by X-ray analysis (see Fig. 4). The results proved that in the crystal structure there is no intramolecular interaction and that the main attractive force is a strong dipole–dipole interaction between the negative fluorine and positive carbon atoms of another pentafluorophenyl ring. On the other hand, it was found that the planes of the carbonyl and the pentafluorophenyl

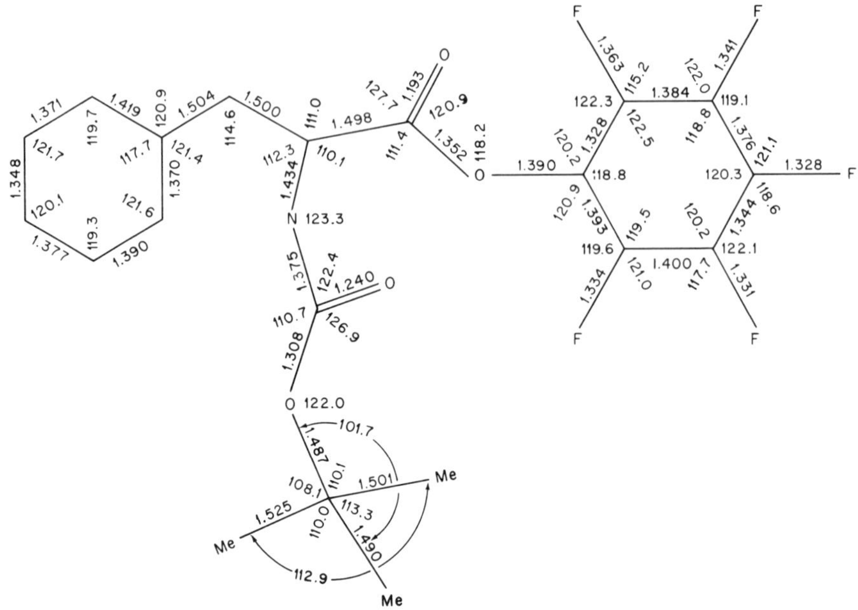

Figure 4. The molecular geometry of *tert*-butyloxycarbonyl-L-phenylalanine pentafluorophenyl ester by X-ray analysis.

groups are nearly perpendicular. We think that this is an important fact to be kept in mind when the rigidity of these types of active esters is discussed. From ir studies in different solvents, it seems likely that the conformation in solution is similar to that in the crystal. The higher activation of an amino acid derivative results in a greater danger of racemization. Kovacs *et al.* (1972) showed that of the commonly used active esters, the pentafluorophenyl exhibit some of the highest rates of racemization. However, from a practical point of view, the ratio of k_{coupling} to $k_{\text{racemization}}$ is of importance, and this ratio is most favorable in the case of pentafluorophenyl esters. Several dipeptide pentafluorophenyl esters could be prepared without racemization (Kisfaludy *et al.*, 1973a) even with sterically hindered amino acids. A possible explanation for racemization-free coupling through a 5(4*H*)-oxazolone intermediate is that the acidic phenol can protonate the oxazolone formed, suppressing the otherwise fast racemization (Kovacs *et al.*, 1969).

On the basis of their chemical and physical properties, we do not classify the pentafluorophenyl esters as overactivated amino acid derivatives. In our opinion, their activity is nearly ideal in providing high reactivity and a favorable k_{coupling} to $k_{\text{racemization}}$ ratio. Pentafluorophenyl ester coupling is performed in the presence of a tertiary base. It is important to note that for rapid completion of the acylation, pH 8 has to be maintained during the entire process. Otherwise, the starting pH will decrease within a few minutes to pH 6–7, due to the release of acidic phenol, and the amine component will be present partly in protonated form, which causes a considerable decrease in rate as the acylation reaction proceeds.

C. Purification

After the completion of acylation, as checked by ninhydrin or fluorescamine, the method of Löw and Kisfaludy (1965) may be used for the removal of excess active esters. In this procedure, the addition of *N,N*-dimethylaminoethylamine to the reaction mixture effects rapid conversion of the excess active ester to the corresponding amide which is readily extracted by aqueous acid due to its basic character (handle principle). This method can be applied advantageously in the early stage of the synthesis when the solubility permits the use of ethyl acetate or chloroform. With increasing length of the peptide chain, the excess of active esters can be removed by various nonsolvents.* The operation is also suitable for removal of the other by-products.

In the deprotection step, hydrogen chloride or bromide in dioxane or ethyl acetate have been used almost exclusively to avoid the possible transacylation previously mentioned. Studies have shown that no trifluoroacetylation occurs in the presence of pentafluorophenyl esters.

D. Examples

Together with simple work-up, the short reaction times, both for coupling and deprotection, considerably decrease the time required for synthesis. Thus, for instance, a synthesis of protected antamanide (Kisfaludy *et al.*, 1974) could be carried out in 510 min in seven cycles (two couplings were carried out with Boc-Pro-Pro-OPfp). The incorporation of one amino acid (or dipeptide segment) required 70–80 min. Nearly the same short periods of time (600 min) were sufficient to synthesize protected angiotensin II (Kisfaludy and Nyéki, 1975) and oxytocin (Kisfaludy and Schön, 1975). It has to be noted that, as with the other three repetitive methods, the main advantage is not only the decrease of time but the ability to control every step, i.e., to purify every intermediate. Because of the individuality of each amino acid and the characteristics of the peptide to be synthesized, the determination of optimal conditions for each step of a synthesis requires

* In the dipeptide stage reactions may alternatively be accelerated by using excess amine component.

Table VI. Rapid Synthesis of Boc-Arg(NO_2)-Val-Tyr(Bzl)-Ile-His(Dnp)-Pro-Ala-ONb Using *tert*-Butyloxycarbonylamino Acid Pentafluorophenyl Esters

Protected peptide[a]	Excess (equiv)	Solvent	Reaction time (min)	Purification[b]
AII(7–8)-peptide	—	Chloroform	20	Citric acid
AII(6–8)-peptide	1.5	Chloroform	30	DMAEA[c]
AII(5–8)-peptide	2.0	Chloroform–DMF (1:1)[d]	20	Ether–hexane 1:9
AII(4–8)-peptide	1.5	Chloroform–DMF (2:1)	15	DMAEA[c]
AII(3–8)-peptide	2.0	Chloroform–DMF (1:1)	20	Ether
AII(2–8)-peptide	3.0	DMF	20	Ethanol

[a] AII, Angiotensin II.
[b] Deprotections were carried out by means of 8N HCl/dioxane.
[c] N,N-Dimethylaminoethylamine.
[d] DMF, Dimethylformamide.

preliminary work. Such efforts pay off particularly when analogs are synthesized, e.g., 1-position analogs of angiotensin II (Kisfaludy and Nyéki, 1977). It has been found advantageous to have a rapid method for the synthesis of angiotensin II(2–8) heptapeptide. The characteristics of this synthesis are shown in Table VI. The overall yield was 72%, which corresponds to 95% average yield per cycle. The time required was 480 min on a 2 mmol scale. With the synthesis of segments of Δ^5-3-ketosteroid isomerase (Kisfaludy *et al.*, 1976b), our method was used up to the tridecapeptide stage in several cases. Beyond that, decreased solubility became prohibitive. Some

Table VII. Protected Peptides Synthesized by the Pentafluorophenyl Ester Method

Protected peptide	Number of cycles	Yield (%)	Total time required (min)	Reference
Antamanid	7	50	510	Kisfaludy *et al.* (1974)
Oxytocin	7	39	600	Kisfaludy and Schön (1975)
Angiotensin II	7	43	600	Kisfaludy and Nyéki (1975)
Δ^5-3-ketosteroid isomerase segments	—	—	—	Kisfaludy *et al.* (1976)
Gastrin(11–17)-peptide	6	39	900	Penke *et al.* (1976)
Angiotensin II analogs	7	61	600	Kisfaludy and Nyeki (1977)
Gastrin(13–17)-peptide	4	45	420	Schön (1977)
Boc-Gly/1-^{14}C/-Trp-Met-Asp-Phe-NH_2	—	—	—	Baláspiri *et al.* (1977)
Calcium-binding protein	—	—	—	Maximov *et al.* (1979)
TRH analogs	2	70	—	Kisfaludy *et al.* (1979)

segments were soluble only in hexamethylphosphoramide, and this insolubility persisted even after deprotection. Insufficient solubility has already been mentioned as a limiting factor in stepwise chain lengthening. The importance of steric factors has also been confirmed in our experiments. The formation of a Val-Val bond in the synthesis of Δ^5-3-ketosteroid isomerase(108–117) at the octapeptide stage required the use of a threefold excess of Boc-Val-OPfp and an unusually long reaction time of 180 min for completion. The biologically active peptides or their segments prepared thus far are summarized in Table VII.

VI. COMPARATIVE STUDIES

There is very little information in the literature that enables a comparison of the efficacy of different methods or protecting groups to be made. A comparison was possible in the case of angiotensin II which was synthesized both by the functional handle and by the pentafluorophenyl ester methods, but in different laboratories. Penke *et al.* (1976) synthesized gastrin(11–17) by the pentafluorophenyl ester method, by the REMA method, and by solid phase and solution syntheses to compare the performances (see Table VIII). Similar comparative studies were carried out with the synthesis of the human gastrin I segments(1–5), (6–13), and (14–17) in which the pentafluorophenyl ester method was found to be the most favorable with respect to both yield and time requirement. Orlowska *et al.* (1976) synthesized protected substance P(5–9) by three different methods, including the *o*-nitrophenyl ester method. With the latter procedure the coupling reactions were not quantitative despite a prolonged reaction time of 72 hr.

Table VIII. Synthesis of Gastrin(11–17) by Different Methods[a]

Method	Time period of one cycle (hours)	Reagent excess (%)	Overall yield (%)	Biological activity (%) (pentagastrin = 100)
Pfp ester[b]	2.5–3	50–100	39	70
REMA[c]	4.5–6	50	10.2	70
Solid phase[d]	5.5–6	150–200	16	65
Conventional[e]	—	—	32	70

[a] Modified from Penke *et al.* (1976).
[b] Pentafluorophenyl ester, repetitive.
[c] Repetitive excess mixed anhydride method.
[d] Merrifield solid-phase synthesis.
[e] Solution synthesis.

VII. CONCLUDING REMARKS

The results obtained with the methods described in this chapter illustrate that there are a variety of efficient and rapid methods available to peptide chemists for syntheses of medium-sized peptides. The ability to monitor the purity of the intermediates is achieved without prolonging the reaction time. In practice, the pentafluorophenyl ester method is less time consuming than the Merrifield solid-phase method. There are of course problems there too. The most serious is the limitation imposed on stepwise chain elongation by insufficient solubility. This is highly dependent on the characteristics of the peptide to be synthesized and applies to every method. A further difficulty encountered is certain side reactions at the amine component, e.g., pyroglutamyl, dioxopiperazine, or cyclic imide formation, that may occur independently of the coupling method used. The ensuing side products are sometimes difficult to detect in the final product and their removal may present problems. Because several of these side reactions are slow, we believe that they can be suppressed to a minimum level by "fast peptide bond formation." The occurrence of other failures such as transacylation or formation of permanently blocked amine component depends on the characteristics of the acylating component.

As for higher peptides, at present the segment condensation seems to be the only way for their synthesis. Each of the four repetitive methods provides the capability to prepare the required segments. As has been mentioned before, a successful synthesis of a large peptide can only be realized by a judicious combination of the existing methods.

ACKNOWLEDGMENT

The author is indebted to Dr. I. Schön for helpful assistance.

REFERENCES

Albertson, N. F. (1962). *Org. React.* **12**, 157–355.

Anderson, G. W., Zimmerman, J. E., and Callahan, F. M. (1967). *J. Am. Chem. Soc.* **89**, 5012–5017.

Baláspiri, L., Penke, B., Pallai, P., Sirokman, F., Varga, L., and Varro, V. (1977). *Acta Phys. Chem.* **23**, 355–359.

Beddell, C. R., Fraser, P. J., Gilbert, D. D., Goodford, P. J., Lowe, L. A., and Wilkinson, S. (1975). *J. Med. Chem.* **18**, 417–423.

Beyerman, H. C., de Leer, E. W. B., and Floor, J. (1973). *Rec. Trav. Chim. Pays-Bas* **92**, 481–492

Beyerman, H. C., Rammeloo, T., Renirie, J. F. M., Syrier, J. L. M., and van Zon, A. (1976). *Rec. Trav. Chim. Pays-Bas* **95**, 143–148.

Bodanszky, M. (1960). *Ann. N.Y. Acad. Sci.* **88**, 655–664.

Bodanszky, M., and Fink, M. L. (1976). *Bioorg. Chem.* **5**, 275–282.

Bodanszky, M., and Tolle, J. C. (1978). *Int. J. Pept. Protein Res.* **10**, 380–384.

Bodanszky, M., Funk, K. W., and Fink, M. L. (1973). *J. Org. Chem.* **38**, 3565–3570.

Bodanszky, M., Fink, M. L., Funk, K. W., Kondo, M., Lin, C. Y., and Bodanszky, A. (1974a). *J. Am. Chem. Soc.* **96**, 2234–2240.

Bodanszky, M., Lin, C. Y., and Said, S. I. (1974b). *Bioorg. Chem.* **3**, 320–323.

Bodanszky, M., Kondo, M., Lin, C. Y., and Sigler, G. F. (1974c). *J. Org. Chem.* **39**, 444–447.

Bodanszky, M., Lin, C. Y., and Yoitakis, A. E. (1976). *Bioorg. Chem.* **5**, 339–350.

Boissonnas, R. A. (1951). *Helv. Chim. Acta* **34**, 874–879.

Burton, J., Fletcher, G. A., and Young, G. T. (1971). *Chem. Commun.* pp. 1057–1058.

Camble, R., Garner, R., and Young, G. T. (1968). *Nature* (*London*) **271**, 247–248.

Coyle, S., and Young, G. T. (1976). *Chem. Commun.* pp. 980–981.

Determann, H., and Wieland, T. (1963). *Justus Liebigs Ann. Chem.* **670**, 136–140.

Fink, M. L., and Bodanszky, M. (1976). *J. Am. Chem. Soc.* **98**, 974–977.

Fletcher, G. A., and Young, G. T. (1972). *J. Chem. Soc., Perkin Trans. 1* p. 1867–1874

Fletcher, G. A., Löw, M., and Young, G. T. (1973). *J. Chem. Soc., Perkin 1* pp. 1162–1164.

Garner, R., and Young, G. T. (1969). *Nature* (*London*) **222**, 177–178.

Garner, R., and Young, G. T. (1971). *J. Chem. Soc. C* pp. 50–53.

Gillessn, D., Schnabel, E., and Meienhofer, J. (1963). *Justus Liebigs Ann. Chem.* **667**, 164–171.

Gosden, A., Macrae, R., and Young, G. T. (1977). *J. Chem. Res.* (*S*) pp. 22–23.

Hirt, J., Kranenburg, P., and Beyerman, H. C. (1979). *Rec. Trav. Chim. Pays-Bas* **98**, 143–154.

Izebond, E., and Beyerman, H. C. (1978). *Recl. Trav. Chim. Pays-Bas* **97**, 1–60.

Kisfaludy, L., and Nyéki, O. (1975). *Acta Chim. Acad. Sci. Hung.* **86**, 343–345.

Kisfaludy, L., and Nyéki, O. (1977). *Hungarian* Patent Prov. NOS. RI-640 and RI-641.

Kisfaludy, L., and Schön, I. (1975). *Acta Chim. Acad. Sci. Hung.* **84**, 227–228.

Kisfaludy, L., Roberts, J. E., Johnson, R. H., Mayers, G. L., and Kovacs, J. (1970). *J. Org. Chem.* **35**, 3563–3565.

Kisfaludy, L., Löw, M., Nyéki, O., Szirtes, T., and Schön, I. (1973a). *Justus Liebigs Ann. Chem.* **1973**, 1421–1429.

Kisfaludy, L., Schön, I., Szirtes, T., Nyéki, O., and Löw, M. (1973b). Hungarian Patent 168, 431.

Kisfaludy, L., Schön, I., Szirtes, T., Nyéki, O., and Löw, M. (1974). *Tetrahedron Lett.* pp. 1785–1786.

Kisfaludy, L., Schön, I., Görög, S., and Rényei, M. (1975). *J. Am. Chem. Soc.* **97**, 5588.

Kisfaludy,L., Löw, M., Argay, Gy., Czugler, M., Kömives, T., Sohár, P., and Darvas,F. (1976a). *In* "Peptides 1976" (A. Loffet, ed.) pp. 55–59. de l'Université de Bruxelles, Belgium.

Kisfaludy, L., Löw, M., Schön, I., Szirtes, T., Nyéki, O., Vezér, Cs., Keresztessy, P., Korenczki, F., and Szirmai, M. (1976b). *In* "Peptides 1976" (A. Loffet, ed.) Editions de l'Université de Bruxelles, Belgium.

Kisfaludy, L., Szirtes, T., Baláspiri, L., Pálosi, É., and Szporny, L. (1979). Hungarian Patent Application. Prov. NOS. RI-717 and RI-718.

Kömives, T., Márton, A. F., Dutka, F., Löw, M., and Kisfaludy, L. (1977). *Z. Naturforsch. Teil B* **32**, 1359–1360.

König, W., and Geiger, R. (1970). *Chem. Ber.* **103**, 788–798.

Kovacs, J., Kisfaludy, L., Ceprini, M. Q., and Johnson, R. H. (1969). *Tetrahedron* **25**, 2555–2573.

Kovacs, J., Mayers, R. H., Johnson, E. R., Cover, E. R., and Ghatak, U. R. (1970). *J. Org. Chem.* **35**, 1810–1815.

Kovacs, J., Davis, E. J., Johnson, R. H., Cortegiano, H., and Roberts, J. (1972). *In* "Chemistry and Biology of Peptides" (J. Meienhofer, ed.) pp. 359–369. Ann Arbor Sci. Publ., Ann Arbor, Michigan.

Kovacs, J., Cover, R. E., Johnson, R. H., Kalas, T. J., Mayers, G. L., and Roberts, J. E. (1973). *J. Org. Chem.* **38**, 2518–2521.

Leister, N. A., and Tarbell, D. S. (1958). *J. Org. Chem.* **23**, 1152–1155.

Löw, M., and Kisfaludy, L. (1965). *Acta Chim. Acad. Sci. Hung.* **44**, 61–66.

Macrae, R., and Young, G. T. (1974). *Chem. Commun.* pp. 446–447.

Macrae, R., and Young, G. T. (1975). *J. Chem. Soc., Perkin Trans, 1* pp. 1185–1191.

Maximov, E. E., Zapevalova, N. P., Mitin, Yu. V., Neubert, K., Jakubke, H. D., Baláspiri, L., Penke, B., Kovacs, K., Lammek, B., and Kupryszevwski, G. (1979). Pept. Proc. Eur. Pept. Symp., 15th, in press.

Meienhofer, J. (1979). *In* "The Peptides" (E. Gross and J. Meienhofer, eds.), Vol. 2. Academic Press, New York.

Ondetti, M. A., and Thomas, P. L. (1965). *J. Am. Chem. Soc.* **87**, 4373–4380.

Orlowska, A., Bankowski, K., and Drabarek, S. (1976). *Rocz. Chem.* **50**, 1701–1707.

Penke, B., Pallai, P., Kovacs, K., and Baláspiri, L. (1976). *In* "Peptides 1976" (ed. Loffet, A.) pp. 101–106. Éditions de l'Université de Bruxelles.

Pinker, T. G., Young, G. T., Elliot, D. F., and Wade, R. (1976). *J. Chem. Soc., Perkin Trans. 1.* pp. 220–228.

Schafer, D. J., Young, G. T., Elliot, D. F., and Wade, R. (1971). *J. Chem. Soc. C* pp. 46–49.

Schnabel, E., Klostermeyer, H., and Berndt, H. (1971). *Justus Liebigs Ann. Chem.* **749**, 90–108

Schön, I. (1977). Dissertation for C. Sci. Degree, Budapest.

Seely, J. H., Ruegg, U., and Rudinger, J. (1973). *In* "Peptides 1972" (H. Hanson and H. D. Jakubke, eds.) pp. 86–89. North Holland Publ., Amsterdam.

Stewart, F. H. C. (1965), *Aust. J. Chem.* **18**, 887–901.

Tilak, M. A. (1970), *Tetrahedron Lett.* pp. 849–854.

Tilak, M. A., Hendricks, M. L., and Wedel, D. S. (1972). *In* "Progress in Peptide Research" (S. Lande ed.), pp. 351–354. Gordon & Breach, New York.

van Zon, A., and Beyerman, H. G. (1973), *Helv. Chim. Acta* **56**, 1729–1740.

van Zon, A., and Beyerman, H. C. (1976), *Helv. Chim. Acta* **59**, 1112–1126.

Vaughan, J. R. (1951), *J. Am. Chem. Soc.* **73**, 3547.

Vaughan, J. R. (1955), *Abstr. Pap., 128th Meet., Am. Chem. Soc.* p. 27.

Vaughan, J. R., and Eichler, J. A. (1953), *J. Am. Chem. Soc.* **75**, 5556–5560.

Wieland, T., and Bernhard, H. (1951), *Justus Liebigs Ann. Chem.* **572**, 190–194.

Yajima, H., Kawaski, K., Okada, Y., Minami, H., Kubo, K., and Yamashita, I. (1968), *Chem. Pharm. Bull.* **16**, 919–928.

Yajima, H., Mizokami, N., Okada, Y., and Kawasaki, K. (1969), *Chem. Pharm. Bull.* **17**, 1958–1962.

Yajima, H., Okada, Y., Kawatani, H., and Mizokami, N. (1969a), *Chem. Pharm. Bull.* **17**, 1229–1236.

Young, G. T. (1973), *In* "Chemistry of Polypeptides" (Katsoyannis, ed.), pp. 43–58. Plenum, New York.

Chapter 7

Partial Synthesis of Peptides and Proteins

ROBERT C. SHEPPARD

THE PEPTIDES, VOLUME 2

ISBN 0-12-304202-X

I. INTRODUCTION

Partial* as opposed to total synthesis would seem to be an ideal strategy in the peptide and protein fields. Much effort is presently expended in the total synthesis of peptide analogues for pharmacological study that differ in amino acid sequence only slightly from their naturally occurring parents. In those cases where the natural peptides or proteins are themselves readily available, partial synthesis could conceivably offer much less laborious synthetic routes, especially to the larger molecules. Likewise, peptides and proteins of the difficult-to-obtain human series often differ at only one residue (as in insulin) or at just a few residues from their more readily available lower-species counterparts and should be accessible by chemical transformation of the latter. Isotopic labeling for physical or metabolic studies by amino acid replacement in natural peptides and proteins is also feasible. These and other potential applications have stimulated significant recent activity in the study of peptide and protein partial synthesis. Most of the work reported concerns modification to a single protein (insulin), but other proteins are also being studied and some general principles, patterns, and problems are emerging.

Two principal strategies for partial synthesis may presently be discerned. In the first, the natural peptide or protein is regarded essentially as a convenient source of peptide fragments, akin to those more conventionally obtained by total synthesis. Reassembly of the polypeptide chain would then

* The description "protein *semisynthesis*" has also found some general acceptance. It has been defined by Smyth (1975) as "combination of two components, the one a fragment of a natural substance, the other a product formed by chemical synthesis." Smyth includes in this definition both covalent and noncovalent (complementation) interactions, as well as combination by nonpeptide covalent bonds (e.g., disulfide bonds, as in hybrid insulins, etc.). This description is too wide for the present discussion which is limited strictly to synthesis through the formation of one or more new *peptide* bonds between natural and synthetic components. Since the term "partial synthesis" has a long history in organic chemistry for such transformation of natural products, its use is retained in the present chapter.

follow the general patterns established for the "fragment condensation" procedure of total synthesis. A range of proteolytic enzymes and some chemical methods are available for the dismemberment of proteins, and powerful chromatographic techniques have been developed for the separation of the often complex mixtures of smaller peptides produced. These chromatographic methods may be operated on a modest preparative scale, and countercurrent distribution provides an alternative, if more cumbersome, technique capable of handling substantial quantities of material. Substitution of synthetic peptides for one or more of the purified natural fragments and reassembly constitutes a general approach to partial synthesis that, in principle, is capable of wide application. However, the later stages in such projected syntheses present formidable problems. A large number of reactive functional groups are usually associated with any significant length of polypeptide chain. These include terminal and side-chain amino (lysine) and carboxyl (aspartic and glutamic acids), disulfide or thiol (cystine, cysteine), imidazole (histidine), guanidino (arginine), indole (tryptophan), and hydroxyl (tyrosine, serine, and threonine) groups. Many of these will require protection before unambiguous reassembly is possible, and this approach to partial synthesis therefore relies heavily on the development of highly specific methods for the protection of individual functional groups in naked, polyfunctional fragments. It is unlikely that the introduction of such protecting groups into complex molecules can be achieved with absolute specificity, and further purification procedures will in general be necessary prior to reassembly. The problems of re-formation of the polypeptide chain are, then, essentially those presently being tackled in total synthesis.

This first strategy of partial synthesis has features in common with those of classical total synthesis, and some words of comparison are in order. The starting point in all conventional peptide syntheses involves the free amino acids which are converted into appropriately protected derivatives at an early stage. These simple, protected amino acids are capable of complete purification and characterization. Assembly of the protected amino acids into short, protected oligopeptides follows laborious but well-developed procedures and does not generally present insurmountable problems. The main difficulties arise in the assembly of these oligopeptide derivatives into larger protected peptides and proteins, and the final removal of all the protecting groups. These are essentially the same problems as those encountered in the partially synthetic approach except that, in the latter case, they are met under less-favorable circumstances. In total synthesis, the points of chain division, the protection or otherwise of individual side chains, and the choice of protecting groups and coupling methods are tactical decisions made at the outset but with considerable flexibility. The partially synthetic approach is much less flexible. The points of chain division are dictated by the cleavage methods available. Protecting groups are usually introduced at a late stage into complex molecules which may be difficult to purity and characterize. The great advantage of the partially synthetic method is a lessening of the labor involved in the initial production of the peptide fragments, but this is gained at the expense of potentially greater difficulty in the reassembly steps. It is not, therefore, to be assumed that this strategy of partial synthesis will necessarily prove more efficient than total synthesis in all cases.

The second principal strategy for partial synthesis involves selective scission or excission of individual amino acids or peptide sequences in the native proteins. In contrast to the foregoing, there is now minimum dismemberment of the native polypeptide, often at just one peptide bond. Fewer recombination steps are therefore required. The main problem now is the severely limited range of methods available for the highly specific chain cleavage which is necessary. This stragegy finds its simplest and most well-explored expression in the stepwise removal and readdition of individual amino acids at the NH_2-termini of peptides and proteins. Apart from this particular application, the approach is unlikely to be a general one since it depends entirely on the presence of peptide bonds in the area of particular chemical or biological interest that are susceptible to specific cleavage. On the other hand, the principle of minimum alteration of the natural structure is an attractive one. Manipulation of the starting peptide or protein during cleavage, separation, and reassembly is minimized. The initial tertiary structure of the protein may be retained during some of these operations, through either noncovalent interactions, or coordination to metal ions, or because of intact disulfide bonds. Retention of the compact folded structure in globular proteins usually results in more favorable stability and solubility properties. By bringing together appropriate COOH- and NH_2-termini, the tertiary structure may facilitate the reassembly process. This last possibility is particularly important potentially, because it opens the way to selective recombination reactions as an alternative to extensive side-chain protection. This strategy of partial synthesis may therefore make better use of the properties of the protein starting material, but it is correspondingly limited in application. Specific resynthesis of cleaved polypeptide chains without the use of protecting groups has been observed in pancreatic trypsin inhibitor (Dyckes *et al.*, 1974, 1978) and in cytochrome c (Corradin and Harbury, 1974); in both cases the tertiary structure of the native protein was almost certainly retained, leading to smooth recombination reactions using only very mildly activated carboxyl derivatives. Resynthesis without the use of protecting groups is also possible in specific cases by the reverse action of proteolytic enzymes (Laskowski, 1978).

II. CHAIN CLEAVAGE OF PEPTIDES AND PROTEINS

A. By Proteolytic Enzymes

General procedures for the cleavage of peptides and proteins by the common proteolytic enzymes are well established and need no discussion here. Often, the sites of action of various enzymes on the particular polypeptide substrate will have been elucidated during the course of structural determination. When this is not so, the specificities of the common proteases are often sufficiently well known to enable prediction of the major points of cleavage, at least for the unfolded (denatured) substrate. Trypsin appears to be the enzyme of first choice for fragmentation reactions. It has very high specificity for cleavage at lysine and arginine residues, although relative rates of cleavage may be much reduced by the presence of adjacent, negatively charged groups (Harris and Offord, 1977). Its action may be restricted to cleavage at arginine by masking of lysine side chains or widened to include cysteine by (irreversible) aminoethylation of sulfydryl groups. An example of its use for the extensive fragmentation of a protein for partial synthesis studies is provided by the work of Offord and his collaborators on lysozyme (Rees and Offord, 1976). Unfolding of the protein was ensured by cleavage of all the disulfide bonds with conversion to *S*-sulfonate derivatives, and all the free amino groups were blocked reversibly with maleic anhydride. Digestion with trypsin (previously treated with the chymotrypsin inhibitor 1-chloro-4-phenyl-3-L-tosylamidobutan-2-one) then afforded a mixture of fragments from which 10 of the 12 possible products could be isolated on a preparative (2–3 gm of digest) scale. A useful feature of procedures of this type is that all the side-chain amino groups are blocked at an early stage, allowing easy differentiation of the newly liberated NH_2-terminal groups in the isolated fragments. A similar study has been carried out on maleylated myoglobin (Borras-Cuesta *et al.*, 1978).

Other reversible blocking groups may be used to restrict the action of trypsin by masking lysine side chains. An attractive procedure is amidination (Section III,A,6) through the action of methyl acetimidate hydrochloride on the protein (Hunter and Ludwig, 1962; Garner and Gurd, 1975). In contrast to analogous guanidation procedures using alkyl isoureas, acetimidylation is reversible and the lysine side chain may be re-formed through the action of strong aqueous ammonia. The positively charged character of the lysine ε-amino group at moderate pH is retained after amidination with minimum alteration of solubility and other properties. Susceptibility to tryptic cleavage appears to be effectively eliminated (Hunter and Ludwig, 1962), as in the related homoarginine derivatives (Weil and Telka, 1957; Shields *et al.*, 1959).

A number of devices may be used to restrict still further the sites of action of trypsin and other proteases. In their native states, many globular proteins are either resistant to the action of proteolytic enzymes or are initially susceptible to cleavage only at specific external loops in the folded polypeptide chain. Thus, highly specific cleavage by enzymes of broader specificity may sometimes be achieved by using the substrate in its native state. Examples include the single cleavages of ribonuclease A by subtilisin (Richards and Vithayathil, 1959) and pepsin (Anfinsen, 1956) and limited proteolysis of *Staphylococcus aureus* nuclease by trypsin (Taniuchi and Anfinsen, 1968). This enhancement of specificity by substrate tertiary structure is also operative in the natural zymogen–enzyme conversion. A very special case is found in the native protease inhibitors (Section V,D) where the very strong, unique association between enzyme and inhibitor may, under appropriate conditions, cause cleavage at just one peptide bond in the active site of the latter. This highly specific cleavage in an important part of the molecule has been used extensively for the partial synthesis of active site-modified analogues.

Enhancement of enzyme specificity may also be achieved by prior binding of the substrate to a high-molecular-weight, insoluble polymer. Thus, both bovine (Gattner, 1975) and camel (Danho *et al.*, 1975) insulins, adsorbed on the strongly acidic cation exchanger sulfoethylcellulose, are cleaved by pepsin at pH 2 exclusively at the B25–B26 phenylalanyl–tyrosyl bond. Both sterically restricted access and electrostatic effects may contribute to this highly specific binding and hydrolysis by a broad specificity enzyme.

A useful adjunct to the selective enzymic cleavage of proteins is subsequent digestion by specific exopeptidases to remove newly exposed terminal residues. This provides an effective method for removal of specific amino acids from within the polypeptide chain. Aminopeptidases have only limited application in this way (Jering and Tschesche, 1974), presumably both because their low specificity may lead to uncontrolled sequential removal of NH_2-terminal residues, and because an efficient stepwise chemical method, the Edman degradation, (Section II,B) is available for this purpose. However, digestion of COOH-terminal residues by carboxypeptidases (for which no equally satisfactory equivalent chemical method apparently exists) has found important application. A particularly powerful combination is digestion with trypsin followed by carboxypeptidase B (Jering and Tschesche, 1974; Sealock and Laskowski, 1969; Gattner *et al.*, 1978); both enzymes are specific for the basic residues lysine and arginine, trypsin cleaving on the carboxyl side and carboxypeptidase B on the amine side of COOH-terminal residues. The overall effect is cleavage with specific excission of a single basic amino acid from the polypeptide chain (Sealock and Laskowski, 1969) (Scheme 1).

In appropriate cases, this same enzyme combination may be used for amino acid replacement and resynthesis of the peptide chain (Section V,E).

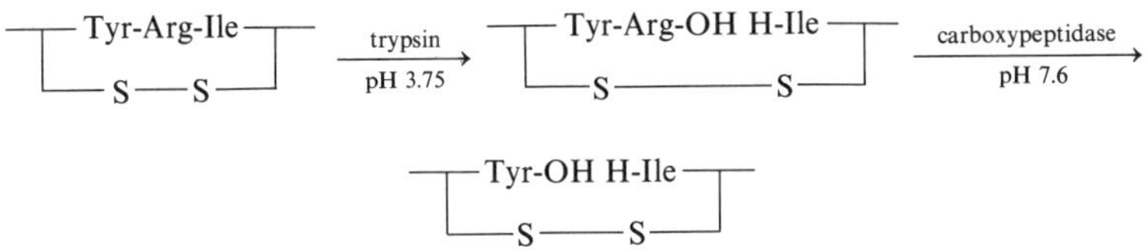

Scheme 1

B. By Chemical Methods

Although a rather substantial number of chemical methods have from time to time been advocated for the cleavage of specific peptide bonds (for a review, see Spande *et al.*, 1970), few have proved of general applicability to complex peptides and proteins. Exceptions are the very important NH_2-terminal degradation of Edman (1950) and the cyanogen bromide-induced cleavage at methionine residues (Gross and Witkop, 1961). Oxidative methods for cleavage at tryptophan (Savige and Fontana, 1977) are showing promise, and a new method for cysteine (Degani and Patchornik, 1974; Jacobson *et al.*, 1973; Stark, 1977) takes place under encouragingly mild conditions. Reagents involved in other suggested procedures have generally proved insufficiently selective in their action on polyfunctional peptides and proteins.

The Edman degradation (Scheme 2) has been extensively applied to the removal of NH_2-terminal residues in partially synthetic studies. Very high

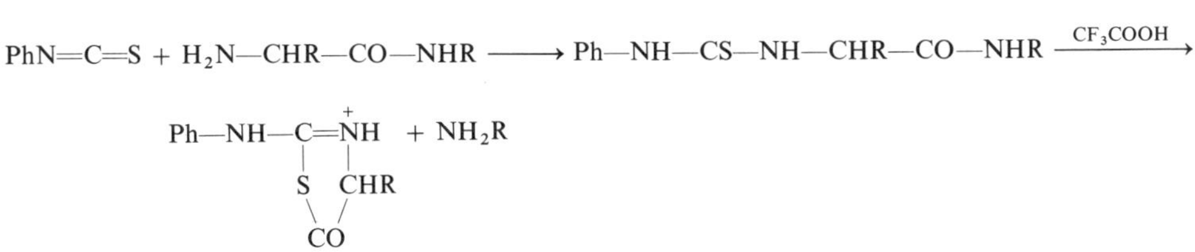

R = peptide chain

Scheme 2

yields may be obtained, particularly if oxygen is excluded. Since phenyl isothiocyanate reacts irreversibly with lysine side chains, protection of the latter is necessary. Selective acylation of α- and ε-amino groups is discussed in Section III. Partial synthesis studies utilizing the Edman degradation include soybean trypsin inhibitor (Kowalski and Laskowski, 1976a,b), phospholipase A_2 (Slotboom and de Haas, 1975), and many examples in the insulin series.

Nonenzymatic transamination has been advocated as a viable alternative to the Edman degradation (Webster and Offord, 1972; Dixon, 1964a,b). It involves exchange of the α-amino group of a peptide or protein with an α-oxo acid (usually glyoxylic acid) [Eq. (1)], and removal of the new terminal α-oxo acyl residue with a bidentate nucleophile (e.g., *o*-phenylenediamine) [Eq. (2)]. The transamination step is specific for α-amino groups, and the advantage of this procedure is that specific protection of lysine side chains is not required; if necessary for subsequent purposes, this can be carried out

$$\mathrm{NH_2{-}CHR{-}CO{-}NHR + OCH{-}CO_2H \longrightarrow \overset{R'}{\overset{|}{OC}}{-}CO{-}NHR + NH_2{-}CH_2{-}CO_2H} \quad (1)$$

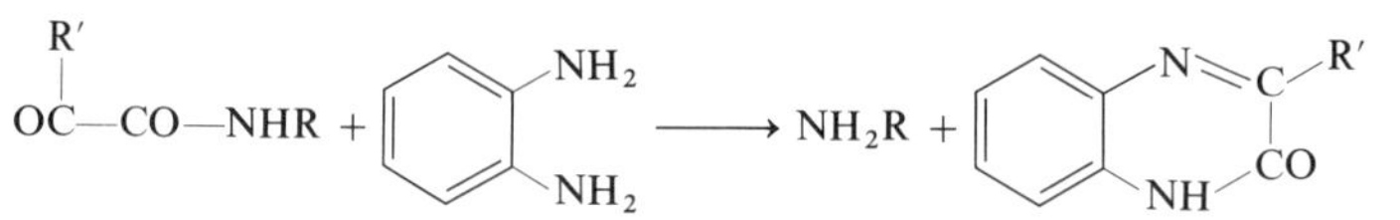
(2)

R = peptide chain

selectively at the α-oxo acylpeptide or protein stage. The transamination reaction has been applied successfully for removal of the terminal isoleucine residue of trypsin (Webster and Offord, 1972), and to cytochrome *c-551* (Dixon and Moret, 1964), carboxypeptidase (Van Heyningen *et al.*, 1968), and insulin (Bünzli and Bosshard, 1971).

Of the chemical methods available for the cleavage of peptides and proteins at specific amino acid residues located within the peptide chain, by far the most efficient appears to be the cyanogen bromide cleavage at methionine (Gross and Witkop, 1961, 1962). It has a firm reputation in protein sequencing studies and is now proving of special value in partial synthesis.

The reaction involves formation of an initial sulfonium salt adduct between the bivalent sulfur atom and cyanogen bromide [Eq. (3)], and nucleophilic displacement of this by the neighboring peptide carbonyl group [Eq. (4)]. The resulting lactone is rapidly hydrolyzed in aqueous acid with cleavage of the peptide chain [Eq. (5)]. Other electrophiles can be used in place of cyanogen bromide, but this reagent has the advantage of volatility, as does the coproduct, methyl thiocyanate. The reaction is usually carried out in 0.1 *N* HCl or 70% aqueous formic acid, preventing reaction of cyanogen bromide with amino groups. The formic acid conditions also presumably ensure unfolding of the protein with exposure of all the methionine residues. Yields are generally good, but examples have been noted of incomplete cleavage of the peptide chain when methionine is followed by either serine or threonine (Schroeder *et al.*, 1969). In these cases the β-hydroxy group

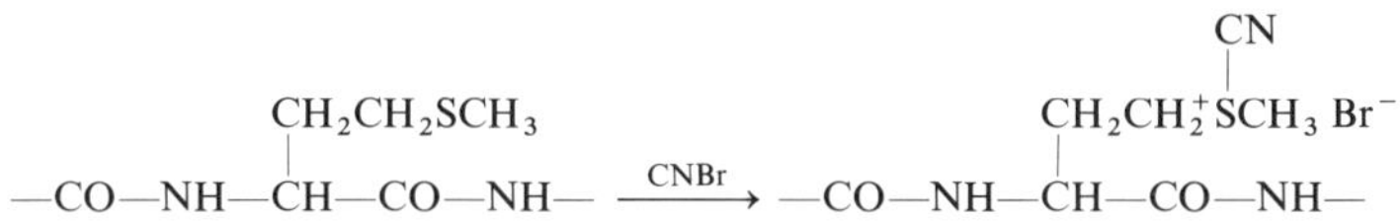
(3)

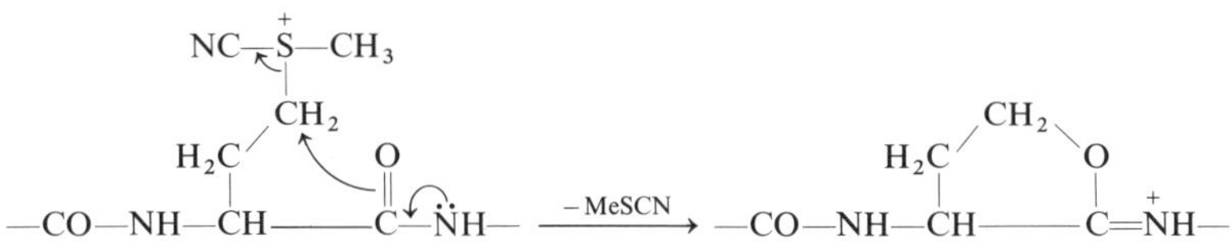
(4)

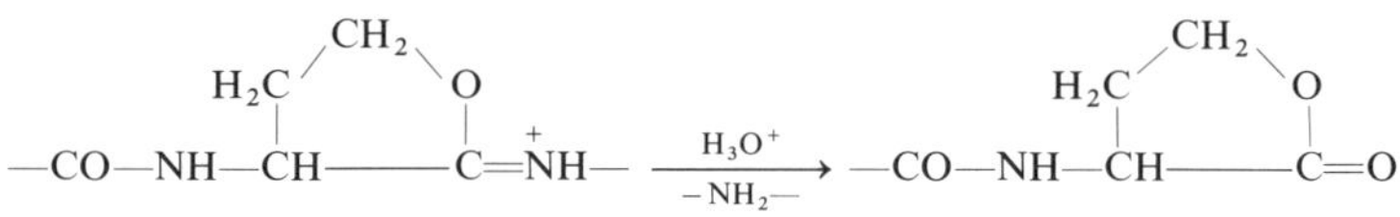
(5)

presumably adds to the iminolactone intermediate giving the oxazoline which hydrolyzes without chain cleavage [Eq. (6)].

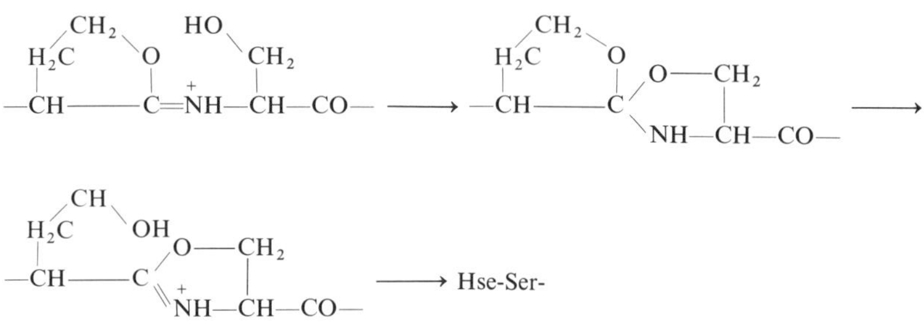
(6)

Methionine is a relatively infrequent amino acid in proteins, and the cyanogen bromide cleavage reaction is therefore well suited to the preparation of large fragments. Cleavage may be limited to individual methionine residues under particular circumstances. Thus, only one of the two residues in chymotrypsin is attacked by alkylating agents under nondenaturing (pH 3) conditions (Koshland *et al.*, 1962), presumably due to steric inaccessibility. Likewise, conditions have been established for limited cleavage of cytochrome *c* (Corradin and Harbury, 1970). In this case, the second methionine residue may be deactivated to attack by cyanogen bromide by coordination of its sulfur atom to heme iron. Methionine sulfoxide residues are unaffected by cyanogen bromide and selective oxidation provides a potentially complementary approach for limiting the cleavage reaction. This selective oxidation of a single methionine residue has been achieved in the case of chymotrypsin (Koshland *et al.*, 1962).

In the cyanogen bromide cleavage reaction the methionine residue is transformed into the lactone of COOH-terminal homoserine. Transformations of this type are common to cleavage reactions at other amino acids, e.g., tyrosine and tryptophan (Spande *et al.*, 1970). If the presence of homoserine is desired or can be tolerated in the partially synthetic analog, then formation of the lactone derivative is highly advantageous. It enables immediate differentiation from all other side-chain carboxyl groups in the peptide fragment and imparts a degree of activation which may be useful in resynthesis reactions (Section IV) without the need for side-chain carboxyl group protection. If the presence of homoserine cannot be tolerated, it can be removed (after alkaline hydrolysis of the lactone ring) by the action of carboxypeptidase A.

Partial synthesis studies have been carried out using cyanogen bromide-cleaved ribonuclease A (Offord, 1972), pancreatic trypsin inhibitor (Dyckes *et al.*, 1974, 1977, 1978), and cytochrome *c* (Corradin and Harbury, 1974; Barstow *et al.*, 1977) and are discussed in Sections IV and V.

Chemical methods for the cleavage of peptides and proteins at tryptophan residues have been reviewed by Savige and Fontana (1977) and Fontana (1972). The reaction is oxidative, employing reagents such as periodate and ozone, and especially brominating agents. Favored reagents are BNPS-skatole (**1**) (Fontana, 1972) and the combination of hydrogen bromide–

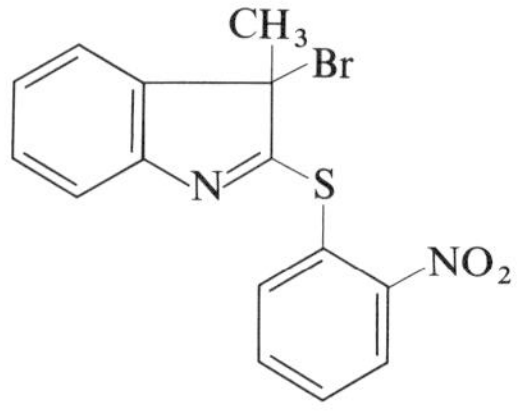

(**1**)

dimethyl sulfoxide (Savige and Fontana, 1977). With both reagents, the tryptophan residue is converted into COOH-terminal oxindolylalanine spirolactone. A number of pathways for the overall reaction can be envisaged; Savige and Fontana favor initial oxidation of the tryptophan residue to oxindolylalanine followed by bromination and lactonization (Scheme 3) for the dimethyl sulfoxide (DMSO) reaction.

Applications to protein cleavage include staphylococcal nuclease (Omenn *et al.*, 1970), A_1-encephalitogenic protein (Bergstrand, 1971; Burnett and Eylor, 1971), cytochrome c (Savige and Fontana, 1977), and myoglobin (Wang *et al.*, 1978). Cleavage yields are bout 60%. Methionine residues are simultaneously converted to methionine sulfoxide; addition of phenol (or tyrosine) to the reaction mixture serves to minimize bromination of tyrosine residues. In cytochrome *c*, tryptophan cleavage was accompanied by partial (BNPS-skatole) or complete (DMSO–HBr) ejection of the heme moiety.

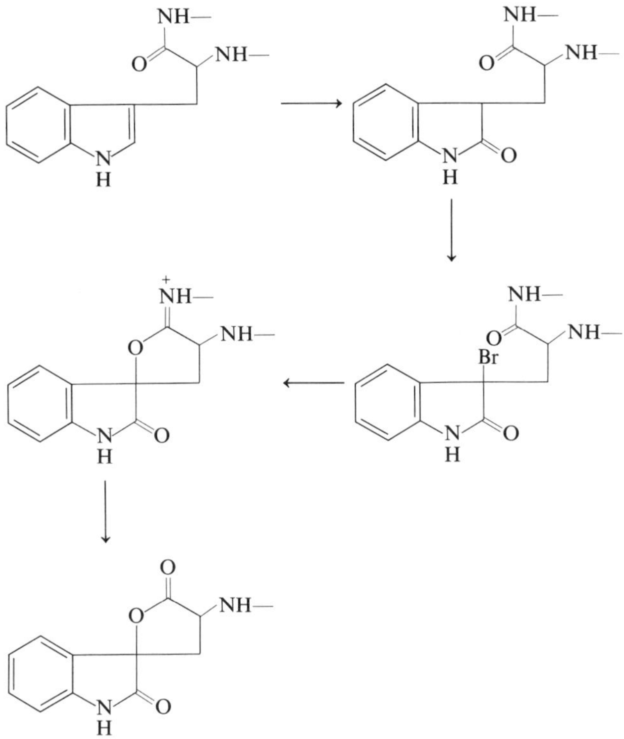

Scheme 3

III. PROTECTION OF FUNCTIONAL GROUPS

A. Terminal and Side-Chain Amino Groups

Of the various functional groups present in proteins and their peptide fragments, free amino groups are the most likely to require protection before unambiguous peptide bond formation can be achieved. Only in those cases mentioned above where specific resynthesis of the peptide chain is sterically directed by the tertiary structure, or where the nature of the protein allows the use of specific enzymatic methods (as in trypsin inhibitors, Section V,E) can amino protection be dispensed with. Generally, various acyl derivatives analogous to those employed in conventional peptide synthesis have been used for amino group protection. Additionally, derivatives of dibasic olefinic acids (maleic, citraconic, and tetrahydrophthalic) previously used in protein structure studies are finding application in partial synthesis. These confer additional negative charge on the protected peptide. Groups which retain the positive charge associated with the protonated amine function (and hence may cause minimal disruption of the secondary structure) are acetimido (from methyl acetimidate) and amidino (from isoureas). The latter are not considered here since the resulting guanidines are not easily revertible to the parent amines. Likewise, phenylthiocarbamoyl derivatives, which have featured in some partial synthesis schemes, are not considered separately here as protected amines, but rather as intermediates in chemical degradation procedures (Section III,B). All the protecting groups considered, when applied to the ε-amino group of lysine, confer resistance to proteolysis by trypsin.

At present it seems that *tert*-butyloxycarbonyl, methylsulfonylethyloxycarbonyl, and acetimidyl derivatives are the most useful in partial synthesis.

1. *tert-Butyloxycarbonyl Derivatives*

This urethane-type protecting group has found extensive application in partial synthesis (Levy and Carpenter, 1967; Geiger *et al.*, 1971; Levy, 1973b; Krail *et al.*, 1975; Rees and Offord, 1976a), notably in the insulin series. It has been universally introduced by means of *tert*-butyloxycarbonyl azide (**2**), but this reagent is now considered to be dangerously explosive (Anonymous, 1976). New alternative reagents are the anhydride (**3**) (Moroder *et al.*,1976) and the oximino derivative (**4**) (Itoh *et al.*, 1975). The group is easily incorporated into peptides and proteins under mildly basic conditions (sodium

$Me_3C{-}O{-}CO{-}N_3$ (**2**)

$Me_3C{-}O{-}CO{-}O{-}CO{-}O{-}CMe_3$ (**3**)

$Me_3C{-}O{-}CO{-}O{-}N{=}C(CN)Ph$ (**4**)

hydrogen carbonate, triethylamine, or *N*-ethylmorpholine), usually in an aqueous or anhydrous polar organic solvent such as dimethylformamide or dimethyl sulfoxide. The azide reagent appears to be highly selective for amino groups in complex molecules, and under appropriate conditions, some selectivity for individual α- and ε-amino functions may be obtained (Section III,A,8). The protecting group is stable to the normal manipulative operations at neutral or moderately alkaline pH but is easily removed by acidolysis, e.g., by trifluoroacetic acid. *tert*-Butyloxycarbonyl groups attached to lysine side chains are appreciably more labile to acid than α-amino derivatives. This acid lability precludes use of *tert*-butyloxycarbonyl derivatives in combination with the Edman procedure (except for just one cycle of the latter) and probably also for cyanogen bromide degradation at methionine residues and other processes carried out at appreciably acid pH.

Mention has been made of the use of the more acid-stable benzyloxycarbonyl derivatives (Offord, 1969; Rees and Offord, 1976a) in partial synthesis.

2. Maleyl, Citraconyl, and Related Acyl Derivatives

Maleyl (**5**) (Butler *et al.*, 1969) and citraconyl (2-methylmaleyl) (**6**) (Dixon and Perham, 1968) derivatives were introduced into protein chemistry as an aid to structural studies but have found substantial application in partial

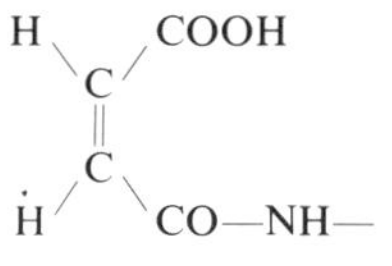

(**5**)

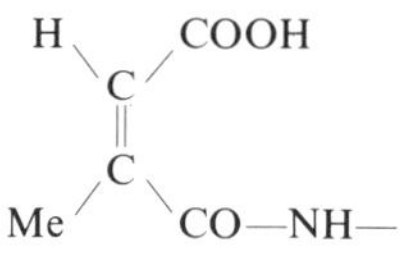

(**6**)

synthesis. At neutral pH and above, they impart additional negative charge to acylated peptides which may facilitate ion-exchange purification (see also di Marchi *et al.*, 1978, for use of tetrahydrophthalyl derivatives for this purpose). Both groups are introduced by reaction of the cyclic anhydrides with the peptide or protein at a low temperature and alkaline pH (Rees and Offord, 1976a). Denaturation by 8 *M* urea may be required for complete acylation of all amino groups (Borrás-Cuesta *et al.*, 1978). Large excesses of the cyclic anhydrides and elevated temperatures are to be avoided as there appears to be some danger of reaction with other amino acid side chains, notably the hydroxyl groups of serine and threonine (Gibbons and Perham, 1970; Rees and Offord, 1976a). Both groups are removed within a few hours in aqueous solution at pH 3.5. Citraconyl derivatives have been considered too labile for easy manipulation (Rees and Offord, 1976a), but this lability may be advantageous when dealing with very sensitive proteins (Habeeb and Atassi, 1970). The maleyl group is stated to be completely resistant to long exposure to liquid hydrogen fluoride and trifluoroacetic acid (Rees and Offord, 1976a), but some loss of citraconyl groups has been observed under the acidic conditions of the Edman degradation (Snell *et al.*, 1978).

Protection of amino groups using maleic and other dibasic acid anhydrides necessarily introduces additional carboxyl groups into the molecule which may themselves require protection. (see discussion by Rees and Offord, 1976a). Esterification of these carboxyl groups serves also to stabilize the amine protecting group.

3. Trifluoroacetyl Derivatives

This base-labile protecting group has found useful application in the insulin series (Borrás and Offord, 1970; Geiger and Obermeier, 1973; Friesen *et al.*, 1978). The thioester (**7**) has been used for the introduction of trifluoroacetyl groups into cytochrome *c* (Fanger and Harbury, 1965) but is inapplicable to disulfide-linked proteins because of disulfide interchange initiated by the liberated thiol. However, the simple phenyl ester **8** (Borrás and Offord, 1970) and even the ethyl ester **9** (Friesen, *et al.*, 1978) have sufficient reactivity. The latter reacts with insulin some three times faster at the A-chain terminus than does *tert*-butyloxycarbonyl azide, and provides a practical route to the monotrifluoroacetyl derivative. Preparation of the analogous B-chain monotrifluoroacetyl derivative has been mentioned

CF_3—COO—SEt	CF_3—COO—Ph	CF_3—COO—Et
(**7**)	(**8**)	(**9**)

(Geiger and Obermeier, 1973). Cleavage of the trifluoroacetyl derivative is achieved by transamidation with dilute ammonia, which is less destructive to insulin at least, than the traditional piperidine treatment (Goldberg and Anfinsen, 1962). The group is stable to strong acids and is compatible with the Edman degradation procedures. The methylsulfonylethyloxycarbonyl group (*q. v.*) has similar characteristics and may be preferred because of its ease of removal.

4. Phthaloyl Derivatives

This protecting group has also been applied successfully in the insulin series. The preferred reagent is *N*-carbethyloxyphthalimide (**10**) which reacts with insulin in dimethylformamide–sodium hydrogen carbonate solution to yield the N^{α}A1,N^{ε}B29 bisphthaloyl derivative (Geiger and Langer, 1975), analogous to that formed with *tert*-butyloxycarbonyl azide and other reagents. The protected derivative is stable to acids and can be subjected to repeated Edman degradation. The phthaloyl groups are removed by hydrazinolysis in 80% phenol (Geiger and Langer, 1975); in the absence of phenol the insulin in destroyed (Geiger and Langer, 1975; Borrás and Offord, 1970).

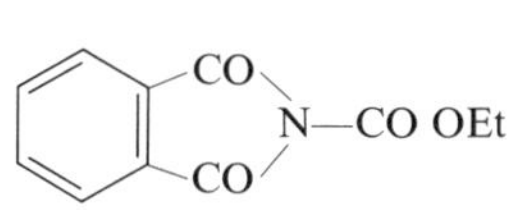

(**10**)

It is likely that difficulties of removal, sensitivity of the phthaloyl ring to opening by aqueous alkali (which inhibits subsequent hydrazinolysis), and insolubility of derivatives will limit further application of this protecting group.

5. Methylsulfonylethyloxycarbonyl Derivatives

This protecting group has grown rapidly in popularity in recent years because of the excellent solubility characteristics of its derivatives, its stability to acids, and its ease of removal by alkali (Tesser and Balvert-Geers, 1975). A number of reagents may be used for introduction of the group, but the *N*-hydroxysuccinimide ester (**11**) seems likely to be the most generally

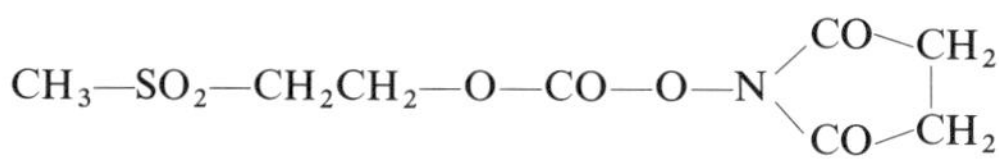

(**11**)

useful. All five amino groups in ACTH(1–24) were readily blocked to give a water-soluble product from which the starting ACTH analog could be recovered with unimpaired lipotropic activity (Boon *et al.*, 1978). Crystallizable, biologically active insulin has also been recovered after conversion to the A1, B29-bis(methylsulfonylethyloxycarbonyl) derivative and deprotection (Geiger and Obermeier, 1975). Cleavage of the protecting group is by brief (15 sec) treatment with dilute alkali. The group is stable to the acidic conditions of both the Edman (Geiger *et al.*, 1975) and cyanogen bromide degradations (Boon *et al.*, 1978) and can be used in combination with *tert*-butyloxycarbonyl protection.

6. Acetimidyl Derivatives

These are unique among reversibly protected amines in that the original basic character of the parent group is retained (or enhanced). The native and protected proteins therefore carry the same distribution of charge at normal pHs, and minimum disruption of tertiary structure, stability, and solubility may be anticipated. This has been amply confirmed in the case of several enzymes [phospholipase A_2 (Slotboom and de Haas, 1975), carboxypeptidase B (Fölsch, 1978), ribonuclease A (Reynolds, 1968)] which give water-soluble, biologically active acetimidylated (amidine) derivatives. The group is introduced by the action of a large excess of methyl acetimidate hydrochloride (**12**) in aqueous alkaline solution (Hunter and Ludwig, 1962) and is removed by alkali, usually ammonium hydroxide (Ludwig and Byrne, 1962). Several proteins have been shown to be stable to the removal conditions

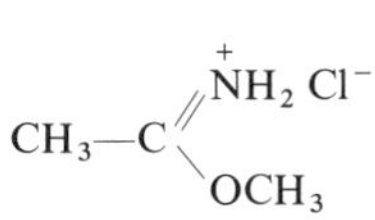

(**12**)

(Ludwig and Byrne, 1961; Harris and Offord, 1977). In other studies, removal has not been considered necessary since the protected and unprotected forms have such similar structures. The group is stable to acids and has been used in conjunction with both the Edman (Slotboom and de Haas, 1975) and the tryptophan (Wang *et al.*, 1978) degradations. Acetimidination of ε-amino groups appears to convey complete resistance to the proteolytic action of trypsin at modified lysine residues (Hunter and Ludwig, 1962).

It should be noted that the acetimido derivatives are only partially protected against acylation, and that as in the case of the guanidino groups of arginine (and homoarginine) residues, side-chain acylation is still possible, especially at elevated pH.

7. Other Amine Protecting Groups

A number of novel protecting groups have been designed to aid in the solution of specific synthetic problems in protein chemistry. In several cases these have made use of established methods of protein degradation for their removal. Good examples are provided by reversible cross-linking agents which have been used successfully to join the A and B chains of insulin. Such groups may have wider application in partial synthesis. Examples are provided by the bis(*tert*-butyloxycarbonylamino)suberyl cross-link, introduced into the A1,B29 amino groups of insulin by the activated ester reagent **13** (Brandenburg *et al.*, 1973b). Removal of the cross-link is achieved by normal acidolysis of the *tert*butyloxycarbonyl groups and one cycle of Edman degradation. Another cross-linking reagent is the carbonyl(bismethionine) activated ester (**14**) (Busse and Carpenter, 1974). This reacts at the same two accessible amino groups of insulin giving the A1,B29-carbonyl(bismethionyl) derivative and is cleaved by a single cyanogen bromide degradation. A cross-linking reagent of different type is bis[2-(succinimidooxycarbonyloxy)ethyl] sulfone (**15**) (Obermeier and Geiger, 1975). This is a bifunctional variant of the methylsulfonylethyloxycardonyl (Msc) protecting group and is cleaved by mildly alkaline treatment.

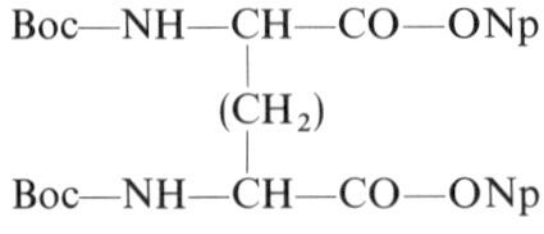

(**13**)

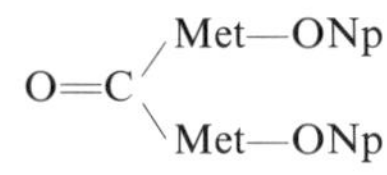

(**14**)

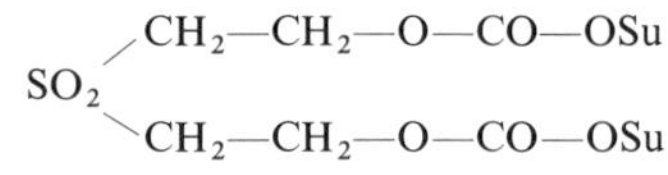

(**15**)

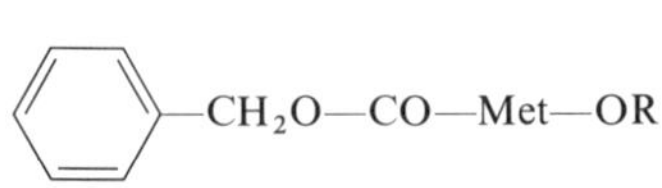

(**16**)

EtO—CO—Met—OR

(**17**)

A range of simple amine protecting groups based on methionine analogous to **14** have been investigated (Saunders and Offord, 1977a,b). These are mostly simple acyl or urethane derivatives of which the benzyloxycarbonyl (**16**) and ethyloxycarbonyl (**17**) activated ester reagents are preferred. Derivatives of the former show slight instability under acidic conditions, but ethyloxycarbonylmethionyl derivatives are completely stable and have been carried through five cycles of Edman degradation without appreciable loss. Removal of the protecting group is by the normal cyanogen bromide reaction.

8. Selectivity in the Introduction of Amine Protecting Groups

Specificity in the partial masking of amino groups in peptides and proteins might be anticipated on two grounds: (i) selectivity due to environmental differences of individual amine functions leading to varying degrees of steric accessibility and (ii) differences in nucleophilic reactivity associated with differing base strengths. The former will clearly depend on the tertiary structure of the polyamine compound as well as on the denaturing or other character of the reaction medium, reagent structure, etc. It cannot easily be discussed in the general case but is clearly to be seen, for example, in the preferential formation of A1,B29 diacyl derivatives in the insulin series. Base strength differences of individual amino groups will also depend on tertiary structure factors, particularly on the polar or nonpolar nature of the immediate environment, the presence of nearby charged groups, etc.; but these may be of relatively minor importance compared to the intrinsic base strength difference of α-amino ($pK_a \sim 8$) and side chain, ε-amino groups ($pK_a \sim 10$). Because these pK_a differences are large, it has sometimes been assumed that selectivity in acylation should be easily achieved merely by appropriate choice of the pH of the reaction medium. Thus, at pH 7 the concentration of free (unprotonated), and hence reactive, α-amine would be 10%, that of free ε-amine only 0.1%. However, this 100-fold equilibrium concentration difference is not necessarily reflected in overall reaction rates for acylation. If the relative nuleophilicity for protons (i.e., basicity) correlates exactly with nucleophilicity for other electrophiles (i.e., Bronsted β-value of unity), then the differences in relative concentration would be completely counterbalanced by the 100-fold increase in kinetic reactivity of the ε-amino group toward acylating reagents. There would thus be no selectivity of the individual amino groups toward the particular electrophile. In general, β is less than one for reactions of representative amines with various acylating agents [e.g., for *p*-nitrophenyl acetate, $\beta = 0.88$ (Fersht, 1977)], giving slight kinetic advantage in favor of α-amine acetylation. Other reagents may have more favorable β-values, but selectivity is clearly not to be assumed just from simple pK_a considerations. It will be affected not only by the nature of the particular acylating agent and nucleophile, but also by the presence of any solvent or cosolvent other than water that may markedly diminish the effective base strength differences between side chain and terminal amino groups.

The need for caution and experimental trial has been emphasized in some specific studies relating to selectivity in the β-melanotropin (Lande, 1970, 1971) and insulin (Friesen *et al.*, 1978) series. The latter authors concluded that the most important factor determining the relative rates for attack at each of the three amino groups, A1 (α), B1 (α), and B29 (ε), is the nature of the acylating reagent. Trifluoroacetylation by ethyl trifluoroacetate gave very largely the A1-monotrifluoroacetyl and then the A1,B1-diacylated derivatives. With *tert*-butyloxycarbonyl azide, the monoacylated product consisted of a mixture of A1(α) and B29(ε) derivatives (ratio 70:30), while the diacylated product was largely the A1,B29 form. Side-chain acylation was thus observed only with the second reagent. Interesting differences were also noted between *p*-nitrophenyl and hydroxysuccinimidyl active ester reagents, the former favouring α-substitution at the A-chain terminus, the latter favouring ε-substitution at lysine B29 in monosubstituted products.

The importance of both peptide and reagent structure in determining α:ε ratios was similarly noted by Lande (1970, 1971). Marked differences were noted in selectivity between *tert*-butyloxycarbonyl azide and a tetrapeptide azide derivative. With the former reagent, conditions were established for N^{α} (pH 7) or N^{ε}-acylation (pyridine–water–triethylamine).

B. Terminal and Side-Chain Carboxyl Groups

Protection of side-chain carboxyl groups of aspartic and glutamic acid residues is generally required only if the natural fragment is to function as the carboxyl component in condensation reactions. When it is to function as the amine component, protection of side-chain and terminal carboxyl groups may not be necessary and may even be disadvantageous by virtue of diminished solubility and later difficulties in deprotection reactions. Even when the naturally derived fragment is to function as carboxyl component, selective activation of the α-carboxyl group may provide an alternative pathway in coupling reactions. This may be achieved enzymatically (Laskowski, 1978) or chemically, particularly as a concomitant of specific chemical cleavage of the polypeptide chain (Dyckes *et al.*, 1977; Harbury, 1978).

Carboxyl groups have been protected invariably as alkyl or aralkyl esters. Methyl esters may be obtained by simple acid-catalyzed esterification, as in the preparation of insulin hexamethyl ester (Mammaerts and Neurath, 1950) using methanolic hydrogen chloride. There is risk of methanolysis of peptide and side chain amide bonds and N $\rightarrow$ O acyl shift at hydroxyl-amino acid residues if the treatment with acid methanol is unduly prolonged. Alternatively, a Lewis acid catalyst such as boron trifluoride may be used (Gattner *et al.*, 1975, and references there cited). This allowed preparation of insulin hexamethyl ester under mild reaction conditions.

Other methylating reagents are diazomethane (Chibnall *et al.*, 1958) and Meerwein's trialkyloxonium tetrafluoroborates. The former requires careful control of reaction conditions for minimization of side reactions. Triethyloxonium tetrafluoroborate may react only sluggishly with protein carboxy groups (Levy, 1973a) even in the presence of denaturing agents (Fölsch, 1974). It is also capable of reacting with a range of other chemical groupings including the phenolic ring of tyrosine, the imidazole of histidine, and under certain conditions even with amide and peptide bonds.

Despite the relative ease of preparation of these simple alkyl esters, they are of limited value in partial synthesis studies. Their removal requires strongly alkaline conditions which have been known for many years to cause rearrangement of aspartyl and glutamyl derivatives in model peptides (Battersby and Robinson, 1955). This rearrangement has been encountered as a serious side reaction in the insulin series, the COOH-terminal asparagine methyl ester being substantially converted to the succinimide on attempted saponification; the latter then hydrolyzes more slowly to a mixture of asparagine and isoasparagine derivatives (Scheme 4) (Obermeier, 1978; Gattner and Schmitt, 1977).

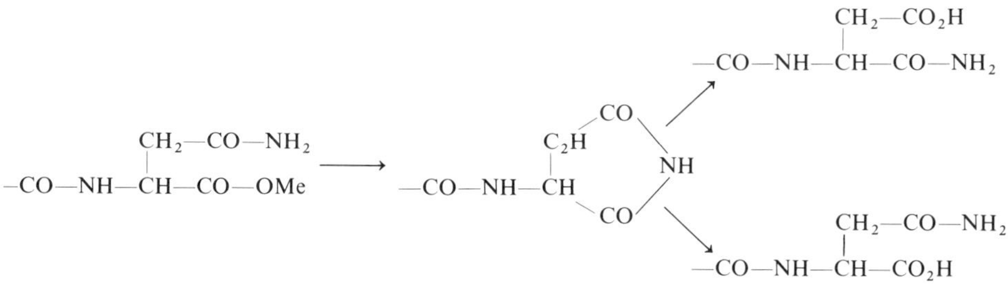

Scheme 4

The use of simple alkyl esters in partial synthesis studies on other proteins will require caution. Because of the dangers attendant on alkaline hydrolysis of methyl and other esters, attention has been devoted to esters cleavable by acidic, nonhydrolytic conditions.

tert-Butyl esters are generally unsuitable because of difficulties in their preparation. Insulin, for example, could not be esterified using *tert*-butyl acetate in the presence of perchloric acid (Obermeier, 1978). Offord and his colleagues (Hayward and Offord, 1971; Rees and Offord, 1976a; Offord *et al.*, 1976) noted that although esterification using diazoalkanes could not be used for the preparation of *tert*-butyl esters, this reaction could be applied in an ingenious way to the preparation of analogous *tert*-amyl esters. Thus it is known that diazo compounds derived from primary aliphatic amines decompose to primary carbonium ions which largely rearrange to the more stable tertiary isomers before further reaction. The diazo compound derived from neopentylamine should therefore give the tertiary carbonium ion **18** as the principle reactive species, capture of which by a carboxylic acid would yield the tertiary amyl ester (Scheme 5). This reaction has been applied to a number of carboxyl-containing peptides. Carbonium ion reagents tend to be unspecific in reactivity and some degree of attack at other nucleophilic centers must be anticipated. The sulfur-containing amino acids, tyrosine, tryptophan, and particularly histidine (*q.v.*), are the most likely residues to be affected.

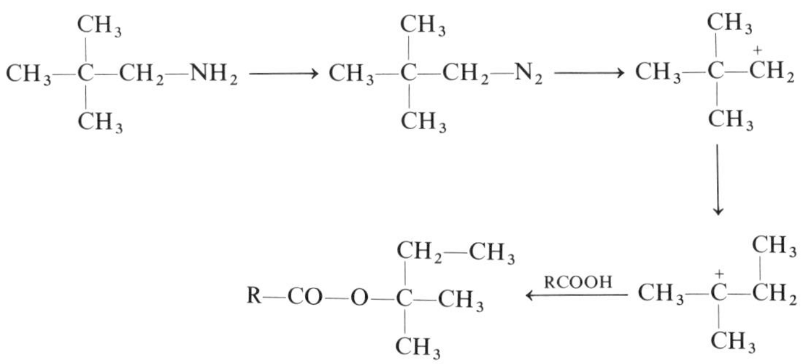

Scheme 5 (**18**)

The terminal *tert*-amyl esters are susceptible to cleavage by trypsin under special conditions. This provides a particularly useful differentiation between side-chain and terminal carboxyl groups for specific protection and activation reactions of tryptic peptides. For general deprotection of *tert*-amyl esters, acidolysis by trifluoroacetic acid at 4°C for 40 min is effective (Rees and Offord, 1976a).

Benzyl esters are also cleaved by anhydrous acids and can be introduced using phenyldiazomethane (Offord, 1969; Offord *et al.*, 1976). Evidence for substantial reaction of this diazo compound with the imidazole ring of histidine residues has been obtained (Offord *et al.*, 1976; Fölsch, 1975), and this amino acid therefore requires side-chain protection. Complete esterification of complex proteins may be difficult to obtain, and the degree of esterification is dependent on the reaction medium Fölsch 1975). The rather vigorous reaction conditions necessary for cleavage of benzyl esters (hydrogen bromide in acetic acid or liquid hydrogen fluoride) are much reduced with the more labile *p*-methoxy derivatives (Offord *et al.*, 1976). These also are accessible by way of the corresponding diazo compound and are cleaved by trifluoroacetic acid somewhat more slowly than *tert*-amyl esters (40–300 *min at* 4°*C*).

Acid-labile diphenylmethyl esters are also of potential value and should be accessible using diphenyldiazomethane. However, only partial esterification of insulin was obtained using this reagent (Obermeier, 1978).

C. Side-Chain Functional Groups

Of the polyfunctional amino acids present in proteins, the sulfur-containing amino acids together with tyrosine, tryptophan, and histidine contain, additional to amino and carboxyl groups, reactive groupings which are likely to interfere in operations of partial synthesis. Involvement of these functional groups can often be minimized through careful choice of reaction conditions, and it is likely that the phenolic group of tyrosine and the indole ring of tryptophan residues may be left unprotected for many operations. Reactions involving large excesses of carbonium ion-forming reagents may be exceptions. Protection of the indole ring of tryptophan by N-formylation was investigated by Rees and Offord (1976a) who concluded that formylation did not show any clear advantages. This and similar conclusions require substantiation in each individual case, but in general, optimization of reagent selectivity is likely to be preferable to attempted widespread introduction of protecting groups. Some functions are, however, too reactive to be left unprotected, and it is probable that the thiol and imidazole groups of cysteine and histidine come into this category.

Cysteine residues occur most frequently as disulfides, and these appear to be adequately stable for many operations. The stability of the natural cysteine pairings in proteins may not be retained in modified derivatives, and extremes of pH and thiol reagents likely to promote disulfide interchange are to be avoided. Much of the partially synthetic work in the insulin series (*q.v.*) has been carried out with intact disulfide bonds, even though these may not represent particularly favored arrangements.

Thiol groups have been protected as *S*-sulfonates in unfolded lysozyme (Rees and Offord, 1976a) and in insulin A-chain (Brandenburg *et al.*, 1975; Weinert *et al.*, 1969). The sulfonate group imparts additional negative charge to the peptide or protein and may aid solubility, which is often poor in reduced thiol-containing peptides. Stability of the *S*-sulfonate group to a variety of partial synthesis reaction conditions has been studied by Rees and Offord (1976a). Anhydrous trifluoroacetic acid (2 hr), anhydrous hydrazine (30 min), and 0.05 *M* ammonia (18 hr) were without effect on a sulfonated test peptide, but the protecting group was rapidly cleaved by mercaptoethanol. *S*-Sulfonated insulin A-chain has been subjected both to Edman degradation (Brandenburg *et al.*, 1975) and to acylation reactions (Weinert *et al.*, 1969). This protecting group would therefore seem to fulfill many of the requirements for cysteine (and cystine) protection.

A potential alternative to *S*-sulfonation is the reaction with ethyl isocyanate to yield *S*-ethylcarbamoyl derivatives. Reaction of the isocyanate with thiol groups is relatively specific (over free amino groups) at low pH, and the protected derivatives are stable in neutral or acidic media. They are rapidly cleaved at pH 9 in the presence of free thiols (e.g., dithiothreitol). Temporary protection of cysteine residues as their *S*-ethylcarbamoyl derivatives proved valuable during purification of a partially synthetic analogue of pancreatic trypsin inhibitor (J. F. Mercer and R. C. Sheppard, unpublished).

Protection of the imidazole ring of histidine residues is not completely solved in conventional synthetic procedures. The free imidazole ring must be expected to participate in both acylation and alkylation (carbonium ion-mediated esterification) procedures, although the former is reversible and may not cause serious permanent damage. Alkylation during diazoalkane esterification has been observed (Offord, 1969; Rees and Offord, 1976a; Fölsch, 1974). In the case of carbonic anhydrase B, esterfication with phenyldiazomethane benzylated the equivalent of 2 of the 11 histidine residues present in the enzyme. Removal of im-benzyl groups requires vigorous reaction conditions, and this side reaction is to be avoided.

Fölsch (1975) has investigated histidine protection by the dinitrophenyl group. Reaction of amine protected carbonic anhydrase with 2,4-dinitrofluorobenzene masked not only the histidine residues but also tyrosine. Since both im- and *O*-dinitrophenyl groups are in principle cleavable by thiolysis, this reagent was regarded as a potential protecting agent for both amino acids. Subsequently, however, Fölsch (1978) reported difficulty in removal of the dinitrophenyl groups from histidine residues. Extensive dinitrophenylation also conferred very unfavorable solubility properties on the protein.

Rees and Offord (1976a) have investigated use of the Weygand-type protecting groups (**19** and **20**) for the histidine imidazole group. Both protected

Z—NH—CH(CF_3)—	Boc—NH—CH(CF_3)—
(**19**)	(**20**)

derivatives were stable under all normal reaction conditions and were cleavable by acids under the conditions expected for benzyloxycarbonyl (liquid HF/0°C/1 hr) and *tert*-butyloxycarbonyl derivatives (trifluoroacetic acid/4°C/40–60 min). The latter would seem to fulfill the requirements for efficient protection, but additional experience on its ease and completeness of introduction into complex histidine-containing peptides is needed.

IV. ACTIVATION AND COUPLING REACTIONS

A. Addition of Single Residues to the NH_2-Termini of Natural Peptides and Proteins

The simplicity, mildness, and general cleanliness of active ester condensations is clearly to be preferred for the stepwise addition of single amino acid residues at the NH_2-termini of N^{ε}-protected proteins and their fragments. This is in line with general experience in peptide synthesis. Freedom from racemization requires that the activated ester be derived from a limited range of N-protected derivatives, notably urethanes (commonly *tert*-butyloxycarbonyl, but possibly also methylsulfonylethyloxycarbonyl and other urethanes), and not from simple acyl derivatives such as trifluoroacetylamino acids (other than glycine and proline). The phenolic or hydroxylic components used include *p*-nitrophenol and 2,4,5-trichlorophenol (Weinert *et al.*, 1969; Krail *et al.*, 1975; Geiger *et al.*, 1975), pentachlorophenol (Shimonishi, 1970), and *N*-hydroxysuccinimide (Garner and Gurd, 1975; Slotboom and de Haas, 1975; Shimonishi, 1970; Krail *et al.*, 1975; di Marchi *et al.*, 1978). Esters of the last-named alcohol have been most popular, probably because of their solubility characteristics and suitability for acylation reactions in both aqueous and organic media. Amino acids incorporated using the activated ester method include the bis(*tert*-butyloxycarbonyl) derivatives of lysine and histidine (Weinert *et al.*, 1969) and *tert*-butyloxycarbonyl-N^{i}-formyltryptophan (Wang *et al.*, 1978). Accessibility of the amine component's amino group is of paramount importance, and denaturing conditions may be required for efficient acylation of native proteins or of derivatives in which the tertiary structure is retained. The trifluoroacetylglycine *N*-hydroxysuccinimide ester reacted efficiently (90%) with acetimidylated myoglobin but not at all with desvaline-1-myoglobin. Coupling with the latter was achieved (70%) in the presence of 8 *M* urea (Wang *et al.*, 1978). In cases where racemization is not a serious risk, activated esters have also been used in peptide couplings. Simple examples include the hydroxysuccinimide ester of *tert*-amyloxycarbonylprolylproline coupled to an elongated insulin A-chain derivative (Shimonishi, 1970) and the *p*-nitrophenyl ester of *tert*-butyloxycarbonylarginylglycine to desglycine A-chain (Weinert *et al.*, 1969). The latter coupling was slow and incomplete. A better route to the arginyl derivative of insulin A-chain involved use of the *N*-carboxy anhydride of arginine as the acylating species (Weinert *et al.*, 1971). *tert*-Butyloxycarbonylamino acid phenylhydrazides (prepared enzymatically using papain) have also been used for single amino acid addition to insulin, the phenylhydrazides being oxidized to phenyldiimides in a preliminary step.

B. Activation and Coupling of Peptide Segments

The coupling of large segments in partial synthesis is more difficult than addition of single residues. All the chemical investigations noted have apparently used either the azide method or a variant of the carbodiimide procedure. In a relatively early investigation, Burton and Lande (1970) coupled the azide, Boc-Ala-Glu(O*t*Bu)-Lys(Boc)-Lys(Boc)-N_3, with porcine β melanotropin in aqueous pyridine. The yield of purified N^{α}-acylated product was only about 4%, but the β-melanotropin contained two unprotected N^{ε}-amino groups. More recently, Wang *et al.* (1978) have reported sequential addition of synthetic octa- and pentapeptide azides and one single amino acid active ester to an acetimidylated segment comprising residues 15–153 of myoglobin. The azides were prepared from fully protected peptide hydrazides by the technique of Honzl and Rudinger (1961). Coupling yields for the active ester condensation were near quantitative, but each peptide azide reacted only to the extent of 25%. The final yield of reconstructed myoglobin protein was 3–5%.

Peptide fragments obtained by tryptic cleavage of amine protected proteins will usually possess COOH-terminal arginine residues. Izumiya *et al.* (1971) have examined the problem of coupling at unprotected COOH-terminal arginine using the formation of the tetrapeptide derivative **21** from two synthetic protected dipeptides as a model system. Reaction mixtures were

$$\text{Z-Phe-Arg-OH} + \text{H-Leu-Asp(OBzl)-OBzl} \longrightarrow \underset{(\mathbf{21})}{\text{Z-Phe-Arg-Leu-Asp(OBzl)OBzl}}$$

examined by thin-layer chromatography and quantitated by ion-exchange separation after hydrogenolytic removal of protecting groups. The best yield of coupled tetrapeptide (66%) was obtained using water-soluble 1-ethyl-3-(3-diethylaminopropyl)carbodiimide as coupling agent in the presence of *N*-hydroxysuccinimide. Absence of *N*-hydroxysuccinimide led to a partially racemized product, and excess of both this reagent and the water-soluble carbodiimide lowered the yield. The mixed anhydride coupling method gave a poor yield of racemized product.

In a somewhat similar study using coupling of natural and synthetic segments related to lysozyme, Rees and Offord (1976b) reported that best results (68% yield) were obtained by prior activation of the carboxyl component using dicyclohexylcarbodiimide in the presence of 1-hydroxybenzotriazole. No racemization was detected. The yield was lower (56%) using the azide method as it was using *in situ* activation, with or without the addition of *N*-hydroxysuccinimide. Racemization was detected in the last two cases.

Dicyclohexylcarbodiimide in combination with *N*-hydroxysuccinimide or 1-hydroxybenzotriazole additives has found extensive application in the insulin series. Weitzel *et al.* (1976) have used dicyclohexylcarbodiimide–*N*-hydroxysuccinimide for the coupling of a range of segments to the COOH-terminus of deoctapeptide(B23–B30) and desnonapeptide(B22–B30) insulins. Terminal amino groups and side-chain carboxyl groups were protected as *tert*-butyloxycarbonyl and methyl ester derivatives, respectively. Since the cleavage of methyl esters presents unresolved problems in the insulin series (Section IV,A), Obermeier and Geiger (1976) and Gattner *et al.* (1978) have investigated similar condensations using carboxyl-unprotected insulin derivatives. The condensing agents were dicyclohexylcarbodiimide and 1-hydroxybenzotriazole. Some selectivity among the six free carboxyl groups is possible, but the yield of the desired purified insulin derivative coupled at the COOH-terminus was very low.

C. Peptide Combination Reactions Using Unprotected Peptide Segments

Mention has been made (see also Section V,B) of coupling reactions carried out in the β-melanotropin (Burton and Lande, 1970) and insulin (Obermeier and Geiger, 1976; Gattner *et al.*, 1978) series in which the naturally derived fragment was devoid of protecting groups. The coupling reactions were therefore ambiguous, and although some selectivity in reaction site may be involved, product mixtures were obtained that required extensive purification. The overall yield in such nonspecific procedures is always likely to be low, and detailed characterization of the product is mandatory. Nevertheless, the advantages of minimum protection, particularly of carboxyl groups, of naturally derived peptide and protein fragments warrant investigation of procedures of this type, if selectivity in coupling sites can be achieved. Such selectivity might be encouraged, e.g., by selective activation of single carboxyl groups (including enzymatic activation), by steric proximity of reacting groups within stable tertiary structures, and by use of massive excesses of one component in bimolecular coupling reactions.

Proteolytic enzymes are specific for derivatives of the α-carboxyl and amino groups of amino acid residues in peptides and proteins. Derivatives of side-chain carboxyl and amino groups are usually totally unaffected. Use of proteolytic enzymes in the reverse manner, i.e., for the synthesis rather than hydrolysis of peptide bonds, is therefore an attractive possibility, providing systems can be found with appropriate thermodynamic equilibria.

Such systems are frequent in nucleic acid chemistry but apparently rare in proteins, where peptide bond hydrolysis is usually greatly favored over resynthesis. The problem has been discussed by Laskowski (1978).

The only examples so far discovered in which equilibria are sufficiently favorable to allow efficient peptide bond synthesis by proteolytic enzymes appear to be the various enzyme inhibitors, notably pancreatic and soybean trypsin inhibitors. Enzymatic transformations of these inhibitors are discussed in Section V.

Nonenzymatic selective activation has been achieved concomitant with selective cleavage of the polypeptide chain. Thus, many of the methods suggested for chemical cleavage of peptides and proteins at specific amino acid residues (Spande *et al.*, 1970) yield fragments in which the particular residue has been modified. Important examples are the cyanogen bromide degradation at methionine (Gross and Witkop, 1961) and oxidative cleavage at tryptophan (Savige and Fontana, 1977). In both cases, peptide fragments are obtained as lactone derivatives (**22** and **23**) from methionine- and tryptophan- containing peptides. Lactone formation serves to distinguish the terminal carboxyl group from all side-chain carboxyl groups elsewhere in the molecule.

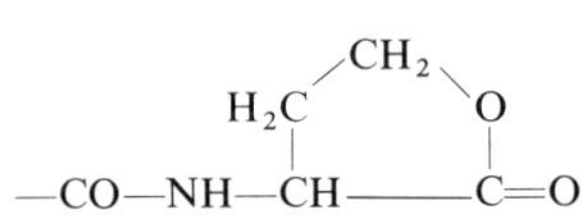

(22)

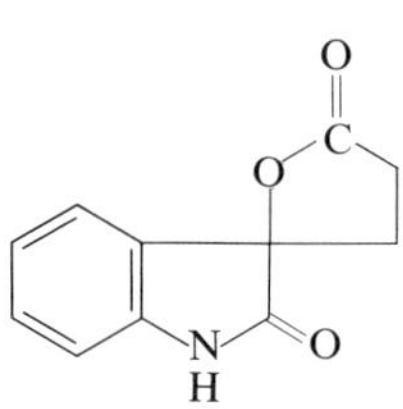

(23)

In the case of methionine-derived peptides, i.e., those containing COOH-terminal homoserine lactone (**22**), this specific labeling of the COOH-terminus has been used in a number of ways. The lactone ring is susceptible to hydrazinolysis so that a route is opened to the generation of carboxyl-unprotected peptides with a COOH-terminal azide group (**24**, Scheme 6) (Offord, 1972). Such azides should be capable of coupling to amine components in resynthesis reactions, but in practice it appears that relactonization of **24** to **22** is a major side reaction, and yields in peptide bond-forming reactions are poor. An alternative is to make direct use of **22** for resynthesis. The homoserine lactone ring in **22** is only weakly activated but is capable of reacting with amine components smoothly and in high yield under particularly favorable circumstances. Such circumstances are to be found in unimolecular combination reactions when the lactone ring and amine function are held in close proximity by the protein tertiary structure. Examples of successful protein resynthesis in which this factor is clearly operative are

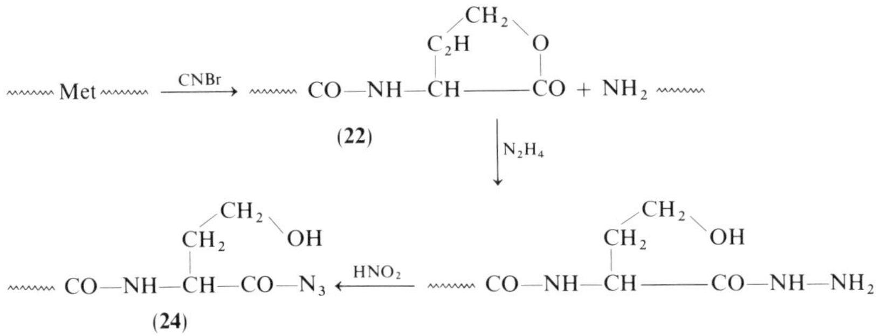

Scheme 6

pancreatic trypsin inhibitor (Dyckes *et al.*, 1974, 1978) and cytochrome *c* (Corradin and Harbury, 1974). Both are discussed in Section V. The cyanogen bromide-cleaved pancreatic trypsin inhibitor system has also been used as a model for bimolecular coupling reactions (Dyckes *et al.*, 1977). It was shown that the homoserine lactone ring was sufficiently reactive to combine in high yield with a model tripeptide ester under denaturing conditions (which slowed competitive intramolecular aminolysis), provided massive concentrations of the latter were used. Reaction of the lactone ring with unprotected side-chain and terminal amino groups in the protein component was thus effectively eliminated.

Of course procedures of these types will be of value only if substitution of homoserine for methionine can be tolerated in the partially synthetic product. No corresponding experiments have apparently been reported in the tryptophan or other series.

V. SELECTED EXAMPLES OF PARTIAL SYNTHESIS OPERATIONS ON NATURAL PEPTIDES AND PROTEINS

A. Insulin

Because of its clinical importance and ready availability in a crystalline, highly purified form, insulin has received by far the greatest study by methods of partial synthesis. Much of the experimental work has been concerned with selective acylation of the terminal (A1, B1) and side-chain (B29) amino groups, with shortening, with amino acid replacement, with lengthening at the NH_2-terminus of one of the insulin chains, or with studies of chain recombination reactions. Some aspects of these studies have been mentioned briefly earlier in this chapter, and in this section only selected examples of more substantial partial synthesis operations are discussed, generally involving substitution of whole peptide sequences rather than individual amino acid residues. For details of aspects not covered here, attention is drawn to a number of review articles dealing with insulin chemistry (Friesen *et al.*, 1977, 1978; Geiger, 1976, 1977; Geiger *et al.*, 1978; Brandenburg *et al.*, 1973a).

The first major partial synthesis in the insulin field, that of the conversion of porcine to human insulin, was reported briefly by Ruttenberg (1972). The hormones from the two species differ only at a single residue, the COOH-terminal alanine of the porcine B-chain being replaced by threonine. Ruttenberg's procedure is illustrated in Scheme 7; it involves protection of all six carboxyl groups by esterification with diazomethane, cleavage of the hexa-

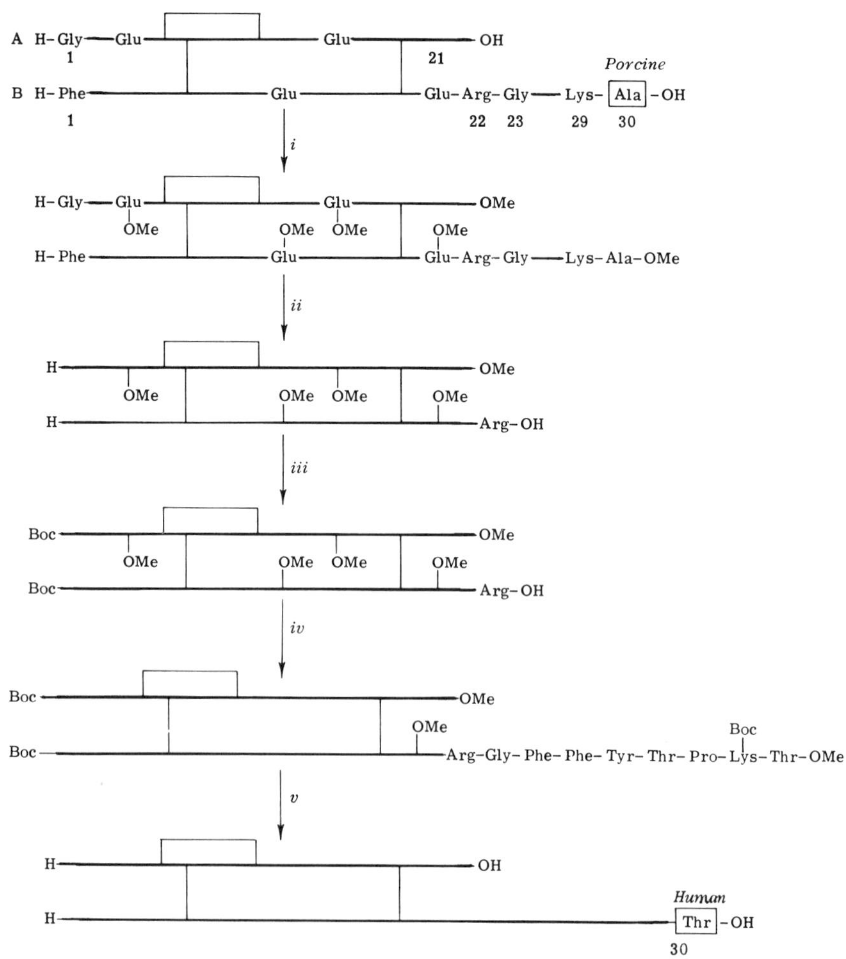

Scheme 7. *i*, CH_2N_2, pH 4.6; *ii*, trypsin; *iii*, Boc-N_3; *iv*, synthetic H-Gly-Phe-Try-Thr-Pro-Lys(Boc)-Thr-OMe, DCCI–HOSu; *v*, "removal of Boc groups" and saponification.

methyl ester with trypsin at arginine B22 (and presumably also at lysine B29), and protection of the A- and B-chain terminal amino groups of the shortened protein by *tert*-butyloxycarbonylation. The resulting protected deoctapeptide(B23–B30) insulin was coupled with a synthetic protected octapeptide corresponding in sequence to the COOH-terminus of human insulin B-chain. The amine and carboxyl protecting groups in the coupled product were cleaved, the latter by alkaline hydrolysis.

This seemingly straightforward procedure attracted early criticism (Jones, 1972) that was soon substantiated by reports from other groups of their inability to obtain the claimed high yield of biologically active material (Shanghai Insulin Research Group, 1973). The main problem appears to be in the alkaline hydrolysis of the final hexamethyl ester derivative. The principal product lacks one carboxyl group of native insulin and has been identified as the succinimide (**25**) derived from the COOH-terminal asparagine residue of the A-chain (Gattner and Schmitt, 1977; Obermeier, 1978). Cyclization of such an asparaginyl methyl ester would be anticipated under alkaline conditions; related observations were made with the more stable asparagine *tert*-butyl ester derivatives (Roeske, 1963). On prolonged alkaline treatment, the imide **25** is hydrolyzed to a mixture of asparagine and isoasparagine derivatives (Scheme 8).

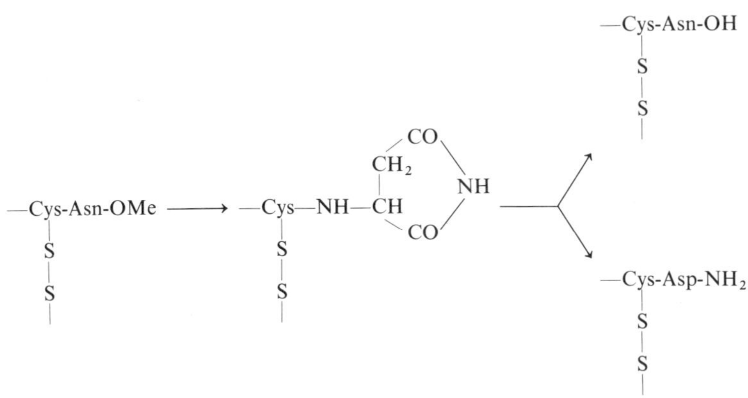

Scheme 8

Notwithstanding its deficiencies, the basic Ruttenberg scheme has been used by others in a variety of partial synthesis-aided structure–function studies. Deoctapeptide insulin methyl ester has been prepared by methanolic hydrogen chloride esterification of insulin and digestion with trypsin. The purified pentamethyl ester was amine protected using *tert*-butyloxycarbonyl azide and coupled using dicyclohexylcarbodiimide–*N*-hydroxysuccinimide to a series of synthetic oligopeptides methyl ester showing variation in the cluster of aromatic amino acids corresponding to residues B24–26 (-Phe-Phe-Tyr-) (Weitzel *et al.*, 1976).

In a further extension, deoctapeptide insulin pentamethyl ester has been digested with carboxypeptidase B to yield desnonapeptide(B22–B30) insulin pentaester. Amine protection, coupling with a range of synthetic tetrapeptide derivatives, and removal of protecting groups gave a series of analogs of despentapeptide(B26–30) insulins with variation of the B22 (arginine) position (Weitzel *et al.*, 1977). Presumably, all these products prepared by methyl ester saponification will be contaminated with the isoasparagine or succinimide derivatives.

These difficulties encountered with methyl esters and the failure to obtain complete esterification of insulin using other reagents (diphenyldiazomethane, *tert*-butyl acetate) giving acid-labile esters, led others to study alternative partially synthetic routes to human insulin. That employing bis-Boc-deoctapeptide(B22–30) insulin *without* protection or selective activation of the six free carboxyl groups (Scheme 9) has already been mentioned. This approach was justified on the grounds that some differential reactivity

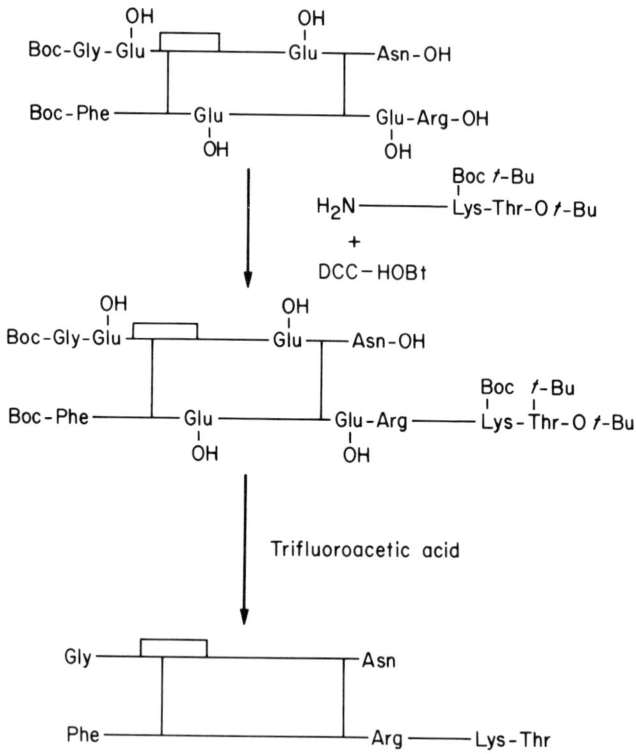

Scheme 9

(favorable or otherwise) might be anticipated because of the different acidities of the A- and B-chain terminal carboxyl groups from those of the four glutamic acid side chains. These intrinsic acidity differences might be further modified by hydrogen bonding and environment effects within the tertiary structure (if this is retained in the deoctapeptide derivative), and to different steric accessibilities (Obermeier and Geiger, 1976). In any event, the yields of insulin activity obtained in the crude condensation product suggested some degree of specificity for the desired reaction at arginine-B22. After extensive purification, 10–15% yields of human insulin were obtained. Four grams of the hormone were prepared in this way (Obermeier, 1978).

The favorable selectivity encountered in the above coupling at terminal arginine (B22) was apparently not repeated when the coupling was performed at phenylalanine (B25) (Scheme 10). In a similar series of experiments, bis-Boc-despentapeptide(B26–30), insulin was coupled with a synthetic protected pentapeptide using dicyclohexylcarbodiimide–1-hydroxybenzotriazole under the same conditions as those of Obermeier and Geiger (1976). In this case the level of biological activity obtained indicated randomized coupling, and the isolated yield of human insulin was less than 1% (Gattner *et al.*, 1978).

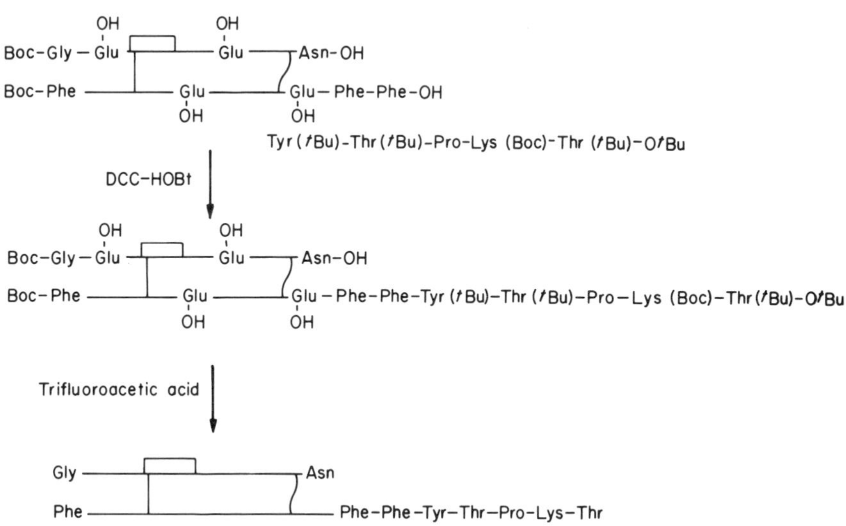

Scheme 10

B. β-Melanotropin (β-MSH)

The conversion of porcine β-MSH (**26**) into the lysine-10 analog (**27**) of human β-MSH (28) (Burton and Lande, 1970) constitutes one of the earliest

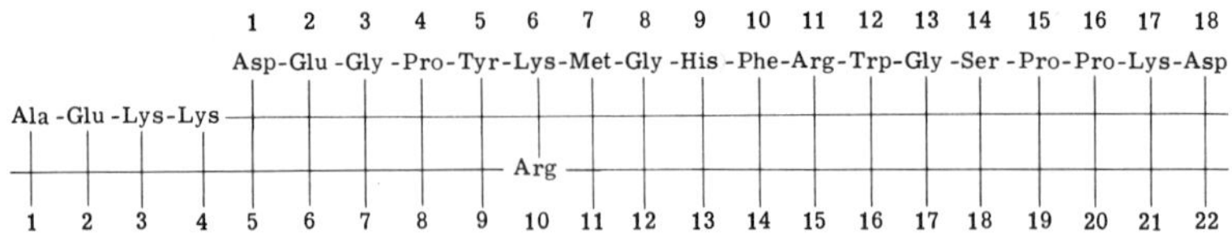

(26)
(27)
(28)

major endeavors in the field. This interconversion requires addition of a tetrapeptide sequence to the terminal amino group of a natural octadecapeptide, the latter containing two additional side-chain amino groups. The approach adopted was that of attempted selective N_α-acylation rather than side-chain protection. The synthetic carboxyl component tetrapeptide was fully protected using *tert*-butyloxycarbonyl and *tert*-butyl ester groups. Coupling was by the azide procedure. A mixture of N^α- and N^ε-acylated products and unchanged β-MSH was obtained from which the desired product could be isolated in only 4% yield after deprotection. Extended reaction times failed to increase the yield of N^α-substituted product, but instead resulted in the formation of polyacyl derivatives and further consumption of β-MSH. The authors concluded that "In spite of a large α–ε-amine p*K* difference, protonation alone is an insufficient deterrent to N^ε-acylation in β-MSH. Apparently high nucleophilicity compensates for low concentration of N^ε-amine species at neutral pH" (cf. discussion Section III,A,B).

C. Myoglobin

This 153-residue protein has been the subject of partial synthesis studies in two laboratories. The molecule contains 19 lysine but only 2 arginine residues, and Borrás-Cuesta *et al.* (1978) have utilized this low arginine concentration in the preparation of three side-chain amine protected fragments spanning the entire sequence. Maleic anhydride failed to react completely with amino groups in the intact protein under a variety of conditions (pH, solvent mixtures, and excess of reagent), and full protection could only be obtained when the heme group was removed from the molecule and the globin reacted with excess maleic anhydride in 8 *M* urea. Hydrolysis with trypsin and fractionation on Sephadex G-50 and DEAE-Sephadex A25 furnished the three expected tryptic peptides together with unidentified components.

Gurd and his colleagues favored acetimidylation for lysine side-chain protection (Garner and Gurd, 1975) and were able to achieve masking of all 19 N^ε-amino groups at the expense of only 60% coverage of the N^α-amine. The protein was then specifically labeled for nuclear magnetic resonance (nmr) studies to the extent of 0.4–0.5 residues per mole by N^{α}-acylation with Boc-[1-^{13}C]glycine *N*-hydroxysuccinimide ester. In an extension of these labeling studies (di Marchi *et al.*, 1978), the crude acetimidylated product was reacted with tetrahydrophthalic anhydride and fractionated, furnishing after cleavage of the tetrahydrophthalyl residue, purified N^{ε} and N^{α},N^{ε}-acetimidylated derivatives. The NH_2-terminal valine residue was removed from the former by one cycle of Edman degradation using the hydrophilic reagent, 3-sulfophenyl isothiocyanate. Use of isotopically labeled trifluoroacetylglycine *N*-hydroxysuccinimide ester in the coupling, then enabled all the protecting groups to be cleaved from the modified protein in a single step.

The N^{α},N^{ε}-peracetimidylated protein was used in more ambitious studies on the removal and replacement of the 14-residue NH_2-terminal sequence (Wang *et al.*, 1978). The BNPS-skatole method of Fontana (1972) was used to cleave the apo derivative at Trp-14. Chain fission was 60–70% complete and only slight incidental cleavage was observed at Trp-7. Inclusion of excess phenol in the reaction mixture prevented modification of tyrosine residues. Sequential coupling reactions were carried out on the purified 15–153 residue fragment using Boc-N^{i}-formyltryptophan *N*-hydroxysuccinimide ester and two protected peptide azides comprising residues 6–13 and 1–5 prepared by the solid-phase method. The yield in each azide coupling reaction was about 25%; the final yield of reconstituted myoglobin sequence as determined by automatic sequence analysis was 3–5%. After reintroduction of the heme moeity, cleavage of acetimidyl groups, and further purification, the isolated product had spectroscopic and circular dichroism properties similar to those of myoglobin and gave a major band on electrophoresis identical to the native protein.

D. Lysozyme

Extensive studies on the preparation of protected fragments of lysozyme and progress in their recombination have been reported by Offord and his colleagues (Offord, 1973; Rees and Offord, 1976a,b). The approach adopted is that of complete protection of side-chain amino and carboxyl groups, and also of thiol and imidazole functions. Maleic anhydride is the preferred reagent for amine protection, diazoalkanes for carboxyl protection, *S*-sulfonate for cysteine thiol groups, and benzyloxycarbonyl or *tert*-butyloxycarbonylamino-2,2,2-trifluoroethyl for the imidazole ring of histidine. Selective deesterification of terminal carboxyl groups (of tryptic peptides) was achieved by the esterolytic action of trypsin. Individual aspects of this work have been discussed earlier in this chapter. Reassembly of the fragments using dicyclohexylcarbodiimide–1-hydroxybenzotriazole has thus far yielded segments corresponding to residues 1–45 and 62–113 of the natural protein. Full details have been published (Rees and Offord, 1976a,b).

E. Trypsin Inhibitors

1. Soybean Trypsin Inhibitor

This protein has proved particularly suitable for single amino acid replacements within the peptide chain. At low pH and enzyme concentrations, the reaction between trypsin and its inhibitor yields an equilibrium mixture consisting of intact and seco-inhibitor with the latter predominating. The seco-inhibitor is apparently cleaved at a single bond between arginine-63 and isoleucine-64, within the active site. Fragmentation does not take place because of the disulfide bond spanning the cleavage point. At higher enzyme concentration and pH the situation is reversed, and at equilibrium most of the inhibitor is in the form of a tight complex with trypsin in which the susceptible peptide bond is now intact. The trypsin–trypsin inhibitor system thus provides a very special and suitable case for use of the reverse action of a proteolytic enzyme for peptide bond synthesis.

Sealock and Laskowski (1969) used these favorable equilibria in an ingenious way to effect amino acid substitution at arginine-63 (Scheme 11). The inhibitor was incubated at pH 3.75 with a low concentration of trypsin,

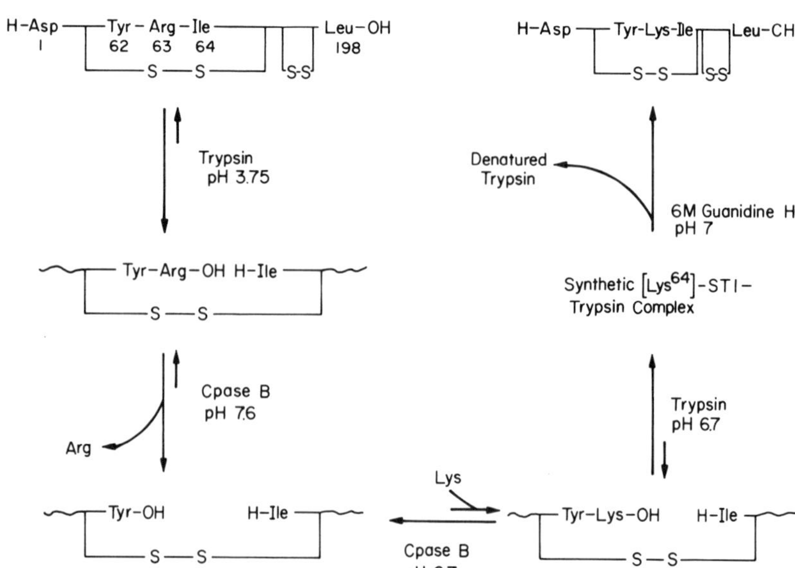

Scheme 11

and after attainment of equilibrium (ratio of cleaved to intact inhibitor ca. 10:1), the pH was raised to 7.6 and carboxypeptidase B added. The newly liberated COOH-terminal arginine residue was rapidly cleaved from the seco-inhibitor. By repeating the trypsin and carboxypeptidase steps the amount of unaltered inhibitor was reduced to less than 2%. The carboxypeptidase-catalyzed equilibrium lies far on the side of hydrolysis, and its reversal requires both a relatively high lysine (or arginine) concentration and coupling with the trypsin resynthesis reaction. Both enzymes are also now used in much larger amounts. After decomposition of the modified inhibitor–trypsin complex, the lysine-63 analogue could be isolated in good yield.

A number of related studies extending this work have been reported. By using appropriate enzymes, the active site arginine-63 has been replaced by tryptophan (Leary and Laskowski, 1973). With prior (irreversible) masking of all amino groups by guanidation and carbamoylation, chemical methods could be used for removal of isoleucine-64, and a number of substitutions were made for this residue (Kowalski and Laskowski, 1976a). Additional residues could also be inserted at this point (Kowalski and Laskowski, 1976b). These and related studies on soybean trypsin inhibitor have been fully reviewed (Kowalski *et al.*, 1974; Kowalski, 1978).

2. *Bovine Pancreatic Trypsin Inhibitor (BPTI)*

This simpler 58-residue trypsin inhibitor has been similarly modified within the peptide chain (Jering and Tschesche, 1974). In this case, selective reduction of one of the three disulfide bridges is a key step in formation of the active site lysine-15-alanine-16 cleaved inhibitor (Scheme 12). After reclosure of the disulfide bridge, trypsin now acts in the reverse manner, resynthesizing the lysine–alanine bond. With the additional aid of carboxypeptidase, coupled with trypsin in the resynthesis step, replacement of the active site lysine by arginine, by phenylalanine, and by tryptophan was achieved.

The success of all these partially synthetic transformations depends critically on the special relationship between the enzyme (trypsin) and its substrate (trypsin inhibitor). This relationship allows preferential single cleavage in a molecule susceptible in principle to widespread proteolysis. It also allows efficient resynthesis under appropriate conditions of enzyme concentration and pH. Clearly these are special examples of protein partial synthesis, but the efficiency and cleanliness of the above transformation is impressive and justifies the search for similar procedures using unprotected peptides and proteins in aqueous solution.

Pancreatic trypsin inhibitor has also been the subject of investigation in the COOH-terminal region (Dyckes *et al.*, 1974, 1977, 1978). This region

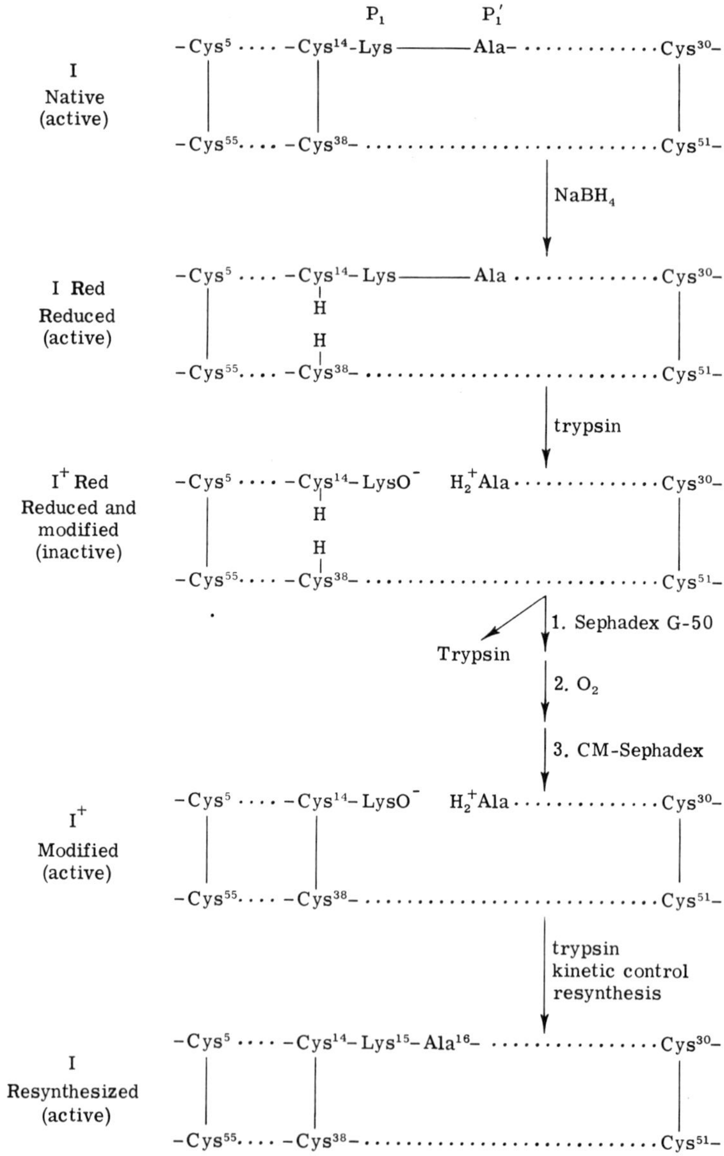
P1
P1'
I
Native
(active)
NaBH4
I Red
Reduced
(active)
trypsin
I+ Red
Reduced and
modified
(inactive)
1. Sephadex G-50
Trypsin
2. O2
3. CM-Sephadex
I+
Modified
(active)
trypsin
kinetic control
resynthesis
I
Resynthesized
(active)

Scheme 12

contains the only methionine residue (at position 52) in the molecule and is smoothly cleaved by cyanogen bromide to yield a 52-homoserine,52,53-seco derivative (**29**: Scheme 13). Fragmentation does not take place because of the intact disulfide bridge spanning the cleavage point. The cleaved molecule retains its biological function, and the tertiary structure is therefore largely retained. This region contains the only significant length of α-helix, including the original methionine residue. Because this tertiary structure holds the new amino group of arginine-53 in close proximity to the homoserine lactone ring, aminolysis of the latter takes place particularly readily with formation of the trypsin inhibitor analogue (**30**). In these particularly favorable steric circumstances, even the weak activation provided by the lactone ring is sufficient to ensure smooth peptide bond formation. Reaction takes place specifically with just one of the several free amino groups present and provides a potentially important route for protein partial synthesis employing cyanogen bromide-type fragmentation and without use of protecting groups. Similar observations were made subsequently with cytochrome *c* (Corradin and Harbury, 1974) and may have some generality for proteins that are cleaved by cyanogen bromide without fragmentation or serious disruption of the tertiary structure. Peptide chain resynthesis in **29** is inhibited when the new NH_2-terminal residue, arginine-53, is removed by aminopeptidase (the next residue in the sequence, threonine 54, would then be held *away* from the homoserine lactone ring) and is markedly slowed

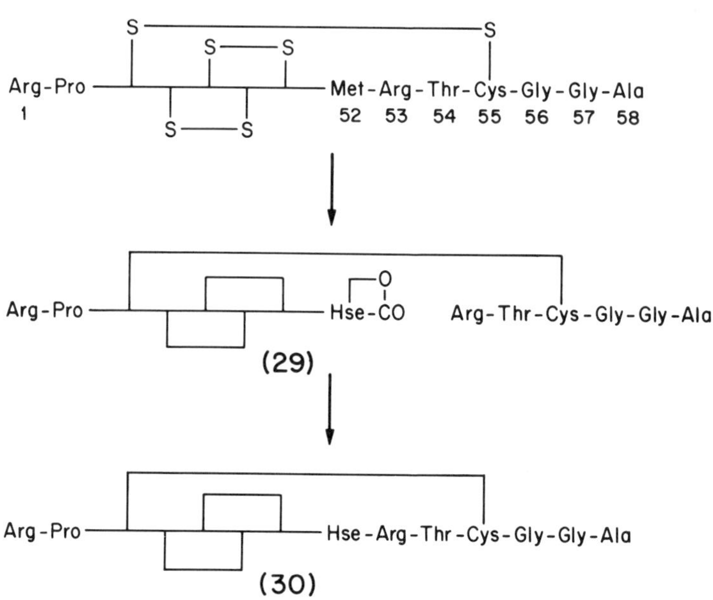

Scheme 13

when **29** is dissolved in a denaturing medium. Under these last circumstances, bimolecular attack on the lactone ring by added peptide amine can compete effectively with the intramolecular resynthesis reaction, provided the concentration of exogenous amine component is high. Thus in a model series of experiments (Dyckes *et al.*, 1977), the seco-lactone (**29**) was reacted (Scheme 14) in dimethyl sulfoxide or aqueous 6 *M* guanidinium chloride solution with a large excess of the tripeptide *tert*-butyl ester, H-Gly-Gly-Ala-O*t*Bu, to give—after reduction of all the disulfide bonds with detachment of the original COOH-terminal hexapeptide and cleavage of the new COOH-terminal protecting group—the reduced 52-homoserine-des(53–55)-trypsin inhibitor homolog (**31**).

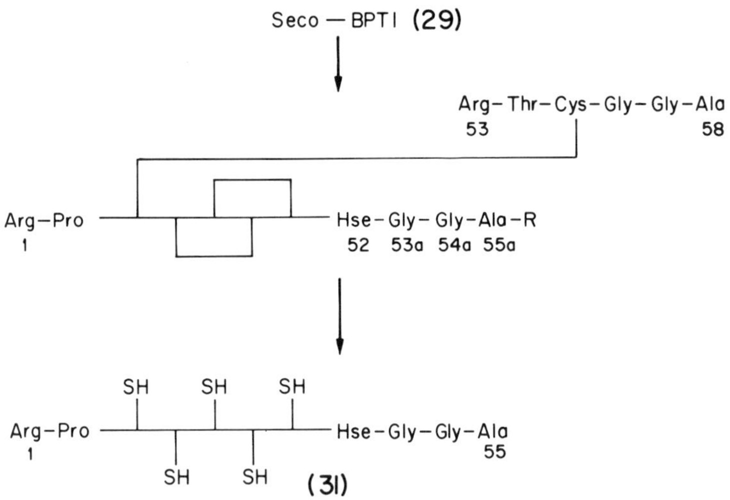

Scheme 14

F. Cytochrome c

This 104-residue protein has been the subject of partial cyanogen bromide cleavage and resynthesis studies. Although the molecule contains two methionine residues, that at position 80 in the polypeptide chain functions as a ligand to the heme iron atom and is therefore presumably partially deactivated to electrophilic attack by cyanogen bromide. Under appropriate conditions, it is possible to react solely at methionine-65 and to isolate the two segments corresponding to residues 1–65 and 66–104 (Corradin and Harbury, 1970). The former residue contains the covalently linked heme moiety. When in the reduced state, the two fragments exhibit a high degree of affinity for one another, and after reoxidation of the complex, a product closely resembling the native oxidized protein is formed (Corradin and Harbury, 1971). However, for an adequate yield of reoxidized product, it was found necessary to keep the reduced complex in solution for 12–24 hours before oxidation. The authors concluded that a slow change in structure was effected during this interval.

With the recognition of the chain resynthesis reaction occurring in cyanogen bromide-treated pancreatic-trypsin inhibitor (see above), it was possible to interpret this change in structure as re-formation of the 65–66 peptide bond in reduced cytochrome (Corradin and Harbury, 1974). Although unlike trypsin inhibitor, cytochrome *c* contains no intrachain disulfide bridges, the coordinate link between methionine-80 and the heme residue, which spans the cleavage point may help to stabilize the appropriate tertiary structure. In the intact protein, methionine-65 is located within a regular α-helical segment, as is the single methionine in pancreatic trypsin inhibitor.

This spontaneous resynthesis reaction has been used to combine 1–65 and 66–104 segments originating from cytochromes of differing species (Harbury, 1978). Barstow *et al.* (1977) also describe a partial synthesis of 65-homoserine cytochrome *c* from the natural heme-containing (1–65)-homoserine lactone fragment and a (66–104)-preparation obtained by solid-phase synthesis. Substantial evidence was obtained for formation of a cytochrome-like product. However, this work is marred by a complete lack of characterization or serious purification of the solid-phase synthesized 39-residue peptide before incorporation. The synthetic methods used were such that truncated and failure sequences, as well as branched chain and other artifactual products, must have been present to a very large extent. It may be argued that the recombination reaction itself constitutes an exceptionally efficient purification procedure, insofar as only those molecules capable of complementing the natural fragment will undergo resynthesis. However the specificity of the complementation process is unknown, and the existence of many sequence variations in natural cytochromes and the ability to combine fragments from different species suggests that this will be by no means absolute. Preparation of cytochrome analogues for study of their biological properties by this method will require rigorous characterization and purification of the synthetic component and incorporation in high yield if the results are to be meaningful.

A partially synthetic route to the (66–104)-fragment of cytochrome has also been reported (Boon *et al.*, 1978). Horse heart cytochrome *c* was converted into its fully amine protected methylsulfonylethyloxycarbonyl derivative, and this was cleaved with cyanogen bromide. The COOH-terminal fragment comprising residues 81–104 was isolated and purified. This side-chain amine protected derivative was extended by addition of a single methionine residue using the *N*-hydroxysuccinimide ester of Boc-methionine and then coupled with a synthetic tetradecapeptide azide comprising residues 66–79 (Scheme 15). This particular route was followed because, very remarkably, the homologous 14-peptide azide terminating in methionine failed completely to react with the natural cytochrome fragment.

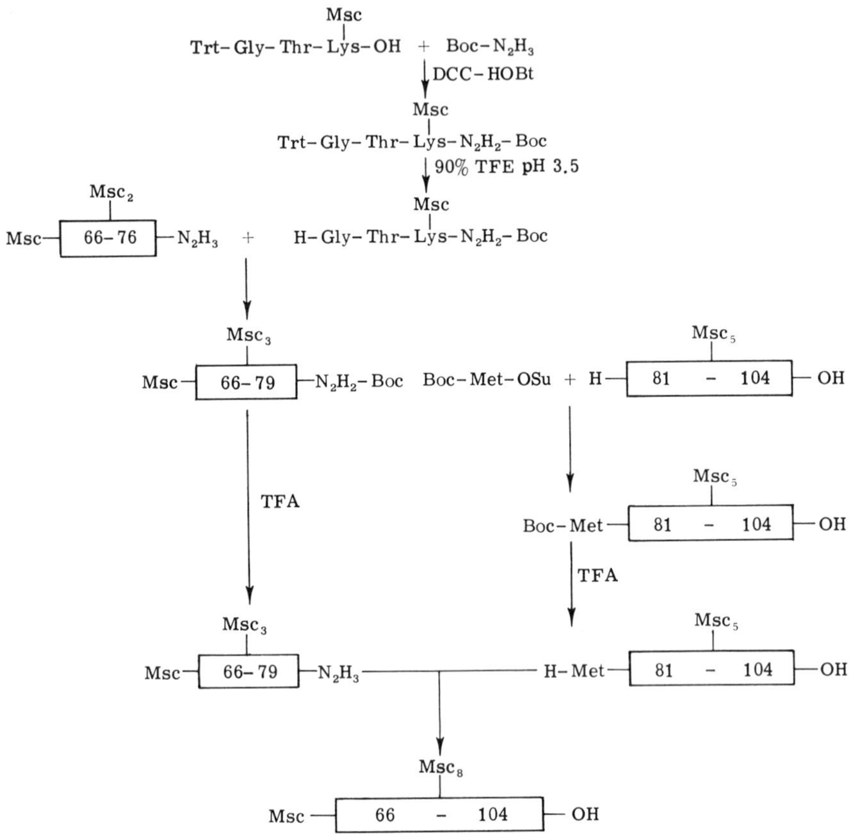

Scheme 15

VI. CONCLUSIONS

The partial synthesis approach to the larger peptides and proteins has made substantial progress in the relatively short period since its inception. Workable methods are available for the protection of side-chain amino groups, and the addition of single residues or short peptide sequences to the NH_2-termini of naturally derived, side-chain amine protected fragments has been achieved in a number of cases. Methods for the protection of side-chain carboxyl groups appear to be less satisfactory. The reagents currently used may give rise to serious side reactions elsewhere in the molecule, the protected products are sometimes insoluble and difficult to purify, and deprotection can cause additional problems. At present, therefore, partial synthesis is most appropriate either for operations at the NH_2-termini of natural peptides or their fragments or at the COOH-termini in those special cases where widespread introduction of protecting groups can be dispensed with through specific enzymatic or chemical carboxyl activation and coupling reactions. In such case, minimal alteration of peptide or protein structure is important and may be crucial for smooth coupling reactions. Complete retention of protein tertiary structure or the initial formation of noncovalent complementing systems prior to covalent combination would appear to provide the most fruitful opportunity for protein analog production through partial synthesis.

REFERENCES

Anfinsen, C. B. (1956). *J. Biol. Chem.* **221**, 405–412.

Anonymous (1976). *Chem. & Eng. News* **54**, 3.

Barstow, L. E., Young, R. S., Yakali, E., Sharp, J. J., O'Brien, J. C., Berman, P. W., and Harbury, H. A. (1977). *Proc. Natl. Acad. Sci. U.S.A.* **74**, 4248–4250.

Battersby, A. R., and Robinson, J. C. (1955). *J. Chem. Soc.* pp. 259–269.

Bergstrand, H. (1971). *Eur. J. Biochem.* **21**, 116–124.

Boon, P. J., Tesser, G. I., and Nivard, R. J. F. (1978). *In* "Semisynthetic Peptides and Proteins" (R. E. Offord and C. Di Bello, eds.), pp. 115–126. Academic Press, New York.

Borrás, F., and Offord, R. E. (1970). *Nature* (*London*) 227, 716–718.

Borrás-Cuesta, F., Pavani, M., Romo, D., and Malebran, P. (1978). *In* "Semisynthetic Peptides and Proteins" (R. E. Offord and C. Di Bellow, eds.), Academic Press, New York.

Brandenburg, D., Büsse, W.-D., Gattner, H.-G., Zahn, H., Wollmer, A., Gliemann, J., and Puls, W. (1973a). *In* "Peptides 1972." (H. Hanson and H. D. Jakubke, eds.) pp. 270–283. North-Holland Publ., Amsterdam.

Brandenburg, D., Schermutzki, W., and Zahn, H. (1973b). *Hoppe-Seyler's Z. Physiol. Chem.* **354**, 1521–1524.

Brandenburg, D., Biela, M., Herbertz, L., and Zahn, H. (1975). *Hoppe-Seyler's Z. Physiol. Chem.* **356**, 961–979.

Bünzli, H. F., and Bosshard, H. R. (1971). *Hoppe-Seyler's Z. Physiol. Chem.* **352**, 1180–1182.

Burnett, P. R., and Eylor, E. H. (1971). *J. Biol. Chem.* **246**, 3425–3430.

Burton, J., and Lande, S. (1970). *J. Am. Chem. Soc.* **92**, 3746–3748.

Busse, W.-D., and Carpenter, F. H. (1974). *J. Am. Chem. Soc.* **96**, 5947–5949.

Butler, P. J. G., Harris, J. I., Hartley, B. S., and Lebermann, R. (1969). *Biochem. J.* **112**, 679–689.

Chibnall, A.-C., Mangan, J.-L., and Rees, M. W. (1958). *Biochem. J.* **68**, 114–118.

Corradin, G., and Harbury, H. A. (1970). *Biochim. Biophts. Acta* **221**, 489–496.

Corradin, G., and Harbury, H. A. (1971). *Proc. Natl. Acad. Sci. U.S.A.* **68**, 3036–3039.

Corradin, G., and Harbury, H. A. (1974). *Biochem. Biophys. Res. Commun.* **61**, 1400–1406.

Danho, W. O., Gattner, H.-G., Nissen, D., and Zahn, H. (1975). *Hoppe-Seyler's Z. Physiol. Chem.* **356**, 1405–1412.

Degani, Y., and Patchornik, A. (1974) *Biochemistry* 13, 1–11.

di Marchi, R. D., Garner, W. H., Wang, C.-C., and Gurd, F. R. N. (1978). *In* "Semisynthetic Peptides and Proteins" (R. E. Offord and C. Di Bello, eds.), pp. 45–58. Academic Press, New York.

Dixon, H. B. F. (1964a). *Biochem. J.* **90**, 2c-3c.

Dixon, H. B. F. (1964b). *Biochem. J.* **92**, 661–666.

Dixon, H. B. F., and Moret, V. (1964). *Biochem. J.* **94**, 433–439.

Dixon, H. B. F., and Perham, R. N. (1968). *Biochem. J.* **109**, 312–314.

Dyckes, D. F., Creighton T. E., and Sheppard, R. C. (1974). *Nature* (*London*) **247**, 202–204.

Dyckes, D. F., Kini, H., and Sheppard, R. C. (1977). *Int. J. Pept. Protein Res.* **9**, 340–348.

Dyckes, D. F., Creighton, T. E., and Sheppard, R. C. (1978). *Int. J. Pept. Protein Res.* **11**, 258–268.

Edman, P. (1950). *Acta Chem. Scand.* **4**, 277–282.

Fanger, M. W., and Harbury, H. A. (1965). *Biochemistry* **4**, 2541–2546.

Fersht, A. R. (1977). *In* "Enzyme Structure and Mechanism," p. 65. Freeman, San Francisco, California.

Fölsch, G. (1974). *Chem. Scr.* **6**, 32–34.

Fölsch, G. (1975). *Chem. Scr.* **8**, 228.

Fölsch, G. (1978). *In* "Semisynthetic Peptides and Proteins" (R. E. Offord and C. Di Bello, eds.), pp. 309–314. Academic Press, New York.

Fontana, A. (1972). *In* "Methods in Enzymology," Vol. 25, (C. H. W. Hirs and S. N. Timasheff, eds.), pp. 419–423. Academic Press, New York.

Friesen, H. J., Brandenburg, D., Diaconescu, C., Gattner, H. G., Naithani, V. K., Nowak, J., Zahn, H., Wood, S. P., and Blundell, T. L. (1977). *In* "Peptides: Proceedings of the Fifth American Peptide Symposium" (M. Goodman and J. Meienhofer, eds.), pp. 136–140. Wiley, New York.

Friesen, H. J., Weimann, J., Nowak, J., and Brandenburg, D. (1978). *In* "Semisynthetic Peptides and Proteins" (R. E. Offord and C. Di Bello, eds.), pp. 161–180. Academic Press, New York.

Garner, W. H., and Gurd, F. R. N. (1975). *Biochem. Biophys. Res. Commun.* **63**, 262–268.

Gattner, H.-G. (1977). *Hoppe-Seyler's Z. Physiol. Chem.* **356**, 1397–1404.

Gattner, H.-G., and Schmitt, E. W. (1977). *Hoppe-Seyler's Z. Physiol. Chem.* **358**, 105–113.

Gattner, H.-G., Schmitt, E. W., and Naithani, V. K. (1975). *Hoppe-Seyler's Z. Physiol. Chem.* **356**, 1465–1467.

Gattner, H.-G., Schmitt, E. W., and Naithani, V. K. (1978). *In* "Semisynthetic Peptides and Proteins" (R. E. Offord and C. Di Bello, eds), pp. 181–192. Academic Press, New York.

Geiger, R. (1976). *Chem. Ztg.* **100**, 111–129.

Geiger, R. (1977). *Med. Chem., Proc. Int. Symp., 5th.*, **1976** pp. 71–78.

Geiger, R., and Langner, D. (1974). *In* "Peptides 1974" (Y. Wolman, ed.) pp. 159–163. Wiley, New York.

Geiger, R., and Obermeier, R. (1973). *Biochem. Biophys. Res. Commun.* **55**, 60–66.

Geiger, R., and Obermeier, R. (1975). *Chem. Ber.* **108**, 2758–2763.

Geiger, R., and Obermeier, R. (1976). *In* "Diabetes Research Today," pp. 25–33. Schattauer Stuttgart.

Geiger, R., Schone, H.-H., and Pfaff, W. (1971). *Hoppe-Seyler's Z. Physiol. Chem.* **352**, 1487–1490.

Geiger, R., Geisen, K., Summ, H.-D., and Langner, D. (1975). *Hoppe-Seyler's Z. Physiol. Chem.* **356**, 1635–1649.

Geiger, R., Teetz, V., Konig, W., and Obermeier, R. (1978). *In* "Semisynthetic Peptides and Proteins" (R. E. Offord and C. Di Bello, eds.), pp. 141–160. Academic Press, New York.
Gibbons, I., and Perham, R. N. (1970). *Biochem. J.* **116**, 843–849.
Goldberg, R. F., and Anfinsen, C. B. (1962). *Biochemistry* **1**, 401–405.
Gross, E., and Witkop, B. (1961). *J. Am. Chem. Soc.* **83**, 1510–1511.
Gross, E., and Witkop, B. (1962). *J. Biol. Chem.* **237**, 1856–1860.
Habeeb, A. F. S. A., and Atassi, M. Z. (1970). *Biochemistry* **9**, 4939–4944.
Honzl, J., and Rudinger, J. (1961). *Collect. Czech. Chem. Commun.* **26**, 2333–2345.
Harbury, H. A. (1978). *In* "Semisynthetic Peptides and Proteins" (R. E. Offord and C. Di Bello, eds.), pp. 73–90 Academic Press, New York.
Harris, D. E., and Offord, R. E. (1977). *Biochem. J.* **161**, 21–25.
Hayward, C. F., and Offord, R. E. (1971). *In* "Peptides 1969" (E. Scoffone, ed.). pp. 116–120. North-Holland Publ., Amsterdam.
Hunter, M. J., and Ludwig, M. (1962). *J. Am. Chem. Soc.* **84**, 3491–3504.
Itoh, M., Haziwara, D., and Kamiya, T. (1975). *Tetrahedron Lett.* pp. 4393–4396.
Izumiya, N., Noda, K., and Anfinsen, C. B. (1971). *Arch. Biochem. Biophys.* **144**, 237–244.
Jacobson, G. R., Schaffer, M. H., Stark, G. R., and Vanaman, T. C. (1973). *J. Biol. Chem.* **248**, 6583–6591.
Jering, H., and Tschesche, H. (1974). *Angew. Chem., Int. Ed., Engl.* **13**, 660–663.
Jones, J. H. (1972). *Amino-Acids, Pept., Proteins* **5**, 330.
Koshland, D. E., Strumeyer, D. H., and Ray, W. J. (1962). *Brookhaven Symp. Biol.* **15**, 101–110.
Kowalski, D. (1978). *In* "Semisynthetic Peptides, and Proteins" (R. E. Offord and C. Di Bello, eds.), pp. 263–282. Academic Press, New York.
Kowalski, D., and Laskowski, M. (1976a). *Biochemistry* **15**, 1300–1309.
Kowalski, D., and Laskowski, M. (1976b). *Biochemistry* **15**, 1309–1315.
Kowalski, D., Leary, T. R., McKee, R. E., Sealock, R. W., Wang, D., and Laskowski, M. (1975). *Proteinase Inhibitors, Bayer-Symp., 5th, 1973* pp. 311–324.
Krail, G., Brandenberg, D., and Zahn, H. (1975). *Hoppe-Seyler's Z. Physiol. Chem.* **356**, 981–996.
Lande, S. (1972). *In* "Peptides: Proceedings of the Second American Symposium" (S. Lande, ed.), pp. 363–367. Gordon & Breach, New York.
Lande, S. (1971). *J. Org. Chem.* **36**, 1267–1270.
Laskowski, M., Jr. (1978). *In* "Semisynthetic Peptides, and Proteins" (R. E. Offord and C. Di Bello, eds.), pp. 255–262. Academic Press, New York.
Leary, T. R., and Laskowski, M. (1973). *Fed. Proc., Fed. Am. Soc. Exp. Biol.* **32**, 465.
Levy, D. (1973a). *Biochim. Biophys. Acta* **310**, 406–415.
Levy, D. (1973b). *Biochim. Biophys. Acta* **328**, 107–113.
Levy, D., and Carpenter, F. H. (1967). *Biochemistry* **6**, 3559–3568.
Ludwig, M., and Byrne, R. (1962). *J. Am. Chem. Soc.* **84**, 4160–4162.
Mammaerts, W. F., and Neurath, H. (1950). *J. Biol. Chem.* **185**, 909–917.
Moroder, L., Hallett, A., Wünsch, E., Keller, O., and Wersin, G. (1976). *Hoppe-Seyler's Z. Physiol. Chem.* **357**, 1651–1653.
Obermeier, R. (1978). *In* "Semisynthetic Peptides and Proteins" (R. E. Offord and C. Di Bello, eds.), pp. 201–212. Academic Press, New York.
Obermeier, R., and Geiger, R. (1975). *Hoppe-Seyler's Z. Physiol. Chem.* **356**, 1631–1634.
Obermeier, R., and Geiger, R. (1976). *Hoppe-Seyler's Z. Physiol. Chem.* **357**, 759–767.
Offord, R. E. (1969). *Nature (London)* **221**, 37–40.
Offord, R. E. (1972). *Biochem. J.* **129**, 499–501.
Offord, R. E. (1972a). *In* "Peptides 1972" (H. Hanson and H. D. Jakubke, eds.), pp. 52–56. North-Holland Publ., Amsterdam.

Offord, R. E., Storey, H. T., Rees, A. R., Hayward, C. F., Johnson, W. H., Pheasey, M. H., and Wightman, D. A. (1976). *Biochem. J.* **159**, 480–486.

Omenn, G. S., Fontana, A., and Anfinsen, C. B. (1970). *J. Biol. Chem.* **245**, 1895–1902.

Rees, A. R., and Offord, R. E. (1976a). *Biochem. J.* **159**, 467–486.

Rees, A. R., and Offord, R. E. (1976b). *Biochem. J.* **159**, 487–493.

Reynolds, J. H. (1968). *Biochemistry* **7**, 3131–3135.

Richard, F. M., and Vithayathil, P. J. (1959). *J. Biol. Chem.* **234**, 1459–1465.

Roeske, R. (1963). *J. Org. Chem.* **28**, 1251–1255.

Ruttenberg, M. A. (1972). *Science* **177**, 623–626.

Saunders, D. J., and Offord, R. E. (1977a). *Biochem. J.* **165**, 479–486.

Saunders, D. J., and Offord, R. E. (1977b). *Hoppe-Seyler's Z. Physiol. Chem.* **358**, 1469–1474.

Savige, W. E., and Fontana, A. (1977). *In* "Methods in Enzymology" (C. H. W. Hirs and S. N. Timasheff, eds.) Vol. 47, pp. 459–469. Academic Press, New York.

Schroeder, W. A., Shelton, J. B., and Shelton, J. R. (1969). *Arch. Biochem. Biophys.* **130**, 551–562.

Sealock, R. W., and Laskowski, M. (1969). *Biochemistry* **8**, 3703–3710.

Shanghai Insulin Research Group (1973). *Sci. Sin.* **16**, 61–70.

Shields, G. B., Hill, R. L., and Smith, E. L. (1959). *J. Biol. Chem.* **234**, 1747–1753.

Shimonishi, Y. (1970). *Bull. Chem. Soc. Jpn.* **43**, 3251–3255.

Slotboom, A. J., and de Haas, G. H. (1975). *Biochemistry* **14**, 5394–5399.

Smyth, D. G. (1975). *Nature (London)* **256**, 699–700.

Snell, C. R., Austen, B. M., Geison, M. J., and Smyth, D. G. (1978). *In* "Semisynthetic Peptides and Proteins" (R. E. Offord and C. Di Bello, eds.), pp. 301–308. Academic Press, New York.

Spande, T. F., Witkop, B., Degani, Y., and Patchornik, A. (1970). *Adv. Protein Chem.* **24**, 98–260.

Stark, G. R. (1977). *In* "Methods in Enzymology" (C. H. W. Hirs and S. N. Timasheff, eds.), Vol. 47, pp. 129–132. Academic Press, New York.

Taniuchi, H., and Anfinsen, C. B. (1968). *J. Biol. Chem.* **243**, 4778–4786.

Tesser, G. I., and Balvert-Geers, I. C. (1975). *Int. J. Pept. Protein Res.* **7**, 295–305.

Van Heyningen, S., Tipton, K. F., and Dixon, H. B. F. (1968). *Biochem. J.* **108**, 508–509.

Wang, C.-C., di Marchi, R., and Gurd, F. R. N. (1978). *In* "Semisynthetic Peptides and Proteins" (R. E. Offord and C. Di Bello, eds.), pp. 59–72. Academic Press, New York.

Webster, D., and Offord, R. E. (1972). *Biochem. J.* **130**, 315–317.

Weil, L., and Telka, M. (1957). *Arch. Biochem. Biophys.* **71**, 473–483.

Weinert, M., Brandenburg, D., and Zahn, H. (1969). *Hoppe-Seyler's Z. Physiol. Chem.* **350**, 1556–1562.

Weinert, M., Kircher, K., Brandenburg, D., and Zahn, H. (1971). *Hoppe-Seyler's Z. Physiol. Chem.* **352**, 719–724.

Weitzel, G., Bauer, F. U., and Eisele, K. (1976). *Hoppe-Seyler's Z. Physiol. Chem.* **357**, 187–200.

Weitzel, G., Bauer, F. U., and Rehe, A. (1977). *Hoppe-Seyler's Z. Physiol. Chem.* **358**, 1573–1582.

Chapter 8

Racemization and Coupling Rates of N^{α}-Protected Amino Acids and Peptide Active Esters: Predictive Potential

JÓZSEF KOVÁCS

THE PEPTIDES, VOLUME 2

ISBN 0-12-304202-X

I. INTRODUCTION

In Volume 1 of "The Peptides," Kemp (1979) reviewed all aspects of racemization in peptide synthesis. Most of the racemization problems described in the literature were studied on model compounds such as the Anderson dipeptide, Z-Gly-Phe-OH, and Young's "dipeptide," Bz-Leu-OH, using different COOH-activating groups and reaction conditions. In this chapter, we discuss studies on racemization and coupling of a large number of N^α-protected amino acid active esters, including representatives of each group (such as aliphatic, aromatic, acidic, basic, sulfur-containing, hydroxyl and heterocyclic amino acids, and N^α-alkyloxycarbonylglycyl-L-amino acid active ester derivatives) under identical conditions. This will allow us to notice "regular" or "irregular" behavior of amino acids and to draw some general conclusions that will help to design better synthetic procedures. In this chapter, N^α-alkyloxycarbonyl-L-amino acid active esters will be referred to as monomers and N^α-alkyloxycarbonylglycyl-L-amino acid active esters will be mentioned as dipeptides. Glycyl dipeptide derivatives were chosen for our study since this allowed the use of polarimetry to follow the kinetics of racemization. A large amount of comparative data were provided both by Kemp and his associates and by our laboratory, and in the following discussion these results will be used primarily.

The kinetics of coupling reactions were investigated in this laboratory mostly in tetrahydrofuran (THF) solution while Kemp's comprehensive studies on N$^{\alpha}$-protected amino acid *p*-nitrophenyl esters were carried out in dimethylformamide (DMF) solution. All these data and conclusions are therefore directly applicable to practical synthetic purposes. The racemization studies in our laboratory were carried out with a tertiary base. Tertiary base is present when the amino component is liberated from its salt or, in specific cases, during the synthesis of sequential polypeptides from peptide active ester salts. For the liberation of the free peptide active ester, a tertiary base is used. Therefore, the correlation of our results with those obtained in the majority of actual practical syntheses presents a problem. Establishment of the necessary correlation between the ability of the tertiary amine and of an amino component to racemize both carboxyl-activated N$^{\alpha}$-protected amino acids and peptides is an important future task. It is assumed that racemization induced by triethylamine (Et_3N) parallels that induced by an amine component during coupling, but data for the latter are not presently available. In the absence of an exact correlation, data based on the results obtained by triethylamine-induced racemization were used, and predictions, which were subsequently proven to be correct, were made concerning the anomalous behavior of some amino acids. For example, Z-Gly-Cys(Bzl)-ONp racemizes, contrary to expectation, largely by enolization and not by the 5(4*H*)-oxazolone mechanism*; this will be discussed in detail later in this chapter. Similar conclusions about enolization were derived from the predictive factors, which are based on the additivity principle.

* Nomenclature according to *Chem. Abstr.* 1977 Index Guide, pp. 836G, 837G. Example: Structure **Ia** is designated as 2-methyl-4-(phenylmethyl)-5(4*H*)-oxazolone in *Chem. Abstr.* **83**, 3633CS (1975) Chemical Substance Index, as a reference to Siemion and Baran (1975). The

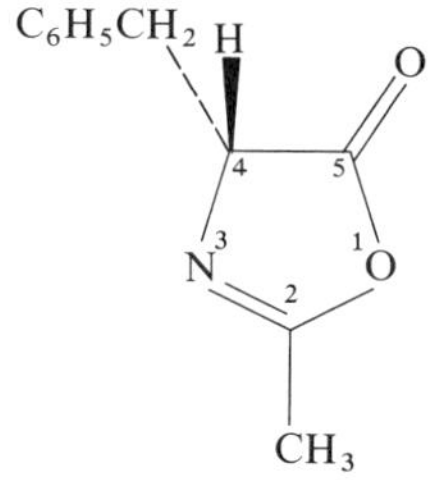

(Derived from *N*$^{\alpha}$-acetyl-L-phenylalanine)

(1a)

term 5(4*H*)-oxazolone replaces all earlier terms (such as azlactone, 4,5-dihydro-1,3-oxazol-5-one, oxazolinone, 1,3-oxazoline-5-one, 2-oxazoline-5-one, etc.) whose use should be discontinued. In this chapter the term 5(4*H*)-oxazolone will be used.

Scheme 1 can be used to derive an equation from which the optical purity of a peptide can be evaluated (Kovacs *et al.*, 1970b). Structure **E** is the carboxyl-activated amino acid or peptide derivative, **E**′ is its enantiomer or diastereomer. Structure **V** is the amine component, **T** is a base such as a tertiary amine (which is added to release **V** from its salt if the salt is used instead of the free amine) or the solvent that can act as a base, k_c is the second-order rate constant for coupling, k_1 and k_2 are the rate constants for the interconversion between **E** and **E**′ with **V** and **T**, respectively, **P** is the optically pure peptide, **P**′ is the undesired diastereomer.

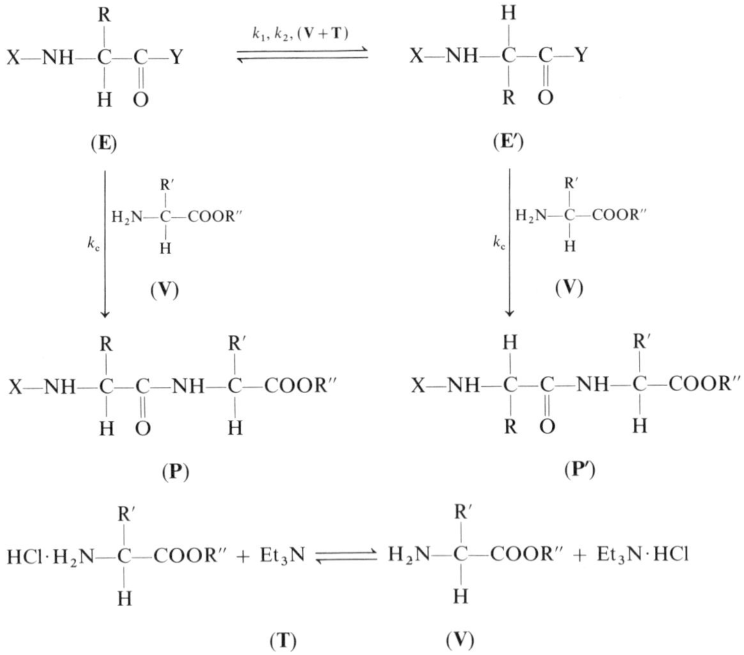

Scheme 1

If the racemization is rapid relative to coupling then obviously the product will be almost completely racemized, assuming that k_c is identical for both **E** and **E**′. This would be the case where an 5(4*H*)-oxazolone racemizes much faster than it couples. If the coupling rate is different for **E** and **E**′, then the concentration ratio of the products will be equal to the ratio of the two coupling rate constants.

However, in most coupling reactions, racemization is much slower than coupling. In this case, assuming that k_c is identical for both **E** and **E**′, and that no tertiary base is present, the ratio of desired to undesired diasteriomer formed at infinite time equals the ratio of rate constants. Thus,

$$\left(\frac{C_{\mathrm{P}}}{C_{\mathrm{P}'}}\right)_{\infty} = \frac{k_{\mathrm{c}}}{k_1} \tag{1}$$

In our model system triethylamine-induced racemization was studied in the absence of an amine component. Assuming that racemization induced by the amine component parallels that measured for triethylamine, then the k_c/k_2 ratio may still indicate the relative extent to which carboxyl-activated N^{α}-protected amino acid or peptide derivatives are susceptible to racemization during coupling.

The ratios in these equations have important preparative applications. The product ratio, $C_{\mathrm{P}}/C_{\mathrm{P}'}$, indicates the optical purity of the peptide formed during coupling. In turn, this product ratio is determined by the ratio of coupling-to-racemization rate constants, k_c/k_{rac}, where $k_{rac} = k_r$ is the total rate constant for racemization. The larger this ratio the smaller the amount of racemization to be expected during coupling. An ideal carboxyl-activated N^{α}-protected amino acid or peptide would be one that does not racemize, or racemizes very slowly, and couples very rapidly.

The coupling rate should be fast enough to complete the reaction within a reasonable time. If larger peptides, e.g., decapeptides, are prepared without isolation of intermediate peptides, as in solid-phase or liquid-phase procedures or peptide synthesis by repetitive methods in solution, it might be considered that at least 99% completion of each coupling reaction is necessary in order to be able to separate the product from contaminating wrong sequences. If however a large peptide, such as a 50–100 peptide, is prepared by these methods without isolation of intermediates, then it might be necessary to require that coupling be up to 99.9% complete, based on the same rationale that Kemp (1979) had previously outlined for racemization. Otherwise, the wrong sequences will result in an unpurifiable mixture of peptides. In the following tables (I–IV), the 99% reaction time will be included to show that only few of the active esters, unless they are used in excess, are suitable for such synthetic purposes. The desired completion time can be calculated by the following equation

$$\frac{1}{C} - \frac{1}{C^0} = k_{\mathrm{c}}t \tag{2}$$

where C^0 is the initial concentration of each reactant and C is the concentration at any time, t. For example, if we want to calculate the time required for 99% completion according to Eq. (2), we obtain Eq. (3), provided that the initial concentration of the active ester,

$$t_{99\%} = \frac{99}{k^{\mathrm{c}}C_{\mathrm{E}}^{0}} \tag{3}$$

C_E^0, is the same as that of the amine component. This equation indicates that the reaction time is inversely proportional to the initial active ester concentration.

It is also possible to decrease the coupling time by increasing the concentration of the active ester relative to that of the amine component. For example, the effect on the 99% coupling time of a 2:1 ratio of active ester to amine component is demonstrated in Table IX. A more detailed discussion of this effect is presented in Section VII.

In the following discussions, amino acids investigated in this laboratory are grouped into tables according to the influence of the side chain, R, of amino acids (**1b**) on racemization and coupling rates. The effect of amino

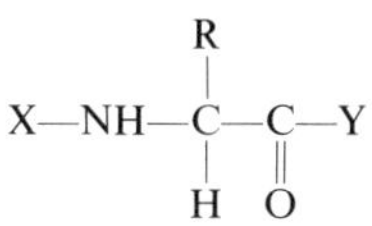

(**1b**)

protecting group, X, which includes N^α-*tert*-butyloxycarbonyl, N^α-benzyloxycarbonyl, N^α-*tert*-butyloxycarbonylglycyl, N^α-benzyloxycarbonylglycyl, as well as the carboxyl activating group, Y, which generally includes *N*-succinimidoyl, pentafluorophenyl, *p*-nitrophenyl, 2,4,5-trichlorophenyl, and pentachlorophenyl, are also compared within each group (Kovacs *et al.*, 1966, 1967a,b, 1969, 1970a,b,c, 1972a,b, 1973, 1975, 1977a,b; 1978, 1979a,b,c; Kisfaludy *et al.*, 1967, 1970; Kupryszew-ky, 1961).

Racemization was carried out in tetrahydrofuran solution with two or three different concentrations of triethylamine, under strictly anhydrous condition, and the second-order rate constants, k_r, are reported. The coupling reactions were also carried out in anhydrous tetrahydrofuran with equimolar amounts of H-Val-OMe. Valine was chosen since it is expected to couple at least an order of magnitude slower than H-Gly-OEt, which therefore permits the use of the ir technique (Conley, 1966). Coupling with H-Val-OMe represents an extreme when compared with H-Gly-OEt, which is the fastest coupling amino acid (Kemp, 1973; Kemp *et al.*, 1974a).

The Tables (I–IV) also show the important k_c/k_r values which indicate the relative extent of racemization that can be expected during coupling. The larger the ratio, the more optically pure the product is expected to be. The k_{r1}/k_{r2} and k_{c1}/k_{c2} values indicate how much faster a benzyloxycarbonyl-protected amino acid active ester racemizes or couples than the corresponding *tert*-butyloxycarbonyl derivatives. The k_{r3}/k_{r1} and k_{c3}/k_{c1} ratios show how much faster the N^α-alkyloxycarbonylglycylamino acid esters (dipeptides) racemize and couple than the corresponding N^α-alkyloxycarbonyl amino acid esters (monomers). The last ratio in Tables I–IV, $(k_{c_1}/k_{r_1})/(k_{c_3}/k_{r_3})$

Table I. Racemization and Coupling Rates in Tetrahydrofuran of Active Ester Derivatives of N^{α}-*tert*-Butyloxycarbonyl- and Benzyloxycarbonylalanine, -Phenyl-Alanine, -Methionine, -Glycyl-Alanine, -Glycyl-Phenyl-Alanine, and -Glycyl-Methionine[a]

Compound	k_r	(k_r)	k_c	(k_c)	$\frac{k_c}{k_r}$	$\frac{(k_c)}{(k_r)}$	t_{99} (hr)	$\frac{k_{r_1}}{k_{r_2}}$	$\frac{k_{c_1}}{k_{c_2}}$	$\frac{k_{r_3}}{k_{r_1}}$	$\frac{k_{c_3}}{k_{c_1}}$	$\frac{k_{c_1}/k_{r_1}}{k_{c_3}/k_{r_3}}$
Boc-Ala-OSu	6.1[b]	(4.7)	4.3[b]	(5.7)	7,063	(12130)	4.9	2.6	1.8			
Z-Ala-OSu	21.9	(20.6)	7.7	(6.7)	3,505	(3252)	2.75					
Boc-Ala-OPfp	3.3	(2.8)	17.1	(17.4)	51,657	(62142)	1.2	2.8	1.1			
Z-Ala-OPfp	9.2	(12.4)	19.3	(18.7)	21,001	(15080)	1.1					
Boc-Ala-ONp	0.4	(0.3)	0.108	(0.1)	2,700	(3333)	195.	3.3	1.4			
Z-Ala-ONp	2.1	(1.3)	0.152	(0.11)	734	(846)	139.					
Z-Gly-Ala-ONp	122	(139)	1.27	(0.95)	104	(68)	16.65			58	8.3	7.0
Boc-Ala-OTcp	0.7	(0.5)	0.29	(0.29)	4,000	(5800)	73.	2.6	1			
Z-Ala-OTcp	1.8	(2.1)	0.30	(0.27)	1,616	(1285)	71.					
Boc-Ala-OPcp	0.3	(0.19)	0.24	(0.47)	7,677	(24736)	89.	2.7	2.1			
Z-Ala-OPcp	0.825	(0.825)	0.506	(0.506)	6,133	(6133)	42.					
Z-Gly-Ala-OPcp	112[b]	(87)	3.2[b]	(4.3)	284	(494)	6.6			135	6.3	21.7
Boc-Phe-OSu	86	(33)	2.2	(4.0)	255	(1210)	9.6	1.6	1.8			
Z-Phe-OSu	139	(146)	4	(4.7)	287	(321)	5.23					
Boc-Phe-OPfp	21	(20)	11.6	(12.2)	5478	(6100)	1.8	3.9	1			
Z-Phe-OPfp	82	(88)	11.9	(13)	1452	(1477)	1.8					
Boc-Phe-ONp	4.8	(2.2)	0.05	(0.07)	98.5	(318)	423.	1.9	1.5			
Z-Phe-ONp	9.2	(9.4)	0.03	(0.08)	34.78	(85)	705.					
Z-Gly-Phe-ONp	187	(984)	0.31	(0.66)	17	(6.7)	68			20	9.7	2.0
Boc-Phe-OTcp	4.6	(3.4)	0.15	(0.17)	334	(506)	136	2.6	1.3			
Z-Phe-OTcp	12	(14.6)	0.20	(0.19)	170	(130)	108					
Boc-Phe-OPcp	1.0	(1.3)	0.33	(0.33)	3284	(2538)	63	3.2	0.9			
Z-Phe-OPcp	3.3	(5.9)	0.29	(0.35)	871	(593)	73					
Z-Gly-Phe-OPcp	431	(615)	2.12	(3.0)	49	(49)	10			130	7.3	17.9
Boc-Gly-Phe-OPcp	404	(141)	—	(0.7)	—	(50)	—	1.06		404[c]		

Table 1 *(Continued)*

Compound	k_r	(k_r)	k_c	(k_c)	$\frac{k_c}{k_r}$	$\frac{(k_c)}{(k_r)}$	t_{99} (hr)	$\frac{k_{r_1}}{k_{r_2}}$	$\frac{k_{c_1}}{k_{c_2}}$	$\frac{k_{r_3}}{k_{r_1}}$	$\frac{k_{c_3}}{k_{c_1}}$	$\frac{k_{c_1}/k_{r_1}}{k_{c_3}/k_{r_3}}$
Boc-Met-OSu	47	(22.4)	2.2	(3.4)	467	(1500)	9.6	1.74	1.18			
Z-Met-OSu	82	(97)	2.6	(3.6)	315	(370)	8.1					
Boc-Met-OPfp	25	(13.5)	11	(9.4)	4494	(7030)	1.9	2.28	1.51			
Z-Met-OPfp	56	(59)	17	(10)	2980	(1724)	1.2			83.6	3.63	22.9
Z-Gly-Met-OPfp	4710	(6146)	61	(85)	130	(138)	0.35					
Boc-Met-ONp	4.9	(1.44)	0.09	(0.06)	188	(394)	235	1.77	0.83			
Z-Met-ONp	8.7	(6.2)	0.08	(0.06)	88	(97)	264			35.6	7.24	4.9
Z-Gly-Met-ONp	308	(656)	0.55	(0.51)	18	(7.8)	38.5					
Boc-Met-OPcp	1.5	(0.9)	0.29	(0.26)	1940	(2865)	73	1.87	1.24			
Z-Met-OPcp	2.8	(3.9)	0.36	(0.27)	1286	(698)	59			112	5.11	21.8
Z-Gly-Met-OPcp	313	(410)	1.84	(2.3)	59	(57)	11.5					

[a] Racemization was carried out with 1, 7, or 35 equivalents of triethylamine. The k_r values are in $10^{-6}\ M^{-1}\ sec^{-1}$. The coupling was carried out with equimolar H-Val-OMe at a concentration of 0.13 M; k_c values are in $10^{-2}\ M^{-1}\ sec^{-1}$. The calculated values are in parenthesis. The k_{r_1}, k_{r_2}, k_{r_3} and k_{c_1}, k_{c_2}, k_{c_3} are rate constants of the N^{α}-benzyloxycarbonyl, N^{α}-*tert*-butyloxycarbonyl, and the corresponding dipeptide active esters respectively. The $t_{99\%}$ is the coupling time for 99% completion.

[b] Boc-Ala-OSu and Z-Gly-Ala-OPcp were investigated at 0.05 M concentration due to their low solubility.

[c] This value is k_{r_3}/k_{r_2}.

Table II. Racemization and Coupling Rates in Tetrahydrofuran of Side-Chain Protected Active Ester Derivatives of N^{α}-*tert*-Butyloxycarbonyl- and Benzyloxycarbonylglutamic Acid, -Aspartic Acid, -Glycyl-Glutamic Acid, and -Glycyl-Aspartic Acid[a]

Compound	k_r	(k_r)	k_c	(k_c)	$\frac{k_c}{k_r}$	$\frac{(k_c)}{(k_r)}$	$t_{99\%}$ (hr)	$\frac{k_{r_1}}{k_{r_2}}$	$\frac{k_{c_1}}{k_{c_2}}$	$\frac{k_{r_3}}{k_{r_1}}$	$\frac{k_{c_3}}{k_{c_1}}$	$\frac{k_{c_1}/k_{r_1}}{k_{c_3}/k_{r_3}}$
Z-Glu(OMe)-OSu	45	(51)	2.7	(3.14)	600	(616)	7.85					
Z-Glu(OMe)-OPfp	32	(31)	11.6	(8.8)	3625	(2838)	1.82					
Boc-Glu(OBzl)-OPfp	19	(11.7)	9.5	(14)	5000	(11966)	2.23					
Z-Glu(OBzl)-OPfp	41	(51)	12.1	(15)	2951	(2941)	1.75	2.2	1.3			
Z-Glu(OMe)-OTcp	6	(5.1)	0.1	(0.12)	170	(235)	212					
Z-Glu(OMe)-ONp	3	(3.3)	0.04	(0.052)	150	(157)	529					
Z-Gly-Glu(OMe)-ONp	246	(347)	0.58	(0.45)	24	(13)	36			82	15	6.2
Z-Glu(OMe)-OPcp	2	(2.1)	0.16	(0.24)	800	(1143)	132					
Z-Gly-Glu(OMe)-OPcp	540	(217)	1.75	(2.0)	32	(92)	12			270	11	26
Boc-Glu(OBzl)-OPcp	1.26	(0.8)	0.28	(0.38)	2222	(4750)	75					
Z-Glu(OBzl)-OPcp	1.6	(3.4)	0.42	(0.4)	2625	(1176)	50	1.3				
Z-Asp(OMe)-OPfp	244	(248)	14.7	(16.9)	605	(681)	1.44					
Z-Asp(OMe)-ONp	27	(26.4)	0.07	(0.1)	27	(37.9)	302					
Z-Asp(OMe)-OTcp	35	(41.3)	0.26	(0.24)	75	(58)	81					
Z-Asp(OMe)-OPcp	17.6	(16.5)	0.74	(0.46)	420	(279)	28.6					
Z-Gly-Asp(OMe)-OPcp	1650	(1730)	1.03	(3.9)	6.24	(22.5)	20.5			94	1.4	67.3

[a] See footnote *a* in Table I.

Table III. Racemization and Coupling Rates in Tetrahydrofuran of Active Ester Derivatives of N^{α}-*tert*-Butyloxycarbonyl- and Benzyloxycarbonylserine, -Tryptophan, -Lysine, -Glycyl-Serine, -Glycyl-Tryptophan, and -Glycyl-Lysine[a]

Compound	k_r	(k_r)	k_c	(k_c)	$\frac{k_c}{k_r}$	$\frac{(k_c)}{(k_r)}$	$t_{99\%}$ (hr)	$\frac{k_{r_1}}{k_{r_2}}$	$\frac{k_{c_1}}{k_{c_1}}$	$\frac{k_{r_3}}{k_{r_1}}$	$\frac{k_{c_3}}{k_{c_1}}$	$\frac{k_{c_1}/k_{r_1}}{k_{c_3}/k_{r_3}}$
Z-Ser-OPfp	297	(359)	19.2	(16)	645	(446)	1.1					
Z-Ser-OPcp	26.3	(23.9)	0.74	(0.66)	281	(275)	27.6					
Z-Gly-Ser-OPcp[b]	846.6	(2512)		(5.6)		(22)				32		
Z-Gly-Ser(Bzl)-ONp	539		0.21		3.71		105.7					
Z-Trp-OPfp	10.7	(11)	12.4	(11.2)	11575	(10182)	1.7					
Z-Trp-OPcp	0.72	(0.73)	0.32	(0.304)	4470	(4164)	66					
Z-Gly-Trp-OPcp	190	(77)	4.5	(2.6)	236	(338)	4.7			263	14.1	19
Z-Lys(Boc)-OPcp	0.6		0.22		3666		96.1					
Z-Gly-Lys(Boc)-OPcp	168		0.76		45.2		27.8			280	3.5	81

[a] See footnote *a* in Table I.

[b] Coupling could not be measured by ir due to its low solubility in THF.

Table IV. Racemization and Coupling Rates in Tetrahydrofuran of Active Ester Derivatives of N^{α}-*tert*-Butyloxycarbonyl- and Benzyloxycarbonylvaline, -Cysteine, -Proline, -Histidine, -Glycyl-Valine, -Glycyl-Cysteine, and -Glycyl-Proline[a]

Compound	k_r	(k_r)	k_c	(k_c)	$\frac{k_c}{k_r}$	$\frac{(k_c)}{(k_r)}$	$t_{99\%}$ (hr)	$\frac{k_{r_1}}{k_{r_2}}$	$\frac{k_{c_1}}{k_{c_2}}$	$\frac{k_{r_3}}{k_{r_1}}$	$\frac{k_{c_3}}{k_{c_1}}$	$\frac{k_{c_1}/k_{r_1}}{k_{c_3}/k_{r_3}}$
Boc-Val-OSu	12	(2.6)	0.045	(0.32)	38	(1230)	470	2.4	3.3			
Z-Val-OSu	28	(11.3)	0.15	(0.34)	53	(300)	141					
Z-Gly-Val-OSu	87	(1188)	1.3	(2.9)	148	(24.6)	16.3			3.1	8.6	0.36
Boc-Val-OPfp	3.6	(1.6)	0.31	(0.22)	873	(1375)	68	2.8	0.6			
Z-Val-OPfp	10	(6.8)	0.19	(0.96)	187	(1409)	111					
Z-Val-ONp	0.7	(0.7)		(0.006)		(82)						
Boc-Val-OPcp	0.2	(0.1)	0.029	(0.025)	1450	(2500)	729	2.3	0.9			
Z-Val-OPcp	0.45	(0.45)	0.026	(0.026)	572	(572)	814					
Z-Gly-Val-OPcp	57	(47)	0.14	(0.2)	25	(46)	151			124	5.4	22.7
Boc-Cys(Bzl)-OSu	888	(1328)	2.68	(8.15)	30.2	(61.4)	7.9	5.49	2.0			
Z-Cys(Bzl)-OSu	4880	(5775)	5.44	(8.7)	11	(15.1)	3.9					
Boc-Cys(Bzl)-OPfp	380	(797)	44.3	(22.7)	1166	(285)	0.5	8.7	0.9			
Z-Cys(Bzl)-OPfp	3300	(3465)	40.4	(24.3)	122	(70)	0.5					
Z-Cys(Bzl)-ODnp	2960	(1650)	18.4	(7.2)	62	(43.6)	1.15					
Z-Cys(Bzl)-ODnp[c]	2900	(1617)	1.73	(0.66)	6	(4.1)	12.2					
Boc-Cys(Bzl)-ONp	46	(85)	0.13	(0.13)	28.3	(15.3)	163	8.6	0.81			
Z-Cys(Bzl)-ONp	394	(370)	0.105	(0.144)	2.65	(3.9)	201					
Z-Gly-Cys(Bzl)-ONp	704	(38,800)	0.6	(1.23)	8.5	(0.32)	35.3			1.8	6	0.31
Boc-Cys(Bzl)-OTcp	95.5	(133)	0.22	(0.32)	23	(24.1)	95	5.11	1.36			
Z-Cys(Bzl)-OTcp	488	(578)	0.3	(0.34)	6.1	(5.9)	71					
Z-Cys(Bzl)-OTcp[c]	80	(44)	0.063	(0.053)	8.0	(12)	336					
Boc-Cys(Bzl)-OPcp	41	(53)	0.52	(0.61)	127	(115)	40.7	10.1	3.3			
Z-Cys(Bzl)-OPcp	414	(231)	1.72	(0.66)	42	(28.6)	12.3					
Z-Gly-Cys(Bzl)-OPcp	1261[b]	(24,300)	5.23[b]	(5.6)	44	(2.3)	4			3	3	0.95
Z-Cys(Bzl)-OTbp[c]	17.2	(9.7)	0.0215	(0.0086)	12.5	(8.9)	984					

Table IV *(Continued)*

Compound	k_r	(k_r)	k_c	(k_c)	$\frac{k_c}{k_r}$	$\frac{(k_c)}{(k_r)}$	$t_{99\%}$ (hr)	$\frac{k_{r_1}}{k_{r_2}}$	$\frac{k_{c_1}}{k_{c_2}}$	$\frac{k_{r_3}}{k_{r_1}}$	$\frac{k_{c_3}}{k_{c_1}}$	$\frac{k_{c_1}/k_{r_1}}{k_{c_3}/k_{r_3}}$
Z-Pro-OPcp	0.00	(0.00)	0.44	(0.25)	∞[d]	(∞)	48					
Z-Gly-Pro-OPcp	0.5		0.686		13720		30.8			∞[d]	1.6	∞[d]
Z-His(Bzl)-OPcp	5600[e]		0.9		1.6		23.5					
Z-His(Z)-OPcp	10[f]		0.23				92					
Z-His(Bzl)-OPcp	Autoracemization for ester concentrations from 0.05 up to 0.15 has a $k_r = 25 \times 10^6$ sec^{-1}.											

[a] See footnote *a* in Table I.

[b] k_r and k_c were investigated at 0.0163 *M* and 0.0184 *M* respectively due to low solubility.

[c] -OTcp and -OTbp are both (2,4,6)substituted phenyl groups; ODnp, 2,6-dinitrophenyl.

[d] No observable racemization in three weeks.

[e] This second-order rate constant is valid up to 0.5 equivalent of triethylamine; beyond 3 equivalents the rate is essentially independent of the base concentrations.

[f] This value is the first-order rate constant with 7 equivalents of triethylamine.

is larger than 1 for most monomers and the corresponding dipeptides. In a few cases this ratio is less than 1, which means that the N^{α}-alkyloxycarbonyl dipeptide active ester gives a higher k_c/k_r ratio than the corresponding monomer. This would indicate that coupling of these N^{α}-alkyloxycarbonyl dipeptide active esters probably will give optically purer products than coupling of the corresponding monomers. These "irregular" amino acids are valine and cysteine, and they are grouped in Table IV with histidine, which racemizes unusually fast, and with proline, which racemizes extremely slowly. In each Table, the $t_{99\%}$ indicates the time necessary for 99% completion of the coupling reaction. The calculated racemization and coupling rate constants (k_c) and (k_r) as well as the (k_c/k_r) ratios are given in parentheses. The method of calculation from the predictive factors will be discussed in Section IX,B together with Kemp's method of calculating the coupling rate constants (Kemp *et al.*, 1974a).

II. RACEMIZATION OF N^{α}-BENZYLOXYCARBONYL- AND N^{α}-tert-BUTYLOXYCARBONYLAMINO ACID ACTIVE ESTERS

A. Effect of Side Chain on the Rate of Racemization by Enolization

The N^{α}-alkyloxycarbonyl-L-amino acid active esters racemize through α-hydrogen abstraction or an enolization mechanism as indicated by Scheme 2.

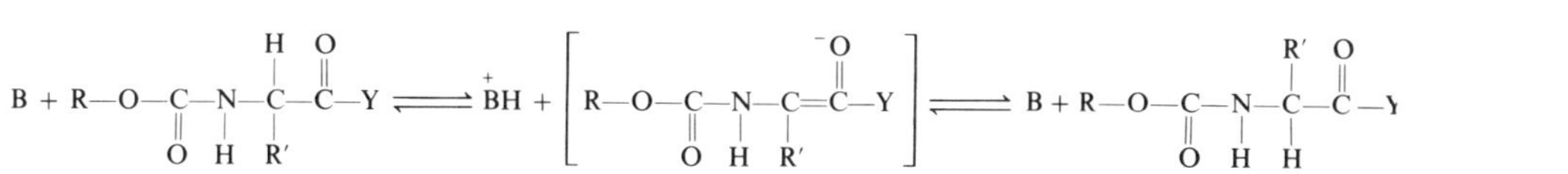

Scheme 2

Table V summarizes the effect of the side chain on the rate of racemization of the N^{α}-benzyloxycarbonylamino acids or side-chain protected derivatives of the frequently used active esters. Alanine was chosen as a standard. It is clear that the side chain has a significant effect on the rate of racemization. In the pentachlorophenyl ester series, N^{im}-benzylhistidine racemizes more than 5600 times faster than alanine and about 12,000 times faster than valine. N^{α}-Benzyloxycarbonyl-*S*-benzylcysteine looses chiral integrity 500 times as fast in this same series and approximately 200 to 300 times as fast in the *N*-succinimidoyl, pentafluorophenyl, *p*-nitrophenyl and 2,4,5-trichlorophenyl ester series. Racemization of all other amino acids in Table V is at least one order of magnitude slower than these two unusually rapidly racemizing

Table V. Comparison of the Side-Chain Effects of N^{α}-Benzyloxycarbonylamino Acid Esters on the Rate of Racemization of Various Active Esters, Relative to Alanine as a Standard

Z-Amino acid	OSu	OPfp	ONp	OTcp	OPcp
His(Z)					44
His(Bzl)					5,600
Cys(Bzl)	223	356	188	271	501
Ser		32			32
Asp(OMe)		26.5	13	19	21
Phe	6.3	9	4.4	6.7	4
Met	3.7	6.1	4.1		3.4
Glu(OMe)	2	3.5	1.4	3.3	2.4
Ala	1	1	1	1	1
Trp		1.2			0.9
Lys(Boc)					0.7
Val	1.3	1.1	0.33		0.55
Pro					0.00

amino acids. At the bottom of Table V are N^{α}-benzyloxycarbonyltryptophan, N^{ε}-*tert*-butyloxycarbonyllysine, valine, and proline, the slowest racemizing amino acids. From these values the order of ease of racemization of amino acids is as follows:

$$\text{His(Bzl)} > \text{Cys(Bzl)} \gg \text{His(Z)}^{*} > \text{Ser} > \text{Asp(OMe)} > \text{Phe} > \text{Met} > \text{Glu(OMe)} > \text{Ala} > \text{Trp} > \text{Lys(Boc)} > \text{Val} \gg \text{Pro} \quad (4)$$

The usefulness of these data for practical peptide synthesis will be discussed later together with the significance of the k_c/k_r values

Barber and Jones (1976) pointed out that the trend in optical lability of amino acids, e.g., Ser(OBzl) > Phe $\simeq$ Met > Ala, corresponds qualitatively to the trend in Taft substituent constants (σ^*) for the side chains. Therefore, the rate of racemization is related to the electron-withdrawing ability of the side chains. However, the N^{α}-protected *S*-benzylcysteine active esters racemize much faster than expected on this basis. They found that the ρ value of + 0.62, from a Hammett plot for the racemization of a series of derivatives of Z-Cys(Bzl)-OPcp with different para-substituents in the *S*-benzyl group (in chloroform by triethylamine), is much higher than expected, and they suggested that the sulfur atom has a special ability to stabilize the negative charge on the α-carbon by overlapping of its *d* orbitals with the α-carbon *p* orbital. They also reported that the racemization of N^{α}-protected cysteine

* Assuming a second-order rate constant.

active esters will be increased with increased electron-withdrawing substitutents on the sulfur.

The racemization of N$^{\alpha}$-protected cysteine active esters proceeds by enolization under either the conditions of coupling or in the presence of a strong tertiary amine. When Z-Cys(Bzl)-OPcp or *p*-nitrophenyl esters were treated with triethylamine in the presence of $C_6H_5—CH_2—^{35}SH$, radioactive sulfur was not incorporated, indicating the absence of β-elimination. On the other hand, dehydroalanine (**2**) yielded 80% of **3** under the same conditions

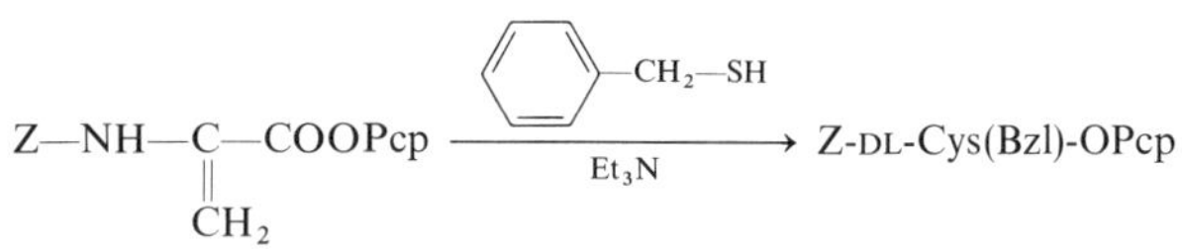

(**2**) (**3**)

(Kovacs *et al.*, 1968, 1970b,c). Most probably, under much more drastic conditions, racemization is practically complete before β-elimination occurs (MacLaren *et al.*, 1958; Bodanszky and Bodanszky, 1967).

In addition, the racemization by enolization of N$^{\alpha}$-protected cysteine active esters in nonpolar solvent proceeds via isoracemization (Mayer and Kovács, 1970; Kovács *et al.*, 1971, 1972a). The racemization rate (k_r) of Z-Cys(Bzl)-OPcp in the presence of Et_3N and a deuterium source ($CH_3—O—D$) in $CHCl_3$ was found to be 1.04×10^{-4} sec^{-1} while the exchange rate $k_e = 5.4 \times 10^{-6}$ sec^{-1} and $k_e/k_r = 0.05$ which, according to Almy and Cram (1969), indicates isoracemization as shown in Scheme 3. The active ester (**4**) and the base form a contact ion pair (**5** and **6**) that racemizes much faster

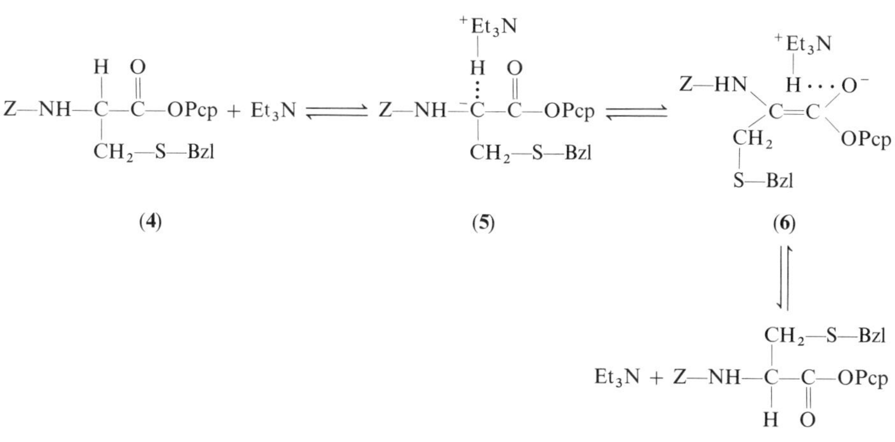

(**4**) (**5**) (**6**) (**7**)

Scheme 3

through a "conducted tour" mechanism than through the formation of a solvent-separated ion pair, which can exchange deuterium with the medium. Isoracemization is probably a general mechanism for amino acid active esters in organic nonpolar solvent: Z-Phe-OPcp also showed an exchange rate value of $k_e/k_r = 0.03$ which indicates isoracemization.

It is important to determine the $k_e/k_r = 1$ value when the extent of racemization is measured by deuterium exchange, otherwise the ensuing values may be too low.

A possible explanation for the extremely slow racemization of proline was proposed by Barber and Jones (1976) as follows: presumably the carbanion formation at the α-carbon can occur only if the charge is delocalized into an enolate. However, enolates of proline active esters (**8**) have unfavorable

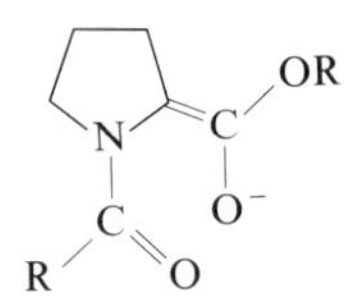

(8)

intramolecular interactions. Molecular models indicate that the *N*-acyl group and the ester oxygen cannot be in one plane as required by structure **8**. Therefore, proline enolization is sterically hindered.

Rapid racemization of histidine was observed during actual synthesis. Windridge and Jorgensen (1970, 1971) detected up to 24% D-histidine during carbodiimide coupling of Boc-His(Bzl)-OH to Tos·H-Glu(OBzl)-OBzl when tributylamine was used to liberate the salt. An intramolecular base catalysis was suggested. Syrier and Beyerman (1974) demonstrated that during the synthesis of thyroliberin by the solid-phase method, 26% D-histidine was formed. Veber (1975) isolated the cyclic acylimidazole derivative (**10**) from the azide (**9**) in neutral or basic solution. Compound **10** is the

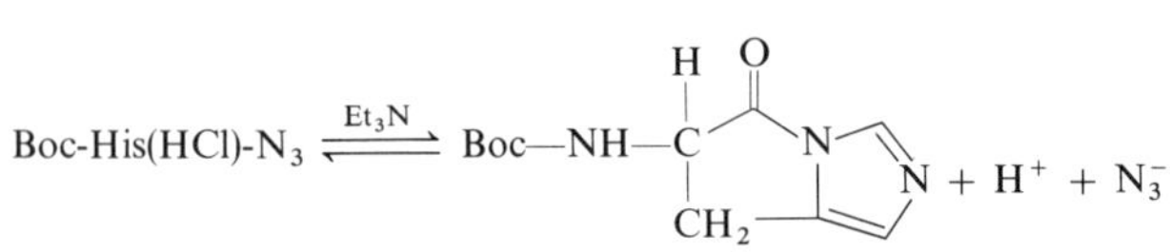

(9) **(10)**

racemizing species and can be detected by ir spectroscopy (1770 cm^{-1}; C=O). Veber suggested that a substituted imidazole in the τ-position does not prevent cyclization and therefore racemization; the safest derivative might prove to be a protecting group in the π-position.

Preliminary racemization rate studies of Z-His(Bzl)-OPcp were carried out both with triethylamine and without base. From the data it appears that there is an intramolecular autoracemization (without triethylamine) and that there is a second-order racemization in the presence of triethylamine, up to about 0.5 equivalents; beyond about three equivalent of triethylamine, the rate is essentially independent of the base concentration. This unusual behavior is under further investigation (see Table IV). The ir spectra show the presence of the active ester only.

Scheme 4 shows a possible explanation for the unusual behavior of histidine in racemization. In one of the possible enol forms (**11**), after α-hydrogen abstraction by triethylamine or another histidine molecule, the imidazole N is sterically close to the negatively charged oxygen. Furthermore, an equilibrium can exist between **11**, $\overset{+}{B}H$, and the protonated imidazole (**12**). Protonated **12** can help to stabilize the enolate intramolecularly.

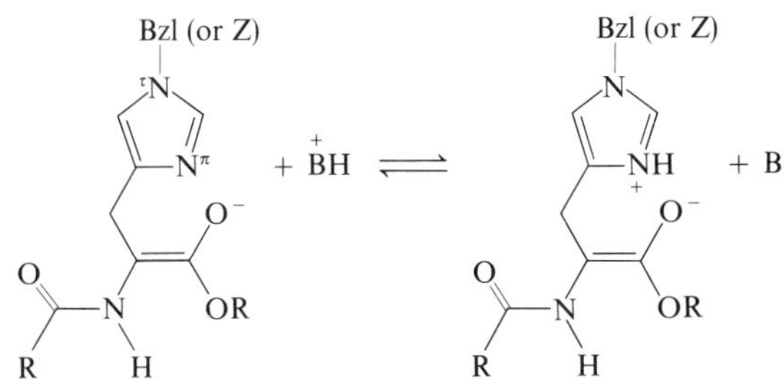

(**11**) (**12**)

Scheme 4

When the τ-nitrogen was protected by the benzyloxycarbonyl group, an electron-withdrawing group, the rate of racemization of Z-His(Z)-OPcp dropped many times and the first-order rate constant value became 10. In Scheme 4 the basicity of **11** decreased; this shifted the equilibrium to the left and decreased the stabilizing effect of **12** on the enolate.

Similar explanations might apply to other amino acids, first of all to those that have heteroatoms adjacent to the β-carbon atom. In the case of serine ($k_r = 26$) for example, the enolate might be stabilized partially by the formation of a hydrogen bond, as indicated by **13**. The stabilizing effect of the

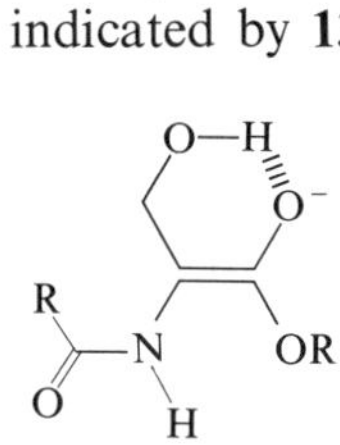

(**13**)

hydrogen bond on the enolate is expected to be less powerful, as indicated by the k_r value. It could be very rewarding to consider the stereochemistry of the planar olefinic enolate in racemization problems.

The rate of racemization of the other amino acids in Table V is mostly parallel to the electron-withdrawing ability of the side chain, with the exception of valine. In the valine case a steric factor, i.e., the bulky isopropyl group, contributes to some extent to the slow racemization rate (Khurgin and Dmitrieva, 1965). The racemization of N^{ε}-*tert*-butyloxycarbonyllysine derivatives is unexpectedly slow, but only preliminary data are presently available.

B. Effect of Active Ester Group on Rate of Racemization

The influence of substituted phenyl ester groups on the racemization rate is mostly parallel with the pK_a value of the corresponding phenol or in general with the electron-withdrawing ability of the ester group (Morawiec *et al.*, 1971). The increasing order of racemizing ability of the active ester groups for N^{α}-benzyloxycarbonylalanine (from Table I) is the following:

$$\begin{array}{ccccccccc} \text{OPcp} & < & \text{OTcp} & < & \text{ONp} & < & \text{OPfp} & < & \text{OSu} \\ 1 & & 2.2 & & 2.5 & & 11.1 & & 26.5 \end{array} \qquad (5)$$

This is the general trend for most amino acids.

The most surprising exception is the pentafluorophenyl ester which racemizes 8 to 22 times faster than the pentachlorophenyl ester, although the pK value of the pentafluorophenol is almost the same as that of pentachlorophenyl (5.5 and 5.2, respectively). The electron-withdrawing ability of the pentafluorophenyl group was investigated in detail by Sheppard (1970) who concluded on the basis of substituent parameters that the pentafluorophenyl group is inductively electron withdrawing, is almost identical to the pentachlorophenyl group, but shows insignificant electron-withdrawing effect by resonance. Kisfaludy *et al.* (1976) investigated in detail the structure of pentafluorophenyl esters of several amino acids, including an X-ray analysis of Boc-Phe-OPfp. They suggested that the high reactivity of the pentafluorophenyl esters in aminolysis is due partly to neighboring group participation by the fluorine atom in the *ortho*-position, which accelerates the collapse of the tetrahedral intermediate in transition state, and partly to the excellent solvation of the leaving $^{-}O—C_6F_5$ group.

We might propose that the enolate (**14**) also becomes stabilized by specific solvent interactions associated with the highly fluorinated ester group. Sheppard (1970) has explained the apparent anomaly that $C_6F_5—COOH$ is a weaker acid than benzoic acid by strong solvation of the carboxylate ion.

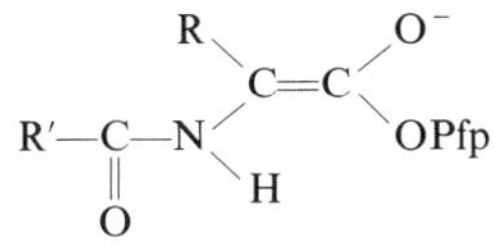

(14)

C. Effect of N^{α}-Benzyloxycarbonyl and N^{α}-*tert*-Butyloxycarbonyl Protecting Groups on Rate of Racemization

The effect of the two amine protecting groups, N^{α}-benzyloxycarbonyl and N^{α}-*tert*-butyloxycarbonyl, on the rate of racemization were studied, and the results are shown in Tables I–IV under the k_{r_1}/k_{r_2} columns. The N^{α}-*tert*-butyloxycarbonyl amino acids racemize about two to three times slower than the corresponding N^{α}-benzyloxycarbonyl amino acids. With cysteine derivatives the difference is larger; for example Boc-Cys(Bzl)-OPcp racemizes 10 times slower than Z-Cys(Bzl)-OPcp.

III. RACEMIZATION OF N^{α}-BENZYLOXYCARBONYL- AND N^{α}-*tert*-BUTYLOXYCARBONYLGLYCYLAMINO ACID ACTIVE ESTERS (DIPEPTIDES)

A. General Consideration

The dipeptide active esters racemize through the 5(4*H*)-oxazolone intermediate, as indicated by Scheme 5. It is understood that the racemization rates reported in the tables are the rates of 5(4*H*)-oxazolone formation. The mechanism of racemization through 5(4*H*)-oxazolone is discussed in detail by Kemp in Volume 1 of "The Peptides." There are two exceptions among the amino acids, i.e., the proline and the cysteine dipeptides, which racemize mainly by the enolization mechanism. Study of the deuterium isotope effect gave supporting evidence for the enolization mechanism in the case of Z-Gly-Cys(Bzl)-ONp, which will be discussed later.

Racemization of the dipeptides is much faster than of the monomers, and this is indicated by the k_{r_3}/k_{r_1} ratio in the tables. This ratio is around 100; if it is much less, as in the case of Z-Gly-Cys(Bzl)-ONp and Z-Gly-Cys-(Bzl)-OPcp where the ratios are 1.8 and 3, respectively, it will most probably indicate deviation from the 5(4*H*)-oxazolone mechanisms. These dipeptides are listed in Table IV.

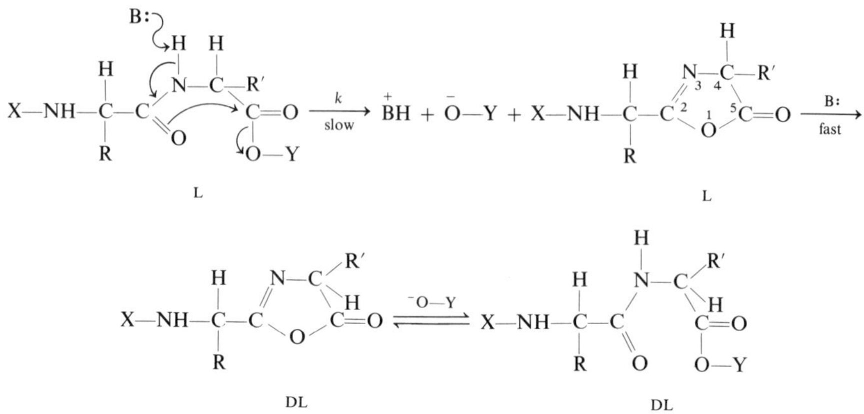

Scheme 5A. Racemization and epimerization through general base catalyzed 5(4*H*)-oxazolone formation in less polar solvent.

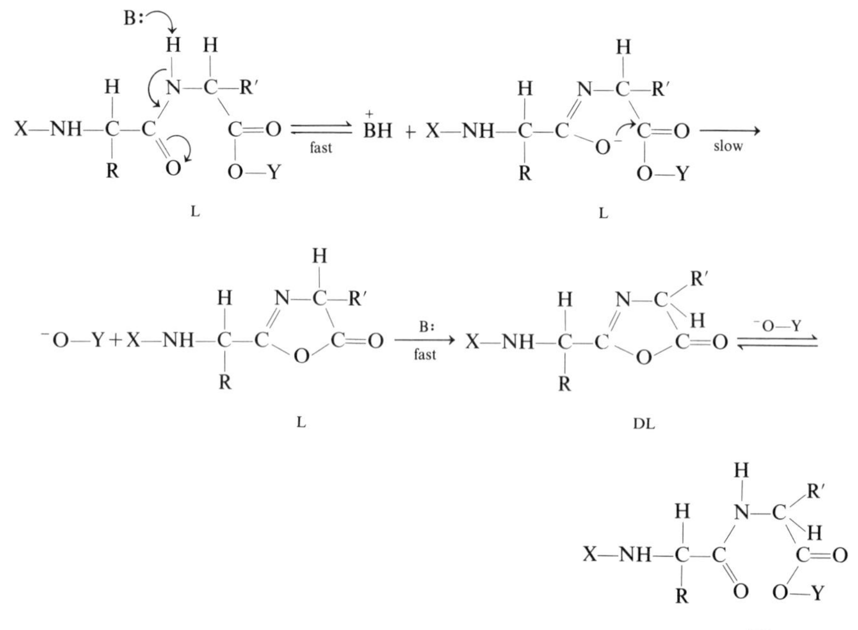

Scheme 5B. Racemization and epimerization through specific base catalyzed 5(4*H*)-oxazolone formation in polar solvent.

B. Effect of 2-Position Side Chain of N^{α}-Protected Dipeptide Active Esters on Racemization Rate

A comparison can be made in the pentachlorophenyl ester series if Z-Gly-Ala-OPcp is again chosen as standard:

$$\underset{14.7}{\text{Asp(OMe)}} > \underset{11.3}{\text{Cys(Bzl)}} > \underset{7.6}{\text{Ser}} > \underset{4.8}{\text{Glu(OMe)}} > \underset{3.8}{\text{Phe}} > \underset{2.8}{\text{Met}} > \underset{1.7}{\text{Trp}} > \underset{1.5}{\text{Lys(Boc)}} > \underset{1}{\text{Ala}} > \underset{0.51}{\text{Val}} \gg \underset{0.005}{\text{Pro}} \qquad (6)$$

As the numbers show, an about 30-fold rate change was observed between aspartic acid β-methyl ester and valine. The effect of the side chain on the racemization rate of dipeptides in tetrahydrofuran solution is far less significant than in the case of monomers for which the mechanism is enolization. The exceptions are proline, which cannot form 5(4*H*)-oxazolone and cysteine, which will be discussed later. According to Goodman and Stueben (1962) and Goodman and Glaser (1970b), the amide oxygen in proline derivatives is prevented from intramolecular attack on the activated function because it is held out of the plane of the ester carbonyl by the rigid five-membered ring. Williams and Young (1964) suggested that a salt of the 5(4*H*)-oxazolone can account for the small amount of racemization occurring in preparation of *p*-nitrobenzoyl-L-proline *p*-nitrophenyl ester by the dicyclohexylcarbodiimide method.

C. Effect of Active Ester Group on Racemization Rate

No significant differences are observed for the various active esters, with the exception of pentafluorophenyl ester, which again increases the rate of racemization about 15 times when compared with *p*-nitrophenyl or penta-chlorophenyl esters of carbobenzoxyglycyl-methionine. There is a poor correlation between the rate of 5(4*H*)-oxazolone ring formation and the electron-withdrawing ability of the substituted phenyl ester group in tetrahydrofuran solution, e.g., Z-Gly-Ala-ONp racemizes faster than the corresponding pentachlorophenyl ester. In the case of pentafluorophenyl esters, the equilibrium between the 5(4*H*)-oxazolone (**15**) and Z-Gly-Met-OPfp probably favors **15** due to the unusual solvation ability of the

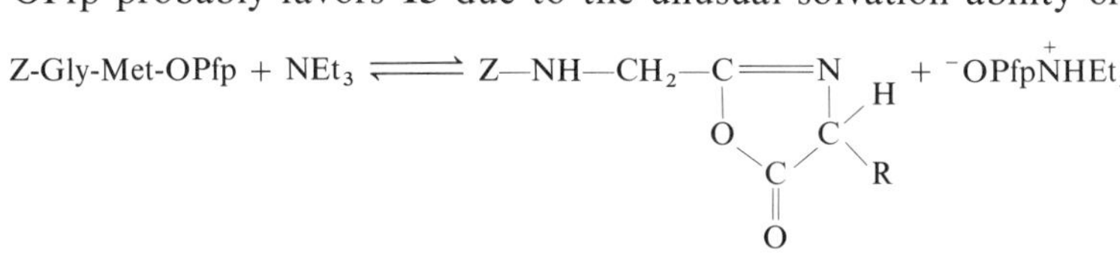

(**15**)

fluorinated pentafluorophenyl leaving group: A strong peak at 5.5 mμ in the ir indicates the presence of a substantial amount of 5(4H)-oxazolone.

D. Ratio of Racemization Rates of N^{α}-Alkyloxycarbonylamino Acid and Corresponding Glycylamino Acid (Dipeptide) Esters

Important conclusions can be drawn from the ratio of the rates of racemization of the amino acid and the dipeptide active esters, k_{r_3}/k_{r_1}. The order of these ratios is given in Eq. (7) for the pentachlorophenyl esters. A similar order was observed for p-nitrophenyl esters. The first in this series is cysteine

Cys(Bzl)	≪ Ser	< Asp(OMe)	< Met	< Val	< Phe	< Ala	< Trp	< Glu(OMe)	< Lys(Boc)
3	32	94	112	124	130	135	263	270	280

≪ Pro (7)

for which the k_{r_3}/k_{r_1} ratio is only 3; this dipeptide does not racemize by the usual 5(4H)-oxazolone mechanism but mostly by enolization, which was unexpected. The supporting evidence will be presented in Section III,E. The enolization mechanism may be operating for the proline dipeptides as well. In the next group, aspartic acid, methionine, valine, phenylalanine, and alanine dipeptides racemize about 100 times faster than the corresponding monomers. Serine is between these values and is suspected to racemize, at least partly, by the enolization mechanism. In the last group are tryptophan, glutamine acid, and lysine, which give k_{r_3}/k_{r_1} ratios of about 300. All dipeptide active esters in the last two groups are expected to racemize by the 5(4H)-oxazolone mechanism.

Whenever the k_{r_3}/k_{r_1} ratio is small, it is reasonable to assume that racemization of those dipeptides will deviate from the 5(4H)-oxazolone mechanism and that they will racemize mostly by enolization. This implies that if the coupling rates of monomers and dipeptides are close, the dipeptide active ester probably will yield the same amount of undesired epimer as will the corresponding monomer active ester. This in turn would mean that in practical peptide synthesis, segment condensation at these amino acids would not be more dangerous than the stepwise method of lengthening the peptide chain, provided the activation step itself is accompanied by minimal racemization. Among the several procedures used to prepare peptide active esters of high optical purity is the "complex" of pentachlorophenol or pentafluorophenol with dicyclohexylcarbodiimide (Kovacs *et al.*, 1967a,b, 1969; Kisfaludy *et al.*, 1967, 1973a,b). Due to "racemization free' and extremely fast coupling, the pentafluorophenyl esters prepared through the "complex" of pentafluorophenol proved to be particularly useful for the preparation of high-molecular-weight, biologically active peptides. Kisfaludy and his co-workers (1972, 1974; Kisfaludy and Schon, 1975) developed a novel, rapid peptide synthesis in solution which is discussed in Chapter 6 of this volume. The "complex" is an isourea derivative (**16**).

$$C_6H_{11}{-}NH{-}\underset{\substack{|\\ OPcP\ (or\ {-}OPfp)}}{C}{=}N{-}C_6H_{11}\cdot 2HOPcp\ (or\ HOPfp)$$

(**16**)

E. Racemization by Enolization Mechanism of Z-Gly-Cys(Bzl)-ONp

The unusually slow racemization of Z-Gly-Cys(Bzl)-ONp relative to Z-Cys(Bzl)-ONp ($k_{r_3}/k_{r_1} = 1.8$), the fact that the ir did not indicate the presence of 5(4*H*)-oxazolone during racemization, and the large deviation of the calculated (k_r) value (38,800) from the experimental k_r(704) (discussion of calculations of rate constants, see Section IX, A), all indicated deviation from the usual 5(4*H*)-oxazolone mechanism. It was concluded that the cysteine dipeptide active esters most probably racemize through α-hydrogen abstraction.

Kemp and Rebek (1970) demonstrated that the kinetic isotope effect can be used to distinguish the enolization from the oxazolone mechanism, provided that the oxazolone formation (k_{cyc}) is rate determining and that the racemization of 5(4*H*)-oxazolone (k_{rac}) is much faster than coupling (k_c). Under these conditions the 5(4*H*)-oxazolone mechanism does not show a deuterium isotope effect, while enolization is expected to give rise to a substantial isotope effect. Kemp and Rebek (1970) used the highly accurate isotope-dilution method (Kemp *et al.*, 1970) and established that racemization observed during coupling of the Anderson peptide by the mixed anhydride method, Kemp's reagent (Kemp and Chien, 1967), or Woodward's reagent (Kemp and Woodward, 1965) proceeds via the 5(4*H*)-oxazolone mechanism while racemization during azide coupling in the presence of triethylamine follows the enolization mechanism.

The α-deuterated active ester, Z-[α-^{2}H]Cys(Bzl)-ONp, when racemized under the same conditions as the undeuterated Z-Cys(Bzl)-ONp (Table IV), gave an isotope effect of $k_H/k_D = 7.5$. During the racemization (94% DL), 36% of the amide hydrogen was exchanged with α-deuterium as detected by mass spectrometry, which somewhat alters the real k_H/k_D value. This isotope effect clearly indicates the enolization mechanism.

The deuterated dipeptide Z-Gly-[α-^{2}H]Cys(Bzl)-ONp (**17**) was prepared from the deuterated monomer by the backing off procedure (Goodman and Stueben, 1959) and showed 77% α-deuterium incorporation by mass spectrometry. Racemization of [α-^{2}H]dipeptide active ester **17** under the same conditions as the undeuterated **17** (Scheme 6) in tetrahydrofuran in the presence of triethylamine showed an isotope effect of $k_H/k_D = 2.5$. This indicates that proton abstraction is the rate-determining step (Kovacs *et al.*, 1977a).

It was necessary to establish that the 5(4*H*)-oxazolone (**18**) of Z-Gly-Cys(Bzl)-OH racemizes much faster than it couples. 5(4*H*)-Oxazolone (**18**) was obtained from the dipeptide (using dicyclohexylcarbodiimide; it is important to use 25% excess of dicyclohexylcarbodiimide, 2 min. cyclization time at 22°C, but −70°C was maintained during the isolation procedure) in 73% optical purity, as determined by conversion to the hydrazide without significant racemization (hydrazine being an α-nucleophile; Goodman and Glaser, 1970a). Alpha-deuterated **18** was similarly prepared. The k_r value for autoracemization of **18** in tetrahydrofuran was 1015×10^{-6} sec^{-1} ($C = 0.1248$ *M*) and that for the α-deuterated **18** was 365×10^{-6} sec^{-1} ($C = 0.1291$ *M*), which indicates an isotope effect of 2.8 to 2.9. The racemization rate is quite slow at −70°C in tetrahydrofuran. This very large temperature coefficient (only 20% decrease in 3 days) supports the observation that, e.g., mixed anhydride coupling at −15°C is practically racemization free for the Anderson dipeptide (Anderson *et al.*, 1967). Racemization is instan-

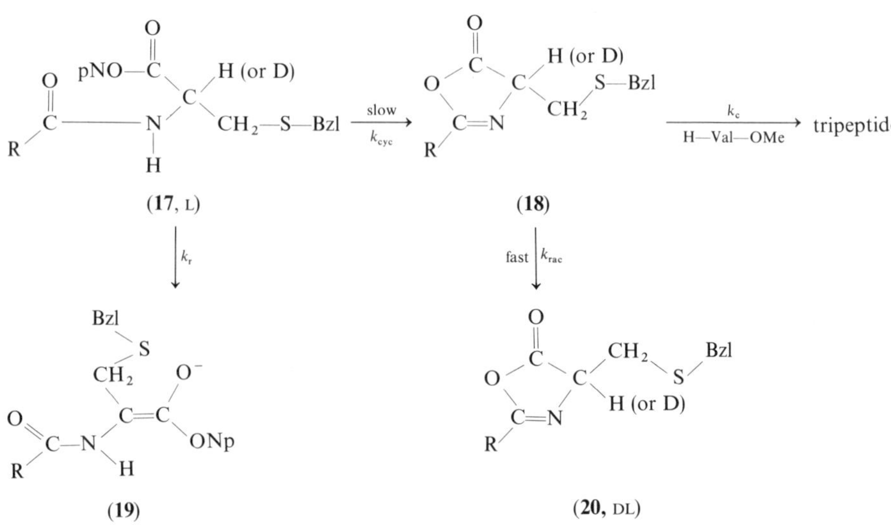

Scheme 6

taneous from triethylamine (measured on Cary 60 spectropolarimeter within 7 sec). Racemization with *N*-methylmorpholine of both the undeuterated **18** and the α-deuterated **18** gave k_r values of 1.1 M^{-1} sec^{-1} and 0.289 M^{-1} sec^{-1}, respectively, which indicates an isotope effect of 3.83. The coupling rate of the 5(4*H*)-oxazolone (**18**) with H-Val-OMe was 8.5×10^{-2} M^{-1} sec^{-1} (ir), i.e., the coupling is much slower than racemization by Et_3N ($k_r \gg k_c$) or even by *N*-methylmorpholine. Racemization during coupling with H-Val-OMe was estimated from data obtained on a Cay 60 spectropolarimeter. The α-values of the 5(4*H*)-oxazolone, H-Val-OMe, and the LL and DL epimers were used for calculation. The epimeric tripeptides, LL and DL, are formed in ratios of 48 and 52, respectively, as determined by high-performance liquid chromatography. The k_r value (0.346 M^{-1} sec^{-1}) was obtained from these measurements and it is still substantially faster than k_c ($k_r = 4k_c$).

It was concluded that Z-Gly-Cys(Bzl)-ONp racemizes largely by the enolization mechanism (Kovacs *et al.*, 1979). The deviation from the 5(4*H*)-oxazolone mechanism is side-chain dependent while the racemization of Z-Gly-Phe-N_3, which is the only known dipeptide derivative which racemizes by enolization, is carboxyl-activation dependent.

It is important to know which racemization mechanism prevails in each case since it may permit us to design synthetic procedures with rational suppression of racemization. Furthermore, the penultimate amino acid is not expected to be affected during racemization via enolization, while it is prone to racemization when the 5(4*H*)-oxazolone mechanism is operating.

A reasonable explanation for the enolization mechanism might be the following: The enolate **19** in Scheme 6 is an olefin and model studies show that the sulfur is close to the negatively charged oxygen. In this conformation it is very likely that the d orbitals of sulfur overlap with the p orbital of oxygen, and by this, substantially increase the stability of the enolate, as indicated by structure **21**. In turn, this quasi five-membered ring will be favored over the 5(4*H*)-oxazolone.

In our opinion the faster racemization of Z-Cys(Bzl)-ONp can be more rationally explained by structure **21** than, as proposed by Barber and Jones (1976), by overlapping the p orbital of the α-carbon and the d orbitals of sulfur, which would form a quasi three-membered ring.

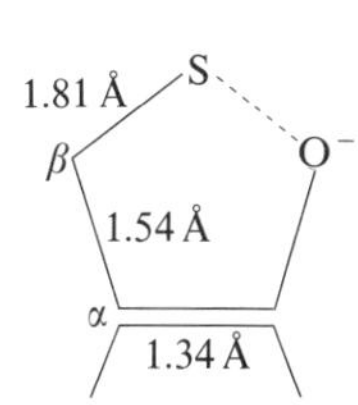

(21)

IV. COUPLING RATES OF N^{α}-ALKYLOXYCARBONYLAMINO ACID ACTIVE ESTERS

A. General Considerations

Since Bodanszky *et al.* (1955) introduced *p*-nitrophenyl esters for peptide synthesis, these esters have been frequently used as models to study both coupling and racemization behavior of the active esters. Kemp (1973) and his co-workers (Kemp *et al.*, 1974a) compared the coupling rates of *p*-nitrophenyl esters of carbobenzoxyglycine, -alanine, -leucine, -proline, -valine, and -phenylalanine in dimethylformamide solution. The coupling amines used in this study were alkyl esters of glycine, alanine, leucine, proline, and valine. This series provides examples for the steric effect of the side chain. The resulting k_c values will be tabulated in Section IX,A (Table X) together with the calculated values.

Kemp also investigated the mechanism of aminolysis of phenyl esters. Their rate of aminolysis depends on the collapse of the reversibly formed tetrahedral intermediate. Three major bond changes are involved in the aminolysis, i.e., a C—N amide bond is formed while C—O ester and N—H amine bonds are broken. In the rate-determining transition state, the acyl carbon and the amide nitrogen are tetrahedral, and the solvent is coordinated with the $—\overset{+}{N}H_2—$as well as with the

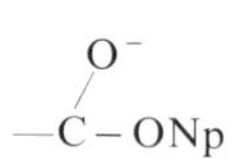

regions. Variation in rate constants reflects steric effects in the transition state. Conformational analysis led to three major conformations for the transition state which are shown below. The first rotamer (**22**) is an *anti* conformation about the

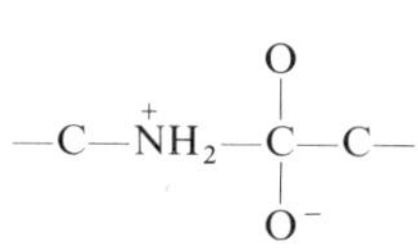

developing C—N bonds, while the two remaining conformations (**23** and **24**) are *gauche* rotamers. (To provide for easy recognition of these conformers, they are built around the cyclohexane skeleton and the nonexisting bonds are dotted lines.) These conformers show the group interactions which are seen at the α-carbons of both the amine and active ester components.

Our study includes amino acids with heteroatoms in the side chains, such as cysteine, serine, glutamic acid, and aspartic acid. In these cases inductive effects of the side chains are expected to influence the rate constants. Further-

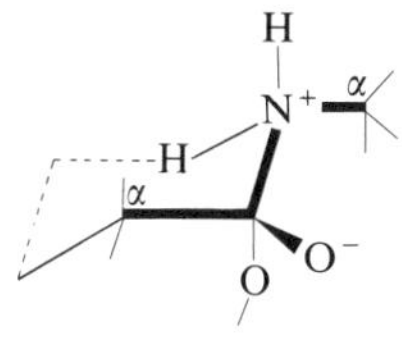

(22)

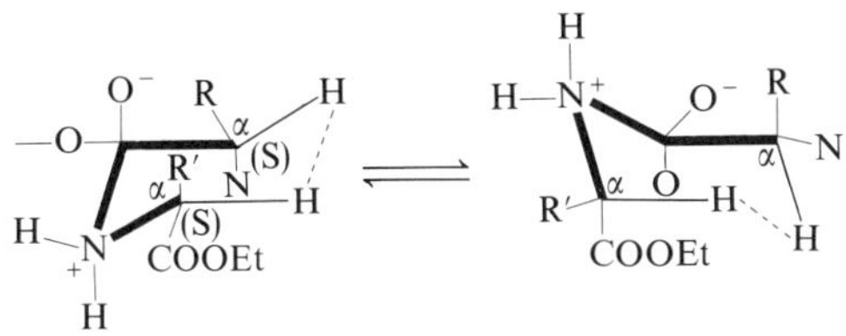

(23) **(24)**

more, the influences of the active ester group as well as of the N^{α}-benzyloxycarbonyl and N^{α}-*tert*-butyloxycarbonyl groups on the coupling rates were also studied.

B. Influence of Side Chain on Coupling Rate

Since the side chain is not directly attached to the carbonyl carbon, the site of the new bond formation, a less significant effect on the rate of coupling is expected than by variation of the active ester groups.

In the N^{α}-benzyloxycarbonyl-protected pentachlorophenyl ester series, the decreasing order of amino acids (from Tables I–IV) is given below [Eq. (8)]. Z-Ala-OPcp was chosen as standard; its coupling constant, k_c, equals $0.506 \times 10^{-2}\ M^{-1}\ \sec^{-1}$.

$$\underset{3.4}{\text{Cys(Bzl)}} > \underset{1.78}{\text{His(Bzl)}} > \underset{1.46}{\text{Ser}} \simeq \underset{1.46}{\text{Asp(OMe)}} > \underset{1}{\text{Ala}} > \underset{0.87}{\text{Pro}} > \underset{0.71}{\text{Met}} > \underset{0.63}{\text{Trp}} > \underset{0.57}{\text{Phe}} > \underset{0.45}{\text{His(Z)}}$$

$$> \underset{0.43}{\text{Lys(Boc)}} > \underset{0.32}{\text{Glu(OMe)}} \gg \underset{0.05}{\text{Val}} \tag{8}$$

In this order, the amino acids which have heteroatoms such as S, O, carboxyl, or imidazole on the β-carbon atom (as in *S*-benzylcysteine, serine, aspartic acid β-methyl ester, and N^{im}-benzylhistidine) all couple faster than the other amino acids, except N^{im}-benzyloxycarbonylhistidine.

A reasonable explanation for N^{im}-benzylhistidine might be the following: Considering the transition states (**22**–**24**), models clearly show that the imidazole N can be in the proper position sterically to act as a base; that is, the proton is transferred intramolecularly from the amine to the imidazole nitrogen as shown below for the *anti* conformation (25). Similar favorable steric situations exist for the *gauche* conformation in intramolecular base catalysis. When the basicity of the imidazole is decreased by protecting it

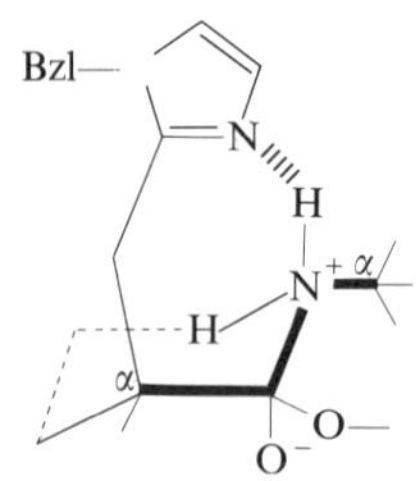

(25)

with a N^{α}-benzyloxycarbonyl group, this intramolecular proton transfer is minimized and the N^{im}-benzyloxycarbonylhistidine active ester couples much slower. Similar considerations could be applied to the other amino acids, such as aspartic acid, but the β-COO- is expected to have a much weaker effect on the coupling rate due to the weaker basicity of the oxygens.

The electron-withdrawing ability of the side chain on the carbonyl carbon is expected to facilitate the coupling, but it should be a weak effect. Valine couples 6 times slower than the next amino acid, N^{α}-benzyloxycarbonyl-glutamic acid γ-methyl ester, but 66 times slower than N^{α}-benzyloxycarbonyl-*S*-benzylcysteine, which is certainly due to steric reasons.

In the pentafluorophenyl ester series, the decreasing order of amino acid coupling rates [Eq. (9)] is practically identical to the pentachlorophenyl ester series, with comparative rates based on alanine. In this series, valine couples 60 times slower then N^{α}-benzyloxycarbonyl-S-benzylcysteine. This is significant and indicates the unusual behavior of pentafluorophenyl esters,

Cys(Bzl) >	Ala $\simeq$	Ser >	Met >	Asp(OMe) >	Trp $\simeq$	Phe $\simeq$	Glu(OMe) $\gg$	Val
2.1	1	0.99	0.88	0.76	0.64	0.62	0.60	0.01

which are the fastest coupling active esters among those frequently used in peptide synthesis. In conclusion, with the exception of valine, the amino acid side chain has a relatively small effect on the coupling rate.

Similar observation can be made when the N^{α}-benzyloxycarbonylamino acid *p*-nitrophenyl esters are coupled with amino acid ethyl esters in dimethylformamide solution (Kemp *et al.*, 1974a). For example, the decreasing order of amino acid coupling rates (with H-Val-OEt) is the following:

(10)

	Gly >	Ala >	Phe >	Leu >	Pro >	Val
In DMF	3.4	2.0	1.0	0.9	0.7	0.1
In THF		0.15	0.03			

C. Influence of Active Ester Group on Coupling Rate

Since the active ester group is directly attached to the reactive center, a very significant influence is exerted. The active esters investigated here are *p*-nitrophenyl, 2,4,5-trichlorophenyl, pentachlorophenyl, pentafluorophenyl, and *N*-succinimidoyl. For these active esters, the rate of coupling is not always parallel to the electron-withdrawing ability of the substituted phenyl ring. It has been mentioned previously that the pentafluorophenyl group has the same electron-withdrawing ability as the pentachlorophenyl (Sheppard, 1970), but the former couples 20 to 73 times faster. Kisfaludy *et al.* (1976) attributed this property of pentafluorophenyl esters to neighboring group participation by the fluorine as well as to excellent solvation of the pentafluorophenyl leaving group. It is worthwhile to compare the aminolysis rates of the pentafluorophenyl esters with those of the *N*-hydroxysuccinimidoyl esters, which were introduced by Anderson *et al.* (1963, 1964) and which are widely used for coupling. These active esters also react rapidly with the amine components. The rapid aminolysis is explained by anchimeric assistance, as indicated in **26**, the transition state using the *anti* conformation.

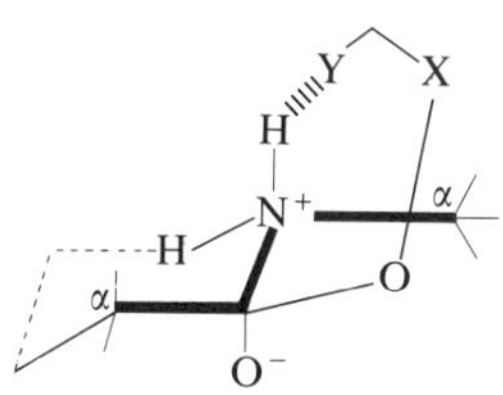

(26)

The active ester group (Y—X) is *N*-succinimidoyl, and the carbonyl oxygen of the *N*-succinimidoyl accepts the proton from the nitrogen and therefore accelerates the collapse of the transition state to the amide (Goodman and Glaser, 1970b; Kemp, 1973, Vol. 1 of "The Peptides"). The order of decreasing reactivity of active esters of N^{α}-benzyloxycarbonyl amino acids is given in Scheme 7. The *p*-nitrophenyl ester was chosen as standard.

	OPfp >	OSu >	OPcp >	OTcp >	ONp
Ala	1287	51	3.4	2	1
Phe	397	133	9.7	6.7	1
Met	212	32.5	4.5		1
Glu(OMe)	290	68	4	2.5	1
Asp(OMe)	210		10.6	3.7	1
Cys(Bzl)	381	51	16	2.9	1

Scheme 7

For all amino acids, compared here in tetrahydrofuran, the pentafluorophenyl esters couple 3 to 25 times faster than the *N*-succinimidoyl esters. In the case of alanine, the pentafluorophenyl ester couples 378 times faster than the pentachlorophenyl ester and 1300 times faster than *p*-nitrophenyl ester! In the case of N^α-benzyloxycarbonyl-*S*-benzylcysteine, the 2,4,6 trichloro- and tribromophenyl esters couple, respectively, 1.6 and 4.7 times slower than *p*-nitrophenyl, and 27 and 80 times slower than the pentachlorophenyl ester. Both esters have substituents in both ortho-positions (Kovács *et al.*, 1970b). If we compare the coupling rates of the 2,4-dinitrophenyl and 2,6-dinitrophenyl esters, again the former is 10 times faster. It seems plausible that steric effects play an important role in the transition state (**26**) if we compare the size of the F, Cl, Br, and NO_2 groups in the two ortho positions. In addition, it is reasonable to assume that in **26** the fluorine (=Y) creates an electron-rich environment and that local dipole interaction will accelerate the aminolysis. Another important effect is the extremely good solvation ability of the leaving pentafluorophenyl group. For Z-Val-ONp k_c is not measured in tetrahydrofuran, but the Z-Val-OPcp couples 20 times slower than Z-Ala-OPcp. However, the coupling constants of the *N*-succinimidoyl and pentafluorophenyl esters with H-Val-OMe are slightly higher than the k_c for Z-Ala-ONp. Steric factors are clearly involved in the unusually slow coupling of valine to valine.

The use of either pentafluorophenyl esters or the "pentafluorophenol–dicyclohexylcarbodiimide complex" (Kovács *et al.*, 1967b; Kisfaludy *et al.*, 1972) seem to be among the best procedures at the present time for the rapid synthesis in solution of medium-sized, biologically active peptides, such as the corrected human ACTH. *Oxytocin, the protected nonapeptide, was obtained in 10 hr with an overall yield of 39% (Kisfaludy and Schön, 1975). Angiotensin (Kisfaludy and Nyéki, 1975), the octapeptide, was obtained in 10 hr in 53% yield and with full biological activity.* The main reason is that the pentafluorophenyl esters are the fastest coupling esters, and furthermore, the coupling is "racemization free" or accompanied by negligible amounts of racemization. The model tetrapeptide, Z-Lys(Boc)-(Lys(Boc))$_2$-Lys(Boc)-NH_2, was prepared by coupling of two dipeptides through azide procedure as well as by use of the "pentafluorophenol–DCC complex"; identical optical activity was observed for both products but 19% racemization occurred when dicyclohexylcarbodiimide was used for coupling. The repetitive excess mixed anhydride method (REMA) (Tilak, 1970; Beyerman, 1972) is rapid; however, fragment condensations are probably accompanied by some racemization and other side reactions.

In the use of the pentafluorophenol–DCC complex, the pentafluorophenol acts as a trapping reagent for 5(4*H*)-oxazolone. Kemp's reagent is racemization free and easy to prepare directly from the acid, but not rapid enough; the rate of aminolysis changes only slightly with solvent, due to the intramolecular base catalysis. The phenyl esters in general couple very much faster in dimethylformamide, dimethyl sulfoxide, or other dipolar aprotic solvents, which are needed to dissolve long peptide chains. The catechol-type active esters couple with similar rates in both polar and nonpolar solvents (Kemp *et al.*, 1974b; Jones, 1974). For example, the rates of coupling of catechol monoester (**27**) with H-Gly-OEt in different solvents can be compared with those of Z-Gly-ONp with H-Gly-OEt (see Table VI) (Kemp *et al.*, 1974a,b).

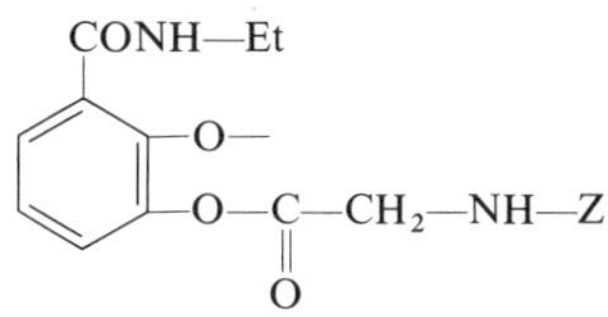

(**27**)

The coupling rates for Z-Gly-Phe-OSu with H-Gly-OEt were determined by Kemp (1973) in different solvents and compared with the coupling for Z-Gly-Phe-OH with H-Gly-OEt through dicyclohexylcarbodiimide. These values are listed in Table VII.

In general, dicyclohexylcarbodiimide couplings are slower in polar solvents, and the rate in dimethylformamide is so slow that it is not practical

Table VI. Coupling Rates of N^{α}-Benzyloxycarbonylglycine Catechol Monoester (**27**) with H-Gly-OEt and Z-Gly-ONp + H-Gly-OEt

Solvent	k_c[a] **27** + H-Gly-OEt	$t_{99\%}$ (hr)[b]	k_c[a] 2-Gly-ONp + H-Gly-OEt	$t_{99\%}$ (hr)[b]
Dioxan	0.25	84	1.16	18
Acetonitrile	1.33	16	1.83	11.5
Dimethylformamide	0.33	64	43.	0.5
Dimethyl sulfoxide	0.33	64	133.	0.15

[a] k_c values are in $10^{-2}\ M^{-1}\ \sec^{-1}$.
[b] For 0.13 *M* solution of both ester and amine.

Table VII. Coupling Rates of Anderson Dipeptide with H-Gly-OEt by OSu and DCC Methods

	Z-Gly-Phe-OSu + H-Gly-OEt		Z-Gly-Phe-OH + H-Gly-OEt + DCC	
Solvent	k_c[a]	$t_{99\%}$ (hr)	k_c[a]	$t_{99\%}$ (hr)[b]
Acetonitrile	61	0.34	6.66	3.2
Dimethylformamide	111	0.19	0.025	845.
Tetrahydrofuran	140	0.15	1.16	18.
Methylene chloride	166	0.13	50	0.42
Toluene	1366	0.015	83	0.25

[a] k_c values are in $10^{-2}\ M^{-1}\ \sec^{-1}$.
[b] For 0.13 *M* solution of both ester and amine.

to use for peptide synthesis. The *N*-succinimidoyl ester couples rapidly; however in the nonpolar solvent, e.g., toluene, the coupling is unusually fast. The calculated k_c value for Z-Gly-Phe-OSu with H-Gly-OEt in tetrahydrofuran is 76 and the actual value, from Table VII, is 140; the calculated k_c for the coupling of Z-Gly-Phe-OPfp with H-Gly-OEt is 228: These values also show that the pentafluorophenyl esters couple faster than the *N*-succinimidoyl esters. Values in toluene can not be compared since no data in this solvent are available.

These data demonstrate the unusual behavior of the pentafluorophenyl esters. We consider that *the pentafluorophenyl ester does not belong in the group of typical active phenyl esters but represent a new type with a different aminolysis mechanism.* The reasons for this can be summarized as follows: a) The pentafluorophenyl ester is the fastest coupling active ester investigated for use in peptide chemistry; b) the pentafluorophenyl esters have unusual high solubility in polar or nonpolar solvent; c) the pentafluorophenyl group is inductively electron-withdrawing and is almost identical with the pentachlorophenyl group, and therefore the inductive effect of former is much less than expected, i.e., "the interaction of the *p*-electrons of the five fluorine and the π-system greatly disturb the electron distribution in the phenyl ring so that it no longer behaves as a typical aromatic system but rather an unusual system that results from p–π interaction (Sheppard, 1970)"; and d) the pentafluorophenol acts as an 5(4*H*)-oxazolone-trapping reagent in the "pentafluorophenol–DCC complex" procedure. This behavior is common for the strongly acidic phenols; acidic environment decreases the rate of racemization of the 5(4*H*)-oxazolone.

The equilibrium between the 5(4*H*)-oxazolone of Z-Gly-Phe-OH and the acidic phenols, as well as the rate of opening of the 5(4*H*)-oxazolone in tetrahydrofuran, were investigated (Kovacs *et al.*, 1969) in both the presence and absence of the strong base, triethylamine. The phenols were *p*-nitro-

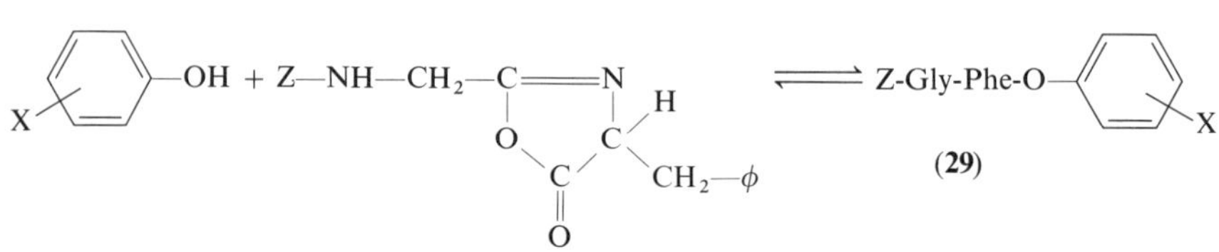

(29)

(28)

phenol, pentachlorophenol, and 2,4-dinitrophenol. With the 1:1:1 ratio of the phenol, 5(4*H*)-oxazolone, and triethylamine, 97%, 52%, and 20% active esters were formed by *p*-nitrophenyl, pentachlorophenyl and 2,4-dinitrophenyl, respectively; the *K* values were 2913, 6.28, and 0.98, respectively. Equilibrium was reached in 15 min in the case of *p*-nitrophenyl, 25 min in the case of 2,4-dinitrophenyl, and 60 min in the case of pentachlorophenyl. In the absence of triethylamine, 15% and 60% active esters were formed by *p*-nitrophenol and 2,4-dinitrophenol; the *K* values were 0.56 and 10.1, respectively. With a 1:1 ratio of pentachlorophenol and 5(4*H*)-oxazolone, equilibrium was not reached within 80 min. With 2 equiv of pentachlorophenol however, practically 100% active ester was formed in 50 min, giving a very large *K* value. This indicates that in neutral solution no 5(4*H*)-oxazolone would be formed from Z-Gly-Phe-OPcp. Values are not available for pentafluorophenol, but it is expected to behave similarly since its p*K* value is very close to that of pentachlorophenol. The order of the rates of 5(4*H*)-oxazolone ring opening is 2,4-dinitrophenol > pentachlorophenol > *p*-nitrophenol. This order is parallel to the acidity of the phenols (p*K* values are 4.1, 5.3, and 7.2, respectively). In the presence of the base this order is reversed ($^{-}$ONp > $^{-}$OPcp > $^{-}$ODnp), which is as expected; that is, the stronger the nucleophile the faster the rate of ring opening. It was concluded that the strongly acidic phenols open the 5(4*H*)-oxazolone much faster than *p*-nitrophenol, and in the presence of excess pentachlorophenol there is no 5(4*H*)-oxazolone present ($K = \infty$). These results suggest a different mechanism for 5(4*H*)-oxazolone ring opening. It was suggested that in a concerted mechanism, the L-oxazolone is protonated by the acidic phenol while another phenol attacks the carbonyl carbon; the driving force for the fast reaction may be attributed to the electron-withdrawing effect of the charged nitrogen. A similar effect occurs when the carboxyl group of Z-Gly-OH, for example, is esterified with pentachlorophenol using dicyclohexylcarbodiimide: The major path to active ester is through direct attack of the more acidic phenols on the acylisourea intermediate. With 2,4-dinitrophenol (pK = 4.1) the reaction proceeds almost exclusively through this path, while with *p*-nitrophenol about 30% of Z-Gly-OH goes through the symmetrical anhydride path. An increase in the acid-to-pentachlorophenol or -pentafluorophenol ratio to 1:3, as in the complexes, will cause practically all active ester formation to proceed through the isourea intermediate. The following data demonstrate that the more acidic phenol will give the more optically pure peptide active ester. When Z-Gly-Phe-ODNp, Z-Gly-Phe-OPcp, and Z-Gly-Phe-ONp were prepared from the dipeptide and the corresponding phenols using dicyclohexylcarbodiimide, the optical purity of the esters was 99%, 90%, and 23%, respectively. Using the "pentachlorophenol–DCC complex" optically pure ester was obtained in 92% yield (Kovács *et al.*, 1969). Similarly optically pure pentafluorophenyl esters were obtained from N^{α}-benzyloxycarbonyl dipeptides of isoleucine (Kisfaludy *et al.*, 1973a).

Rebek and Feitler (1973, 1974) determined that dicyclohexylcarbodiimide-mediated coupling in solution proceeds through *O*-acylisourea, while in solid-phase synthesis the acylating agent is the symmetrical anhydride.

The kinetics of 5(4*H*)-oxazolone ring opening and racemization was extensively studied by Goodman and co-workers—including the ring opening by *p*-nitrophenol—and the results were summarized by Goodman and Glaser (1970b). A pertinent point here is the result of Goodman and Levine (1964) that the equilibrium between Bzl-Phe-ONp and the 5(4*H*)-oxazolone in dioxan in the presence of tributylamine is far in favor of the active ester ($K = 17$). This is in agreement with our findings.

A final point in connection with the high reactivity of the pentafluorophenyl esters is that the frequently mentioned "overactivation" of peptides does not necessarily lead to excessive racemization.

D. Influence of N^{α}-Protecting Group on Coupling Rate

The N^{α}-benzyloxycarbonyl and N^{α}-*tert*-butyloxycarbonyl protecting group effects were determined and the results, $k_{c_Z}/k_{c_{Boc}}$, are summarized in Table VIII. The values indicate that the smallest effect is obtained for the pentafluorophenyl esters; but generally, the N^{α}-*tert*-butyloxycarbonyl group decreases the coupling rate. The maximum effect is 3.3 times. In several cases however, the N^{α}-*tert*-butyloxycarbonyl derivatives are coupled somewhat faster. Since the N^{α}-benzyloxycarbonyl and N^{α}-*tert*-butyloxycarbonyl groups are two atoms away from the reaction center, their effect is not significant. It seems that the effect is steric rather than electronic.

Table VIII. Influence of N^{α}-Benzyloxycarbonyl and N^{α}-*tert*-Butyloxycarbonyl Groups on the Coupling Rate ($k_{c_Z}/k_{c_{Boc}}$)

Compound	OSu	OPfp	ONp	OTcp	OPcp
Ala	1.8	1.1	1.4	1	2.1
Phe	1.8	1	1.5	1.3	0.9
Met	1.18	1.5	0.8		1.24
Glu(Bzl)		1.3			
Val	3.3	0.6			0.9
Cys(Bzl)	2.0	0.9	0.8	1.36	3.3

V. COUPLING RATES OF GLYCYLAMINO ACID (DIPEPTIDE) ACTIVE ESTERS

Most of our investigations were carried out with *p*-nitrophenyl and pentachlorophenyl esters and some data are available for pentafluorophenyl and *N*-succinimidoyl esters as well.

A. Effect of 2-Position Side Chain on Rate of Coupling

In the N^{α}-benzyloxycarbonyl-protected pentachlorophenyl ester series, the decreasing order of glycylamino acids are given in Eq. (11). Alanine was chosen as standard with a coupling rate constant $k_c = 3.2 \times 10^{-2}\,M^{-1}\,\mathrm{sec}^{-1}$. The maximum difference is between valine and cysteine, the former

$$\underset{1.63}{\mathrm{Cys(Bzl)}} > \underset{1.4}{\mathrm{Trp}} > \underset{1}{\mathrm{Ala}} > \underset{0.66}{\mathrm{Phe}} > \underset{0.57}{\mathrm{Met}} \simeq \underset{0.54}{\mathrm{Glu(OMe)}} > \underset{0.32}{\mathrm{Asp(OMe)}} > \underset{0.23}{\mathrm{Lys(Boc)}} \simeq \underset{0.23}{\mathrm{Pro}} > \underset{0.04}{\mathrm{Val}} \tag{11}$$

couples 37 times slower. In general the side chain has a small effect on the coupling rate, due to the distance from the reaction center. Similar conclusions can be drawn for the *p*-nitrophenyl esters from the data in Tables I–IV.

B. Effect of Active Ester Group on Coupling Rate

Significant differences in coupling rates are found among the pentafluorophenyl, *N*-succinimidoyl, pentachlorophenyl, and *p*-nitrophenyl esters. In the case of methionine the decreasing order is shown in Eq. (12). The penta-

$$\underset{61}{\mathrm{OPfp}} \gg \underset{1.84}{\mathrm{OPcp}} > \underset{0.55}{\mathrm{ONp}} \tag{12}$$

fluorophenyl ester couples 111 times faster than the *p*-nitrophenyl ester and 33 times faster than the pentachlorophenyl ester. In general, the pentachlorophenyl esters couple about 4 times faster than the *p*-nitrophenyl esters. The unusual behavior of the pentafluorophenyl ester is again clearly observable. In the case of valine, the *N*-succinimidoyl ester couples 10 times faster than the pentachlorophenyl ester.

C. Comparison of Coupling Rates of Dipeptides with Monomers

The dipeptides usually couple faster than the corresponding monomers, as shown in the pentachlorophenyl ester series below by the k_{c_3}/k_{c_1} values (taken from Tables I–IV and arranged in decreasing order). The N^{α}-protecting groups are N^{α}-benzyloxycarbonyl for both monomers and dimers. There is

$$\underset{14}{\mathrm{Trp}} > \underset{11}{\mathrm{Glu(OMe)}} > \underset{7.3}{\mathrm{Phe}} > \underset{6.3}{\mathrm{Ala}} > \underset{5.4}{\mathrm{Val}} > \underset{5.1}{\mathrm{Met}} > \underset{3.5}{\mathrm{Lys(Boc)}} > \underset{3}{\mathrm{Cys(Bzl)}} > \underset{1.7}{\mathrm{Pro}} > \underset{1.4}{\mathrm{Asp(OMe)}} \tag{13}$$

no clear structure–coupling rate correlation between the monomer and dipeptide. Similar observation can be made for the *p*-nitrophenyl esters; again the dipeptides couple 6 to 15 times faster than the corresponding monomers.

The faster a dipeptide active ester coupling proceeds relative to the 5(4*H*)-oxazolone ring formation, the more optically pure is the product obtained. This will be discussed in connection with k_c/k_r values in Section VI.

VI. RATIO OF COUPLING TO RACEMIZATION RATE CONSTANTS

Comparison of the data shows that the decreasing order of racemization rate constants for the active esters studied is not always the same as the decreasing order of the coupling rate constants; this indicates that *the activity of an active ester is not strictly parallel to its ability to racemize.* This is best illustrated by the ratio of coupling to racemization. In our model system triethylamine-induced racemization was studied in the absence of the amine component. Assuming that racemization induced by the amine component would parallel that measured for triethylamine, the k_c/k_r ratios would indicate the relative extent of racemization that could be expected during coupling. The larger this ratio, the smaller the amount of racemization expected during coupling. There are experimental findings which show that such a parallelism may exist between triethylamine and the amine components. For example, the k_c/k_r ratios for N^{im}-benzylhistidine active esters are very small, which indicates extensive racemization during coupling. As mentioned before, Windridge and Jorgensen (1970, 1971) and Syrier and Beyerman (1974) observed unusually high racemization (50%) during Z-His(Bzl)-OH couplings. At the other extreme, proline, whose k_c/k_r value is extremely large, is used for racemization-free segment coupling.

A. Comparison of k_c/k_r Values for N^{α}-Benzyloxycarbonylamino Acid Active Esters. Effect of Side Chain, Active Ester Group, and N^{α}-Protecting Group on the k_c/k_r Values

The decreasing order of k_c/k_r values for amino acids in the pentachlorophenyl ester series (Kovacs *et al.*, 1973) is given in Eq. (14). This

–OPcp:	Pro >>	Ala >	Trp >	Lys(Boc) >	Met >	Phe >	Glu(OMe) >	Val >
Z:	∞	6133	4470	3666	1286	871	800	572
Boc:		7677			1940	3284	2222	1450

–OPcp:	Asp(OMe) >	Ser >	His(Z)* >	Cys(Bzl) >	His(Bzl)
Z:	420	281		42	1.6
Boc:				127	

(14)

* Assuming a second-order rate constant.

order is not identical to the reverse order for racemization since these ratios also depend on the coupling rates. The order of these ratios will indicate the vulnerability of the amino acids to racemization by the enolization mechanism during coupling. It is well known that N^{im}-benzylhistidine, *S*-benzylcysteine, or serine, for example, are easily racemized amino acids. Few data are available for other amino acids. Rebek and Feitler (1975) determined the racemization of Boc-Phe-OH during coupling with H-Gly-OMe by the dicyclohexylcarbodiimide method in CH_2Cl_2. Isotope dilution indicated that racemization falls in the 0.01 to 0.1% range and that it is linearly increased with concentration.

The order of the k_c/k_r values for the pentafluorophenyl esters is similar [Eq. (15)]; however, these ratios are much larger numbers, indicating that

–OPfp:	Ala >	Trp >	Glu(OMe) >	Met >	Phe >	Ser >	Asp(OMe) >	Cys(Bzl)
Z:	21,000	11,600	3,590	2,980	1,452	645	605	122
Boc:	51,600		5,000	4,500	5,500			1,166

(15)

much less racemization is expected during coupling. These values are especially indicative if the amine component is liberated from its salt with a strong tertiary base. Equations (16) to (18) illustrate the order of the k_c/k_r values for the *N*-hydroxysuccinimide esters, the 2,4,5-trichlorophenyl esters

—OSu:	Ala >	Glu(OMe) >	Met >	Phe >	Val >	Cys(Bzl)
Z:	3500	600	315	287	53	11
Boc:	7100		467	255		30

(16)

(note the much smaller numbers), and the *p*-nitrophenyl esters, respectively.

—OTcp:	Ala >	Phe ≃	Glu(OMe) >	Asp(OMe) >	Cys(Bzl)
Z:	1616	170	170	75	6.1
Boc:	4000	334			23

(17)

—ONp:	Ala >	Glu(OMe) >	Met >	Phe >	Asp(OMe) >	Cys(Bzl)
Z:	734	150	88	35	27	2.7
Boc:	2700		188	99		28

(18)

The faster the coupling, the smaller the amount of racemization through enolization. The order of esters in terms of the k_c/k_r values is not the same as the order of racemization or coupling. For alanine and *S*-benzylcysteine, the order of k_c/k_r values is given in Eq. (19).

—Ala:	OPfp >	OPcp >	OSu >	OTcp >	ONp
Z:	21,000	6,100	3,500	1,600	730
Boc:	51,600	7,600	7,000	5,000	2,700

—Cys(Bzl):	OPfp >	OPcp >	OSu >	OTcp >	ONp
Z:	122	42	11	6.1	2.7
Boc:	1,160	127	30	23	28

(19)

The order is the same but the numbers between the series for the two amino acids differ widely, indicating that *S*-benzylcysteine is about 200 to 300 times more vulnerable to racemization during coupling.

This order thus indicates which ester is advantageous to use for coupling. The figures for Z-Val-OPcp and Z-Val-OPfp are unusual, since the former gives a 3 times larger k_c/k_r value (572 and 181) than the latter.

Comparing the effect of the N^α-benzyloxycarbonyl and *tert*-butyloxycarbonyl groups, it is clear that the latter provides substantially better protection against racemization than the former. This is significant in the case of *S*-benzylcysteine, pentafluorophenyl, and *p*-nitrophenyl esters: The N^α-*tert*-butyloxycarbonyl group gives an order of magnitude better protecttion than the N^α-benzyloxycarbonyl group [Eq. (19)].

B. Comparison of k_c/k_r Values of Glycylamino Acid (Dipeptide) Active Esters. Effect of Side Chain, Active Ester Group, and N^α-Protecting Group on k_c/k_r Values

Significant work has been done with the Anderson and Young dipeptides to determine the exact amount of racemization during coupling with H-Gly-OEt, using most of the carboxyl-activating groups, including the active esters. Particularly important, because of its high accuracy, are the results obtained by Kemp's isotope dilution method. Kemp summarized these results in Vol. 1 of "The Peptides." In all these tests H-Gly-OEt was used as amine component; the rate of coupling was at least 10 to 40 times faster than in the coupling with H-Val-OMe. For example, coupling of Z-Phe-ONp with H-Gly-OEt in dimethylformamide was $43.8 \times 10^{-2}\ M^{-1}\ \text{sec}^{-1}$ and with H-Val-OEt only $1 \times 10^{-2}\ M^{-1}\ \text{sec}^{-1}$ (Kemp *et al.*, 1974a). It is therefore expected that the extent of racemization will also be 10 to 40 times larger. Besides the C-activated peptides of phenylalanine, leucine, valine, alanine, and isoleucine, only a few other amino acids have been subjected to racemization studies. From the available data (see Kemp, "The Peptides," Vol. 1; Table 6), it seems reasonable to assume that there is at least qualitative correlation between the k_c/k_r values obtained by triethylamine and the extent of racemization during peptide coupling conditions, provided the mechanism of racemization is the same.

The effect of the side chain on the k_c/k_r value is summarized by giving the decreasing order for both the pentachlorophenyl ester series [Eq. (20)] and

$$\text{Z-Gly-A-OPcp:}\quad \underset{13{,}720}{\text{Pro}} \gg \underset{284}{\text{Ala}} > \underset{236}{\text{Trp}} > \underset{59}{\text{Met}} > \underset{49}{\text{Phe}} > \underset{45}{\text{Lys(Boc)}} \simeq \underset{44}{\text{Cys(Bzl)}} > \underset{32}{\text{Glu(OMe)}} > \underset{25}{\text{Val}} > \underset{6.2}{\text{Asp(OMe)}} \tag{20}$$

the *p*-nitrophenyl ester series [Eq. (21)]. Comparing the effect of active

Z-Gly-A-ONp:	Ala >	Glu(OMe) >	Met >	Phe >	Cys(Bzl) >	Ser(Bzl)	(21)
	104	24	18	17	8.5	3.7	

esters on k_c/k_r, the order for methionine and valine is given in Eq. (22).

	OSu ≃	OPfp >	OPcp >	ONp	
Z-Gly-Met-	—	130	59	18	(22)
Z-Gly-Val-	148	—	25	—	

The ratios in Eqs. (20)–(22) indicate how vulnerable to racemization the dipeptides are during coupling. The numbers will *suggest which amino acid should be used for segment condensation as well as which ester should be the carboxyl-activating group.* After proline and glycine, the next choice is alanine or tryptophan pentachlorophenyl or alanine *p*-nitrophenyl ester, if only these two active esters are considered. It has been shown that pentachlorophenyl esters couple "racemization free" in the Young test (Kovacs *et al.*, 1969) or during coupling of Z-Gly-Ala-OPcp with H-Leu-OBzl (Kisfaludy *et al.*, 1973a). *Particularly vulnerable are aspartic acid, cysteine, and serine peptides.* The ratios in Eq. (22) indicate that the best choice for active ester will be *N*-succinimidoyl or pentafluorophenyl.

However, comparing the k_c/k_r values of dipeptides to monomers, $(k_{c_1}/k_{r_1})/(k_{c_3}/k_{r_3})$, the following interesting conclusion can be made (Scheme 8) for the N^{α}-benzyloxycarbonyl derivatives. These ratios show how much better

	Ala	Phe	Met	Glu(OMe)	Asp(OMe)	Val	Cys(Bzl)	Pro
ONp	7	2	5	6.2			0.31	
OPcp	22	18	22	26	67	23	0.55	∞
OPfp			23					
OSu						0.36		

Scheme 8

it is to use the monomer active ester than the dipeptide active ester to minimize racemization in coupling. Unlike all other amino acids, cysteine and valine give ratios less than 1, which indicates that it is probably safer to couple the dipeptide active ester than the monomer. However, this has not been established experimentally as yet. An alternative conclusion could be that in these exceptional cases, racemization proceeds via enolization rather than 5(4*H*)-oxazolone formation. This was established for Z-Gly-Cys(Bzl)-ONp using the deuterium isotope effect (Kovács *et al.*, 1977a, 1978) and is most probably valid for proline, which cannot form 5(4*H*)-oxazolone.

An interesting case is phenylalanine, for which this ratio is 2; although there is no direct evidence that Z-Gly-Phe-ONp partially racemizes through enolization, Kemp and Rebek (1970) have established that the corresponding azide does. Coupling of Boc-Phe-OH with H-Gly-OEt using dicyclohexylcarbodiimide gave 0.07% racemization (in 0.2 *M* solution in CH_2Cl_2; Rebek and Feitler, 1975); and the coupling of Z-Gly-Phe-ONp with H-Gly-OEt yielded 0.6% racemate (Kemp, 1973). Although these values are not directly comparable, they indicate that the racemization of monomer is not even an order of magnitude smaller.

Most interesting is a comparison of the following two reports: Rebek and Feitler (1975) observed that racemization by enolization increased with increased reactant concentrations while Kemp (1973) concluded, from a kinetic study of coupling Z-Gly-Phe-ONp with H-Gly-OEt in CH_3CN or dimethylformamide, that the fraction of product racemized by the 5(4*H*)-oxazolone mechanism should be and indeed is inversely dependent on the reactant concentration.

There are several reports on the extent of racemization during coupling of X-A-Phe-Y with H-Val-OR or of Tfa-Val-Y with H-Val-OR when racemization is very extensive (summarized by Kemp, "The Peptides," Vol. 1, Table 6). In our study the k_c/k_r values are low for N^{α}-alkyloxycarbonylglycylvaline active esters, indicating more racemization than one would expect for other amino acids, with larger k_c/k_r values.

VII. RELATION OF COUPLING TIME TO CONCENTRATION

A. General Considerations

It has been pointed out that if larger peptides are prepared without isolation of the intermediate, as in solid-phase synthesis or in repetitive methods of peptide synthesis in solution, coupling must be close to quantitative in order to obtain predominantly the desired product rather than wrong sequences. For practical purposes the coupling reaction must be completed within a reasonable time, and consequently only rapidly reacting carboxyl-activated amino acid derivatives should be used for such syntheses. However, if it should be necessary to use slower-coupling derivatives, then the coupling times can be shortened by increasing the concentrations. If the initial concentrations of both components have the same value, C^0, then the time, t_x, required to complete any fraction, x, of the coupling is inversely proportional to C^0, as seen from Eq. (23).

$$t_x = \frac{x}{(1 - x)k_c C^0} \tag{23}$$

Thus for 99% completion, the time required is $99/(k_c C^0)$, as was seen in Eq. 5. Table IX includes the 99% coupling times calculated for various amino

Table IX. 99% Coupling Time for H-Val-OMe in Tetrahydrofuran at Two Concentrations (for $C^0_{ester} = C^0_{Val}$ and $C^0_{ester} = 0.26\ M = 2C^0_{Val}$)

Compound	k_c	$t_{99\%}$ for 0.01 M (hr)	$t_{99\%}$ for 0.13 M (hr)	$t_{99\%}$ for two equiv. ester (hr)
Boc-Ala-OSu[a]	4.3[a]	64	4.9	0.19
Z-Ala-OSu	7.7	36	2.75	0.109
Boc-Ala-OPfp	17.1	16	1.2	0.05
Z-Ala-OPfp	19.3	14.25	1.1	0.043
Boc-Ala-ONp	0.108	2546	195	7.75
Z-Ala-ONp	0.152	1809	139	5.51
Z-Gly-Ala-ONp	1.27	216.54	16.65	0.66
Boc-Ala-OTcp	0.29	948	73	2.9
Z-Ala-OTcp	0.30	917	70.5	2.80
Boc-Ala-OPcp	0.24	1146	89	3.5
Z-Ala-OPcp	0.506	543	42	1.65
Z-Gly-Ala-OPcp	3.2[a]	86	6.6[b]	0.26
Boc-Phe-OSu	2.2	125	9.6	0.38
Z-Phe-OSu	4	69	5.23	0.209
Boc-Phe-OPfp	11.6	24	1.8	0.07
Z-Phe-OPfp	11.9	23	1.80	0.07
Boc-Phe-ONp	0.05	5500	423	16.74
Z-Phe-ONp	0.03	9166	705	27.9
Z-Gly-Phe-ONp	0.31	887	68	2.7
Boc-Phe-OTcp	0.15	1833	136	5.6
Z-Phe-OTcp	0.20	1375	106	4.19
Boc-Phe-OPcp	0.33	833	63	2.5
Z-Phe-OPcp	0.29	948	73	2.89
Z-Gly-Phe-OPcp	2.12	129	10.00	0.39
Z-Pro-OPcp	0.44	625	48	1.90
Z-Gly-Pro-OPcp	0.75	367	28	1.11
Z-His(Bzl)-OPcp	0.9	306	23.5	0.93
Z-His(Z)-OPcp	0.23	1196	92	3.64
Boc-Val-OSu	0.045	6111	470	18.6
Z-Val-OSu	0.15	1833	141	5.58
Z-Gly-Val-OSu	1.3	211	16.3	0.64
Boc-Val-OPfp	0.31	887	68	2.7
Z-Val-OPfp	0.19	1447	111	4.4
Boc-Val-OPcp	0.029	9483	729	28.9
Z-Val-OPcp	0.026	10577	814	32
Z-Gly-Val-OPcp	0.14	1964	151	5.97
Boc-Met-OSu	2.2	125	9.6	0.38
Z-Met-OSu	2.6	106	8.1	0.32
Boc-Met-OPfp	11	25	1.9	0.08
Z-Met-OPfp	17	16.17	1.2	0.049
Z-Gly-Met-OPfp	61	4.50	0.35	0.0137

Table IX (*Continued*)

Compound	k_c	$t_{99\%}$ for 0.01 *M* (hr)	$t_{99\%}$ for 0.13 *M* (hr)	$t_{99\%}$ for two equiv. ester (hr)
Boc-Met-ONp	0.09	3055	235	9.3
Z-Met-ONp	0.08	3438	264	10.46
Z-Gly-Met-ONp	0.55	500	38.5	1.5
Boc-Met-OPcp	0.29	948	73	2.9
Z-Met-OPcp	0.36	764	59	2.32
Z-Gly-Met-OPcp	1.84	149	11.5	0.46
Z-Glu(OMe)-OSu	2.7	102	7.85	0.31
Z-Glu(OMe)-OPfp	11.6	23.7	1.82	0.072
Boc-Glu(OBzl)-OPfp	9.5	29	2.23	0.09
Z-Glu(OBzl)-OPfp	12.1	22.7	1.75	0.069
Z-Glu(OMe)-OTcp	0.1	2750	212	8.37
Z-Glu(OMe)-ONp	0.04	6875	529	21
Z-Gly-Glu(OMe)-ONp	0.58	474	36	1.44
Z-Gly(OMe)-OPcp	0.16	1719	132	5.23
Z-Gly-Glu(OMe)-OPcp	1.75	157	12	0.48
Boc-Gly(OBzl)-OPcp	0.28	982	75	3
Z-Glu(OBzl)-OPcp	0.42	655	50	1.99
Z-Asp(OMe)-OPfp	14.7	18.70	1.44	0.057
Z-Asp(OMe)-ONp	0.07	3929	302	11.95
Z-Asp(OMe)-OTcp	0.26	1058	81	3.22
Z-Asp(OMe)-OPcp	0.74	372	28.6	1.13
Z-Gly-Asp(OMe)-OPcp	1.03	267	20.5	0.81
Boc-Cys(Bzl)-OSu	2.68	103	7.9	0.3
Z-Cys(Bzl)-OSu	5.44	51	3.90	0.15
Boc-Cys(Bzl)-OPfp	44.3	6.2	0.5	0.02
Z-Cys(Bzl)-OPfp	40.4	6.81	0.5	0.021
Boc-Cys(Bzl)-ONp	0.13	2115	163	6.44
Z-Cys(Bzl)-ONp	0.105	2619	201	7.97
Z-Gly-Cys(Bzl)-ONp	0.6	458	35	1.40
Z-Cys(Bzl)-ODnp(2,6)	18.4	15	1.15	0.045
Z-Cys(Bzl)-ODnp(2,6)	1.73	159	12.2	0.5
Boc-Cys(Bzl)-OTcp	0.22	1250	95	3.8
Z-Cys(Bzl)-OTcp	0.3	916	71	2.8
Z-Cys(Bzl)-OTcp(2,4,6)	0.063	4365	336	13
Boc-Cys(Bzl)-OPcp	0.52	529	40.7	1.6
Z-Cys(Bzl)-OPcp	1.72	160	12.3	0.49
Z-Gly-Cys(Bzl)-OPcp	5.23[c]	53	4[d]	0.16
Z-Cys(Bzl)-OTbp(2,4,6)	0.0215	12790	984	39
Z-Pro-OPcp	0.44	625	48	1.9
Z-Gly-Pro-OPcp	0.75	367	28	1.1
Z-His(Bzl)-OPcp	0.9	306	23.5	0.9
Z-His(Z)-OPcp	0.23	1196	92	3.6

Table IX (*Continued*)

Compound	k_c	$t_{99\%}$ for 0.01 M (hr)	$t_{99\%}$ for 0.13 M (hr)	$t_{99\%}$ for two equiv. ester (hr)
Z-Ser-OPfp	19.2	14	1.1	0.04
Z-Ser-OPcp	0.74	372	27.6	1.1
Z-Gly-Ser(Bzl)-ONp	0.21	1310	105	4
Z-Trp-OPfp	12.4	22	1.7	0.07
Z-Trp-OPcp	0.32	860	66	2.6
Z-Gly-Trp-OPcp	4.5	61	4.7	0.19
Z-Lys(Boc)-OPcp	0.22	1250	96	3.8
Z-Gly-Lys(Boc)-OPcp	0.76	362	27.8	1.1

[a] k_c is investigated at 0.05 M.
[b] Experimentally, a 0.05 M conc. was used due to low solubility.
[c] k_c is investigated at 0.0184 M.
[d] Experimentally, a 0.0184 M conc. was used due to low solubility.

acid and peptide active esters with equimolar H-Val-OMe in tetrahydrofuran solution for $C^0 = 0.01\ M$ and for $C^0 = 0.13\ M$. The required times for the larger concentration are all 13 times smaller than for the smaller concentration.

When larger protected segments are coupled, their solubility may be too poor to allow sufficiently high molar concentration to apply this method effectively. It is possible that new protecting groups could be found that would increase the solubility of the larger fragments, or possibly peptide synthesis could be done in aqueous solution; this latter solution however, would require a new development, with new protecting and activating groups.

A much simpler way to reduce the coupling time of one component (say the less soluble one) is to use an excess of the other component. This produces rather striking results, as shown by Kovács *et al.* (1978a,b, 1979). If C^0 is the initial concentration of the less concentrated component, and nC^0 that of the more concentrated, then T, the ratio of the time required for a fraction, x, of the limiting component to be coupled as compared to the time required when both initial concentrations are the same (C^0) is given by

$$T = \frac{(1 - x)}{x(n - 1)} \ln\left[\frac{(n - x)}{n(1 - x)}\right] \tag{24}$$

For example, if $n = 2$, and $x = 0.99$, then $T = 0.0396$, showing that the time required for 99% coupling of the limiting component is less than 4% of the time required when the initial concentrations are equal. This is illustrated in

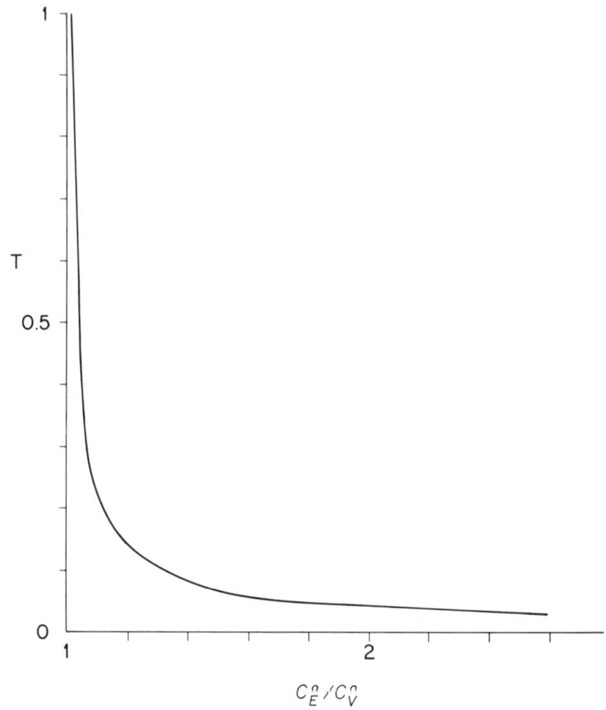

Figure 1. Concentration ratio dependence of 99% coupling time (T).

Table IX which includes the 99% coupling times for active ester and amine when both concentrations are equal to 0.13 *M*, and also for 0.13 *M* amine reacting with 0.26 *M* active ester. The latter times are all about 25 times smaller. Figure 1 shows how *T* varies with *n* for 99% completion. It is seen that a considerable reduction in coupling time is caused by even a modest excess in concentration.

B. Racemization and Concentration

Kemp (1973) provided experimental evidence to demonstrate this important factor in practical peptide synthesis; for example, coupling of Z-Gly-Phe-ONp with H-Gly-OEt in CH_3CN at 0.1 *M* and 0.4 *M* concentrations gave 0.25% and 0.13% racemate, respectively. For the coupling of Boc-Ala-OH with H-Gly-OMe by dicyclohexylcarbodiimide, however, Rebek and Feitler (1975) observed increased racemization with increased concentrations.

C. Special Problems in Preparing Sequential Polypeptides by Polycondensation of Peptide Active Esters

Racemization is a serious problem during the synthesis of sequential polypeptides from peptide active ester salts. Most frequently tripeptides or higher peptide active ester salts, such as HCl · H-A-B-C-OPcp, are used for polycondensation in dimethylformamide solution, and a tertiary base is necessarily used to liberate the free amine (Kovács *et al.*, 1966, 1972c,d, 1977b). The inverted amino acid residue is permanently incorporated into the polypeptide chain, and such closely related epimers cannot be separated.

The following considerations are important for preparation of such sequential polypeptides with maximum optical purity. (a) The COOH-terminal amino acid should be chosen on the basis of k_c/k_r values. This can be done as follows: If a tripeptide of 3 different amino acids is chosen for the required sequence, then A-B-C, B-C-A, and C-A-B combinations will also give identical polypeptides, with the exception of the NH_2- and COOH-termini, as

$$\begin{array}{l} \text{A-B-C} \longrightarrow \text{A-B-C-A-B-C-A-B-C-A-B-C-} \cdots \\ \text{B-C-A} \longrightarrow \cdots\text{-B-C-A-B-C-A-B-C-A-} \cdots \\ \text{C-A-B} \longrightarrow \cdots \text{C-A-B-C-A-B-C-A-B-} \cdots \end{array} \quad (25)$$

indicated in Eq. (25). The physical properties, such as ORD and conformation, were considered to be practically identical, since the effect of the terminal groups was expected to be negligible if the molecular weights were sufficiently high. (b) The carboxyl-activating group should be an active ester with the largest possible k_c value in order to minimize the coupling time. (c) Since both coupling time and racemization are inversely proportional to the initial active ester concentration under these conditions, the concentration of the tripeptide active ester should be as high as possible. At the end of the reaction the molar concentration of the active ester species will be very low, and racemization is more extensive at the end of polycondensation (see Section VII,B). The following result will illustrate some of these points: Polycondensation of H-Gly-Gly-Phe-OPcp in dimethylformamide gave a polypeptide which contained 5% inverted phenylalanine residues when compared with the polymer obtained by polymerization of H-Phe-Gly-Gly-OPcp (Kovács *et al.*, 1972a) which cannot racemize.

VIII. RACEMIZATION OF PENULTIMATE AMINO ACID RESIDUE

It was observed by Bergman and Zervas (1929) that during 5(4*H*)-oxazolone formation of phenylalanyl-tyrosine, the phenylalanine residue was also racemized. Neuberger (1948) explained this by assuming an equilibrium between the 5(4*H*)-oxazolone (**30**) and the enamine (**31**). Racemization of the penultimate amino acid was investigated later by Weygand *et al.* (1966) and Dzieduszycka *et al.* (1973). A significant amount of epimer formation was observed with model peptides.

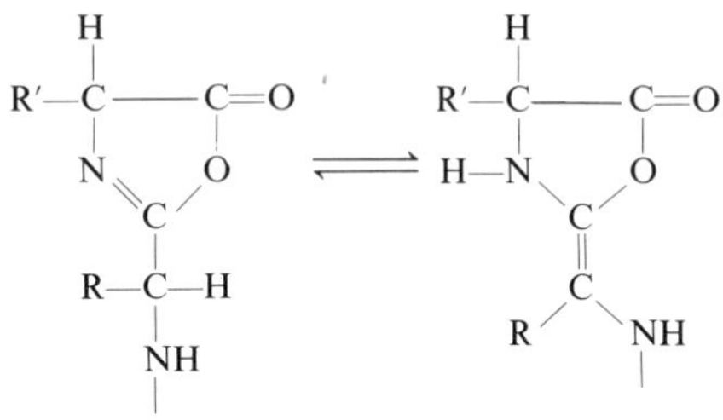

(**30**) (**31**)

The triethylamine-catalyzed racemization of Z-Phe-Gly-OPcp in both tetrahydrofuran and dimethylformamide were studied in our laboratory. In tetrahydrofuran, the racemization constant $k_r = 4.71 \times 10^{-6}\ M^{-1}\ \sec^{-1}$ (7 equiv of triethylamine with ester concentration equalling 0.0084 *M*; in 200 hours about 11.7% was racemized); this indicates that racemization is somewhat faster than the enolization of Z-Phe-OPcp ($k_r = 3.3$). In dimethylformamide, $k_r = 78 \times 10^{-6}\ M^{-1}\ \sec^{-1}$, while the racemization rate (k_r) of Z-Gly-Phe-OPcp in dimethylformamide is 4870, which is 62 times faster than the former.

The Z-Cys(Bzl)-Gly-OPcp racemization rate (k_r) in dimethylformamide is $43 \times 10^{-6}\ M^{-1}\ \sec^{-1}$, which is 5 times slower than that for Boc-Cys(Bzl)-OPcp. On the other hand, when the last amino acid is not activated and cysteine is in a peptide bond, as in Z-Cys(Bzl)-Gly-OEt, no racemization was detectable with triethylamine in tetrahydrofuran in 7 days.

These limited data indicate that racemization of the penultimate amino acid is due to 5(4*H*)-oxazolone formation and also that the rate of racemization of the penultimate amino acid is of the same order of magnitude as that of the C-activated monomer.

IX. PREDICTION OF COUPLING AND RACEMIZATION RATE CONSTANTS

A. Kemp's Method of Calculating Coupling Rates

At the 11th European Peptide Symposium in 1971, Kemp reported (Kemp, 1973; Kemp *et al.*, 1974a) a method for calculating the rate constants of the coupling of N^{α}-alkyloxycarbonylamino acid *p*-nitrophenyl esters with amino acid esters in dimethylformamide solution. The following equation was used for the calculations:

$$k_{A-B} = \frac{k_{A-Gly}k_{Gly-B}}{k_{Gly-Gly}} \tag{26}$$

where k_{A-B} is the second-order rate constant for the coupling of Z-A-ONp with H-B-OEt. The calculated and experimentally determined rate constants are given in Table X, together with the time required to complete 99% of the coupling. This series includes coupling of benzyloxycarbonyl-protected glycine, alanine, leucine, proline, valine, and phenylalanine *p*-nitrophenyl esters with glycine, alanine, leucine, proline, and valine ethyl esters, and provide examples for nearly all steric effects on both reactants. The data in Table X show that the calculated values are in very good agreement with the experimental values, with the exception of the coupling to H-Pro-OEt. The deviations from the experimental values are a measure of the side chain–side chain interaction for the transition state of the coupling. For coupling with H-Val-OEt, the calculated values by our approach are also included and both are in good agreement. Kemp provided conformational analysis for a tetrahedral transition state; the collapse of this transition state is the rate determining step, as we have discussed in Section IV. The anomalous behavior of proline was clearly explained by steric considerations.

B. Application of the Additivity Principle for Prediction of Rate Constants

Recognition of the additivity principle for prediction of properties in organic chemistry goes back to the 1820s when Liebig recognized the homologous series. Later, van't Hoff applied the additivity principle to the prediction of optical rotation, known as the principle of optical superposition. In our work, the additivity principle (Rouvray, 1973) is used to predict racemization and coupling rate constants.

It is postulated that log k is the sum of contributions from each reactant, each component of each reactant, and the solvent; this is indicated in Eqs. (27) and (28), where k is the experimental rate constant

$$\log k = \log f_1 + \log f_2 + \cdots + \log f_i \tag{27}$$

$$k = f_1 \times f_2 \times \cdots \times f_i \tag{28}$$

and f_1, f_2, etc. are the contribution of the reactants, their substituents, or the solvent.

Z-Ala-OPcp was arbitrarily chosen as standard. The racemization reaction was carried out in tetrahydrofuran in the presence of triethylamine at 23°C ($k_r = 0.825 \times 10^{-6}\ M^{-1}\ \text{sec}^{-1}$), the coupling was with H-Val-OMe in tetrahydrofuran at 23°C ($k_c = 0.506 \times 10^{-2}\ M^{-1}\ \text{sec}^{-1}$).

Table X. Rate of Coupling of N^{α}-Benzyloxycarbonyl- and *tert*-Butyloxycarbonylamino Acid *p*-Nitrophenyl Ester Derivatives with Amino Acid Esters in Dimethylformamide[a]

Active ester	Amino acid ester	k_c	(k_c)	$t_{99\%}$ (hr)
Z-Gly-ONp	H-Gly-OEt	43.8		0.48
Boc-Gly-ONp	H-Gly-OEt	38.5		0.55
Z-Ala-ONp	H-Gly-OEt	27.8		0.76
Z-Val-ONp	H-Gly-OEt	2.1		10.1
Z-Leu-ONp	H-Gly-OEt	18.6		1.13
Boc-Leu-ONp	H-Gly-OEt	16.1		1.31
Z-Pro-ONp	H-Gly-OEt	15		1.41
Z-Phe-ONp	H-Gly-OEt	23.5		0.9
Z-Asn-ONp	H-Gly-OEt	12		1.76
Z-Gly-ONp	H-Ala-O*t*Bu	14		1.51
Z-Gly-ONp	H-Ala-OEt	10.1		2.1
Boc-Gly-ONp	H-Ala-OEt	9.6	(8.5)	2.2
Z-Ala-ONp	H-Ala-O*t*Bu	11.1	(9.0)	1.9
Z-Ala-ONp	H-Ala-OEt	7.3	(6.16)	2.9
Z-Val-ONp	H-Ala-O*t*Bu	0.43	(0.46)	49.2
Z-Leu-ONp	H-Ala-O*t*Bu	3.6	(4.16)	5.9
Boc-Leu-ONp	H-Ala-O*t*Bu	3.16	(3.5)	6.7
Z-Pro-ONp	H-Ala-O*t*Bu	2.2	(3.33)	9.6
Z-Phe-ONp	H-Ala-O*t*Bu	4.1	(5.16)	5.16
Z-Gly-ONp	H-Val-OEt	3.4		6.22
Z-Ala-ONp	H-Val-OEt	2.0	(2.16)(2.25)[b]	10.6
Z-Val-ONp	H-Val-OEt	0.1	(0.165)(0.085)[b]	211
Z-Leu-ONp	H-Val-OEt	0.9	(1.46)(0.93)[b]	23.5
Z-Pro-ONp	H-Val-OEt	0.7	(1.18)(0.83)[b]	30.2
Z-Phe-ONp	H-Val-OEt	1.0	(1.83)(1.16)[b]	21.1
Z-Gly-ONp	H-Leu-O*t*Bu	8.1		2.6
Z-Gly-ONp	H-Leu-OEt	4.7		4.5
Z-Ala-ONp	H-Leu-O*t*Bu	5.8	(5.16)	3.6
Z-Ala-ONp	H-Leu-OEt	3.16	(3.0)	6.7
Z-Val-ONp	H-Leu-OEt	0.19	(0.23)	111.6
Z-Leu-ONp	H-Leu-OEt	1.75	(2.0)	11.9
Z-Pro-ONp	H-Leu-OEt	1.1	(1.61)	19.2
Z-Phe-ONp	H-Leu-OEt	2.16	(2.5)	9.8
Z-Gly-ONp	H-Pro-OEt	11.4		1.8
Z-Ala-ONp	H-Pro-OEt	2.7	(7.33)	7.8
Z-Val-ONp	H-Pro-OEt	0.25	(0.55)	84.6
Z-Leu-ONp	H-Pro-OEt	0.85	(5.0)	24.9
Z-Pro-ONp	H-Pro-OEt	0.22	(4.0)	96.1
Z-Phe-ONp	H-Pro-OEt	2.31		9.1
Z-Gly-ONp	H-Phe-OEt	1.66		12.7
Z-Phe-ONp	H-Phe-OEt	0.76	(0.9)	27.8

[a] Kemp *et al.*, 1974a. The k_c values are in $10^{-2}\ M^{-1}\ \text{sec}^{-1}$. The calculated values are in parentheses. The $t_{99\%}$ is for 0.13 *M* solution for both reactants.

[b] The calculated (k_c) values from the predictive factors (Kovács *et al.*, 1975).

The relative contribution of the side chains of amino acids can be calculated from the ratio of racemization rate constants of, for example, Z-A-OPcp and Z-Ala-OPcp, since from Eq. (29), all the contributory factors would cancel out except those due to the side chains.

$$\frac{k_{\text{Z-Met-OPcp}}}{k_{\text{Z-Ala-OPcp}}} = \frac{k_{\text{Met}}}{k_{\text{Ala}}} = F_{\text{Met}} \tag{29}$$

In this way the relative contributions of N^{α}-protecting groups, amino acid side chains, carboxyl-activating groups, and solvents for coupling as well as for racemization were obtained from the available data. The factors (F) in Table XI are the "best" average values, obtained by pooling the available estimates from different active ester. In general, the percent average deviation in the factor is $\pm 25\%$.

Calculation of k_r for Z-Gly-Phe-OPcp. in dimethylformamide with triethylamine is demonstrated in Eq. (30). The calculated k_r equals

$$\begin{aligned} k_r &= F_Z\, F_{\text{-Gly-}}\, F_{\text{Phe}}\, F_{\text{OPcp}}\, F_{\text{DMF}}\, F_{\text{Et}_3\text{N}}\, (0.825 \times 10^{-6}\ M^{-1}\ \text{sec}^{-1}) \\ k_r &= (1)\ (105)\ (7.1)\ \ (1)\ \ (10)\ \ (1)\ \ (0.825 \times 10^{-6}\ M^{-1}\ \text{sec}^{-1}) \end{aligned} \tag{30}$$

$6150 \times 10^{-6}\ M^{-1}\ \text{sec}^{-1}$ and the experimental k_r equals $4870 \times 10^{-6}\ M^{-1}\ \text{sec}^{-1}$. The latter value is actually the rate of triethylamine-catalyzed 5(4*H*)-oxazolone formation.

Equation (31) illustrates the calculations of the coupling rate constant of Z-Met-ONp with H-Val-OMe in tetrahydrofuran solution.

$$\begin{aligned} k_c &= F_Z\ F_{\text{Met}}\ F_{\text{ONp}}\ F_{\text{THF}}\ F_{\text{Val-OMe}}\ (0.506 \times 10^{-2}\ M^{-1}\ \text{sec}^{-1}) \\ k_c &= (1)\ (0.94)\ (0.22)\ \ (1)\ \ \ (1)\ \ \ (0.506 \times 10^{-2}\ M^{-1}\ \text{sec}^{-1}) \end{aligned} \tag{31}$$

The calculated k_c equals $0.06 \times 10^{-2}\ M\ \text{sec}^{-1}$ and the experimental k_c equals $0.08 \times 10^{-2}\ M^{-1}\ \text{sec}^{-1}$.

In Tables I–IV and X the calculated k_r and k_c values are included and, with a few exceptions, the results show good agreement with the experimental values. These exceptional amino acids are listed in Table IV and include Z-Gly-Cys(Bzl)-ONp and pentachlorophenyl esters. For these esters, the calculated k_r values are 55 and 19.2 times larger, respectively, then the ex-experimental values. Since the theory behind the rate constant calculations is a form of linear free energy relationship (Chapman and Shorter, 1972), large deviation from the calculated values indicate deviation from the usual reaction mechanism. It was concluded that Z-Gly-Cys(Bzl)-ONp and Z-Gly-Cys(Bzl)-OPcp do not racemize through the 5(4*H*)-oxazolone mechanism but through the enolization mechanism. This conclusion was supported later, experimentally, for the *p*-nitrophenyl ester, using the kinetic isotope effect as discussed previously.

Table XI. Predictive Factors

Amino acid side chain	F_r	F_c
Ala	1	1
Phe	7.1	0.7
Trp	0.89	0.6
Cys(Bzl)	280	1.3
Met	5.69	0.94
Val	0.552	0.0514
Pro	0.000	0.87
Ser	29	1.3
Ser(Ac)		4.4
Ser(*t*Bu)		0.47
Asp(OMe)	20	0.9
Asp(Bzl)	22	1.4
Glu(OMe)	2.5	0.47
Glu(OBzl)	4.1	0.8
Asn		0.45
Gln		0.58
Gly		1.5
Ileu		0.053
Leu		0.56
Lys(Boc)	0.72	0.44
Lys(Z)		0.48
Tyr		0.89
Tyr(Z)		0.83
His(Bzl)	763	1.76
His(Z)	44	0.45
Activating ester	F_r	F_c
OSu	25	13
OPfp	15	37
ODnp	7.2	11
ODnp(2,6)	7	1
OTcp	2.5	0.53
OTcp(2,4,6)	0.19	0.08
ONp	1.6	0.22
OPcp	1	1
N^{α}-Protecting group	F_r	F_c
Z	1	1
Boc	0.23	0.94
Bz	1500	
Amino acid residue	F_r	F_c
-Gly-	105	8.8

Table XI *(Continued)*

Solvent	F_r	F_c
Tetrahydrofuran	1	1
Ethylacetate	2.5	0.32
Chloroform	2.2	0.2
Dioxane	0.43	0.92
Dimethylformamide	8.9	15
Dimethyl sulfoxide	29	17
Coupling amine[a]		F_c
Val-		1
Gly-		19
Ala-		3.6
Phe-		1.2
Leu-		1.7
Ser-		0.77
Met-		0.58
Lys(Z)-		0.95
Racemizing base	F_r	
Triethylamine	1	
N-Methylmorpholine	0.08	

[a] The F_c is approximately the same for -OMe, -OEt, or -OBzl but twice the value for -O*t*Bu.

Similarly, anomalous behavior was observed for Z-Gly-Val-OSu, for which the calculated k_r value is 13.6 times larger than the experimental value; furthermore, this dipeptide racemizes only 3 times faster than Z-Val-OSu. The k_r value is usually 10 to 100 times larger for the dipeptide, and indeed Z-Gly-Val-OPcp racemizes 124 times faster than Z-Val-OPcp. It is suggested again that Z-Gly-Val-OSu probably racemizes by enolization.

Besides these specific cases, whenever the additivity principle is operable, large numbers of peptide coupling and racemization rate constants can be estimated, with only few rate constants determined experimentally.

X. SUMMARY AND PROGNOSIS

From this chapter, it can be seen that quantitative work in synthetic peptide chemistry can show possible solutions for long-unresolved problems. Kinetic studies of coupling reactions of amino acids show large differences in coupling rates; therefore, sufficient time should always be used in order to complete the coupling reaction. When large peptides are synthesized, it is desirable to achieve 99 to 99.9% completion in order to avoid large numbers of wrong sequences. Since the completion time is concentration dependent, coupling of large peptide fragments require much longer coupling times or new "supersolvents" to obtain sufficiently large molar concentrations. Racemization is always a problem in synthetic peptide chemistry. Kinetic studies of tertiary base-catalyzed racemizations of amino acid and dipeptide active esters led, for example, to the recognition of the enolization mechanism for the racemization of both Z-Cys(Bzl)-ONp and Z-Gly-Cys(Bzl)-ONp, which in turn can provide a basis for improved synthetic procedures. Furthermore, using the additivity principle for predicting race mization, as well as coupling rates, can extend our knowledge without additional experimental work.

It is important to establish a correlation between the ability of the tertiary amine and the amino component to racemize carboxyl-activated amino acid and peptide derivatives. A simple, rapid procedure is needed to detect small amounts of diastereomers: The high-performance liquid chromatography is the most promising and convenient method for this purpose.

In view of the general failure of coupling of large peptide segments in acceptable yield, studies on coupling as well as racemization rates should be extended to this almost completely unknown field. For coupling of larger peptide segments, much faster coupling reagents than the frequently used azide are needed. At the same time racemization should not be increased; that is, the new reagents should have a large k_c/k_r ratio.

ACKNOWLEDGMENT

The author is grateful to K. Y. Hui who collected data and was helpful in preparing this manuscript. Discussion with Dr. James E. Brady regarding the feasibility of structure 21 in connection with the racemization by enolization mechanism was helpful and greatfully appreciated. Frequent discussion and cooperation with Dr. Richard E. Cover and Dr. Eugene M. Holleran on kinetic problems contributed significantly to the results presented in this chapter.

REFERENCES

Almy, J., and Cram, D. J. (1969). *J. Am. Chem. Soc.* **91**, 4459–4468.

Anderson, G. W., Zimmerman, J. E., and Callahan, F. M. (1963). *J. Am. Chem. Soc.* **85**, 3039.

Anderson, G. W., Zimmerman, J. E., and Callahan, F. M. (1964). *J. Am. Chem. Soc.* **86**, 1839–1842

Anderson, G. W., Zimmerman, J. E., and Callahan, F. M. (1967). *J. Am. Chem. Soc.* **89**, 5012–5017

Barber, M., and Jones, J. H. (1976). *In* "Peptides 1976" (A. Loffet, ed.) pp. 109–116. Editions de l'Université de Bruxelles, Belgium.

Bergmann, M., and Zervas, L. (1929). *Biochem. Z.* **203**, 280

Beyerman, H. C. (1972). *In* "Chemistry and Biology of Peptides" (J. Meienhofer, ed.), pp. 351–357. Ann Arbor Sci. Publ., Ann Arbor Michigan.

Bodanszky, M., and Bodanszky, A. (1967). *Chem. Commun.* pp. 591–593.

Bodanszky, M., Szelke, M., Tömörkeny, E., and Weisz, E. (1955). *Chem. Ind. (London)* p. 1517.

Camble, R., Garner, R., and Young, G. T. (1969). *J. Chem. Soc. C* pp. 1911–1916.

Chapman, N. B., and Shorter, J. (1972). "Advances in Linear Free Energy Relationship." Plenum, New York.

Conley, R. T. (1966). "Infrared Spectroscopy," Allyn & Bacon, Boston, Massachusetts.

Dzieduszycka, M., Smulkowsky, M., and Taschner, E. (1973). *In* "Peptides 1972" (H. Hanson and H. D. Jakubke, eds.) pp. 103–107. North Holland Publ., Amsterdam.

Goodman, M., and Levine, L. (1964). *J. Am. Chem. Soc.* **86**, 2918–2922.

Goodman, M., and Glaser C. (1970a). *J. Org. Chem.* **35**, 1954–1962.

Goodman, M., and Glaser C. (1970). *In* "Peptides: Chemistry and Biochemistry" (B. Weinstein and S. Lande, eds.) pp. 267–335. Dekker, N.Y.

Goodman, M., and Stueben, K. C. (1959). *J. Am. Chem. Soc.* **81**, 3980–3983.

Goodman, M., and Stueben, K. C. (1962). *J. Org. Chem.* **27**, 3409–3416.

Jones, J. H. (1974). *Chem. Ind. (London)* **21**, 723–727.

Kemp, D. S. (1973). *In* "Peptides 1971" (H. Nesvadba, ed.) pp. 1–19. North-Holland Publ., Amsterdam.

Kemp, D. S., and Chien, S. W. (1967). *J. Am. Chem. Soc.* **89**, 2743–2745.

Kemp, D. S., and Rebek, J. (1970). *J. Am. Chem. Soc.* **92**, 5792–5793.

Kemp, D. S., Wang, S. W., Busby, G., III, and Hugel, G. (1970). *J. Am. Chem. Soc.* **92**, 1043–1055.

Kemp, D. S., and Woodward, R. B. (1965). *Tetrahedron* **21**, 3019.

Kemp, D. S., Choong, S. L. H., and Pekaar, J. (1974a). *J. Org. Chem.* **39**, 3841–3847.

Kemp, D. S., Wang, S. W., Rebek, J., Mollan, R. C., Banguer, C., and Subramanyam, G. (1974b). *Tetrahedron* **30**, 3955–3967.

Khurgin, Y. I., and Dmitrieva, M. G. (1965). *Tetrahedron* **21**, 2305–2312.

Kisfaludy, L., and Nyéki, O. (1975). *Acta Chim. Acad. Sci. Hung.* **86**, 343–345.

Kisfaludy, L., and Schön, I. (1975) *Acta Chem. Acad. Sci. Hung.* **84**, 227–228.

Kisfaludy, L., Ceprini, M. Q., Rakoczy, B., and Kovacs, J. (1967). *In* "Peptides 1966" (H. C. Beyerman, A. vande Linde, and W. Massen vanden Brink, eds.) pp. 25–27. North Holland Publ., Amsterdam.

Kisfaludy, L., Roberts, J., Johnson, R. H., Mayers, G. L., and Kovacs, J. (1970). *J. Org. Chem.* **35**, 3563–3565.

Kisfaludy, L., Löw, M., Szirtes, T., Schön, I., Sarközi, M., Bajusz, S., Turan, A., Juhász, A., Beke, R., Graf, L., and Medzihradszky, K. (1972). *In* "Chemistry and Biology of Peptides" (J. Meienhofer, ed.), pp. 299–303. Ann Arbor Sci. Publ. Ann Arbor Michigan.

Kisfaludy, L., Löw, M., Nyéki, O., Szirtes, T., and Schön, I. (1973a). *Justus Liebigs Ann. Chem.* **1973**, 1421–1429.

Kisfaludy, L., Nyéki, O., and Szirtes, T. (1973b) *In* "Peptides 1971" (H. Nesvabda, ed.) pp. 54–59. North Holland Publ., Amsterdam.

Kisfaludy, L., Schön, I., Szirtes, T., Nyéki, O., and Löw, M. (1974). *Tetrahedron Lett.* pp. 1785–1786.

Kisfaludy, L., Löw, M., Argay, G., Czugler, M., Kömives, T., Sohár, P., and Darvas, F. (1976).

In "Peptides 1976" (A. Loffet, ed.) pp. 55–60. Editions de l'Universite de Bruxelles, Belgium.
Kovács, J., Hsieh, Y., Holleran, E. M., and Ting, Y. F. (1979). Submitted for publication.
Kovács, J., Giannotti, R., and Kapoor, A. (1966). *J. Am. Chem. Soc.* **88**, 2282–2292.
Kovács, J., Kisfaludy, L., and Ceprini, M. Q. (1967a). *In* "Peptides 1966" (H. C. Beyerman, A. van de Linde, and W. Massen vanden Brink, eds.) pp. 23–24. North Holland Publ., Amsterdam.
Kovács, J., Kisfaludy, L., and Ceprini, M. Q. (1967b). *J. Am. Chem. Soc.* **89**, 183–184.
Kovács, J., Mayers, G. L., Johnson, R. H., and Ghatak, U. R. (1968). *Chem. Commun.* 1066–1067.
Kovács, J., Kisfaludy, L., Ceprini, M. Q., and Johnson, R. H. (1969). *Tetrahedron* **25**, 2555–2573.
Kovács, J., Mayers, G. L., Johnson, R. H., Cover, R. E., and Ghatak, U. R. (1970a). *Chem. Commun*, pp. 53–54.
Kovács, J., Mayers, G. L., Johnson, R. H., Cover, R. E., and Ghatak, U. R. (1970b). *J. Org. Chem.* **35**, 1810–1815.
Kovács, J., Mayers, G. L., Johnson, R. H., and Ghatak, U. R. (1970c). *In* "Peptides: Chemistry and Biochemistry" (B. Weinstein and S. Lande, eds.) pp. 337–345. Dekker, New York.
Kovács, J., Cortegiano, H., Cover, R. E., and Mayers, G. L. (1971). *J. Am. Chem. Soc.* **93**, 1541–1543.
Kovács, J., Mayers, G. L., Johnson, R. H., Giannotti, R., Cortegiano, H., and Roberts, J. (1972a). *In* "Progress in Peptide Research" (S. Lande, ed.), pp. 185–193. Gordon & Breach, New York.
Kovács, J., Davis, E. J., Johnson, R. H., Cortegiano, H., and Roberts, J. (1972b). *In* "Chemistry and Biology of Peptides" (J. Meienhofer, ed.), pp. 359–364. Ann Arbor Sci. Pub., Ann Arbor, Michigan.
Kovács, J., Kapoor, A., Ghatak, U. R., Mayers, G. L., Giannasio, V. R., Giannotti, R., Senyk, G. Nitecki, D. E., and Goodman, J. W. (1972c). *Biochemistry* **11**, 1953–1958.
Kovács, J., Kalita, C., and Ghatak, U. R. (1972d). *J. Org. Chem.* **37**, 30–34.
Kovács, J., Cover, R. E., Johnson, R. H., Kalas, T., Mayers, G. L., and Roberts, J. (1973). *J. Org. Chem.* **38**, 2518–2521.
Kovács, J., Cover, R. E., Jham, G., Hsieh, Y., and Kalas, T. (1975). *In* "Peptides: Chemistry, Structure and Biology" (R. Walter and J. Meienhofer, eds.), pp. 317–324. Ann Arbor Sci. Publ., Ann Arbor, Michigan.
Kovács, J., Hsieh, Y., Hui K., and Kim, S. (1977a). *In* "Peptides, Proceedings of the Fifth American Peptide Symposium," (M. Goodman and J. Meienhofer, eds.), p. 461–464. Wiley, New York.
Kovács, J., Kalita, C., and Cervi, F. (1977b). *173rd Meet., Am. Chem. Soc. 1977* Abst. organic p. 226.
Kovács, J., Holleran, E. M., and Ting, Y. F. (1979a). In preparation.
Kovács, J., Holleran, E. M., and Hui, K. Y. (1979b). In preparation.
Kovács, J., and Centrone, D. A. (1979c). In preparation.
Kupryszewski, G. (1961). *Rocz. Chem.* **35**, 1533.
MacLaren, J. A., Savige, W. E., and Swan, J. M. (1958). *Aust. J. Chem.* **11**, 345–359.
Mayers, G. L., and Kovacs, J. (1970). *Chem. Commun.* pp. 1145–1146.
Morawiec, J., Konopinska, D., and Siemion, I. Z., (1971). *Rocz. Chem.* **45**, 771–779.
Neuberger, A. (1948). *Adv. Protein Chem.* **4**, 297–383.
Rebek, J., and Feitler, D. (1973). *J. Am. Chem. Soc.* **95**, 4052–4053.
Rebek, J., and Feitler, D. (1974). *J. Am. Chem. Soc.* **96**, 1606–1607.
Rebek, J., and Feitler, D. (1975). *Int. J. Pept. Protein Res.* **7**, 167–169.

Rouvray, D. H. (1973). *Chem. Technol.* **3**, 379–384.
Sheppard, W. A. (1970). *J. Am. Chem. Soc.* **92**, 5419–5422.
Siemion, I. Z., and Baran, G. (1975). *Bull. Acad. Pol. Sci., Ser. Sci. Chim.* **23**, 317–322.
Syrier, J. L., and Beyerman, H. C. (1974). *Recl. Trav. Chim. Pays-Bas* **93**, 117–120 and 256–257.
Tilak, M. A. (1970). *Tetrahedron Lett.* pp. 849–854.
Veber, D. F. (1975). *In* "Peptides: Chemistry, Structure and Biology" (R. Walter, and J. Meienhofer, eds.), pp. 307–316. Ann Arbor Sci. Publ., Ann Arbor, Michigan.
Weygand, F., Prox, A., and König, W. (1966). *Chem. Ber.* **99**, 1446–1450.
Williams, M. W. and Young, G. T. (1964). *J. Chem. Soc.* 3701–3708.
Windridge, G. C., and Jorgensen, E. C. (1970). *Intra-Sci. Chem. Rep.* **5**, 375–380.
Windridge, G. C., and Jorgensen, E. C. (1971). *J. Am. Chem. Soc.* **93**, 6318–6319.

Author Index

Numbers in italics refer to the pages on which the complete references are listed.

C

E

F

G

H

I

J

M

N

R

S

T

Y

Z

Subject Index

B

C

D

G

H

N

O

P

R

T

U